# *When Did Plate Tectonics Begin on Planet Earth?*

edited by

Kent C. Condie
Department of Earth and Environmental Science
New Mexico Institute of Mining and Technology
Socorro, New Mexico 87801, USA

Victoria Pease
Department of Geology and Geochemistry
Stockholm University
Stockholm SE-106 91, Sweden

THE
GEOLOGICAL
SOCIETY
OF AMERICA®

Special Paper 440

3300 Penrose Place, P.O. Box 9140 ▪ Boulder, Colorado 80301-9140 USA

2008

Published by The Geological Society of America, Inc.
3300 Penrose Place, P.O. Box 9140, Boulder, Colorado 80301-9140, USA
www.geosociety.org

Printed in U.S.A.

GSA Books Science Editors: Marion E. Bickford and Donald I. Siegel

**Library of Congress Cataloging-in-Publication Data**

When did plate tectonics begin on planet Earth? / edited by Kent C. Condie, Victoria Pease.
p. cm. -- (Geological Society of America special paper ; 440)
Includes bibliographic references.
ISBN-13: 978-0-8137-2440-9 (pbk.)
1. Plate tectonics--History. 2. Earth--Crust. I. Condie, Kent C. II. Pease, Victoria.

QE511.4.W44 2008
551.1'36--dc22

2008014665

Cover image courtesy of Dennis Tasa.

10 9 8 7 6 5 4 3 2 1

# Contents

*Preface* . . . . . . . . . . v
Kent C. Condie and Victoria Pease

**Isotopic and geochemical constraints**

1. ***A review of the isotopic and trace element evidence for mantle and crustal processes in the Hadean and Archean: Implications for the onset of plate tectonic subduction*** . . . . . . . . . . 1
Steven B. Shirey, Balz S. Kamber, Martin J. Whitehouse, Paul A. Mueller, and Asish R. Basu

2. ***A trace element perspective on Archean crust formation and on the presence or absence of Archean subduction*** . . . . . . . . . . 31
Stephen Foley

3. ***An overview of the lithological and geochemical characteristics of the Mesoarchean (ca. 3075 Ma) Ivisaartoq greenstone belt, southern West Greenland*** . . . . . . . . . . 51
Ali Polat, Robert Frei, Peter W.U. Appel, Brian Fryer, Yıldırım Dilek, and Juan C. Ordóñez-Calderón

4. ***Stratigraphy, geochemistry, and depositional environments of Mesoarchean sedimentary units in western Superior Province: Implications for generation of early crust*** . . . . . . . . . . 77
Philip Fralick, Pete Hollings, and David King

**Constraints from metamorphism and mineralization**

5. ***Characteristic thermal regimes of plate tectonics and their metamorphic imprint throughout Earth history: When did Earth first adopt a plate tectonics mode of behavior?*** . . . . . . . . . . 97
Michael Brown

6. ***Evidence for modern-style subduction to 3.1 Ga: A plateau–adakite–gold (diamond) association*** . . . . . . . . . . 129
D.A. Wyman, C. O'Neill, and J.A. Ayer

**Mantle geodynamic constraints**

7. ***Effects of a warmer mantle on the characteristics of Archean passive margins*** . . . . . . . . . . 149
Andrew Hynes

8. ***Tectonics of early Earth: Some geodynamic considerations*** . . . . . . . . . . 157
Jeroen van Hunen, Peter E. van Keken, Andrew Hynes, and Geoffrey F. Davies

### Crustal geodynamic and geophysical constraints

*9. The Late Archean Abitibi-Opatica terrane, Superior Province: A modified oceanic plateau* . . . 173
Keith Benn and Jean-François Moyen

*10. When did plate tectonics begin? Evidence from the orogenic record* . . . . . . . . . . . . . . . . . . . . . . 199
Victoria Pease, John Percival, Hugh Smithies, Gary Stevens, and Martin Van Kranendonk

*11. Seismic images of Paleoproterozoic microplate boundaries in the Fennoscandian Shield* . . . . . 229
Annakaisa Korja and Pekka J. Heikkinen

*12. Plate tectonics on early Earth? Weighing the paleomagnetic evidence* . . . . . . . . . . . . . . . . . . . . 249
David A.D. Evans and Sergei A. Pisarevsky

### Synthesis

*13. Modern-style plate tectonics began in Neoproterozoic time: An alternative interpretation of Earth's tectonic history* . . . . . . . . . . . . . . . . . . . . . . . . . . . . . . . . . . . . . . . . . . . . . . . . . . . . . . . . 265
Robert J. Stern

*14. When did plate tectonics begin? Evidence from the geologic record* . . . . . . . . . . . . . . . . . . . . . . 281
Kent C. Condie and Alfred Kröner

# *Preface*

Planet Earth is unique in many ways from the other silicate-metal planets in the solar system. In addition to oceans and an atmosphere that contains oxygen, Earth is the only planet that exhibits plate tectonics. Plate tectonics appears to be the result of convective cooling of the mantle, although the gravitational pull of subducted slabs probably drives plate motions (Conrad and Lithgrow-Bertelloni, 2002). Plate tectonics includes sites of plate formation at ocean ridges and sites of recycling into the mantle, which today are at subduction zones. Although geodynamic models support coherent plates throughout Earth history, just when and how plates became negatively buoyant is not yet clear. This is partly due to our uncertainty about the early thermal history of the mantle. Was the mantle hotter than at present, as generally assumed, or was it the same temperature (or even lower) due to enhanced rates of convective cooling during the early Archean? This leads to the question of how we track modern-style plate tectonics back in time.

To address this problem, a Penrose Conference was held in Lander, Wyoming, June 14–18, 2006, and was attended by ~70 scientists, representing a wide spectrum of specialties in the earth sciences (Condie et al., 2006). Although most participants agreed that plate tectonics involves horizontal plate motions on the Earth's surface and includes sites of plate formation and sites of recycling into the mantle, there was considerable discussion on how best to track plate tectonics back in time. Part of the disagreement was related to the likelihood that plate tectonics has changed with time, and especially the mechanism of recycling lithosphere into the mantle may have evolved. Not all investigators accept that modern plate tectonics operated in the Archean (Davies, 1992; Hamilton, 1998; Bleeker, 2002; Stern, 2005). One way to partially bypass this issue is to consider "modern-style" subduction, i.e.—steep subduction—and "pre-modern-style" subduction, which may have been different. For instance, during the magma ocean stage at ca. 4.5 Ga, subduction may have been symmetric, as we observe today on lava lakes. During the Hadean, but after the magma ocean had crystallized, some form of eclogite-driven delamination of the oceanic lithosphere may have been the principal return mechanism of plate material to the mantle.

In this volume, we include a cross section of papers from the Lander Conference, which address the question of when "modern-style" plate tectonics began on planet Earth. In chapter 1, Steve Shirey and others focus on isotopic and trace element evidence from the mantle and crust that indicates recycling of near-surface rocks into the deep mantle. They show quite convincingly that the geochemical signals indicate subduction-altered slabs have been recycled into the deep mantle beginning at or before 3.8 Ga. Thick mantle keels that have subduction imprints begin to appear at about 3.5 Ga. Collectively, the isotopic and trace element results indicate extraction of continental crust from the mantle by plate tectonic processes from 3.8–3.5 Ga onwards.

Stephen Foley's contribution on the geochemistry of tonalite-trondhjemite-granodiorites (TTGs) is directly relevant to when plate tectonics began: early (>3.8 Ga) continental crust is dominated by TTG gneisses, some of which are thought to have formed in subduction zones. Combining major and trace elements with experimental petrology, Foley argues that most Archean TTGs form by melting of garnet amphibolite of broadly basaltic composition. He suggests that the lack of large volumes of TTGs formed by melting of garnet-free amphibolite in the early Archean indicates that the Archean mantle was only marginally hotter than the present-day mantle and was consequently dominated by low-$SiO_2$ melt production. This prevented the generation of voluminous felsic crust and requires only a slightly modified plate tectonic scenario: marginally hotter average geotherms and substantially hotter subduction geotherms to enable melting of garnet amphibolite in the late Archaean.

Ali Polat and others review the lithological and geochemical characteristics of the Mesoarchean Ivisaartoq greenstone belt of southern West Greenland and compare them to other Archean greenstone belts in the region. The geochemical signatures of the least altered rocks are similar to Phanerozoic forearc/backarc ophiolites and intra-oceanic island arcs, suggesting that the Ivisaartoq greenstone belt represents a relic of dismembered Mesoarchean suprasubduction zone oceanic crust. This supports the onset of modern-style plate tectonics processes by ca. 3075 Ma.

Philip Fralick and colleagues document the stratigraphy and geochemistry of Mesoarchean sedimentary rocks in the Western Superior Province in Canada. They argue that the chemical and depositional environment associated with these rocks is analogous to the modern Kerguelen Plateau. Consequently, they suggest that models requiring fast, or numerous, spreading ridges to increase mantle heat dissipation are unlikely in the Mesoarchean. They conclude that mantle plume processes dominated the Earth's tectonic regime between 3.0 and 2.8 Ga. Given these different interpretations (Polat et al. and Fralick et al.) for Mesoarchean time, the onset of modern-style plate tectonics was either diachronous or requires better age constraints. Clearly, more work needs to be done in order to resolve this apparent contradiction.

In chapter 5, Mike Brown proposes that a duality of thermal regimes is the hallmark of modern plate tectonics, and a duality of metamorphic belts is the characteristic imprint of plate tectonics in the rock record. He suggests that the appearance of both ultra-high temperature and ultra-high pressure metamorphic belts after the Archean records the onset of a Proterozoic plate tectonic regime, which evolved during the Neoproterozoic into the modern plate tectonic regime. He further points out that during amalgamation of supercontinents, extreme metamorphism occurs along the sutures.

In chapter 6, Wyman and others propose that the formation of diamonds in the Archean requires geotherms similar to modern ones and probably reflects the presence of cool mantle roots beneath the continents. Stretching of continents underlain by these thickened roots during Archean plate tectonics would yield passive margins similar to modern passive margins. Hence, development of significant passive margins may have occurred through rifting of continents underlain by cool mantle roots. Widespread subcontinental melting associated with rifting of continents may have been a significant contributor to development of the thick, cool lithospheric roots during the Archean.

In chapter 7, Andrew Hynes discusses some of the geological consequences of a hotter Archean mantle. He shows that with a 200 °C increase in mantle temperature during the Archean, continental lithosphere would ride high with its upper surface near sea level. Hence, many submarine Archean greenstone successions could have formed along continental margins during extension over warmer mantle. The warmer mantle also results in greatly reduced subsidence along the rifted continental margins, thus decreasing the space available for passive-margin sediments, and hence, may account for the scarcity of passive-margin sequences in the Archean geological record. In chapter 8, Jeroen van Hunen and others suggest, from geodynamic modeling, that the mantle temperature has dropped 200–300 °C since the Archean. They conclude that the viability of Archean plate tectonics is limited by the availability of plate driving forces and lithospheric strength. Alternative mechanisms of cooling include magma ocean plate tectonics, diapirism, independent dynamics of crust and underlying lithospheric mantle, and large-scale (Venus-type) mantle overturns.

Next, we have four chapters that address geodynamic or geophysical constraints for when plate tectonics began. In chapter 9, Keith Benn and Jean-François Moyen propose a testable geodynamic model for the late Archean Abitibi greenstone belt in the Superior province in Canada. These authors use a combination of geological, structural, geochronological, and geochemical data to evaluate a plate tectonic model involving subduction of an ocean basin beneath an oceanic plateau at 2.7 Ga. The proposed model explains the presence of plume-type and subduction-type geochemical signatures in the volcanics and in the TTG plutons, and requires a single period of Archean plate convergence and subduction that lasted for about 35 My.

In chapter 10, Victoria Pease and colleagues combine geological, geochemical, and geochronological data from three Archean orogens to argue for the existence of modern-style plate tectonics during the mid- to late Archean. They review geological and geochemical evidence from the Pilbara craton for continental rifting at 3.2 Ga, development of an oceanic subduction-related arc complex at 3.12 Ga, and terrane accretion at 3.07 Ga. The authors also present thermobarometric evidence from high-grade rocks in the Barberton granite-greenstone terrane in the Kaapvaal craton and make a case for a paired metamorphic belt from this Archean composite terrane. Lastly, they summarize the accretionary and collisional assembly of the Superior craton at 2.7 Ga, which has been the subject of numerous studies, most of which call upon a plate tectonic interpretation.

In chapter 11, Annakaisa Korja and Pekka Heikkinen discuss deep seismic reflection data from the Fennoscandian Shield from which they suggest the Svecofennian lithosphere (1.9–1.8 Ga) is composed largely of accreted terranes. A variety of tectonic settings are recognized, including fossil subducted slabs, large transform faults, and various accreted arcs and basins. The results support the use of seismic reflection profiles in tracking plate tectonics into the Precambrian, and clearly suggest the operation of modern-style plate tectonics by 1.9 Ga.

Chapter 12, by David Evans and Sergei Pisarevsky, addresses the question of when plate tectonics began by assessing the relative motion between rigid plates in the Precambrian. There are statistically significant differences in apparent polar wander path lengths and/or changes in paleolatitude when comparing well-dated paleomagnetic poles of different cratons as old as Meso-Paleoproterozoic. However, even the most reliable pre-1880 Ma data are unable to distinguish between differential motion and true polar wander. Lateral motion alone is not unique to modern-style plate tectonics, but when combined with other indicators of cratonic rigidity (e.g., extensive dyke swarms), the authors suggest that a recognizable form of plate kinematics was typical throughout the last ca. 1900 million years of Earth history. This will be a fertile research topic in the coming years.

The final two contributions are by Bob Stern, and Kent Condie and Alfred Kröner. These papers synthesize a broad range of evidence, from planetary analogues and geodynamic modeling, to Earth's preserved geologic record. Bob Stern adopts a provocative stance, arguing for a non-uniformitarian approach to the question, "When did plate tectonics begin on Earth?". He advocates a pre-Neoproterozoic "proto-plate tectonic" mechanism that lacks plate subduction. This period in Earth's history is followed by the development of modern-style (subduction-driven) plate tectonics in the Neoproterozoic. On the other hand, Condie and Kröner consider that while any single piece of evidence may be explained by an alternative tectonic mechanism, many are explained by modern-style plate tectonic processes and that, in fact, the existence of modern-style plate tectonic processes since the late Archean best explains the co-existing combined wealth of evidence preserved in Earth's geologic record. Due to our increasing knowledge of even a limited rock record, it seems likely that in the future we will be able to advocate an even older beginning to modern-style plate tectonics on Earth.

With the renewed interest in how far back in time we can track modern-style plate tectonics, we might ask where we should go from here. Such a challenge should bring about renewed efforts to understand if and how the Archean differs from later times in Earth's history. Despite the outstanding questions about the Archean, a wealth of geologic, geochemical, structural, volcanologic, and geodynamic results seem to demand plate tectonics in the late Archean. Yes, there are some differences with modern-style plate tectonics. However, most of these can be explained by a hotter mantle, which results in a weaker lithosphere. From the papers in this volume and the discussions at the Lander Penrose Conference, it is necessary to understand the differences in Archean dynamic processes and tectonic regimes in order to explain the "unusual" tectonic and geochemical signatures preserved in the Archean geological record, rather than to abandon plate tectonics, which so well explains so much observational data. Discussion about the best criteria to identify ancient plate tectonics should be encouraged. Once the list of the best indicators of plate tectonic activity matures, these indicators should be catalogued and mapped on a 4D global crustal GIS (geographic information system) that is easily accessible to earth scientists (R.J. Stern, personal commun., 2007).

We would like to extend our thanks to those who donated their time and energy to reviewing the manuscripts submitted to this volume. Their willingness to participate in the review process contributes to maintaining a high scientific standard through constructive criticism and feedback. Reviewers include Jean Bedard, Ken Collerson, Alan Collins, Chris Fedo, Ron Frost, Richard Goldfarb, Jack Hillhouse, Rob Kerrich, Reiner Klemd, Jan Kramers, S. Labrosse, Hervé Martin, Al Hofmann, Eldridge Moores, Jean-François Moyen, Simon Peacock, Hugh Rollinson, Erik Scherer, Hugh Smithies, Robert

Stern, Phil Thurston, Arie van der Velden, Peter van Thienen, and Walter Mooney as well as several anonymous reviewers.

Kent C. Condie
Department of Earth and Environmental Science
New Mexico Institute of Mining and Technology
Socorro, New Mexico 87801, USA
kcondie@nmt.edu

Victoria Pease
Department of Geology and Geochemistry
Stockholm University
Stockholm SE-106 91, Sweden
vicky.pease@geo.su.se

## REFERENCES CITED

Bleeker, W., 2002, Archean tectonics: A review, with illustrations from the Slave craton: Geological Society [London] Special Publication, v. 199, p. 151–181.

Condie, K.C., Kröner, A., and Stern, R.J., 2006, When Did Plate Tectonics Begin? Penrose Conference Report: GSA Today, v. 16, no. 10, p. 40–41.

Conrad, C.P., and Lithgrow-Bertelloni, C., 2002, How mantle slabs drive plate tectonics: Science, v. 298, p. 207–209.

Davies, G.F., 1992, On the emergence of plate tectonics: Geology, v. 20, p. 963–966.

Hamilton, W.B., 1998, Archean magmatism and deformation were not products of plate tectonics: Precambrian Research, v. 91, p. 143–179.

The Geological Society of America
Special Paper 440
2008

# *A review of the isotopic and trace element evidence for mantle and crustal processes in the Hadean and Archean: Implications for the onset of plate tectonic subduction*

**Steven B. Shirey***
*Department of Terrestrial Magnetism, Carnegie Institution of Washington, 5241 Broad Branch Road NW, Washington, D.C. 20015, USA*

**Balz S. Kamber***
*Department of Earth Sciences, Laurentian University, Sudbury, Ontario P3E 2C6, Canada*

**Martin J. Whitehouse***
*Laboratory for Isotope Geology, Swedish Museum of Natural History, Box 50007, Stockholm S-105 05, Sweden*

**Paul A. Mueller***
*Department of Geological Sciences, Box 112120-241, Williamson Hall, University of Florida, Gainesville, Florida 32611, USA*

**Asish R. Basu***
*Earth and Environmental Sciences, 227 Hutchinson Hall, University of Rochester, Rochester, New York 14627, USA*

## ABSTRACT

**Considerable geochemical evidence supports initiation of plate tectonics on Earth shortly after the end of the Hadean. Nb/Th and Th/U of mafic-ultramafic rocks from the depleted upper mantle began to change from 7 to 18.2 and 4.2 to 2.6 (respectively) at 3.6 Ga. This signals the appearance of subduction-altered slabs in general mantle circulation from subduction initiated by 3.9 Ga. Juvenile crustal rocks began to show derivation from progressively depleted mantle with typical igneous $\varepsilon_{Nd}$:$\varepsilon_{Hf}$ = 1:2 after 3.6 Ga. Cratons with stable mantle keels that have subduction imprints began to appear by at least 3.5 Ga. These changes all suggest that extraction of continental crust by plate tectonic processes was progressively depleting the mantle from 3.6 Ga onwards. Neo-archean subduction appears largely analogous to present subduction except in being able to produce large cratons with thick mantle keels. The earliest Eoarchean juvenile rocks and Hadean zircons have isotopic compositions that reflect the integrated effects of separation of an early enriched reservoir and fractionation of Ca-silicate and Mg-silicate perovskite from the terrestrial magma oceans associated with Earth accretion and Moon formation, superposed on subsequent crustal processes. Hadean zircons most likely were derived from a continent-absent, mafic to ultramafic protocrust that was**

*Corresponding author, Shirey: shirey@dtm.ciw.edu; Kamber: bkamber@laurentian.ca; Whitehouse: martin.whitehouse@nrm.se; Mueller: mueller@geology.ufl.edu; Basu: abasu@earth.rochester.edu

Shirey, S.B., Kamber, B.S., Whitehouse, M.J., Mueller, P.A., and Basu, A.R., 2008, A review of the isotopic and trace element evidence for mantle and crustal processes in the Hadean and Archean: Implications for the onset of plate tectonic subduction, *in* Condie, K.C., and Pease, V., eds., When Did Plate Tectonics Begin on Planet Earth?: Geological Society of America Special Paper 440, p. 1–29, doi: 10.1130/2008.2440(01). For permission to copy, contact editing@geosociety.org. 

**multiply remelted between 4.4 and 4.0 Ga under wet conditions to produce evolved felsic rocks. If the protocrust was produced by global mantle overturn at ca. 4.4 Ga, then the transition to plate tectonics resulted from radioactive decay-driven mantle heating. Alternatively, if the protocrust was produced by typical mantle convection, then the transition to plate tectonics resulted from cooling to the extent that large lithospheric plates stabilized.**

**Keywords:** subduction, plate tectonics, mantle, crust, Hadean, Archean, radiogenic isotopes.

## INTRODUCTION

The surface of Earth is composed of geodynamically stable continental cratons, orogenically active continental and oceanic margins, and freshly created ocean floor that compositionally reflects 4.5 billion years of geological evolution. While present plate tectonic processes (summed up as the "Wilson cycle"), driven by subduction and its consequent return flow, adequately explain much of the surface topography and composition of the crust and lithosphere, it is not known how plate tectonics in the Precambrian was different from today or even whether a particular time in Earth's geological history can be recognized as the time when plate tectonics started in its present form. The oceanic sedimentary record, the location of earthquake foci, and magnetic anomaly patterns on the ocean floor are the key evidence that show how plate tectonics operates today. The pre-Mesozoic record has been fragmentally compressed into the continents or reassimilated in to the mantle and is lost to simple, direct inspection or measurement. The isotope geochemistry of old continental and young oceanic rocks, however, can provide insight into the record of global lithospheric recycling in a manner that, if consistent with slab subduction and plume/ridge return flow, could reveal when plate tectonics began. Even for data obtained on recent igneous rocks, this isotopic record appears to date from earliest Earth history as a consequence of the continuity of Earth's tectonic activity, its global extent, and the ongoing nature of crustal differentiation. While these isotopic data will not directly correlate spatially with plate boundaries, they will record the time-integrated geochemical effects of plate tectonic processes. This manuscript reviews chiefly the isotopic evidence plus some supporting trace element evidence that relate specifically to the onset of plate tectonics on Earth.

## EVIDENCE FOR PLATE TECTONICS FROM CRUST-MANTLE RECYCLING AND MANTLE DEPLETION

### Mantle Heterogeneity and Its Implications for Early Earth

Melting at the linear volcanic chains that form at mid-ocean ridges and at the central volcanic complexes of oceanic islands provides a geochemical probe of upper-mantle and in some cases lower-mantle compositions. Radiogenic isotopic data of Pb and Sr for these zero-age basalts have long been known to display isotopic variations (heterogeneity) that can only be explained by isolation of their mantle sources for hundreds of millions to several billion years (e.g., Gast et al., 1964; Sun and Hanson, 1975a, 1975b; Sun, 1980). In the 1980s these variations were systematized into end-member mantle components or reservoirs (e.g., EMI, EMII, HIMU, DMM, FOZO; see Hofmann, 1997, 2003, for more recent reviews) to express their geochemical similarities and thus the combined petrogenetic histories of portions of mantle (e.g., Zindler and Hart, 1986). Implicit in these isotopic mantle reservoirs was the recycling of surface materials: pelagic sediments to explain EMI, continental sediments to explain EMII, and altered oceanic crust to explain HIMU (Hofmann and White, 1982; White and Hofmann, 1982; Zindler and Hart, 1986). Concurrently it was recognized that oceanic mantle isotopic compositions represented by both mid-ocean ridges and ocean islands are not truly randomly distributed geographically (Hart, 1984; Hawkesworth et al., 1986; Shirey et al., 1987) and in some cases define patterns that can be related to the opening of ocean basins and the movement of continental terranes with their attendant mantle keels (Hawkesworth et al., 1986; Luais and Hawkesworth, 2002). Continuing efforts have served to refine the identity of the end members by adding new isotopic systems (Hf and Os; Salters and Hart, 1991; Hauri and Hart, 1993), to clarify the extent of the compositional effects by pairing them with trace element and stable isotopic data (e.g., O; Eiler et al., 2000) and to better define the petrology of the recycled materials by using major elements (high-silica components; Hauri, 1996). The most viable current model to explain mantle heterogeneity focuses on the erosion of ancient continental mantle keel components and the subduction of oceanic lithosphere with its included volatiles, sediments, and seawater-altered peridotite and basalt. The basic subduction aspects of this geochemical model were presciently advocated by Ringwood (1991). More recent geophysical studies using tomographic inversion and shear wave splitting basically corroborate all aspects of this model: slab subduction into and through the transition zone (van der Hilst et al., 1997), deep-seated return flow in plumes (Montelli et al., 2004), and sublithospheric flow directed around mantle keels (Behn et al., 2004).

The isotopic expression of these processes is perhaps best illustrated using the Pb isotopic compositions of mid-oceanic-ridge and oceanic-island basalts (Fig. 1). The Pb paradoxes presented by these data have been discussed by numerous authors (e.g., Hofmann, 2003; Hart and Gaetani, 2006) and are not the

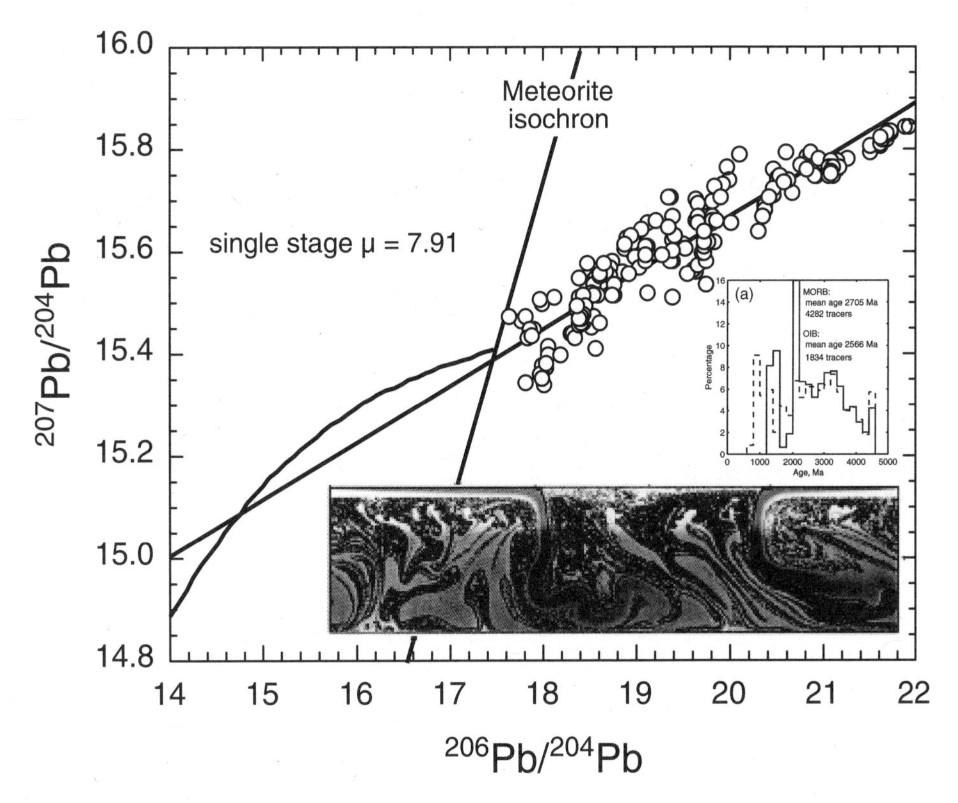


Figure 1. Representative worldwide ocean island basalt (OIB) common Pb isotope compositions (compilation of Murphy et al., 2003) relative to the positions of the meteorite isochron and a single-stage mantle evolution curve (μ = 7.91). Note that a regression line through the OIB array has a slope corresponding to ca. 1.6–2.0 Ga and also intersects the single-stage growth curve at a similar date. Convection model after Davies (2002) in insets shows how old tracers would be distributed in the convecting upper mantle that will be sampled by MORB and OIB and could explain Mesoproterozoic ages.

subject of discussion here. What is of importance is the average 2 Ga slope of the $^{207}Pb/^{204}Pb$ versus $^{206}Pb/^{204}Pb$ array and the scatter of data within the array. This array represents the mixing of the many isolated mantle sources that were melted to generate these oceanic basalts. The parent radionuclide of $^{207}Pb$, $^{235}U$, has a short enough half-life that by 1.5 billion years ago roughly 75% of it had decayed. Thus, the bulk slope of the oceanic array and higher-angle excursions of subsets of data above this array have long been interpreted as reflecting a mixing history with portions of the mantle that inherited U/Pb variations from before the earliest Mesoproterozoic when more $^{235}U$ was actively decaying. Because the *average* slope of this array is Paleoproterozoic and because the array was at least partially derived from mantle sources with geologically recently established U/Pb variations (which are only capable of affecting $^{206}Pb/^{204}Pb$ variability), then it is most probable that these trends reflect mixing of end members of Mesoarchean, Eoarchean, or perhaps even Hadean ages. (Note that the International Commission on Stratigraphy [ICS] has revised the boundaries and nomenclature of the Precambrian to be Eoarchean >3600 Ma, Paleoarchean 3200–3600 Ma, Mesoarchean 2800–3200 Ma, Neoarchean 2500–2800 Ma, Paleoproterozoic 1600–2500 Ma, Mesoproterozoic 1000–1600 Ma, and Neoproterozoic 542–1000 Ma. See Gradstein et al., 2005. The term Hadean was left undefined but is widely accepted to refer to the time before Earth's oldest rocks, ca. >4000 Ma.)

The continents, though, provide a ready storehouse of Paleoarchean to Neoarchean material that could be recycled as sediment, and for this reason, there is always potential for recent incorporation of ancient, isolated continental crust or ancient incorporation of juvenile continental crust as the old mixing end members of the oceanic array. Detailed studies of the trace element and isotopic composition of oceanic basalts have shown though that continental material only appears in the EM signature found in a few oceanic-island basalts and some Indian Ocean MORBs (mid-oceanic-ridge basalts) (Hofmann, 1997, 2003; Rehkamper and Hofmann, 1997; Sims and DePaolo, 1997). Therefore, the chief cause of the apparent ancient ages on the Pb isotope array of oceanic basalts must be the recycling of altered oceanic lithosphere (and perhaps to a lesser degree, some continental lithosphere) into the upper mantle beginning at least in the Neoarchean to Paleoproterozoic. For these components, radiogenic ingrowth occurs during residence in the upper mantle before they are returned nearer to the surface to be sampled in the mantle sources of ocean islands and, in more diluted fashion, ocean ridges.

## Geodynamic Aspects of Recycling

Accepting from the modern evidence in oceanic basalts that oceanic lithosphere recycling occurs and that it can carry older components at least to the transition zone, it is important for the question of plate tectonic initiation to understand just when in the Proterozoic, Archean, or Hadean recycling of the oceanic lithosphere started. The effect of this process over time is to produce a "marble-cake" mantle (Allegre and Turcotte, 1986), an idea that can be traced through the literature in discussions of veined oceanic mantle (Sun and Hanson, 1975b; Hanson, 1977), pyroxenite veins in orogenic lherzolite and harzburgite (Allegre and Turcotte, 1986), and the pyroxenitic-eclogitic component in oceanic-island basalts (Hirschmann and Stolper, 1996; Kogiso et al., 2003; Sobolev et al., 2005). Note, however, that in most samples of veined mantle, the veins have been interpreted as melt, not stretched lithosphere (e.g., Bodinier and Godard, 2003). Nonetheless, pyroxenitic components are thought to be ubiquitously distributed and are preferentially sampled by melting at low extents as shown by studies of posterosional volcanism at hot spots, off-axis seamounts (Zindler et al., 1984), and magmatism at propagating rifts.

Mantle convection calculations (Christensen and Hofmann, 1994; Davies, 2002; Huang and Davies, 2007) show that a small percentage of tracers as old as 3.6–4 Ga, when introduced from the top, do survive whole-mantle convection for nearly the age of Earth (Fig. 1, insets). These models roughly reproduce a Paleoproterozoic Pb-Pb age for the oceanic array, albeit one that can be too old to match its true 1.8 Ga age without starting the process later (Christensen and Hofmann, 1994; Davies, 2002) or reducing lower mantle viscosity (Davies, 2002). Recent three-dimensional models that involve rapid early mantle overturn, and that can be scaled to Earth, more closely approximate the 1.8 Ga age (Huang and Davies, 2007). Thus, the Pb-Pb array for modern oceanic rocks, the prime example of mantle heterogeneity, is consistent with recycling of oceanic lithosphere back into mantle convective flow by some mechanism perhaps as early as the Eoarchean. This process could be plate subduction, but the geochemical effects of recycling do not unambiguously demand this. Note, though, that this evidence is independent of the subduction signatures seen within continental rocks themselves, as discussed below.

The current mode by which Earth releases 90% of its deep-seated internal heat (from accretion, inner core crystallization, and radioactive decay) is through mantle convection coupled to hydrothermal circulation at the ocean ridges (Davies, 1999). Continents provide 24% of the surface heat, but this comes chiefly from decay of radionuclides in the crust. Only 12% of Earth's heat emanates from the lithospheric mantle beneath the continents because the lithospheric mantle has low heat production and it must move by conduction. So the effect of continents is to insulate the asthenospheric mantle beneath them (Gurnis, 1988; Lowman and Jarvis, 1999) and to limit the surface area over which the oceanic convective + hydrothermal heat-loss engine can operate. Recent work on the interaction of conduction and convection between ocean basins and continents (Lenardic et al., 2005; Lenardic, 2006) extrapolates these processes back through the Eoarchean and does not preclude extension into the Hadean. The percentage of surface area covered by continental crust had a minimal effect on global mantle heat flow and perhaps even contributed to its increase (Lenardic et al., 2005). Raising the internal temperature of the mantle with insulating continents would have increased convective velocities in the whole mantle because of the mantle's temperature-dependent viscosity. Earth's mantle is near its solidus, and too much early continental area would have triggered widespread mantle melting in the Eoarchean, for which there is no direct evidence (Lenardic, 2006). Thus, Lenardic et al. (2005) concludes that progressive growth of the continental crust (McCulloch and Bennett, 1994; Kramers and Tolstikhin, 1997; Collerson and Kamber, 1999) offers a better explanation for the heat loss mechanisms of the mantle through time than does the early formation of areally extensive early continental crust.

A corollary of the above arguments is that an oceanic-lithospheric, mantle-convective heat loss model may have been as applicable to the convecting mantle on early Earth as it is to today's mantle. Although the nature of convection in the Hadean is poorly understood (van Hunen et al., this volume) and depends critically on the heat distribution of Earth (e.g., whether there was a mantle overturn after the magma ocean that was generated by a Mars-sized impact; see Kamber, 2007; Kramers, 2007; and discussion below), the Eoarchean mantle would have eventually started to convect. When it did, the early existence of a global ocean (Marty and Yokochi, 2006; Kramers, 2007) means that heat transfer by hydrothermal circulation would have been as important then as it is today if not more so. It would have accompanied oceanic volcanism, and it would have altered the composition of oceanic lithosphere, which was likely to have been mostly composed of harzburgite, komatiite, and basalt (Takahashi and Brearley, 1990). It also would have been available for lowering the solidus of altered komatiite and basalt and for recycling into melting mantle sources. What is typically expected to have been different on Eoarchean to Hadean Earth was a hotter mantle, an atmosphere that was less oxidizing, and lithosphere that was bombarded more frequently by meteorites. A hotter mantle existed in the Archean because Earth is now and was then cooling, radioactive decay produced more heat in the past (Davies, 1999), and komatiite petrogenesis supports higher temperatures estimated to range from 100 °C (Grove and Parman, 2004) to 300 °C (Nisbet et al., 1993) hotter depending on source water content (Arndt et al., 1998). The atmosphere was less oxidizing because there is clear evidence for a rise in the oxygen content in the Proterozoic (e.g., Bekker et al., 2004). Meteorite bombardment was more frequent as shown by the lunar cratering record (e.g., Koeberl, 2006).

Another major difference expected between the Archean and younger Earth models would be the lack of topographically emergent stable continents of sizeable area because such continental masses only could have been preserved when they were developed concomitantly with thick, stable, depleted keels. Without extensive early sizeable continents then, the question of when plate tectonics started on Earth and its form can be reduced to the question of how the different conditions affected whether the style of lithosphere creation was at ridges and whether its destruction was in subduction zones or whether other geometries would be viable. While the isotopic data cannot directly constrain the tectonic form of oceanic lithosphere creation and destruction in the Eoarchean and Hadean, this paper will evaluate whether it is consistent with a plate tectonic mode for the oceanic regime. Since, in the absence of continents, an early basalt-capped, oceanic lithosphere would have entirely covered Earth, its stability and longevity become important considerations (see discussion below). Rheological modeling (Korenaga, 2006) and Pb isotopic data on the oldest sediments derived from basaltic precursors (Kamber, 2007) support a stable, near-surface, long-lived reservoir of basalt. The need to release Earth's heat by convection with hydrothermal circulation (Lenardic, 2006), however, seems critically limited by a stable, early depleted oceanic lithosphere (Davies, 2006). Though reconciling such contrasting models remains challenging, the continually evolving geochemical databases discussed below offer significant insight into the geodynamic evolution of Earth, including the initiation of plate tectonics.

## Evolution of Depleted Mantle Trace Element Compositions

An important aim of trace element studies of ancient, mantle-derived rocks is to reconstruct the chemical evolution of the depleted mantle through time. In this regard, the most useful chemical elements are those that are heavily concentrated in continental crust, through the isolation of which the mantle becomes ultimately and severely depleted. It is important to remember that continental crust cannot form directly by mantle melting. Rather, the chemical inventory of modern arc-type continental crust reflects a step in a complex chain of processes beginning with the formation, alteration, and hydration of oceanic lithosphere. Eventual subduction, metamorphism, and associated devolatilization of such lithosphere leads to fluid-induced melting of the suprasubduction zone mantle (Tatsumi and Kogiso, 1997). The resulting melt undergoes differentiation and eventually forms sialic crust as well as ultramafic cumulates that are returned to the mantle (Muentener et al., 2001). This succession of processes ultimately imparts a characteristic geochemical fingerprint on continental crust. In a diagram where elements are arranged according to their relative incompatibility in mantle melting, it is evident (Fig. 2) that certain elements appear more concentrated than expected, while others are less abundant than could be expected from their incompatibility during upper-mantle, MORB-style melting (Hofmann, 1988). The most widely discussed element that is overenriched in arc-type crust is Pb (Miller et al., 1994), which shows a prominent positive spike in the N-MORB (normal mid-oceanic-ridge basalt) normalized trace element plot (Fig. 2). The most conspicuous group of elements that are less abundant than could be expected from their incompatibility are Ti, Nb, and Ta. A detailed discussion of how exactly the arc geochemical pattern develops is beyond the purpose of this treatment, but to a first order, the overly enriched elements are those that are particularly soluble in the metamorphic fluids expelled from the oceanic slab upon subduction, while the underabundant elements are much less fluid-soluble and are preferentially retained in the slab, mainly in Ti minerals.

It has been known for some time (Jochum et al., 1991) that the gradual depletion of highly incompatible elements of oppo-

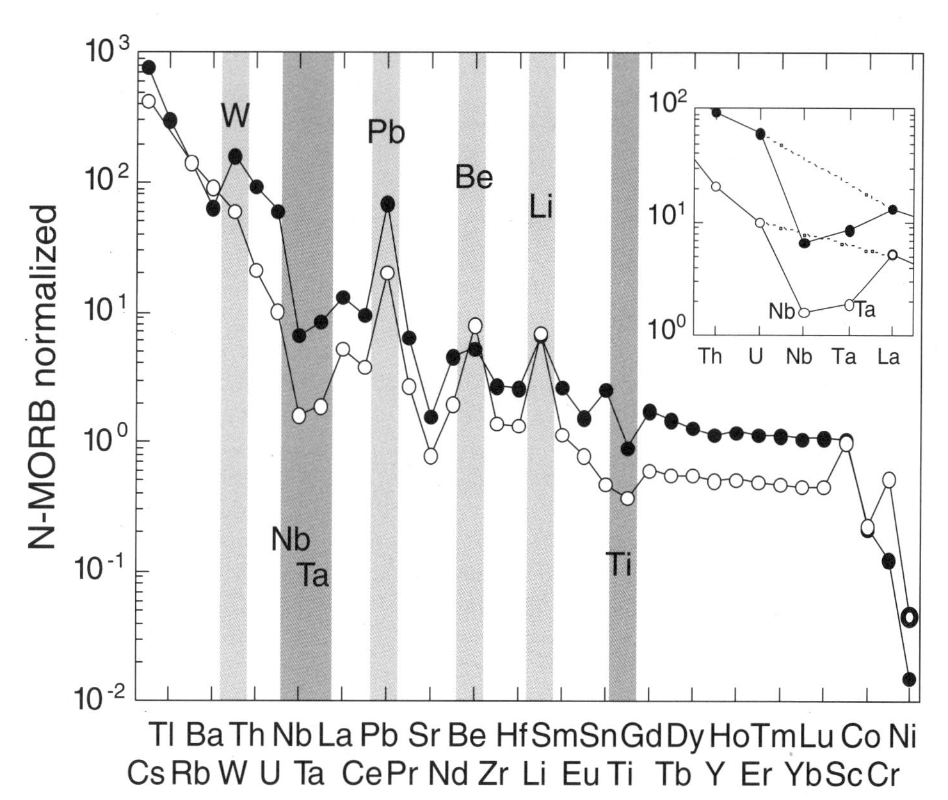


Figure 2. Full trace element patterns of modern continental sediment composite (solid circles; Kamber et al., 2005a) and average sediment from the 3.7 Ga Isua greenstone belt (open circles; Bolhar et al., 2004). Data are normalized to N-MORB (normal mid-ocean ridge basalt), and elements are arranged according to relative incompatibility (see Kamber et al., 2002, 2005a, for further information). In mid-ocean-ridge-style mantle melting, the liquid is most enriched for elements on the left-hand side of the plot and least enriched for the most compatible elements plotting at the very right. Note that this trend is also generally true for suprasubduction zone melting, leading to the pronounced exponentially decaying trend seen on this plot. However, suprasubduction zone melting is characterized by more complex processes that lead to the overenrichment of fluid-mobile elements (W, Pb, Be, and Li; highlighted with light gray backgrounds) and depletion in the refractory Nb, Ta, and Ti (highlighted with dark gray backgrounds). Importantly, the extent of the Nb deficit is greater than that of Ta, resulting in a characteristic positive step in the pattern highlighted on the inset of the figure. All these fingerprints of arc-type melting are already evident in the 3.7 Ga sediment composite.

site behavior during the fluid-induced melting that accompanies subduction (e.g., those that are conserved versus those that are fugitive during slab dehydration) could be used to reconstruct the amount of continental crust present through time. To accomplish this, two conditions must be met: that the crust form in a similar way as crust today and that fossil mantle melts be found that truthfully reflect the inventory of these elements in a rather uniformly depleted portion of the mantle. With respect to the first condition, there is excellent evidence from the trace element geochemistry of Eoarchean granitoids (Nutman et al., 1999; Kamber et al., 2002) and the oldest Eoarchean terrestrial clastic sediments (Bolhar et al., 2005) that the arc geochemical fingerprint was already established by 3.7–3.8 Ga. Remarkably, despite their long, complex geological histories, these Eoarchean rocks not only show the modern arc-characteristic depletion in the relatively immobile Ti-Nb-Ta (Fig. 2) but also display the overenrichment in the much less conservative elements Pb, Li, Be, and W. The consistency of this overenrichment with that seen in more immobile elements suggests that it is not due entirely to alteration. Indeed, the magnitude of the positive Pb spike is as large in Eoarchean sediment as in modern alluvial arc-dominated continental sediment (Fig. 2).

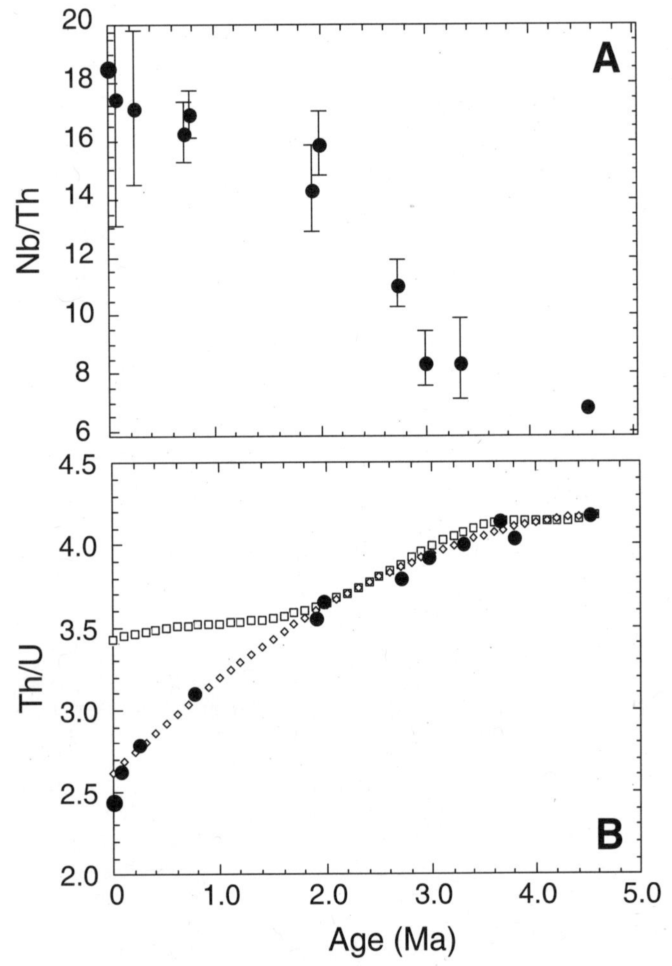


Figure 3. (A) Reconstruction of the depleted mantle Nb/Th (after Collerson and Kamber, 1999; see Kamber et al., 2003b, for chondritic values and input parameters). Note that the Hadean bulk silicate Earth Nb/Th is assumed to be slightly subchondritic due to either Nb incorporation into the core or Hadean silicate Earth differentiation. Importantly, the modern Nb/Th of the depleted mantle is more than twice as high as it was in the Paleoarchean. The increase in the ratio is due to extraction of continental crust. (B) Reconstruction of the depleted mantle Th/U (after Collerson and Kamber, 1999). Extraction of continental crust alone predicts a modern MORB Th/U of ~3.4 because the continental crust Th/U of ~4.5 is only slightly higher than chondrite (4.0–4.2). However, there is a marked deviation some time after 2.0 Ga, of the observed (solid black circles, from high-Mg basalts with juvenile Nd isotopes) from the modeled (open squares) trend due to formation of crust with constant Th/U. A better fit can be obtained (open diamonds) with preferential U recycling via subduction after the first significant atmospheric oxygenation at ca. 2.2 Ga.

In terms of the second condition, many studies have shown that chemically diverse mantle source areas have existed at least since the Neoarchean (Kerrich et al., 1999). For this reason, it is necessary to limit this type of trace element investigation to those basalts that can independently be shown to have been derived from the depleted portion of the mantle. $^{147}Sm$-$^{143}Nd$ isotope systematics are most commonly used as a screening tool to identify the purest melts from the most depleted mantle based on the most radiogenic initial $^{143}Nd/^{144}Nd$ composition (Collerson and Kamber, 1999).

In such basalts, the element pair Nb-Th (niobium-thorium) has proven very insightful for reconstruction of depleted mantle chemistry. Both elements are similarly and strongly incompatible (hence plotting close together and far to the left in Fig. 2) as well as relatively immobile during metamorphism and alteration. As a result, the ratio of Nb to Th in basalts of great geological antiquity reflects, to a first order, the Nb/Th in the depleted mantle melt source (Jochum et al., 1991). Figure 3A shows that the Nb/Th of modern MORB is substantially higher than that of chondritic meteorites. This increase in Nb/Th cannot be attributed to the earliest terrestrial mantle depletion event recorded by $^{146}Sm/^{142}Nd$ (see below) because it is not evident in early Precambrian basalts. Rather, the increase in the depleted mantle Nb/Th was initially slow (until ca. 2.9 Ga), then proceeded rapidly until ca. 1.8 Ga, since which time it has slowed again (Collerson and Kamber, 1999; Kamber et al., 2003b). The gradual sigmoidal increase in Nb/Th is readily explained by the progressively increasing separation of arc-type continental crust, with its relative Nb deficit and is prima facie evidence against the notion of constant continental volume (Armstrong, 1981; see below).

A related observation can be made by comparing the abundances of Nb and Ta. It can be seen in the inset of Figure 2 that although both elements are relatively depleted in average continental crust, Ta is noticeably less so. This is expressed in the positive slope from Nb to Ta (Fig. 2, inset) in the otherwise exponentially decaying trend of a typical continental trace element plot. This qualitative statement can be quantified via the Nb/Ta, which in continental crust is ~11 (Kamber et al., 2005b). This is 40% lower than in chondritic meteorites (~18; Münker et al., 2003). The exact properties of Nb and Ta that cause this effect during arc magmatism are a matter of ongoing debate, but what is beyond any doubt is that this process was operative in the Eoarchean, because the same rocks that show characteristic Pb, Li, Be, and W enrichment and Ti, Nb, and Ta depletion also show a strongly subchondritic Nb/Ta. In the case of the oldest known clastic metasedimentary rocks, the average Nb/Ta is indistinguishable within error from average continental crust. A bulk silicate Earth mass balance issue arises because the vast majority of mantle melts and continental rocks have Nb/Ta lower than chondrites. While some of the missing Nb probably was sequestered into the core (Wade and Wood, 2001), the relatively constant depleted mantle Nb/Ta through time (Kamber et al., 2003b) requires the long-term storage of eclogitic oceanic slabs in the mantle (Rudnick et al., 2000).

Another important ratio to define the evolution of the mantle is Th/U. Judging from the high incompatibility of these elements (Fig. 2) and the slightly higher incompatibility of Th, the extraction of continental crust from the mantle would be expected to lead to a very modest reduction of Th/U in the depleted mantle from the chondritic value of ~4.0–4.2 to ~3.75. Pb isotopic systematics of modern MORB allow an independent estimate of the time-integrated Th/U of ~3.4 for the depleted mantle (Kramers and Tolstikhin, 1997). However, the present-day measured MORB Th/U, both directly as concentrations and indirectly from U-series systematics (Galer and O'Nions, 1985), is 2.6, much lower than the time-integrated ratio. Furthermore, as the data on mafic to ultramafic rocks (Fig. 3B) (Collerson and Kamber, 1999) and the deep-seated parental magmas to kimberlites (Zartman and Richardson, 2005) show, the Th/U of the depleted upper mantle has been on a steady decline since ca. 3.6 Ga. This observation constitutes the second terrestrial Pb isotope paradox and is explained by preferential recycling of continental U at subduction zones (Elliott et al., 1999). This process, whereby oxidized U (+6) is lost during weathering and incorporated into oceanic sediment and hydrated oceanic lithosphere through valence reduction (+4) and returned to the mantle with subducting slabs, became progressively more effective as the atmosphere became more oxidizing by ca. 2.2 Ga (Fig. 3B) (Bekker et al., 2004). But it had evidently been developing since 3.6 Ga due to the slightly higher incompatibility of Th relative to U. Subduction is the chief mechanism responsible for the preferential recycling of U relative to Th into the mantle. Uranium recycling into the mantle is also the reason why it is not possible to directly estimate the continental crust volume versus time curve from Nb/U systematics (e.g., Sylvester et al., 1997). As pointed out by Collerson and Kamber (1999), the Nb/U of the MORB-source mantle was higher two billion years ago than it is today, not because more continental crust existed but because of U recycling.

Observations of depleted mantle Nb-Ta-Th-U systematics show that after the earliest mantle depletion events, recorded by $^{142}Nd/^{144}Nd$ (see below), gradual extraction and isolation of continental crust and continuous recycling of slabs of oceanic lithosphere progressively changed the abundances of these elements in the depleted mantle. Suprasubduction zone melting and slab subduction are the only known geological processes that are capable of separating these similarly incompatible elements in the recorded fashion. The near-chondritic Nb/Th of Eo- and Mesoarchean basalts with slightly superchondritic initial $^{143}Nd/^{144}Nd$ values clearly argues against a voluminous Eoarchean continental crust. It is important to realize that, unlike long-lived radiogenic tracers, Nb/Th would also record the temporary extraction of voluminous short-lived, juvenile continental crust, for which there is again no evidence. However, Nb/Th and Th/U place no constraints on the amount of oceanic lithosphere (including plateaus) that could have coexisted with nascent continental crust. Oceanic lithosphere is less visible to these element ratios, because it forms by processes that are not nearly as effective at discriminating between these elements. Nonetheless, it seems quite capable of recording the chemical effects in the mantle of the slab return flow, the main mechanism by which recycling is accomplished. Hence, while Nb-Th-Ta-U strongly argue for the initiation of subduction and the start of formation of continental crust at ca. 3.9 Ga (Kamber et al., 2002; and see below), the possibility remains that voluminous long-lived mafic crust existed during the earliest (pre–3.9 Ga) part of Earth history.

## PLATE TECTONIC IMPLICATIONS OF ARCHEAN CRUSTAL EVOLUTION

### Earliest Mantle Isotopic Signatures from Juvenile Crust

The "juvenile" component of Archean crust includes either those rocks that are derived directly from the mantle (e.g., komatiites, basalts, andesites, diorites) or those rocks that form from crustal sources rapidly enough (e.g., <50 m.y.) to be considered to come indirectly from the mantle (e.g., tonalities, trondhjemites, and some granodiorites whose isotopic systems do not reveal a prolonged continental crustal history). Collectively these rocks have long been used to track the evolution of the mantle in the Archean (e.g., Peterman, 1979) because they are the best ancient analogues to modern oceanic mantle-derived rocks. Many are recognized to have had suprachondritic initial isotopic compositions in the long-lived $^{147}Sm$-$^{143}Nd$ system (based on the decay of $^{147}Sm$ to $^{143}Nd$) throughout the Neoarchean to the Paleoarchean (e.g., Shirey and Hanson, 1986; Shirey, 1991; Nägler and Kramers, 1998; Bennett, 2003). It is important to recognize that Sm-Nd isotopic data are obtained on whole rocks and that all the oldest terrestrial rocks are severely and multiply metamorphosed in the crust. This has led to a debate in the literature (e.g., Arndt and Goldstein, 1987; Moorbath et al., 1997; Bennett and Nutman, 1998; Kamber et al., 1998) on the ability of a whole rock to retain its igneous $^{147}Sm/^{144}Nd$ unchanged during metamorphic recrystallization and has cast doubt on the veracity of the highest $\varepsilon_{Nd}$ measured, especially in the oldest rocks. Nonetheless, the Nd isotopic signatures of the oldest rocks clearly reflect a depleted origin, regardless of their absolute initial value. This has previously been interpreted as the result of earlier Hadean depletion events that resulted from the removal of an enriched (e.g., low-

Sm/Nd) component of oceanic (Carlson and Shirey, 1988; Chase and Patchett, 1988) or continental (Armstrong, 1981, 1991; McCulloch and Bennett, 1994; Harrison et al., 2005) crustal affinity. The timing of the prior depletion event(s), the extent of depletion, and the composition of the sequestered enriched components (continental versus oceanic) are important because they are keys to constraining the range of Hadean tectonic processes.

Additional insight into the nature of early mantle depletion processes has been gained from the documentation of variations in the Lu-Hf isotopic system (based on the decay of $^{176}$Lu to $^{176}$Hf) and in the short-lived $^{146}$Sm-$^{142}$Nd system (based on the decay of $^{146}$Sm to $^{142}$Nd). The Lu-Hf system has the advantage of zircon as its chief host mineral. Zircon, besides being resistant to isotopic resetting, carries most of the Hf in a whole rock. With a low Lu/Hf, it has a measurable $^{176}$Hf/$^{177}$Hf that requires little correction to establish the initial value, and can be dated independently with the U-Pb system (see detailed discussion below). The $^{146}$Sm-$^{142}$Nd system has the advantage of being able to record time-averaged Sm/Nd that predates much later metamorphic impacts on Sm/Nd that lead to inaccurate initial $^{143}$Nd/$^{144}$Nd. Initially, studies of Hf isotopes in zircons and whole rocks appeared to support the existence of the early depleted mantle reservoirs evident in the $^{147}$Sm-$^{143}$Nd isotopic system from whole rocks. This is because they were compatible with systematic, early separation of continental and/or oceanic crustal components. For example, initial Hf isotopic composition of juvenile crustal rocks, expressed as $\varepsilon_{Hf}$, was about twice that seen with the Nd system (e.g., $\varepsilon_{Hf} \approx 2 \times \varepsilon_{Nd}$)—a value close to the slope of the modern Nd-Hf array (Vervoort et al., 1996; Vervoort and Blichert-Toft, 1999; Bennett, 2003). This simple picture was subject to the accuracy of the poorly determined $^{176}$Lu decay constant. When the early Hf data are recalculated with the 4% lower decay constant proposed by Nir-El and Lavi (1998) and subsequently confirmed by others (Scherer et al., 2001; Söderlund et al., 2004; Amelin, 2005), the Hf-Nd correspondence disappeared for the oldest terrestrial rocks (cf. Bennett, 2003) because the recalculated initial Hf ratios were now much less radiogenic and thus much lower relative to Nd-based mantle evolution models (e.g., Kramers, 2001).

$^{146}$Sm-$^{142}$Nd isotope system studies aimed at constraining the timing of early mantle depletion also corroborated the creation of depleted reservoirs in the first 200 m.y. of Earth history, but suggested that they were sampled rarely and unevenly (McCulloch and Bennett, 1993; Bennett et al., 2007b; Caro et al., 2003, 2006; Boyet et al., 2003; Sharma and Chen, 2004) perhaps because of dilution of the depleted source signatures by mantle-mixing processes (e.g., Bennett et al., 2007b). The recent discovery (Fig. 4A) that all chondrites and most eucrites have an average 20 ppm lower $^{142}$Nd/$^{144}$Nd than the average of all measured terrestrial igneous rocks (Boyet and Carlson, 2005, 2006; Andreasen and Sharma, 2006) suggests that if a chondritic model for Earth's composition at its accretion is valid (Bennett et al., 2007; Carlson et al., 2007), then an enriched component must have been removed from Earth's upper mantle before the production of any of the Hadean or Archean crust preserved today as rocks or zircons (Fig. 4B) (Boyet and Carlson, 2005, 2006). Taken together, both the refined $^{176}$Lu decay constant and the meteorite $^{146}$Sm-$^{142}$Nd data require that the Nd-Hf systematics and the identity of Neoarchean to Paleoarchean juvenile rocks be reconsidered for their Hadean tectonic implications.

Geochemically, the average depletion reflected in the initial $^{143}$Nd/$^{144}$Nd isotopic compositions of Eoarchean to Paleoarchean juvenile crustal rocks of +2.5 ($\varepsilon_{Nd}$) is best explained by separating an early enriched reservoir (EER) from the newly accreted, still molten, mantle within the first 30 m.y. of Earth history. The lack of negative $\varepsilon_{Nd}$ values reported for Eoarchean crustal rocks suggests this EER was not continental crust but rather enriched mafic to ultramafic material that was segregated at the base of the mantle (perhaps in the D″ layer) or in the lower mantle (Fig. 4) (Boyet and Carlson, 2005, 2006). In this model, the entire upper mantle would have contained both a 20 ppm average $^{142}$Nd/$^{144}$Nd anomaly and an $\varepsilon_{Nd}$ = +2.5 (250 ppm) $^{143}$Nd/$^{144}$Nd anomaly resulting from evolution subsequent to the early separation of the EER. As such, it would be a viable source for all Hadean and Eoarchean juvenile crustal rocks.

The timing of the separation of the EER, its putative composition, and the formation of the Moon are closely connected (e.g. Boyet and Carlson, 2005) and important to constrain the composition of the Hadean mantle. Geochemical research on the Moon and Hadean Earth is active, and the understanding is evolving rapidly. Recent work on the Hf-W system in lunar metals shows a remarkable homogeneity and an $^{182}$W/$^{184}$W isotopic composition identical to the Earth's mantle (Touboul et al., 2007). This shows that the Moon formed after core formation on the Earth, further strengthens the compositional similarities between the Moon and the Earth's silicate mantle, and indicates that the Moon must have formed after $^{182}$Hf had decayed away, or 62 m.y. after formation of the Solar System (Brandon, 2007; Touboul et al., 2007). The oldest lunar samples (4.2–4.4 Ga) have similar suprachondritic signatures in both their initial $^{142}$Nd/$^{144}$Nd and $^{143}$Nd/$^{144}$Nd isotopic compositions (Boyet and Carlson, 2005), whereas younger lunar basalts (3.1–3.9 Ga) record chondritic initial $^{142}$Nd/$^{144}$Nd, but both supra- and subchondritic initial $^{143}$Nd/$^{144}$Nd values (Rankenburg et al., 2006). These datasets imply that lunar basalt sources were in equilibrium some 200 m.y. after formation of the Moon (Rankenburg et al., 2006). EER formation from the postaccretion terrestrial magma ocean better fits the lunar data (Boyet and Carlson, 2005; Rankenburg et al., 2006) chiefly because of the highly variable Nd isotopic compositions of differentiated lunar rocks and the newly refined later age for formation of the Moon, but it would require the EER to remain separated during the moon-forming, giant impact. Formation of an EER from the Moon-forming, giant-impact-generated magma ocean seems less feasible because of the extreme Sm/Nd required to explain subsequent terrestrial mantle compositions (Boyet and Carlson, 2005) and the requirement that it produce suprachondritic $^{142}$Nd/$^{144}$Nd and $^{143}$Nd/$^{144}$Nd in the oldest lunar rocks.

Nonetheless, the EER is modeled to be enriched in the light rare earth elements (LREEs) and depleted in the heavy rare earth

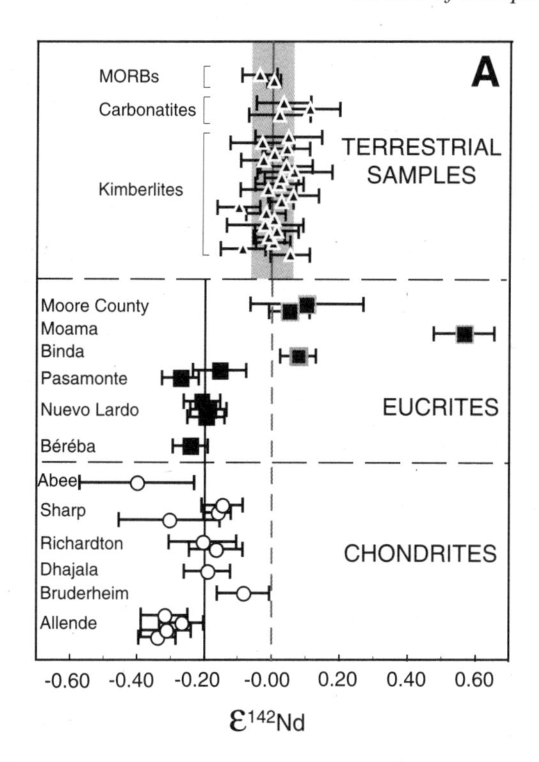


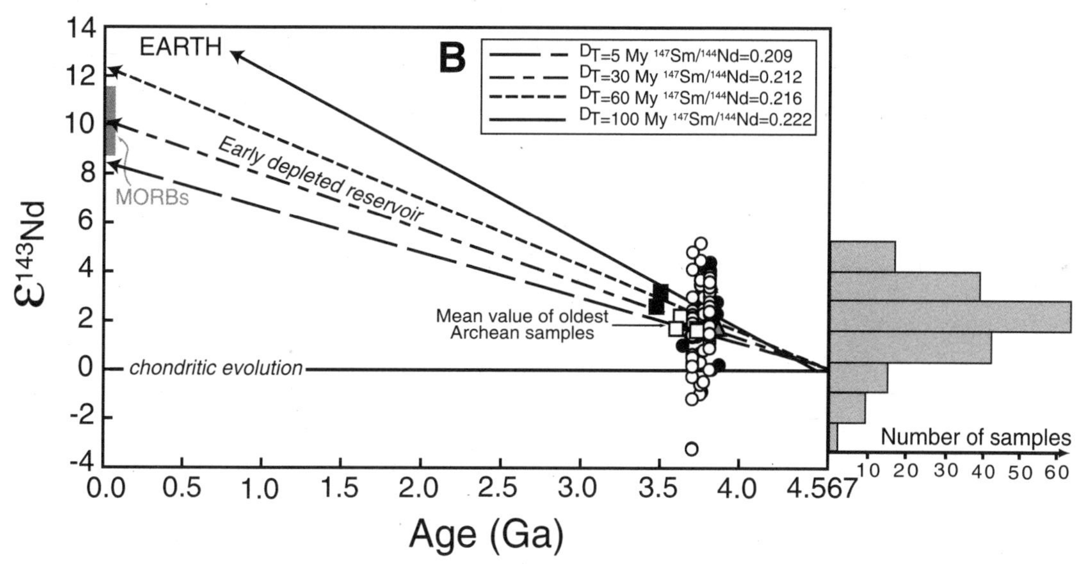


Figure 4. Composition of Earth and evolution of the mantle as constrained by $^{146}$Sm-$^{142}$Nd data. (A) Comparison of meteorites to terrestrial samples. (B) Model of the growth of an early depleted reservoir to explain the depletion of the Archean mantle source of some juvenile crustal rocks. After Boyet and Carlson (2005, 2006), reproduced from *Science* with permission.

elements (HREEs), so its separation also leads to an upper mantle with a radiogenic Hf isotopic composition that grows to an $\varepsilon_{HF}$ of +4 by 3 Ga. Thus, separation of the EER produces an early depleted reservoir that becomes the starting composition for the Hadean upper mantle and thus the base composition for the genesis of Archean juvenile crust. Excursions above and below in Nd and Hf isotopic composition are the isotopic signal that reflects the nature of the Hadean and Archean tectonic processes necessary to develop spatially separate geochemical reservoirs. Consequently, Eoarchean to Paleoarchean juvenile crustal rocks that have higher $\varepsilon_{Nd}$ than the average mantle growth model (Figs. 4B and 5A) result from (1) isolated, extreme LREE depletions of mantle sources (e.g., McCulloch and Bennett, 1994) and/or (2) secondary metamorphic perturbations of their $^{147}$Sm/$^{144}$Nd (Arndt and Goldstein, 1987), which leads to anomalously high calculated $^{143}$Nd/$^{144}$Nd for a given age (Moorbath et al., 1997). Apparently juvenile crustal rocks with lower $\varepsilon_{Nd}$, i.e., below the average mantle evolution model, result from either separate LREE-enriched sources or a short, but discernable, involvement of some source component(s) with low-Sm/Nd material (e.g., not totally juvenile crust).

The extension of these arguments to include the Lu-Hf data from Hadean and Eoarchean rocks and zircons, however, requires a more complicated scenario. A canon of isotope geology and igneous petrogenesis is that melting processes producing basalts from the mantle or producing felsic rocks from mafic crust fractionate Sm/Nd and Lu/Hf in a corresponding way; the melts have low Sm/Nd and Lu/Hf and the residues have higher Sm/Nd and Lu/Hf, which evolve with time to higher $\varepsilon_{Nd}$ and $\varepsilon_{Hf}$ isotopic compositions. Another canon is that the fractionation is more extreme for crustal melting because the Sm/Nd and Lu/Hf in the felsic melts relative to the basalts are much lower than for the ultramafic to mafic melting that produces the original mafic crust. Ideally then, the Hf isotopic composition of whole rocks and especially zircons should be able to support the interpretations based on the $^{143}Nd/^{144}Nd$ isotopic data. But with recent revisions to the Lu decay constant (Nir-El and Lavi, 1998; Scherer et al., 2001; Söderlund et al., 2004; Amelin, 2005), Eoarchean to Paleoarchean juvenile crustal rocks do not appear to have been derived from a depleted mantle reservoir in the Lu-Hf isotopic system that corresponds to the depleted mantle required by the $^{147}Sm$-$^{144}Nd$ isotopic system (Figs. 5A and 5B). Rather, they derive from one that is chondritic to slightly LREE-enriched (e.g., $\varepsilon_{Hf} = 0$ to $-4$; Fig. 5B) (Bennett, 2003; Caro et al., 2005; Bennett et al., 2007; Kramers, 2007), thus violating the basic Nd-Hf canons (whole-rock Sm/Nd perturbations accompanying alteration notwithstanding; see discussion above) and presenting a major challenge for understanding tectonic and petrogenetic processes for early Earth.

On the basis of the evidence presented above, the formation of either basaltic or granitic protocrusts and their separation from the mantle should form residual upper-mantle reservoirs that on remelting produce rocks that have a positive $\varepsilon_{Nd}$ and even more positive $\varepsilon_{Hf}$. Furthermore, the effects should be more pronounced if granitic crust was involved. Therefore, despite the existence of Hadean zircons, scenarios that invoke the Hadean formation and sequestering of volumetrically sizeable continental crust (e.g., Armstrong, 1981, 1991; Harrison et al., 2005) are inconsistent with the lack of Archean rocks derived from mantle reservoirs that had a high $\varepsilon_{Hf}$ and a high $\varepsilon_{Nd}$ (Figs. 5A, 5B, and 6). In addition, sizeable amounts of basaltic or granitic crustal components could not have been recycled into the upper-mantle sources of juvenile Archean crustal rocks because this process would have produced a similarly more negative $\varepsilon_{Nd}$ (e.g., evidence of LREE-enriched crustal material) compared to $\varepsilon_{Hf}$, which again is not seen in the data. Therefore, the oldest Archean juvenile crustal rocks can be explained by the high-pressure segregation of a phase that fractionates Sm/Nd differently from Lu/Hf, such as Mg-silicate and Ca-silicate perovskite (Figs. 5A, 5B, and 6). Early work on Mg-silicate and Ca-silicate perovskite partitioning (Kato et al., 1988a, 1988b) showed that substantial fractionation (>10%) of either phase from a magma ocean would lead to nonchondritic trace element ratios for many elements unless the perovskite phases were homogenized by mantle convection (Kato et al., 1988b; Drake et al., 1993). More recent work corroborates this effect and sets a new limit for Ca-silicate perovskite fractionation of less than 8%–11% (Corgne et al., 2005; Liebske et al., 2005). Caro et al. (2005) recently recognized that even at these low percentages, Ca-silicate perovskite crystallization, in particular, substantially lowers the Lu/Hf while raising the Sm/Nd of the resulting melt. This could have produced a resultant mantle with chondritic or bulk silicate Earth $\varepsilon_{Hf}$, but suprachondritic or depleted $\varepsilon_{Nd}$. The likely magma ocean candidate in which such fractionation may have occurred was the one created by the Moon-forming giant impact (Canup and Asphaug, 2001), which occurred after the separation of the EER (but see uncertainties in timing of EER separation, discussed above).

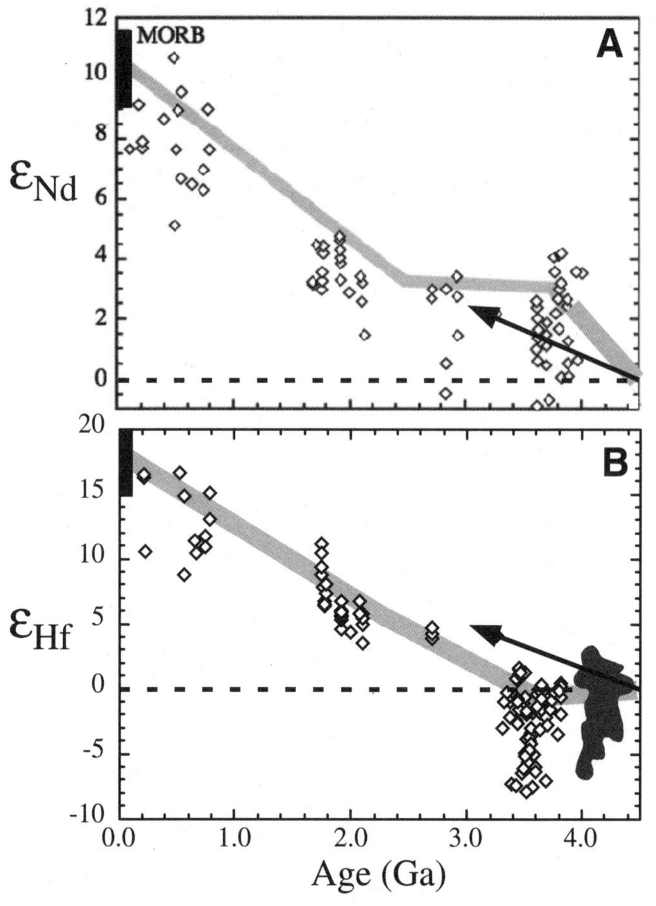


Figure 5. Nd isotopic composition (A) and Hf isotopic composition (B) of Archean to recent mantle-derived rocks and juvenile granitoids (both shown as open diamonds and the oldest samples on B containing some granitoids with older crustal components) versus age. Hadean zircons are the dark gray field in B. Range of MORB shown by black bars. Both A and B have been modified from Bennett (2003), Figures 1 and 3 therein (respectively), and the $\varepsilon_{Hf}$ in B has been recalculated using the decay constant proposed by Scherer et al. (2001) and the chondritic composition for $^{176}Hf/^{177}Hf$ and $^{176}Lu/^{177}Hf$ given in Blichert-Toft and Albarede (1997). The dashed line represents chondritic evolution, the thick gray line the evolution of the mantle sources for juvenile rocks, and the arrow the growth of the early depleted reservoir (Boyet and Carlson, 2005). Primary data sources for Nd are Baadsgaard et al. (1986), Jacobsen and Dymek (1988), Collerson et al. (1991), Bennett et al. (1993), Bowring and Housh (1995), Moorbath et al. (1997), Vervoort and Blichert-Toft (1999), and the compilation of Shirey (1991). Primary data sources for Hf are Salters (1996), Amelin et al. (1999, 2000), Vervoort and Blichert-Toft (1999), and Harrison et al. (2005).

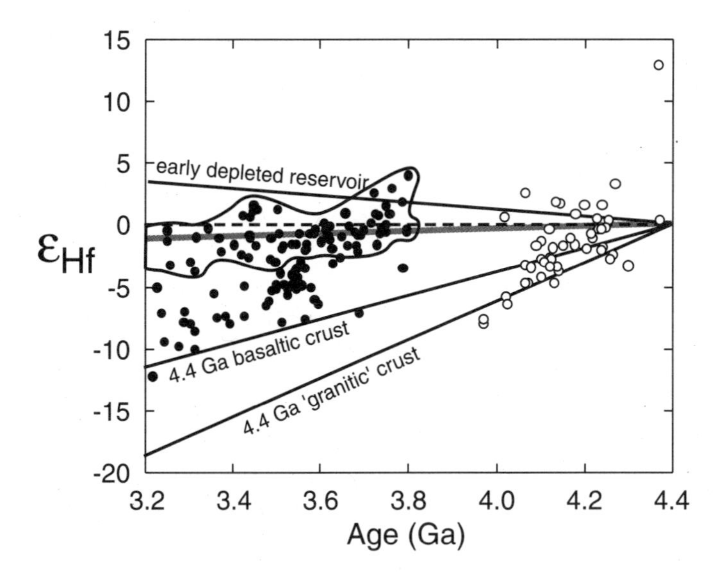


Figure 6. Initial Hf isotope systematics of Hadean and Eoarchean zircon after Kramers (2007) and Kamber (2007) with original data sources given in the caption to Figure 5. $\varepsilon_{Hf}$ was calculated from a present-day chondritic $^{176}Hf/^{177}Hf$ of 0.282772 and $^{176}Lu/^{177}Hf$ of 0.0332 with a $^{176}Lu$ decay constant of $1.865 \times 10^{-11}$ yr$^{-1}$. Early depleted reservoir evolution (separation at 4.5 Ga) after Boyet and Carlson (2005) projects to modern N-MORB Hf isotope composition. Two types of early Hadean (4.4 Ga) crustal reservoirs were modeled to separate from chondritic composition with $^{176}Lu/^{177}Hf$ of 0.021 (basaltic) and 0.012 ("granitic"). Note that all initial Hf isotope compositions of Eoarchean zircons from the Acasta, southern West Greenland, Barberton, North China, Yangtze, and the western Superior localities fall into the fields defined by the early depleted reservoir and the *mafic* Hadean crust. Also, the bulk of the depleted juvenile whole-rock data (inside balloon) straddles estimates of mantle evolution (gray line) following Ca-silicate and Mg-silicate perovskite fractionation from the magma ocean generated by the Mars-sized impact (Caro et al., 2005).

## The Oldest Sialic Rocks: Evidence for the Transition to Plate Tectonics in the Archean

The only samples presently available to us from the Hadean eon are ancient detrital zircons preserved in substantially younger sedimentary rocks, key evidence from which is discussed below. Although there may be significant petrogenetic differences between the host rocks for the Hadean zircons and the first preserved terrestrial rocks from the Eoarchean, the latter may still be used to provide some clues about the nature of Earth's earliest tectonic regime and its transition to a style we might loosely term "plate tectonics." Particularly important among these earliest rocks are the extensive 3.6–3.8 Ga tonalite-trondhjemite-granodiorite gneisses and supracrustal rocks of southern West Greenland, the Pb isotopic systematics of which have been used to propose a single-plate model for the Hadean (Kamber et al., 2003a, 2005b). This approach is based on the much faster decay of $^{235}U$ compared to $^{238}U$, which results in more rapid production of radiogenic $^{207}Pb$ relative to $^{206}Pb$ in early Earth, defining the characteristic shape of the $^{207}Pb/^{204}Pb$ versus $^{206}Pb/^{204}Pb$ growth curve shown in Figure 7 (see also Stacey and Kramers, 1975; Kramers and Tolstikhin, 1997).

When plotted in this coordinate space, Eoarchean samples from southern West Greenland define a broad array between two ca. 3.7 Ga reference isochrons. The lower margin of this array is defined by 3.65–3.75 Ga tonalite-trondhjemite-granodiorite (Amîtsoq) gneisses from the Godthåbsfjord region, while the upper margin is defined by chemical sediments (banded iron formation and chert, as well as galenas) of the 3.7–3.8 Ga Isua greenstone belt. Clastic metasedimentary rocks from the Isua greenstone belt, together with some amphibolitic gneisses and ca. 3.8 Ga tonalite-trondhjemite-granodiorite gneisses from south of the Isua greenstone belt, scatter between these two boundaries. The lower trend intersects the Pb growth curve at essentially the same age as the age of the regression line itself, indicating that these rocks were derived from a typical mantle Pb isotope reservoir, which has evolved predictably to the present day. In contrast, appropriate age regression lines through other sample sets intersect the growth curve of this reservoir at ages that are unrealistically young. For example, in the case of >3.7 Ga banded iron formations and cherts, this intersection occurs at <3.4 Ga (Fig. 7). The more radiogenic (specifically, elevated $^{207}Pb$) composition of these samples requires long-term (~500 m.y.) isolation of their Pb isotope source reservoir with a time-averaged $\mu$ ($^{238}U/^{204}Pb$ extrapolated from the present) value of 10.5, considerably higher than that typical of convecting mantle ($\mu \approx 8$).

In order to physically achieve this isolation in a reservoir that ultimately is accessible to surface weathering, Kamber et al. (2003a) proposed that this reservoir might have represented a stable crustal lid of basaltic composition that existed for most of the Hadean. In this case, earliest Earth could have been a single-plate planet without active convection and subduction processes (Kamber, 2007; Kramers, 2007). Additional supporting evidence for the crustal lid hypothesis comes from solar rare gas composition of the deep mantle sampled in plumes, which requires a prolonged period of surface exposure (Tolstikhin and Hofmann, 2005). The presence of a stagnant lid in the Hadean is by no means certain (e.g., Davies, 2006), and it would in any case be stable only in the absence of convection as a terrestrial heat loss mechanism. There is also the potential of impacts to break up the crust, and the crustal reprocessing evident in the Hadean zircon record must be explained. Both argue for some local recycling of the lid (see below). Nonetheless, if a stagnant lid occurred, the obvious implication would be that subduction tectonics, which in effect defines the tectonic regime of a multiple-plate planet, did not operate during the Hadean.

The Pb isotopic data provide further insights into the actual transition from a possible single-plate crustal lid to a multiple-plate

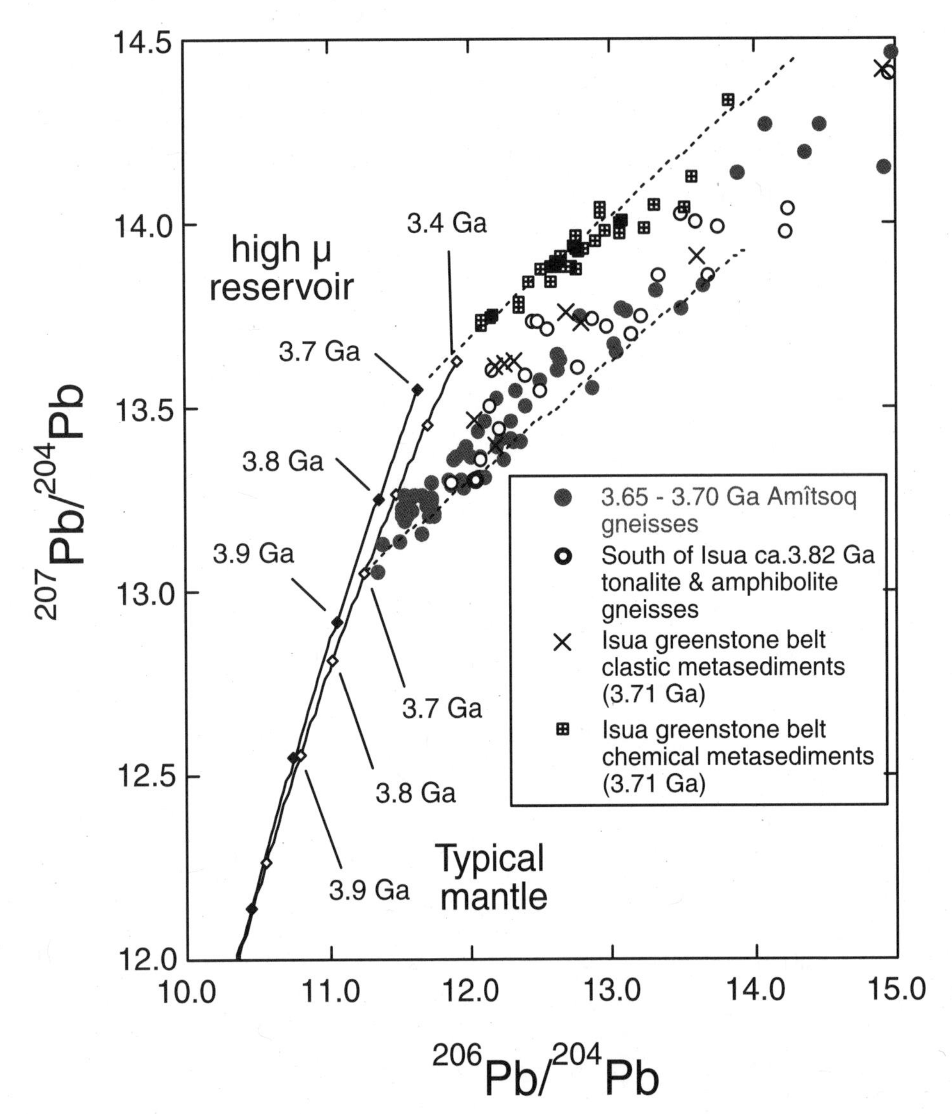


Figure 7. Pb isotope systematics of Eoarchean rocks of southern West Greenland (modified from Kamber et al., 2003a). A 3.65 Ga regression line through Godthabsfjord Amîtsoq gneisses intersects the typical mantle evolution curve (open diamonds; after Kramers and Tolstikhin, 1997) at the expected age of ca. 3.67 Ga. By contrast, 3.71 Ga chemical metasediments intersect the same mantle evolution line at a much younger, implausible date of <3.4 Ga, implying that their Pb had previously evolved in a high-U/Pb environment for several hundred million years. This is approximated with the "high-μ reservoir" growth curve (solid diamonds; see Kamber et al., 2003a, for details of growth curve). This required separation from the mantle at 4.3 Ga with a μ of 10.5. Eoarchean clastic metasediments incorporated Pb from both sources. Ca. 3.82 Ga tonalite and amphibolite gneisses from south of Isua also have an affinity with the high-μ reservoir.

system. Of particular significance are the ca. 3.8 Ga tonalite-trondhjemite-granodiorite (TTG) gneisses from south of Isua. Near this area, remnants of Earth's oldest ophiolite have been discovered (Furnes et al., 2007). More importantly, these rocks have typical trace element characteristics expected of magmatic rocks produced above a subduction zone (Kamber et al., 2002), such as enrichment in fluid-mobile incompatible elements (B, Li, Pb, U) and depletion in Nb and the heavy rare earth elements. These depletions can be explained by melting of garnet amphibolite (e.g., Foley, this volume). The TTGs also contain a Pb isotopic memory of the high-μ Hadean reservoir. This dichotomy can be explained if these earliest preserved TTGs represent some of the first true subduction-related rocks in which the fluid-mobile, but incompatible, element Pb was derived from the initial subduction and melting of the high-μ basaltic reservoir. If so, the inherent instability of the early basaltic crustal lid following initiation of global subduction would have resulted in its rapid recycling into the mantle within 100–150 m.y., so that no discernable isotopic trace of its former existence was imparted to the 3.65 Ga Amîtsoq TTG gneisses.

An inevitable consequence of voluminous TTG magmatism marking the onset of subduction tectonics between 3.75 and 3.65 Ga is the formation of cratonic nuclei capable of preserving some vestige of the Hadean crust. In this way, the high-μ signature developed during the lifetime of the Hadean crustal lid was tapped at a much later time and is evident in long-recognized (Oversby, 1975) high-μ cratons that are formed largely of Meso- to Neoarchean rocks, i.e., the North Atlantic, Slave, Wyoming, Yilgarn, and Zimbabwe cratons. In a similar way, physical relicts of the Hadean protocrust, most notably 4.0–4.4 Ga zircon from the Jack Hills, were eventually liberated by unroofing and erosion. The significance of these zircon grains for evolution of earliest Earth is the subject of great debate as discussed below.

## Arc-Like Juvenile Crustal Growth Starting in the Mesoarchean-Paleoarchean

Nd and Hf isotopic data for rocks of direct mantle derivation and juvenile crustal rocks from the Mesoarchean onwards define a trend toward increasing $\varepsilon_{Nd}$ and $\varepsilon_{Hf}$ as the rocks get younger, culminating in isotopic compositions that approach the composition of modern MORB (Figs. 5A and 5B). During this time, the $\varepsilon_{Nd}$ and $\varepsilon_{HF}$ take on the roughly one-to-two correspondence expected from the normal igneous fractionation of $^{147}Sm/^{144}Nd$ versus $^{176}Lu/^{177}Hf$ during mantle and crustal melting. This suggests that crustal formation processes at the top of the mantle (e.g., low pressure and not associated with a magma ocean as hypothesized for the Hadean) are dominating the composition of the progressively depleting mantle. The complementary nature of the trace element composition of continental crust compared to the depletions estimated for the MORB source (Hofmann, 1988; Workman and Hart, 2005), the very survival of ancient continental crust in cratons, and the increasing areal accumulations of continental crust through time (Hurley and Rand, 1969) have long led to the idea that irreversible continental crustal extraction has caused a progressive depletion of the upper mantle through time that can be traced by changes in its isotopic compositions in the $^{147}Sm$-$^{143}Nd$ and Lu-Hf systems.

Nb/Th and Th/U of mantle-derived rocks (Figs. 3A and 3B) echo the beginning of this major, unidirectional change in mantle composition near the same time as, or perhaps slightly before, the change seen in the Nd-Hf isotopic curves. Nb versus Th is a compelling tracer of slab-wedge interaction in subduction zones, whereas U versus Th is a compelling tracer of preferential incorporation of U into the altered oceanic crustal slab (as discussed above). The Nb-Th-U data trace the injection of oceanic slabs into the deeper mantle circulation, and this is an important adjunct of the arc crustal growth process and a strong indicator of plate subduction. Thus the Nb-Th-U data support the onset of continental crustal growth by the conventional arc process in an entirely complementary way to the Nd-Hf curves; the former records the long-term effects of slab additions to the mantle, whereas the latter records the long-term effects of crustal melts extracted from the mantle.

Arc models of continental crustal growth have long been advocated by studies of Pb isotopes in crustal hydrothermal ore deposits (Stacey and Kramers, 1975; Zartman and Doe, 1981; Zartman and Haines, 1988) and rare earth elements in continental sediments (Taylor et al., 1981; Taylor and McLennan, 1995). The crust itself also carries a record of the onset of the arc process in the change of the common Pb isotopic composition of galena and feldspar during the Mesoproterozoic, which requires an increase in time-averaged U/Pb of the crustal sources of these low-U/Pb minerals (Stacey and Kramers, 1975; Tera, 1982, 2003). The Re-Os system is one radiogenic isotopic system that is a relatively poor recorder of this process. The high Os content of the mantle makes it much less sensitive to the Re increases resulting from the introduction of basaltic components added during slab recycling than the other isotope systems in which lithophile trace element daughter elements (e.g., Sr, Nd, Pb, and Hf) are greatly enriched in the crust relative to the mantle. This is perhaps the reason that return plume flow as seen in oceanic-island basalts only starts to show significant enrichments relative to the mantle after the Paleoproterozoic (Shirey and Walker, 1998) or Neoarchean (Bennett, 2003).

Nd and Hf isotopic data on the oldest Paleoarchean and Eoarchean rocks are difficult to interpret in the simple context of arc-growth/mantle-depletion/crustal-recycling models. The $\varepsilon_{Nd}$ of the oldest juvenile rocks have long appeared extraordinarily high, which supported an apparent constancy of $\varepsilon_{Nd}$ throughout the Mesoarchean (Fig. 5A). This feature in Nd isotopic data was ascribed (e.g., DePaolo, 1983; Chase and Patchett, 1988; McCulloch and Bennett, 1994) to inputs of crustal material with low Sm/Nd back into the depleted mantle in the period before the progressive rise in the $\varepsilon_{Nd}$ of the depleted mantle after the Neoarchean. While such crustal recycling may have happened, the Nd isotopic data are no longer especially good evidence for it. Nd isotopic data are obtained on whole rocks, which can be subject to age uncertainties and to metamorphic perturbations of their Sm/Nd (see above). Both features can lead to anomalously high estimates of initial $\varepsilon_{Nd}$. The $\varepsilon_{Nd}$ of +3 to +6 on 3.8 Ga rocks cannot be easily explained by the early separation of enriched material (whether it is continental crust or the EER) because the high Sm/Nd needed would lead to ultradepleted mantle reservoirs at younger ages, which are not seen. Furthermore, a similar constancy in the Mesoarchean of Nb/Th, Th/U, and the ratio of $\varepsilon_{Hf}$ to $\varepsilon_{Nd}$ is missing. This suggests that straightforward slab return flow into the general mantle circulation is likely not the cause of constant $\varepsilon_{Nd}$ evolution in the Mesoarchean. For the late Neoarchean and after though, both the Nd and Hf evolution curves show an ever increasing positive $\varepsilon_{Nd}$ and $\varepsilon_{Hf}$ coupled with an increase in Nb/Th and decrease in Th/U. This occurs during and just postdating the addition of large volumes of continental crust on Earth. This temporal connection and the well-established complementary nature of the isotopic composition and trace element content of the continental crust to depleted upper mantle suggest that permanent extraction of continental crust progressively depleted the upper mantle since at least the Neoarchean.

## Examples from Mesoarchean to Neoarchean Terranes

The arc model clearly is not the only way the continental crust has been hypothesized to form. Lateral accretion of oceanic plateaus (Boher et al., 1992), mantle plume upwellings (Hollings et al., 1999; Bedard, 2006; Fralick et al., this volume), subduction into differentiated oceanic plateaus (Benn and Moyen, this volume), large mantle overturns (Stein and Hofmann, 1994), and vertical differentiation due to density downwellings (Zegers and van Keken, 2001; Bedard, 2006; van Hunen et al., this volume) all have been suggested. Apparent Neoarchean crustal growth so rapid as to lead to "un-subduction-like" crustal growth rates (e.g., Reymer and Schubert, 1984) was taken as indirect justification for some of these nonsubduction crustal growth models. Since then, it has become clear that in some terranes, such as the western Superior province, these high growth rates resulted from incomplete sampling of magmatic rocks for the U-Pb zircon age record. More thorough sampling and dating reveals development of some Neoarchean continental crustal terranes over longer periods than previously inferred, approaching modern subduction rates.

So, at present, the arc model of crustal growth by plate tectonic processes is directly supported by observations of Mesoarchean to Neoarchean terranes (e.g., western Superior province, Canada; the Pilbara craton, Australia; the Dharwar craton, India; and the North Atlantic craton) that have geological evidence that they were formed by or closely associated with subduction or accretion by subduction (Krogstad et al., 1989; Smithies et al., 2005; Percival et al., 2006b; Polat et al., this volume; Wyman et al., this volume). The western Superior province in particular, provides the most striking such case. Geological/tectonic syntheses of the western Superior province (Card, 1990; Stott, 1997; Percival et al., 2004, 2006b; Pease et al., this volume) recognize three distinct terrane types that repeat in five linear crustal domains: (1) old remnants of continental crust, (2) oceanic domains, and (3) metasedimentary belts separating continental and oceanic domains (Fig. 8). The old remnants of continental crust were created chiefly in the Mesoarchean and must have existed independently as continental nuclei rafted into position in the Neoarchean entrained in younger oceanic lithosphere. Geological and geophysical observations strongly favor a plate tectonic model of building the western Superior province by operation of a southward-younging succession of subduction zones: (1) the presence of clear continental and oceanic domains indicating a Neoarchean plate tectonic Wilson cycle (Percival et al., 2006b); (2) juvenile mantle-derived magmas with source enrichments due to the introduction of fluids in the mantle wedge (Shirey and Hanson, 1984; Stern et al., 1989; Stern and Hanson, 1992); (3) calc-alkaline granitoid batholiths whose dimensions are comparable to those in modern continental arcs such as Sierra Nevada, Patagonia, and British Columbia (Percival et al., 2006b); (4) long strike-slip fault systems (Percival et al., 2006a) similar to those in California, British Columbia, and Alaska; and (5) shallowly dipping crustal and mantle seismic reflectors that trace relict subduction zones (van der Velden and Cook, 2005) and show fossil slabs preserved near the Moho (Musacchio et al., 2004). Attached to this subduction terrane is the most seismically well-defined mantle root of any craton (Grand, 1987; van der Lee and Nolet, 1997; Goes and van der Lee, 2002; van der Lee and Frederiksen, 2005).

Cratons such as the Superior province show that as far back as 3.1 Ga, plate tectonics strikingly similar to that operating on present-day Earth not only existed but resulted in a voluminous peak in global crustal production. This voluminous peak in crustal production ca. 2.7–2.9 Ga (McCulloch and Bennett, 1994; Condie, 1998) occurred as the mantle evolved to higher $\varepsilon_{Nd}$, $\varepsilon_{Hf}$, and Nb/Th (Figs. 3A, 5A, and 5B) from 2.8 Ga onwards. Presumably this change was made larger by the irreversibility of removing crust from the mantle and stabilizing it with depleted mantle keels. Nonetheless, the evidence that peaks in crustal production were produced by subduction seems to argue against plume models or mantle overturns (e.g., Stein and Hofmann, 1994) for direct production of large crustal volumes in the Mesoarchean because it is not evident how they would produce crustal rocks with clear subduction-like geochemical features.

## ZIRCON CONSTRAINTS ON ARCHEAN CRUSTAL RECYCLING

Understanding the impact of crustal recycling on the development of the modern mantle and crustal reservoirs requires accurate descriptions of their Hadean and Eoarchean counterparts. As noted above, these relationships have been primarily explored using whole-rock abundances of isotopes in the U-Pb, Sm-Nd, Lu-Hf, and Rb-Sr systems in Archean rocks, which are susceptible to change during alteration and metamorphism leading to uncertainty when extrapolating observed parameters to initial parameters (e.g., Vervoort et al., 1996; Moorbath et al., 1997). More recently, however, direct measurements of the Hf, O, and trace element abundances in individual zircon grains have been used to provide additional constraints on the Hadean-Eoarchean crust-mantle system that are more robust than measurements in any whole-rock system (Vervoort et al., 1996; Amelin et al., 2000; Machado and Simonetti, 2001; Griffin et al., 2002; Zheng et al., 2004; Cavosie et al., 2005; Davis et al., 2005; Halpin et al., 2005; Harrison et al., 2005; Nemchin et al., 2006).

### Trace Element and O Isotopic Systematics of Zircon

Studies of O isotopic and trace element variations in zircons have been interpreted to varying degrees to indicate the presence of differentiated, sialic crust and a liquid hydrosphere (e.g., Peck et al., 2001; Wilde et al., 2001; Cavosie et al., 2005) on Hadean Earth (see discussion below). Although the trace element abundances in individual zircons do not uniquely constrain the bulk composition of the rocks from which they were derived (e.g., Hoskin and Ireland, 2000; Whitehouse and Kamber, 2002; Coogan and Hinton, 2006), some of the most ancient zircons apparently retain primitive (mantle-like) O isotopic signatures (e.g., Cavosie et al., 2005; Nemchin et al., 2006). As noted by Valley et al. (2006), however, the retention of these relatively primitive ratios is best documented when O isotopic analyses can be spatially paired with U-Pb data (e.g., via ion probe). Using a spatially resolved data set helps reduce effects related to subsequent alteration and metamorphism, which can dilute the original signal via mixing at later times (e.g., Valley et al., 2006). The ability of individual zircons or specific domains within individual zircons to consistently preserve their original O isotopic composition in conjunction with zircon's well-established ability to preserve U-Pb isotopic systematics, despite even granulite facies metamorphism, provides important evidence that the Lu-Hf system in zircon can also preserve original signals especially in nonmetamict, low-U, low-Th zircons.

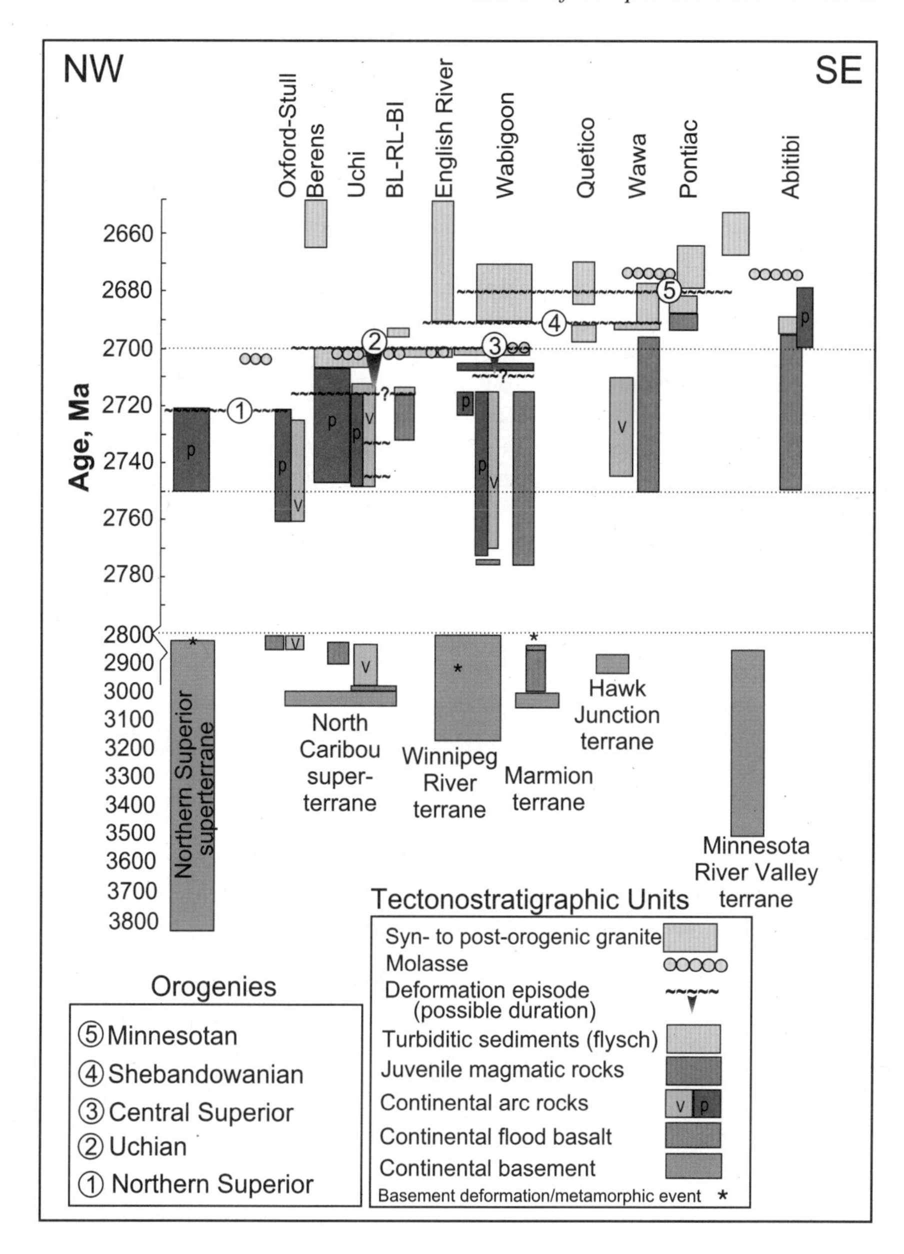


Figure 8. Repeating geomagmatic cycles of crustal evolution are evidence of subduction-accretion of the western Superior province, Canada, from oceanic and continental fragments (horizontally oriented names just above the legend). Vertically oriented names at top of figure designate the various terranes of the Superior province. Lithologic makeup of each terrane is shown directly under the terrane name. Note that the same sequence of tectonostratigraphic units is repeated five times at successively younger ages toward the southeast: continental arc rocks (v–volcanic, p—plutonic); juvenile magmatic rocks; sediments; and postorogenic granites. After Percival et al. (2006b), with permission of *Canadian Journal of Earth Sciences*.

## Lu-Hf Systematics of Zircon and Their Utility in Studies of Crustal Recycling

While the whole-rock Sm-Nd system has been a staple of Archean crustal studies for many years, the Lu-Hf system (in zircon) is rapidly becoming recognized as at least a valuable complement to the traditional Sm-Nd measurements for a number of reasons: (1) The Lu-Hf system has a shorter half-life that provides better temporal resolution; (2) Hf (a high field strength element) is relatively immobile during metamorphic and/or metasomatic processes; (3) Hf can be readily measured in very robust mineral reservoirs characterized by both high Hf and low Lu/Hf (i.e., zircon and baddeleyite) with no chemical preparation in the case of laser ablation and smaller potential errors of extrapolation to initial values; (4) the concordancy of the U-Pb data for an individual zircon or part thereof can be used as a guide to the extent to which the Hf isotopic systematics in an individual zircon (or part of a zircon) have been disturbed; and (5) quartz-saturated mafic rocks may contain zircon or if undersaturated, baddeleyite, providing a significant advantage over extrapolating the relatively high and more readily disturbed whole-rock Sm/Nd over billions of years to obtain initial ratios for mafic compositions. This is particularly true for Archean rocks because the differences in isotopic composition between models for chondritic and depleted sources become progressively smaller with age, regardless of the parameters chosen (e.g., Patchett et al., 2004). Although these differences appear larger for the more rapidly decaying Lu-Hf system, it is still critically important to measure the Hf isotopic system in the most robust and carefully characterized reservoirs. Consequently, the Lu-Hf system in zircon offers a better opportunity to obtain reliable initial isotopic ratios throughout the Hadean and Eoarchean. This system, therefore, is better able to distinguish the extent of juvenile and recycled components in Archean crust even though zircons typically form in evolved, felsic magmas that may have incorporated Hf from preexisting crust.

### Zircon Hf Evidence from the Oldest Zircons for Hf Isotopic Evolution in the Hadean and Eoarchean

At present there are only a limited number of studies in which U-Pb ages and Lu-Hf data have been reported for individual Eoarchean and Hadean zircons and even fewer that report O isotopic and trace element data for the same grains or parts thereof (e.g., Amelin et al., 1999; Cavosie et al., 2005; Harrison et al., 2005). Collectively these studies have focused on defining the extent of Hf isotopic variability by developing time-$\varepsilon_{Hf}$ databases for individual rocks or regions. For example, Amelin et al. (2000), Harrison et al. (2005), Bennett et al. (2007a), and Griffin et al. (2004) have reported $\varepsilon_{Hf}$ values for 2.5–4.3 Ga zircons from the Archean cratons of Australia. Reported values for $\varepsilon_{Hf}$ at 4.2 Ga range from +15 to –7 and imply the early development of reservoirs both higher ($^{176}Lu/^{177}Hf > 0.1$) and lower ($^{176}Lu/^{177}Hf < 0.01$) than modern depleted mantle (~0.04) and continental crust (~0.01). The least extreme Hf compositions from the oldest Hadean zircons appear to derive from a range of sources that include those consistent with the upper mantle after global differentiation of the EER (Fig. 6; slightly positive $\varepsilon_{Hf}$), whereas the more extreme compositions may reflect not-yet-mixed, short-lived, depleted and enriched reservoirs (e.g., Kamber, 2007). Many other younger Hadean zircons must have been derived from enriched sources (negative $\varepsilon_{Hf}$; Fig. 6) (Harrison et al., 2005; Kramers, 2007). In the case of the zircon derived from the very depleted reservoir noted in Harrison et al. (2005), the extrapolated value exceeds the modern depleted mantle value by more than an order of magnitude. Assuming that the reported Hf values are correct for the depleted reservoir, this early, "ultradepleted" reservoir must have been severely and rapidly altered by recycling and mixing with less depleted or even enriched in order to produce the more moderate values that characterize the modern depleted mantle (e.g., Salters and Stracke, 2004).

Harrison et al. (2005) proposed that the mixing of an early ultradepleted reservoir occurred with the enriched reservoir represented by $\varepsilon_{Hf}$ values <0, which these authors interpreted to be differentiated sialic crust based on Ti crystallization temperatures and felsic mineral inclusions found in some zircons. Though these arguments are strong with regard to the existence of a sialic crustal composition, they do not speak directly to the relative volumes of such materials in the Hadean, nor do they provide insight into how these sialic materials may have been recycled. In modern Earth, recycling of differentiated compositions is highly concentrated at convergent plate boundaries in which subduction of sialic detritus and/or tectonic erosion occurs. For either of these mechanisms to operate, however, much larger volumes of oceanic lithosphere are concomitantly recycled (e.g., Chase and Patchett, 1988). For Hadean Earth, we have no firm understanding of whether terrestrial geodynamics included plates, and if so, were plate boundary interactions similar to those of today? We do know, however, that differentiated rocks at mid-ocean ridges do produce zircon with some compositional similarities to zircons produced in convergent margin settings (e.g., Hoskin and Ireland, 2000; Coogan and Hinton, 2006), although several cycles of remelting may be needed to produce the closest similarities to continental zircons (Grimes et al., 2007).

A clear understanding of the geodynamic and geochemical implications of reported Hf isotopic compositions in Hadean zircons from Australia will ultimately require verification from other localities as well as evidence from other systems in ancient rocks that are consistent with these data (e.g., Kramers, 2007). To date, the very enriched and depleted reservoirs implied by the data of Harrison et al. (2005) have not been confirmed. Bennett et al. (2007), however, report $^{142-143}Nd/^{144}Nd$ from 3.6 to 3.9 Ga rocks from Greenland and Australia and $^{176}Hf/^{177}Hf$ from their zircons that show evidence for extraction from a reservoir with strong, early depletion in the Sm-Nd system, but with a chondritic signature in the Lu-Hf system (as recorded in zircon). A first-order mass balance between sialic crust and depleted mantle cannot account for this "discrepancy" and strongly suggests that a pairing of less fractionated mafic crust (e.g., stronger LREE fractionation than HREE fractionation) and depleted mantle may be more realistic for early Earth (Kramers, 2007). Dominance of a mafic differentiate rather than a sialic differentiate as the primary complement to the early depleted mantle would not, however, preclude the development of strongly LREE-depleted mantle sources, as seen in lunar samples. The (relatively) moderate $^{143}Nd/^{144}Nd$ and $^{176}Hf/^{177}Hf$ of the modern MORB source, the petrogenetically reasonable relationship between its Sm-Nd and Lu-Hf systems as evidenced by the Hf-Nd array of MORB (e.g., Vervoort and Blichert-Toft, 1999), and the lack of evidence from the mantle keels of ancient continental nuclei in the form of either xenoliths or younger magmatic rock compositions (see below), however, suggest that any ultradepleted reservoirs were short-lived. The formation of these depleted reservoirs, regardless of the degree of depletion, is undoubtedly tied to melt extraction. Whether this melt extraction occurred in a plate tectonic scenario (i.e., melting at ridges) or was related to formation of a magma ocean (Kramers, 2007) is not yet resolvable.

## CONTINENTAL LITHOSPHERIC MANTLE WITH SUBDUCTION IMPRINTS

Continental cratons are hallmark features of Earth because of their high topographic surface, old geologic age, sialic composition, seismic stability, fast P-wave velocity, and high proportion of mineral deposits. Ancient cratonic regions also are characterized by lithospheric mantle keels that are similar in age to the overlying crust and are thought to stabilize cratons over time.

Thus the process of making ancient continental crust requires the production of a full continental lithospheric section (termed the "continental tectosphere" by Jordan, 1981) with an attendant mantle keel, in some cases more than 150 km thick. Although present subduction produces some depleted mantle, no modern subduction zone seems capable of depleting the mantle to the extent that it is seen in the Archean. Furthermore, Meso- to Neoarchean subduction terranes with mantle keels (see above) provide evidence from samples of the mantle keel itself (e.g., kimberlite-borne xenolith suites) that some variant of subduction was the key to forming continental cratons. These samples of peridotite, eclogite, and diamond are important to the debate on the onset of plate tectonics on Earth because they come from the only portion of the mantle isolated from plate tectonic convection since the Meso- to Neoarchean.

## Evidence from Mantle Xenoliths

Recent reviews (Pearson et al., 2003; Carlson et al., 2005) cover the geochemical aspects of the two main types of xenoliths in kimberlites: peridotites (harzburgites, lherzolites, and wehrlites) and eclogites (the high-pressure metamorphic equivalent of basalt). Most studies have focused on the Kaapvaal craton of southern Africa, although detailed studies have been completed also on the Siberian, Tanzanian, and Slave cratons. Peridotites, although they make up the bulk of the continental mantle keel by volume (Schulze, 1989), carry ambiguous subduction signatures. They are characterized by strong depletion in melt components (e.g., Fe, Ca, Al) but also secondary enrichments in Si and the large ion lithophile trace elements. The latter is taken as primary evidence of the process of mantle metasomatism. The depletion is understood to be a primary feature of the peridotites due to high extents of melting and has been dated by Re-Os model age systematics (Carlson et al., 2005) as Neoarchean to Mesoarchean in the Kaapvaal. The depletion could be related to melting in the mantle wedge, especially if Archean subduction zones were hotter than and perhaps as wet as those today. Such differences, though, are not clear, as the debate over wet (subduction-related; Grove et al., 1999; Parman et al., 2004) versus dry (plume-related; Arndt et al., 1998) komatiites shows. Si enrichment in cratonic peridotites leads to high modal orthopyroxene contents (Boyd et al., 1997; Boyd, 1999) and has been linked to subduction through the partial melting of eclogite. Eclogites melt to form tonalitic magmas that react with the bulk peridotite, leaving it Si-enriched (e.g., Kelemen et al., 1998). In principle, large ion lithophile trace element enrichments in the peridotite would be a good indicator of subduction zone fluids except that they are nearly completely overprinted at much younger times by the high trace element content of the kimberlitic host magma. Although there are no direct age constraints on when Si enrichment occurs, it likely follows closely after the depletion for two reasons: This is when eclogite would be available to melt, and Si enrichment is not related to the very young trace element enrichments imparted by the host kimberlitic magma because kimberlites are silica-undersaturated. Thus, cratonic mantle peridotites carry evidence consistent with, but not requiring, Mesoarchean subduction.

On the other hand, eclogites do carry clear subduction signatures. The most direct of these signatures is oxygen isotopic composition. Eclogites from Roberts Victor were discovered in the 1970s to have both anomalously light and heavy $\delta^{18}O$ (MacGregor and Manton, 1986) similar to that seen in ophiolites. The recent summary of models for the formation of mantle eclogite (Pearson et al., 2003; Jacob, 2004) shows that this is a ubiquitous feature of mantle eclogites (Fig. 9). Eclogites, therefore, are not simply high-pressure magmas formed below or within the lithosphere, because temperatures at those depths would be in excess of 1100 °C and would not permit O isotopes to retain their low-temperature, fractionated compositions. Rather, eclogites carry a surface geochemical imprint of seawater alteration in the hydrothermal systems that interact with oceanic lithosphere to continental lithospheric mantle keel depths. Also consistent with the former residence of eclogite xenoliths at Earth's surface is the common occurrence of coesite and reduced carbon in the form of graphite or diamond. Eclogites, like the peridotites, are subject to infiltration from the host kimberlitic magma, thus requiring very careful treatment of clean mineral separates that can be used to reconstruct whole-rock compositions. Trace element contents of whole-rock eclogites reconstructed from their constituent minerals are roughly consistent with the patterns seen in basalt or gabbro from the depleted oceanic lithosphere (Jacob et al., 1994; Jacob, 2004). Radiogenic isotopes (U-Pb, Sm-Nd, and Re-Os) establish eclogite ages as Neoarchean to Mesoarchean (Jacob

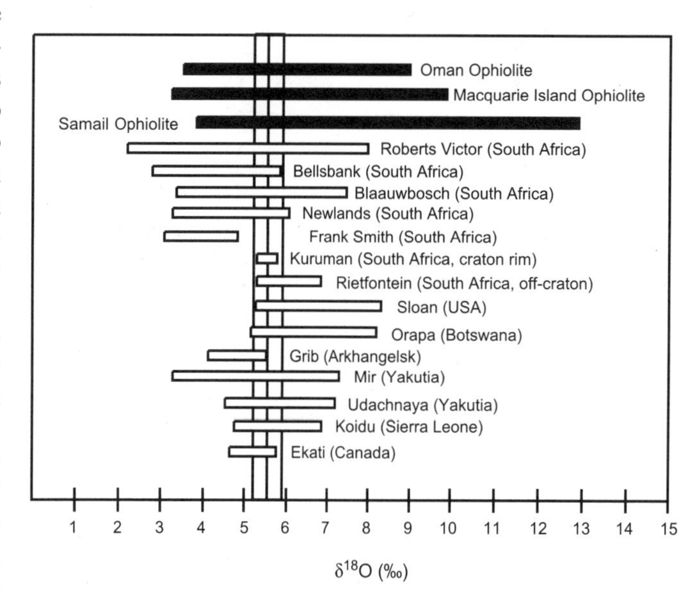


Figure 9. Oxygen isotopic composition ($\delta^{18}O$) of eclogite xenoliths (white bars) compared to that of ophiolites (black bars) and isotopically unaltered mantle (white vertical field). Note that most eclogite suites fall well outside the range for the mantle, similar to ophiolites, which is widely accepted as due to low-temperature seafloor alteration. After Jacob (2004), which cites the original data sources.

and Jagoutz, 1994; Pearson et al., 1995; Shirey et al., 2001; Barth et al., 2002; Menzies et al., 2003; Jacob, 2004), proving that they cannot be pieces of modern oceanic lithosphere and implying subduction by that time. Furthermore, the high Os and low Re content of eclogite xenoliths are typical of basaltic komatiites to komatiite rather than basalt (Shirey et al., 2001; Menzies et al., 2003). Komatiite or basaltic komatiite is an expected product of melting in a hotter mantle either at ridges (Takahashi and Brearley, 1990) or in plumes (Herzberg, 1999; Arndt, 2003), thus providing another link between eclogite and the Archean oceanic lithosphere.

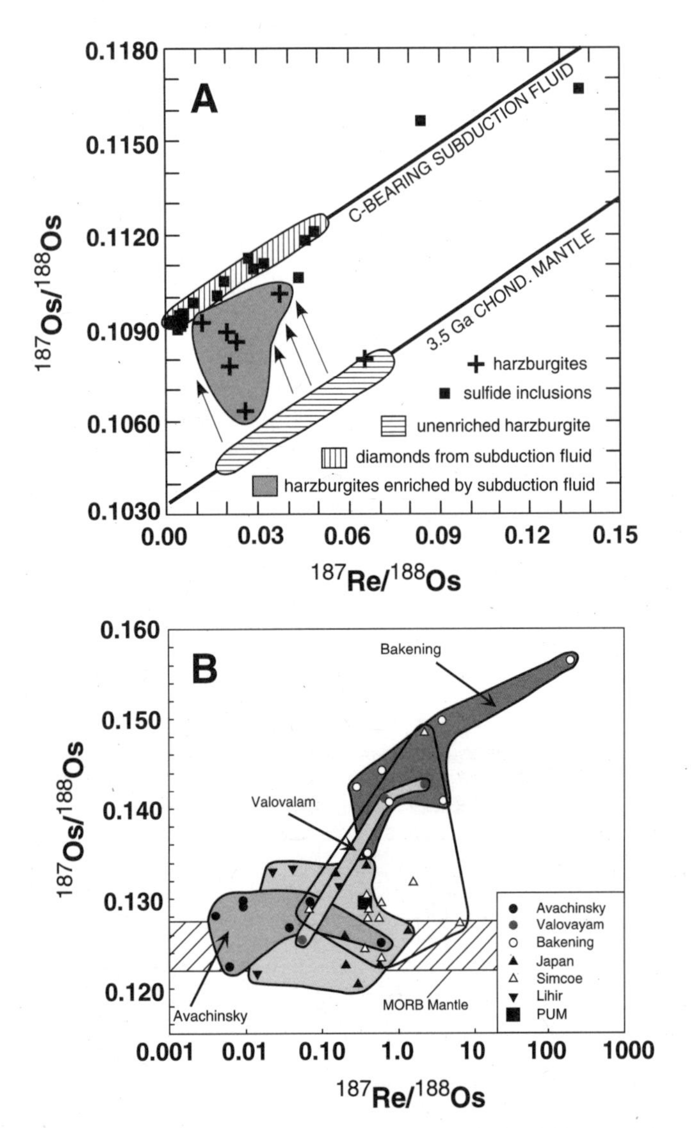


Figure 10. Comparison of Re-Os isotopic systematics for 3.52 Ga sulfide inclusions in diamonds and harzburgites from the Panda kimberlite pipe, Slave craton, Canada (A), versus recently erupted Kamchatka peridotites from the mantle wedge (B) (adapted from Widom et al., 2003; Westerlund et al., 2006). Horizontally ruled field and 3.5 Ga reference line in A show the expected position for harzburgites if they had chondritic Re-Os systematics. Note that only one harzburgite falls in this field; the other harzburgites have distinctly elevated Os isotopic compositions (within gray field) toward that of the diamonds that they host (vertically ruled field and reference line for C-bearing fluid). Kamchatka peridotites (Bakening, Valovalam, Avachinsky) and others from arc settings (Japan, Simcoe, Lihir) show similar enrichments in $^{187}Os/^{188}Os$ (PUM—primitive undepleted mantle).

## Evidence from Diamonds and Their Inclusions

Macrodiamonds from the continental lithospheric mantle carry two types of inclusion suites that have been amenable to trace element and isotopic analysis: silicates (garnet, olivine, orthopyroxene, and clinopyroxene) and sulfides (pyrrhotite, pentlandite, or chalcopyrite). Silicate inclusions have been divided into p-type (peridotitic, including both harzburgitic and lherzolitic compositions) and e-type (eclogitic) parageneses based on garnet (Cr, Al) and pyroxene (Na, Mg, Fe) compositions, whereas sulfide inclusions have been subdivided into p-type versus e-type based on the Ni content of their sulfide. Sm-Nd and Rb-Sr isotopes from silicate inclusions from both suites have allowed dating of their diamond hosts (see review of Pearson and Shirey, 1999) as Archean (e.g., Richardson et al., 1984) or Proterozoic (e.g., Richardson, 1986; Richardson et al., 1990; Richardson and Harris, 1997). The Sm-Nd isotopic data on silicate inclusions have been difficult to relate directly to subduction either because the use of model age systematics assumes the extraction of their host from a chondritic mantle reservoir or because the Sm-Nd isochrons with negative initial $\varepsilon_{Nd}$ compositions more typically reflect a lithospheric history of metasomatism with low-Sm/Nd fluids prior to diamond growth.

Sulfide inclusions analyzed for their Re-Os isotopic systematics provide much clearer age and petrogenetic links to subduction (Shirey et al., 2004). P-type sulfides have proven to be rare in the Kaapvaal craton. A remarkable suite of p-type sulfides from diamonds in the Panda kimberlite pipe, Slave craton, Canada, however, provides a Paleoarchean (3.52 Ga) isochron age and an elevated initial Os isotopic composition that is most readily explained by fluid enrichment in a mantle wedge (Fig. 10) (Westerlund et al., 2006). E-type sulfides from the Kimberley pool of the Kaapvaal craton show sources similarly enriched in Os isotopic composition that also are interpreted to reflect subduction-related enrichments (Richardson et al., 2001). Sulfur isotopic compositions of eclogitic sulfides from Orapa on the Zimbabwe craton show isotopically light values of $\delta^{34}S$ (Eldridge et al., 1991) and unusual Pb isotopic compositions (Rudnick et al., 1993) that have long been linked to low temperature, and thus implicate surficial sulfur. Mass-independent fractionations in $\Delta^{33}S$ (Farquhar et al., 2002a, 2002b) are further evidence that the sulfur was volcanogenic and obtained its distinctive mass-independent fractionation signature via photocatalyzed reactions in a low-$pO_2$ atmosphere that existed prior to 2.3 Ga (Farquhar et al., 2002a, 2002b). Subduction is by far the most plausible way to incorporate such sulfur. For the Kaapvaal craton in particular, an indirect but perhaps more extensive link to subduction is made by the association between diamondiferous eclogite xenoliths and eclogitic diamonds, the overlap in Neoarchean to Mesoarchean ages of both diamonds and eclogites, and the widespread occurrence across the craton of these sulfide inclusions (Shirey et al., 2001).

The trace element and isotopic compositions of macrodiamonds that host inclusions have been recently reviewed (Pearson et al., 2003). Briefly, macrodiamonds can be distinguished on the basis of the paragenesis of their inclusions as described above into peridotitic and eclogitic types. The p-type diamonds have C and N isotopic compositions centered roughly around mantle values ($\delta^{13}C$ = –6‰ to 0‰ and $\delta^{15}N$ = –10‰ to 0‰, except for one locality; see Fig. 11), whereas the e-type diamonds overlap these compositions but also carry a subpopulation that tails off to isotopically light carbon ($\delta^{13}C$ = –22‰) and heavy nitrogen ($\delta^{15}N$ = +7‰). The cause of these isotopic differences in carbon and nitrogen relative to the mantle is by no means clear, and it has been argued from mass balance considerations and the composition of isotopic end members that they result from intra-mantle fractionation (Cartigny et al., 1998, 2001). However, note that the position of some eclogitic diamonds in carbon and nitrogen composition is consistent with having incorporated some of the isotopic signatures of organic-rich sediments (Fig. 11) (Pearson et al., 2003).

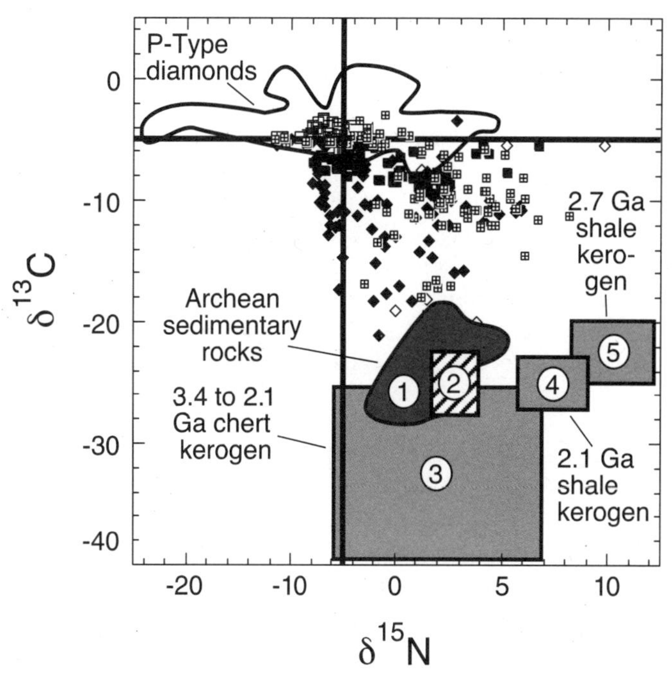


Figure 11. Carbon and nitrogen isotopic composition of macroscopic diamonds mined from kimberlite compared to the mantle (solid lines) and various supracrustal rocks and organic components. Separate symbols are e-type diamonds; p-type diamonds are within a field. Most of the diamond isotopic data are from the work of Cartigny (2005) with references as given in Pearson et al. (2003; see caption to Fig. 57 therein, after which this figure was modified). Composition of Precambrian and Phanerozoic sedimentary rocks (1) taken from Navon (1999) and the references therein. Modern subducting sediment compositions (2) from Sadofsky and Bebout (2004). Fields for 2.1–3.4 Ga chert kerogen (3), 2.1 Ga shale kerogen (4), and 2.7 Ga shale kerogen (5) estimated from the work of Jia and Kerrich (2004) and Marty and Dauphas (2003).

## HADEAN AND ARCHEAN OCEANIC LITHOSPHERE-HYDROSPHERE: EVIDENCE FOR EARLY WATER

The presence of liquid water at the surface of early Earth is important in the plate tectonic debate because of water's role in producing silicic magmas typical of continents through wet melting of mantle peridotite and basaltic crust. Detailed consideration of the ultimate source of terrestrial water is beyond the scope of this paper. Furthermore, sources as diverse as chondrites, comets, hydrous planetesimals, hydrous minerals, and gas-absorbing grains recently have been proposed by the cosmochemical and astrophysical community (Abe et al., 2000; Morbidelli et al., 2000; Kramers, 2003; Drake, 2005; Marty and Yokochi, 2006). Experimental data on water solubility and partitioning in mantle minerals and various geophysical arguments and measurements in nominally anhydrous minerals indicate a bulk Earth (including its surface water) content of 350–500 ppm water (Marty and Yokochi, 2006). Consideration of this water originating from chondritic sources requires the early Earth water content to be much higher, as high as 50 times the current ocean masses of water (Abe et al., 2000). The current estimate of much lower bulk Earth water content is possible due to the water lost during terrestrial evolution and the Moon-forming event.

Recent high-precision $^{142}Nd$ measurements in primitive meteorites (Boyet and Carlson, 2005) indicate terrestrial differentiation within 30–40 m.y. after the solar system condensation event. It is interesting to speculate how much water was retained by Earth after the early differentiation episode and the Moon-forming event, although it is generally believed that some volatiles were added via asteroidal and interplanetary dust input during later accretion. The relative abundances of the radiogenic, nucleogenic, and fissiogenic noble gas isotopes and, in particular, the Xe isotope record as measured in present-day MORB and plume sources are consistent with extensive loss of volatiles in the first 100 m.y. after Earth formation. The Xe isotope constraints also imply that during the Hadean the rate of loss of volatile elements was at least one order of magnitude higher than at present (Yokochi and Marty, 2005). These data also imply that Hadean Earth was more thermally active.

Elevated $\delta^{18}O$ in excess of typical mantle values observed in 4.1–4.3 Ga detrital zircons in Western Australia (Mojzsis et al., 2001; Peck et al., 2001; Wilde et al., 2001) imply water-rock interaction in crustal processes at least as early as these well-dated zircon ages indicate. One of these studies (Mojzsis et al., 2001) suggested that the elevated values were evidence of generation of granitic melts in a suprasubduction zone environment and thus were evidence for plate tectonic processes analogous to those of the present day operating in the Hadean. Subsequent studies have cast doubt on the veracity of the elevated values in unambiguously magmatic, undisturbed zircon (Nemchin et al., 2006) and/or questioned whether the hydrous granite origin for such zircon is a necessary conclusion from O data (e.g., Whitehouse and Kamber, 2002; Coogan and Hinton, 2006). Alternative suggestions to explain the elevated $\delta^{18}O$ values include burial of hydrated metabasalts to depth by volcanic resurfacing (Kamber et al., 2005b) or fractionation resulting from carbonation of basalts in a $CO_2$-rich Hadean atmosphere (Coogan and Hinton, 2006).

In order to evaluate the possible existence of an early liquid water ocean on Earth, Kramers (2003, 2007) considered the relative abundances, with respect to carbonaceous chondrites, of the volatile elements H, C, N, Cl, Br, I, Ne, Ar, Kr, and Xe that are presently concentrated in the outer Earth reservoirs such as atmosphere, hydrosphere, and sediments, and concluded that their elevated abundances including overabundances of H and Cl are best explained by the presence of liquid water at the surface early in Earth history.

## MAFIC-ULTRAMAFIC COMPOSITION OF THE EARLIEST RECYCLED CRUST

Consideration of the type of crust that was available to have been recycled on earliest Earth is critical to the debate on when plate tectonics started. It is central to whether we can say that plate tectonics operated from formation of the very first crust and constrains how early plate tectonics must have worked. Estimates for the composition of early crust range from silicic (e.g., granitic; Armstrong, 1981, 1991; Mojzsis et al., 2001; Harrison et al., 2005), to mafic-ultramafic (e.g., basaltic to komatiitic; Carlson and Shirey, 1988; Chase and Patchett, 1988; Takahashi and Brearley, 1990; Galer and Goldstein, 1991; Kamber, 2007; Kramers, 2007). A granitic crust implies differentiation, access of water to melt source regions, buoyancy, and a recycling mechanism such as erosion and sedimentation that can lead to entrainment of continental material into the oceanic lithosphere and eventually the mantle. A mafic to ultramafic crust implies direct mantle derivation, little first-stage differentiation, no necessary access of water to the mantle source, and a bulk recycling mechanism involving density such as lithospheric slabs or eclogitic blobs.

Armstrong (1981, 1991) was the main proponent of early, complete, and large-volume crustal differentiation and recycling. His model hinges on the continental freeboard argument of Wise (1974) extrapolated back into the Hadean, and on the mechanisms of continental erosion and sediment recycling at subduction zones, and it draws parallels to the early differentiation that occurred on the Moon and the terrestrial planets (Armstrong, 1981, 1991). Recent work on the oldest terrestrial zircons that show evidence of crustal reprocessing (e.g., Fig. 6) (Amelin et al., 2000; Harrison et al., 2005), rather cool magmatic temperatures (Watson and Harrison, 2005), and heavier-than-mantle $\delta^{18}O$ (Mojzsis et al., 2001, the arguments of Nemchin et al., 2006, above notwithstanding; Cavosie et al., 2005) has revived the idea of an early granitic crust and thus the early crustal growth models of Armstrong (e.g., Harrison et al., 2005). Because sediment subduction is inherent in the Armstrong model, its acceptance would argue that plate tectonics starts with the earliest differentiation of Earth, whereas its refutation would suggest plate tectonics starts at some later time. These questions are central to this volume.

While the separation of an early continental crust does explain some of the isotopic and geochemical features of the oldest zircons mentioned above, it is inconsistent with many features of Hadean-Eoarchean Earth and its juvenile rock record. Its key premise of continental freeboard (Wise, 1974) has dubious applicability to continents that had not been cratonized with thick mantle keels (e.g., the Hadean to Eoarchean). When freeboard arguments are reexamined, they preclude large volumes of continental crust in the Hadean and Archean (Hynes, 2001). As discussed in detail above ("Earliest Mantle Isotopic Signatures from Juvenile Crust"), separation of large volumes of continental crust would produce depletions in the Nd and Hf isotopic composition of the depleted mantle such that $\varepsilon_{Hf}$ would be about twice $\varepsilon_{Nd}$. Uncertainties about the pristine nature of the Nd and Hf isotopic compositions on the earliest rocks aside, this relationship is not seen for earliest Earth but rather after the Paleoarchean (Figs. 5A and 5B). Erosion and sedimentation is not an effective way to recycle continental crust without the high elevations to speed erosion. Not only are high elevations unlikely in the Hadean, but they should have left larger amounts of residual Hadean crust preserved at the surface than the detrital zircons occurring in the few small outcrop belts that currently exist. Careful examination of the oldest sediments (3.5–3.8 Ga) shows that they only contain zircons that are 100–300 m.y. older than the deposition age and no Hadean zircons (Fig. 12) (Nutman, 2001). This pattern precludes widespread existence of voluminous continental crust. The continent age versus surface area map first published by Hurley and Rand (1969) shows *preserved* crust, not crustal growth. Armstrong was able to mimic the histogram of crustal age versus crustal area preserved by appealing to more rapid ocean-floor creation, subduction, and mantle mixing in the Archean—all driven by higher Archean temperatures (Armstrong, 1981). Because subduction is widely accepted as a process to form continental crust, it not clear from the Armstrong model how or why subduction should destroy crust in the Hadean yet create it from the Neoarchean onward. Finally, implicit in the mechanism of continental crustal recycling of Armstrong is the necessity to recycle much larger amounts of oceanic lithosphere to incorporate the sediment into the mantle. Thus, the idea of a voluminous early continental or granitic crust that was fully recycled is not tenable.

Instead, the early crust on Earth was likely to have been mostly mafic to ultramafic (e.g., Kamber, 2007; Kramers, 2007), and its melting and recycling must have dominated geochemical processes in the Hadean. The Hadean zircon record itself is consistent with this scenario and does not provide unequivocal evidence for a dominantly granitic crust. In addition to being found in continental rocks, zircon also occurs today in oceanic settings where silica-saturated to -oversaturated melts are produced from silica-undersaturated parents: pegmatitic patches or plagiogranites that crystallize within gabbros, dacites that form as extreme differentiates of mid-oceanic-ridge magma chambers, and rhyolites or trondhjemites generated by differentiation or melting of oceanic-island basalt. The last petrogenetic setting, where hydrothermally altered basalt can be buried to the depths of wet

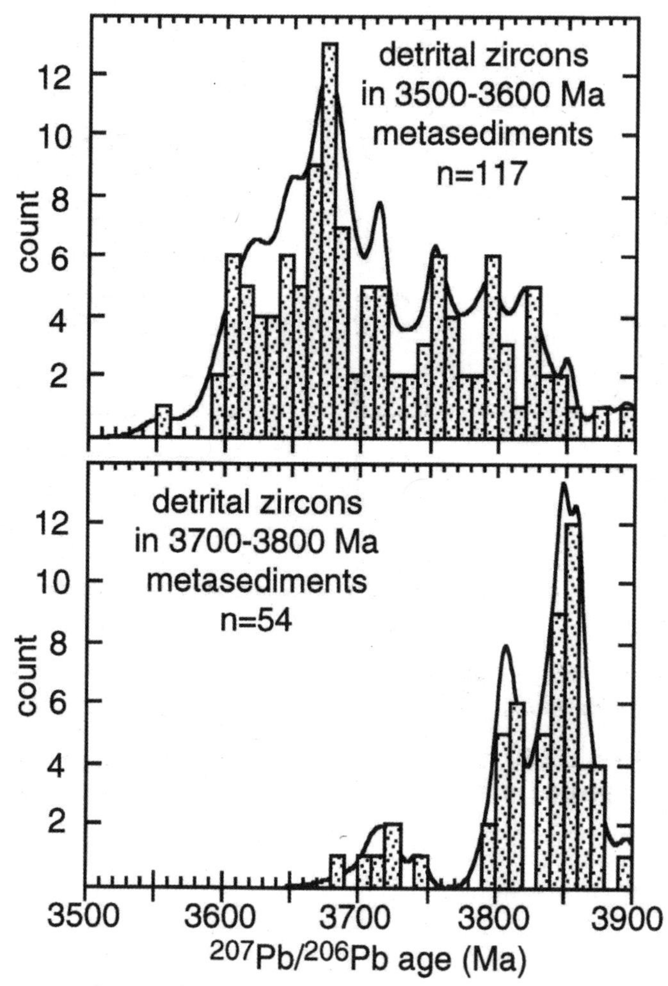


Figure 12. Histograms of detrital zircon in Mesoarchean sediments plotted versus $^{207}Pb/^{206}Pb$ age (with relative probability curves in background) showing a lack of grains older than 3.9 Ga. After Nutman (2001), with permission from *Precambrian Research*.

melting (e.g., Marsh et al., 1991) by the repeated addition of new basalt in shield-building stages, is interesting because it may be the most akin to Hadean conditions. The composition of some Hadean zircon does resemble the composition of modern oceanic zircons (Coogan and Hinton, 2006), and there is a population of least altered Hadean zircon that displays magmatic, mantle-like $\delta^{18}O$ (Nemchin et al., 2006). Many other Hadean zircons display inclusions of crustal minerals (e.g., quartz, K-feldspar) and much heavier $\delta^{18}O$ (Maas et al., 1992; Peck et al., 2001; Wilde et al., 2001; Cavosie et al., 2005), but these could have an explanation in the repeated remelting of newly formed silica-oversaturated parents, again under wet conditions. Thus, we interpret the Hadean zircon record chiefly as evidence of remelting of mafic to ultramafic crust under wet melting conditions that would have produced silica-oversaturated trondhjemitic to tonalitic melts from the first stage and ever more siliceous and granitic melts in subsequent stages. These nascent sialic components apparently lacked a well-developed, depleted mantle keel to preserve them and disaggregated rapidly, but remained near the surface to survive recycling. Such multiple processing could be areally restricted and shallow, leaving small terranes whose zircons were eventually collected into the restricted occurrences seen today. Each silica-oversaturated melt is a product of small degrees of partial melting, so rather than representing large volumes of granitic crust, the zircons may instead represent many tens to hundreds of times the volume of reprocessed mafic to ultramafic crust (e.g., Kamber, 2007). Thus these zircons reveal more about the hydrothermal/anatectic processes accompanying melting and alteration in the shallow lithosphere than about the early tectonic processes on Earth involving deeper portions of the upper mantle and clearly do not dictate continental crust as the dominant component of any EER.

## ONSET OF PLATE SUBDUCTION AND CONTINENT FORMATION

Although this paper has focused chiefly on the isotopic data, the question of when plate tectonics started on Earth and its form require an integration of these data with additional evidence from geophysics, trace elements, and petrology. Taken together, the evidence produces a self-consistent model for Earth's geologic history and a specific starting point for plate tectonics at ca. 3.9 Ga. This is not inconsistent with specific terrane studies (e.g., Benn and Moyen, this volume; Polat et al., this volume), key lithological associations (e.g., Foley, this volume; Wyman et al., this volume), and geologic syntheses (e.g., Brown, this volume; Condie and Kröner, this volume; Pease et al., this volume) that have plate tectonics operating by the late Mesoarchean or Neoarchean. Rather, it is apparent from the global isotopic and trace element record that plate tectonics, even restricted in extent, likely started earlier. Prior to ca. 3.9 Ga, earliest Earth underwent global dynamic and differentiation processes that were not plate tectonic, but that need to be considered to interpret the later rock isotopic record's implications for the start of plate tectonics: the separation of the core, the deep sequestering of the EER and the consequent formation of the early depleted reservoir, the giant impact to form the Moon, and the formation of an impact-generated magma ocean from which some perovskite fractionated (Table 1). Magma oceans inevitably convect rapidly and cool quickly, so the mantle would have solidified and the oceans condensed back to liquid water within a few thousand to at most a few million years. Kramers (2007) has proposed that the magma ocean would have frozen along its solidus from the bottom of the lower mantle upward, leading to a solidus-determined mantle that would have been gravitationally unstable. A mantle overturn would have ensued, producing large volumes of mafic to ultramafic crust above a stable, nonconvecting mantle that led to a "quiescent" Hadean period (Kramers, 2007). After the quiescent period, the accumulated heat from perhaps 400 m.y. of radioactive decay would initiate convection where the thermal gradient was greatest—from the top down. Throughout the quiescent

TABLE 1. SUGGESTED SUMMARY OF EARLY EARTH EVENTS

| Time (Ga) | Event | Evidence |
|---|---|---|
| 4.57 | Accretion | Short- and long-lived isotopic systems in meteorites |
| 4.57–4.51 | Core formation | Composition, Hf-W |
| | First magma ocean | Modeling, energetics |
| | Separation of early enriched reservoir (EER) | Sm-Nd, composition, Hf-W, Sm-Nd |
| 4.51–4.4 | Giant impact to form Moon | Composition, modeling |
| | Second magma ocean | Modeling |
| | Perovskite fractionation | Lu-Hf, Sm-Nd |
| | Mantle overturn? | Hypothesis |
| 4.4–3.9 | Oceanic lithosphere (noncontinental) | Hadean zircons, O, Lu-Hf |
| 3.9 | Onset of subduction, and first continental crust | Arc-like rocks, ophiolites |
| 3.5 | First mantle keels and permanent crust extraction | Nb-Th, Sm-Nd, samples |
| 2.5–3.1 | Terranes with direct subduction imprints | Preserved arcs, geophysics |

*Note:* This table is compiled from the literature discussed in the text. Ages are meant to serve as guides, not firm boundaries, since future research will produce new age constraints. The short- and long-lived radioisotopic systems that have been applied to meteorites include: Al-Mg ($^{26}Al \rightarrow ^{26}Mg$), Mn-Cr ($^{53}Mn \rightarrow ^{53}Cr$), Pd-Ag ($^{107}Pd \rightarrow ^{107}Ag$), Rb-Sr ($^{87}Rb \rightarrow ^{87}Sr$), Sm-Nd ($^{146}Sm \rightarrow ^{142}Nd$; $^{147}Sm \rightarrow ^{143}Nd$), Lu-Hf ($^{176}Lu \rightarrow ^{176}Hf$), Hf-W($^{182}Hf \rightarrow ^{182}W$), U-Pb ($^{235}U \rightarrow ^{207}Pb$; $^{238}U \rightarrow ^{205}Pb$).

period, there would have been no continents and Earth's surface would have been dominated by some form of stable, nonconvecting, oceanic lithosphere (Kamber, 2007; Kramers, 2007). This model has the attractions of being able to rapidly remove an excess of terrestrial heat, of a long period where little new crust of any kind was produced, and of providing a surficial reservoir in which to store the high-U/Pb protoliths of the first Archean sediments (e.g., Fig. 7).

The stability of this earliest oceanic lithosphere, however, is an open question because whether Earth's mantle underwent such a convective overturn is not known. Also, it is unclear how a quiescent, stable oceanic lithosphere would have withstood the late, heavy meteorite bombardment or been able to produce Hadean zircons throughout the period from 4.4 to 4.0 Ga without being continually remelted. If, instead, there was no mantle overturn, a more "uniformitarian" scenario can be envisioned where the oceanic lithosphere was produced throughout the Hadean by rapid mantle convection occurring in both ridge and plume modes. Crustal differentiation occurred, producing the early zircon record, but it must have happened around dispersed and recyclable ocean islands or plateaus where water could access the thickening basaltic-komatiitic piles and produce silica-saturated, relatively low-temperature melting. Recycling of these terranes into the mantle would not have been complete because the mafic sources to the earliest sediments were prevented from recycling into the mantle and zircons crystallized from these silicic rocks survived. This period (4.3–3.9 Ga) would have been marked by convective vigor in the mantle and intense meteorite bombardment. Both processes worked to break up the lithosphere, preventing the differentiation from occurring long enough in one localized setting such that the mantle could not form a large enough mass of depleted keel to resist being recycled. Recycling did not occur by slab subduction, but by some other form that was density-driven from the bottom (e.g., Bedard, 2006). This period of dominant recycling would have included only nascent crustal differentiation; there would have been no continent formation or recognizable plate tectonics as marked by the slab subduction geometry observed today.

With the meteorite bombardment tailing off by 3.6 Ga, the oceanic lithosphere could have been maintained in large lithospheric plates with a long enough residence at the surface to cool and become slabs that could sink by subduction. This was the beginning of plate tectonics and the formation of continents. The sequestering of Nb, Ta, and Ti in subducting slabs and their anomalously lower abundance in arc rocks and continental crustal materials is the hallmark signature of slab subduction, modern plate tectonics, and continental growth by the subduction process. Ratios involving Nb in the depleted mantle from which crust was extracted clearly start to change at 3.5 Ga, signaling that the plate tectonic process already had started to make preservable continental crust. The presence of subduction imprints within the continental lithospheric mantle as far back as 3.5 Ga and within the crust as far back as 3.8 Ga support this model. Continental crust only began to be preserved on Paleoarchean to Mesoarchean Earth because deep mantle keels are more effectively produced in Archean subduction settings where water-fluxed, localized melting can achieve crustal differentiation, residue removal, a high degree of mantle depletion, and advective thickening.

The geophysical, isotopic, trace element, and petrologic synthesis presented here allows us to propose that plate tectonics started on Earth ca. 3.6–3.9 Ga and that it is marked by the onset of crustal preservation. It is important to identify what conditions changed in Hadean Earth to allow this change. In the quiescent Hadean model, it would have been the accumulation of radioactive heat in the cool-bottom, postoverturn, nonconvecting, mantle, which took some 400 m.y. to build up after the giant impact that formed the Moon. In the active "uniformitarian" Hadean model, a loss of heat would have, after 400 m.y., permitted silicic crust to begin to stabilize in large enough masses to serve as centers of crustal production and storage. In both models, termination of the late heavy meteorite bombardment would have permitted the survival of larger lithospheric plates than was possible before this time.

## CONCLUSIONS

Regardless of initial conditions, diverse lines of evidence point to the start of plate tectonics on Earth shortly after the end of the Hadean. The present upper mantle retains old heterogeneities, some of which could date from subduction in the early Eoarchean, although some of them also could be Eoarchean material subducted more recently. Nb/Th and Th/U of well-characterized mafic to ultramafic rocks though time, representing good estimates of the depleted upper mantle, begin to change from initial values (Nb/Th = 7; Th/U = 4.2) at 3.6 Ga in a unidirectional fashion toward present values (Nb/Th = 18.2; Th/U = 2.6). This signals the appearance of subduction-altered slabs in general mantle circulation from subduction initiated at 3.9 Ga. Juvenile crustal rocks also begin to show that they were derived from progressively depleted mantle after 3.6 Ga from sources that have a typical igneous fractionation of $\varepsilon_{Nd}:\varepsilon_{Hf} = 1:2$. This, and the start of the appearance of cratons with stable mantle keels that have subduction imprints at 3.5 Ga, strongly suggest that continental crustal extraction was progressively depleting the mantle from 3.6 Ga onwards. By the Neoarchean, subduction had taken on an appearance similar to present subduction and was capable of producing large cratons with thick mantle keels.

The Hadean zircons precede the subduction processes evident in the earliest Eoarchean juvenile rocks. Their compositions are incompatible with derivation from a voluminous continental crust but instead reflect the integrated effects of separation of the EER and fractionation of Ca-silicate and Mg-silicate perovskite from the magma ocean generated by the Mars-sized impact. This requires the Hadean zircons to come from a continent-absent, mafic to ultramafic protocrust that was recycled in transitory, small-scale fashion that would not have reflected a global plate tectonic system. If the mafic-ultramafic protocrust was a product of a magma ocean overturn after the giant impact, then a stable, cool-bottom mantle might have resulted and the transition to plate tectonics could have been the result of mantle heating from radioactive decay. If, alternatively, the mafic-ultramafic protocrust was produced by traditional mantle convection with no global overturn, then the transition to plate tectonics might have marked the cooling of the mantle to the point where large lithospheric plates could have been formed on a global scale. Cessation of the late heavy meteorite bombardment also contributed to creating an environment that allowed lithospheric plates to stabilize.

## ACKNOWLEDGMENTS

The careful editorial handling of Vicky Pease and Kent Condie is appreciated. Bob Stern, Alfred Kröner, and Kent Condie are thanked for originally convening the Penrose Conference "When Did Plate Tectonics Start?" in Lander, Wyoming, June 2006, which led to the review and synthesis presented here. The manuscript was improved by the careful reviews of Al Hofmann, Jan Kramers, and Erik Scherer. Discussions with Al Brandon, Rick Carlson, Janne Blichert-Toft, Vickie Bennett, and Jeff Vervoort were helpful in preparing the original text.

## REFERENCES CITED

Abe, Y., Ohtani, E., Okuchi, T., Righter, K., and Drake, M.J., 2000, Water in the early Earth, *in* Righter, K., and Canup, R.M., eds., Origin of the Earth and Moon: Tucson, University of Arizona Press, p. 413–433.

Allegre, C.J., and Turcotte, D.L., 1986, Implications of a two-component marble-cake mantle: Nature, v. 323, p. 123–127, doi: 10.1038/323123a0.

Amelin, Y., 2005, Meteorite phosphates show constant $^{176}$Lu decay rate since 4557 million years ago: Science, v. 310, p. 839–841, doi: 10.1126/science.1117919.

Amelin, Y., Lee, D.-C., Halliday, A.N., and Pidgeon, R.T., 1999, Nature of the Earth's earliest crust from hafnium isotopes in single detrital zircons: Nature, v. 399, p. 252–255, doi: 10.1038/20426.

Amelin, Y., Lee, D.C., and Halliday, A.N., 2000, Early-middle Archaean crustal evolution deduced from Lu-Hf and U-Pb isotopic studies of single zircon grains: Geochimica et Cosmochimica Acta, v. 64, p. 4205–4225, doi: 10.1016/S0016-7037(00)00493-2.

Andreasen, R., and Sharma, M., 2006, Solar nebula heterogeneity in p-process samarium and neodymium isotopes: Science, v. 314, p. 806–809, doi: 10.1126/science.1131708.

Armstrong, R.L., 1981, Radiogenic isotopes: The case for crustal recycling on a near-steady-state no-continental-growth Earth: Royal Society of London Philosophical Transactions, ser. A, v. 301, p. 443–472, doi: 10.1098/rsta.1981.0122.

Armstrong, R.L., 1991, The persistent myth of crustal growth: Australian Journal of Earth Sciences, v. 38, p. 613–630, doi: 10.1080/08120099108727995.

Arndt, N.T., 2003, Komatiites, kimberlites, and boninites: Journal of Geophysical Research, v. 108, no. B6, p. 2293, doi: 10.1029/2002JB002157.

Arndt, N.T., and Goldstein, S.L., 1987, Use and abuse of crust-formation ages: Geology, v. 15, p. 893–895, doi: 10.1130/0091-7613(1987)15<893:UAAOCA>2.0.CO;2.

Arndt, N.T., Ginibre, C., Chauvel, C., Albarede, F., Cheadle, M., Herzberg, C., Jenner, G., and Lahaye, Y., 1998, Were komatiites wet?: Geology, v. 26, p. 739–742, doi: 10.1130/0091-7613(1998)026<0739:WKW>2.3.CO;2.

Baadsgaard, H., Nutman, A.P., Rosing, M., Bridgwater, D., and Longstaffe, F.J., 1986, Alteration and metamorphism of Amitsoq gneisses from the Isukasia area, West Greenland: Recommendations for isotope studies of the early crust: Geochimica et Cosmochimica Acta, v. 50, p. 2165–2172, doi: 10.1016/0016-7037(86)90071-2.

Barth, M.G., Rudnick, R.L., Carlson, R.W., Horn, I., and McDonough, W.F., 2002, Re-Os and U-Pb geochronological constraints on the eclogite-tonalite connection in the Archean Man Shield, West Africa: Precambrian Research, v. 118, p. 267–283, doi: 10.1016/S0301-9268(02)00111-0.

Bedard, J.H., 2006, A catalytic delamination-driven model for coupled genesis of Archaean crust and sub-continental lithospheric mantle: Geochimica et Cosmochimica Acta, v. 70, p. 1188–1214, doi: 10.1016/j.gca.2005.11.008.

Behn, M.D., Conrad, C.P., and Silver, P.G., 2004, Detection of upper mantle flow associated with the African superplume: Earth and Planetary Science Letters, v. 224, p. 259–274, doi: 10.1016/j.epsl.2004.05.026.

Bekker, A., Holland, H.D., Wang, P.L., Rumble, D., Stein, H.J., Hannah, J.L., Coetzee, L.L., and Beukes, N.J., 2004, Dating the rise of atmospheric oxygen: Nature, v. 427, p. 117–120, doi: 10.1038/nature02260.

Benn, K., and Moyen, J.-F., 2008, this volume, The Late Archean Abitibi-Opatica terrane, Superior province: A modified oceanic plateau, *in* Condie, K.C., and Pease, V., eds., When did plate tectonics begin on planet Earth?: Geological Society of America Special Paper 440, doi: 10.1130/2008.2440(09).

Bennett, V.C., 2003, Compositional evolution of the mantle, *in* Carlson, R.W., ed., Treatise on geochemistry, volume 2: The mantle: New York, Elsevier, p. 493–519.

Bennett, V.C., and Nutman, A.P., 1998, Extreme Nd isotope heterogeneity in the Early Archean—Fact or fiction? Case histories from northern Canada and West Greenland: Comment: Chemical Geology, v. 148, p. 213–217, doi: 10.1016/S0009-2541(98)00031-X.

Bennett, V.C., Nutman, A.P., and McCulloch, M.T., 1993, Nd isotopic evidence for transient, highly depleted mantle reservoirs in the early history of the Earth: Earth and Planetary Science Letters, v. 119, p. 299–317, doi: 10.1016/0012-821X(93)90140-5.

Bennett, V.C., Brandon, A.D., Heiss, J., and Nutman, A.P., 2007a, Coupled $^{142}$Nd, $^{143}$Nd, and $^{176}$Hf isotopic data from 3.6–3.9 Ga rocks: New constraints on the timing and composition of early terrestrial chemical reservoirs: 38th Lunar Science Conference, Houston, Texas March 12–16.

Bennett, V.C., Brandon, A.D., and Nutman, A.P., 2007b, Coupled $^{142}$Nd-$^{143}$Nd isotopic evidence for Hadean mantle dynamics: Science, v. 318, p. 1907–1909, doi:10.1126/science.1145928.

Blichert-Toft, J., and Albarede, F., 1997, The Lu-Hf isotope geochemistry of chondrites and the evolution of the mantle-crust system: Earth and Planetary Science Letters, v. 148, p. 243–258, doi: 10.1016/S0012-821X(97)00040-X.

Bodinier, J.-L., and Godard, M., 2003, Orogenic, ophiolitic, and abyssal peridotites, *in* Carlson, R.W., ed., Treatise on geochemistry, volume 2: The mantle: New York, Elsevier, p. 103–170.

Boher, M., Abouchami, W., Michard, A., Albarede, F., and Arndt, N.T., 1992, Crustal growth in West Africa at 2.1 Ga: Journal of Geophysical Research, v. 97, p. 345–369.

Bolhar, R., Kamber, B.S., Moorbath, S., Fedo, C.M., and Whitehouse, M.J., 2004, Characterisation of Early Archaean chemical sediments by trace element signatures: Earth and Planetary Science Letters, v. 222, p. 43–60, doi: 10.1016/j.epsl.2004.02.016.

Bolhar, R., Kamber, B.S., Moorbath, S., Whitehouse, M.J., and Collerson, K.D., 2005, Chemical characterization of Earth's most ancient clastic metasediments from the Isua Greenstone Belt, southern West Greenland: Geochimica et Cosmochimica Acta, v. 69, p. 1555–1573, doi: 10.1016/j.gca.2004.09.023.

Bowring, S.A., and Housh, T., 1995, The Earth's early evolution: Science, v. 269, p. 1535–1540, doi: 10.1126/science.7667634.

Boyd, F.R., 1999, The origin of cratonic peridotites: A major-element approach, *in* Snyder, G.A., et al., eds., Planetary petrology and geochemistry: The Lawrence A. Taylor 60th Birthday Volume: Columbia, Maryland, Bellwether Publishing, p. 5–14.

Boyd, F.R., Pokhilenko, N.P., Pearson, D.G., Mertzman, S.A., Sobolev, N.V., and Finger, L.W., 1997, Composition of the Siberian cratonic mantle: Evidence from Udachnaya peridotite xenoliths: Contributions to Mineralogy and Petrology, v. 128, p. 228–246, doi: 10.1007/s004100050305.

Boyet, M., and Carlson, R.W., 2005, $^{142}$Nd evidence for early (>4.53 Ga) global differentiation of the silicate Earth: Science, v. 309, p. 576–581, doi: 10.1126/science.1113634.

Boyet, M., and Carlson, R.W., 2006, A new geochemical model for the Earth's mantle inferred from $^{146}$Sm-$^{142}$Nd systematics: Earth and Planetary Science Letters, v. 250, p. 254–268, doi: 10.1016/j.epsl.2006.07.046.

Boyet, M., Blichert-Toft, J., Rosing, M., Storey, M., Telouk, P., and Albarede, F., 2003, $^{142}$Nd evidence for early Earth differentiation: Earth and Planetary Science Letters, v. 214, p. 427–442, doi: 10.1016/S0012-821X(03)00423-0.

Brandon, A., 2007, Planetary science: A younger Moon: Nature, v. 450, p. 1169–1170, doi:10.1038/4501169a.

Brown, M., 2008, this volume, Characteristic thermal regimes of plate tectonics and their metamorphic imprint throughout Earth history: When did Earth first adopt a plate tectonics mode of behavior?, *in* Condie, K.C., and Pease, V., eds., When did plate tectonics begin on planet Earth?: Geological Society of America Special Paper 440, doi: 10.1130/2008.2440(05).

Canup, R.M., and Asphaug, E., 2001, Origin of the Moon in a giant impact near the end of the Earth's formation: Nature, v. 412, p. 708–712, doi: 10.1038/35089010.

Card, K.D., 1990, A review of the Superior province of the Canadian Shield, a product of Archean accretion: Precambrian Research, v. 48, p. 99–156, doi: 10.1016/0301-9268(90)90059-Y.

Carlson, R.W., and Shirey, S.B., 1988, Magma oceans, ocean ridges, and continental crust: Relative roles in mantle differentiation: Lunar and Planetary Institute Contribution 681, p. 13–14.

Carlson, R.W., Pearson, D.G., and James, D.E., 2005, Physical, chemical and chronological characteristics of continental mantle: Reviews of Geophysics, v. 43, RG1001, doi: 10.1029/2004RG000156.

Carlson, R.W., Boyet, M., and Horan, M., 2007, Chondrite barium, neodymium, and samarium isotopic heterogeneity and early Earth differentiation: Science, v. 316, p. 1175–1178, doi:10.1126/science.1140189.

Caro, G., Bourdon, B., Birck, J.-L., and Moorbath, S., 2003, $^{146}$Sm-$^{142}$Nd evidence from Isua metamorphosed sediments for early differentiation of the Earth's mantle: Nature, v. 423, p. 428–432, doi: 10.1038/nature01668.

Caro, G., Bourdon, B., Wood, B.J., and Corgne, A., 2005, Trace-element fractionation in Hadean mantle generated by melt segregation from a magma ocean: Nature, v. 436, p. 246–249, doi: 10.1038/nature03827.

Caro, G., Bourdon, B., Birck, J.L., and Moorbath, S., 2006, High-precision $^{142}$Nd/$^{144}$Nd measurements in terrestrial rocks: Constraints on the early differentiation of the Earth's mantle: Geochimica et Cosmochimica Acta, v. 70, p. 164–191, doi: 10.1016/j.gca.2005.08.015.

Cartigny, P., 2005, Stable isotopes and the origin of diamond: Elements, v. 1, p. 79–84.

Cartigny, P., Harris, J.W., and Javoy, M., 1998, Eclogitic diamond formation at Jwaneng: No room for a recycled component: Science, v. 280, p. 1421–1424, doi: 10.1126/science.280.5368.1421.

Cartigny, P., Harris, J.W., and Javoy, M., 2001, Diamond genesis, mantle fractionations and mantle nitrogen content: A study of $\delta^{13}$C-N concentrations in diamonds: Earth and Planetary Science Letters, v. 185, p. 85–98, doi: 10.1016/S0012-821X(00)00357-5.

Cavosie, A.J., Valley, J.W., and Wilde, S.A., 2005, Magmatic $\delta^{18}$O in 4400–3900 Ma detrital zircons: A record of the alteration and recycling of crust in the Early Archean: Earth and Planetary Science Letters, v. 235, p. 663–681, doi: 10.1016/j.epsl.2005.04.028.

Chase, C.G., and Patchett, P.J., 1988, Stored mafic/ultramafic crust and Early Archean mantle depletion: Earth and Planetary Science Letters, v. 91, p. 66–72, doi: 10.1016/0012-821X(88)90151-3.

Christensen, U.R., and Hofmann, A.W., 1994, Segregation of subducted oceanic crust in the convecting mantle: Journal of Geophysical Research, v. 99, p. 19,867–19,884, doi: 10.1029/93JB03403.

Collerson, K.D., and Kamber, B.S., 1999, Evolution of the continents and the atmosphere inferred from Th-U-Nb systematics of the depleted mantle: Science, v. 283, p. 1519–1522, doi: 10.1126/science.283.5407.1519.

Collerson, K.D., Campbell, L.M., Weaver, B.L., and Palacz, Z.A., 1991, Evidence for extreme mantle fractionation in Early Archaean ultramafic rocks from northern Labrador: Nature, v. 349, p. 209–214, doi: 10.1038/349209a0.

Condie, K.C., 1998, Episodic continental growth and supercontinents: A mantle avalanche connection?: Earth and Planetary Science Letters, v. 163, p. 97–108, doi: 10.1016/S0012-821X(98)00178-2.

Condie, K.C., and Kröner, A., 2008, this volume, When did plate tectonics begin? Evidence from the geologic record, *in* Condie, K.C., and Pease, V., eds., When did plate tectonics begin on planet Earth?: Geological Society of America Special Paper 440, doi: 10.1130/2008.2440(14).

Coogan, L.A., and Hinton, R.W., 2006, Do the trace element compositions of detrital zircons require Hadean continental crust?: Geology, v. 34, p. 633–636, doi: 10.1130/G22737.1.

Corgne, A., Liebske, C., Wood, B.J., Rubie, D.C., and Frost, D.J., 2005, Silicate perovskite-melt partitioning of trace elements and geochemical signature of a deep perovskitic reservoir: Geochimica et Cosmochimica Acta, v. 69, p. 485–496, doi: 10.1016/j.gca.2004.06.041.

Davies, G.F., 1999, Dynamic Earth: Cambridge, UK, Cambridge University Press, 458 p.

Davies, G.F., 2002, Stirring geochemistry in mantle convection models with stiff plates and slabs: Geochimica et Cosmochimica Acta, v. 66, p. 3125–3142, doi: 10.1016/S0016-7037(02)00915-8.

Davies, G.F., 2006, Gravitational depletion of the early Earth's upper mantle and the viability of early plate tectonics: Earth and Planetary Science Letters, v. 243, p. 376–382, doi: 10.1016/j.epsl.2006.01.053.

Davis, D.W., Amelin, Y., Nowell, G.M., and Parrish, R.R., 2005, Hf isotopes in zircon from the western Superior province, Canada: Implications for Archean crustal development and evolution of the depleted mantle reservoir: Precambrian Research, v. 140, p. 132–156, doi: 10.1016/j.precamres.2005.07.005.

Debaille, V., Brandon, A.D., Yin, Q.Z., and Jacobsen, B., 2007, Coupled $^{142}$Nd-$^{143}$Nd evidence for a protracted magma ocean on Mars: Nature, v. 450, p. 525–528, doi:10.1038/nature06317.

DePaolo, D.J., 1983, The mean life of continents: Estimates of continent recycling rates from Nd and Hf isotopic data and implications for mantle structure: Geophysical Research Letters, v. 10, p. 705–708.

Drake, M.J., 2005, Origin of water in the terrestrial planets: Meteoritics and Planetary Science, v. 40, p. 519–527.

Drake, M.J., McFarlane, E.A., Gasparik, T., Rubie, D.C., and Agee, C.B., 1993, Mg-perovskite/silicate melt and majorite garnet/silicate melt partition coefficients in the system CaO-MgO-$SiO_2$ at high temperatures and pressures: Journal of Geophysical Research, v. 98, p. 5427–5431.

Eiler, J.M., Schiano, P., Kitchen, N., and Stolper, E.M., 2000, Oxygen-isotope evidence for recycled crust in the sources of mid-ocean-ridge basalts: Nature, v. 403, p. 530–534, doi: 10.1038/35000553.

Eldridge, C.S., Compston, W., Williams, I.S., Harris, J.W., and Bristow, J.W., 1991, Isotope evidence for the involvement of recycled sediments in diamond formation: Nature, v. 353, p. 649–653, doi: 10.1038/353649a0.

Elliott, T., Zindler, A., and Bourdon, B., 1999, Exploring the kappa conundrum: The role of recycling in the lead isotope evolution of the mantle: Earth and Planetary Science Letters, v. 169, p. 129–145, doi: 10.1016/S0012-821X(99)00077-1.

Farquhar, J., Wing, B., McKeegan, K.D., and Harris, J.W., 2002a, Insight into crust-mantle coupling from anomalous $\Delta^{33}$S of sulfide inclusions in diamonds, *in* Abstracts of the 12th annual V.M. Goldschmidt conference: Geochimica et Cosmochimica Acta, v. 66, p. 225.

Farquhar, J., Wing, B.A., McKeegan, K.D., Harris, J.W., Cartigny, P., and Thiemens, M.H., 2002b, Mass-independent sulfur of inclusions in diamond and sulfur recycling on early Earth: Science, v. 298, p. 2369–2372, doi: 10.1126/science.1078617.

Foley, S., 2008, this volume, A trace element perspective on Archean crust formation and on the presence or absence of Archean subduction, *in* Condie, K.C., and Pease, V., eds., When did plate tectonics begin on planet Earth?: Geological Society of America Special Paper 440, doi: 10.1130/2008.2440(02).

Fralick, P., Hollings, P., and King, D., 2008, this volume, Stratigraphy, geochemistry, and depositional environments of Mesoarchean sedimentary units in western Superior province: Implications for generation of early crust, *in* Condie, K.C., and Pease, V., eds., When did plate tectonics begin on planet Earth?: Geological Society of America Special Paper 440, doi: 10.1130/2008.2440(04).

Furnes, H., de Wit, M.J., Staudigel, H., Rosing, M., and Muehlenbachs, K., 2007, A vestige of the Earth's oldest ophiolite: Science, v. 315, p. 1704–1707, doi: 10.1126/science.1139170.

Galer, S.J.G., and Goldstein, S.L., 1991, Early mantle differentiation and its thermal consequences: Geochimica et Cosmochimica Acta, v. 55, p. 227–239, doi: 10.1016/0016-7037(91)90413-Y.

Galer, S.J.G., and O'Nions, R.K., 1985, Residence time of uranium, thorium and lead in the mantle with implications for mantle convection: Nature, v. 316, p. 778–782, doi: 10.1038/316778a0.

Gast, P.W., Tilton, G.R., and Hedge, C., 1964, Isotopic composition of lead and strontium from Ascension and Gough islands: Science, v. 145, p. 1181–1185, doi: 10.1126/science.145.3637.1181.

Goes, S., and van der Lee, S., 2002, Thermal structure of the North American uppermost mantle inferred from seismic tomography: Journal of Geophysical Research, v. 107, no. B3, p. 2050, doi: 10.1029/2000JB000049.

Gradstein, F.M., Ogg, J.G., and Smith, A.G., 2005, A geologic scale 2004: New York, Cambridge University Press, 610 p.

Grand, S.P., 1987, Tomographic inversion for shear velocity beneath the North American plate: Journal of Geophysical Research, v. 92, p. 14,065–14,090.

Griffin, W.L., Wang, X., Jackson, S.E., Pearson, N.J., O'Reilly, S.Y., Xu, X., and Zhou, X., 2002, Zircon chemistry and magma mixing, SE China: In situ analysis of Hf isotopes, Tonglu and Pingtan igneous complexes, *in* Wiebe, R.A., ed., Magmatic processes: A special issue in honor of R.H. Vernon: 15th Australian Geological Convention, Sydney: Lithos, v. 61, p. 237–269.

Griffin, W.L., Belousova, E.A., Shee, S.R., Pearson, N.J., and O'Reilly, S.Y., 2004, Archean crustal evolution in the northern Yilgarn Craton: U-Pb and Hf-isotope evidence from detrital zircons, *in* Van Kranendonk, M.J., ed., Archaean tectonics, volume 2: 4th International Archean Symposium, Perth, Western Australia, Australia, September 24–28, 2001: Precambrian Research, v. 131, p. 231–282.

Grimes, C.B., John, B.E., Keleman, P.B., Mazdab, F.K., Wooden, J.L., Cheadle, M.J., Hanghøj, K., and Schwartz, J.J., 2007, Trace element chemistry of zircons from oceanic crust: A method for distinguishing detrital zircon provenance: Geology, v. 35, p. 643 646, doi:10.1130/G23603A.1.

Grove, T.L., and Parman, S.W., 2004, Thermal evolution of the Earth as recorded by komatiites: Earth and Planetary Science Letters, v. 219, p. 173–187, doi: 10.1016/S0012-821X(04)00002-0.

Grove, T.L., Parman, S.W., and Dann, J.C., 1999, Conditions of magma generation for Archean komatiites from the Barberton Mountain Land, South Africa: Geochemical Society Special Publication 6, p. 155–167.

Gurnis, M., 1988, Large-scale mantle convection and the aggregation and dispersal of supercontinents: Nature, v. 332, p. 695–699, doi: 10.1038/332695a0.

Halpin, J.A., Gerakiteys, C.L., Clarke, G.L., Belousova, E.A., and Griffin, W.L., 2005, In situ U-Pb geochronology and Hf isotope analyses of the Rayner Complex, East Antarctica: Contributions to Mineralogy and Petrology, v. 148, p. 689–706, doi: 10.1007/s00410-004-0627-6.

Hanson, G.N., 1977, Geochemical evolution of the suboceanic mantle: Geological Society [London] Journal, v. 134, p. 1–19.

Harrison, T.M., Blichert-Toft, J., Mueller, W., Albarede, F., Holden, P., and Mojzsis, S.J., 2005, Heterogeneous Hadean hafnium: Evidence of continental crust at 4.4 to 4.5 Ga: Science, v. 310, p. 1947–1950, doi: 10.1126/science.1117926.

Hart, S.R., 1984, A large-scale isotope anomaly in the Southern Hemisphere mantle: Nature, v. 309, p. 753–757, doi: 10.1038/309753a0.

Hart, S.R., and Gaetani, G.A., 2006, Mantle Pb paradoxes: The sulfide solution: Contributions to Mineralogy and Petrology, v. 152, p. 295–308, doi: 10.1007/s00410-006-0108-1.

Hauri, E.H., 1996, Major-element variability in the Hawaiian mantle plume: Nature, v. 382, p. 415–419, doi: 10.1038/382415a0.

Hauri, E.H., and Hart, S.R., 1993, Re-Os isotope systematics of HIMU and EMII oceanic island basalts from the South Pacific Ocean: Earth and Planetary Science Letters, v. 114, p. 353–371, doi: 10.1016/0012-821X(93)90036-9.

Hawkesworth, C.J., Mantovani, M.S.M., Taylor, P.N., and Palacz, Z., 1986, Evidence from the Parana of south Brazil for a continental contribution to Dupal basalts: Nature, v. 322, p. 356–359, doi: 10.1038/322356a0.

Hiess, J. Nutman, A.P., Bennett, V.C., and Holden, P., 2008, Ti-in-zircon thermometry applied to contrasting Archean metamorphic and igneous systems: Chemical Geology, v. 247, p. 323–338, doi:10.1016/j.chemgeo.2007.10.012.

Herzberg, C., 1999, Phase equilibrium constraints on the formation of cratonic mantle: Geochemical Society Special Publication 6, p. 241–257.

Hirschmann, M.M., and Stolper, E.M., 1996, A possible role for garnet pyroxenite in the origin of the "garnet signature" in MORB: Contributions to Mineralogy and Petrology, v. 124, p. 185–208, doi: 10.1007/s004100050184.

Hofmann, A.W., 1988, Chemical differentiation of the Earth: The relationship between mantle, continental crust, and oceanic crust: Earth and Planetary Science Letters, v. 90, p. 297–314, doi: 10.1016/0012-821X(88)90132-X.

Hofmann, A.W., 1997, Mantle geochemistry: The message from oceanic volcanism: Nature, v. 385, p. 219–229, doi: 10.1038/385219a0.

Hofmann, A.W., 2003, Sampling mantle heterogeneity through oceanic basalts: Isotopes and trace elements, *in* Carlson, R.W., ed., Treatise on geochemistry, volume 2: The mantle: New York, Elsevier, p. 61–101.

Hofmann, A.W., and White, W.M., 1982, Mantle plumes from ancient oceanic crust: Earth and Planetary Science Letters, v. 57, p. 421–436, doi: 10.1016/0012-821X(82)90161-3.

Hollings, P., Wyman, D., and Kerrich, R., 1999, Komatiite-basalt-rhyolite volcanic associations in northern Superior province greenstone belts: Significance of plume-arc interaction in the generation of the proto-continental Superior province: Lithos, v. 46, p. 137–161, doi: 10.1016/S0024-4937(98)00058-9.

Hoskin, P.W.O., and Ireland, T.R., 2000, Rare earth element chemistry of zircon and its use as a provenance indicator: Geology, v. 28, p. 627–630, doi: 10 .1130/0091-7613(2000)28<627:REECOZ>2.0.CO;2.

Huang, J., and Davies, G.F., 2007, Stirring in three-dimensional mantle convection models and implications for geochemistry: Passive tracers: Geochemistry, Geophysics, Geosystems, v. 8, p. 17.

Hurley, P.M., and Rand, J.R., 1969, Pre-drift continental nuclei: Science, v. 164, p. 1229–1242, doi: 10.1126/science.164.3885.1229.

Hynes, A., 2001, Freeboard revisited: Continental growth, crustal thickness change and Earth's thermal efficiency: Earth and Planetary Science Letters, v. 185, p. 161–172, doi: 10.1016/S0012-821X(00)00368-X.

Jacob, D.E., 2004, Nature and origin of eclogite xenoliths from kimberlites, *in* Mitchell, R.H., et al., eds., 8th International Kimberlite Conference, Victoria, BC: Lithos, v. 77, p. 295–316.

Jacob, D.E., and Jagoutz, E., 1994, A diamond-graphite bearing eclogitic xenoliths from Roberts Victor (South Africa): Indication for petrogenesis from Pb-, Nd- and Sr-isotopes, *in* Meyer, H.O.A., and Leonardos, O.H., eds., Kimberlites, related rocks and mantle xenoliths: Proceedings of the Fifth Kimberlite Conference, CPRM, p. 304–317.

Jacob, D.E., Jagoutz, E., Lowry, D., Mattey, D., and Kudrjavtseva, G., 1994, Diamondiferous eclogites from Siberia: Remnants of Archean ocean crust: Geochimica et Cosmochimica Acta, v. 58, p. 5191–5207, doi: 10.1016/0016-7037(94)90304-2.

Jacobsen, S.B., and Dymek, R.F., 1988, Nd and Sr isotope systematics of clastic metasediments from Isua, West Greenland: Identification of pre–3.8 Ga differentiated crustal components: Journal of Geophysical Research, v. 93, p. 338–354.

Jia, Y., and Kerrich, R., 2004, Nitrogen 15–enriched Precambrian kerogen and hydrothermal systems: Geochemistry, Geophysics, Geosystems, v. 5, Q07005, doi: 10.1029/2004GC000716.

Jochum, K.P., Arndt, N.T., and Hofmann, A.W., 1991, Nb-Th-La in komatiites and basalts: Constraints on komatiite petrogenesis and mantle evolution: Earth and Planetary Science Letters, v. 107, p. 272–289, doi: 10.1016/0012-821X(91)90076-T.

Jordan, T.H., 1981, Continents as a chemical boundary layer: Royal Society of London Philosophical Transactions, ser. A, v. 301, p. 359–373.

Kamber, B.S., 2007, The enigma of the terrestrial protocrust: Evidence for its former existence and the importance of its complete disappearance, *in* Van Kranendonk, M.J., Smithies, H., and Bennett, V., eds., Earth's Oldest Rocks: Amsterdam, Elsevier, Developments in Precambrian Geology, v. 15, p. 75–89.

Kamber, B.S., Moorbath, S., and Whitehouse, M.J., 1998, Extreme Nd-isotope heterogeneity in the Early Archaean—fact or fiction? Case histories from northern Canada and West Greenland: Reply: Chemical Geology, v. 148, p. 219–224, doi: 10.1016/S0009-2541(98)00032-1.

Kamber, B.S., Ewart, A., Collerson, K.D., Bruce, M.C., and McDonald, G.D., 2002, Fluid-mobile trace element constraints on the role of slab melting and implications for Archaean crustal growth models: Contributions to Mineralogy and Petrology, v. 144, p. 38–56.

Kamber, B.S., Collerson, K.D., Moorbath, S., and Whitehouse, M.J., 2003a, Inheritance of Early Archaean Pb-isotope variability from long-lived Hadean protocrust: Contributions to Mineralogy and Petrology, v. 145, p. 25–46.

Kamber, B.S., Greig, A., Schoenberg, R., and Collerson, K.D., 2003b, A refined solution to Earth's hidden niobium: Implications for evolution of continental crust and depth of core formation: Precambrian Research, v. 126, p. 289–308, doi: 10.1016/S0301-9268(03)00100-1.

Kamber, B.S., Greig, A., and Collerson, K.D., 2005a, A new estimate for the composition of weathered young upper continental crust from alluvial sediments, Queensland, Australia: Geochimica et Cosmochimica Acta, v. 69, p. 1041–1058, doi: 10.1016/j.gca.2004.08.020.

Kamber, B.S., Whitehouse, M.J., Bolhar, R., and Moorbath, S., 2005b, Volcanic resurfacing and the early terrestrial crust: Zircon U-Pb and REE constraints from the Isua Greenstone Belt, southern West Greenland: Earth and Planetary Science Letters, v. 240, p. 276–290, doi: 10.1016/j.epsl.2005.09.037.

Kato, T., Ringwood, A.E., and Irifune, T., 1988a, Constraints on element partition coefficients between $MgSiO_3$ perovskite and liquid determined by direct measurements: Earth and Planetary Science Letters, v. 90, p. 65–68, doi: 10.1016/0012-821X(88)90111-2.

Kato, T., Ringwood, A.E., and Irifune, T., 1988b, Experimental determination of element partitioning between silicate perovskites, garnets and liquids: Constraints on early differentiation of the mantle: Earth and Planetary Science Letters, v. 89, p. 123–145, doi: 10.1016/0012-821X(88)90038-6.

Kelemen, P.B., Hart, S.R., and Bernstein, S., 1998, Silica enrichment in the continental upper mantle via melt/rock reaction: Earth and Planetary Science Letters, v. 164, p. 387–406, doi: 10.1016/S0012-821X(98)00233-7.

Kerrich, R., Polat, A., Wyman, D., and Hollings, P., 1999, Trace element systematics of Mg-, to Fe-tholeiitic basalt suites of the Superior province: Implications for Archean mantle reservoirs and greenstone belt genesis: Lithos, v. 46, p. 163–187, doi: 10.1016/S0024-4937(98)00059-0.

Koeberl, C., 2006, Impact processes on the early Earth: Elements, v. 2, p. 211–216.

Kogiso, T., Hirschmann, M.M., and Frost, D.J., 2003, High-pressure partial melting of garnet pyroxenite: Possible mafic lithologies in the source of ocean island basalts: Earth and Planetary Science Letters, v. 216, p. 603–617, doi: 10.1016/S0012-821X(03)00538-7.

Korenaga, J., 2006, Archean geodynamics and the thermal evolution of Earth, *in* Benn, K., et al., eds., Archean geodynamics and environments: American Geophysical Union Geophysical Monograph 164, p. 7–32.

Kramers, J., 2001, The smile of the Cheshire Cat: Science, v. 293, p. 619–620, doi: 10.1126/science.1063288.

Kramers, J.D., 2003, Volatile element abundance patterns and an early liquid water ocean on Earth, *in* Appel, P.W.U., et al., eds., Early Archean processes and the Isua greenstone belt, West Greenland: Isua multidisciplinary project workshop, Berlin, FDR, January 17–20, 2002: Precambrian Research, v. 126, p. 379–394.

Kramers, J.D., 2007, Hierarchical Earth accretion and the Hadean eon: Geological Society [London] Journal, v. 164, p. 3–17, doi: 10.1144/0016-76492006-028.

Kramers, J.D., and Tolstikhin, I.N., 1997, Two terrestrial lead isotope paradoxes, forward transport modelling, core formation and the history of the continental crust, *in* Hawkesworth, C., and Arndt, N.T., eds., Highlights of the Goldschmidt meeting, in honor of A.W. Hofmann: Goldschmidt meeting, Heidelberg, FRG, March 31–April 4, 1996: Lithos, v. 139, p. 75–110.

Krogstad, E.J., Balakrishnan, S., Mukhopadhyay, D.K., Rajamani, V., and Hanson, G.N., 1989, Plate tectonics 2.5 billion years ago: Evidence at Kolar, South India: Science, v. 243, p. 1337–1340.

Lenardic, A., 2006, Continental growth and the Archean paradox, *in* Benn, K., et al., eds., Archean geodynamics and environments: American Geophysical Union Geophysical Monograph 164, p. 33–45.

Lenardic, A., Moresi, L.N., Jellinek, A.M., and Manga, M., 2005, Continental insulation, mantle cooling, and the surface area of oceans and continents: Earth and Planetary Science Letters, v. 234, p. 317–333, doi: 10.1016/j.epsl.2005.01.038.

Liebske, C., Corgne, A., Frost, D.J., Rubie, D.C., and Wood, B.J., 2005, Compositional effects on element partitioning between Mg-silicate perovskite and silicate melts: Contributions to Mineralogy and Petrology, v. 149, p. 113–128, doi: 10.1007/s00410-004-0641-8.

Lowman, J.P., and Jarvis, G.T., 1999, Effects of mantle heat source distribution on supercontinent stability: Journal of Geophysical Research, v. 104, p. 12,733–12,747, doi: 10.1029/1999JB900108.

Luais, B., and Hawkesworth, C.J., 2002, Pb isotope variations in Archaean time and possible links to the sources of certain Mesozoic-Recent basalts, *in* Fowler, C.M.R., et al., eds., The early Earth: Physical, chemical and biological development: Geological Society [London] Special Publication 199, p. 105–124.

Maas, R., Kinny, P.D., Williams, I.S., Froude, D.O., and Compston, W., 1992, The Earth's oldest known crust: A geochronological and geochemical study of 3900–4200 Ma old detrital zircons from Mt. Narryer and Jack Hills, Western Australia: Geochimica et Cosmochimica Acta, v. 56, p. 1281–1300, doi: 10.1016/0016-7037(92)90062-N.

MacGregor, I.D., and Manton, W.I., 1986, Roberts Victor eclogites: Ancient oceanic crust: Journal of Geophysical Research, B, Solid Earth and Planets, v. 91, p. 14,063–14,079.

Machado, N., and Simonetti, A., 2001, U-Pb dating and Hf isotopic composition of zircon by laser ablation-MC-ICP-MS, *in* Sylvester, P.J., ed., Laser-ablation-ICPMS in the earth sciences: Principles and applications: Toronto, MAC Short Course Handbook, v. 29, p. 121–146.

Marsh, B.D., Gunnarsson, B., Congdon, R., and Carmody, R., 1991, Hawaiian basalt and Icelandic rhyolite: Indicators of differentiation and partial melting: International Journal of Earth Sciences, v. 80, p. 481–510.

Marty, B., and Dauphas, N., 2003, The nitrogen record of crust-mantle interaction and mantle convection from Archean to present: Earth and Planetary Science Letters, v. 206, p. 397–410, doi: 10.1016/S0012-821X(02)01108-1.

Marty, B., and Yokochi, R., 2006, Water in the early Earth: Water in nominally anhydrous minerals, in Keppler, H. and Smyth, J.R., ed., Water in Nominally Anhydrous Minerals: Reviews in Mineralogy and Geochemistry 62, p. 421–450.

McCulloch, M.T., and Bennett, V.C., 1993, Evolution of the early Earth: Constraints from $^{143}$Nd-$^{142}$Nd isotopic systematics, *in* Campbell, I.H., Maruyama, S., and McCulloch, M.T., eds., The Evolving Earth: Lithos, v. 30, p. 237–255.

McCulloch, M.T., and Bennett, V.C., 1994, Progressive growth of the Earth's continental crust and depleted mantle: Geochemical constraints: Geochimica et Cosmochimica Acta, v. 58, p. 4717–4738, doi: 10.1016/0016-7037(94)90203-8.

Menzies, A.H., Carlson, R.W., Shirey, S.B., and Gurney, J.J., 2003, Re-Os systematics of diamond-bearing eclogites from Newlands kimberlite: Lithos, v. 71, p. 323–336, doi: 10.1016/S0024-4937(03)00119-1.

Miller, D.M., Goldstein, S.L., and Langmuir, C.H., 1994, Cerium/lead and lead isotope ratios in arc magmas and the enrichment of lead in the continents: Nature, v. 368, p. 514–520, doi: 10.1038/368514a0.

Mojzsis, S.J., Harrison, T.M., and Pidgeon, R.T., 2001, Oxygen-isotope evidence from ancient zircons for liquid water at the Earth's surface 4300 Myr ago: Nature, v. 409, p. 178–181, doi: 10.1038/35051557.

Montelli, R., Nolet, G., Dahlen, F.A., Masters, G., Engdahl, E.R., and Hung, S.-H., 2004, Finite-frequency tomography reveals a variety of plumes in the mantle: Science, v. 303, p. 338–343, doi: 10.1126/science.1092485.

Moorbath, S., Whitehouse, M.J., and Kamber, B.S., 1997, Extreme Nd-isotope heterogeneity in the Early Archaean—Fact or fiction? Case histories from northern Canada and West Greenland: Chemical Geology, v. 135, p. 213–231, doi: 10.1016/S0009-2541(96)00117-9.

Morbidelli, A., Chambers, J., Lunine, J.I., Petit, J.M., Robert, F., Valsecchi, G.B., and Cyr, K.E., 2000, Source regions and timescales for the delivery of water to Earth: Meteoritics and Planetary Science, v. 35, p. 1309–1320.

Muentener, O., Keleman, P.B., and Grove, T.L., 2001, The role of $H_2O$ during crystallization of primitive arc magmas under uppermost mantle conditions and genesis of igneous pyroxenites: An experimental study: Contributions to Mineralogy and Petrology, v. 141, p. 643–658.

Münker, C., Pfander, J.A., Weyer, S., Buchl, A., Kleine, T., and Mezger, K., 2003, Evolution of planetary cores and the Earth-Moon system from Nb/Ta systematics: Science, v. 301, p. 84–87, doi: 10.1126/science.1084662.

Murphy, D.T., Kamber, B.S., and Collerson, K.D., 2003, A refined solution to the first terrestrial Pb-isotope paradox: Journal of Petrology, v. 44, p. 39–53, doi: 10.1093/petrology/44.1.39.

Musacchio, G., White, D.J., Asudeh, I., and Thomson, C.J., 2004, Lithospheric structure and composition of the Archean western Superior province from seismic refraction/wide-angle reflection and gravity modeling: Journal of Geophysical Research, v. 109, B03304, doi: 10.1029/2003JB002427.

Nägler, T.F., and Kramers, J.D., 1998, Nd isotopic evolution of the upper mantle during the Precambrian: Models, data and the uncertainty of both: Precambrian Research, v. 91, p. 233–252, doi: 10.1016/S0301-9268(98)00051-5.

Navon, O., 1999, Diamond formation in the Earth's mantle, *in* Gurney, J.J., Pascoe, M.D., and Richardson, S.H. et al., eds., Proceedings of the 7th International Kimberlite Conference, p. 584–604.

Nemchin, A.A., Pidgeon, R.T., and Whitehouse, M.J., 2006, Re-evaluation of the origin and evolution of >4.2 Ga zircons from the Jack Hills metasedimentary rocks: Earth and Planetary Science Letters, v. 244, p. 218–233, doi: 10.1016/j.epsl.2006.01.054.

Nir-El, Y., and Lavi, N., 1998, Measurement of the half-life of $^{176}$Lu: Applied Radiation and Isotopes, v. 49, p. 1653–1655, doi: 10.1016/S0969-8043(97)10007-0.

Nisbet, E.G., Cheadle, M.J., Arndt, N.T., and Bickle, M.J., 1993, Constraining the potential temperature of the Archaean mantle: A review of the evidence from komatiites, *in* Campbell, I.H., et al., eds., The evolving Earth: Lithos, v. 30, p. 291–307.

Nutman, A.P., 2001, On the scarcity of >3900 Ma detrital zircons in ≥3500 Ma metasediments: Precambrian Research, v. 105, p. 93–114, doi: 10.1016/S0301-9268(00)00106-6.

Nutman, A.P., Bennett, V.C., Friend, C.R.L., and Norman, M.D., 1999, Meta-igneous (non-gneissic) tonalites and quartz-diorites from an extensive ca. 3800 Ma terrain south of the Isua supracrustal belt, southern West Greenland: Constraints on early crust formation: Contributions to Mineralogy and Petrology, v. 137, p. 364–388, doi: 10.1007/s004100050556.

Oversby, V.M., 1975, Lead isotopic systematics and ages of Archaean acid intrusives in the Kalgoorlie-Norseman area, Western Australia: Geochimica et Cosmochimica Acta, v. 39, p. 1107–1125, doi: 10.1016/0016-7037(75)90053-8.

Parman, S.W., Grove, T.L., Dann, J.C., and de Wit, M.J., 2004, A subduction origin for komatiites and cratonic lithospheric mantle: South African Journal of Geology, v. 107, p. 107–118, doi: 10.2113/107.1-2.107.

Patchett, P.J., Vervoort, J.D., Soderlund, U., and Salters, V.J.M., 2004, Lu-Hf and Sm-Nd isotopic systematics in chondrites and their constraints on the Lu-Hf properties of the Earth: Earth and Planetary Science Letters, v. 222, p. 29–41, doi: 10.1016/j.epsl.2004.02.030.

Pearson, D.G., and Shirey, S.B., 1999, Isotopic dating of diamonds, *in* Lambert, D.D., and Ruiz, J., eds., Application of radiogenic isotopes to ore deposit research and exploration: Reviews in economic geology: Boulder, Colorado, Society of Economic Geologists, p. 143–171.

Pearson, D.G., Snyder, G.A., Shirey, S.B., Taylor, L.A., Carlson, R.W., and Sobolev, N.V., 1995, Archaean Re-Os age for Siberian eclogites and constraints on Archaean tectonics: Nature, v. 374, p. 711–713, doi: 10.1038/374711a0.

Pearson, D.G., Canil, D., and Shirey, S.B., 2003, Mantle samples included in volcanic rocks: Xenoliths and diamonds, *in* Carlson, R.W., ed., Treatise on geochemistry, volume 2: The mantle: New York, Elsevier, p. 171–277.

Pease, V., Percival, J., Smithies, H., Stevens, G., and Van Kranendonk, M., 2008, this volume, When did plate tectonics begin? Evidence from the orogenic record, *in* Condie, K.C., and Pease, V., eds., When did plate tectonics begin on planet Earth?: Geological Society of America Special Paper 440, doi: 10.1130/2008.2440(10).

Peck, W.H., Valley, J.W., Wilde, S.A., and Graham, C.M., 2001, Oxygen isotope ratios and rare earth elements in 3.3 to 4.4 Ga zircons: Ion microprobe evidence for high $\delta^{18}O$ continental crust and oceans in the Early Archean: Geochimica et Cosmochimica Acta, v. 65, p. 4215–4229, doi: 10.1016/S0016-7037(01)00711-6.

Percival, J.A., McNicoll, V.J., Brown, J.L., Whalen, J.B., and Percival, J.A., 2004, Convergent margin tectonics, central Wabigoon subprovince, Superior province, Canada: Precambrian Research, v. 132, p. 213–244, doi: 10.1016/j.precamres.2003.12.016.

Percival, J.A., McNicoll, V., and Bailes, A.H., 2006a, Strike-slip juxtaposition of ca. 2.72 Ga juvenile arc and >2.98 Ga continent margin sequences and its implications for Archean terrane accretion, western Superior province, Canada: Canadian Journal of Earth Sciences, v. 43, p. 895–927, doi: 10.1139/E06-039.

Percival, J.A., Sanborn-Barrie, M., Skulski, T., Stott, G.M., Helmstaedt, H., and White, D.J., 2006b, Tectonic evolution of the western Superior province from NATMAP and lithoprobe studies: Canadian Journal of Earth Sciences, v. 43, p. 1085–1117, doi: 10.1139/E06-062.

Peterman, Z.E., 1979, Strontium isotope geochemistry of Late Archean to Late Cretaceous tonalites and trondhjemites, *in* Barker, F., ed., Tonalites, trondhjemites, and related rocks: IGCP Project 092: Amsterdam, Elsevier, p. 133–147.

Polat, A., Frei, R., Appel, P.W.U., Fryer, B., Dilek, Y., and Ordóñez-Calderón, J.C., 2008, this volume, An overview of the lithological and geochemical characteristics of the Mesoarchean (ca. 3075 Ma) Ivisaartoq greenstone belt, southern West Greenland, *in* Condie, K.C., and Pease, V., eds., When did plate tectonics begin on planet Earth?: Geological Society of America Special Paper 440, doi: 10.1130/2008.2440(03).

Rankenburg, K., Brandon, A.D., and Neal, C.R., 2006, Neodymium isotope evidence for a chondritic composition for the Moon: Science, v. 312, p. 1369–1372, doi: 10.1126/science.1126114.

Rehkamper, M., and Hofmann, A.W., 1997, Recycled ocean crust and sediment in Indian Ocean MORB: Earth and Planetary Science Letters, v. 147, p. 93–106, doi: 10.1016/S0012-821X(97)00009-5.

Reymer, A., and Schubert, G., 1984, Phanerozoic addition rates to the continental crust and crustal growth: Tectonics, v. 3, p. 63–77.

Richardson, S.H., 1986, Latter-day origin of diamonds of eclogitic paragenesis: Nature, v. 322, p. 623–626, doi: 10.1038/322623a0.

Richardson, S.H., and Harris, J.W., 1997, Antiquity of peridotitic diamonds from the Siberian craton: Earth and Planetary Science Letters, v. 151, p. 271–277, doi: 10.1016/S0012-821X(97)81853-5.

Richardson, S.H., Gurney, J.J., Erlank, A.J., and Harris, J.W., 1984, Origin of diamonds in old enriched mantle: Nature, v. 310, p. 198–202, doi: 10.1038/310198a0.

Richardson, S.H., Erlank, A.J., Harris, J.W., and Hart, S.R., 1990, Eclogitic diamonds of Proterozoic age from Cretaceous kimberlites: Nature, v. 346, p. 54–56, doi: 10.1038/346054a0.

Richardson, S.H., Shirey, S.B., Harris, J.W., and Carlson, R.W., 2001, Archean subduction recorded by Re-Os isotopes in eclogitic sulfide inclusions in Kimberley diamonds: Earth and Planetary Science Letters, v. 191, p. 257–266, doi: 10.1016/S0012-821X(01)00419-8.

Ringwood, A.E., 1991, Phase transformations and their bearing on the constitution and dynamics of the mantle: Geochimica et Cosmochimica Acta, v. 55, p. 2083–2110, doi: 10.1016/0016-7037(91)90090-R.

Rudnick, R.L., Eldridge, C.S., and Bulanova, G.P., 1993, Diamond growth history from in situ measurement of Pb and S isotopic compositions of sulphide inclusions: Geology, v. 21, p. 13–16, doi: 10.1130/0091-7613(1993)021<0013:DGHFIS>2.3.CO;2.

Rudnick, R.L., Barth, M., Horn, I., and McDonough, W.F., 2000, Rutile-bearing refractory eclogites: Missing link between continents and depleted mantle: Science, v. 287, p. 278–281, doi: 10.1126/science.287.5451.278.

Sadofsky, S.J., and Bebout, G.E., 2004, Nitrogen geochemistry of subducting sediments: New results from the Izu-Bonin-Mariana margin and insights regarding global nitrogen subduction: Geochemistry, Geophysics, Geosystems, v. 5, doi: 10.1029/2003GC000543.

Salters, V.J.M., 1996, The generation of mid-ocean ridge basalts from the Hf and Nd isotope perspective: Earth and Planetary Science Letters, v. 141, p. 109–123, doi: 10.1016/0012-821X(96)00070-2.

Salters, V.J.M., and Hart, S.R., 1991, The mantle sources of ocean ridges, islands and arcs: The Hf-isotope connection: Earth and Planetary Science Letters, v. 104, p. 364–380, doi: 10.1016/0012-821X(91)90216-5.

Salters, V.J.M., and Stracke, A., 2004, Composition of the depleted mantle: Geochemistry, Geophysics, Geosystems, v. 5, Q05004, doi: 10.1029/2003GC000597.

Scherer, E., Muenker, C., and Mezger, K., 2001, Calibration of the lutetium-hafnium clock: Science, v. 293, p. 683–687, doi: 10.1126/science.1061372.

Schulze, D.J., 1989, Constraints on the abundance of eclogite in the upper mantle: Journal of Geophysical Research, v. 94, p. 4205–4212.

Sharma, M., and Chen, C., 2004, Neodymium isotope fractionation in the mass spectrometer and the issue of $^{142}$Nd anomalies in Early Archean rocks, *in* Mojzsis, S.J., ed., The first billion years: Selected papers presented at the 13th V.M. Goldschmidt conference, Kurashiki, Japan, September 7–12, 2003: Precambrian Research, v. 135, p. 315–329.

Shirey, S.B., 1991, The Rb-Sr, Sm-Nd and Re-Os isotopic systems: A summary and comparison of their applications to the cosmochronology and geochronology of igneous rocks, *in* Heaman, L., and Ludden, J.N., eds., Short course handbook on application of radiogenic isotopic systems to problems in geology: Toronto, Mineralogical Association of Canada, p. 103–166.

Shirey, S.B., and Hanson, G.N., 1984, Mantle-derived Archaean monzodiorites and trachyandesites: Nature, v. 310, p. 222–224, doi: 10.1038/310222a0.

Shirey, S.B., and Hanson, G.N., 1986, Mantle heterogeneity and crustal recycling in Archean granite-greenstone belts: Evidence from Nd isotopes and trace elements in the Rainy Lake area, Superior province, Ontario, Canada: Geochimica et Cosmochimica Acta, v. 50, p. 2631–2651, doi: 10.1016/0016-7037(86)90215-2.

Shirey, S.B., and Walker, R.J., 1998, The Re-Os isotope system in cosmochemistry and high-temperature geochemistry: Annual Review of Earth and Planetary Sciences, v. 26, p. 423–500, doi: 10.1146/annurev.earth.26.1.423.

Shirey, S.B., Bender, J.F., and Langmuir, C.H., 1987, Three-component isotopic heterogeneity near the oceanographer transform, Mid-Atlantic Ridge: Nature, v. 325, p. 217–223, doi: 10.1038/325217a0.

Shirey, S.B., Carlson, R.W., Richardson, S.H., Menzies, A.H., Gurney, J.J., Pearson, D.G., Harris, J.W., and Wiechert, U., 2001, Archean emplacement of eclogitic components into the lithospheric mantle during formation of the Kaapvaal Craton: Geophysical Research Letters, v. 28, p. 2509–2512, doi: 10.1029/2000GL012589.

Shirey, S.B., Richardson, S.H., and Harris, J.W., 2004, Integrated models of diamond formation and craton evolution: Lithos, v. 77, p. 923–944, doi: 10.1016/j.lithos.2004.04.018.

Sims, K.W.W., and DePaolo, D.J., 1997, Inferences about magma sources from incompatible element concentration ratios in oceanic basalts: Geochimica et Cosmochimica Acta, v. 61, p. 765–784, doi: 10.1016/S0016-7037(96)00372-9.

Smithies, R.H., Champion, D.C., Van Kranendonk, M.J., Howard, H.M., and Hickman, A.H., 2005, Modern-style subduction processes in the Mesoarchaean: Geochemical evidence from the 3.12 Ga Whundo intra-oceanic arc: Earth and Planetary Science Letters, v. 231, p. 221–237, doi: 10.1016/j.epsl.2004.12.026.

Sobolev, A.V., Hofmann, A.W., Sobolev, S.V., and Nikogosian, I.K., 2005, An olivine-free mantle source of Hawaiian shield basalts: Nature, v. 434, p. 590–597, doi: 10.1038/nature03411.

Söderlund, U., Patchett, P.J., Vervoort, J.D., and Isachsen, C.E., 2004, The $^{176}$Lu decay constant determined by Lu-Hf and U-Pb isotope systematics of Precambrian mafic intrusions: Earth and Planetary Science Letters, v. 219, p. 311–324, doi: 10.1016/S0012-821X(04)00012-3.

Stacey, J.S., and Kramers, J.D., 1975, Approximation of terrestrial lead isotope evolution by a two-stage model: Earth and Planetary Science Letters, v. 26, p. 207–221, doi: 10.1016/0012-821X(75)90088-6.

Stein, M., and Hofmann, A.W., 1994, Mantle plumes and episodic crustal growth: Nature, v. 372, p. 63–68, doi: 10.1038/372063a0.

Stern, R.A., and Hanson, G.N., 1992, Origin of Archean lamprophyre dykes, Superior province, Canada: Rare earth element and Nd-Sr isotopic evidence: Contributions to Mineralogy and Petrology, v. 111, p. 515–526, doi: 10.1007/BF00320906.

Stern, R.A., Hanson, G.N., and Shirey, S.B., 1989, Petrogenesis of mantle-derived, LILE-enriched Archean monzodiorites and trachyandesites (sanukitoids) in southwestern Superior province: Canadian Journal of Earth Sciences, v. 26, p. 1688–1712.

Stott, G.M., 1997, The Superior province, Canada, *in* de Wit, M.J., and Ashwal, L.D., eds., Greenstone belts: New York, Oxford, Clarendon Press, p. 480–557.

Sun, S.S., 1980, Lead isotopic study of young volcanic rocks from mid-ocean ridges, ocean islands, and island arcs: Royal Society of London Philosophical Transactions, v. 297, p. 409–445, doi: 10.1098/rsta.1980.0224.

Sun, S.S., and Hanson, G., 1975a, Evolution of the mantle: Geochemical evidence from alkali basalt: Geology, v. 3, p. 297–302, doi: 10.1130/0091-7613(1975)3<297:EOTMGE>2.0.CO;2.

Sun, S.S., and Hanson, G.N., 1975b, Origin of Ross Island basanitoids and limitations upon the heterogeneity of mantle sources for alkali basalts and nephelinites: Contributions to Mineralogy and Petrology, v. 52, p. 77–106, doi: 10.1007/BF00395006.

Sylvester, P.J., Campbell, I.H., and Bowyer, D.A., 1997, Niobium/uranium evidence for early formation of the continental crust: Science, v. 275, p. 521–523, doi: 10.1126/science.275.5299.521.

Takahashi, E., and Brearley, M., 1990, Speculations on the Archean mantle: Missing link between komatiite and depleted garnet peridotite, *in* Dingwell, D.B., ed., Special section on silicate melts and mantle petrogenesis (in memory of Christopher M. Scarfe): American Geophysical Union 1989 Spring Meeting Symposium on Silicate Melts and Mantle Petrogenesis, Baltimore, Maryland, May 8–12, 1989: Journal of Geophysical Research, v. 95, p. 15,941–15,954.

Tatsumi, Y., and Kogiso, T., 1997, Trace element transport during dehydration processes in the subducted oceanic crust, 2: Origin of chemical and physical characteristics in arc magmatism: Earth and Planetary Science Letters, v. 148, p. 207–221, doi: 10.1016/S0012-821X(97)00019-8.

Taylor, S.R., and McLennan, S.M., 1995, The geochemical evolution of the continental crust: Reviews of Geophysics, v. 33, p. 241–265, doi: 10.1029/95RG00262.

Taylor, S.R., McLennan, S.M., Armstrong, R.L., and Tarney, J., 1981, The composition and evolution of the continental crust: Rare earth element evidence from sedimentary rocks: Royal Society of London Philosophical Transactions, ser. A, v. 301, p. 381–399, doi: 10.1098/rsta.1981.0119.

Tera, F., 1982, An outline of Earth's evolution as inferred from lead isotopes: Yearbook of the Carnegie Institution of Washington, v. 81, p. 539–543.

Tera, F., 2003, A lead isotope method for the accurate dating of disturbed geologic systems: Numerical demonstrations, some applications and implications: Geochimica et Cosmochimica Acta, v. 67, p. 3687–3716, doi: 10.1016/S0016-7037(03)00132-7.

Tolstikhin, I., and Hofmann, A.W., 2005, Early crust on top of the Earth's core: Physics of the Earth and Planetary Interiors, v. 148, p. 109–130, doi: 10.1016/j.pepi.2004.05.011.

Touboul, M., Kleine, T., Bourdon, B., Palme, H., and Weiler, R., 2007, Late formation and prolonged differentiation of the Moon inferred from W isotopes in lumar metals: Nature, v. 450, p. 1206–1209, doi:10.1038/nature06428.

Valley, J.W., Cavosie, A.J., Fu, B., Peck, W.H., and Wilde, S.A., 2006, Comment on "Heterogeneous Hadean hafnium: Evidence of continental crust at 4.4 to 4.5 Ga": Science, v. 312, p. 1139, doi: 10.1126/science.1125301.

van der Hilst, R.D., Widiyantoro, S., and Engdahl, E.R., 1997, Evidence for deep mantle circulation from global tomography: Nature, v. 386, p. 578–584, doi: 10.1038/386578a0.

van der Lee, S., and Frederiksen, A., 2005, Surface wave tomography applied to the North American upper mantle: American Geophysical Union Geophysical Monograph 157, p. 67–80.

van der Lee, S., and Nolet, G., 1997, Upper mantle S-velocity structure of North America: Journal of Geophysical Research, v. 102, p. 22,815–22,838, doi: 10.1029/97JB01168.

van der Velden, A.J., and Cook, F.A., 2005, Relict subduction zones in Canada: Journal of Geophysical Research, v. 110, B08403, doi: 10.1029/2004JB003333.

van Hunen, J., van Keken, P.E., Hynes, A., and Davies, G.F., 2008, this volume, Tectonics of early Earth: Some geodynamic considerations, *in* Condie, K.C., and Pease, V., eds., When did plate tectonics begin on planet Earth?: Geological Society of America Special Paper 440, doi: 10.1130/2008.2440(08).

Vervoort, J.D., and Blichert-Toft, J., 1999, Evolution of the depleted mantle: Hf isotope evidence from juvenile rocks through time: Geochimica et Cosmochimica Acta, v. 63, p. 533–556, doi: 10.1016/S0016-7037(98)00274-9.

Vervoort, J.D., Patchett, P.J., Gehrels, G.E., and Nutman, A.P., 1996, Constraints on early Earth differentiation from hafnium and neodymium isotopes: Nature, v. 379, p. 624–627, doi: 10.1038/379624a0.

Wade, J., and Wood, B.J., 2001, The Earth's "missing" niobium may be in the core: Nature, v. 409, p. 75–78, doi: 10.1038/35051064.

Watson, E.B., and Harrison, T.M., 2005, Zircon thermometer reveals minimum melting conditions on earliest Earth: Science, v. 308, p. 841–844, doi: 10.1126/science.1110873.

Westerlund, K.J., Shirey, S.B., Richardson, S.H., Carlson, R.W., Gurney, J.J., and Harris, J.W., 2006, A subduction origin for Early Archean peridotitic diamonds and harzburgites from the Panda kimberlite, Slave craton: Implications from Re-Os isotope systematics: Contributions to Mineralogy and Petrology, v. 152, p. 275–294, doi: 10.1007/s00410-006-0101-8.

White, W.M., and Hofmann, A.W., 1982, Sr and Nd isotope geochemistry of oceanic basalts and mantle evolution: Nature, v. 296, p. 821–825, doi: 10.1038/296821a0.

Whitehouse, M.J., and Kamber, B.S., 2002, On the overabundance of light rare earth elements in terrestrial zircons and its implication for Earth's earliest magmatic differentiation: Earth and Planetary Science Letters, v. 204, p. 333–346, doi: 10.1016/S0012-821X(02)01000-2.

Widom, E., Kepezhinskas, P., Defant, M., Horan, M.F., Brandon, A.D., and Neal, C.R., 2003, The nature of metasomatism in the sub-arc mantle wedge: Evidence from Re-Os isotopes in Kamchatka peridotite xenoliths: Chemical Geology, v. 196, p. 283–306, doi: 10.1016/S0009-2541(02)00417-5.

Wilde, S.A., Valley, J.W., Peck, W.H., and Graham, C.M., 2001, Evidence from detrital zircons for the existence of continental crust and oceans on the Earth 4.4 Gyr ago: Nature, v. 409, p. 175–178, doi: 10.1038/35051550.

Wise, D.U., 1974, Continental margins, freeboard and the volumes of continents and oceans through time, *in* Nairn, A.E.M., and Stehli, F.G., eds., The geology of continental margins: New York, Springer, p. 45–58.

Workman, R.K., and Hart, S.R., 2005, Major and trace element composition of the depleted MORB mantle (DMM): Earth and Planetary Science Letters, v. 231, p. 53–72, doi: 10.1016/j.epsl.2004.12.005.

Wyman, D.A., O'Neill, C., and Ayer, J.A., 2008, this volume, Evidence for modern-style subduction to 3.1 Ga: A plateau–adakite–gold (diamond) association, *in* Condie, K.C., and Pease, V., eds., When did plate tectonics begin on planet Earth?: Geological Society of America Special Paper 440, doi: 10.1130/2008.2440(06).

Yokochi, R., and Marty, B., 2005, Geochemical constraints on mantle dynamics in the Hadean: Earth and Planetary Science Letters, v. 238, p. 17–30, doi: 10.1016/j.epsl.2005.07.020.

Zartman, R.E., and Doe, B.R., 1981, Plumbotectonics—The model: Tectonophysics, v. 75, p. 125–162.

Zartman, R.E., and Haines, S., 1988, The plumbotectonic model for Pb isotopic systematics among major terrestrial reservoirs—A case for bi-directional transport: Geochimica et Cosmochimica Acta, v. 52, p. 1327–1339, doi: 10.1016/0016-7037(88)90204-9.

Zartman, R.E., and Richardson, S.H., 2005, Evidence from kimberlitic zircon for a decreasing mantle Th/U since the Archean: Chemical Geology, v. 220, p. 263–283, doi: 10.1016/j.chemgeo.2005.04.003.

Zegers, T.E., and van Keken, P.E., 2001, Middle Archean continent formation by crustal delamination: Geology, v. 29, p. 1083–1086, doi: 10.1130/0091-7613(2001)029<1083:MACFBC>2.0.CO;2.

Zheng, J., Griffin, W.L., O'Reilly, S.Y., Lu, F., Wang, C., Zhang, M., Wang, F., and Li, H., 2004, 3.6 Ga lower crust in central China: New evidence on the assembly of the North China Craton: Geology, v. 32, p. 229–232, doi: 10.1130/G20133.1.

Zindler, A., and Hart, S.R., 1986, Chemical geodynamics: Annual Review of Earth and Planetary Sciences, v. 14, p. 493–571, doi: 10.1146/annurev.ea.14.050186.002425.

Zindler, A., Staudigel, H., and Batiza, R., 1984, Isotope and trace element geochemistry of young Pacific seamounts: Implications for the scale of upper mantle heterogeneity: Earth and Planetary Science Letters, v. 70, p. 175–195, doi: 10.1016/0012-821X(84)90004-9.

Manuscript Accepted by the Society 14 August 2007

Printed in the USA

The Geological Society of America
Special Paper 440
2008

# *A trace element perspective on Archean crust formation and on the presence or absence of Archean subduction*

**Stephen Foley***
*Institute of Geosciences, University of Mainz, Becherweg 21, 55099 Mainz, Germany*

## ABSTRACT

**The early continental crust is dominated by high-grade gneisses with the composition of sodic granites (the tonalite-trondhjemite-granodiorite or TTG suite) that date as far back as >3800 Ma. These are considered by many to be formed in subduction zones, and so have a critical role in the discussion about when plate tectonics may have begun. Trace elements can be used to learn about the identity of minerals in the source rocks during melting, but only indirectly to infer tectonic environments. The integrated results from experimental petrology and major and trace element geochemistry of the TTG suite indicate that most of them formed by melting of garnet amphibolites of broadly basaltic composition. This can explain low Nb/Ta coupled with high Zr/Sm, as well as low concentrations of HREEs. Melting of eclogite probably increased in importance in the Late Archean, as shown by an increase in Nb/Ta. Melting of garnet amphibolite can be achieved either in subduction zones at appropriate geotherms or in the lower reaches of thick basaltic crust. Water contents must be much higher than in the original basalts, indicating hydrothermal alteration at near-surface conditions. Thus, the thick crust scenario requires volcanic piling to deeply bury hydrothermally altered basalts, and also delamination of underlying thick cumulates. If subduction occurred in the Archean (implying the operation of plate tectonics), then with a slightly higher average mantle temperature, subduction geotherms would have been disproportionately hotter than today. However, there is no evidence for large volumes of TTG gneisses formed by melting of garnet-free amphibolites in the Early Archean, which constrains the average mantle temperature to be only marginally hotter than on modern Earth. In the Early Archean, melting of more magnesian volcanics and cumulates produced low-$SiO_2$ melts and prevented the production of voluminous *continental* crust with >55 wt% $SiO_2$. Given current trace element evidence, the most likely scenario for Archean tectonics is a slightly modified plate tectonics with only marginally hotter average geotherms, but substantially hotter subduction geotherms, enabling melting of garnet amphibolites in the Late Archean.**

**Keywords:** crust formation, trace elements, subduction, Archean.

*foley@uni-mainz.de

Foley, S., 2008, A trace element perspective on Archean crust formation and on the presence or absence of Archean subduction, *in* Condie, K.C., and Pease, V., eds., When Did Plate Tectonics Begin on Planet Earth?: Geological Society of America Special Paper 440, p. 31–50, doi: 10.1130/2008.2440(02). For permission to copy, contact editing@geosociety.org. 

## 1. INTRODUCTION

Participants at the Penrose Conference held at Lander, Wyoming, in the summer of 2006 debated the time of the onset of plate tectonics. Four days considering the evidence from diverse aspects of geology brought no unanimity of opinion, but most favored the beginning of plate tectonics sometime during the Archean (4.0–3.0 Ga). Although there is no universally accepted definition of plate tectonics, it can nevertheless be characterized by a list of criteria that geologists expect to see when plate tectonics operate. The most commonly cited examples are the creation, lateral movement, and destruction of rigid lithospheric plates, the operation of subduction and seafloor spreading, the generation of arcs, continuous formation and recycling of crust, and a combination of compressional, extensional, and strike-slip deformation.

The main area of application of the trace element analysis of rocks to this complex of problems lies in the recognition of subduction and the mechanism of formation of new continental crust. Most models for Archean continental crust consider a large proportion of it to be broadly of sodic granitoid composition (TTG = tonalite-trondhjemite-granodiorite; Condie, 1981; Martin, 1994), which is generally explained by melting of basaltic source material during subduction of ocean crust, and thus is taken to imply the operation of plate tectonics. TTG gneisses were formed at least as early as 3.8 Ga in West Greenland (Nutman et al., 1999), and a large peak in TTG production occurred between 2.9 and 2.6 Ga; both of these could be taken to indicate the operation of subduction at these times (Condie, 1986; Martin and Moyen, 2002).

However, since subduction is critical in deciding whether plate tectonics operated, it is essential to establish whether igneous rocks really were formed at subduction zones, and so we must take a closer look at the criteria used to recognize a subduction association of igneous rocks. A catalogue of geochemical criteria for recognizing a subduction association in igneous rocks has been drawn up over the years, including enrichment in the most incompatible large ion lithophile elements (Rb, Cs, Ba, K, Th, U), a relative depletion in high field strength elements (Ti, Nb, Ta, and to a lesser extent Zr and Hf), and lower concentrations of the heavy rare earth elements than in mid-oceanic-ridge basalts (Pearce and Norry, 1979; Pearce, 1983; Foley and Wheller, 1990). However, two factors compromise the apparent simplicity of this approach. Firstly, we do not have a complete understanding of the complex and diverse processes that lead to particular geochemical signatures in subduction zones on modern Earth. Without a robust set of criteria to identify subduction-related igneous rocks on modern Earth, how can they be applied satisfactorily to the Archean? Secondly, many Archean rocks, particularly the TTG gneisses, are strongly metamorphosed, so that some of the trace element information is modified greatly, and our understanding of trace element behavior during metamorphism is still in its infancy. Thus, the assignment of a geochemical "subduction signature" to Archean TTG gneisses is often little more than a vague association. Here, I restrict my attention to assessing the use of trace elements in the assignment of continental crust formation to subduction zone processes. I concentrate on the TTG gneisses and do not consider the volumetrically less significant greenstone belts, sanukitoids, and potassic granites, which are the other main components of the Archean rock record (Condie, 1981; Martin et al., 2005).

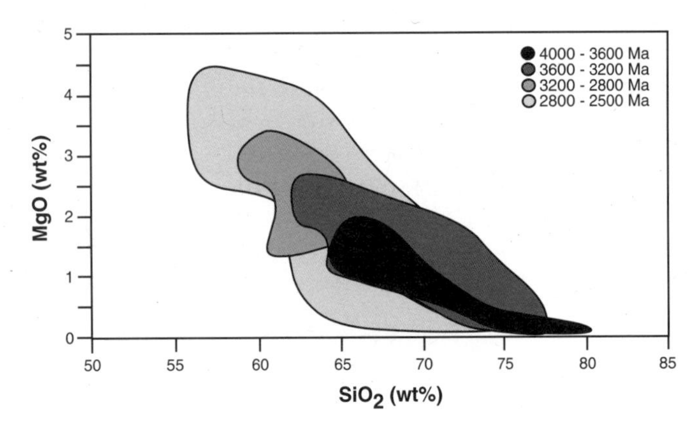


Figure 1. Changes in the major element compositions of tonalite-trondhjemite-granodiorite (TTG) gneisses and plutons as a function of time during the Archean, illustrated by time slices 4000–3600 Ma (n = 35), 3600–3200 Ma (n = 57), 3200–2800 Ma (n = 137), and 2800–2500 Ma (n = 149). There is a gradual trend toward intermediate $SiO_2$ (55–65 wt%) and higher MgO (2.5–4.5 wt%), but a sudden increase indicative of a process change is lacking.

## 2. THE COMPOSITION OF THE EARLY CONTINENTAL CRUST

The greater part of the Archean crust is made up principally of rocks of sodic granitoid composition, many of which have undergone upper amphibolite or granulite facies metamorphism (Tarney et al., 1979; Percival, 1994). The protoliths of most high-grade deformed gneisses are believed to be igneous based on the similarity of compositions to little-deformed Late Archean intrusives, and on the dissimilarity of their whole-rock chemical compositions to sediments of any age (Tarney, 1976). The high Na/K ratios are reflected in low modal abundances of K-feldspar, so that they generally classify as tonalites, trondhjemites, granodiorites, and quartz-diorites and are therefore commonly referred to as TTG (Jahn et al., 1981; Martin, 1994). Silica contents range from 55 to 80 wt% (Fig. 1), averaging around 68–70 wt% (Condie, 2005; Martin et al., 2005). Estimates of the composition of the bulk continental crust and the bulk Archean crust (Rudnick and Fountain, 1995) are both considerably less silicic.

Recent summaries of available TTG analyses have investigated whether there are systematic changes in the composition of TTG gneisses as a function of time (Martin and Moyen, 2002; Martin et al., 2005; Condie, 2005). This is an important question because cooling curves indicate that there should have been more change in heat flow during the Archean than in all the time since the end of the Archean.

Following the time scale of Gradstein et al. (2004), the Archean is conveniently divided into four time slices, 4.0–3.6 Ga, 3.6–3.2 Ga, 3.2–2.8 Ga, and 2.8–2.5 Ga, from which it can be seen that Archean gneisses progress smoothly from lower MgO and higher $SiO_2$ contents in the Early Archean to higher MgO and lower $SiO_2$ contents in the Late Archean (Fig. 1). The average MgO content increases from 0.89 to 1.52 wt% and corresponds to an increase in Mg# (100 × Mg/(Mg + Fe)) from 36.4 to 44.4. The average concentrations of other major elements not shown in Figure 1 include increases in $TiO_2$ (0.33–0.45 wt%), $Al_2O_3$ (15.0–15.6 wt%), and $CaO/Na_2O$ but no systematic change in $Na_2O$ or $K_2O$. These differences may indicate changes in the petrological mechanism of TTG magma production, which would carry over into tectonic scenarios for their origin.

The trace element patterns of TTG gneisses are shown compared to patterns for the bulk continental crust in Figure 2. The most significant differences between average TTGs and the average continental crust are the lower concentrations of the heavy rare earth elements in the TTG pattern, which has been taken to indicate the presence of residual garnet in the source (Tarney et al., 1979; Martin, 1987). Similarities are seen in the general negative slope of the pattern, in the troughs in the pattern at Nb, Ta, and Ti, and in the low Nb/Ta ratio. Temporal developments in the trace element patterns of TTG gneisses are investigated in Figure 3, in which patterns for gneisses with different ages are plotted in different panels. Despite the limitations on database size it brings, attention here is restricted to trace element concentrations measured by ICP-MS (inductively coupled plasma–mass spectrometry) to ensure comparability and consistency of critical element ratios such as Nb/Ta that are otherwise frequently analyzed with different techniques (e.g., XRF and INAA; Luais and Hawkesworth, 1994; Martin et al., 1997). It is immediately apparent that the diversity of rock patterns at individual localities is often greater than any temporal effect, and this diversity is only slightly accentuated by the effects of crystal fractionation, which are generally low and would tend to increase the proportion of trondhjemites relative to tonalites (Tarney et al., 1979; Martin and Moyen, 2002; Bedard, 2006b). Nevertheless, some tendencies can be recognized, such as an overall trend toward higher concentrations of many incompatible trace elements in the Late Archean, and particularly low concentrations at the left of the diagram (Ba to Ta) in the earliest Archean. There is often a more marked trough at Nb-Ta in Late Archean rocks due to higher increases in Ba, Th, U, and La than in Nb and Ta. No systematic change in heavy rare earth element (HREE) concentrations is apparent that may indicate a change in the role of garnet in the residue.

Due to uncertainties about the behavior of trace elements during metamorphism, not all of these trace element signals can be given equal value when considering the question of whether they constitute a subduction signature. However, the high field strength elements and heavy rare earth elements are considered to be the least mobile in geological processes up to temperatures of 700 °C and thus the most dependable in metamorphic rocks at the grades considered here (Peucat et al., 1996; Jacob and Foley, 1999). Carefully measured concentrations of the high field strength elements Nb, Ta, and Zr, and critical ratios between these and the rare earth elements will therefore be given more credence than the full trace element pattern in the arguments that follow.

## 3. EXPERIMENTAL CONSTRAINTS ON THE FORMATION OF SODIC GRANITES

Several experimental studies of the melting relations of basaltic starting material at pressures between 1 and 3 GPa have shown that melt compositions correspond broadly to those of the high-grade Archean gneisses (Fig. 4). These are experiments in which the solidus assemblages range from amphibolite, through garnet amphibolite to eclogite (Rushmer, 1991; Winther and Newton, 1991; Wolf and Wyllie, 1994; Rapp et al., 1991). The

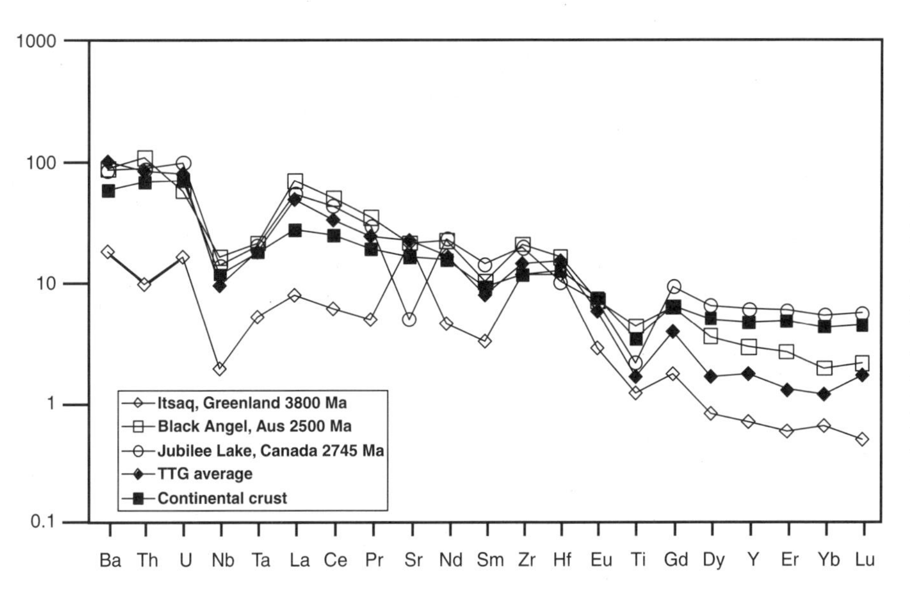


Figure 2. Trace element patterns of TTG gneisses compared to the bulk continental crust, illustrating the difference between early continental crust and time-integrated continental crust. Most characteristics of the bulk crust are seen in the Archean, including low Nb/Ta and high Zr/Sm, but the HREEs are consistently lower in most Archean gneisses; only rare Late Archean gneisses are similar. The Early Archean Itsaq gneisses have much lower concentrations of trace elements than the TTG gneiss average. Data sources: Rudnick and Fountain (1995), Sage et al. (1996), Nutman et al. (1999), and Kamber et al. (2002).

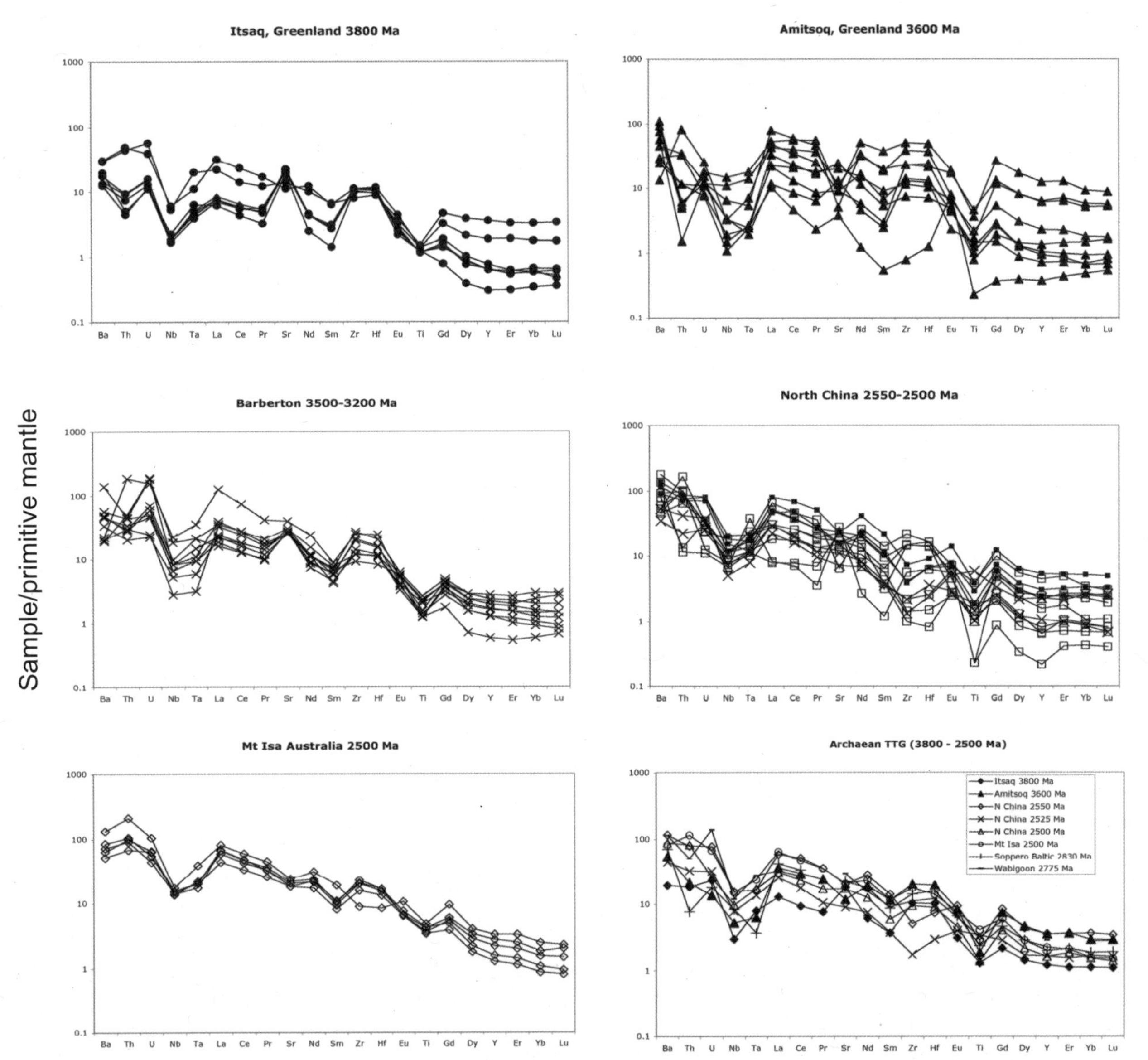


Figure 3. Temporal developments in the trace element patterns of TTG gneisses during the Archean (normalized to primitive mantle using values of McDonough and Sun, 1995). Variation at individual localities are generally larger than between localities, but nevertheless trends emerge: Concentrations of earliest Archean Itsaq gneisses are unusually low, and the more marked trough at Nb is characteristic of the Late Archean. Average patterns for the localities are compared in the last panel. Data sources: Öhlander et al. (1987), Nutman et al. (1999), Whalen et al. (2002), Kamber et al. (2002), Liu et al. (2004), and Kleinhanns et al. (2003).

experiments on amphibolites have been reviewed recently by Moyen and Stevens (2006), who included rock compositions more mafic than basalts in their consideration (although only three of 23 compositions studied so far have >10 wt% MgO). The tendency of the experimental melt compositions to plot toward Na-rich compositions may be underestimated owing to Na loss during analysis of glasses in the experimental charges (Moyen and Stevens, 2006). Melts of source rocks with basaltic compositions thus appear to be restricted to trondhjemite and tonalite.

Although the experimental results show that the major element compositions of TTG gneisses may be produced by partial melting of basaltic material, they do not distinguish clearly between melting as amphibolite, garnet amphibolite, or eclogite. Melt compositions vary as a function of pressure, temperature, water content, and the exact composition of the protolith, so that subdivisions of the TTG suite can be recognized: Tonalitic melts are favored by higher temperatures and thus degrees of melting, lower pressures, and more water, whereas trondhjemites are favored by higher pressures, lower temperatures, and lower water contents (Winther and Newton, 1991; Rapp and Watson, 1995). Very low degrees of melting may produce high-K granites that do not correspond to the TTG suite, but it is in any case questionable that these low-degree melts could be extracted from their source regions (Rapp, 1995; Rapp and Watson, 1995). Modeling

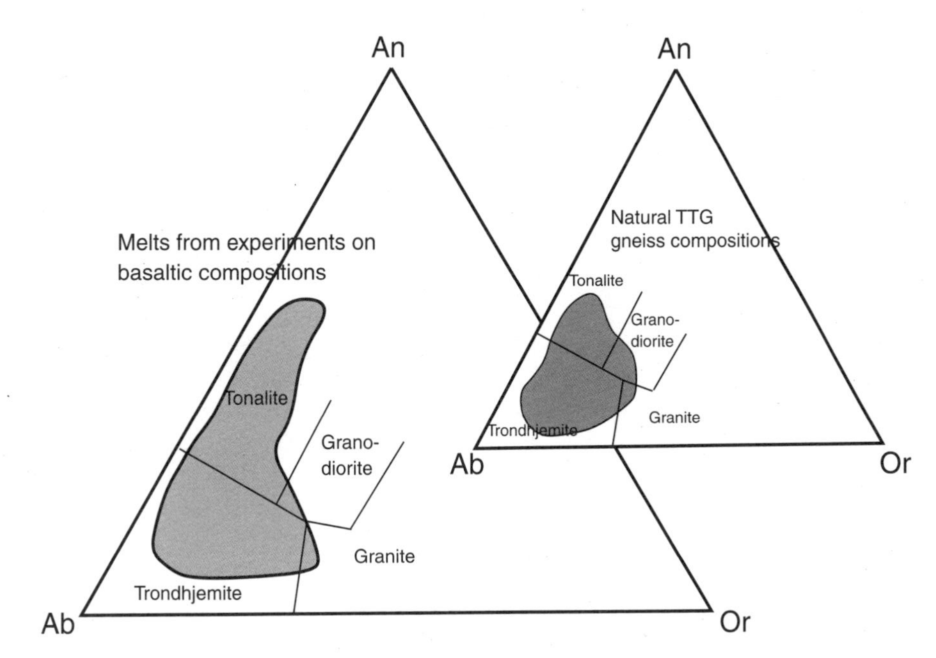


Figure 4. Normative feldspar contents (An-Ab-Or) of TTG gneisses compared to analyzed glass compositions from experiments on basaltic compositions. Granite and granodiorite compositions are avoided almost entirely by the experimental melts. Post-Archean granitoids are dominantly granite and granodiorite with subordinate tonalite and trondhjemite. Experimental data summarized from Rushmer (1991), Winther and Newton (1991), Rapp et al. (1991), and Rapp and Watson (1995).

by Moyen and Stevens (2006) predicts that these granitic melts would not be produced where the protolith is quartz-free, and that granodioritic melts will be restricted to pressures below 10 kbar from quartz-bearing source rocks. Other than these general tendencies, the origin of tonalitic and trondhjemitic melts cannot be assigned to amphibolite, garnet amphibolite, or eclogite melting on the basis of major element compositions alone.

Although these experiments provide only general constraints on the major element compositions of melts, they are useful in constraining the stability ranges of residual phases, which in turn are important in determining the behavior of trace elements. The stability of garnet is pressure-dependent and has its minimum at 0.8–1.2 GPa, regardless of the type of basalt being melted (Rapp and Watson, 1995), and moves to higher pressures in more MgO-rich compositions (1.2–1.6 GPa at 17.5 wt%; Foley et al., 2003). Rutile is restricted to pressures higher than 1.4–1.5 GPa (Moyen and Stevens, 2006). It should be noted here that the discussion of whether garnet amphibolite or eclogite represents the source of TTG melt compositions (Foley et al., 2002; Rapp et al., 2003) addresses the residual mineral assemblage during melting and not the protolith. Eclogite as an unmelted source rock would melt by dehydration melting of paragonite or phengite at higher pressures and much lower temperatures than garnet amphibolites (Moyen and Stevens, 2006). Eclogitic mineralogy results from higher-degree melting of garnet amphibolites following the elimination of amphibole from the residue, which may occur at <20% melting at pressures below 1.5 GPa but at >30% melting at higher pressures (Rapp and Watson, 1995; Rapp, 1995). Rapp et al. (2003) conclude that only high-degree melts above the amphibole-out temperature are relevant for TTG melts, as these degrees of melting are required for mobility of the melts (Rapp, 1995). However, this conclusion neglects the strong effect of higher water contents reducing the viscosity and so increasing the separability of lower-degree melts (Holtz et al., 1996). This may mean that low-degree melting should not be dismissed as a possibility in TTG genesis.

## 4. A SUMMARY AND CRITIQUE OF TRACE ELEMENT PARTITIONING DURING CONTINENTAL CRUST FORMATION

The exact conditions of origin of the early continental crust can be constrained further by trace elements, provided that the partitioning of the elements is well understood and partition coefficients are appropriately chosen from the plethora of coefficients now available. Trace elements can help to distinguish between melts in equilibrium with amphibolite and eclogite residues, to constrain the presence or absence of garnet, and even to make implications about the water contents in melts through the effects on trace element partitioning between plagioclase and melts. In this section, recent partitioning data relevant to the melting of amphibolites and eclogites are summarized, following which the application to TTG formation is considered (section 5). Partition coefficients are noted $D_i$ for an element i in keeping with internationally accepted terminology (Takahashi and Irvine, 1981; Beattie et al., 1993).

Numerous trace element partitioning studies have been carried out, both in experiments and on natural rocks (Klein et al., 1997; Barth et al., 2002a; van Westrenen et al., 1999; Bedard, 2006a), and attempts have been made to parameterize partitioning as a function of temperature, pressure, and bulk composition for individual minerals (Wood and Blundy, 1997; van Westrenen et al., 2001). However, for specific applications such as the melting of amphibolites and eclogites, the most appropriate choice of coefficients and applications of parameterizations has to be made. Numerous studies have shown that the availability and dimensions of crystal sites is the most important control on mineral/

melt partitioning (Blundy and Wood, 1994; Green, 1994; Lundstrom et al., 1998; Bottazzi et al., 1999). The importance of nonidealities in melts affecting the activities of trace element species may increase toward more silica-rich compositions (Linnen and Keppler, 1997; Schmidt et al., 2006).

### 4.1. Trace Element Partitioning Between Amphibolite Minerals and Melts

Mineral assemblages in basaltic compositions at solidus temperatures of 650–800 °C vary as a function of pressure but are simpler than in lower-temperature parts of the amphibolite facies. The volumetrically most important minerals are calcic amphibole and plagioclase, joined by garnet toward higher pressures and more aluminum-rich compositions. Accessory phases may include ilmenite, apatite, and epidote, but these are unlikely to persist into the melting interval and remain as residual phases that affect trace element partitioning. The melting reaction of amphibole is incongruent, producing the anhydrous phases clinopyroxene and orthopyroxene, which coexist with amphibole during the initial stages of melting but will replace amphibole completely at higher degrees of melting (Rapp et al., 1991; Rapp and Watson, 1995; Moyen and Stevens, 2006).

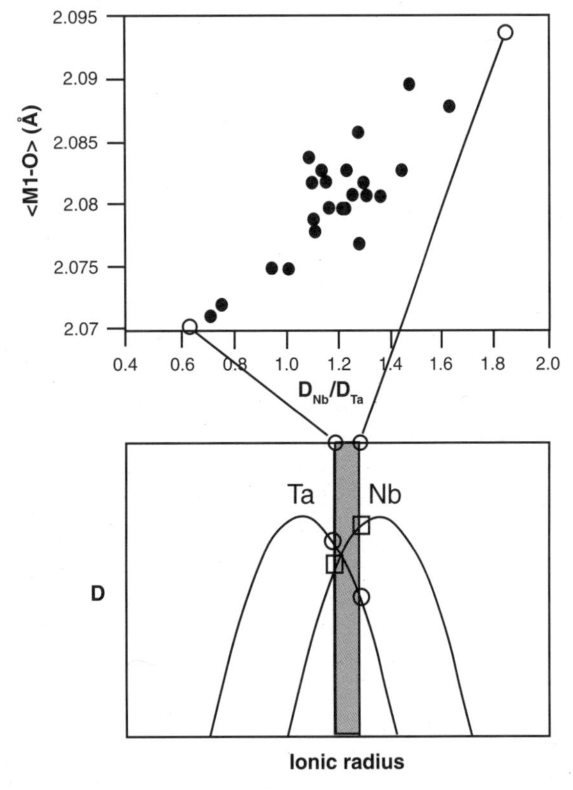


Figure 5. Variation of the partition coefficient ratio for amphibole/melt pairs in experiments with the average M1 polyhedron size as illustrated by the average bond length between the center of the M1 polyhedron and neighboring oxygens (Tiepolo et al., 2000a). The increase in Nb/Ta with increasing polyhedron size can be explained by a larger ionic radius for Nb than Ta, as illustrated in the lower diagram. The optimum site size for Nb is larger than the largest size offered by amphibole in this experimental series, whereas the optimum size for Ta is smaller. The difference in ionic radii of Nb and Ta can be estimated as 1.5 pm from these results (Tiepolo et al., 2000a).

Trace element partitioning between calcic amphibole and melts has been summarized in a series of papers by Tiepolo and coworkers (Bottazzi et al., 1999; Tiepolo et al., 2000a, 2000b, 2001). Their experiments filled a gap in $SiO_2$ contents of melts from previous studies (LaTourette et al., 1995; Klein et al., 1997; Brenan et al., 1995) and also greatly improved the understanding of trace element partitioning by presenting single-crystal structure refinements of amphiboles from each of the experimental charges. The structure refinement results were combined with elastic strain modeling (Blundy and Wood, 1994) and existing crystal chemical knowledge of amphiboles to make conclusions about the exact site occupancies of individual trace elements. Important chemical parameters such as Ti contents, Mg# (100 × Mg/(Mg + Fe)), and $Al_2O_3$ contents were varied deliberately in the experimental series to refine understanding of partitioning as a function of crystal chemistry. Important findings resulting from this systematic study included the recognition that Ti and HREEs are distributed between more than one site, the off-center positioning of trace elements within coordination polyhedra (Oberti et al., 2000; Bottazzi et al., 1999), and the difference in ionic radii for Nb and Ta (Tiepolo et al., 2000a), explaining their fractionation in geochemical processes. I concentrate here on results for the high field strength elements (HFSEs) Nb, Ta, Zr, and Hf because of their importance for the assessment of troughs in trace element patterns (Fig. 3) of melts coexisting with amphibole-bearing as opposed to amphibole-free residual assemblages.

Although the trace HFSEs Nb, Ta, Zr, and Hf are generally treated as a single geochemical group, it could be shown that they are not taken up on the same crystallographic sites in calcic amphiboles, but follow Ti, which partitions between two or three sites (Oberti et al., 2000). The three octahedral sites are not equivalent; Nb and Ta preferentially occupy the M1 site, where they assist Ti in the electronic charge balancing of dehydrogenation (substitution of $OH^-$ by $O^{2-}$, which is coupled by replacement of $Mg^{2+}$ and $Fe^{2+}$ by the higher-charged HFSEs; Tiepolo et al., 2000a). In contrast, Zr and Hf are sited preferentially on the M2 and M3 sites, which are farther from the OH sites and therefore do not contribute meaningfully to the dehydrogenation mechanism. This means that variations in Nb/Zr can be expected and are observed (Tiepolo et al., 2001), and that troughs may appear in the trace element pattern for Nb and Ta, but not for Zr and Hf, where amphibole has an effect.

The variation in Mg# and $TiO_2$ contents in the experimental program enabled a difference in ionic radii for Nb and Ta to be demonstrated. The measured average distance between the M1 polyhedron center and neighboring oxygens was found to correlate well with the ratio $D_{Nb}/D_{Ta}$ (Fig. 5). This cation site size is greatest where $D_{Nb}/D_{Ta}$ of the amphibole was highest, indicating that the ionic radius of Nb must be greater than that of Ta. The

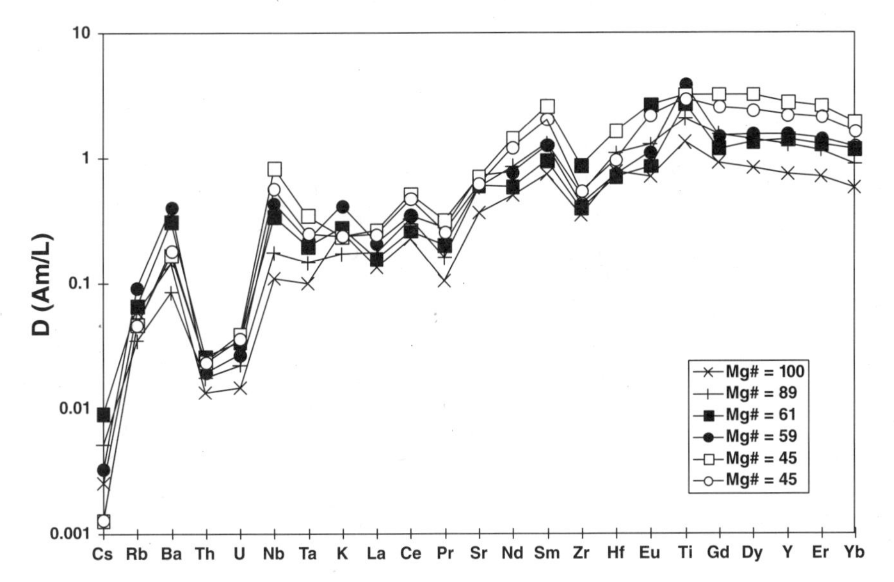


Figure 6. Variation in amphibole/melt partition coefficients for trace elements as a function of amphibole composition (Mg#), compiled from the results of Tiepolo et al. (2000a, 2000b, 2001). Lower-Mg# amphiboles have higher Nb/Ta and La/Nb, resulting in the opposite tendencies in coexisting melts, whereas high-Mg# amphiboles such as those in peridotites cannot impart these characteristics to coexisting melts (Foley et al., 2002). Amphiboles coexisting with average TTG melts (Mg# 40–45) have intermediate Mg# of 60–70.

resulting difference in partition coefficient implies a difference in ionic radius of ~1.5 pm (or 0.015 Å; Fig. 5) (Tiepolo et al., 2000a). The values of $D_{Nb}$ and $D_{Ta}$ depend most strongly on $TiO_2$ concentrations and Mg# of the amphibole, so that an empirical equation predicting $D_{Nb}$ and $D_{Ta}$ for application to any amphibole for which a major element composition analyzed by electron microprobe was available could be calibrated (Tiepolo et al., 2000a). From this experimental program, the dependence of amphibole/melt trace element partitioning on Mg# of the bulk composition can be illustrated (Fig. 6), and a preferred partition coefficient set for calcic amphibole/melt pairs applicable to the melting of amphibolites can be drawn up (Table 1).

The applicability of these values to amphibolites in equilibrium with melts of TTG composition has been questioned recently on the basis that TTG melts are more $SiO_2$-rich and that the $D_{Nb}$ values found by Klein et al. (1997) were partly lower (0.18–0.39 versus 0.44). However, this will be more a question of the water contents, and thus degree of dehydrogenation of the amphiboles, than of the overall $SiO_2$ contents of the melts, so that the $D_{Nb}$ and $D_{Ta}$ given in Table 1, for which 0.86–1.25 of the two "OH sites" are occupied by OH (Tiepolo et al., 2000a), should apply to natural conditions. $D_{Ti}$ can be used as a guide; wherever $D_{Ti}$ is >3, high $D_{Nb}$ and $D_{Ta}$ will result. Much higher amphibole/melt $D_{Nb}$ has been measured in rhyolites (0.98–6.7; Ewart and Griffin, 1994).

There are remarkably few experimental data on plagioclase/melt pairs, and this has been reviewed recently by Bedard (2006a). The most complete experimental set is by Bindeman et al. (1998), which constitutes a reanalysis of the earlier experimental charges of Drake and Weill (1975). These show a preference for the LREEs over HREEs, although the absolute D values are on the order of 0.1. The most prominent features are a high $D_{Sr}$, and a variable $D_{Eu}$ that depends on the oxidation state of Eu in the system. However, the Bindeman et al. (1998) values differ considerably from those measured or predicted by others (Blundy and Wood, 1994; Dunn and Sen, 1994; Norman et al., 2005; Bedard, 2006a), which show lower values for many elements such as the REEs. $D_{Sr}$ may be indirectly affected by the water content of the system (Foley et al., 2002), because $D_{Sr}$ is inversely proportional to the An content of the plagioclase, and An has been shown to increase with increasing $H_2O$ in the melt (Panjasawatwong et al., 1995; Takagi et al., 2005). $D_{Sr}$ values of 3–5 may thus be more realistic for melting of amphibolites than the lower values measured in Hawaiian lavas (Norman et al., 2005). Preferred D values for plagioclase/liquid pairs are given in Table 1.

TABLE 1. PARTITION COEFFICIENTS FOR AMPHIBOLE/MELT AND PLAGIOCLASE/MELT PAIRS SUITABLE FORMELTING OF AMPHIBOLITE (Mg# 60)

| | D (amph/liquid) | D (plg/liquid) |
|---|---|---|
| Ba | 0.40 | 0.2 |
| Th | 0.019 | 0.1 |
| U | 0.027 | 0.1 |
| Nb | 0.44 | 0.005 |
| Ta | 0.22 | 0.005 |
| La | 0.21 | 0.05 |
| Ce | 0.35 | 0.044 |
| Pr | 0.28 | 0.17 |
| Sr | 0.72 | 1.84 |
| Nd | 0.77 | 0.035 |
| Sm | 1.26 | 0.035 |
| Zr | 0.43 | 0.0005 |
| Hf | 0.75 | 0.0005 |
| Eu | 1.11 | 0.15 |
| Gd | 1.5 | 0.02 |
| Dy | 1.55 | 0.018 |
| Y | 1.56 | 0.005 |
| Er | 1.43 | 0.012 |
| Yb | 1.23 | 0.01 |

*Note:* Amphibole (amph) data compiled from Tiepolo et al. (2000a, 2000b, 2001); plagioclase (plg) values are estimated based on Norman et al. (2005) and Dunn and Sen (1994).

TABLE 2. RECOMMENDED TRACE ELEMENT PARTITION COEFFICIENTS FOR ECLOGITIC MINERALS

| | Garnet* | Cpx* | Rutile† | Opx§ |
|---|---|---|---|---|
| Ba | 0.0004 | 0.009 | 0.0043 | 0.00017 |
| Th | 0.001 | 0.020 | | 0.00011 |
| U | 0.002 | 0.031 | 0.02 | 0.00023 |
| Nb | 0.0045 | 0.017 | 102 | 0.00038 |
| Ta | 0.0045 | 0.017 | 170 | 0.00040 |
| La | 0.0032 | 0.116 | 0.0055 | 0.00010 |
| Ce | 0.0138 | 0.198 | 0.006 | 0.00014 |
| Pr | 0.0519 | 0.317 | 0.007 | 0.0005 |
| Sr | 0.0072 | 0.093 | 0.06 | 0.00025 |
| Nd | 0.169 | 0.478 | 0.008 | 0.0010 |
| Sm | 0.965 | 0.856 | 0.009 | 0.0033 |
| Zr | 0.517 | 0.229 | 4.24 | 0.0064 |
| Hf | 0.350 | 0.457 | 5.32 | 0.008 |
| Eu | 1.80 | 1.04 | 0.01 | 0.004 |
| Gd | 3.13 | 1.23 | 0.01 | 0.009 |
| Dy | 7.52 | 1.56 | 0.01 | 0.018 |
| Y | 9.30 | 1.63 | 0.007 | 0.028 |
| Er | 12.9 | 1.74 | 0.012 | 0.029 |
| Yb | 17.2 | 1.77 | 0.012 | 0.059 |
| Lu | 18.6 | 1.76 | 0.012 | 0.070 |

*Barth et al. (2002a).
†Foley et al. (2000, 2002).
§Eggins et al. (1998); Kennedy et al. (1993).

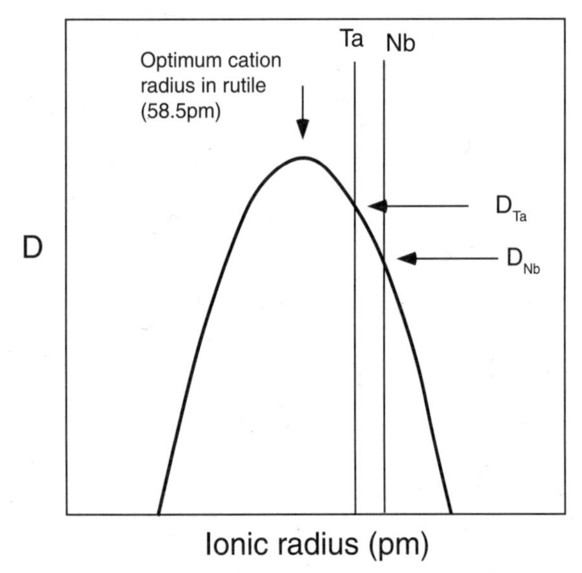


Figure 7. Ionic radii of Nb and Ta are larger than the polyhedron size offered by the rutile structure. The smaller ionic radius of Ta is therefore favored relative to Nb, resulting in low Nb/Ta in the mineral rutile, which will be expressed as high Nb/Ta in coexisting melts. This will be a general rule for Ti-bearing oxide minerals, all of which take up Nb and Ta in sites smaller than the ionic radius of tantalum, whereas silicate minerals take up Ta and Nb on larger sites.

### 4.2. Trace Element Partitioning Between Eclogite Minerals and Melts

The mineralogy of metabasite eclogites is relatively simple at solidus temperatures, consisting principally of omphacitic clinopyroxene and pyrope-grossular-rich garnet, with minor rutile and possibly orthopyroxene, depending on the exact bulk composition. This reduces the number of phases that need to be considered for the distribution of trace elements, but also increases the complexity of the crystal chemistry, particularly in clinopyroxene and garnet. Partition coefficients for augitic clinopyroxenes found in basalts cannot be simply applied, because the uptake of the jadeite molecule leads to splitting of both M1 and M2 sites, resulting in essentially four subsites (Carpenter, 1980; Smyth and Bish, 1988) between which the incompatible trace elements will be distributed. The number of experimental studies of garnet/melt trace element partitioning has increased recently (Klein et al., 2000; Green et al., 2000; van Westrenen et al., 1999, 2000, 2001; Klemme et al., 2002), but applications are complicated by complex activity-composition relations in garnets, which are less well characterized for pyrope-grossular solid solutions than for Fe-bearing garnets. Cation-oxygen bond lengths vary more along the pyrope-grossular join than along any other garnet solid-solution series and in a nonlinear way (Merli et al., 1995). Atomistic calculations predict higher trace element uptake in intermediate garnets (van Westrenen et al., 2003), but these results do not agree with crystal chemical measurements at present (Quatieri et al., 2004). Until these are better understood, preferably by studies that incorporate crystal structural refinements as for calcic amphiboles (see above), the few direct experimental determinations on natural systems involving melts of TTG-relevant compositions are preferred (Klein et al., 2000; Barth et al., 2002a). Partition coefficients for the melting of eclogites are given in Table 2. These are preferred to more recent data that predict much lower D values for many elements based on water-free experiments at much higher temperatures (Klemme et al., 2002; Pertermann et al., 2004).

The partitioning of the HFSEs depends critically on the presence or absence of rutile. The edge-sharing $TiO_2$ octahedra that make up the structure of rutile offer only one crystal site with cation-oxygen bond lengths between 56.5 and 59.7 pm (Smyth and Bish, 1988). This means that the optimal cation radius of ~58 pm for inclusion in rutile is smaller than all the main trace elements considered here. Nb and Ta are closest and thus will have the highest rutile/melt partition coefficients. Furthermore, given the difference in cation radius mentioned above, $D_{Ta}$ must be systematically higher than $D_{Nb}$ (Fig. 7), and indeed recent studies have shown the $D_{Nb}/D_{Ta}$ ratio to be between 0.6 and 0.7 (Schmidt et al., 2004). There are suggestions that chemical complexing of Nb and Ta in melts with high $SiO_2$ contents (>65 wt% $SiO_2$) may cause this difference to be reduced, but recent systematic studies show that this will not change the conclusion that $D_{Nb}$ will be appreciably higher than $D_{Ta}$ for application to eclogite melting and the production of TTG-like melt compositions.

Hf and Zr have larger cationic radii than Ta and Nb (by ~8 pm), but are much smaller than other elements in the full trace element patterns of Figures 2 and 3. This leads to a three-tier grouping of partition coefficients with $D_{Nb}$ and $D_{Ta}$ from ~80 to 300, $D_{Zr}$ and $D_{Hf}$ from ~4 to 6, and other elements another two orders of magnitude smaller (Foley et al., 2000). The trend of $D_{Nb} < D_{Ta}$ must apply equally to other titanium oxide minerals, contrasting with the pattern found for calcic amphiboles.

## 5. TRACE ELEMENT CONSTRAINTS ON EARLY CRUST FORMATION

As noted above, experimental melting studies have shown that the major element compositions represented by TTG gneisses correspond to melting of basaltic source material but cannot distinguish adequately between melting in the form of amphibolite or eclogite. Trace elements should help because of potentially large fractionations between melts and residual minerals, but results are controversial, with both amphibolite and eclogite being proposed to explain trace element patterns.

The plot of Nb/Ta against Zr/Sm (Fig. 8) was designed by Foley et al. (2002) to test the applicability of current amphibole partition coefficients to this question. The high Nb/Ta of amphiboles should result in low Nb/Ta in coexisting melts, and thus a position in the lower half of the diagram, whereas the decoupling of Nb from Zr means that no trough in the trace element pattern is present at Zr. Since Sm is adjacent to Zr in this pattern, melts coexisting with calcic amphiboles will have high Zr/Sm and thus be in the lower right quadrant of Figure 8. Because the Nb/Ta fractionation by amphibole is a function of the Mg# and $TiO_2$ contents of the amphibole, it is instructive to consider how these compositional parameters vary in natural amphiboles. Figure 9 shows that amphiboles in peridotites have consistently high Mg#, although variable $TiO_2$ contents, because amphibole is a volumetrically minor phase whose Mg# is buffered at high values by the coexisting silicate phases. In contrast, amphibolites of basaltic bulk composition have much higher amphibole modes, so that the chemistry of their amphiboles reflects that of the whole rock. Only the amphibolites contain amphiboles with low $TiO_2$ and low Mg#, which can fractionate Nb from Ta as modeled in Figure 8B (Foley et al., 2002), and this will be more effective the lower the Mg# is.

Despite considerable scatter, the generalization can be made that TTG gneisses are in the lower right quadrant, whereas modern island arc basalts tend to plot in the lower left quadrant (Foley et al., 2002), corresponding to a minor negative anomaly for Zr. Figure 8A is updated from Foley et al. (2002) to show the

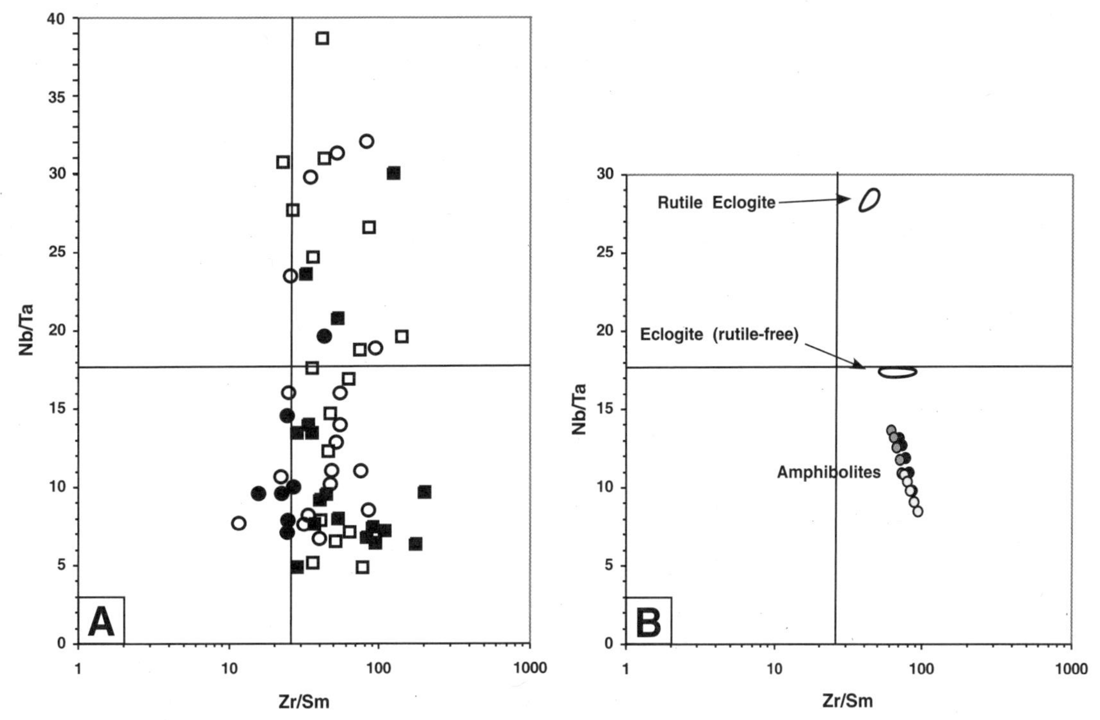


Figure 8. Nb/Ta versus Zr/Sm diagrams. (A) Signatures of TTG gneisses in the same time slices as in Figure 1 (4000–3600 Ma solid squares; 3600–3200 Ma solid circles; 3200–2800 Ma open squares; 2800–2500 Ma open circles). The average and most individual values for rocks older than 3200 Ma plot in the lower right quadrant, whereas more data points (although still the minority) in the Late Archean plot in the upper right quadrant. (B) Melting models for garnet amphibolites plot in the lower right quadrant with low Nb/Ta and high Zr/Sm (batch melting models; reproduced from Foley et al., 2002). Melting of rutile-bearing eclogites lie in the upper right quadrant, whereas in the absence of rutile, eclogite does not appreciably fractionate Nb from Ta. Full trace element patterns are given in Figure 10.

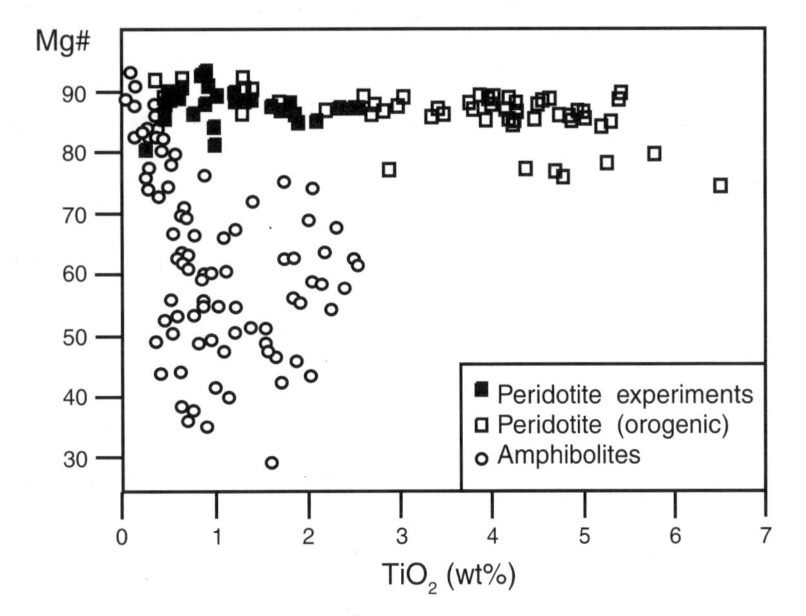


Figure 9. Mg# versus $TiO_2$ in natural amphiboles from amphibolites (open circles) and in peridotites from high-pressure experiments (solid squares) and from orogenic bodies (open squares). Fractionation of Nb from Ta can only be controlled by amphiboles with low Mg# and low $TiO_2$ contents such as those found in amphibolites (Tiepolo et al., 2000a). Peridotitic amphiboles are buffered to high Mg# and so cannot cause low Nb/Ta in coexisting melts.

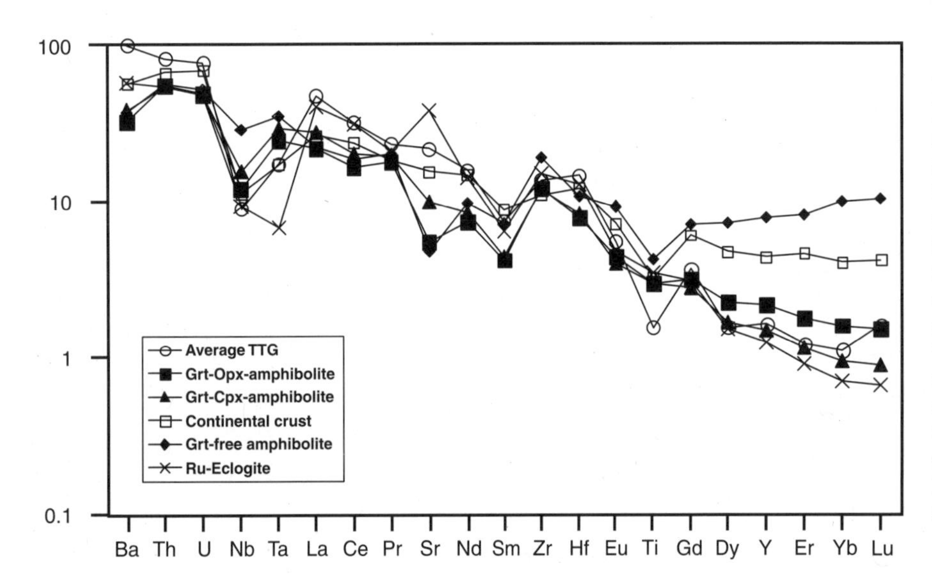


Figure 10. Trace element pattern for melts of basaltic compositions with differing residual mineral assemblages (normalized to primitive mantle values of McDonough and Sun, 1995). Calculations use partition coefficients listed in Tables 1 and 2. The low Nb/Ta ratio can be reproduced by melting of amphibolites, but not by rutile-eclogites. A rutile-free eclogite does not appreciably fractionate Nb from Ta and so also does not fit the patterns of most TTG gneisses. Garnet-free amphibolite produces melts with much higher concentrations of HREEs; these characteristics are not seen in the TTG rock record. Modal mineralogy of residua are taken from experiments on melting of basaltic compositions (Rapp et al., 1991; Wolf and Wyllie, 1994).

same time slices as in Figure 1, restricting attention to ICP-MS analyses. This shows an increasing number of TTG gneisses outside the lower right quadrant toward the Late Archean; a similar conclusion was reached by Condie (2005, Fig. 7 therein), on the basis of averaged analyses for individual locations.

The position of melts of amphibolite in the lower right quadrant (Fig. 8B) will apply as long as amphibole remains in the residue. The full trace element pattern calculated for melts coexisting with several pyroxene-bearing garnet amphibolites are shown in Figure 10; these residual mineralogies are taken from experiments with 10%–30% melt (Rapp et al., 1991; Wolf and Wyllie, 1994) with TTG-like major element compositions. The average TTG pattern follows the low Nb/Ta and high Zr/Sm expected to be due to amphibole control, but not the high Nb/Ta that must result from rutile-eclogite (Fig. 10) because of the crystal site size (Fig. 7). Garnet appears to be required in the residue to explain the low concentrations of HREEs (Fig. 10) (Martin, 1994; Foley et al., 2002; Moyen and Stevens, 2006). The most aberrant fit in Figure 10 other than Ta is for Sr, which Foley et al. (2002) explained as a function of the $H_2O$ content: Plagioclase is richer in anorthite in water-bearing conditions (Panjasawatwong et al., 1995; Takagi et al., 2005), and $D_{Sr}$ is lower in plagioclase with high anorthite content (Blundy and Wood, 1994; Bedard, 2006a). Thus, an increase in the amount of water present during melting stabilizes more Ca-rich plagioclase, which is not able to hold back so much Sr, resulting in more Sr being taken up in the melt.

However, Rapp et al. (2003) preferred melting of eclogite to explain the trace element patterns of TTG gneisses, criticizing the conclusions of Foley et al. (2002) on three counts:

(1) inappropriate choice of rutile/melt partition coefficients, (2) requirement for low degrees of partial melting, and (3) the choice of basaltic starting material with chondritic Nb/Ta ratio for their melting model (Fig. 8B). The first two of these are answered here, whereas the third remains as an important and unresolved problem that is central to interpretations of whether plate tectonics operated during the Archean. For the first point, Rapp et al. (2003) recommended lower values of $D_{Nb}$ (55–60) and $D_{Ta}$ (85–90) based on the earlier experimental work of Green and Pearson (1987), overlooking the fact that Foley et al. (2002) used more recent, higher values (102 and 170; Foley et al., 2000, Table 1 therein), which are supported by subsequent studies (Schmidt et al., 2004). The second objection about the degree of partial melting does not apply: On the original diagram by Foley et al. (2002), the two models of fractional melting with low degrees of melting are especially noticeable because they are the only ones that plot outside the lower right quadrant. However, *all* batch melting models (reproduced here in Fig. 8B) and some fractional melting models plot in the lower right quadrant and do not depend greatly on the degree of melting; even high degrees of melting plot in the lower right quadrant. Furthermore, all melting models use garnet amphibolite residual mineralogy as guided by experiments with measured mineral modes with 10%–30% melting in which TTG-like melts were analyzed (Rapp et al., 1991; Wolf and Wyllie, 1994).

The third objection is especially interesting for the purposes of this paper. Rapp et al. (2003) note that melts of eclogite will plot in the lower right quadrant provided that the protolith already had a low Nb/Ta ratio, which they assign to crustal imbricate thrust stacking and tectonic accretion *of subduction-related terranes*. This scenario will not be greatly affected by any differences in opinion about the correct choice of $D_{Nb}$ for rutile, nor by fractional crystallization. They thus assign the low Nb/Ta of TTG gneisses to a *previous* subduction signature. This will be evaluated further in section 6.

## 6. TRACE ELEMENTS AND THE ARCHEAN SUBDUCTION SIGNATURE

Summarizing the above evidence and its application to the trace element patterns seen in TTG gneisses (Fig. 3), it can be concluded that melting of garnet amphibolite played a major role in the formation of the sodic granitoid material throughout the Archean. Melting of eclogites may well have played a role if Nb/Ta was minimized previously by other processes (Rapp et al., 2003), but these processes remain to be specified. In a multistage process of melting, crystallization of these melts, and subsequent remelting of the resulting igneous rocks, the sources of the final-stage melts may well have low Nb/Ta.

The relevance of the trace element evidence to the theme of this volume comes through the question of whether subduction is needed to explain the trace element budgets of the early continental crust. A subduction-related genesis of the TTG suite is generally assumed or invoked (Burke et al., 1976; Tarney et al., 1976; Windley, 1977; Condie, 1986; Martin, 1986; de Wit, 1998; Martin et al., 2005; Samsonov et al., 2005). However, we need to be careful to critically examine whether claims of subduction processes operating in the Archean are due to simple application of the uniformitarian principle, which would *assume* that petrological explanations on modern Earth with plate tectonics must necessarily apply to all periods of Earth history. This assumption surfaces again specifically in the interpretation of Rapp et al. (2003), who assign the subchondritic Nb/Ta of their eclogite source for TTG gneisses to previous arc processes.

However, experimental petrological studies for major and trace element composition of melts constrain directly only the minerals present in the melting assemblage, and the temperature and pressure at which melting occurred. The geodynamic interpretation should be based on whether the solidi of relevant assemblages can be crossed by geodynamically reasonable geothermal gradients at the appropriate pressure-temperature conditions. Constraints on tectonic setting, although frequently drawn by geochemists from petrological results, usually on the basis of the HFSEs in the case of arc or subduction processes, are at best indirect and are a further step removed from the evidence. Our understanding of the petrological melting mechanisms and the consequent trace element patterns is now much better than our grasp of the tectonic scenarios in which this occurred. To evaluate the robustness of conclusions about tectonic settings, the following must be considered: (1) how much trust we can put in the "subduction signature", (2) factors that may differ between subduction in the Archean and on modern Earth, and (3) alternative tectonic scenarios that may satisfy the petrological and geochemical evidence.

### 6.1. The Robustness of the Geochemical Signature of Subduction

The recognition of subduction takes a key place in the theory of plate tectonics. The explanation that oceanic plates are returned to the mantle was an essential counterpart to seafloor spreading (Pitman and Heirtzler, 1966) and relied on geophysical evidence, i.e., the interpretation of the downward-sloping pattern defined by earthquake foci as outlining the subducting slab (Oliver and Isacks, 1967). There is little geophysical evidence for the Archean, with the possible exception of paleomagnetic data suggesting that plates were moving as rigid bodies (Borradaile et al., 2003; Bleeker, 2003; Evans and Pisarevsky, this volume). There are also a number of geochemical indicators missing that have helped greatly to constrain what is subducted and how long it takes in modern systems, such as Be and U-Th series isotopes (Turner et al., 1997). Nevertheless, seismic reflectors from Late Archean cratonic terranes have been interpreted as representing subducting slabs trapped in time (Calvert et al., 1995).

The recognition of a trace element "subduction signature" as it is generally accepted for application to igneous rocks of the geological past relies basically on three features: (1) an enrichment in strongly incompatible elements, particularly the large ion lithophile elements (LILEs), (2) lower concentrations of the less incompatible elements, particularly the HREEs, compared to mid-oceanic-ridge basalts, and (3) much lower concentrations of the trace HFSEs than would be expected from their position on the incompatibility scale, resulting in troughs in the pattern, often referred to as "negative anomalies" (Pearce and Norry, 1979; Pearce, 1983; Thirlwall et al., 1994; Gamble et al., 1996). These features are so widely accepted as characteristic of subduction that it is often overlooked that the reason, i.e., the underlying mechanism, for these features is still debated. There is unanimity that material transfer from the subducting slab to the mantle wedge and so the magma source region is necessary (Ringwood, 1982; Tatsumi, 1989; McCulloch and Gamble, 1991), but whether this takes the form of hydrous fluids, silicate melts, or supercritical fluids is uncertain (Stalder et al., 2001; Kessel et al., 2005). The last of these appears to be gaining credence, particularly for the transfer of LILEs, but the explanation of the HFSE signature remains problematical. The anomalously low concentrations of Nb, Ta, and Ti, and to a lesser extent Zr and Hf, prompt comparison with the three-tier partitioning pattern of rutile (Foley et al., 2000), which holds also for rutile/fluid pairs (Stalder et al., 1998), and indeed rutile has long been held responsible by some (e.g., Nicholls and Ringwood, 1973). However, filtering of HFSEs by amphibole from either melts (Ionov and Hofmann, 1995) or fluids (Foley et al., 2002) or fractionation by clinopyroxene (Keppler, 1996) have also been suggested.

Added to the problem that the causal mechanisms for the geochemical subduction signature are not well understood comes the fact that many TTG gneisses are high-grade metamorphic rocks, meaning that the original trace element signature is unlikely to be undisturbed. Pearce et al. (2005) have recently assessed trace element behavior in subduction zones, and suggest the use of specific trace elements such as Ba and Th that are mobile in the presence of a fluid. However, high-temperature metamorphic reactions will commonly involve fluid or even melt loss (White and Powell, 2002), so that the more incompatible trace elements are unlikely to closely represent the original rocks. In their consideration of eclogite xenoliths, Jacob and Foley (1999) restricted their attention to HFSEs and HREEs, considering the LILEs to bear no resemblance to the concentrations in the protolith.

### 6.2. Differences between "Subduction Signatures" during the Archean and on Modern Earth

Overlooking the problems concerning the mechanisms responsible for the modern subduction signature, and accepting that the TTG gneisses are formed at subduction zones, then Figure 8 contains a simple demonstration that modern and Archean subduction-related processes differ. The much-debated closest analogy to TTG formation on modern Earth is that of adakites (Martin, 1999; Smithies, 2000). These fall together with TTG gneisses in the lower right quadrant of Figure 8A, whereas island arc basalts are characterized by subchondritic Zr/Sm. This difference may be due to the control of HFSEs by melting during the Archean, whereas modern subduction is characterized by fluid transfer, in which Nb/Zr fractionation is not so extreme (Stalder et al., 1998). This implies important differences in the geothermal gradient between low gradients that pass through the blueschist facies in modern subduction (Clarke et al., 1997; Fryer et al., 1999) and higher gradients that avoid blueschist and possibly also eclogite facies before reaching melting temperatures.

These differences in geothermal gradient have important ramifications for the development of trace element budgets during metamorphism, irrespective of whether subduction operates. The partitioning of trace elements in high-grade metamorphic processes depends on the prograde path followed: During a metamorphic reaction, elements will be redistributed between the new phases or partly lost to a fluid phase (if formed). This gives the minerals on the low-temperature side of a reaction a different determining factor from that of the residual minerals in a melting reaction; the trace elements will be shared out according to the relative mineral/mineral partition coefficients, which can result in trace element patterns in minerals that differ strongly from those in igneous systems. For modern subduction zones, hydrous phases such as phengite have been shown to be important in the control of LILEs (Zack et al., 2001), and great attention has recently been paid to high-pressure, low-temperature stabilities of minerals such as phengite, lawsonite, and glaucophane (Pawley, 1994; Domanik and Holloway, 1996; Poli and Schmidt, 1998; Forneris and Holloway, 2003), which may be avoided completely at much higher geothermal gradients. Here, minerals such as chlorite, epidote, amphibole, plagioclase, and garnet will exert more control, and a smaller set of hydrous minerals will be involved. Little is known about the differences in behavior of trace elements between many of these minerals.

### 6.3. Alternative Tectonic Scenarios for Early Crustal Formation

"Where nobody knows anything, there is no point in changing your mind." Bertrand Russell (1952)

Most scientists support the idea that plate tectonics started sometime between 4.0 and 3.0 Ga (Shirey et al., this volume; Wyman et al., this volume; Condie and Kröner, this volume), although many seem quite happy to consider the question of what preceded plate tectonics to be thoroughly intractable. If subduction operated, then it must have differed from that on modern Earth, so that the options for the Archean are either a modification of modern subduction processes, and thus plate tectonics, or some kind of pre-plate tectonic scenario. The main problem is our inability to judge the initial heat content and thus the cooling rate of Earth, and much of this uncertainty is related to poorly constrained scenarios for magma ocean cooling and crystallization. Oxygen isotopes on 4.4 Ga Jack Hills zircons, however, are interpreted to indicate surface water just 150 million years after formation of Earth (Peck et al., 2001).

TTG gneisses present a different temporal record from that of other petrological sources of information such as metamorphic belts, the eclogite xenolith record, and the subcratonic mantle. The earliest continental crust dates from 3.8 to 4.0 Ga, whereas the other sources appear to start between 3.5 and 3.3 Ga. Nevertheless, it is pertinent to consider the information these other rocks give us about the early operation of plate tectonics. Regional-scale metamorphic belts can offer evidence for plate tectonics when two contrasting *P-T* regimes are found in close proximity (Brown, 2006, this volume). The earliest demonstration to date is for 3.23 Ga in the Barberton region of South Africa (Moyen et al., 2006; Pease et al., this volume). The oxygen and sulfur isotope signatures of eclogite xenoliths demonstrate that they were earlier close to the surface and exchanged with ocean water or atmosphere (Jacob et al., 1994; Farquhar et al., 2002; Jacob, 2004), and they have therefore been widely interpreted as samples of subducted ocean crust. Many of their ages fall between 3.3 and 2.6 Ga (Jacob and Foley, 1999; Barth et al., 2002b; Jacob, 2004), whereas older examples are still unknown. Peridotite xenoliths of the subcratonic mantle date as far back as 3.5 Ga using the relatively inexact Re depletion ages (Pearson, 1999). Sulfide inclusions in diamond, which should give more dependable ages according to Alard et al. (2002), give similar ages as far back as 3.52 Ga (Westerlund et al., 2006).

Despite claims that formation of crust and subcontinental lithospheric mantle occurred during the Archean (Davis et al., 2003; Bedard, 2006b), the earliest TTGs (>3.5 Ga) appear unaccompanied by these other features. The conclusion is that good evidence for plate tectonics in terms of orogenic metamorphism, eclogites, and stable mantle lithosphere appears restricted to later than 3.5–3.3 Ga, and plate tectonics must be considered uncertain before that. The first 500–700 million years of TTG gneisses lack counterparts in terms of mantle lithosphere or orogenic events expected of plate tectonics.

Returning to the geochemical characteristics of TTG gneisses and how they might help to resolve this question, the question is how does the low Nb/Ta originate? This can be either generated by direct partial melting of amphibolite or inherited from a source already having low Nb/Ta due to an earlier process. In the latter case, subduction is presumed but may not necessarily apply. The distinction between amphibolites and eclogites is potentially important for tectonic scenarios because it raises the question of the water content: Recycling of surface basalts by subduction easily explains the transfer of water to great depths, and the accumulation of hydrothermally altered basalt piles away from subduction zones may also lead to melting of basaltic material in the form of amphibolites. I now consider early plate tectonic and subduction-free scenarios for the opportunities they present for amphibolite and eclogite melting.

### *6.3.1. Archean Plate Tectonics*

It was established on the basis of trace element signatures (Figs. 8 and 10) that if Archean subduction existed, then the geochemical characteristics of subduction-related igneous rocks were different from those on modern Earth (see also Smithies, 2000). Melting must have been more widespread if mantle temperatures were higher, although by no means ubiquitous, and slabs melted as garnet amphibolite to form TTG-like melts. However, it is not easy to draw conclusions about the overall geotherms from this information. It is often supposed that given higher mantle temperatures, plates would be smaller and maybe move quicker (Hynes, this volume), allowing greater heat loss through a longer ridge system. Thus the average age of oceanic crust at subduction would be younger, perhaps 20–30 Ma instead of the current average of 56 Ma (Hargraves, 1986; Karsten et al., 1996). If true, then subducting crust would have been warmer and so would be more prone to melt, explaining TTG formation at subduction zones. The oceanic geotherm away from subduction zones may not, however, have been much higher than today, meaning that the subduction geotherm would be disproportionately hotter in comparison to today (Fig. 11).

The next problem is oceanic plate thickness. Assuming similar rates of seafloor spreading, melting would begin at deeper levels in a hotter upper mantle, and the integrated melt volume would cause a much thicker crust than today. How much thicker depends on the temperature profile, with most estimates ranging between 20 km and >40 km (Sleep and Windley, 1982; Bickle, 1986;

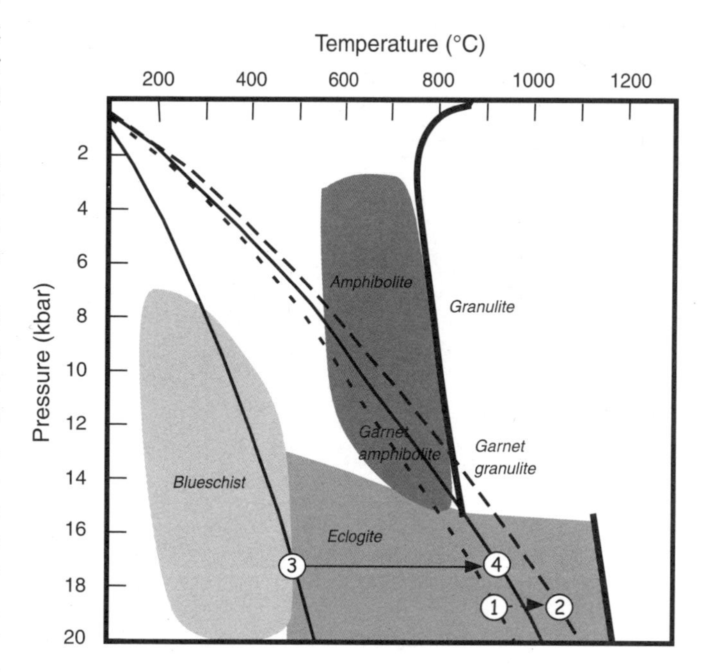


Figure 11. Variation of oceanic and subduction geotherms through time. Slightly higher heat loss from the mantle during the Archean may have been achieved through longer ridges and thus smaller plates with a lower average age at subduction. Subducting plates were thus hotter, resulting in much higher subduction geotherms (4 versus 3) despite average oceanic geotherms being only marginally hotter (2 versus 1). Geotherms 2 and 4 cut the solidus of garnet amphibolite (thick solid line), resulting in much commoner melting of subducting crust during the Archean than in modern Earth. Dry eclogite melts at higher temperatures that are not reached along any geotherm at <20 kbar.

Bickle et al., 1994; Kent et al., 1996). Because their crustal thicknesses are similar to these estimates, oceanic plateaux on modern Earth have been used as proxies for Archean crustal processes (Saunders et al., 1996; Kent et al., 1996). These analogies should be taken seriously, but not too literally, as differences cited (e.g., HFSE anomalies in Archean rocks; Karsten et al., 1996) need not negate the analogy for a process in the Archean, which differs in detail from *all* environments on modern Earth. The greater degree of melting involved means that the overall composition of the crust would have been more magnesian and may have been picritic to komatiitic. This would, however, have differentiated internally into basaltic upper parts and ultramafic cumulates (Farnetani et al., 1996; Foley et al., 2003). Cumulates—consisting of dunites, pyroxenites, and melanonorites—and gabbros would form the lower crust with up to a third of the total thickness (Neal et al., 1997), and gabbros would dominate the middle sections.

We can now consider the fate of various parts of the Archean oceanic crust, assuming subduction at a younger age, on average.

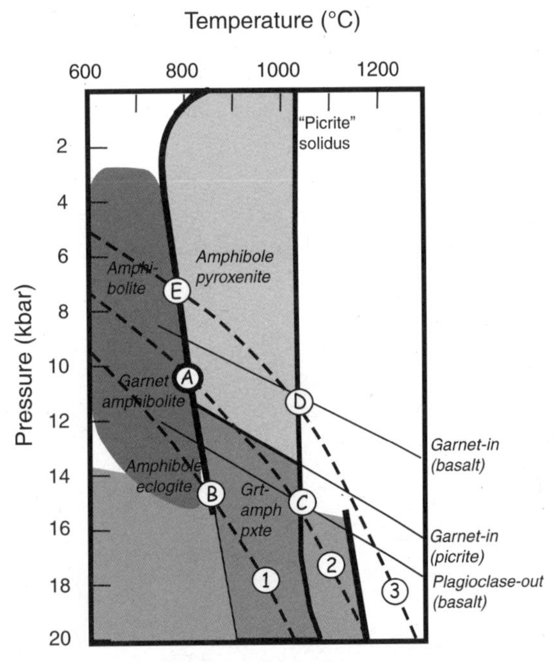


Figure 12. Comparison of possible Archean geotherms (dashed lines 1, 2, and 3) and the melting curves of water-bearing basalts and picrites (heavy solid lines). Shading indicates rock types at the solidus immediately to the right. Picrites melt as amphibole-bearing pyroxenite at pressures below 13–14 kbar and contain garnet at higher pressures. Melting can occur on the elevated geotherms 2 and 3, and produces basaltic melts (Foley et al., 2003). Metabasalts could melt as amphibole eclogite along geotherm 1 at B, as garnet amphibolite along geotherm 2 at A, or as garnet-free amphibolite along geotherm 3 at E. Trace element patterns in TTG gneisses (Fig. 3) argue against abundant melt production at E in the Early Archean, which would lead to a tendency for HREE-richer melts in the Early Archean. Thus, Early Archean geotherms are unlikely to have been as hot as geotherm 3.

Zircons dating from before the rock record appear to demonstrate the existence of surface water (Peck et al., 2001), so that the phenomenon of hydrothermal alteration of the uppermost oceanic crust should apply throughout the rock record. Those parts of the upper basaltic layers that had been hydrothermally altered could melt as garnet amphibolite to give rise to TTG-like melts, fitting their major and trace element characteristics (Figs. 4 and 8) (Moyen and Stevens, 2006). If the lower crust melted, it would not have produced TTG-like melts, as ultramafic pyroxene- or amphibole-rich rocks produce basaltic or even silica-undersaturated melts (Foley et al., 2003; Kogiso and Hirschmann, 2006). This led Foley et al. (2003) to propose that the large volumes of TTG gneisses dating from the end of the Archean may be due to the hindrance of subduction of the upper parts of the oceanic crust much earlier than 3.3 Ga. Before this, crust more than ~25 km thick may not have been subductable as a unit (Abbott and Mooney, 1995), so that the cumulate-rich lower parts delaminated and melted, but without extensive production of silica-rich melts. If this had not been the case, and geotherms were still hotter in the Early Archean, then immense volumes of silicic plutons should have been produced by melting of hydrothermally altered basaltic crust between 4.0 and 3.5 Ga, many with trace element characteristics attesting to melting of garnet-free amphibolites, particularly earlier in time (Fig. 12). However, all these would have to have been recycled into the mantle, as very few candidates are observed in the rock record. This may form an important argument against the existence of large volumes of *continental* crust in the Early Archean (cf. Armstrong, 1991).

Eclogite xenoliths in kimberlites sample the cratonic mantle, and their chemical compositions can be used to crudely calibrate the geotherms of the Late Archean if plate tectonics operated. They bear many geochemical characteristics of hydrothermally altered oceanic crust; most were originally gabbros as indicated by oxygen isotopes, and coupled Sr and Eu anomalies (Jacob et al., 1994; Rollinson, 1997; Jacob and Foley, 1999; Barth et al., 2002b; Jacob, 2004), whereas a few have high $\delta^{18}O$, indicating low-temperature alteration (volcanics). Those of volcanic origin are picritic in composition at an age of 2.6 Ga (Jacob and Foley, 1999) and are geochemically similar to eclogites as old as 3.3 Ga. However, no eclogites as old as the earliest TTG gneisses are known that could constrain the oceanic components of plate tectonics.

The attribution of low Nb/Ta in TTGs to melting of eclogitic source material that already possessed low Nb/Ta due to previous tectonic thickening of arc crust (Rapp et al., 2003) thus equates to a proposal for subduction processes in the earliest Archean, probably before 4.0 Ga. This would be in keeping with suggestions that plate tectonics already operated by 4.4 Ga. This cannot be discounted, but widens the time window between the proposed start of plate tectonics and our real evidence for it from other geological constraints (see also Shirey et al., this volume). There are two reasonable mechanisms to cause low Nb/Ta (discounting a strongly subchondritic Nb/Ta in silicate Earth), which are (1) previous melting of amphibolite (Foley et al., 2002) or (2) melting of an upper-mantle reservoir much more depleted than the present MORB reservoir (Bennett et al., 1993; Foley et al., 2003). The first of these could explain low Nb/Ta in just one or two episodes because of the correct sense of active fractionation by crystal sites in the amphibole lattice (see above). The second mechanism appeals to lowering of Nb/Ta by the difference in incompatibility in peridotite minerals, which is only slight (Eggins et al., 1998): The severe depletion of a restricted layer of upper mantle needed would presumably require several melting cycles, which would probably require an important role for water in fluxing the later episodes of melting. The Nb/Ta fractionation may favor the action of amphibolite melting, and the progression from amphibolite to eclogite melting with time (Fig. 12) is in keeping with the later appearance of high Nb/Ta in TTG gneisses (Fig. 8).

### *6.3.2. Pre-Plate Tectonic Scenarios*

Pre-plate tectonic magma genesis may be limited to pre-Archean times from which we have almost no evidence in the rock record, or it may be needed to explain the first 500 million years or so of TTG history. Bedard (2006b) has recently suggested that TTG genesis as late as 2.7 Ga was caused by a catalytic delamination process related to positions above mantle plumes. Ever since the crystallization of the magma ocean (assuming this existed), some kind of crust must have been present at Earth's surface, but differentiating between oceanic and continental crust may first become meaningful at much later stages. As long as melting assemblages are restricted either to mantle peridotite itself, to pyroxenite streaks or blocks within it, or to delaminated ultramafic cumulates from the lower crust, no melt with $SiO_2$ contents above 55 wt% would have been formed, and so no continents could have formed. They require melting of quartz-bearing eclogite or amphibolites (Foley et al., 2002, 2003; Moyen and Stevens, 2006; Kogiso and Hirschmann, 2006), noting that the trace element models appear to require the latter at an earlier stage than the former.

So *where* did melting occur on an Earth before plate tectonics? An important difference from the modern world is that the two locations that now produce the most abundant magmatism, the mid-ocean ridges and the subduction zones, would not exist to provide the locations for melting. At mid-ocean ridges, voluminous melt production is allowed by the divergence of the plates permitting decompression melting at depths of less than ~45 km where the geotherm defined by upwelling mantle crosses the solidus (McKenzie and Bickle, 1988). At subduction zones, melting is permitted by water returned to the mantle. Taken together, these two locations are estimated to produce at least 88% of magmatism (Schmincke, 2004; Jicha et al., 2006) on modern Earth.

Without seafloor spreading and subduction, basaltic volcanism would accumulate at the surface and could probably only be returned to the mantle by delamination of the lower crust (Condie, 1986; Foley et al., 2003). The amount of crust produced in a given period depends largely on the temperature of the mantle, which is one of the great unknowns for early Earth history. It is commonly assumed that mantle temperatures decreased gradually at rates subparallel to radioactive decay curves (Abbott et al., 1994), which would mean a considerably hotter Early Archean mantle and consequently much higher rates of crustal production. However, there is now considerable evidence for episodic production of Archean continents (Condie, 1989; Bedard, 2006b), which is possibly due to mantle turnover events or at least strongly nonlinear development of temperature profiles.

It is debatable how rapidly heat was lost by turbulent convection following solidification of the magma ocean; thorough cooling to temperatures comparable to the Late Archean before 4.0 Ga is possible, and would be appealing in view of the trace element evidence presented above. If Earth had been much hotter, then it would be difficult to avoid considerable melting of delaminated or subducted basaltic crust, which would lead to the production of huge volumes of TTGs at relatively low pressures below the stability of garnet at ~10 kbar (Fig. 12) (Moyen and Stevens, 2006). Such a crust would have trace element characteristics that are not typical of TTG gneisses, with higher HREEs and low Sr. As this is not seen, it would have to be argued that it has been almost completely recycled into the mantle, but the volume of recycled crust must be limited to much less than this in order to prevent very different radiogenic isotope signatures from those seen.

If we assume that mantle temperature was similar to or slightly higher than that today but seafloor spreading did not operate, then melting would be restricted to local areas such as ocean islands today. The melts would be picritic, because melting depths would be deeper than under mid-ocean ridges today where tholeiitic basaltic rocks dominate. However, the melting conditions of modern mid-ocean ridges would not occur, because the conditions of extension and plate divergence allowing passage of the upper mantle so close to the surface would be prevented. Also, subduction would not be present for flux melting of peridotite by transport of water into the mantle, and so the overall rate of magma production may have been much lower even if the mantle was slightly hotter.

The production of continental crust (meaning crust with $SiO_2$ contents >60 wt%) would require delamination of the lower crust in some way. The most $SiO_2$-rich melts from peridotite are boninites with ~54–55 wt% $SiO_2$, but these require both elevated water contents and low pressures (<5 kbar; van der Laan et al., 1987), which are discounted by the presence of the thicker crust in the same way as are primary tholeiitic basalts. Without subduction and seafloor spreading, large volcanic piles would build up associated with central volcanoes, and these may eventually become so thick that their lower parts would be delaminated and may melt on returning to the mantle. Initially, these lower parts would be pyroxenites and would not produce continental crust-style melts, but eventually continued volcanism would press the earliest basaltic upper levels so deep that they would melt. Many of the rocks would have been hydrothermally altered at the surface and so would contain considerable water and would melt as amphibolites. When this stage is reached, the potential to produce silicic plutons is realized for the first time. These will be tonalitic to trondhjemitic in composition (see section 3), and the proportion of trondhjemites may be higher due to additional production by fractionation: Trondhjemitic rocks can be produced by extensive fractionation of basaltic parent melts, whereas tonalites cannot (Martin and Sigmarsson, 2005). It can be expected that the return flux of hydrothermally altered basalt to the lower crust and thus to melting conditions is much lower in the volcanic piling scenario than in the subduction scenario, which may better explain the missing large volumes of TTGs with chemical signature corresponding to low-pressure melting of garnet-free amphibolite (Fig. 12).

In summary, the requirement for TTG genesis by melting of garnet amphibolite can be accommodated in petrogenetic scenarios either with modified plate tectonics or without it. The question of how to avoid the production of large volumes of TTG magmas at low pressures in the Archean with trace elements that are not seen may be a more useful constraint in the long term.

## ACKNOWLEDGMENTS

Work on Archean crust formation is supported by the Geocycles Cluster at Mainz, which is part of the "Wissen Schafft Zukunft" program of the State of Rheinland-Pfalz. This paper is Geocycles Publication 45. The arguments about trace element behavior used here stem largely from the amphibole trace element partitioning project run in cooperation with Massimo Tiepolo, Riccardo Vannucci, Roberta Oberti, and Alberto Zanetti in Pavia, to whom I am greatly indebted. Reviews by Hervé Martin and Jean-Francois Moyen and the editorial comments of Vicky Pease have helped to improve the paper.

## REFERENCES CITED

Abbott, D., and Mooney, W., 1995, The structural and geochemical evolution of the continental crust: Support for the oceanic plateau model of continental growth: Reviews of Geophysics, v. 33, p. 231–242.

Abbott, D., Burgess, L., Longhi, J., and Smith, W., 1994, An empirical thermal history of the Earth's upper mantle: Journal of Geophysical Research, v. 99, p. 13,835–13,850, doi: 10.1029/94JB00112.

Alard, O., Griffin, W.L., Pearson, N.J., Lorand, J.P., and O'Reilly, S.Y., 2002, New insights into the Re-Os systematics of sub-continental lithospheric mantle from in situ analysis of sulphides: Earth and Planetary Science Letters, v. 203, p. 651–663, doi: 10.1016/S0012-821X(02)00799-9.

Armstrong, R.L., 1991, The persistent myth of crustal growth: Australian Journal of Earth Sciences, v. 38, p. 613–630, doi: 10.1080/08120099108727995.

Barth, M.G., Foley, S.F., and Horn, I., 2002a, Partial melting in Archean subduction zones: Constraints from experimentally determined trace element partition coefficients between eclogitic minerals and tonalitic melts under upper mantle conditions: Precambrian Research, v. 113, p. 323–340, doi: 10.1016/S0301-9268(01)00216-9.

Barth, M.G., Rudnick, R.L., Carlson, R.W., Horn, I., and McDonough, W.F., 2002b, Re-Os and U-Pb geochronological constraints on the eclogite-tonalite connection in the Archean Man Shield, West Africa: Precambrian Research, v. 118, p. 267–283, doi: 10.1016/S0301-9268(02)00111-0.

Beattie, P., Drake, M., Jones, J., Leeman, W., Longhi, J., McKay, G., Nielsen, R., Palme, H., Shaw, D., Takahashi, E., and Watson, B., 1993, Terminology for trace-element partitioning: Geochimica et Cosmochimica Acta, v. 57, p. 1605–1606, doi: 10.1016/0016-7037(93)90015-O.

Bedard, J.H., 2006a, Trace element partitioning in plagioclase feldspar: Geochimica et Cosmochimica Acta, v. 70, p. 3717–3742, doi: 10.1016/j.gca.2006.05.003.

Bedard, J.H., 2006b, A catalytic delamination-driven model for coupled genesis of Archean crust and sub-continental lithospheric mantle: Geochimica et Cosmochimica Acta, v. 70, p. 1188–1214, doi: 10.1016/j.gca.2005.11.008.

Bennett, V.C., Nutman, A.P., and McCulloch, M.T., 1993, Nd isotopic evidence for transient, highly depleted mantle reservoirs in the early history of the Earth: Earth and Planetary Science Letters, v. 119, p. 299–317, doi: 10.1016/0012-821X(93)90140-5.

Bickle, M.J., 1986, Implications of melting for stabilization of the lithosphere and heat loss in the Archaean: Earth and Planetary Science Letters, v. 80, p. 314–324, doi: 10.1016/0012-821X(86)90113-5.

Bickle, M.J., Nisbet, E., and Martin, A., 1994, Archean greenstone belts are not oceanic crust: Journal of Geology, v. 102, p. 121–138.

Bindeman, I.N., Davis, A.M., and Drake, M.J., 1998, Ion microprobe study of plagioclase-basalt partition experiments at natural concentration levels of trace elements: Geochimica et Cosmochimica Acta, v. 62, p. 1175–1193, doi: 10.1016/S0016-7037(98)00047-7.

Bleeker, W., 2003, The Late Archean record: A puzzle in ca. 35 pieces: Lithos, v. 71, p. 99–134, doi: 10.1016/j.lithos.2003.07.003.

Blundy, J., and Wood, B., 1994, Prediction of crystal-melt partition coefficients from elastic moduli: Nature, v. 372, p. 452–454, doi: 10.1038/372452a0.

Borradaile, G.J., Lemmetty, T.J., and Werner, T., 2003, Apparent polar wander paths and the close of Late Archean crustal transpression, northern Ontario: Journal of Geophysical Research, v. 108, article 2402.

Bottazzi, P., Tiepolo, M., Vannucci, R., Zanetti, A., Brumm, R., Foley, S.F., and Oberti, R., 1999, Distinct site preferences for heavy and light REE in amphibole and the prediction of D-Amph/L(REE): Contributions to Mineralogy and Petrology, v. 137, p. 36–45, doi: 10.1007/s004100050580.

Brenan, J., Shaw, H., Ryerson, F., and Phinney, D., 1995, Experimental determination of trace-element partitioning between pargasite and a synthetic hydrous andesitic melt: Earth and Planetary Science Letters, v. 135, p. 1–11, doi: 10.1016/0012-821X(95)00139-4.

Brown, M., 2006, Duality of thermal regimes is the distinctive characteristic of plate tectonics since the Neoarchean: Geology, v. 34, p. 961–964, doi: 10.1130/G22853A.1.

Brown, M., 2008, this volume, Characteristic thermal regimes of plate tectonics and their metamorphic imprint throughout Earth history: When did Earth first adopt a plate tectonics mode of behavior?, *in* Condie, K.C., and Pease, V., eds., When did plate tectonics begin on planet Earth?: Geological Society of America Special Paper 440, doi: 10.1130/2008.2440(05).

Burke, K., Dewey, J.F., and Kidd, W.S.F., 1976, Dominance of horizontal movements, arc and microcontinental collisions during the later permobile regime, *in* Windley, B.F., ed., The early history of the Earth: Chichester, J. Wiley and Sons, p. 113–129.

Calvert, A., Sawyer, E., Davis, W., and Ludden, J., 1995, Archean subduction inferred from seismic images of a mantle suture in the Superior province: Nature, v. 375, p. 670–674, doi: 10.1038/375670a0.

Carpenter, M.A., 1980, Mechanisms of exsolution in sodic pyroxenes: Contributions to Mineralogy and Petrology, v. 71, p. 289–300, doi: 10.1007/BF00371671.

Clarke, G.L., Aitchison, J.C., and Cluzel, D., 1997, Eclogites and blueschists of the Pam Peninsula, NE New Caledonia: A reappraisal: Journal of Petrology, v. 38, p. 843–876, doi: 10.1093/petrology/38.7.843.

Condie, K.C., 1981, Archean greenstone belts: Amsterdam, Elsevier, 434 p.

Condie, K.C., 1986, Origin and early growth rate of continents: Precambrian Research, v. 32, p. 261–278, doi: 10.1016/0301-9268(86)90032-X.

Condie, K.C., 1989, Plate tectonic and crustal evolution: Oxford, UK, Pergamon Press, 476 p.

Condie, K.C., 2005, TTGs and adakites: Are they both slab melts?: Lithos, v. 80, p. 33–44, doi: 10.1016/j.lithos.2003.11.001.

Condie, K.C., and Kröner, A., 2008, this volume, When did plate tectonics begin? Evidence from the geologic record, *in* Condie, K.C., and Pease, V., eds., When did plate tectonics begin on planet Earth?: Geological Society of America Special Paper 440, doi: 10.1130/2008.2440(14).

Davis, W.J., Jones, A.G., Bleeker, W., and Grutter, H., 2003, Lithosphere development in the Slave craton: A linked crustal and mantle perspective: Lithos, v. 71, p. 575–589, doi: 10.1016/S0024-4937(03)00131-2.

de Wit, M.J., 1998, On Archean granites, greenstones, cratons and tectonics: Does the evidence demand a verdict?: Precambrian Research, v. 91, p. 181–226, doi: 10.1016/S0301-9268(98)00043-6.

Domanik, K.J., and Holloway, J.R., 1996, The stability and composition of phengitic muscovite and associated phases from 5.5 to 11 GPa: Implications for deeply subducted sediments: Geochimica et Cosmochimica Acta, v. 60, p. 4133–4150, doi: 10.1016/S0016-7037(96)00241-4.

Drake, M.J., and Weill, D., 1975, Partition of Sr, Ba, Ca, Y, $Eu^{2+}$, $Eu^{3+}$ and other REE between plagioclase feldspar and magmatic liquid: An experimental study: Geochimica et Cosmochimica Acta, v. 39, p. 689–712, doi: 10.1016/0016-7037(75)90011-3.

Dunn, T., and Sen, C., 1994, Mineral/matrix partition coefficients for orthopyroxene, plagioclase, and olivine in basaltic to andesitic systems: A combined analytical and experimental study: Geochimica et Cosmochimica Acta, v. 58, p. 717–733, doi: 10.1016/0016-7037(94)90501-0.

Eggins, S., Rudnick, R., and McDonough, W., 1998, The composition of peridotites and their minerals: A laser-ablation ICP-MS study: Earth and Planetary Science Letters, v. 154, p. 53–71, doi: 10.1016/S0012-821X(97)00195-7.

Evans, D.A.D., and Pisarevsky, S.A., 2008, this volume, Plate tectonics on early Earth? Weighing the paleomagnetic evidence, *in* Condie, K.C., and Pease, V., eds., When did plate tectonics begin on planet Earth?: Geological Society of America Special Paper 440, doi: 10.1130/2008.2440(12).

Ewart, A., and Griffin, W.L., 1994, Application of proton-microprobe data to trace-element partitioning in volcanic rocks: Chemical Geology, v. 117, p. 251–284, doi: 10.1016/0009-2541(94)90131-7.

Farnetani, C.G., Richards, M.A., and Ghiorso, M.S., 1996, Petrological models of magma evolution and deep crustal structure beneath hotspots and flood basalt provinces: Earth and Planetary Science Letters, v. 143, p. 81–94, doi: 10.1016/0012-821X(96)00138-0.

Farquhar, J., Wing, B.A., McKeegan, K.D., Harris, J.W., Cartigny, P., and Thiemens, M.H., 2002, Mass-independent sulfur of inclusions in diamond and sulfur recycling on early Earth: Science, v. 298, p. 2369–2372, doi: 10.1126/science.1078617.

Foley, S.F., and Wheller, G.E., 1990, Parallels in the origin of the geochemical signatures of island arc volcanics and continental potassic igneous rocks: The role of residual titanates: Chemical Geology, v. 85, p. 1–18, doi: 10.1016/0009-2541(90)90120-V.

Foley, S.F., Barth, M.G., and Jenner, G.A., 2000, Rutile/melt partition coefficients for trace elements and an assessment of the influence of rutile on the trace element characteristics of subduction zone magmas: Geochimica et Cosmochimica Acta, v. 64, p. 933–938, doi: 10.1016/S0016-7037(99)00355-5.

Foley, S.F., Tiepolo, M., and Vannucci, R., 2002, Growth of early continental crust controlled by melting of amphibolite in subduction zones: Nature, v. 417, p. 837–840, doi: 10.1038/nature00799.

Foley, S.F., Buhre, S., and Jacob, D.E., 2003, Evolution of the Archaean crust by delamination and shallow subduction: Nature, v. 421, p. 249–252.

Forneris, J.F., and Holloway, J.R., 2003, Phase equilibria in subducting basaltic crust: Implications for $H_2O$ release from the slab: Earth and Planetary Science Letters, v. 214, p. 187–201, doi: 10.1016/S0012-821X(03)00305-4.

Fryer, P., Wheat, C.G., and Mottl, M.J., 1999, Mariana blueschist mud volcanism: Implications for conditions within the subduction zone: Geology, v. 27, p. 103–106, doi: 10.1130/0091-7613(1999)027<0103:MBMVIF>2.3.CO;2.

Gamble, J., Woodhead, J., Wright, I., and Smith, I., 1996, Basalt and sediment geochemistry and magma petrogenesis in a transect from oceanic island arc to rifted continental margin arc: The Kermadec Hikurangi margin, SW Pacific: Journal of Petrology, v. 37, p. 1523–1546, doi: 10.1093/petrology/37.6.1523.

Gradstein, F.M., Ogg, J.G., Smith, A.G., Bleeker, W., and Lourens, L.J., 2004, A new geologic time scale, with special reference to Precambrian and Neogene: Episodes, v. 27, p. 83–100.

Green, T.H., 1994, Experimental studies of trace-element partitioning applicable to igneous petrogenesis—Sedona 16 years later: Chemical Geology, v. 117, p. 1–36, doi: 10.1016/0009-2541(94)90119-8.

Green, T.H., and Pearson, N.J., 1987, An experimental study of Nb and Ta partitioning between Ti-rich minerals and silicate liquids at high pressure and temperature: Geochimica et Cosmochimica Acta, v. 51, p. 55–62, doi: 10.1016/0016-7037(87)90006-8.

Green, T.H., Blundy, J.D., Adam, J., and Yaxley, G.M., 2000, SIMS determination of trace element partition coefficients between garnet, clinopyroxene and hydrous basaltic liquids at 2–7.5 GPa and 1080–1200 degrees C: Lithos, v. 53, p. 165–187, doi: 10.1016/S0024-4937(00)00023-2.

Hargraves, R., 1986, Faster spreading or greater ridge length in the Archean?: Geology, v. 14, p. 750–752, doi: 10.1130/0091-7613(1986)14<750:FSOGRL>2.0.CO;2.

Holtz, F., Scaillet, B., Behrens, H., Schulze, F., and Pichavant, M., 1996, Water contents of felsic melts: Application to the rheological properties of granitic magmas: Transactions of the Royal Society of Edinburgh, Earth Sciences, v. 87, p. 57–64.

Hynes, A., 2008, this volume, Effects of a warmer mantle on the characteristics of Archean passive margins, *in* Condie, K.C., and Pease, V., eds., When did plate tectonics begin on planet Earth?: Geological Society of America Special Paper 440, doi: 10.1130/2008.2440(07).

Ionov, D., and Hofmann, A., 1995, Nb-Ta-rich mantle amphiboles and micas: Implications for subduction-related metasomatic trace element fractionations: Earth and Planetary Science Letters, v. 131, p. 341–356, doi: 10.1016/0012-821X(95)00037-D.

Jacob, D.E., 2004, Nature and origin of eclogite xenoliths from kimberlites: Lithos, v. 77, p. 295–316, doi: 10.1016/j.lithos.2004.03.038.

Jacob, D.E., and Foley, S.F., 1999, Evidence for Archean ocean crust with low high field strength element signature from diamondiferous eclogite xenoliths: Lithos, v. 48, p. 317–336, doi: 10.1016/S0024-4937(99)00034-1.

Jacob, D., Jagoutz, E., Lowry, D., Mattey, D., and Kudrjavtseva, G., 1994, Diamondiferous eclogites from Siberia: Remnants of Archean oceanic crust: Geochimica et Cosmochimica Acta, v. 58, p. 5191–5207, doi: 10.1016/0016-7037(94)90304-2.

Jahn, B.M., Glikson, A.Y., Peucat, J.-J., and Hickman, A.H., 1981, REE geochemistry and isotopic data of Archaean silicic volcanics and granitoids from the Pilbara block, Western Australia: Geochimica et Cosmochimica Acta, v. 45, p. 1633–1652.

Jicha, B.R., Scholl, D.W., Singer, B.S., Yogodzinski, G.M., and Kay, S.M., 2006, Revised age of Aleutian island arc formation implies high rate of magma production: Geology, v. 34, p. 661–664, doi: 10.1130/G22433.1.

Kamber, B.S., Ewart, A., Collerson, K.D., Bruce, M.C., and McDonald, G.D., 2002, Fluid-mobile trace element constraints on the role of slab melting and implications for Archaean crustal growth models: Contributions to Mineralogy and Petrology, v. 144, p. 38–56.

Karsten, J.L., Klein, E.M., and Sherman, S.B., 1996, Subduction zone geochemical characteristics in ocean ridge basalts from the southern Chile Ridge: Implications of modern ridge subduction systems for the Archean: Lithos, v. 37, p. 143–161, doi: 10.1016/0024-4937(95)00034-8.

Kennedy, A.K., Lofgren, G.E., and Wasserburg, G.J., 1993, An experimental study of trace-element partitioning between olivine, orthopyroxene and melt in chondrules—equilibrium values and kinetic effects: Earth and Planetary Science Letters, v. 115, p. 177–195.

Kent, R.W., Hardarson, B.S., Saunders, A.D., and Storey, M., 1996, Plateaux ancient and modern: Geochemical and sedimentological perspectives on Archaean oceanic magmatism: Lithos, v. 37, p. 129–142, doi: 10.1016/0024-4937(95)00033-X.

Keppler, H., 1996, Constraints from partitioning experiments on the composition of subduction-zone fluids: Nature, v. 380, p. 237–240, doi: 10.1038/380237a0.

Kessel, R., Ulmer, P., Pettke, T., Schmidt, M.W., and Thompson, A.B., 2005, The water-basalt system at 4 to 6 GPa: Phase relations and second critical endpoint in a K-free eclogite at 700 to 1400 degrees C: Earth and Planetary Science Letters, v. 237, p. 873–892, doi: 10.1016/j.epsl.2005.06.018.

Klein, M., Stosch, H.G., and Seck, H.A., 1997, Partitioning of high field strength and rare earth elements between amphibole and quartz-dioritic to tonalitic melts: An experimental study: Chemical Geology, v. 138, p. 257–271, doi: 10.1016/S0009-2541(97)00019-3.

Klein, M., Stosch, H.G., Seck, H.A., and Shimizu, N., 2000, Experimental partitioning of high field strength and rare earth elements between clinopyroxene and garnet in andesitic to tonalitic systems: Geochimica et Cosmochimica Acta, v. 64, p. 99–115, doi: 10.1016/S0016-7037(99)00178-7.

Kleinhanns, I.C., Kramers, J.D., and Kamber, B.S., 2003, Importance of water for Archaean granitoid petrology: A comparative study of TTG and potassic granitoids from Barberton Mountain Land, South Africa: Contributions to Mineralogy and Petrology, v. 145, p. 377–389, doi: 10.1007/s00410-003-0459-9.

Klemme, S., Blundy, J.D., and Wood, B.J., 2002, Experimental constraints on major and trace element partitioning during partial melting of eclogite: Geochimica et Cosmochimica Acta, v. 66, p. 3109–3123, doi: 10.1016/S0016-7037(02)00859-1.

Kogiso, T., and Hirschmann, M.M., 2006, Partial melting experiments of bimineralic eclogite and the role of recycled mafic oceanic crust in the genesis of ocean island basalts: Earth and Planetary Science Letters, v. 249, p. 188–199, doi: 10.1016/j.epsl.2006.07.016.

LaTourette, T., Hervig, R., and Holloway, J., 1995, Trace element partitioning between amphibole, phlogopite, and basanite melt: Earth and Planetary Science Letters, v. 135, p. 13–30, doi: 10.1016/0012-821X(95)00146-4.

Linnen, R., and Keppler, H., 1997, Columbite solubility in granitic melts: Consequences for the enrichment and fractionation of Nb and Ta in the Earth's crust: Contributions to Mineralogy and Petrology, v. 128, p. 213–227, doi: 10.1007/s004100050304.

Liu, S.W., Pan, Y.M., Xie, Q.L., Zhang, J., and Li, Q.G., 2004, Archean geodynamics in the Central Zone, North China Craton: Constraints from geochemistry of two contrasting series of granitoids in the Fuping and Wutai complexes: Precambrian Research, v. 130, p. 229–249, doi: 10.1016/j.precamres.2003.12.001.

Luais, B., and Hawkesworth, C.J., 1994, The generation of continental crust: An integrated study of crust-forming processes in the Archean of Zimbabwe: Journal of Petrology, v. 35, p. 43–93.

Lundstrom, C.C., Shaw, H.F., Ryerson, F.J., Williams, Q., and Gill, J., 1998, Crystal chemical control of clinopyroxene-melt partitioning in the Di-Ab-An system: Implications for elemental fractionations in the depleted mantle: Geochimica et Cosmochimica Acta, v. 62, p. 2849–2862, doi: 10.1016/S0016-7037(98)00197-5.

Martin, E., and Sigmarsson, O., 2005, Trondhjemitic and granitic melts formed by fractional crystallization of an olivine tholeiite from Reykjanes Peninsula, Iceland: Geological Magazine, v. 142, p. 651–658, doi: 10.1017/S0016756805001160.

Martin, H., 1986, Effect of steeper Archean geothermal gradient on geochemistry of subduction zone magmas: Geology, v. 14, p. 753–756, doi: 10.1130/0091-7613(1986)14<753:EOSAGG>2.0.CO;2.

Martin, H., 1987, Petrogenesis of Archaean trondhjemites, tonalites and granodiorites from eastern Finland: Major and trace element geochemistry: Journal of Petrology, v. 28, p. 921–953.

Martin, H., 1994, The Archaean grey gneisses and the genesis of the continental crust, *in* Condie, K.C., ed., Archean crustal evolution: Amsterdam, Elsevier, p. 205–259.

Martin, H., 1999, Adakitic magmas: Modern analogues of Archaean granitoids: Lithos, v. 46, p. 411–429, doi: 10.1016/S0024-4937(98)00076-0.

Martin, H., and Moyen, J.-F., 2002, Secular changes in tonalite-trondhjemite-granodiorite composition as markers of the progressive cooling of Earth: Geology, v. 30, p. 319–322, doi: 10.1130/0091-7613(2002)030<0319:SCITTG>2.0.CO;2.

Martin, H., Peucat, J.J., Sabate, P., and Cunha, J.C., 1997, Crustal evolution in the Early Archaean of South America: Example of the Sete Voltas Massif, Bahia State, Brazil: Precambrian Research, v. 82, p. 35–62, doi: 10.1016/S0301-9268(96)00054-X.

Martin, H., Smithies, R.H., Rapp, R.P., Moyen, J.-F., and Champion, D., 2005, An overview of adakite, tonalite-trondhjemite-granodiorite (TTG), and sanukitoid: Relationships and some implications for crustal evolution: Lithos, v. 79, p. 1–24, doi: 10.1016/j.lithos.2004.04.048.

McCulloch, M., and Gamble, J., 1991, Geochemical and geodynamical constraints on subduction zone magmatism: Earth and Planetary Science Letters, v. 102, p. 358–374, doi: 10.1016/0012-821X(91)90029-H.

McDonough, W.F., and Sun, S.-S., 1995, The composition of the Earth: Chemical Geology, v. 120, p. 223–253, doi: 10.1016/0009-2541(94)00140-4.

McKenzie, D., and Bickle, M., 1988, The volume and composition of melt generated by extension of the lithosphere: Journal of Petrology, v. 29, p. 625–679.

Merli, M., Callegari, A., Cannillo, E., Caucia, F., Leona, M., Oberti, R., and Ungaretti, L., 1995, Crystal-chemical complexity in natural garnets: Structural constraints on chemical variability: European Journal of Mineralogy, v. 7, p. 1239–1249.

Moyen, J.-F., and Stevens, G., 2006, Experimental constraints on TTG petrogenesis: Implications for Archean geodynamics, *in* Benn, K., et al., eds., Archean geodynamics and environments: American Geophysical Union Geophysical Monograph 164, p. 149–175.

Moyen, J.-F., Stevens, G., and Kisters, A., 2006, Record of mid-Archaean subduction from metamorphism in the Barberton terrain, South Africa: Nature, v. 442, p. 559–562, doi: 10.1038/nature04972.

Neal, C.R., Mahoney, J.J., Kroenke, L.W., Duncan, R.A., and Petterson, M.G., 1997, The Ontong-Java Plateau, *in* Mahoney, J.J., and Coffin, M.F., eds., Large igneous provinces: Continental, oceanic, and planetary flood volcanism: American Geophysical Union Geophysical Monograph 100, p. 183–216.

Nicholls, I.A., and Ringwood, A.E., 1973, Effect of water on olivine stability in tholeiites and the production of silica-saturated magmas in the island-arc environment: Journal of Geology, v. 81, p. 285–300.

Norman, M., Garcia, M.O., and Pietruszka, A.J., 2005, Trace-element distribution coefficients for pyroxenes, plagioclase, and olivine in evolved tholeiites from the 1955 eruption of Kilauea Volcano, Hawai'i, and petrogenesis of differentiated rift-zone lavas: American Mineralogist, v. 90, p. 888–899, doi: 10.2138/am.2005.1780.

Nutman, A.P., Bennett, V.C., Friend, C.R.L., and Norman, M.D., 1999, Meta-igneous (non-gneissic) tonalites and quartz-diorites from an extensive ca. 3800 Ma terrain south of the Isua supracrustal belt, southern West Greenland: Constraints on early crust formation: Contributions to Mineralogy and Petrology, v. 137, p. 364–388, doi: 10.1007/s004100050556.

Oberti, R., Vannucci, R., Zanetti, A., Tiepolo, M., and Brumm, R.C., 2000, A crystal chemical re-evaluation of amphibole/melt and amphibole/clinopyroxene D-Ti values in petrogenetic studies: American Mineralogist, v. 85, p. 407–419.

Öhlander, B., Skiöld, T., Hamilton, P.J., and Claesson, L.A., 1987, The western border of the Archaean province of the Baltic Shield: Evidence from northern Sweden: Contributions to Mineralogy and Petrology, v. 95, p. 437–450, doi: 10.1007/BF00402204.

Oliver, J., and Isacks, B., 1967, Deep earthquake zones, anomalous structures in the upper mantle, and the lithosphere: Journal of Geophysical Research, v. 72, p. 4259–4275.

Panjasawatwong, Y., Danyushevsky, L.V., Crawford, A.J., and Harris, K.L., 1995, An experimental study of the effects of melt composition on plagioclase-melt equilibria at 5 and 10 kbar: Implications for the origin of magmatic high-An plagioclase: Contributions to Mineralogy and Petrology, v. 118, p. 420–432, doi: 10.1007/s004100050024.

Pawley, A.R., 1994, The pressure and temperature stability limits of lawsonite: Implications for $H_2O$ recycling in subduction zones: Contributions to Mineralogy and Petrology, v. 118, p. 99–108, doi: 10.1007/BF00310614.

Pearce, J.A., 1983, Role of sub-continental lithosphere in magma genesis at active continental margins, *in* Hawkesworth, C.J., and Norry, M.J., eds., Continental basalts and mantle xenoliths: Nantwich, Shiva, p. 230–249.

Pearce, J.A., and Norry, M., 1979, Petrogenetic implications of Ti, Zr, Y and Nb variations in volcanic rocks: Contributions to Mineralogy and Petrology, v. 69, p. 33–47, doi: 10.1007/BF00375192.

Pearce, J.A., Stern, R.J., Bloomer, S.H., and Fryer, P., 2005, Geochemical mapping of the Mariana arc-basin system: Implications for the nature and distribution of subduction components: Geochemistry, Geophysics, Geosystems, v. 6, doi:10.1029/2004GC000895.

Pearson, D.G., 1999, The age of continental roots: Lithos, v. 48, p. 171–194, doi: 10.1016/S0024-4937(99)00026-2.

Pease, V., Percival, J., Smithies, H., Stevens, G., and Van Kranendonk, M., 2008, this volume, When did plate tectonics begin? Evidence from the orogenic record, *in* Condie, K.C., and Pease, V., eds., When did plate tectonics begin on planet Earth?: Geological Society of America Special Paper 440, doi: 10.1130/2008.2440(10).

Peck, W.H., Valley, J.W., Wilde, S.A., and Graham, C.M., 2001, Oxygen isotope ratios and rare earth elements in 3.3 to 4.4 Ga zircons: Ion microprobe evidence for high $\delta^{18}O$ continental crust and oceans in the Early Archean: Geochimica et Cosmochimica Acta, v. 65, p. 4215–4229, doi: 10.1016/S0016-7037(01)00711-6.

Percival, J.A., 1994, Archean high-grade metamorphism, *in* Condie, K.C., ed., Archaean crustal evolution: Amsterdam, Elsevier, p. 357–410.

Pertermann, M., Hirschmann, M.M., Hametner, K., Gunther, D., and Schmidt, M.W., 2004, Experimental determination of trace element partitioning between garnet and silica-rich liquid during anhydrous partial melting of MORB-like eclogite: Geochemistry, Geophysics, Geosystems, v. 5, doi:10.1029/2003GC000638.

Peucat, J.J., Capdevilla, R., Drareni, A., Choukroune, P., Fanning, C.M., Bernard Griffiths, J., and Foucarde, S., 1996, Major and trace element geochemistry and isotope (Sr, Nd, Pb, O) systematics of an Archaean basement involved in a 2.0 Ga very high-temperature (1000 °C) metamorphic event: In Ouzzal Massif, Hoggar, Algeria: Journal of Metamorphic Geology, v. 14, p. 667–692, doi: 10.1111/j.1525-1314.1996.00054.x.

Pitman, W.C., and Heirtzler, J.R., 1966, Magnetic anomalies over the Pacific-Antarctic Ridge: Science, v. 154, p. 1164–1171, doi: 10.1126/science.154.3753.1164.

Poli, S., and Schmidt, M.W., 1998, The high-pressure stability of zoisite and phase relationships of zoisite-bearing assemblages: Contributions to Mineralogy and Petrology, v. 130, p. 162–175, doi: 10.1007/s004100050357.

Quartieri, S., Dalconi, M.C., Boscherini, F., Oberti, R., and D'Acapito, F., 2004, Changes in the local coordination of trace rare earth elements in garnets by high-energy XAFS: New data on dysprosium: Physics and Chemistry of Minerals, v. 31, p. 162–167, doi: 10.1007/s00269-003-0377-4.

Rapp, R.P., 1995, Amphibole-out phase boundary in partially melted metabasalt, its control over liquid fraction and composition, and source permeability: Journal of Geophysical Research-Solid Earth, v. 100, p. 15,601–15,610, doi: 10.1029/95JB00913.
Rapp, R.P., and Watson, E.B., 1995, Dehydration melting of metabasalt at 8–32 kbar: Implications for continental growth and crust-mantle recycling: Journal of Petrology, v. 36, p. 891–931.
Rapp, R.P., Watson, E., and Miller, C., 1991, Partial melting of amphibolite/eclogite and the origin of Archean trondhjemites and tonalites: Precambrian Research, v. 51, p. 1–25, doi: 10.1016/0301-9268(91)90092-O.
Rapp, R.P., Shimizu, N., and Norman, M.D., 2003, Growth of early continental crust by partial melting of eclogite: Nature, v. 425, p. 605–609, doi: 10.1038/nature02031.
Ringwood, A., 1982, Phase transformations and differentiation in subducted lithosphere: Implications for mantle dynamics, basalt petrogenesis, and crustal evolution: Journal of Geology, v. 90, p. 611–643.
Rollinson, H., 1997, Eclogite xenoliths in west African kimberlites as residues from Achaean granitoid crust formation: Nature, v. 389, p. 173–176, doi: 10.1038/38266.
Rudnick, R., and Fountain, D., 1995, Nature and composition of the continental crust: A lower crustal perspective: Reviews of Geophysics, v. 33, p. 267–309, doi: 10.1029/95RG01302.
Rushmer, T., 1991, Partial melting of two amphibolites: Contrasting experimental results under fluid-absent conditions: Contributions to Mineralogy and Petrology, v. 107, p. 41–59, doi: 10.1007/BF00311184.
Russell, B., 1952, Dictionary of mind, matter and morals: New York, Philosophical Library, 290 p.
Sage, R.P., Lightfoot, P.C., and Doherty, W., 1996, Geochemical characteristics of granitoid rocks from within the Archean Michipicoten Greenstone Belt, Wawa subprovince, Superior province, Canada: Implications for source regions and tectonic evolution: Precambrian Research, v. 76, p. 155–190, doi: 10.1016/0301-9268(95)00021-6.
Samsonov, A.V., Bogina, M.M., Bibikova, E.V., Petrova, A.Y., and Shchipansky, A.A., 2005, The relationship between adakitic, calc-alkaline volcanic rocks and TTGs: Implications for the tectonic setting of the Karelian greenstone belts: Lithos, v. 79, p. 83–106, doi: 10.1016/j.lithos.2004.04.051.
Saunders, A.D., Tarney, J., Kerr, A.C., and Kent, R.W., 1996, The formation and fate of large oceanic igneous provinces: Lithos, v. 37, p. 81–95, doi: 10.1016/0024-4937(95)00030-5.
Schmidt, M.W., Dardon, A., Chazot, G., and Vannucci, R., 2004, The dependence of Nb and Ta rutile-melt partitioning on melt composition and Nb/Ta fractionation during subduction processes: Earth and Planetary Science Letters, v. 226, p. 415–432, doi: 10.1016/j.epsl.2004.08.010.
Schmidt, M.W., Connolly, J.A.D., Gunther, D., and Bogaerts, M., 2006, Element partitioning: The role of melt structure and composition: Science, v. 312, p. 1646–1650, doi: 10.1126/science.1126690.
Schmincke, H.-U., 2004, Volcanism: Berlin, Springer, 324 p.
Shirey, S.B., Kamber, B.S., Whitehouse, M.J., Mueller, P.A., and Basu, A.R., 2008, this volume, A review of the isotopic and trace element evidence for mantle and crustal processes in the Hadean and Archean: Implications for the onset of plate tectonic subduction, *in* Condie, K.C., and Pease, V., eds., When did plate tectonics begin on planet Earth?: Geological Society of America Special Paper 440, doi: 10.1130/2008.2440(01).
Sleep, N., and Windley, B., 1982, Archean plate tectonics: Constraints and inferences: Journal of Geology, v. 90, p. 363–379.
Smithies, R.H., 2000, The Archaean tonalite-trondhjemite-granodiorite (TTG) series is not an analogue of Cenozoic adakite: Earth and Planetary Science Letters, v. 182, p. 115–125, doi: 10.1016/S0012-821X(00)00236-3.
Smyth, J.R., and Bish, D.L., 1988, Crystal structures and cation sites of the rock-forming minerals: Boston, Allen and Unwin, 297 p.
Stalder, R., Foley, S.F., Brey, G.P., and Horn, I., 1998, Mineral aqueous fluid partitioning of trace elements at 900–1200 degrees C and 3.0–5.7 GPa: New experimental data for garnet, clinopyroxene, and rutile, and implications for mantle metasomatism: Geochimica et Cosmochimica Acta, v. 62, p. 1781–1801, doi: 10.1016/S0016-7037(98)00101-X.
Stalder, R., Ulmer, P., Thompson, A.B., and Gunther, D., 2001, High pressure fluids in the system $MgO-SiO_2-H_2O$ under upper mantle conditions: Contributions to Mineralogy and Petrology, v. 140, p. 607–618.
Takagi, D., Sato, H., and Nakagawa, M., 2005, Experimental study of a low-alkali tholeiite at 1–5 kbar: Optimal condition for the crystallization of high-An plagioclase in hydrous arc tholeiite: Contributions to Mineralogy and Petrology, v. 149, p. 527–540, doi: 10.1007/s00410-005-0666-7.
Takahashi, E., and Irvine, T.N., 1981, Stoichiometric control of crystal/liquid single-component partition coefficients: Geochimica et Cosmochimica Acta, v. 45, p. 1181–1185, doi: 10.1016/0016-7037(81)90141-1.
Tarney, J., 1976, Geochemistry of Archaean high-grade gneisses, with implications as to the origin and evolution of the Precambrian crust, *in* Windley, B.F., ed., The early history of the Earth: Chichester, J. Wiley and Sons, p. 405–417.
Tarney, J., Dalziel, I.W.D., and de Wit, M.J., 1976, Marginal basin "Rocas Verdes" complex from southern Chile: A model for Archaean greenstone belt formation, *in* Windley, B.F., ed., The early history of the Earth: Chichester, J. Wiley and Sons, p. 131–146.
Tarney, J., Weaver, B., and Drury, S.A., 1979, Geochemistry of Archean trondhjemitic and tonalitic gneisses from Scotland and East Greenland, *in* Barker, F., ed., Trondhjemites, dacites, and related rocks: Amsterdam, Elsevier, p. 275–299.
Tatsumi, Y., 1989, Migration of fluid phases and genesis of basalt magmas in subduction zones: Journal of Geophysical Research, v. 94, p. 4697–4707.
Thirlwall, M., Smith, T., Graham, A., Theodorou, N., Hollings, P., Davidson, J., and Arculus, R., 1994, High field strength element anomalies in arc lavas: Source or process?: Journal of Petrology, v. 35, p. 819–838.
Tiepolo, M., Vannucci, R., Oberti, R., Foley, S., Bottazzi, P., and Zanetti, A., 2000a, Nb and Ta incorporation and fractionation in titanian pargasite and kaersutite: Crystal-chemical constraints and implications for natural systems: Earth and Planetary Science Letters, v. 176, p. 185–201, doi: 10.1016/S0012-821X(00)00004-2.
Tiepolo, M., Vannucci, R., Bottazzi, P., Oberti, R., Zanetti, A., and Foley, S.F., 2000b, Partitioning of REE, Y, Th, U and Pb between pargasite, kaersutite and basanite to trachyte melts: Implications for percolated and veined mantle: Geochemistry, Geophysics, Geosystems, v. 1, doi: 10.1029/2000GC000064.
Tiepolo, M., Bottazzi, P., Foley, S.F., Oberti, R., Vannucci, R., and Zanetti, A., 2001, Fractionation of Nb and Ta from Zr and Hf at mantle depths: The role of titanian pargasite and kaersutite: Journal of Petrology, v. 42, p. 221–232, doi: 10.1093/petrology/42.1.221.
Turner, S., Hawkesworth, C., Rogers, N., Bartlett, J., Worthington, T., Hergt, J., Pearce, J., and Smith, I., 1997, $^{238}U-^{230}Th$ disequilibria, magma petrogenesis, and flux rates beneath the depleted Tonga-Kermadec island arc: Geochimica et Cosmochimica Acta, v. 61, p. 4855–4884, doi: 10.1016/S0016-7037(97)00281-0.
van der Laan, S.R., Flower, M.F.J., and Koster van Groos, A.F., 1987, Experimental evidence for the origin of boninites: Near-liquidus phase relations to 7.5 kbar, *in* Crawford, A.J., ed., Boninites and related rocks: London, Unwin-Hyman, p. 112–147.
van Westrenen, W., Blundy, J., and Wood, B., 1999, Crystal-chemical controls on trace element partitioning between garnet and anhydrous silicate melt: American Mineralogist, v. 84, p. 838–847.
van Westrenen, W., Blundy, J.D., and Wood, B.J., 2000, Effect of $Fe^{2+}$ on garnet-melt trace element partitioning: Experiments in FCMAS and quantification of crystal-chemical controls in natural systems: Lithos, v. 53, p. 189–201, doi: 10.1016/S0024-4937(00)00024-4.
van Westrenen, W., Wood, B.J., and Blundy, J.D., 2001, A predictive thermodynamic model of garnet-melt trace element partitioning: Contributions to Mineralogy and Petrology, v. 142, p. 219–234.
van Westrenen, W., Allan, N.L., Blundy, J.D., Lavrentiev, M.Y., Lucas, B.R., and Purton, J.A., 2003, Trace element incorporation into pyrope-grossular solid solutions: An atomistic simulation study: Physics and Chemistry of Minerals, v. 30, p. 217–229, doi: 10.1007/s00269-003-0307-5.
Westerlund, K.J., Shirey, S.B., Richardson, S.H., Carlson, R.W., Gurney, J.J., and Harris, J.W., 2006, A subduction wedge origin for Paleoarchean peridotitic diamonds and harzburgites from the Panda kimberlite, Slave craton: Evidence from Re-Os isotope systematics: Contributions to Mineralogy and Petrology, v. 152, p. 275–294, doi: 10.1007/s00410-006-0101-8.
Whalen, J.B., Percival, J.A., McNicoll, V.J., and Longstaffe, F.J., 2002, A mainly crustal origin for tonalitic granitoid rocks, Superior province, Canada: Implications for Late Archean tectonomagmatic processes: Journal of Petrology, v. 43, p. 1551–1570, doi: 10.1093/petrology/43.8.1551.
White, R.W., and Powell, R., 2002, Melt loss and the preservation of granulite facies mineral assemblages: Journal of Metamorphic Geology, v. 20, p. 621–632, doi: 10.1046/j.1525-1314.2002.00206.x.
Windley, B.F., 1977, The evolving continents: London, J. Wiley and Sons, 385 p.
Winther, K.T., and Newton, R.C., 1991, Experimental melting of hydrous low-K tholeiite: Evidence on the origin of Archaean cratons: Bulletin of the Geological Society of Denmark, v. 39, p. 213–228.

Wolf, M., and Wyllie, P., 1994, Dehydration-melting of amphibolite at 10 kbar: The effects of temperature and time: Contributions to Mineralogy and Petrology, v. 115, p. 369–383, doi: 10.1007/BF00320972.

Wood, B.J., and Blundy, J.D., 1997, A predictive model for rare earth element partitioning between clinopyroxene and anhydrous silicate melt: Contributions to Mineralogy and Petrology, v. 129, p. 166–181, doi: 10.1007/s004100050330.

Wyman, D.A., O'Neill, C., and Ayer, J.A., 2008, this volume, Evidence for modern-style subduction to 3.1 Ga: A plateau–adakite–gold (diamond) association, *in* Condie, K.C., and Pease, V., eds., When did plate tectonics begin on planet Earth?: Geological Society of America Special Paper 440, doi: 10.1130/2008.2440(06).

Zack, T., Rivers, T., and Foley, S.F., 2001, Cs-Rb-Ba systematics in phengite and amphibole: An assessment of fluid mobility at 2.0 GPa in eclogites from Trescolmen, central Alps: Contributions to Mineralogy and Petrology, v. 140, p. 651–669.

MANUSCRIPT ACCEPTED BY THE SOCIETY 14 AUGUST 2007

Printed in the USA

The Geological Society of America
Special Paper 440
2008

# *An overview of the lithological and geochemical characteristics of the Mesoarchean (ca. 3075 Ma) Ivisaartoq greenstone belt, southern West Greenland*

**Ali Polat***
*Department of Earth and Environmental Sciences, University of Windsor, Windsor, Ontario N9B 3P4, Canada*

**Robert Frei**
*Geological Institute, University of Copenhagen, Øster Voldgade 10, DK-1350, Copenhagen, Denmark* and *Nordic Center for Earth Evolution, NordCEE, Denmark*

**Peter W.U. Appel**
*Geological Survey of Denmark and Greenland, Geocenter Copenhagen, Øster Voldgade 10, DK-1350, Copenhagen, Denmark*

**Brian Fryer**
*Great Lakes Institute for Environmental Research, University of Windsor, Windsor, Ontario N9B 3P4, Canadà*

**Yıldırım Dilek**
*Department of Geology, Miami University, Oxford, Ohio 45056, USA*

**Juan C. Ordóñez-Calderón**
*Department of Earth and Environmental Sciences, University of Windsor, Windsor, Ontario N9B 3P4, Canada*

## ABSTRACT

**In this review we summarize the major lithological and geochemical characteristics of the Mesoarchean (ca. 3075 Ma) Ivisaartoq greenstone belt, Nuuk region, southern West Greenland. In addition, the geological characteristics of the Ivisaartoq greenstone belt are compared with those of other Archean greenstone belts in the area. The Ivisaartoq greenstone belt is the largest Mesoarchean supracrustal lithotectonic assemblage in the Nuuk region. The belt contains well-preserved primary magmatic structures including pillow lavas, volcanic breccias, and cumulate (picrite) layers. It also includes variably deformed gabbroic to dioritic dikes, actinolite schists, serpentinites, siliciclastic sediments, and minor cherts. The Ivisaartoq rocks underwent at least two stages of postmagmatic metamorphic alteration, including seafloor hydrothermal alteration and syn- to post-tectonic calc-silicate metasomatism, between 3075 and 2961 Ma. The trace element systematics of the least altered rocks are consistent with a subduction zone geodynamic setting. On the basis of lithological similarities**

*polat@uwindsor.ca

Polat, A., Frei, R., Appel, P.W.U., Fryer, B., Dilek, Y., and Ordóñez-Calderón, J.C., 2008, An overview of the lithological and geochemical characteristics of the Mesoarchean (ca. 3075 Ma) Ivisaartoq greenstone belt, southern West Greenland, *in* Condie, K.C., and Pease, V., eds., When Did Plate Tectonics Begin on Planet Earth?: Geological Society of America Special Paper 440, p. 51–76, doi: 10.1130/2008.2440(03). For permission to copy, contact editing@geosociety.org. 

**between the Ivisaartoq greenstone belt and Phanerozoic forearc/backarc ophiolites, and intra-oceanic island arcs, we suggest that the Ivisaartoq greenstone belt represents a relic of dismembered Mesoarchean suprasubduction zone oceanic crust.**

**The Sm-Nd isotope system appears to have remained relatively undisturbed in picrites, tholeiitic pillow lavas, gabbros, and diorites. As a group, picrites have more depleted initial Nd isotopic signatures ($\varepsilon_{Nd}$ = +4.2 to +5.0) than gabbros, diorites, and tholeiitic basalts ($\varepsilon_{Nd}$ = +0.3 to +3.1), consistent with a strongly depleted mantle source. In some areas gabbros include up to 15 cm long white inclusions (xenoliths). These inclusions are composed primarily (>90%) of Ca-plagioclase and are interpreted as anorthositic cumulates of the lower oceanic crust brought to the surface by upwelling gabbroic magmas. Alternatively, the inclusions may represent the xenoliths from older (>3075 Ma) anorthositic crust onto which the Ivisaartoq magmas were emplaced as an autochthonous sequence. However, no geological evidence has been found for such older anorthositic crust in the region. The anorthositic cumulates have significantly higher initial $\varepsilon_{Nd}$ values (+4.8 to +6.0) than the surrounding gabbroic matrix (+2.3 to +2.8), suggesting two different mantle sources for these rocks.**

**Keywords:** Archean, greenstone belt, oceanic crust, pillow basalt, anorthosite, ocelli, isotope.

## INTRODUCTION

The distribution of rock types and the internal structure of many Archean greenstone belts suggest that they are the products of multiple geological processes, such as tectonism, magmatism, sedimentation, and metamorphism, operating over different spatial and temporal scales (Condie, 1981, 2005; Goodwin, 1991; Eriksson et al., 1994; Corcoran and Dostal, 2001; Sandeman et al., 2004; Smithies et al., 2005a, b; Kerrich and Polat, 2006). Plate tectonic models have been proposed to account for the geological characteristics of Archean greenstone-granitoid terranes since the 1970s (Bridgwater et al., 1974; Windley, 1976; Condie, 1981, 2005; Card, 1990; Eriksson et al., 1994; Mueller et al., 1996; de Wit, 1998; Kusky and Polat, 1999; Percival et al., 2006a, and references therein). The seismic and structural characteristics of many Archean greenstone belts are comparable to those of lithotectonic assemblages occurring in Phanerozoic convergent plate boundaries (Kusky and Polat, 1999; Kerrich and Polat, 2006; Percival et al., 2006a, 2006b; van der Velden et al., 2006). In addition, greenstone belts from 3.8 to 2.5 Ga include rare volcanic rock types reported from Phanerozoic convergent margins, such as boninites, picrites, adakites, Mg-andesites, and Nb-enriched basalts (Polat and Kerrich, 2006, and references therein).

The question of whether Archean greenstone belts represent the fragments of ancient oceanic crust or alternatively are the remnants of continental flood basalts remains controversial (Bickle et al., 1994; Kusky, 2004; Polat and Kerrich, 2006). Based on the criteria of xenocrystic zircons and geochemical contamination trends of their mafic-ultramafic lavas, some greenstone belts may be considered intra-oceanic in origin, whereas some others may have formed from mantle magmas erupted through continental crust (Nisbet and Fowler, 1983; Polat et al., 1998, 2002, 2006; Arndt et al., 2001; Bleeker, 2002; Condie, 2005).

Ophiolites and their fragments in Phanerozoic orogenic belts are likely to represent the closest Phanerozoic analogues of Archean greenstone belts (Sylvester et al., 1997; Kusky and Polat, 1999; Dilek, 2003; Kusky, 2004; Şengör and Natal'in, 2004). The majority of Phanerozoic ophiolites appear to have formed in suprasubduction zone geodynamic settings (Dewey, 2003; Hawkins, 2003; Pearce, 2003, and references therein). Accordingly, the recognition of Archean oceanic crust can be done most effectively through comparative studies of the well-established lithological and geochemical characteristics for oceanic crustal fragments occurring in modern suprasubduction zone environments and accretionary complexes.

In general, the geochemical characteristics of modern volcanic rocks from different tectonic settings are distinct in terms of their rare earth element (REE), large ion lithophile element (LILE), and high field strength element (HFSE) systematics (Sun and McDonough, 1989; Hawkesworth et al., 1993; Pearce, 2003; Hofmann, 2004). In this study we assume that similar, uniformitarian, geochemical behavior also prevailed in the Archean, given that certain groups of elements will behave consistently in petrogenetic processes, including source composition, residual mineralogy, partial melting, hybridization, and crustal recycling throughout Earth's history (see Polat and Kerrich, 2006; Foley, this volume; Shirey et al., this volume). However, thermal conditions in a given particular geodynamic setting were likely different in the Archean, because the Archean mantle had higher geothermal gradients than its modern counterpart (see Fyfe, 1978; Bickle, 1986). Therefore, conditions of partial melting under the influence of a particular geodynamic regime likely differed in the Archean. Accordingly, in this contribution the trace element systematics of Archean volcanic rocks are used in conjunction with field characteristics to constrain their geodynamic setting.

Notwithstanding two major phases of ductile deformation and amphibolite facies metamorphism (Friend et al., 1981; Hall, 1981; Chadwick, 1985, 1986, 1990; Appel, 1997; Polat et al., 2007), pillow structures, volcanic breccias, and cumulate and ocellar (eye-shaped solidified immiscible liquids) textures have been well preserved in low-strain domains of the Mesoarchean (ca. 3075 Ma) Ivisaartoq greenstone belt in southern West Greenland (Figs. 1 and 2). The presence of the primary magmatic features in the Ivisaartoq greenstone belt provides a unique opportunity to study Mesoarchean petrogenetic and geodynamic processes. In this review we summarize the recent field and geochemical studies of the Ivisaartoq greenstone belt (Polat et al., 2007, 2008). Specifically, we address the significance of trace element data to constrain the geodynamic origin of the belt. Given the fact that all rocks in the Ivisaartoq greenstone belt and surrounding region have been metamorphosed (Nutman and Friend, 2007), the prefix "meta" will be taken implicit.

## REGIONAL GEOLOGY AND FIELD CHARACTERISTICS

The Ivisaartoq greenstone belt is the largest Mesoarchean supracrustal assemblage in southern West Greenland (Figs. 1 and 2) (Hall and Friend, 1979; Brewer et al., 1984; Chadwick, 1985, 1986, 1990; Friend and Nutman, 2005a). It is located in the central part of the inner Nuuk region (Fig. 1A). The belt occurs within the recently recognized Mesoarchean (ca. 3075–2950 Ma) Kapisilik tectonic terrane (Friend and Nutman, 2005a), which is tectonically bounded by the Eoarchean Isukasia terrane (3600–3800 Ma) to the north, and the Eoarchean Færingehavn and the Neoarchean Tre Brødre terranes to the south and west, respectively (Fig. 1A) (Friend and Nutman, 2005a). Geochronological, structural, and metamorphic studies indicate that the Kapisilik and Isukasia terranes were juxtaposed and metamorphosed by 2950 Ma (Friend and Nutman, 2005a). Field relationships indicate that the Isukasia terrane is structurally overlain by the Kapisilik terrane to the south; and the Kapisilik terrane is in turn structurally overlain by the Færingehavn and Tre Brødre terranes to the south-southwest (Fig. 1B).

The precise age of the volcanic and intrusive (gabbro, diorite) rocks in the Ivisaartoq greenstone belt is unknown. Siliceous volcaniclastic sedimentary rocks have yielded an average U-Pb zircon age of 3075 ± 15 Ma (Friend and Nutman, 2005a; Polat et al., 2007), constraining the maximum age of the belt. The Ivisaartoq greenstone belt is intruded by weakly deformed 2961 ± 12 Ma granites to the north, constraining the minimum age of the belt (Chadwick, 1990; Friend and Nutman, 2005a). The Ivisaartoq sequence is truncated by an up to 2 m thick mylonite zone to the south, separating the belt from an association of leucogabbros and anorthosites. These leucogabbros and anorthosites are intruded by 2963 ± 8 Ma tonalites and granodiorites (now gneisses). On the basis of field observations and zircon ages, Friend and Nutman (2005a) interpreted the mylonite zone as a post–2960 Ma structure deforming the Kapisilik terrane. The leucogabbro and anorthosite association is lithologically and structurally similar to those found in the Fiskenaesset region of southern West Greenland, and is interpreted as intrusive into the Ivisaartoq greenstone belt (Chadwick, 1990).

The Ivisaartoq greenstone belt is composed mainly of mafic to ultramafic volcanic rocks, gabbros, minor diorites, and serpentinites (Figs. 2, 3, and 4) (Hall, 1981; Chadwick, 1985, 1986, 1990; Polat et al., 2007). Volcanic rocks consist dominantly of deformed tholeiitic pillow basalts and ultramafic lava flows (Fig. 4) (Chadwick, 1990; Polat et al., 2007). Sedimentary rocks constitute a volumetrically minor component of the belt (Figs. 2 and 3). There are minor, up to several-meters-thick lenses of siliciclastic sedimentary rocks in the upper unit (Fig. 5A). Siliceous pyrite-bearing rocks were interpreted as metamorphosed cherts (Chadwick, 1990) (Fig. 5B). Contacts between volcanic and sedimentary rocks are sharp (Fig. 5B).

Chadwick has subdivided the Ivisaartoq greenstone belt into a lower and an upper amphibolite unit (Fig. 3) (Chadwick, 1985, 1986, 1990). These units are separated by a layer (up to 50 m thick) of magnetite-rich ultramafic schists, called the "magnetic marker" (Fig. 3) (Chadwick, 1986, 1990). Hydrothermal alteration of the "magnetic marker" and volcanic rocks in its vicinity resulted in the formation of calc-silicate rocks hosting strata-bound scheelite mineralization (Appel, 1994, 1997). The intensity of deformation appears to increase toward the boundary between the two amphibolite units (Fig. 5C), suggesting that they are tectonically juxtaposed.

Pillow basalts are characterized by well-preserved core and rim structures (Figs. 4A, 4B, and 4C). The least deformed pillow basalts have concentric cracks filled mainly with quartz, and display way-up directions (Fig. 4B). Pillow cores are mineralogically zoned (Figs. 4A and 4C). Many inner pillow cores display drainage cavities at the center, which are either empty or filled with quartz (Fig. 4C). The pillow cores often display ocellar texture consisting chiefly of white ellipsoidal (eye-shaped) millimeter- to centimeter-sized ocelli set in a dark green finer-grained mafic matrix (Figs. 5D and 5E). The contacts between ocelli and matrix are sharp. In many cores the ocelli-matrix texture has been partly to completely replaced by a calc-silicate metasomatic assemblage (Fig. 5E). Pillow rims often display silica alteration; some pillows have been completely silicified. Some pillows are composed predominantly of actinolite, consistent with an ultramafic protolith. Primary magmatic textures, such as clinopyroxene cumulates, are locally preserved in ultramafic flows of low-strain domains (Fig. 4E). With increasing intensity of deformation, clinopyroxene cumulates grade into actinolite schists.

Gabbros and minor diorites occur as sills and dikes, one to several tens of meters thick, in pillow basalts (Fig. 4D). They also occur sporadically between pillow basalts and ultramafic flows. Chilled margins between pillow basalts and gabbroic dikes are preserved in a few locations. Primary igneous textures and minerals are locally preserved in low-strain domains (Fig. 4). Some gabbros contain deformed anorthositic inclusions (xenoliths) up to 15 cm long (Figs. 5F and 6A). Like pillow basalts, gabbros and

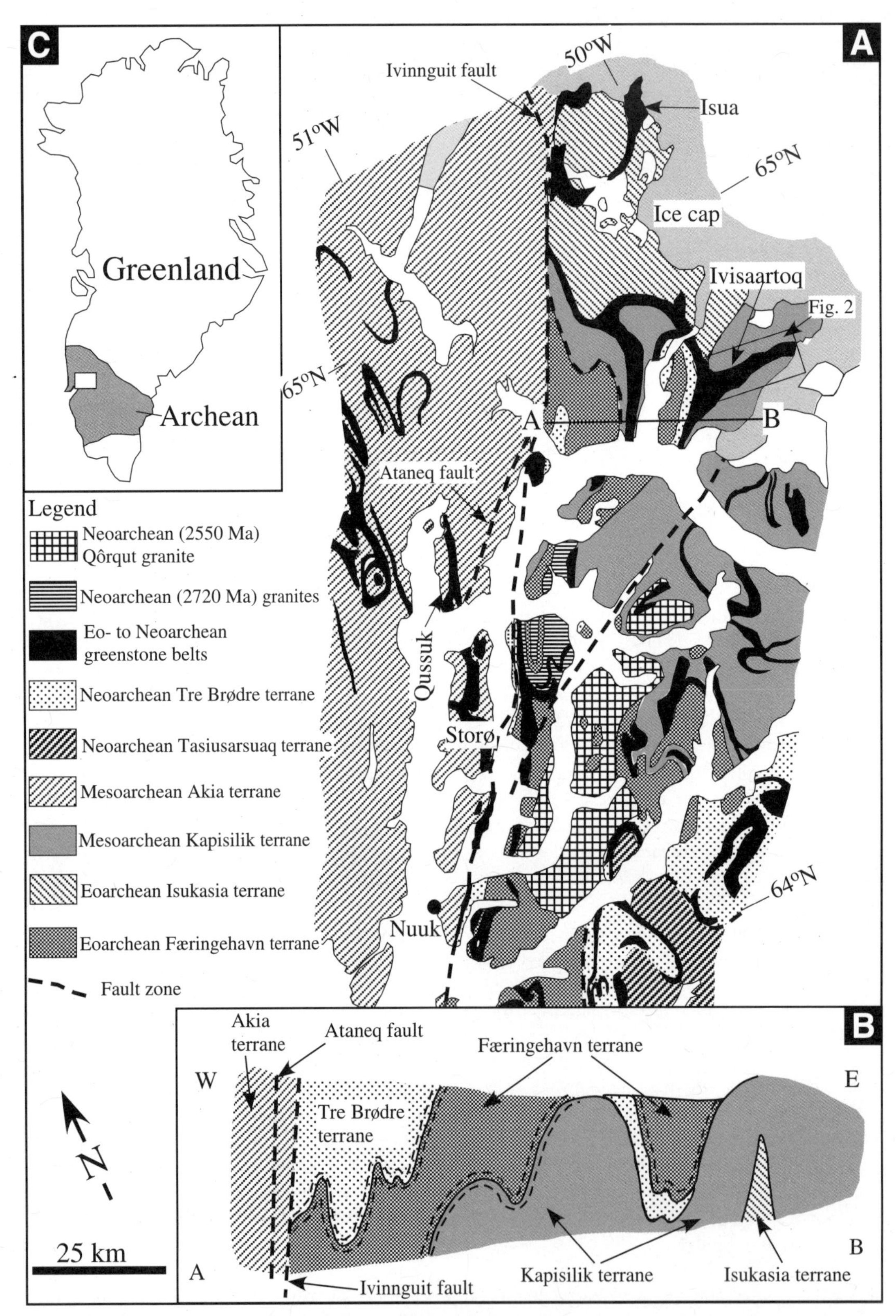


Figure 1. (A) A simplified geological map of the Nuuk region, showing the Eo- to Neoarchean tectonic terranes and the locations of the Isua, Ivisaartoq, and Qussuk greenstone belts. Also shown are the locations of line A–B and Figure 2. (B) A simplified geological cross section through line A–B; vertical and horizontal scales are exaggerated for illustration purposes. The map is modified from Friend and Nutman (2005a, 2005b) and Nutman and Friend (2007). The cross section is modified from Friend and Nutman (2005a). Because of recent revisions in both terrane boundaries and terminology, there are some discrepancies between new and old terrane boundaries and names. (C) Study area (rectangle) and the boundaries of Archean terrane in Greenland.

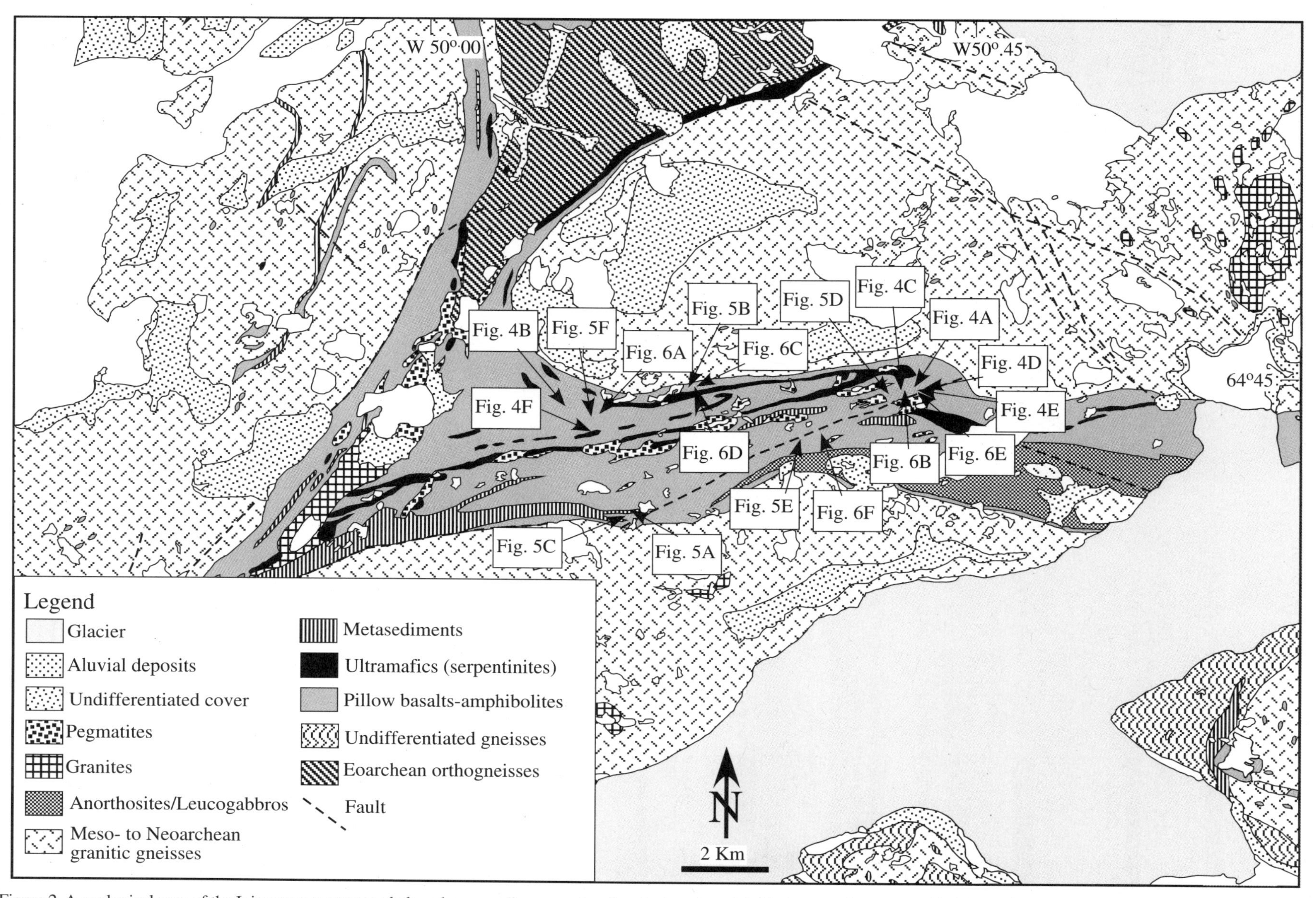


Figure 2. A geological map of the Ivisaartoq greenstone belt and surrounding area, showing the location of field photographs presented in Figures 4, 5, and 6. Modified from Chadwick and Coe (1988).

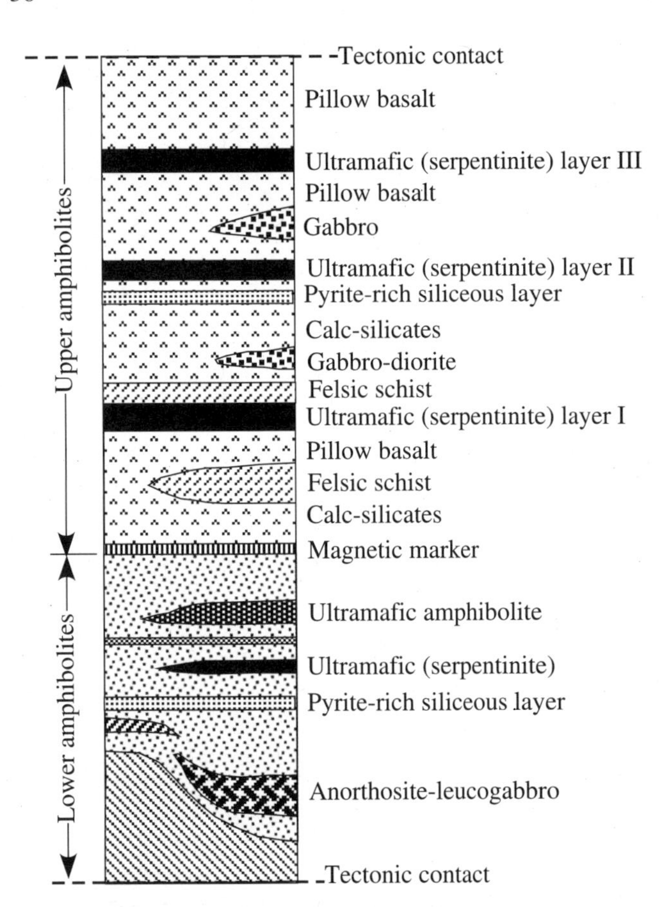


Figure 3. A simplified tectonostratigraphic column of the Ivisaartoq greenstone belt. Modified from Chadwick (1986, 1990).

diorites underwent calc-silicate metasomatic alteration mainly along fractures and pillow basalt contacts (Fig. 6B).

Ultramafic rocks (now serpentinites) are exposed discontinuously as three major layers throughout the sequence (Figs. 2, 3, and 4F) (Chadwick, 1986, 1990). On the basis of field relationships, Chadwick (1986, 1990) suggested that the protoliths of the serpentinites intruded as sills into submarine lavas. In many outcrops, they are in tectonic contact with pillow basalts and gabbros (Fig. 4F).

## PETROGRAPHY

The mineralogical characteristics of different rock types are summarized in Table 1. Cumulates are composed primarily of altered clinopyroxene phenocrysts (Fig. 4E; Table 1). Ultramafic schists are composed dominantly of actinolite (Table 1). Gabbros and diorites are composed mainly of hornblende + plagioclase ± epidote ± quartz. White inclusions (xenoliths) in gabbros have an assemblage of plagioclase (90%) + hornblende (5%–10%) + quartz (0%–5%), consistent with an anorthositic composition (Fig. 7; Table 1). Because of extensive recrystallization, magmatic plagioclase is rarely present. Amphiboles in the anorthositic inclusions typically have dark green to blue green pleochroism and range from subhedral to euhedral.

Ocelli in the pillow cores consist mainly of plagioclase (30%–50%) + quartz (30%–40%) + amphibole (10%–20%) ± epidote (0%–5%). No internal structure has been observed in the ocelli. The darker matrix (outer pillow core) surrounding the ocelli is made of hornblende (50%–60%) + plagioclase (20%–30%) + quartz (10%–20%) ± epidote (0%–5%) ± titanite (0%–5%) (Table 1). The pillow rims are composed of fine-grained hornblende + quartz + plagioclase ± epidote ± biotite.

The magnetic marker is composed primarily of actinolite + olivine (metamorphic) + diopside + magnetite ± plagioclase ± epidote ± garnet ± vesuvianite ± scapolite ± calcite ± titanite ± scheelite. The stage I metasomatic assemblage is composed predominantly of epidote (now mostly diopside) + quartz + plagioclase ± hornblende ± scapolite (Table 1). The stage II metasomatic assemblage consists mainly of diopside + garnet + amphibole + plagioclase + quartz ± vesuvianite ± scapolite ± epidote ± titanite ± calcite ± scheelite (Table 1). Amphibolites (foliated pillows) are composed mainly of hornblende + plagioclase + quartz ± diopside ± epidote ± titanite ± sulfide.

## ANALYTICAL METHODS

Details of analytical methods are given in Polat et al. (2007, 2008). Thus, only a summary of the analytical methods is presented here. All whole-rock samples were powdered using an agate mill in the Department of Earth and Environmental Sciences, University of Windsor, Canada. Major elements and some trace elements (Zr, Sc, Ni) were determined by Thermo Jarrel-Ash ENVIRO II ICP at ACTLABS in Ancaster, Canada. Samples were mixed with a flux of lithium metaborate and lithium tetraborate, and fused in an induction furnace. Totals of major element oxides are 100 ± 1 wt%, and the analytical precisions are 1%–2%. Mg-numbers (%) were calculated as the molecular ratio of Mg/(Mg + $Fe^{2+}$), where $Fe^{2+}$ is assumed to be 90% of total Fe.

Samples were analyzed for REEs, HFSEs, LILEs, and transition metals (Co, Cr, and V) by a high-sensitivity Thermo Elemental X7 ICP-MS in the Great Lakes Institute for Environmental Research (GLIER), University of Windsor, Canada. Wet chemical procedures were conducted under clean lab conditions, and all acids were distilled twice.

Whole-rock Pb and Sm-Nd isotope analyses were carried out on a VG Sector 54-IT TIMS in the Geological Institute, University of Copenhagen, Denmark. A mixed $^{150}$Nd-$^{149}$Sm spike was added to the REE aliquot beforehand. Chemical separation of REEs was carried out on conventional cation exchange columns, followed by an Sm-Nd separation using HDEHP-coated beads (BIO-RAD) charged in 6 ml quartz glass columns. Neodymium ratios were normalized to $^{146}$Nd/$^{144}$Nd = 0.7219. The mean value for our internal JM Nd standard (referenced against La Jolla) during the period of measurement was 0.511098 for $^{143}$Nd/$^{144}$Nd, with a 2σ external reproducibility of ±0.000011 (seven measurements). Procedural blanks run during the period of these analyses

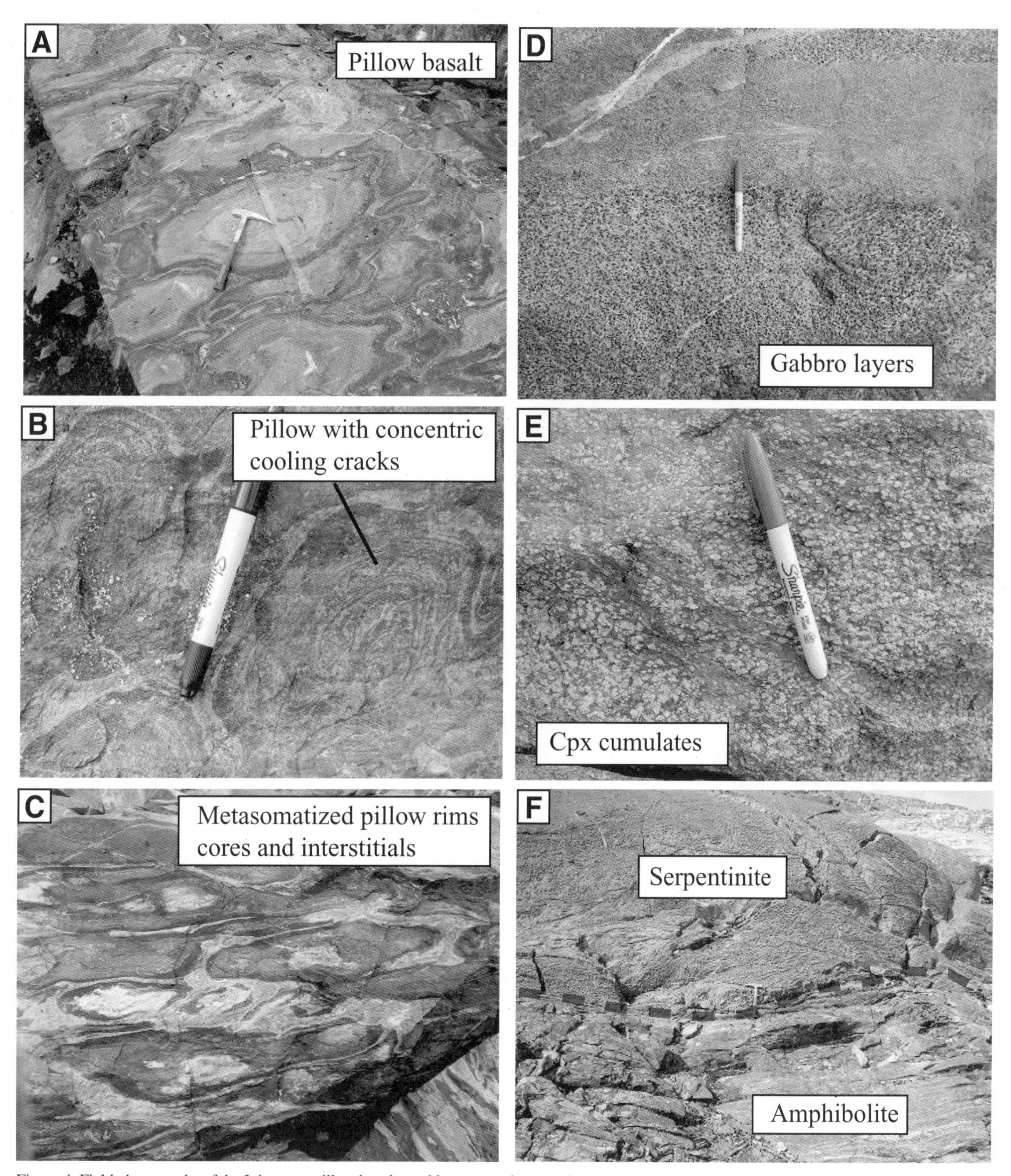


Figure 4. Field photographs of the Ivisaartoq pillow basalts, gabbros, cumulates, and serpentinites. (A) Compositionally zoned pillow basalts, recording the formation of the stage I metasomatic assemblage during seafloor hydrothermal alteration. (B) Pillow basalts with concentric cooling fractures, filled with plagioclase and quartz. (C) Pillow cores, rims, and interstitials replaced by the stage I metasomatic assemblage. (D) Gabbro with well-preserved fine- and coarse-grained layers. (E) Clinopyroxene-bearing cumulate. (F) Deformed serpentinite in amphibolites (deformed pillow basalts).

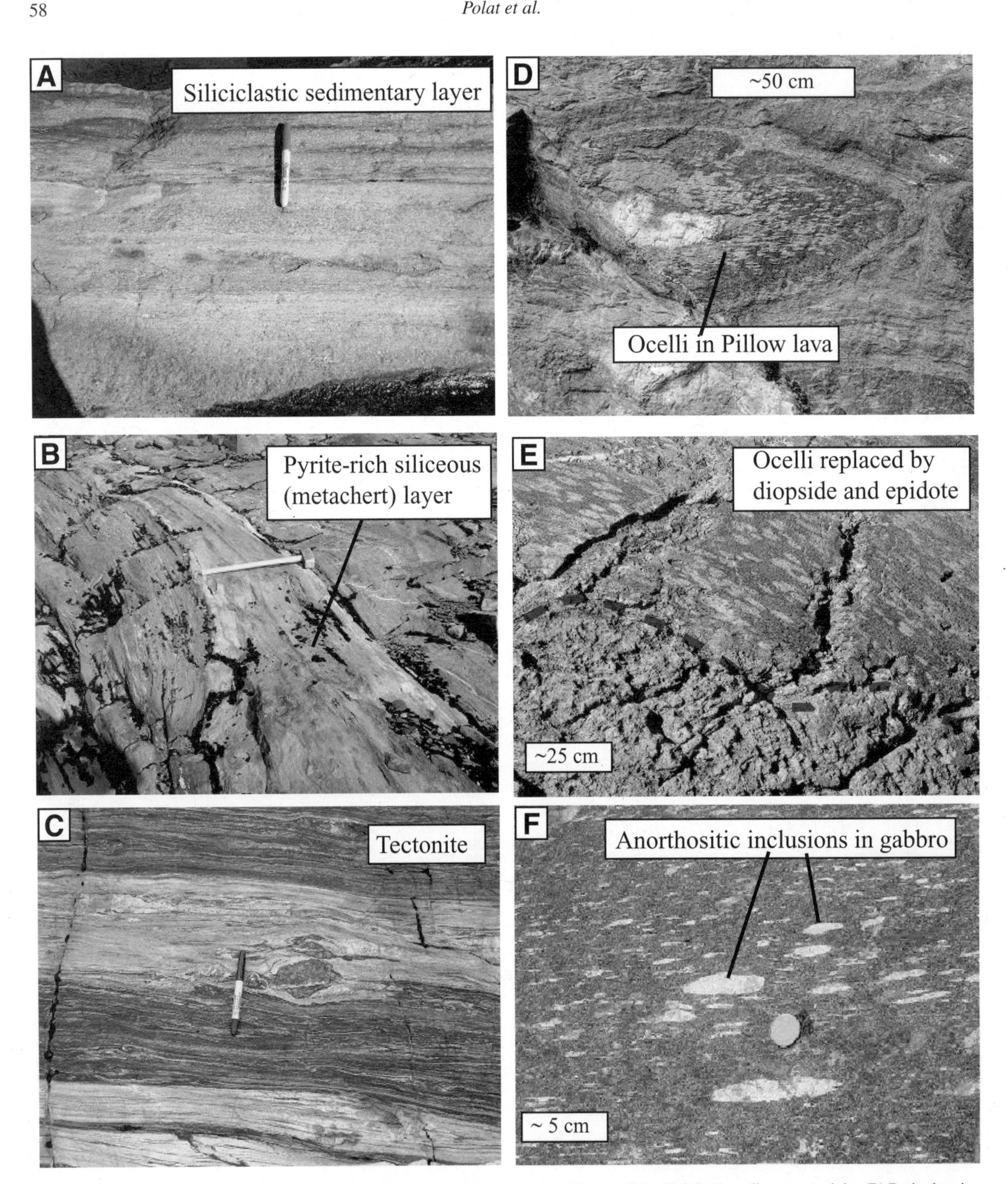


Figure 5. Field photographs of the Ivisaartoq rocks. (A) Strongly deformed rock with a possible siliciclastic sedimentary origin. (B) Pyrite-bearing siliceous (metachert) sedimentary layer within amphibolites. Contacts are typically sharp. (C) Tectonite with the stage II calc-silicate assemblage near the lower and upper amphibolite contact. (D) Pillow basalts with drainage cavity–filling quartz and ocelli in the outer core. (E) Ocelli in an outer pillow core partly replaced by the stage I calc-silicate assemblage. (F) Flattened centimeter-sized anorthositic inclusions in gabbros.

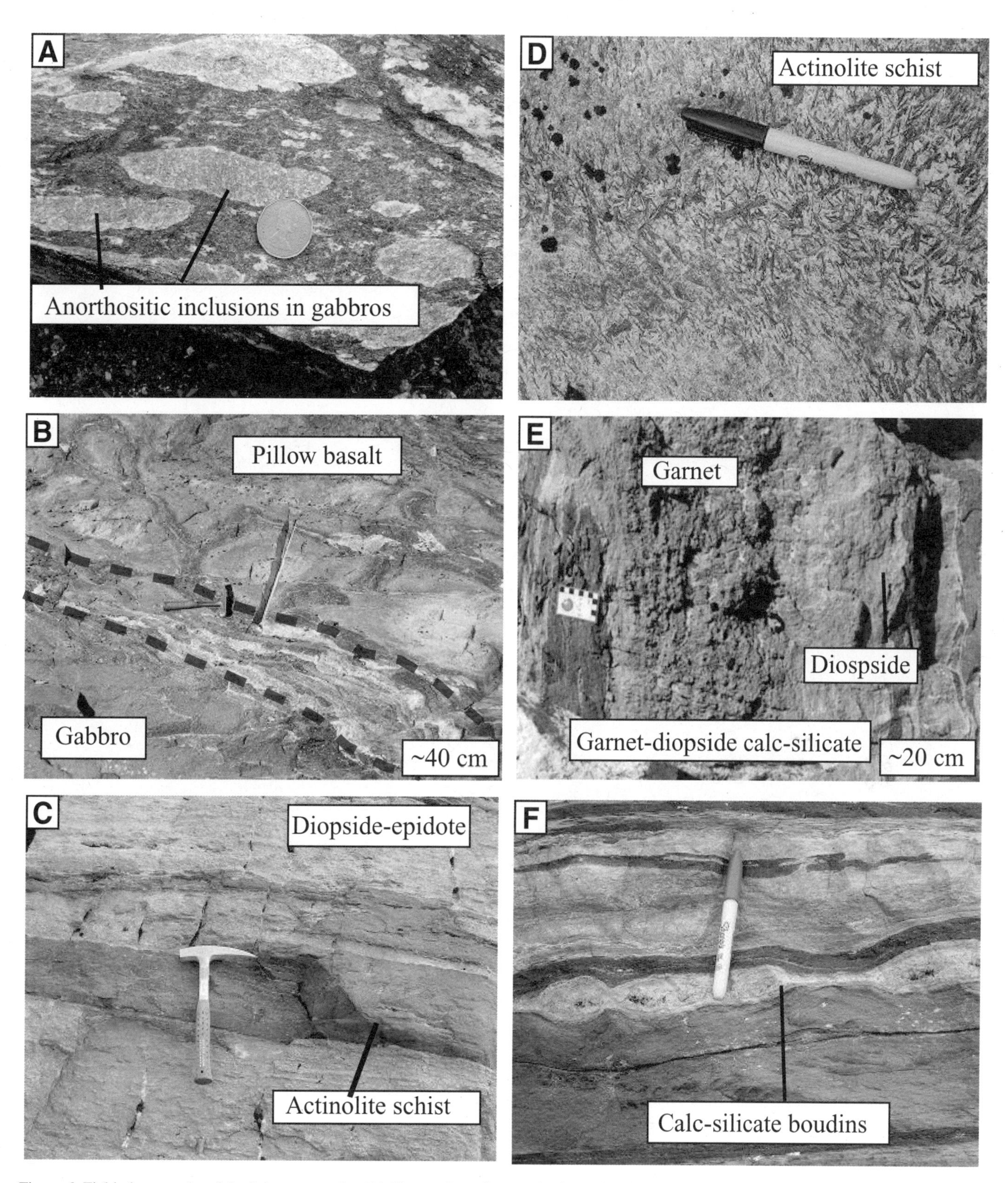


Figure 6. Field photographs of the Ivisaartoq rocks. (A) Flattened centimeter-sized anorthositic inclusions (xenoliths) in gabbros. (B) Stage I calc-silicate alteration at a pillow basalt–gabbro contact and in pillow cores. (C) Stage II calc-silicate metasomatic assemblage, replacing actinolite schists. (D) Actinolite schist (dark) replaced by a massive layered stage II calc-silicate rock assemblage. (E) A diopside + garnet + hornblende + quartz ± epidote vein. (F) Boudins of the stage II calc-silicate assemblage in banded amphibolite.

TABLE 1. MINERALOGICAL COMPOSITIONS OF THE IVISAARTOQ ROCKS

| Lithology | Mineral assemblage |
|---|---|
| Cumulate | actinolite + clinopyroxene ± plagioclase ± quartz |
| Actinolite schist | actinolite + diopside ± plagioclase ± quartz |
| Inner pillow core | diopside + plagiocalse + quartz + epidote ± amphibole ± sulphide ± titanite |
| Outer pillow core | hornblende + plagioclase + quartz ± diopside ± epidote ± titanite |
| Pillow rim | hornblende + quartz + plagioclase + epidote ± biotite ± titanite |
| Gabbro | hornblende + plagioclase ± epidote ± quartz |
| Diorite | hornblende + plagioclase ± epidote ± quartz ± biotite |
| Amphibolite | hornblende + plagioclase + quartz ± diopside ± epidote ± titanite ± sulphide |
| Anorthosite inclusions in gabbros | plagioclase + hornblende ± quartz |
| Ocelli in pillows | plagioclase + quartz + amphibole ± epidote |
| Calc-silicate stage I | diopside + quartz + plagioclase + epidote ± hornblende ± scapolite |
| Calc-silicate stage II | diopside + garnet + amphibole + plagioclase + quartz ± vesuvianite ± scapolite ± epidote ± titanite ± calcite ± scheelite |
| Magnetic marker | actinolite + olivine + diopside + magnetite ± plagioclase ± epidote ± garnet ± vesuvianite ± scapolite ± calcite ± titanite ± scheelite |

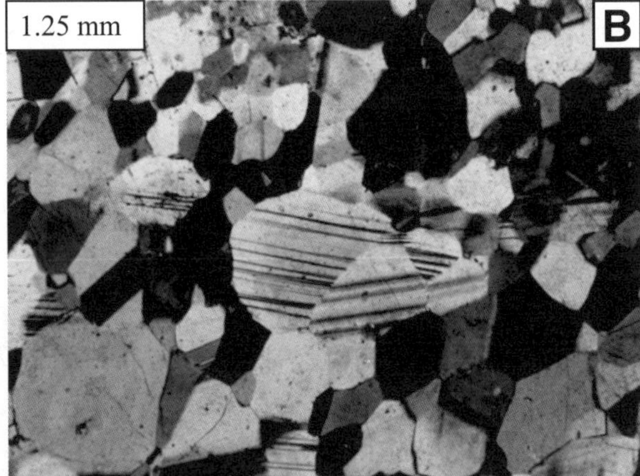


Figure 7. Photomicrographs of anorthositic inclusions in gabbros (see Fig. 6A), showing recrystallized plagioclase. Both A and B are composed predominantly of plagioclase (>90%) and hornblende, with minor (<5%) quartz. Hornblende appears to have a metamorphic origin, but an igneous origin cannot be ruled out.

show insignificant blank levels of ~5 pg Sm and ~12 pg Nd. Precisions for concentration analysis are ~0.5% for Sm and Nd. Initial $\varepsilon_{Nd}$ values were calculated at 3075 Ma from U-Pb zircon ages obtained from volcaniclastic rocks (Friend and Nutman, 2005a).

## GEOCHEMISTRY

### Major and Trace Elements

The summary of major and trace element compositions and significant ratios for the Ivisaartoq rocks are presented in Table 2. Despite their simple mineralogical composition, actinolite schists display large variations in $Al_2O_3$ (4.0–15.2 wt%), Cr (1325–12,500 ppm), Ni (430–1250 ppm), V (100–550 ppm), $TiO_2$ (0.10–0.67 wt%), Nb (0.16–2.68 ppm), and Y (2.6–21.6 ppm) (Table 2). They have moderate variations in $SiO_2$ (44–55 wt%), MgO (15.5–25.6 wt%), $Fe_2O_3$ (6.8–12.4 wt%), Zr (11–25 ppm), and Co (68–107 ppm). The abundances of REEs (e.g., La = 0.23–80 ppm) are extremely scattered (Table 2). On a chondrite-normalized diagram, they have the following characteristics: (1) variably depleted to strongly enriched LREE patterns, (2) slightly depleted to enriched HREE patterns, (3) negative to positive Eu anomalies, and (4) positive Ce anomalies (Fig. 8A; Table 2). On a primitive mantle–normalized diagram, they have the following significant features: (1) variably negative Nb anomalies and (2) negative to positive Ti and Zr anomalies (Fig. 8B; Table 2).

Cumulates are compositionally uniform at 22.3–23.5 wt% MgO, 48.7–50.3 wt% $SiO_2$, 9.2–10.4 wt% $Fe_2O_3$, 0.27–0.28 wt% $TiO_2$, 6.3–7.1 wt% $Al_2O_3$, 9.6–10.0 wt% CaO, and Mg-number = 82–83 (Table 2). They display minor variation in Ni (730–830 ppm), Cr (1575–1670 ppm), Co (80–84 ppm), and V (100–170 ppm) abundances. They have subchondritic Nb/Ta (11.5–16.9) and Zr/Y (1.8–2.1) ratios, and slightly superchondritic

TABLE 2. SUMMARY OF MAJOR (wt%), TRACE (ppm) ELEMENT CONCENTRATIONS, AND SIGNIFICANT ELEMENT RATIOS FOR THE IVISAARTOQ ROCKS

| Major element concentration (wt%) | Actinolite schist | Cumulates | Pillow lavas | Gabbros | Diorites | Anorthositic inclusions | Gabbroic matrix |
|---|---|---|---|---|---|---|---|
| $SiO_2$ | 44.2–55.4 | 48.7–50.3 | 47.7–55.7 | 47.6–51.0 | 55.2–57.1 | 47.4–49.0 | 46.9–50.4 |
| $TiO_2$ | 0.10–0.67 | 0.27–0.28 | 0.37–0.75 | 0.50–1.00 | 0.64–1.14 | 0.04–0.08 | 0.73–1.08 |
| $Al_2O_3$ | 4.0–15.2 | 6.3–7.1 | 8.7–15.2 | 13.6–16.0 | 15.0–16.9 | 29.1–30.3 | 15.8–16.4 |
| $Fe_2O_3$ | 6.8–12.4 | 9.2–10.4 | 7.7–13.9 | 9.0–13.2 | 6.2–8.0 | 2.4–3.2 | 10.7–13.0 |
| MgO | 15.5–25.6 | 22.3–23.5 | 4.5–18.8 | 7.8–14.1 | 3.8–7.6 | 1.1–1.6 | 8.1–8.6 |
| CaO | 4.5–12.4 | 9.6–10.0 | 8.5–17.2 | 8.3–12.3 | 9.2–12.0 | 14.0–15.9 | 10.6–11.1 |
| Mg-number | 72.7–86.7 | 81.7–83.2 | 53.6–76.9 | 53.9–75.6 | 50.6–70.1 | 43.6–49.5 | 56.6–60.6 |
| **Trace element concentration (ppm)** | | | | | | | |
| Cr | 1325–12,700 | 1575–1670 | 62–5700 | 230–1060 | 180–1030 | 6–11 | 210–240 |
| Co | 68–107 | 80–84 | 48–96 | 45–61 | 36–48 | 6–9 | 45–50 |
| Ni | 430–1520 | 730–830 | 125–705 | 87–230 | 60–120 | <2 | 90–140 |
| Sc | 12–50 | 24–26 | 26.5–43.4 | 33–41 | 35–48 | 1–3 | 38–42 |
| V | 100–550 | 100–170 | 136–485 | 190–475 | 230–300 | 14–21 | 240–284 |
| Nb | 0.16–2.68 | 0.69–0.77 | 0.09–1.61 | 0.10–1.77 | 1.33–3.06 | 0.10–0.21 | 1.45–2.15 |
| Zr | 11–25 | 12.5–15.6 | 22.3–42.4 | 28.2–48.5 | 36.3–68.4 | 3.6–21.1 | 48–57 |
| Th | 0.06–0.30 | 0.22–0.36 | 0.37–0.79 | 0.22–0.74 | 0.15–0.36 | 0.04–0.12 | 0.33–0.38 |
| Y | 2.6–21.6 | 7.0–8.4 | 9.4–15.2 | 13.6–19.8 | 9.5–22.2 | 1.3–2.2 | 18–20 |
| La | 0.23–80.0 | 1.59–1.83 | 2.25–3.56 | 2.21–7.98 | 2.44–6.63 | 0.68–0.98 | 2.67–3.09 |
| Nd | 0.63–54.0 | 2.35–2.85 | 3.66–5.26 | 4.12–9.47 | 4.50–10.30 | 0.61–1.00 | 6.38–7.25 |
| Sm | 0.26–7.42 | 0.69–0.87 | 1.10–1.90 | 1.26–2.47 | 1.34–2.76 | 0.15–0.24 | 2.04–2.36 |
| Gd | 0.42–6.62 | 0.98–1.22 | 1.52–2.47 | 1.83–2.94 | 1.70–3.47 | 0.19–0.32 | 2.90–3.21 |
| Yb | 0.34–2.44 | 0.80–1.00 | 1.08–1.69 | 1.49–2.20 | 1.08–2.58 | 0.16–0.21 | 2.07–2.29 |
| **Significant element ratios** | | | | | | | |
| $La/Yb_{cn}$ | 0.47–23.5 | 1.27–1.60 | 1.09–2.19 | 0.80–3.30 | 1.05–2.05 | 2.80–3.32 | 0.92–1.00 |
| $La/Sm_{cn}$ | 0.64–7.74 | 1.50–1.90 | 0.97–2.31 | 0.90–2.30 | 0.89–1.89 | 2.60–3.35 | 0.92–0.97 |
| $Gd/Yb_{cn}$ | 0.84–1.91 | 0.97–1.02 | 1.00–1.20 | 1.00–1.40 | 1.10–1.30 | 0.94–1.28 | 1.12–1.16 |
| Eu/Eu* | 0.77–2.16 | 0.61–0.81 | 0.63–1.05 | 0.70–1.00 | 0.89–1.07 | 4.1–6.5 | 0.84–0.94 |
| $Al_2O_3/TiO_2$ | 23–38 | 23.5–25.6 | 20.3–26.3 | 15.8–27.0 | 14.8–23.5 | 390–673 | 15–22 |
| Nb/Ta | 4.7–17.4 | 11.5–16.9 | 13.3–16.3 | 8.7–16.7 | 12.6–16.0 | n.d. | 12.7–14.6 |
| Y/Ho | 3.9–72.0 | 25.5–28.5 | 25.7–28.7 | 24.5–27.0 | 23.5–27.1 | 29–32 | 24–26 |
| Zr/Y | 1.1–6.0 | 1.8–2.1 | 2.0–2.9 | 1.2–2.7 | 1.2–3.8 | 1.8–9.8 | 2.6–2.9 |
| Ti/Zr | 59–163 | 103–128 | 76–116 | 94–123 | 95–106 | 13–90 | 17–23 |
| Zr/Zr* | 0.1–1.9 | 0.66–0.78 | 0.52–0.94 | 0.50–0.90 | 0.86–1.00 | 0.50–3.05 | 0.86–0.95 |
| Nb/Nb* | 0.01–0.96 | 0.31–0.41 | 0.28–0.63 | 0.30–0.80 | 0.30–0.90 | 0.11–0.16 | 0.53–0.82 |
| Ti/Ti* | 0.25–1.78 | 0.64–0.81 | 0.64–0.82 | 0.60–0.93 | 0.80–1.00 | 0.40–0.64 | 0.69–0.91 |

*Note:* cn—chondrite normalized.

$Al_2O_3/TiO_2$ (24–26) ratios (Table 2). On chondrite- and primitive mantle–normalized diagrams, they have the following trace element characteristics: (1) moderately enriched LREE patterns, (2) flat HREE patterns, and (3) negative Eu, Nb, Zr, and Ti anomalies (Figs. 9A and 9E; Table 2).

Pillow lavas are tholeiitic basalts, with a variable composition (Table 2). The ratios of Ti/Zr (76–116) and Zr/Y (2.03–2.94) range from subchondritic to slightly superchondritic values. $Al_2O_3/TiO_2$ (20–26) ratios are slightly superchondritic. The ratios of Nb/Ta (13.3–16.3) and Y/Ho (25.7–28.7) tend to be subchondritic. In addition, they have the following trace element characteristics: (1) flat to moderately enriched LREE patterns, (2) flat to slightly enriched HREE patterns, and (3) negative Nb, Zr, and Ti anomalies (Figs. 9B and 9F; Table 2).

The major element compositions of gabbros and diorites are similar to those of pillow lavas (Table 2). There are large variations in Ni (121–234 ppm), Cr (246–1060 ppm), and REEs (e.g., La = 2.2–8.0 ppm), and moderate variations in Co (52–61 ppm), V (186–264 ppm), Zr (28–44 ppm), and Y (14–20 ppm). Mg-numbers vary between 54 and 76 (Table 2). The ratios of $Al_2O_3/TiO_2$ (16–27), Zr/Y (2.1–2.7), and Ti/Zr (94–121) extend from subchondritic to superchondritic values (see Sun and McDonough, 1989). Nb/Ta (8.7–16.7) and Y/Ho (23.5–27.1) ratios are subchondritic. In addition, they have the following trace element characteristics: (1) slightly depleted to moderately enriched LREE patterns, (2) flat to slightly enriched HREE patterns, and (3) variably large negative Nb and Ti anomalies (Figs. 9C, 9D, 9G, and 9H; Table 2). Diorites have more evolved geochemical

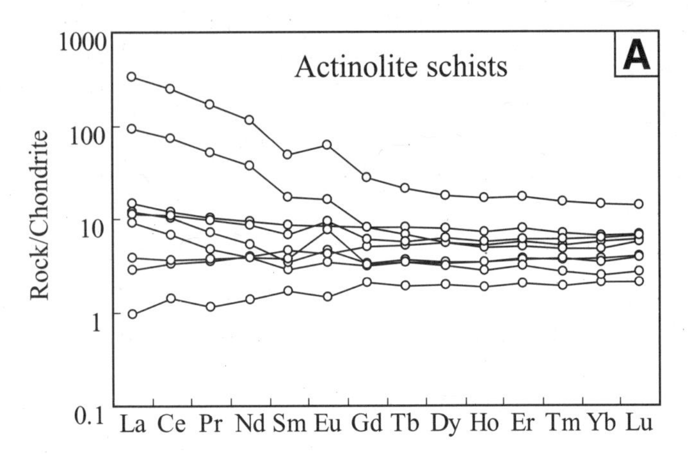


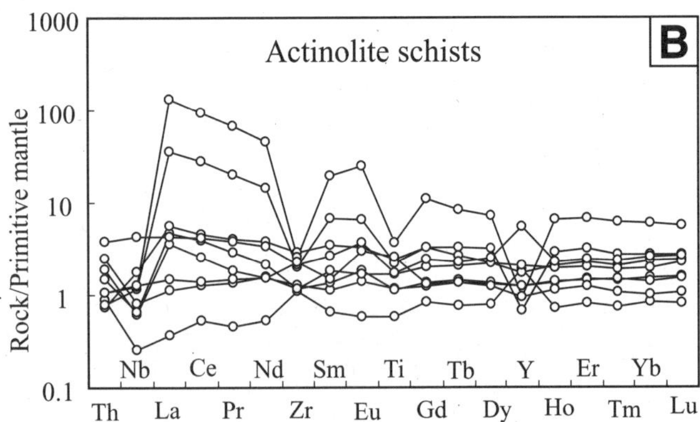


Figure 8. Chondrite-normalized REE patterns (A) and primitive mantle–normalized trace element patterns (B) for actinolite schists. Noncoherent patterns in primitive mantle–normalized diagram reflect the mobility of LREEs. Given that samples with very different REE patterns (and concentrations) have similar Th, Nb, Zr, and Ti concentrations, we suggest that these elements were less mobile than REEs. Data are from Polat et al. (2008). Chondrite normalization values are from Sun and McDonough (1989), and primitive mantle normalization values are from Hofmann (1988).

compositions (i.e., higher $SiO_2$, but lower MgO, $Fe_2O_3$, Ni, and Cr concentrations and Mg-numbers) than gabbros (Table 2). The chondrite- and primitive mantle–normalized trace element patterns of diorites are similar to those of gabbros (Fig. 9).

The anorthositic inclusions (xenoliths) in gabbros have high concentrations of $Al_2O_3$ (29–30 wt%) and CaO (14–16 wt%) (Table 2). They have extremely low Ni (<2 ppm), Cr (5.6–10.9 ppm), Co (6.4–9.4 ppm), Sc (<2 ppm), V (14–21 ppm), and $TiO_2$ (0.04–0.08 wt%). Mg-numbers range between 44 and 50. In addition, they display very low HFSE (Nb = 0.10–0.21 ppm; Y = 1.3–2.2 ppm) and REE (La = 0.68–0.98 ppm; Yb = 0.16–0.21 ppm) concentrations (Table 2). $Al_2O_3/TiO_2$ (390–673) ratios are extremely high. On primitive mantle– and chondrite-normalized diagrams, they have the following significant features: (1) moderately enriched LREE patterns, (2) flat to slightly fractionated HREE patterns, (3) large positive Eu anomalies, and (4) negative Nb and Ti anomalies (Fig. 10).

The gabbroic matrix surrounding the anorthositic inclusions has higher MgO, $Fe_2O_3$, $TiO_2$, Sc, Ni, Cr, and Co, but lower CaO, $Al_2O_3$, and Sr contents than the inclusions (Table 2). In addition, the matrix is characterized by higher absolute concentrations of REEs and HFSEs than the inclusions (Table 2). In comparison to the inclusions, the matrix has less fractionated LREE patterns, and smaller Nb and Ti anomalies (Fig. 10; Table 2).

### Neodymium Isotopes

Regression of the Sm-Nd isotope data for picritic cumulates and their more deformed counterpart, actinolite schists, yields an errorchron age of 3092 ± 260 Ma (MSWD = 97) (Fig. 11A). This age, within uncertainties, is in good agreement with the 3075 ± 15 Ma U-Pb zircon age of the spatially associated siliceous volcaniclastic sedimentary rocks (see Friend and Nutman, 2005a). Large uncertainty in the errorchron age is likely due to large scatter in the data. Cumulates have a narrow range of initial $\varepsilon_{Nd}$ values (+4.2 to +5.0), whereas actinolite schists display large variations (+2.5 to +9.5) (Table 3). Strongly sheared actinolite schists, with very large $\varepsilon_{Nd}$ values (+8.3 and +9.5), have much higher LREE concentrations (e.g., Nd = 16–43 ppm) and more fractionated LREE patterns, compared to the rest of the samples in the group (Figs. 8 and 11; Table 2). In addition, these samples display the lowest $^{147}Sm/^{144}Nd$ (0.088–0.089) ratios in the group (Table 3).

Pillow lavas, gabbros, and diorites define an errorchron age of 3069 ± 220 Ma (MSWD = 80) (Fig. 11B). This age, within uncertainties, agrees well with the 3075 ± 15 Ma U-Pb zircon age of siliceous volcaniclastic sedimentary rocks (Friend and Nutman, 2005a; Polat et al., 2007). The large uncertainty in the errorchron ages reflects the narrow compositional range and large scatter in the data points. The initial $\varepsilon_{Nd}$ values in pillow lavas (+1.1 to +3.1) overlap with, but extend to higher values than, gabbros and diorites (+0.3 to +2.9) (Table 3). Samples with higher MgO content tend to have greater initial $\varepsilon_{Nd}$ values. All rock types have similar range of $^{147}Sm/^{144}Nd$ (Table 3).

The anorthositic inclusions have much higher initial $\varepsilon_{Nd}$ values than the surrounding mafic matrix ($\varepsilon_{Nd}$ = +4.8 to +6.0 versus +2.3 to +2.8) (Table 4). The Nd isotopic compositions of the inclusions are comparable to those of clinopyroxene cumulates (Tables 3 and 4). The initial $\varepsilon_{Nd}$ (+2.3 to +2.8) isotopic composition of the matrix overlaps with, but extends to higher values than, gabbros (+0.3 to +2.9), which are devoid of anorthositic inclusions, and diorites (+1.0 to +1.5).

## DISCUSSION

### Postmagmatic Alteration, Element Mobility, and Modification of Isotopic Composition

The Ivisaartoq greenstone belt underwent at least two stages of metamorphic alteration prior to the intrusion of 2961 ± 12 Ma granitoids, resulting in the formation of widespread calc-silicate metasomatic mineral assemblages (Figs. 4, 5, and 6) (Polat et al., 2007, 2008). The stage I metasomatic assemblage appears

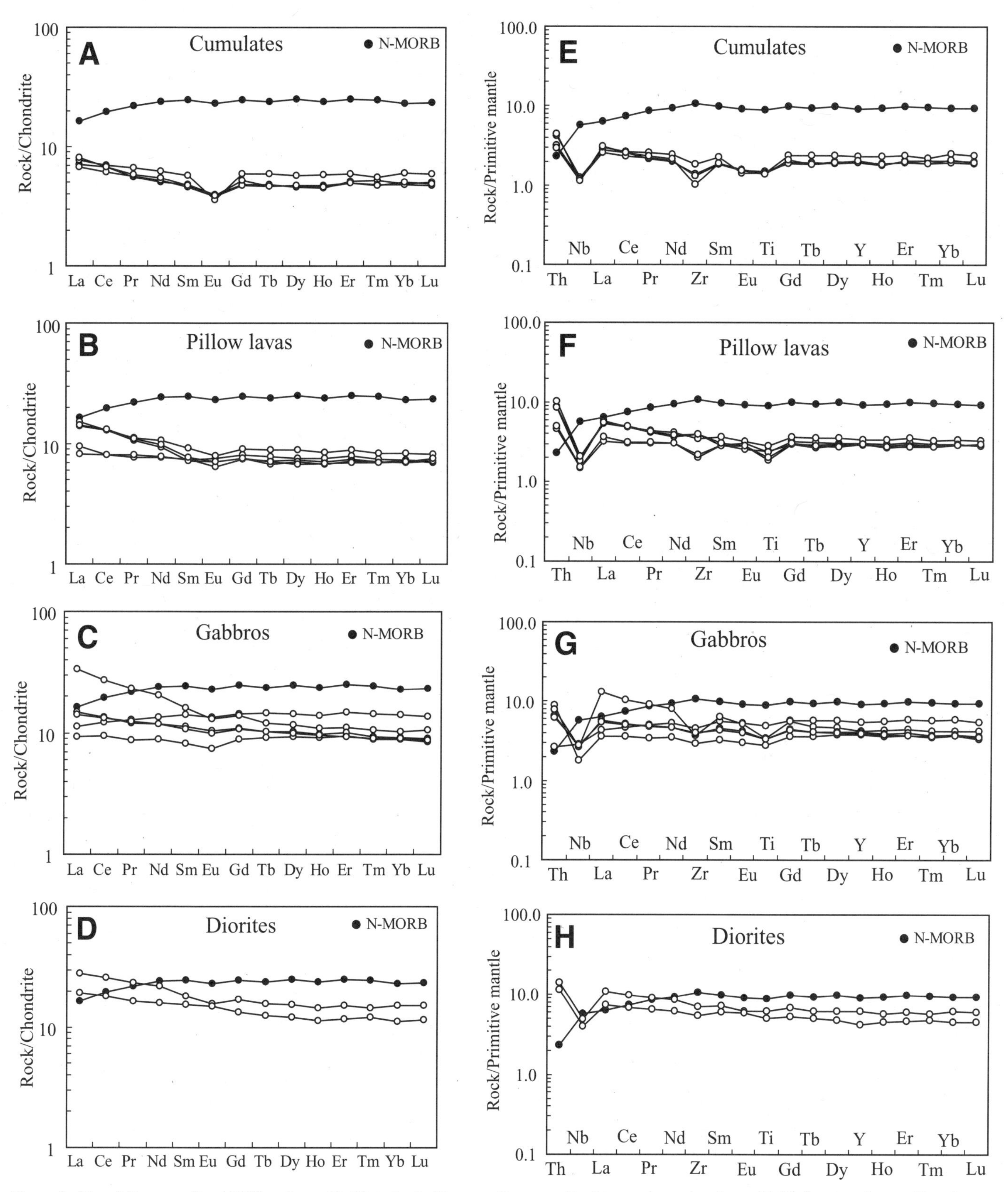


Figure 9. Chondrite-normalized REE patterns (A-D) and primitive mantle–normalized trace element patterns (E–H) for cumulates, pillow lavas, gabbros, and diorites. Data are from Polat et al. (2007, 2008). Chondrite normalization values are from Sun and McDonough (1989), and primitive mantle normalization values are from Hofmann (1988). N-MORB—normal mid-oceanic-ridge basalt.

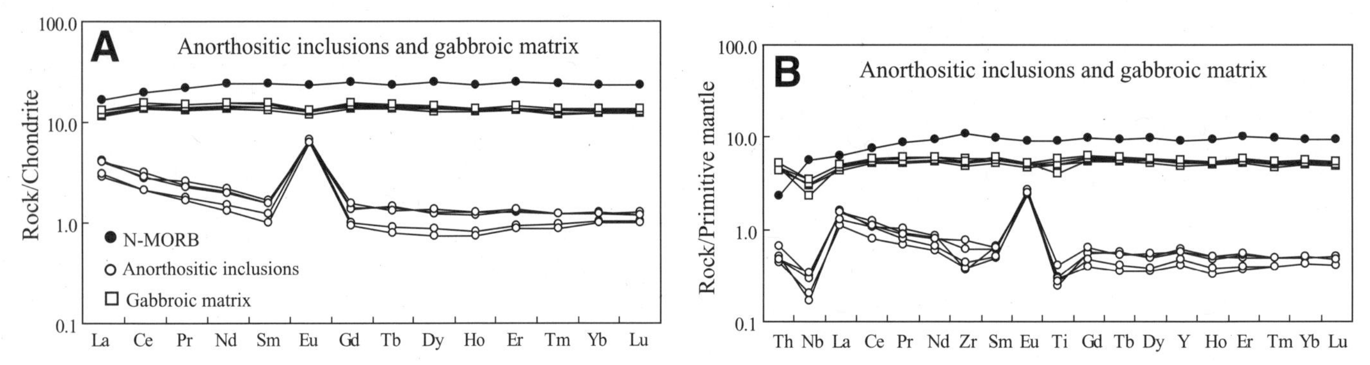


Figure 10. Chondrite-normalized REE patterns (A) and primitive mantle–normalized trace element patterns (B) for inclusions (xenoliths) in gabbros (see Fig. 6A). Data are from Polat et al. (2008). Chondrite normalization values are from Sun and McDonough (1989), and primitive mantle normalization values are from Hofmann (1988). N-MORB—normal mid-oceanic-ridge basalt.

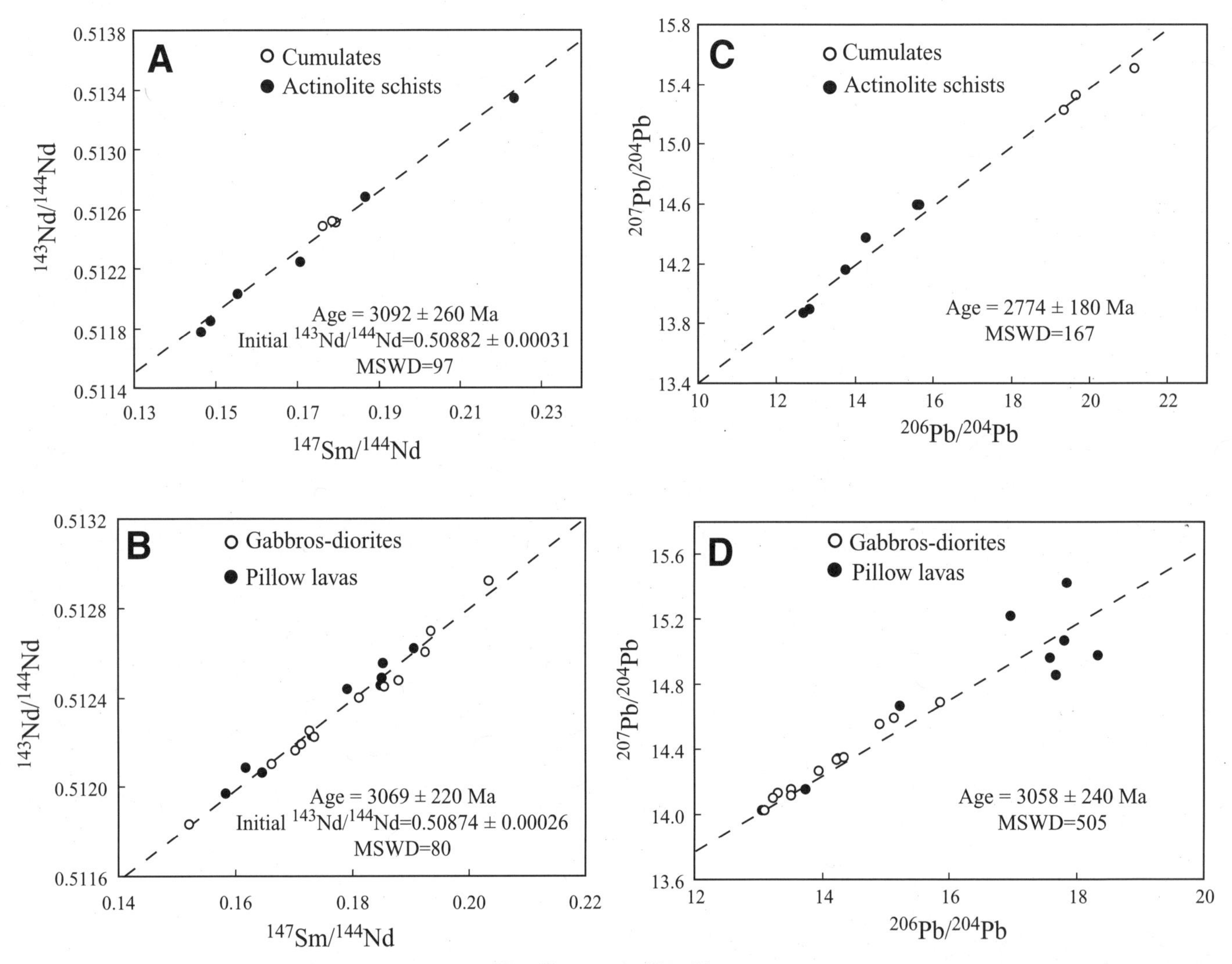


Figure 11. $^{143}Nd/^{144}Nd$ versus $^{147}Sm/^{144}Nd$ (A and B) and $^{207}Pb/^{204}Pb$ versus $^{206}Pb/^{204}Pb$ (C and D) errorchron diagrams for cumulates, actinolite schists, pillow lavas, gabbros, and diorites. Data are from Polat et al. (2007, 2008). The Isoplot program of Ludwig (2003) was used for age and initial $^{143}Nd/^{144}Nd$ ratio calculations.

TABLE 3. Sm-Nd ISOTOPE COMPOSITIONS OF THE IVISAARTOQ ACTINOLITE SCHISTS, CUMULATES, PILLOW LAVAS, GABBROS, AND DIORITES

| Sample # | Rock type | $^{143}Nd/^{144}Nd$ | ±2σ | Nd (ppm) | $^{147}Sm/^{144}Nd$ | Sm (ppm) | $\varepsilon_{Nd}$ (3075 Ma) | Sm/Nd | North | West |
|---|---|---|---|---|---|---|---|---|---|---|
| 485426 | Actinolite schist | 0.512027 | 11 | 1.82 | 0.1557 | 0.469 | 4.4 | 0.26 | 64°44.903′ | 49°53.261′ |
| 485430 | Actinolite schist | 0.512244 | 9 | 3.45 | 0.1709 | 0.974 | 2.5 | 0.28 | 64°44.900′ | 49°53.466′ |
| 485431 | Actinolite schist | 0.513339 | 28 | 0.72 | 0.2236 | 0.267 | 3.1 | 0.37 | 64°44.900′ | 49°53.466′ |
| 485433 | Actinolite schist | 0.511848 | 11 | 3.67 | 0.1489 | 0.902 | 3.6 | 0.25 | 64°44.940′ | 49°53.402′ |
| 485434 | Actinolite schist | 0.510883 | 13 | 16.3 | 0.0894 | 2.413 | 8.3 | 0.15 | 64°44.940′ | 49°53.402′ |
| 485435 | Actinolite schist | 0.511767 | 11 | 3.59 | 0.1465 | 0.869 | 2.9 | 0.24 | 64°44.940′ | 49°53.402′ |
| 485436 | Actinolite schist | 0.510923 | 9 | 42.6 | 0.0884 | 6.218 | 9.5 | 0.15 | 64°44.943′ | 49°53.379′ |
| 485437 | Actinolite schist | 0.512678 | 11 | 1.53 | 0.1869 | 0.471 | 4.7 | 0.31 | 64°44.962′ | 49°53.313′ |
| 485473 | Cpx cumulate | 0.512504 | 10 | 2.07 | 0.1795 | 0.615 | 4.2 | 0.30 | 64°44.763′ | 49°51.318′ |
| 485474 | Cpx cumulate | 0.512512 | 11 | 2.44 | 0.1787 | 0.719 | 4.7 | 0.30 | 64°44.763′ | 49°51.318′ |
| 485475 | Cpx cumulate | 0.512479 | 12 | 2.77 | 0.1764 | 0.807 | 5.0 | 0.29 | 64°44.763′ | 49°51.318′ |
| 485481 | Pillow basalt | 0.512086 | 11 | 4.56 | 0.1618 | 1.219 | 3.1 | 0.27 | 64°44.515′ | 49°53.299′ |
| 485482 | Pillow basalt | 0.511967 | 10 | 4.34 | 0.1584 | 1.135 | 2.1 | 0.26 | 64°44.515′ | 49°53.361′ |
| 485486 | Pillow basalt | 0.512060 | 13 | 3.95 | 0.1646 | 1.075 | 1.4 | 0.27 | 64°44.438′ | 49°54.399′ |
| 485414 | Pillow basalt | 0.512617 | 14 | 5.67 | 0.1906 | 1.786 | 2.0 | 0.31 | 64°44.055′ | 49°56.286′ |
| 485418 | Pillow basalt | 0.512549 | 10 | 3.13 | 0.1853 | 0.958 | 2.8 | 0.31 | 64°44.375′ | 49°56.130′ |
| 485420 | Pillow basalt | 0.512226 | 12 | 4.59 | 0.1732 | 1.313 | 1.3 | 0.29 | 64°44.334′ | 49°56.183′ |
| 485422 | Pillow basalt | 0.512438 | 9 | 3.88 | 0.1792 | 1.149 | 3.0 | 0.30 | 64°44.361′ | 49°56.678′ |
| 485468 | Pillow basalt | 0.512453 | 12 | 5.19 | 0.1848 | 1.586 | 1.1 | 0.31 | 64°44.798′ | 49°51.544′ |
| 485469 | Pillow basalt | 0.512483 | 12 | 4.86 | 0.1850 | 1.486 | 1.6 | 0.31 | 64°44.812′ | 49°51.477′ |
| 485428 | Gabbro | 0.512922 | 9 | 6.0 | 0.204 | 2.025 | 2.9 | 0.34 | 64°44.464′ | 49°56.311′ |
| 485432 | Gabbro | 0.512472 | 10 | 5.72 | 0.1878 | 1.776 | 0.3 | 0.31 | 64°44.940′ | 49°53.402′ |
| 485438 | Gabbro | 0.511830 | 12 | 8.28 | 0.1521 | 2.080 | 1.9 | 0.25 | 64°44.962′ | 49°53.313′ |
| 485467 | Gabbro | 0.512601 | 13 | 3.97 | 0.1925 | 1.263 | 0.9 | 0.32 | 64°44.941′ | 49°51.831′ |
| 485472 | Gabbro | 0.512449 | 6 | 4.99 | 0.1854 | 1.528 | 0.8 | 0.31 | 64°44.767′ | 49°51.576′ |
| 485476 | Gabbro | 0.512395 | 11 | 5.49 | 0.1813 | 1.643 | 1.4 | 0.30 | 64°44.815′ | 49°51.046′ |
| 485477 | Gabbro | 0.512248 | 11 | 5.49 | 0.1727 | 1.565 | 1.9 | 0.29 | 64°44.767′ | 49°51.576′ |
| 499730-A | Gabbro | 0.512693 | 5 | 6.94 | 0.1935 | 2.218 | 2.4 | 0.32 | 64°44.492′ | 49°56.613′ |
| 499739 | Diorite | 0.512219 | 5 | 4.64 | 0.1735 | 1.332 | 1.0 | 0.29 | 64°44.742′ | 49°54.082′ |
| 485483 | Diorite | 0.512098 | 10 | 9.97 | 0.1663 | 2.739 | 1.5 | 0.27 | 64°44.450′ | 49°53.799′ |
| 485429 | Diorite | 0.512158 | 10 | 5.57 | 0.1702 | 1.566 | 1.1 | 0.28 | 64°44.776′ | 49°53.759′ |
| 485484 | Diorite | 0.512188 | 11 | 6.84 | 0.1713 | 1.937 | 1.28 | 0.28 | 64°44.452′ | 49°53.789′ |

Note: Table is from Polat et al. (2008); all initial $\varepsilon_{Nd}$ ages calculated at 3075 Ma yielded by U-Pb zircon analyses (see Friend and Nutman, 2005a).

TABLE 4. Sm-Nd ISOTOPE COMPOSITIONS OF THE ANORTHOSITIC INCLUSIONS AND SURROUNDING GABBROIC MATRIX IN THE IVISAARTOQ GREENSTONE BELT

| Sample # | Rock type | $^{143}Nd/^{144}Nd$ | ±2σ | Nd (ppm) | $^{147}Sm/^{144}Nd$ | Sm (ppm) | $\varepsilon_{Nd}$ (3075 Ma) | North | West |
|---|---|---|---|---|---|---|---|---|---|
| 499728-A1 | Anorthositic inclusion | 0.511941 | 6 | 0.850 | 0.1505 | 0.2114 | 4.77 | 64° 44.492′ | 49° 56.613′ |
| 499729-A1 | Anorthositic inclusion | 0.511935 | 7 | 0.980 | 0.1474 | 0.2388 | 5.85 | 64° 44.492′ | 49° 56.613′ |
| 499731-A1 | Anorthositic inclusion | 0.511802 | 7 | 0.578 | 0.1406 | 0.1343 | 5.99 | 64° 44.492′ | 49° 56.613′ |
| 499731-B1 | Anorthositic inclusion | 0.511811 | 6 | 0.615 | 0.1420 | 0.1444 | 5.57 | 64° 44.492′ | 49° 56.613′ |
| 499731-C1 | Anorthositic inclusion | 0.512015 | 6 | 0.871 | 0.1532 | 0.2204 | 5.14 | 64° 44.492′ | 49° 56.613′ |
| 499728-A2 | Gabbroic matrix | 0.512699 | 6 | 6.732 | 0.1929 | 2.146 | 2.69 | 64° 44.492′ | 49° 56.613′ |
| 499729-A2 | Gabbroic matrix | 0.512687 | 6 | 6.612 | 0.1929 | 2.107 | 2.49 | 64° 44.492′ | 49° 56.613′ |
| 499731-A2 | Gabbroic matrix | 0.512732 | 5 | 6.936 | 0.1943 | 2.227 | 2.79 | 64° 44.492′ | 49° 56.613′ |
| 499731-B2 | Gabbroic matrix | 0.512713 | 4 | 7.269 | 0.1934 | 2.323 | 2.81 | 64° 44.492′ | 49° 56.613′ |
| 499731-C2 | Gabbroic matrix | 0.512703 | 4 | 7.223 | 0.1942 | 2.317 | 2.30 | 64° 44.492′ | 49° 56.613′ |

*Note:* Table is from Polat et al. (2008); all initial $\varepsilon_{Nd}$ ages calculated at 3075 Ma yielded by U-Pb zircon analyses (see Friend and Nutman, 2005a).

to have formed during seafloor hydrothermal alteration under greenschist to lower-amphibolite facies metamorphic conditions (Polat et al., 2007). The stage II metasomatic assemblage was formed during a regional tectonothermal metamorphic event under middle- to upper-amphibolite facies metamorphic conditions (Appel, 1997; Polat et al., 2008).

Many pillow basalts are mineralogically and chemically zoned (Figs. 4A and 4C) (Polat et al., 2007). The rims have higher contents of $Fe_2O_3$, MgO, MnO, and $K_2O$, whereas the inner and outer cores possess higher concentrations of CaO, and $Na_2O$ and $SiO_2$, respectively, consistent with the mobility of these elements during postmagmatic alteration. Similarly, large variations in Ba, Sr, Pb, Rb, Cs, Li, U, Zn, and Cu contents between pillow cores and rims are consistent with the mobility of these elements. Compared with the less altered cores, the rims have lower LREE abundances and $La/Sm_{cn}$ ratios, indicating the loss of these elements. In contrast to the above elements, $Al_2O_3$, $TiO_2$, Th, Zr, Y, Cr, Ni, Co, Ga, and HREEs display minor variations between the cores and rims, suggesting that these elements were relatively immobile during Mesoarchean seafloor hydrothermal alteration. Likewise, REEs and HFSEs (Ti, Nb, Ta, Zr, Y) in gabbros, diorites, pillow lavas, and cumulates display coherent primitive mantle– and chondrite-normalized patterns (Fig. 9), indicating that these elements were also relatively immobile during postmagmatic alteration.

Cumulates with relic clinopyroxene phenocrysts are characterized by more coherent trace element patterns and narrower ranges of many major and trace elements than their more deformed actinolite schist counterparts. In addition, they have a narrow range of Sm/Nd (0.29–0.30) ratios and initial $\varepsilon_{Nd}$ values (+4.2 to +5.0) (Table 3), consistent with the Sm-Nd system remaining closed. Given the preservation of primary minerals and texture in cumulates, the initial $\varepsilon_{Nd}$ values in these rocks likely reflect the near-primary magmatic composition (Fig. 4E; Table 3).

In contrast to those in cumulates, many elements in actinolite schists display large variations (Fig. 8; Table 2). Despite the fact that actinolite schist samples analyzed for this study came from the least metasomatized outcrops, they still display a large spread in Sm and Nd concentrations, and in isotopic ratios (Tables 2 and 3). Additionally, these samples have large positive to negative Eu and Ce anomalies (Fig. 8; Table 2). Even the samples that were collected from the same outcrop display significant variations in the isotopic ratios. In summary, these geochemical characteristics are consistent with the mobility of Sm and Nd in actinolite schists during postmagmatic alteration.

The initial $\varepsilon_{Nd}$ values in gabbros and diorites (+0.3 to +2.9) overlap with, but extend to lower values than, the pillow lavas (+1.1 to +3.1). We suggest that these initial $\varepsilon_{Nd}$ values likely reflect the near-primary magmatic compositions for the following reasons: (1) Both groups plot colinearly on a $^{143}Nd/^{144}Nd$ versus $^{147}Sm/^{144}Nd$ diagram, yielding a Mesoarchean age (3069 ± 220 Ma; MSWD = 80); (2) there are no covariations between the initial $\varepsilon_{Nd}$ values and $La/Sm_{cn}$ ratios within each group; and (3) there are no correlations between $\varepsilon_{Nd}$ and mobile elements (e.g., LILEs).

**Geodynamic Setting**

Despite the recent advances in understanding the origin of Archean greenstone-granitoid terranes, the question regarding the operation of modern-style plate tectonics in early Earth remains still controversial (see Stern, 2005; Bédard, 2006; Kerrich and Polat, 2006; Ernst, 2007). Both uniformitarian and nonuniformitarian geodynamic models have been proposed to explain the origin of Archean greenstone belts. Phanerozoic-style plate tectonic models have been constructed for Archean greenstone-granitoid terranes since the 1970s. These models were based primarily on field relationships, in which multiple lithotectonic terranes, each having distinct lithological associations, ages, structural history, and metamorphic grade(s) were juxtaposed along regional-scale structures, and therefore were interpreted to have been accreted allochthonously (see Kusky and Polat, 1999; Friend and Nutman, 2005a; Kerrich and Polat, 2006; Nutman and Friend, 2007). Similarly, based on uniformitarian principles, the geochemical characteristics of Archean volcanic rocks have been used to interpret their geodynamic settings. Tectonomagmatic settings proposed for Archean greenstone belts include oceanic and continental arcs, plume-derived intra-oceanic (plateau) and -continental flood basalt provinces, mid-ocean ridges, backarcs, and forearcs (see Kerrich and Polat, 2006, for a review).

On the Th/Yb versus Nb/Yb variation diagram of Pearce and Peate (1995), the Mesoarchean Ivisaartoq volcanic rocks plot in the field of modern subduction-derived magmas (Figs. 12A and 12B). Similarly, on a Ti/Zr versus Zr discrimination diagram (Knittel and Oles, 1994), the Ivisaartoq lavas plot in the field of modern island arc basalts (Fig. 12C). In the following section, we use field observations and normalized trace elements patterns to evaluate various possible geodynamic settings, including the island arc geodynamic setting as suggested by the Th/Yb-Nb/Yb and Ti/Zr-Zr variations, for the Ivisaartoq greenstone belt.

Given the presence of pillow structures and evidence for seafloor hydrothermal alteration, the Ivisaartoq greenstone belt appears to have originated in an oceanic rather than a continental environment. There is no field evidence (e.g., basal conglomerates, normal faulting) indicating that the Ivisaartoq greenstone belt was deposited on older continental basement as an autochthonous continental flood basalt province (e.g., the Deccan, Karoo, and Parana basalts). The presence of gabbroic and dioritic dikes and siliciclastic sedimentary rocks are inconsistent with a typical oceanic plateau (e.g., Ontong Java, Caribbean) setting. No komatiitic flows, the origin of which has typically been attributed to mantle plumes (Storey et al., 1991; Herzberg, 1992; Condie, 2004; Ernst et al., 2004), have been documented in the belt to suggest a mantle plume origin. Most plume-derived intra-oceanic (e.g., OIB) and intra-continental (e.g., rifts) volcanic sequences tend to contain alkaline volcanic rocks; such rocks have not been documented in the Ivisaartoq

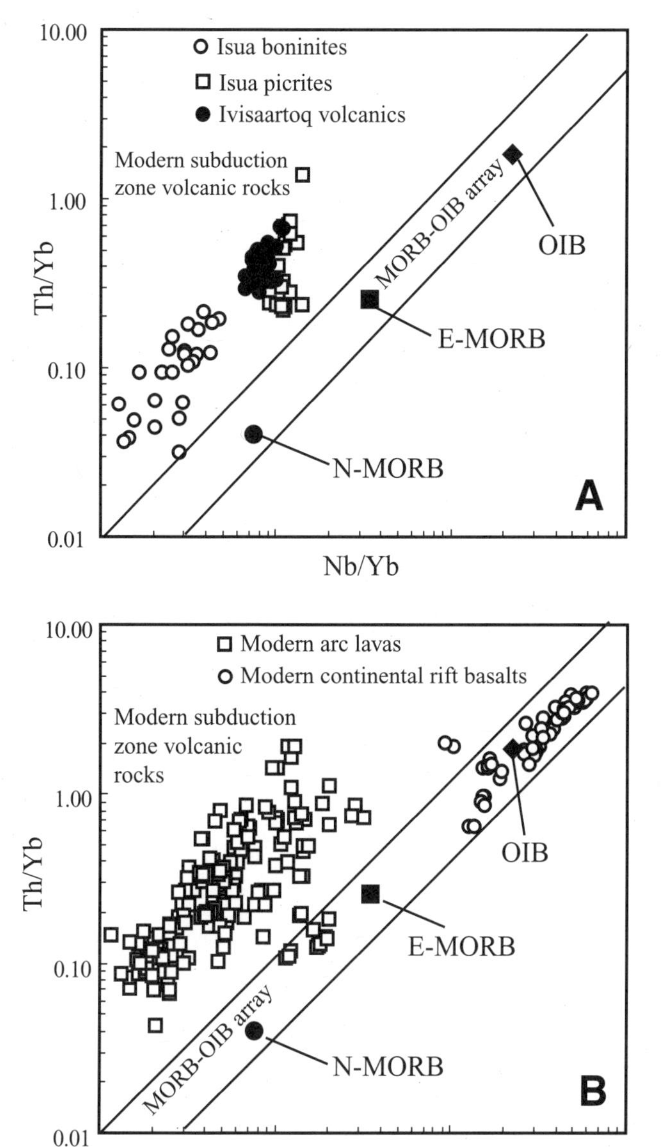


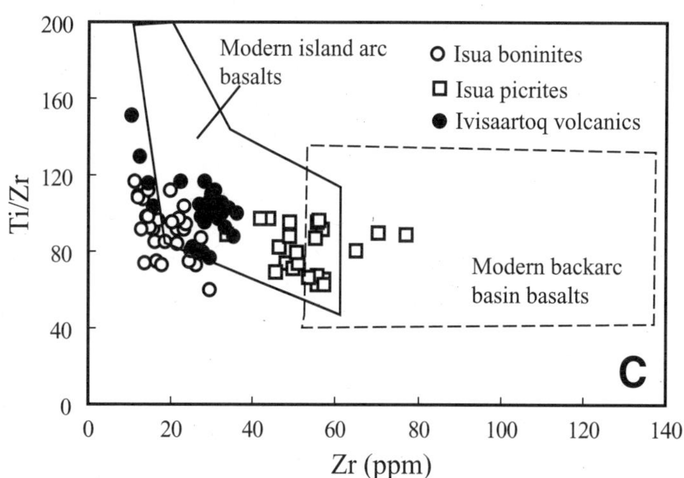


Figure 12. (A) Th/Yb versus Nb/Yb variation diagrams for the 3800–3700 Ma Isua boninites and picrites, and 3075 Ma Ivisaartoq mafic to ultramafic volcanic rocks, suggesting that all Eo- to Neoarchean southern West Greenland volcanic rocks plot in the field of modern subduction zone volcanic rocks. (B) Th/Yb versus Nb/Yb variation diagram for modern subduction zone and intra-continental rift volcanic rocks. None of the Greenland volcanic rocks plot in the field of intra-continental rift volcanic rocks. Data for Isua volcanic rocks are from Polat et al. (2002) and Polat and Hofmann (2003). Data for modern subduction and intra-continental rift rocks are from Class et al. (1994), Pearce et al. (1995), Peate et al. (1997), Polat et al. (1997), Ewart et al. (1998), Jung and Masberg (1998), Woodhead et al. (1998), Yilmaz and Polat (1998), and Schuth et al. (2004). Diagrams modified after Pearce and Peate (1995). N-MORB (normal mid-oceanic-ridge basalt), E-MORB (enriched mid-oceanic-ridge basalt), and OIB (oceanic-island basalt) values are from Sun and McDonough (1989). (C) Ti/Zr versus Zr variation diagram for the Isua and Ivisaartoq. The majority of the Ivisaartoq volcanic rocks plot in the field of modern island arc basalts. Fields for modern backarc basalts and island arc basalts are from Knittel and Oles (1994).

greenstone belt. Sylvester et al. (2006) interpreted the geochemical composition of the "magmatic marker" as representing ultramafic alkaline lavas derived from a mantle plume by a small degree (<5%) of partial melting. Given the presence of extensive calc-silicate metasomatic alteration, intensive deformation, and a large amount of metamorphic magnetite (~10%) in the magnetic marker, the geochemistry of this unit should be interpreted with great caution. The primitive mantle–normalized trace element patterns of the Ivisaartoq lavas are inconsistent with a typical mid-ocean tectonic setting (Fig. 9). Finally, on the Th/Yb versus Nb/Yb variation diagram, none of the Ivisaartoq samples plot in the fields of modern plume-derived oceanic-island basalts (OIBs), continental rift basalts, and mid-oceanic-ridge basalts (MORBs) (Figs. 12A and 12B).

Although the trace element systematics of the Ivisaartoq rocks are consistent with subduction zone geochemical signatures (cf. Saunders et al., 1991; Hawkesworth et al., 1993; Pearce and Peate, 1995), crustal contamination cannot be completely ruled out to account for the origin of the negative Nb anomalies in these rocks. Accordingly, the origin of the Ivisaartoq rocks can be attributed to the following geodynamic settings: (1) a Red Sea–type proto-ocean setting (Fig. 13A), (2) a Rocas Verdes–type (southern Andes) continental arc rift or backarc basin (Fig. 13B) (see Tarney et al., 1976; Stern and de Wit, 1997), and (3) a Western Pacific–type oceanic arc-backarc-forearc system (e.g., Izu-Bonin-Mariana, Solomon–New Hebrides, Tonga-Kermadec-Lau) (Fig. 13C) (see Pearce and Peate, 1995; Hawkins, 2003; Schuth et al., 2004).

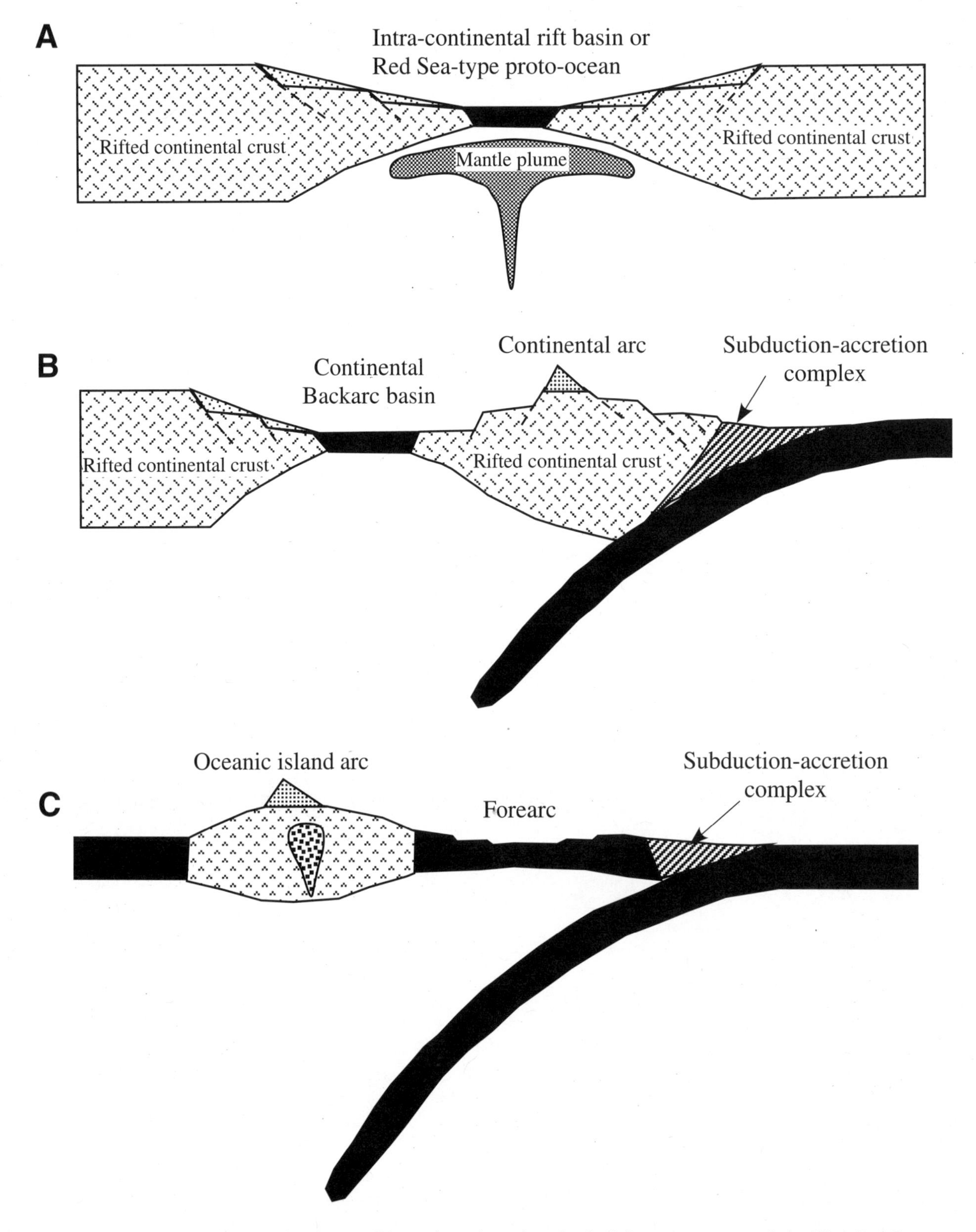


Figure 13. Cartoon diagrams showing possible geodynamic settings for the Ivisaartoq greenstone belt. (A) A Red Sea–type proto-ocean resulting from continental rifting where plume-lithosphere interaction could have resulted in negative Nb anomalies. (B) A Rocas Verdes–type continental backarc basin where a combination of crustal assimilation and subduction zone processes would have been the cause of the negative Nb anomalies. (C) An oceanic arc-forearc setting where the negative Nb anomalies resulted from subduction zone petrogenetic processes.

High MgO (14–24 wt%), Ni (600–850 ppm), and Cr (1600–1700 ppm) concentrations, and Mg-numbers (70–83) in clinopyroxene cumulates and high-Mg pillow lavas are consistent with an island arc picritic composition. Tertiary island arc picrites have been documented in the Solomon and New Hebrides (Vanuatu arc) oceanic island arcs (see Schuth et al., 2004, and references therein). In the Solomon Islands, picrites occur only in New Georgia Island above the subducting Woodlark spreading center. In the New Hebrides subduction system, the overriding plate is currently undergoing extension east of the Vanuatu arc, forming a suprasubduction oceanic crust within the North Fiji Basin (Hawkins, 2003). As a corollary, given that the geochemical and field characteristics of the Ivisaartoq rocks are comparable to those of modern intra-oceanic island arcs, we suggest that the Ivisaartoq greenstone belt originated as Mesoarchean intra-oceanic forearc or backarc oceanic crust (Fig. 13C).

### Petrogenesis and Mantle Source Characteristics

The positive initial $\varepsilon_{Nd}$ values (e.g., +4.2 to +5.0 in clinopyroxene cumulates; +4.8 to +6.0 in anorthositic inclusions; +1.1 to +3.1 in pillow basalts) require long-term depleted upper mantle sources. Given the near-flat HREE patterns in these lithologies, melting occurred at <80 km (cf. Hirschmann and Stolper, 1996). The depleted Nd isotopic signatures and low LREE and HFSE (Nb, Ta, Zr, Ti, and Y) abundances indicate that the mantle source region had experienced significant melt extraction prior to 3075 Ma.

Picritic cumulates and anorthositic inclusions plot above the estimated evolution curve of the depleted mantle, whereas the majority of pillow lavas, gabbros, and diorites plot below the predicted depleted mantle evolution curve (Polat et al., 2008). Large variations in the initial $\varepsilon_{Nd}$ values (+0.3 to +3.1) may reflect either mantle source heterogeneity or crustal contamination. Contamination of the Ivisaartoq rocks by continental crust during magma ascent, rather than contamination of their source regions by subducted crustal material, may be ruled out on the basis of the following observations: (1) The association of pillow basalts, gabbros, and sulfide-rich siliceous volcaniclastic sedimentary rocks is expected to have formed in an oceanic rather than a continental setting; (2) there is no field evidence indicating that the Ivisaartoq greenstone belt was deposited on older continental crust; (3) there are no covariations between $\varepsilon_{Nd}$ and abundances of contamination-sensitive elements or their ratios (e.g., $SiO_2$, Ni, Cr, Th, Zr, $La/Sm_{cn}$) within each volcanic group; and (4) there are no correlations between $^{147}Sm/^{144}Nd$ and initial $\varepsilon_{Nd}$ values, which, according to Vervoort and Blichert-Toft (1999), is a robust criterion for identifying crustal contamination (Tables 3 and 4). Accordingly, the lower initial $\varepsilon_{Nd}$ values (<+2.0) likely indicate an Nd-enriched component in the source region, rather than crustal contamination. We therefore suggest that an enriched component was added to the mantle wedge in variable proportions by recycling of older continental material, with superchondritic Nd/Sm ratios (cf. Shirey and Hanson, 1986; Polat and Kerrich, 2002; Shirey et al., this volume).

The LREE-enriched patterns of the Ivisaartoq rocks, however, imply that the depleted subarc mantle source must have been metasomatized shortly before or during partial melting that took place at ca. 3075 Ma. Hydrous fluids and/or melts derived from either subducted altered oceanic crust or sediments, with subchondritic Nb/La, Nb/Th, Sm/Nd, and Ti/Gd ratios, were probably the main cause of the metasomatism, generating the LREE-enriched and HFSE-depleted trace element patterns in the Ivisaartoq rocks.

### Origin of the Ocelli in Pillow Lavas

On the basis of field and petrographic observations, Polat et al. (2007) proposed that the felsic ocelli and surrounding mafic matrix in the Mesoarchean Ivisaartoq pillow lavas (Figs. 5D and 5E) may have derived from immiscible dacitic and basaltic liquids, respectively. However, in the absence of detailed major element, trace element, and radiogenic isotopic data from the ocelli and matrix separates, other petrogenetic processes cannot be ruled out. There are six possible explanations for the presence of ocelli in the Ivisaartoq pillow lavas: (1) They may be deformed amygdules; (2) they could be xenoliths derived from the older tonalite-trondhjemite-granodiorite (TTG) gneisses (e.g., Isukasia terrane); (3) they could have been derived from melting of TTG gneiss (e.g., Isukasia terrane) by mafic magmatism; (4) they may represent the dacitic end member of mixed magmas in which the surrounding matrix constitutes the basaltic end member; (5) the felsic ocelli and mafic matrix might have been related to each other by crystal-liquid differentiation; and (6) they may represent globules of immiscible dacitic liquids. The first two models can be negated on the basis of the following arguments: (1) The ocelli are composed of polycrystalline feldspar, and quartz, with minor amount of hornblende and biotite. In general, amygdules tend to be monomineralic (e.g., quartz, calcite, zeolite). (2) The ocelli are finer-grained than the gneisses in the Ivisaartoq region (see Friend and Nutman, 2005a, and references therein). A comprehensive geochemical dataset from both the ocelli and matrix separates is required to test the last four models.

### Origin of the Anorthositic Inclusions in Gabbros

Petrographic observations, such as the presence of relic reaction rims and smaller pieces of peeled anorthosites at the inclusion-matrix contacts, suggest that the ocellar anorthositic inclusions had already been solidified before they were incorporated into gabbroic magma. The smaller anorthositic inclusions, aligned parallel to the inclusion-matrix contacts (Fig. 6A), might have resulted from the thermal erosion of the larger ones during their transportation to shallower depths. Both the inclusions and matrix were deformed and recrystallized under amphibolite facies metamorphic conditions before the intrusion of 2961 ± 12 Ma granitoids. On the basis of existing geochemical data and field relationships, we suggest that the anorthositic inclusions were carried, as crustal xenoliths, from the lower oceanic crust to the shallower depths by upwelling magmas. It is possible that the anorthosite-leucogabbro association exposed to the south of the Ivisaartoq greenstone belt (Fig. 2) could also have been the source of the inclusions in gabbros. Chadwick (1990) interpreted the anorthosite-leucogabbro association as intrusive into the Ivisaartoq greenstone belt (Chadwick, 1990). In contrast, recent studies (Friend and Nutman, 2005a) indicate that the contact between the greenstone belt and anorthosite-leucogabbro association is characterized by a 2-m-thick mylonitic shear zone. If the anorthosite-leucogabbro association belongs to the orthogneisses of the ca. 3000 Ma Kapisilik tectonic terrane, this association could not have been the source of the inclusions in the ca. 3075 Ma Ivisaartoq greenstone belt. Alternatively, the anorthositic inclusions might have derived from an unknown source of anorthosite body that might have existed in the older continental crust in the area, assuming that the belt was emplaced onto the preexisting crust as an autochthonous sequence. However, there is no field evidence for the existence of older (>3075 Ma) anorthosites in the region.

Low abundances of MgO, Ni, Cr, Co, and Sc in the anorthositic inclusions are consistent with olivine, clinopyroxene, and/or orthopyroxene fractionation. Depletion of Nb, relative to Th and La, and near-flat HREE patterns are consistent with a subduction zone geodynamic setting and a shallow mantle source (Fig. 10). The gabbroic matrix shares the negative Nb anomalies (Fig. 10; Table 2). The initial $\varepsilon_{Nd}$ values of the anorthositic inclusions are much larger than those of the gabbroic matrix (+4.8 to +6.0 versus +2.3 to +2.8), indicating two different mantle sources. On the other hand, the anorthositic inclusions are isotopically comparable to picrites (clinopyroxene cumulates) (Tables 3 and 4), suggesting a petrogenetic link between the two rock types. It is likely that picrites and anorthosites were derived from the same parental magma through clinopyroxene and plagioclase fractionation, respectively. Given their low viscosity, picritic magmas could easily have reached the surface to form the ultramafic sills and/or flows. In contrast, anorthositic magmas might have been too viscous to reach to the surface; instead, they might have crystallized within the lower oceanic crust to form layered anorthosites.

## Comparison of the Ivisaartoq Greenstone Belt with Other Belts in Southern West Greenland

In this section, we summarize the main geological and geochemical characteristics of the Eoarchan (3800–3700 Ma) Isua and Mesoarchean (ca. 3075 Ma) Ivisaartoq and Qussuk (ca. 3071 Ma) greenstone belts, SW Greenland (Fig. 1). All these greenstone belts occur within orthogneisses with TTG compositions (Bridgwater et al., 1974; Nutman et al., 1996; Garde, 2007; Nutman and Friend, 2007). Notwithstanding intense deformation and amphibolite facies metamorphism, the presence of well-preserved primary magmatic structures and near-primary geochemical signatures in the Isua, Ivisaartoq, and Qussuk greenstone belts provide important information on the nature of Archean oceanic crust and geodynamic processes (Nutman et al., 2002; Polat, 2005; Polat and Hofmann, 2003; Furnes et al., 2007; Garde, 2007).

The Eoarchean (3800–3700 Ma) Isua greenstone belt occurs in orthogneisses of the recently recognized Isukasia tectonic terrane (Fig. 1A) (Friend and Nutman, 2005a). The belt contains the oldest rocks deposited on the surface of Earth, comprising volcanic, volcaniclastic, clastic, and chemical sedimentary rocks (Rosing et al., 1996; Appel et al., 1998; Nutman et al., 2002; Fedo, 2000; Myers, 2001; Frei and Polat, 2007). Polymictic conglomerates and pelites, mainly staurolite-mica schists, can be traced for some kilometers along strike. Banded iron formations and cherts are interbedded with volcanic rocks. In the low-strain domains, the variably deformed pillow basalts are intercalated with ultramafic units. Despite polyphase deformation and metamorphism, some low-strain zones display a wealth of well-preserved primary volcanic structures such as pillow lavas, minor debris flows, and pillow breccias. The Isua volcanic rocks are characterized by tectonically juxtaposed boninitic and island arc picritic compositions (Fig. 14) (Polat and Hofmann, 2003). Given that Cenozoic boninites and picrites tend to form in restricted intra-oceanic suprasubduction zone settings (e.g., Izu-Bonin-Marina subduction, Solomon Islands), it is plausible that the Isua volcanic rocks were also erupted in a similar geodynamic setting. These two geochemically distinct sequences were likely to have been juxtaposed as a consequence of Phanerozoic-style plate tectonic processes operating in early Earth. Recently, a relic sheeted dike complex has been recognized in the picritic to basaltic pillows (Furnes et al., 2007). Compositionally, the sheeted dikes are comparable to the spatially associated pillow basalts, indicating a cogenetic relationship between the two rock types. The Isua and Ivisaartoq picrites have similar trace element characteristics (Figs. 12 and 14), suggesting a similar geodynamic origin.

The 3071 Ma Qussuk greenstone belt (informal name) is located in the Akia terrane (Fig. 1). Like the other tectonic terranes in the Nuuk region, the Akia terrane underwent polyphase deformation and multiple events of metamorphism and magmatism (Garde, 2007, and references therein). The belt was subjected to amphibolite-grade metamorphism at ca. 2980 Ma. The geochemical and geochronological characteristics of supracrustal rocks exposed in the Qussuk Peninsula have recently been investigated by Garde (2007). The amphibolites and associated gabbros have LREE-depleted trace element patterns, with minor negative Nb anomalies. Mafic to intermediate pyroclastic rocks display LREE-enriched patterns with large negative anomalies. According to Garde (2007), the Qussuk volcanic and volcaniclastic rocks originated in a convergent plate margin at ca. 3071 Ma and were intruded by slab-derived granitoids ca. 3060 Ma, forming an intra-oceanic island arc system. Volcanic and intrusive events were followed by thrusting and folding at ca. 3030 Ma. Despite occurring in two different tectonic terranes, both the ca. 3071 Ma Qussuk (Akia terrane) and the ca. 3075 Ma Ivisaartoq (Kapisilik terrane) volcanic associations formed at similar times and geodynamic settings.

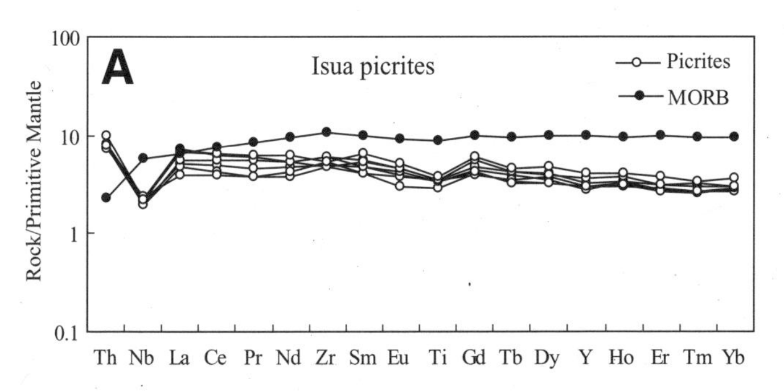


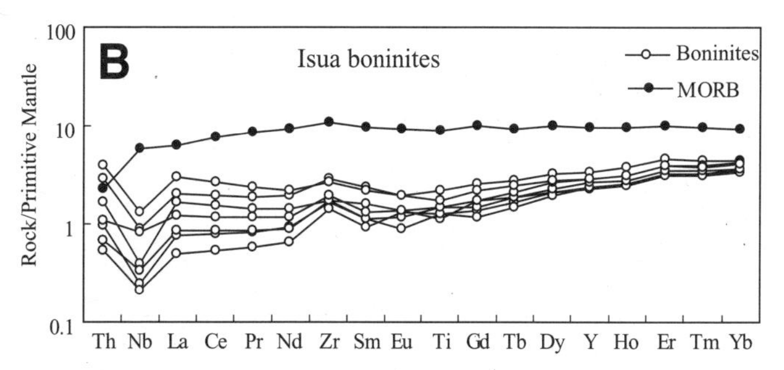


Figure 14. Primitive mantle–normalized trace element patterns for the Eoarchean Isua greenstone belt volcanic rocks. Data for picrites (A) and boninites (B) are from Polat and Hofmann (2003) and Polat et al. (2002), respectively. Primitive mantle normalization values are from Hofmann (1988).

## FORMATION OF THE IVISAARTOQ GREENSTONE BELT AS A MESOARCHEAN FOREARC OCEANIC CRUST

Field relationships and geochemical data indicate that all volcanic and intrusive rock types in the Ivisaartoq greenstone belt are part of the same lithotectonic assemblage, sharing a common history of magmatism, deformation, and metamorphism (Chadwick, 1985, 1990; Polat et al., 2007, 2008). The association of pillow basalts, gabbros, and sulfide-rich siliceous volcaniclastic sedimentary rocks in the belt suggests an intra-oceanic depositional environment.

The large initial $\varepsilon_{Nd}$ (e.g., +2 to +6) isotopic values in the Mesoarchean Ivisaartoq rocks indicate a long-term LREE-depleted ($Sm/Nd_{cn} > 1$) mantle source, similar to the source of modern normal mid-oceanic-ridge basalts (N-MORBs) (see Hofmann, 2004). However, the majority of the least altered samples have LREE-enriched ($La/Sm_{cn} > 1$; $Sm/Nd_{cn} < 1$) but Nb-depleted, relative to Th and La, trace element patterns (Figs. 9 and 10), consistent with a subduction zone geodynamic setting (Figs. 13B and 13C). Accordingly, we propose a two-stage evolutionary geodynamic model to explain the geological characteristics of the Ivisaartoq greenstone belt (Figs. 15A, B). In the first stage, the mantle source of the Ivisaartoq rocks had originated as a suboceanic depleted upper mantle, like the source of present-day N-MORBs. The second stage marks the development of an intra-oceanic subduction system. Following the initiation of an intra-oceanic subduction zone along either a mid-ocean ridge or a transform fault, the mantle source of the Ivisaartoq rocks was converted to a subarc mantle wedge (cf. Casey and Dewey, 1984; Dilek and Flower, 2003). Hydrous fluids and/or melts originating from the subducted slab metasomatized the subarc mantle wedge, resulting in LREE-enriched and HFSE-depleted, relative to Th and LREE, patterns (Figs. 9 and 10).

The forearc region of the overriding plate may have undergone a significant extension in response to slab rollback, resulting in a large degree of partial melting of the hydrated upper mantle wedge at shallow depths (Fig. 15). Such a high degree of partial melting is expected to have resulted in the formation of a large magma chamber (Figs. 15A, B). Extensive partial melting beneath the Ivisaartoq forearc may have generated a thick (>20 km) oceanic crust (cf. Sleep and Windley, 1982). Such an intact oceanic crust might have been composed of two major crustal sections: (1) a lower layer of anorthosites and leucogabbros, and (2) an upper layer of basaltic to picritic flows, gabbroic to dioritic dikes, and dunitic to wehrlitic sills (Fig. 15B).

## IMPLICATIONS FOR THE ORIGIN OF ARCHEAN ANORTHOSITES

The major and trace element characteristics of the anorthositic inclusions (xenoliths) in the Ivisaartoq gabbros are comparable to those of Meso- to Neoarchean anorthosite complexes in southern West Greenland (Windley et al., 1973; Weaver et al., 1981; Ashwal and Myers, 1994; Dymek and Owens, 2001). Like the Buksefjorden, Nordland, and Fiskenaesset anorthosites, the Ivisaartoq counterparts have LREE-enriched chondrite-normalized patterns with large positive Eu anomalies (see Weaver et al., 1981; Dymek and Owens, 2001), suggesting a similar petrogenetic process. However, the Ivisaartoq anorthositic inclusions have flat to less fractionated HREE patterns compared to the Buksefjorden, Nordland, and Fiskenaesset anorthosites, indicating a shallower, garnet-free mantle source region.

If the anorthosite-leucogabbro association to the south of the Ivisaartoq greenstone belt was originally intrusive into the lower amphibolites (Figs. 2 and 3) as suggested by Chadwick (1990), it might initially have been thicker. There are two main reasons why this would have been the case. First, given the record of several generations of deformation in the region (Chadwick, 1990; Friend and Nutman, 2005a), finding an intact, thicker anorthosite-leucogabbro association is unlikely. Second, if Archean oceanic crust was thicker due to potentially higher mantle temperatures (Sleep and Windley, 1982; McKenzie and Bickle, 1988), then it is possible that only the upper basaltic crustal section was peeled off and accreted while the lower anorthosite-leucogabbro section was subducted. Notwithstanding these problems, partial sections of 2800–3000 Ma anorthosite-leucogabbro associations have been identified throughout the Archean terranes of southern West Greenland (Windley, 1970; Windley et al., 1973, 1981; Escher and Myers,

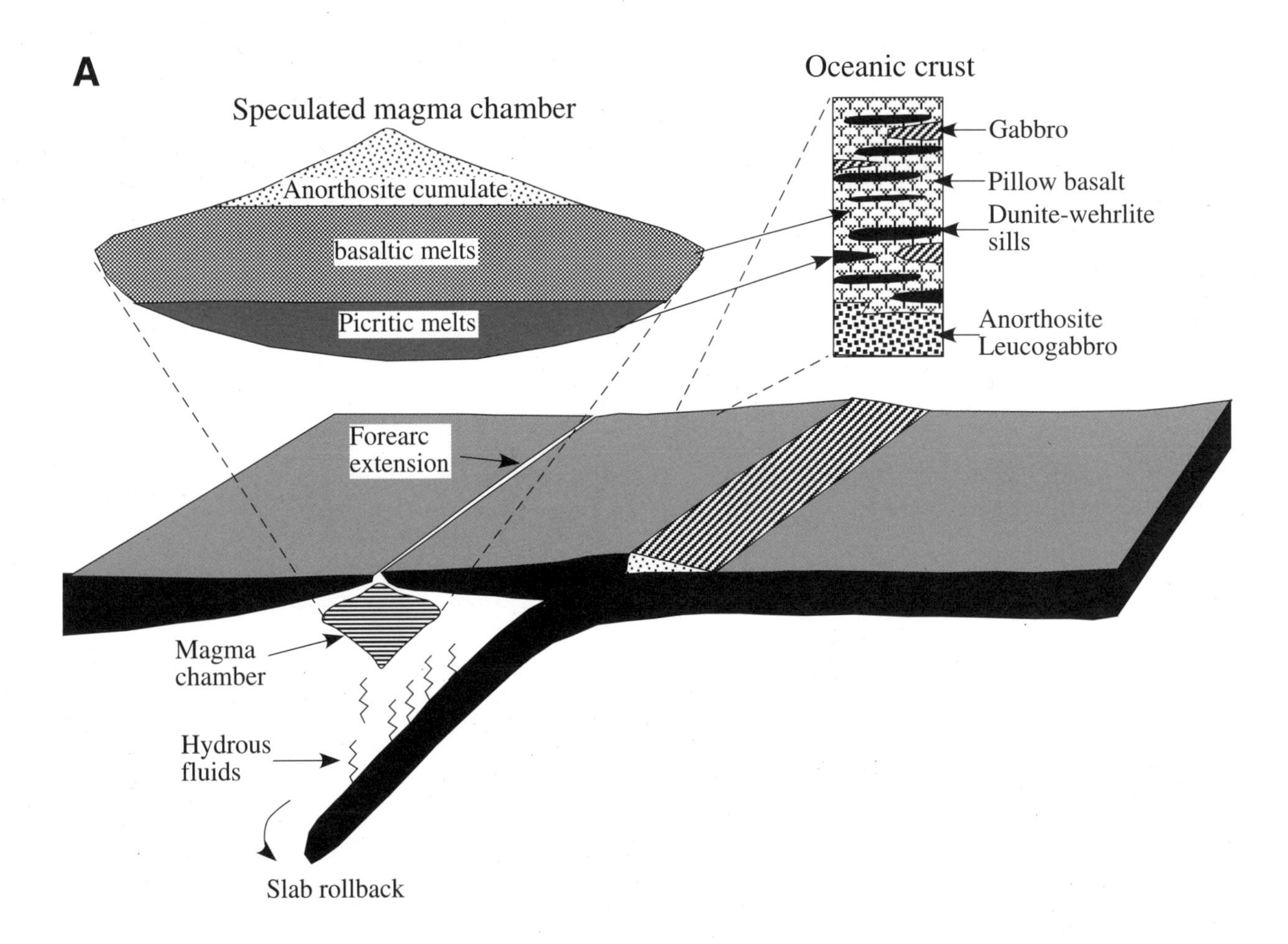


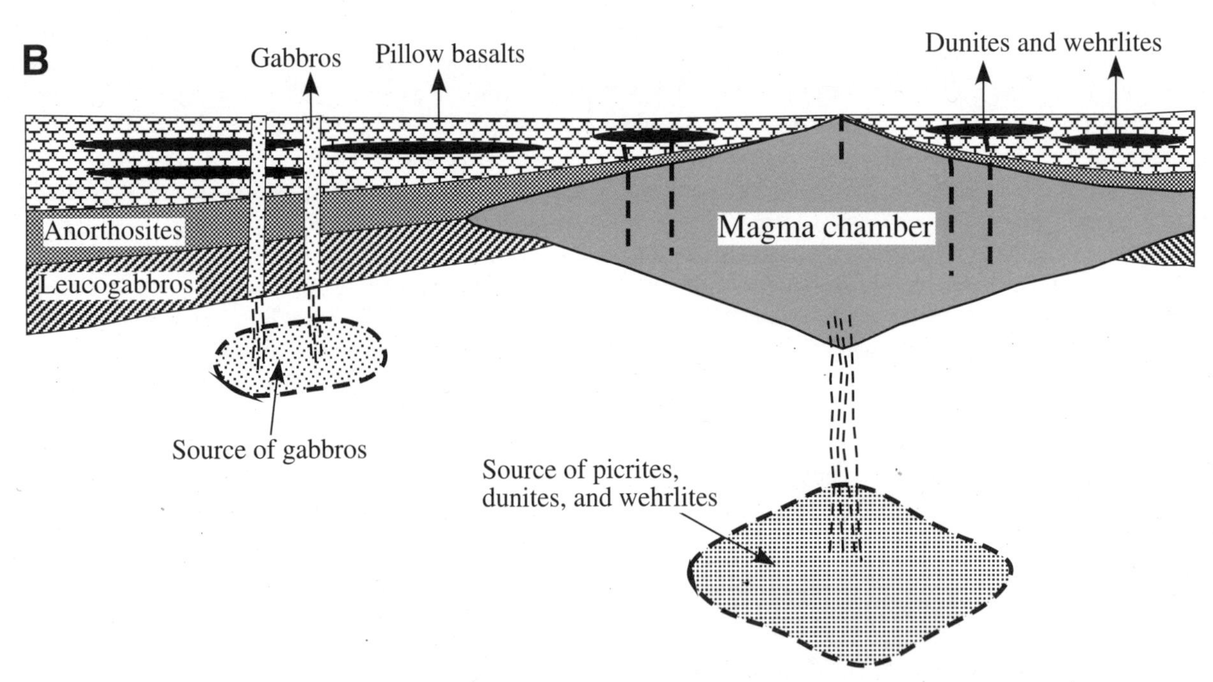


Figure 15. A simplified possible geodynamic and petrologic model for the Ivisaartoq greenstone belt and Mesoarchean forearc oceanic crust. (A) Extension of the forearc oceanic crust and formation of a magma chamber beneath the spreading center. Stratigraphic section shows different rock types in the forearc oceanic crust and their possible melt sources. (B) A speculated cross section through the forearc oceanic crust and magma chamber, showing the mantle sources of various rock types.

1975; Myers, 1985; Ashwal and Myers, 1994; Owens and Dymek, 1997).

The petrogenetic origin of Archean anorthosites remains unresolved (Weaver et al., 1981, 1982; Pinney et al., 1988; Ashwal and Myers, 1994; Dymek and Owens, 2001). In the best-studied Fiskenaesset complex, the anorthosites and leucogabbros appear to have intruded into the overlying greenstone sequences (Escher and Myers, 1975; Ashwal and Myers, 1994). Geochemical studies suggest that the Fiskenaesset anorthosite complex is genetically related to spatially associated mafic to ultramafic volcanic rocks by fractional crystallization (Weaver et al., 1981, 1982; Peck and Valley, 1996) and is derived from a long-term depleted mantle source (Ashwal et al., 1989). The anorthositic inclusions in the Ivisaartoq belt might have been related to the ultramafic lithologies in the belt through fractional crystallization.

Greenstone-anorthosite associations in southern West Greenland (e.g., Fiskenaesset complex) are interpreted as remnant Archean oceanic crust (ophiolite?) in several studies (Myers, 1985; Windley et al., 1981; Weaver et al., 1982; Ashwal and Myers, 1994). On the basis of geological similarities between the Ivisaartoq and Fiskenaesset greenstone belts, we interpret the Ivisaartoq greenstone belt as a relic of an Archean forearc oceanic crust. We do not propose that all Archean anorthosites formed in an arc-forearc tectonic setting. Like basaltic counterparts, Archean anorthosites, depending on their geological and geochemical characteristics, might have formed in diverse geodynamic settings, including in mid-ocean ridges, forearcs, backarcs, plume-derived oceanic plateaus, and intra-continental rifts.

## ACKNOWLEDGMENTS

The authors thank Victoria Pease and Kent Condie for organizing this Geological Society of America Special Paper, and for the invitation to contribute a paper. R.J. Stern and an anonymous reviewer are acknowledged for their comprehensive critiques, which have resulted in significant improvements to the paper. This is a contribution of NSERC (Natural Sciences and Engineering Research Council) grant 250926 to Polat. Frei is supported by FNU (Forskningsrådet for Natur of Univers) grant 21-01-0492 56493. Fieldwork was supported by the Bureau of Minerals and Petroleum in Nuuk and the Geological Survey of Denmark and Greenland (GEUS). Polat thanks Tekin Demir for helping his family during fieldwork in Greenland.

## REFERENCES CITED

Appel, P.W.U., 1994, Stratabound scheelite in altered Archean komatiites, West Greenland: Mineralium Deposita, v. 29, p. 341–352.

Appel, P.W.U., 1997, High bromine and low Cl/Br ratios in hydrothermally altered Archean komatiitic rocks, West Greenland: Precambrian Research, v. 82, p. 177–189, doi: 10.1016/S0301-9268(96)00046-0.

Appel, P.W.U., Fedo, C.M., Moorbath, S., and Myers, J.S., 1998, Recognizable primary volcanic and sedimentary features in a low-strain domain of the highly deformed, oldest known (3.7–3.8 Gyr) greenstone belt, Isua, West Greenland: Terra Nova, v. 10, p. 57–62, doi: 10.1046/j.1365-3121.1998.00162.x.

Arndt, N.T., Bruzak, G., and Reischmann, T., 2001, The oldest continental and oceanic plateaus-geochemistry of basalts and komatiites of the Pilbara Craton, Australia, *in* Buchan, K.L., and Ernst, R.E., eds., Mantle plumes: Their identification through time: Geological Society of America Special Paper 352, p. 1–30.

Ashwal, L.D., and Myers, J., 1994, Archean anorthosites, *in* Condie, K.C., ed., Archean crustal evolution: Amsterdam, Elsevier, p. 315–355.

Ashwal, L.D., Jacobsen, S.B., Myers, J.S., Kalsbeek, F., and Goldstein, S.J., 1989, Sm-Nd age of the Fiskenaesset Anorthosite Complex, West Greenland: Earth and Planetary Science Letters, v. 91, p. 261–270, doi: 10.1016/0012-821X(89)90002-2.

Bédard, J.H., 2006, A catalytic delamination-driven model for coupled gneisses of Archean crust and sub-continental lithospheric mantle: Geochimica et Cosmochimica Acta, v. 70, p. 1188–1214, doi: 10.1016/j.gca.2005.11.008.

Bickle, M.J., 1986, Implications of melting for stabilization of lithosphere and heat loss in the Archean: Earth and Planetary Science Letters, v. 80, p. 314–324, doi: 10.1016/0012-821X(86)90113-5.

Bickle, M.J., Nisbet, E.G., and Martin, A., 1994, Archean greenstone belts are not oceanic crust: Journal of Geology, v. 102, p. 121–138.

Bleeker, W., 2002, Archean tectonics: A review, with illustrations from the Slave craton, *in* Fowler, M.R., et al., eds., The early Earth: Physical, chemical and biological development: Geological Society [London] Special Publication 199, p. 151–181.

Brewer, M.A., Chadwick, B., Coe, K., and Park, J.F.W., 1984, Further field investigations in the Ivisaartoq region of southern West Greenland: Rapport Grønlands Geologiske Undersøgelse, v. 120, p. 55–67.

Bridgwater, D., McGregor, V.R., and Myers, J.S., 1974, Horizontal tectonic regime in the Archean of Greenland and its implications for early crustal thickening: Precambrian Research, v. 1, p. 179–197, doi: 10.1016/0301-9268(74)90009-6.

Card, K.D., 1990, A review of the Superior Province of the Canadian Shield, a product of Archean accretion: Precambrian Research, v. 48, p. 99–156.

Casey, J.F., and Dewey, J.F., 1984, Initiation of subduction zones along transform and accreting plate boundaries, triple-junction evolution, and forearc spreading centres—Implications for ophiolitic geology and obduction, *in* Gass, I.G., et al., eds., Ophiolites and oceanic lithosphere: Geological Society [London] Special Publication 13, p. 269–290.

Chadwick, B., 1985, Contrasting styles of tectonism and magmatism in the Late Archean crustal evolution of the northeastern part of the Ivisaartoq region, inner Godthåbsfjord, southern West Greenland: Precambrian Research, v. 27, p. 215–238, doi: 10.1016/0301-9268(85)90013-0.

Chadwick, B., 1986, Malene stratigraphy and Late Archean structure: New data from Ivisaartoq, inner Godthåbsfjord, southern West Greenland: Rapport Grønlands Geologiske Undersøgelse, v. 130, p. 74–85.

Chadwick, B., 1990, The stratigraphy of a sheet of supracrustal rocks within high-grade orthogneisses and its bearing on Late Archean structure in southern West Greenland: Geological Society [London] Journal, v. 147, p. 639–652.

Chadwick, B., and Coe, K., 1988, Geological map of Greenland, Ivisaartoq sheet 64 V. Nord: Copenhagen, Geological Survey of Greenland, scale 1:100,000.

Class, C., Altherr, R., Volker, F., Eberz, G., and McCulloch, M.T., 1994, Geochemistry of Pliocene to Quaternary alkali basalts from the Huri Hills, northern Kenya: Chemical Geology, v. 113, p. 1–22, doi: 10.1016/0009-2541(94)90002-7.

Condie, K.C., 1981, Archean greenstone belts: Amsterdam, Elsevier, 1981, 435 p.

Condie, K.C., 2004, Precambrian superplume events, *in* Eriksson, P.G., et al., eds., The Precambrian Earth: Tempos and events: Amsterdam, Elsevier, p. 163–173.

Condie, K.C., 2005, Earth as an evolving planetary system: Amsterdam, Elsevier, 447 p.

Corcoran, P.L., and Dostal, J., 2001, Development of an ancient back-arc basin overlying continental crust: The Archean Peltier Formation, Northwest Territories, Canada: Journal of Geology, v. 109, p. 329–348, doi: 10.1086/319976.

Dewey, J., 2003, Ophiolites and lost oceans: Rifts, ridges, arcs, and/or scraping, *in* Dilek, Y., and Newcomb, S., eds., Ophiolite concept and the evolution of geological thought: Geological Society of America Special Paper 373, p. 153–158.

de Wit, M.J., 1998, On Archean granites, greenstones, cratons, and tectonics: Does the evidence demand a verdict?: Precambrian Research, v. 91, p. 181–226.

Dilek, Y., 2003, Ophiolite concept and its evolution, *in* Dilek, Y., and Newcomb, S., eds., Ophiolite concept and the evolution of geological thought: Geological Society of America Special Paper 373, p. 1–15.

Dilek, Y., and Flower, M.F.J., 2003, Arc-trench rollback and forearc accretion, 2: A model template for ophiolites in Albania, Cyprus, and Oman, *in* Dilek,

Y., and Robinson, P.T., eds., Ophiolites in Earth history: Geological Society [London] Special Publication 218, p. 43–68.

Dymek, R.F., and Owens, B.R., 2001, Chemical assembly of Archean anorthosites from amphibolite- and granulite-facies terranes, SW Greenland: Contributions to Mineralogy and Petrology, v. 141, p. 513–528.

Eriksson, K.A., Krapez, B., and Fralick, P.W., 1994, Sedimentology of Archean greenstone belts: Signatures and tectonic evolution: Earth-Science Reviews, v. 37, p. 1–88, doi: 10.1016/0012-8252(94)90025-6.

Ernst, R.E., Buchan, K.L., and Prokoph, A., 2004, Large igneous province record through time, *in* Eriksson, P.G., et al., eds., The Precambrian Earth: Tempos and events: Amsterdam, Elsevier, p. 173–180.

Ernst, W.G., 2007, Speculations on evolution of the terrestrial lithosphere-asthenosphere system—Plumes and plates: Gondwana Research, v. 11, p. 38–49, doi: 10.1016/j.gr.2006.02.007.

Escher, J.C., and Myers, J.S., 1975, New evidence concerning the original relationships of early Precambrian volcanics and anorthosites in the Fiskenaesset region, southern West Greenland: Rapport Grønlands Geologiske Undersøgelse, v. 75, p. 72–76.

Ewart, A., Collerson, K.D., Regelous, M., Wendt, J.I., and Niu, Y., 1998, Geochemical evolution within the Tonga-Kermadec-Lau arc-back-arc systems: The role of varying mantle wedge composition in space and time: Journal of Petrology, v. 39, p. 331–368, doi: 10.1093/petrology/39.3.331.

Fedo, C.M., 2000, Setting and origin for problematic rocks from the >3.7 Ga Isua greenstone belt, southern West Greenland: Earth's oldest coarse clastic sediments: Precambrian Research, v. 101, p. 69–78, doi: 10.1016/S0301-9268(99)00100-X.

Foley, S., 2008, this volume, A trace element perspective on Archean crust formation and on the presence or absence of Archean subduction, *in* Condie, K.C., and Pease, V., eds., When did plate tectonics begin on planet Earth?: Geological Society of America Special Paper 440, doi: 10.1130/2008.2440(02).

Frei, R., and Polat, A., 2007, Source heterogeneity for the major components of ~3.7 Ga banded iron formations (Isua Greenstone Belt, Western Greenland): Tracing the nature of interacting water masses in BIF formation: Earth and Planetary Science Letters, v. 253, p. 266–281, doi: 10.1016/j.epsl.2006.10.033.

Friend, C.R.L., and Nutman, A.P., 2005a, New pieces to the Archean jigsaw puzzle in the Nuuk region, southern West Greenland: Steps in transforming a simple insight into a complex regional tectonothermal model: Geological Society [London] Journal, v. 162, p. 147–162, doi: 10.1144/0016-764903-161.

Friend, C.R.L., and Nutman, A.P., 2005b, Complex 3670–3500 Ma orogenic episodes superimposed on juvenile crust accreted between 3850 and 3690 Ma, Itsaq gneiss complex, southern West Greenland: Journal of Geology, v. 113, p. 375–397, doi: 10.1086/430239.

Friend, C.R.L., Hall, R.P., and Hughes, D.J., 1981, The geochemistry of the Malene (mid-Archean) ultramafic-mafic amphibolite suite, southern West Greenland, *in* Glover, J.E., and Groves, D.E., eds., Archean geology: Perth, Geological Society of Australia Special Publication 7, p. 301–312.

Furnes, H., de Wit, M., Staudigel, H., Rosing, M., and Muehlenbachs, K., 2007, A vestige of Earth's oldest ophiolite: Science, v. 315, p. 1704–1707, doi: 10.1126/science.1139170.

Fyfe, W.S., 1978, The evolution of the Earth's crust: Modern plate tectonics to ancient hot spot tectonics?: Chemical Geology, v. 23, p. 89–114, doi: 10.1016/0009-2541(78)90068-2.

Garde, A.A., 2007, A mid-Archean island arc complex in eastern Akia terrane, Godthåbsfjord, southern West Greenland: Geological Society [London] Journal, v. 164, p. 565–579, doi: 10.1144/0016-76492005-107.

Goodwin, M.A., 1991, Precambrian geology: The dynamic evolution of the continental crust: London, Academic Press, 666 p.

Hall, R.P., 1981, The tholeiitic and komatiitic affinity of Malene metavolcanic amphibolites from Ivisaartoq, southern West Greenland: Rapport Grønlands Geologiske Undersøgelse, v. 97, 20 p.

Hall, R.P., and Friend, C.R.L., 1979, Structural evolution of the Archean rocks in Ivisaartoq and neighboring inner Godthåbsfjord region, southern West Greenland: Geology, v. 7, p. 311–315, doi: 10.1130/0091-7613(1979)7<311:SEOTAR>2.0.CO;2.

Hawkesworth, C.J., Gallanger, K., Hergt, J.M., and McDetmott, F., 1993, Mantle and slab contributions in arc magmas: Annual Review of Earth and Planetary Sciences, v. 21, p. 175–204.

Hawkins, J.W., 2003, Geology of supra-subduction zones—Implications for the origin of ophiolites, *in* Dilek, Y., and Newcomb, S., eds., Ophiolite concept and the evolution of geological thought: Geological Society of America Special Paper 373, p. 227–268.

Herzberg, C., 1992, Depth and degree of melting of komatiites: Journal of Geophysical Research, v. 97, p. 4521–4540.

Hirschmann, M.M., and Stolper, E.M., 1996, A possible role for garnet pyroxenite in the origin of "garnet signature" in MORB: Contributions to Mineralogy and Petrology, v. 124, p. 185–208, doi: 10.1007/s004100050184.

Hofmann, A.W., 1988, Chemical differentiation of the Earth: The relationships between mantle, continental crust, and oceanic crust: Earth and Planetary Science Letters, v. 90, p. 297–314, doi: 10.1016/0012-821X(88)90132-X.

Hofmann, A.W., 2004, Sampling mantle heterogeneity through oceanic basalts: Isotopes and trace elements, *in* Carson, R.W., ed., Treatise on geochemistry, volume 2: The mantle and core: New York, Elsevier, p. 61–101.

Jung, S., and Masberg, P., 1998, Major- and trace-element systematics and isotope geochemistry of Cenozoic mafic volcanic rocks from the Vogelsberg (central Germany): Constraints of the origin of continental alkaline and tholeiitic basalts and their mantle sources: Journal of Volcanology and Geothermal Research, v. 86, p. 151–177, doi: 10.1016/S0377-0273(98)00087-0.

Kerrich, R., and Polat, A., 2006, Archean greenstone-tonalite duality: Thermochemical mantle convection models or plate tectonics in the early Earth global dynamics?: Tectonophysics, v. 415, p. 141–165, doi: 10.1016/j.tecto.2005.12.004.

Knittel, U., and Oles, D., 1994, Basaltic volcanism associated with extensional tectonics in the Taiwan-Luzon island arc: Evidence for non-depleted sources and subduction enrichment, *in* Smellie, J.L., ed., Volcanism associated with extension at consuming plate margins: Geological Society [London] Special Publication 81, p. 77–93.

Kusky, T.M., 2004, Introduction, *in* Kusky, T.M., ed., Precambrian ophiolites and related rocks: Amsterdam, Elsevier, p. 1–34.

Kusky, T.M., and Polat, A., 1999, Growth of granite-greenstone terranes at convergent margins, and stabilization of Archean cratons: Tectonophysics, v. 305, p. 43–73, doi: 10.1016/S0040-1951(99)00014-1.

Ludwig, K., 2003, User's manual for Isoplot/Ex version 3.0: A geochronological toolkit for Microsoft Excel: Berkeley Geochronology Center Special Publication 1a.

McKenzie, D.P., and Bickle, M.J., 1988, The volume and composition of melt generated by extension of the lithosphere: Journal of Petrology, v. 29, p. 625–679.

Mueller, W.U., Daigneault, R., Mortensen, J.K., and Chown, E.H., 1996, Archean terrane docking: Upper crustal collision tectonics, Abitibi greenstone belt, Quebec, Canada: Tectonophysics, v. 265, p. 127–150, doi: 10.1016/S0040-1951(96)00149-7.

Myers, J.S., 1985, Stratigraphy and structure of the Fiskenaesset Complex, West Greenland: Grønlands Geologiske Undersøgelse Bulletin 150, 72 p.

Myers, J.S., 2001, Protoliths of the 3.7–3.8 Ga Isua greenstone belt, West Greenland: Precambrian Research, v. 105, p. 129–141, doi: 10.1016/S0301-9268(00)00108-X.

Nisbet, E.G., and Fowler, C.M.R., 1983, Model for Archean plate tectonics: Geology, v. 11, p. 376–379.

Nutman, A.P., and Friend, C.R.L., 2007, Adjacent terranes with ca. 2715 and 2650 Ma high-pressure metamorphic assemblages in the Nuuk region of the North Atlantic Craton, southern West Greenland: Complexities of Neoarchean collisional orogeny: Precambrian Research, v. 155, p. 159–203, doi: 10.1016/j.precamres.2006.12.009.

Nutman, A.P., McGregor, V.R., Friend, C.R.L., Bennett, V.C., and Kinny, P.D., 1996, The Itsaq Gneiss Complex of southern West Greenland: The world's most extensive record of early crustal evolution (3900–3600 Ma): Precambrian Research, v. 78, p. 1–39, doi: 10.1016/0301-9268(95)00066-6.

Nutman, A.P., Friend, C.R.L., and Bennett, V.C., 2002, Evidence for 3650–3600 Ma assembly of the northern end of the Itsaq Gneiss Complex, Greenland: Implications for Early Archean tectonics: Tectonics, v. 21, 1005, doi: 10.1029/2000TC001203.

Owens, B.E., and Dymek, R.F., 1997, Comparative petrology of Archean anorthosites in amphibolite and granulite facies terranes, SW Greenland: Contributions to Mineralogy and Petrology, v. 128, p. 371–384, doi: 10.1007/s004100050315.

Pearce, J.A., 2003, Supra-subduction zone ophiolites: The search for modern analogues, *in* Dilek, Y., and Newcomb, S., eds., Ophiolite Concept and the Evolution of Geological Thought: Geological Society of America Special Paper 373, p. 269–293.

Pearce, J.A., and Peate, D.W., 1995, Tectonic implications of the composition of volcanic arc magmas: Annual Review of Earth and Planetary Sciences, v. 23, p. 251–285, doi: 10.1146/annurev.ea.23.050195.001343.

Pearce, J.A., Baker, P.E., Harvey, P.K., and Luff, I.W., 1995, Geochemical evidence for subduction fluxes, mantle melting and fractional crystallization beneath South Sandwich island arc: Journal of Petrology, v. 36, p. 1073–1109.

Peate, D.W., Pearce, J.A., Hawkesworth, C.J., Colley, H., Edwards, C.M.H., and Hirso, K., 1997, Geochemical variations in Vanuatu arc lavas: The role of subducted material and a variable mantle wedge composition: Journal of Petrology, v. 38, p. 1331–1358, doi: 10.1093/petrology/38.10.1331.

Peck, W.H., and Valley, J.W., 1996, The Fiskenaesset Anorthosite Complex: Stable isotope evidence for shallow emplacement into Archean oceanic crust: Geology, v. 24, p. 523–526, doi: 10.1130/0091-7613(1996)024<0523:TFACSI>2.3.CO;2.

Percival, J.A., Sanborn-Barrie, M., Skulski, T., Sott, G.M., Helmstaedt, H., and White, D.J., 2006a, Tectonic evolution of the western Superior province from NATMAP and Lithoprobe studies: Canadian Journal of Earth Sciences, v. 43, p. 1085–1117, doi: 10.1139/E06-062.

Percival, J.A., McNicoll, V., and Bailes, A.H., 2006b, Strike-slip juxtaposition of ca. 2.72 Ga juvenile arc and >2.98 Ga continental margin sequences and its implications for Archean terrane accretion, western Superior province, Canada: Canadian Journal of Earth Sciences, v. 43, p. 895–927, doi: 10.1139/E06-039.

Pinney, W.C., Morrison, D.A., and Maczuga, D.E., 1988, Anorthosites and related megacrystic units in the evolution of Archean crust: Journal of Petrology, v. 29, p. 1283–1323.

Polat, A., and Hofmann, A.W., 2003, Alteration and geochemical patterns in the 3.7–3.8 Ga Isua greenstone belt, West Greenland: Precambrian Research, v. 126, p. 197–218, doi: 10.1016/S0301-9268(03)00095-0.

Polat, A., and Kerrich, R., 2002, Nd-isotope systematics of ~2.7 Ga adakites, magnesian andesites, and arc basalts, Superior province, Canada: Evidence for shallow crustal recycling at Archean subduction zones: Earth and Planetary Science Letters, v. 202, p. 345–360, doi: 10.1016/S0012-821X(02)00806-3.

Polat, A., and Kerrich, R., 2006, Reading the geochemical fingerprints of Archean hot subduction volcanic rocks: Evidence for accretion and crustal recycling in a mobile tectonic regime, *in* Benn, K., et al., eds., Archean geodynamics and environments: American Geophysical Union Geophysical Monograph 164, p. 189–213.

Polat, A., Kerrich, R., and Casey, J.F., 1997, Geochemistry of Quaternary basalts erupted along the east Anatolian and Dead Sea fault zones of southern Turkey: Implications for mantle sources: Lithos, v. 40, p. 55–68, doi: 10.1016/S0024-4937(96)00027-8.

Polat, A., Kerrich, R., and Wyman, D.A., 1998, The Late Archean Schreiber-Hemlo and White River–Dayohessarah greenstone belts, Superior province: Collages of oceanic plateaus, oceanic arcs, and subduction-accretion complexes: Tectonophysics, v. 289, p. 295–326, doi: 10.1016/S0040-1951(98)00002-X.

Polat, A., Hofmann, A.W., and Rosing, M.T., 2002, Boninite-like volcanic rocks in the 3.7–3.8 Ga Isua greenstone belt, West Greenland: Geochemical evidence for intra-oceanic subduction zone processes in the early Earth: Chemical Geology, v. 184, p. 231–254, doi: 10.1016/S0009-2541(01)00363-1.

Polat, A., Li, J., Fryer, B., Kusky, T., Gagnon, J., and Zhang, S., 2006, Geochemical characteristics of the Neoarchean (2800–2700 Ma) Taishan Greenstone Belt, North China Craton: Evidence for plume-craton interaction: Chemical Geology, v. 230, p. 60–87, doi: 10.1016/j.chemgeo.2005.11.012.

Polat, A., Appel, P.W.U., Frei, R., Pan, Y., Dilek, Y., Ordonez-Caldeeron, J.C., Fryer, B., and Raith, J.G., 2007, Field and geochemical characteristics of the Mesoarchean (~3075 Ma) Ivisaartoq greenstone belt, southern West Greenland: Evidence for seafloor hydrothermal alteration in a supra-subduction oceanic crust: Gondwana Research, v. 11, p. 69–91, doi: 10.1016/j.gr.2006.02.004.

Polat, A., Frei, R., Appel, P.W.U., Dilek, Y., Fryer, B., Ordóñez-Calderón, J.C., and Yang, Z., 2008, The origin and compositions of Mesoarchean oceanic crust: Evidence from the 3075 Ma Ivisaartoq greenstone belt, SW Greenland: Lithos, v. 100, p. 293–321, doi:10.1016/j.lithos.2007.06.021.

Rosing, M.T., Rose, N.M., Bridgwater, D., and Thomsen, H.S., 1996, Earliest part of Earth's stratigraphic record: A reappraisal of the >3.7 Ga Isua (Greenland) supracrustal sequence: Geology, v. 24, p. 43–46, doi: 10.1130/0091-7613(1996)024<0043:EPOESS>2.3.CO;2.

Sandeman, H.A., Hanmer, S., Davis, W.J., Ryan, J.J., and Peterson, T.D., 2004, Neoarchean volcanic rocks, Central Hearne supracrustal belt, Western Churchill province, Canada: Geochemical and isotopic evidence supporting intra-oceanic, supra-subduction zone extension: Precambrian Research, v. 134, p. 113–141, doi: 10.1016/j.precamres.2004.03.014.

Saunders, A.D., Norry, M.J., and Tarney, J., 1991, Fluid influence on the trace element compositions of subduction zone magmas: Royal Society of London Philosophical Transactions, ser. A, v. 335, p. 377–392, doi: 10.1098/rsta.1991.0053.

Schuth, S., Rohrbach, A., Münker, C., Ballhaus, C., Garbe-Schönberg, D., and Qopoto, C., 2004, Geochemical constraints on the petrogenesis of arc picrites and basalts, New Georgia Group, Solomon Islands: Contributions to Mineralogy and Petrology, v. 148, p. 288–304, doi: 10.1007/s00410-004-0604-0.

Şengör, A.M.C., and Natal'in, B.A., 2004, Phanerozoic analogues of Archean oceanic basement fragments: Altaid ophiolites and ophirags, *in* Kusky, T.M., ed., Precambrian ophiolites and related rocks: Amsterdam, Elsevier, p. 675–726.

Shirey, S.B., and Hanson, G.N., 1986, Mantle heterogeneity and crustal recycling in Archean granite-greenstone belts: Evidence from Nd-isotopes and trace elements in the Rainy Lake area, Superior province, Ontario, Canada: Geochimica et Cosmochimica Acta, v. 50, p. 2631–2651, doi: 10.1016/0016-7037(86)90215-2.

Shirey, S.B., Kamber, B.S., Whitehouse, M.J., Mueller, P.A., and Basu, A.R., 2008, this volume, A review of the isotopic and trace element evidence for mantle and crustal processes in the Hadean and Archean: Implications for the onset of plate tectonic subduction, *in* Condie, K.C., and Pease, V., eds., When did plate tectonics begin on planet Earth?: Geological Society of America Special Paper 440, doi: 10.1130/2008.2440(01).

Sleep, N.H., and Windley, B.F., 1982, Archean plate tectonics: Constrains and inferences: Journal of Geology, v. 90, p. 363–379.

Smithies, R.H., Champion, D.C., Van Kranendonk, M.J., Howard, H.M., and Hickman, A.H., 2005a, Modern-style subduction processes in the Mesoarchean: Geochemical evidence from the 3.12 Ga Whundo intra-oceanic arc: Earth and Planetary Science Letters, v. 231, p. 221–237, doi: 10.1016/j.epsl.2004.12.026.

Smithies, R.H., Van Kranendonk, M.J., and Champion, D.C., 2005b, It started with a plume—Early Archean basaltic proto-continental crust: Earth and Planetary Science Letters, v. 238, p. 284–297, doi: 10.1016/j.epsl.2005.07.023.

Stern, C.R., and de Wit, M.J., 1997, The Rocas Verdes "Greenstone Belt," southernmost South America, *in* de Wit, M., and Ashwal, L., eds., Tectonic evolution of greenstone belts: Oxford, UK, Oxford University Press, p. 55–90.

Stern, R.J., 2005, Evidence from ophiolites, blueschists, and ultrahigh-pressure metamorphic terranes that the modern episode of subduction tectonics began in Neoproterozoic time: Geology, v. 33, p. 557–560, doi: 10.1130/G21365.1.

Storey, M., Mahoney, J.J., Kroenke, L.W., and Saunders, A.D., 1991, Are oceanic plateaus sites of komatiite formation?: Geology, v. 19, p. 376–379, doi: 10.1130/0091-7613(1991)019<0376:AOPSOK>2.3.CO;2.

Sun, S.S., and McDonough, W.F., 1989, Chemical and isotopic systematics of oceanic basalts: Implications for mantle composition and processes, *in* Saunders, A.D., and Norry, M.J., eds., Magmatism in the ocean basins: Geological Society [London] Special Publication 42, p. 313–345.

Sylvester, P.J., Harper, G.D., Byerly, R.G., and Thurston, P.C., 1997, Volcanic aspects, *in* de Wit, M., and Ashwal, L., eds., Tectonic evolution of greenstone belts: Oxford, UK, Oxford University Press, p. 55–90.

Sylvester, P.J., Mader, M.M., and Myers, J.S., 2006, Ultramafic alkaline magmas (meymechites) from the mid-Archean Ivisaartoq greenstone belt, Southwest Greenland: Geochimica et Cosmochimica Acta, v. 70, Supplement 1, p. A633, doi: 10.1016/j.gca.2006.06.1175.

Tarney, J., Dalziel, I., and de Wit, M.J., 1976, Marginal basin Rocas Verdes complex from southern Chile: A model for Archean greenstone belt formation, *in* Windley, B.F., ed., The early history of the Earth: London, Wiley Publishing, p. 131–146.

van der Velden, A.J., Cook, F.A., Drummond, B.J., and Goleby, B.R., 2006, Reflections of the Neoarchean: A global perspective, *in* Benn, K., et al., eds., Archean geodynamics and environments: American Geophysical Union Geophysical Monograph 164, p. 255–265.

Vervoort, J.D., and Blichert-Toft, J., 1999, Evolution of the depleted mantle: Hf isotope evidence from juvenile rocks through time: Geochimica et Cosmochimica Acta, v. 63, p. 533–556, doi: 10.1016/S0016-7037(98)00274-9.

Weaver, B.L., Tarney, J., and Windley, B., 1981, Geochemistry and petrogenesis of the Fiskenaesset anorthosite complex, southern West Greenland: Nature and parent magma: Geochimica et Cosmochimica Acta, v. 45, p. 711–725, doi: 10.1016/0016-7037(81)90044-2.

Weaver, B.L., Tarney, J., Windley, B., and Leake, B.E., 1982, Geochemistry and petrogenesis of Archean metavolcanic amphibolites from Fiskenaesset, SW Greenland: Geochimica et Cosmochimica Acta, v. 46, p. 2203–2215, doi: 10.1016/0016-7037(82)90195-8.

Windley, B.F., 1970, The stratigraphy of the Fiskenaesset complex: Rapport Grønlands Geologiske Undersøgelse, v. 35, p. 19–21.

Windley, B.F., 1976, New tectonic models for the evolution of Archean continents and oceans, *in* Windley, B., ed., The early history of the Earth: London, Wiley, p. 105–112.

Windley, B.F., Herd, R.K., and Bowden, A.A., 1973, The Fiskenaesset Complex, West Greenland, part 1: Copenhagen, Grønlands Geologiske Undersøgelse Bulletin 106, 80 p.

Windley, B.F., Bishop, F.C., and Smith, J.V., 1981, Metamorphosed layered igneous complexes in Archean granulite-gneiss belts: Annual Review of Earth and Planetary Sciences, v. 9, p. 175–198, doi: 10.1146/annurev.ea.09.050181.001135.

Woodhead, J.D., Eggins, S.M., and Johnson, R.W., 1998, Magma genesis in the New Britain island arc: Further insights into melting and mass transfer processes: Journal of Petrology, v. 39, p. 1641–1668, doi: 10.1093/petrology/39.9.1641.

Yilmaz, Y., and Polat, A., 1998, Geology and evolution of the Thrace volcanism, Turkey: Acta Vulcanologica, v. 10, p. 293–303.

Manuscript Accepted by the Society 14 August 2007

The Geological Society of America
Special Paper 440
2008

# *Stratigraphy, geochemistry, and depositional environments of Mesoarchean sedimentary units in western Superior Province: Implications for generation of early crust*

**Philip Fralick**
**Pete Hollings***
**David King**
*Department of Geology, Lakehead University, Thunder Bay, Ontario P7B 5E1, Canada*

## ABSTRACT

**The Steep Rock Group, Little Falls assemblage, Finlayson greenstone belt, and Lumby Lake greenstone belt form a once-continuous Mesoarchean terrane at the southern margin of Wabigoon subprovince, Canadian Shield. Synchronous with eruptive activity, oceanic plateau basalts in this region were intruded by tonalitic batholiths, which fed intermediate to felsic volcanic cones and associated mass-flow sediment aprons. A sedimentary succession capping the volcanic pile records stream-induced channel incision into upraised basalt and tonalite, feeding sediment to a delta complex. Base-level rise caused the incised channels to backfill, stromatolitic carbonates to develop in shallow-marine areas, and finally, deposition of iron formation to dominate. A Mesoarchean terrane was also examined in the Wallace Lake area of Manitoba. Here a kilometer-thick, transgressive succession records a transition from fluvial to delta-front to prodelta turbiditic environments, with eventual drowning and siliciclastic sediment starvation leading to dominance by carbonate and iron formation deposition. Sedimentary rocks that cap the oceanic Cretaceous to Holocene Kerguelen Plateau are a direct analogue for the depositional environments, sediment source rocks, and drowning events that define the stratigraphy of the Mesoarchean successions studied here. The Steep Rock–Lumby oceanic plateau formed and actively grew during an interval of 230 m.y. This sequence produces serious problems for Mesoarchean tectonic models in which increased heat dissipation produces fast spreading or numerous spreading ridges. A Mesoarchean, mantle plume–dominated heat dissipation process is in much better agreement with the data presented here.**

**Keywords:** Archean sedimentology, greenstone belt, Archean stratigraphy, mantle plume.

*Corresponding author: peter.hollings@lakeheadu.ca

Fralick, P., Hollings, P., and King, D., 2008, Stratigraphy, geochemistry, and depositional environments of Mesoarchean sedimentary units in western Superior Province: Implications for generation of early crust, *in* Condie, K.C., and Pease, V., eds., When Did Plate Tectonics Begin on Planet Earth?: Geological Society of America Special Paper 440, p. 77–96, doi: 10.1130/2008.2440(04). For permission to copy, contact editing@geosociety.org. 

## INTRODUCTION

Mesoarchean volcanic-sedimentary successions of the Superior Province contain stromatolitic carbonate- and/or quartz arenite–bearing sedimentary rocks (Thurston and Chivers, 1990) associated with thick sequences of tholeiitic basalt that commonly interlayer with minor proportions of komatiitic flows (Hollings and Wyman, 1999; Hollings et al., 1999) or, very rarely, ultramafic pyroclastics (Schaefer and Morton, 1991). These ca. 2900–3000 Ma assemblages are present in the northern half of Superior Province as scattered greenstone belts, usually of limited extent. They have been interpreted to reflect platformal to rift settings (Thurston and Chivers, 1990; Tomlinson et al., 1996), oceanic plateaus (Hollings et al., 1999), or plume-arc interaction (Wyman and Hollings, 1998). Although similar sedimentary assemblages in other, smaller shield areas around the world have received considerable study (see summaries in Eriksson et al., 1994, 1997), aside from work carried out in the Steep Rock area by Wilks (1986) and Wilks and Nisbet (1988), the Mesoarchean sedimentary rocks of the northern Superior Province have not been subjected to a detailed appraisal.

Extensive investigations have recently focused on associated volcanic rocks. Tomlinson et al. (1999) proposed that the mantle source of the Mesoarchean volcanic assemblages in the Steep Rock and Lumby Lake areas was geochemically similar to the source of basaltic rocks that form the Cretaceous Ontong-Java oceanic plateau and that eruption probably occurred in a continental setting following rifting of the crust. Wyman and Hollings (1998) proposed that most of volcanic units in the Lumby Lake belt have an oceanic plateau affinity. However, intercalated calc-alkalic volcanic rocks and coeval intrusive tonalite-trondhjemite-granodiorite (TTG) suites attest to formation of the terrane not by plateau accretion, rather by plume-arc interaction during approach and subduction of a plume-fed ridge (Wyman and Hollings, 1998; Hollings and Wyman, 1999; Hollings et al., 1999). In their model, movement of the plume under the terrane then caused rifting and possible breakup. Sedimentary successions are present at several stratigraphic levels in these assemblages and thus are capable of providing additional information on the crust-forming mechanisms.

The Mesoarchean successions provide information on the crustal and mantle processes of early Earth that built the foundations on which the Superior Province developed during later, Neoarchean collisional events (Langford and Morin, 1976; Thurston et al., 1991; Stott, 1997). The scattered remnants of Mesoarchean rocks combined with Nd isotopic studies indicating that extensive areas of Neoarchean volcanic and plutonic rocks erupted or intruded through older crust (Tomlinson et al., 2000; Tomlinson, 2000) imply that northwestern Superior Province is largely underlain by Mesoarchean basement. However, positive $\varepsilon_{Nd}$ values reported for Lumby Lake rocks suggest no assimilation of basement material (Hollings and Wyman, 1999). Understanding the genesis of this basement complex will provide insight into the processes that formed the first cratonic nucleus of the largest present-day fragment of Archean crust.

The use of depositional systems, provenance, and basin architectural studies to determine geodynamic setting and factors that caused subsidence is well established (Busby and Ingersoll, 1995). These techniques have been applied to four Mesoarchean successions in the western Superior Province. Most of the outcrops examined have a metamorphic grade of greenschist facies, but changes in orientation due to strain make paleocurrent evaluation unreliable. Therefore, drainage direction was inferred, where possible, from provenance established from clast lithologies, sandstone geochemistry, and detrital zircon geochronology.

Three of the successions studied (Little Falls assemblage, Finlayson greenstone belt, and Lumby Lake greenstone belt) occur along a disrupted band near the southern margin of Wabigoon subprovince, proximal to the Steep Rock Group (Fig. 1). The fourth (Wallace Lake area) occurs ~330 km to the northwest, near the southern margin of Uchi subprovince. The areas are separated by the western Wabigoon subprovince, the Winnipeg River subprovince, and a region of Neoarchean, deep-water sediments, the English River subprovince. The similar ages and lithologies present in these two Mesoarchean areas permit comparison of their developmental histories.

## STRATIGRAPHY AND SEDIMENTOLOGY

### Steep Rock Group

The Steep Rock Group, Little Falls assemblage, Finlayson greenstone belt, and Lumby Lake greenstone belt form a once-continuous succession of volcanic, sedimentary, and intrusive rocks that has been partly dissected by several major faults (Fralick and King, 1996; King, 1998; Wyman and Hollings, 1998; Tomlinson et al., 1999) (Fig. 1). The Steep Rock Group, as formally defined by Wilks and Nisbet (1988), consists of five formations. The 0–15-m-thick basal Wagita Formation is composed of discontinuous lenses of pebble-cobble conglomerate and sandstone that occupy paleochannels (Wilks and Nisbet, 1988) eroded into the 3003 ± 5 Ma Marmion Batholith (Davis and Jackson, 1988; also 3003 ± 3 Ma, Tomlinson et al., 1999; and 3001.6 ± 1.7 Ma, Tomlinson et al., 2003), a tonalitic body cut by numerous gabbroic dikes (Stone et al., 1992). The clasts are dominated by tonalite with lesser amounts of felsic and mafic volcanic rock fragments. Six detrital zircons from sandstones have U-Pb zircon ages of ca. 2999 Ma (Davis, 1993). The Wagita Formation is conformably overlain by the up to 500-m-thick Mosher Carbonate, which consists of stromatolitic ankerite, dolomite, and calcite, and ribbon chert-carbonate. This unit is conformably succeeded by up to 400 m of iron formation overlain by a 50–400-m-thick, high-Mg, predominantly pyroclastic komatiite assemblage, the Dismal Ashrock (Wilks and Nisbet, 1988). A komatiitic lapilli tuff from this unit has a maximum age of 2780.4 ± 1.4 Ma based on the youngest inherited zircons (Tomlinson et al., 2003) (Fig. 2).

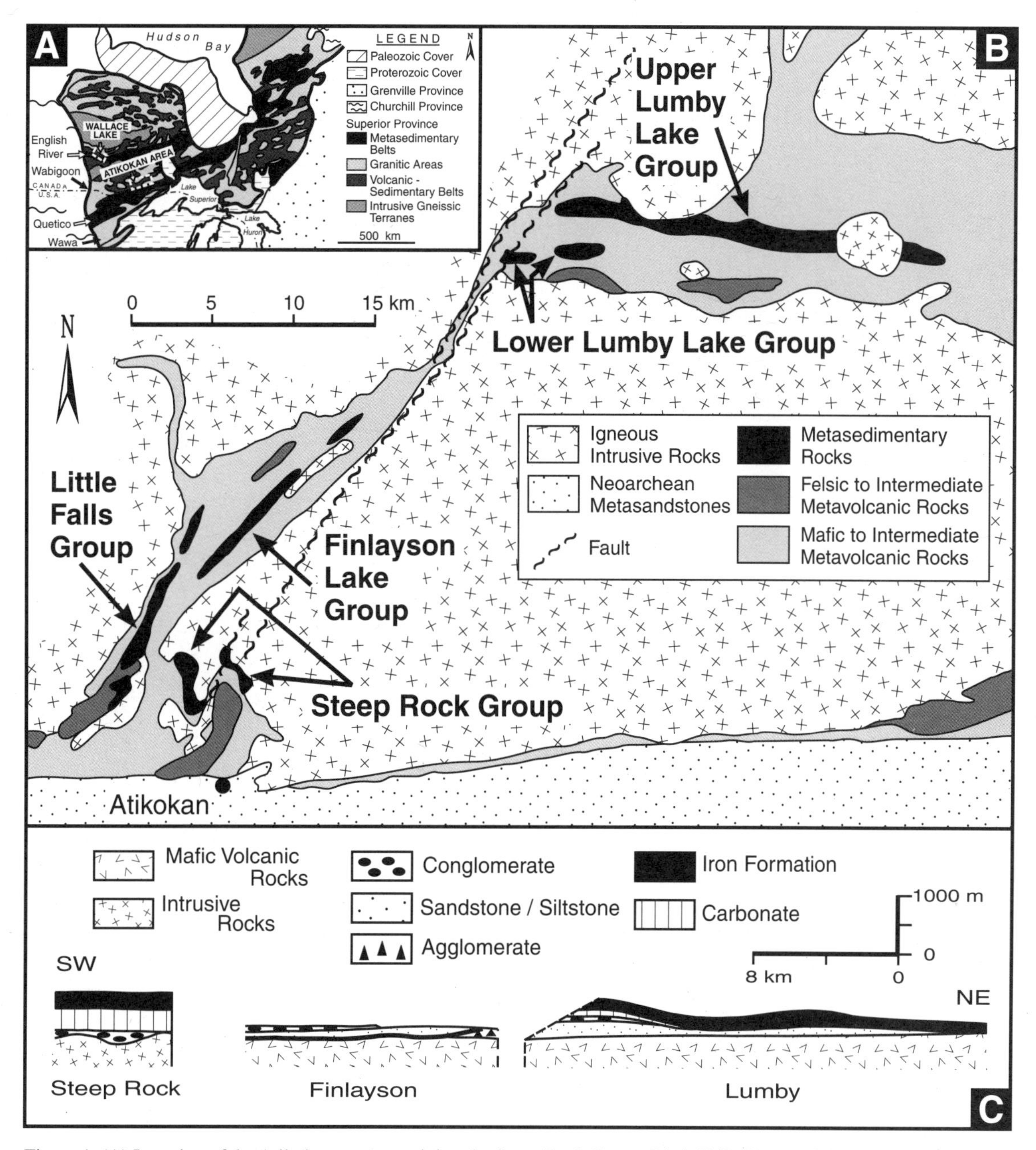


Figure 1. (A) Location of the Atikokan area containing the Steep Rock Group, Little Falls Group, Finlayson greenstone belt, and Lumby Lake greenstone belt, and the Wallace Lake area containing the Conley and Overload Bay Formations. These two areas are separated by Neoarchean sediments of the English River subprovince, which represents a closed remnant ocean basin. (B) Geology of the Atikokan area showing the locations of the sedimentary units and the fault dissection of the greenstone belts (modified from Stone et al., 2001). (C) Correlation of generalized stratigraphic successions in the Steep Rock Group, Finlayson Lake Group, and Upper Lumby Lake Group.

## Little Falls Group

The Little Falls area lies to the immediate west of the Steep Rock Group. Stone et al. (1992) outlined a synclinal axis running down the center of the Little Falls area. This places the basalts, which dominate the area, under a felsic, volcanic tuff breccia-sandstone unit (Little Falls Group; Figs. 1 and 2). Single-grain, U-Pb age determinations on six zircons from resedimented sandstones derived from, and laterally traceable into, the felsic pyroclastic pile have an age of 2996.9 ± 0.8 Ma (age determined by D. Davis, reported in King, 1998). This is considered to be a depositional age as the mass-flow sandstones are traceable along strike as the lateral equivalents of the pyroclastic eruptive units. The underlying mafic volcanic rocks are

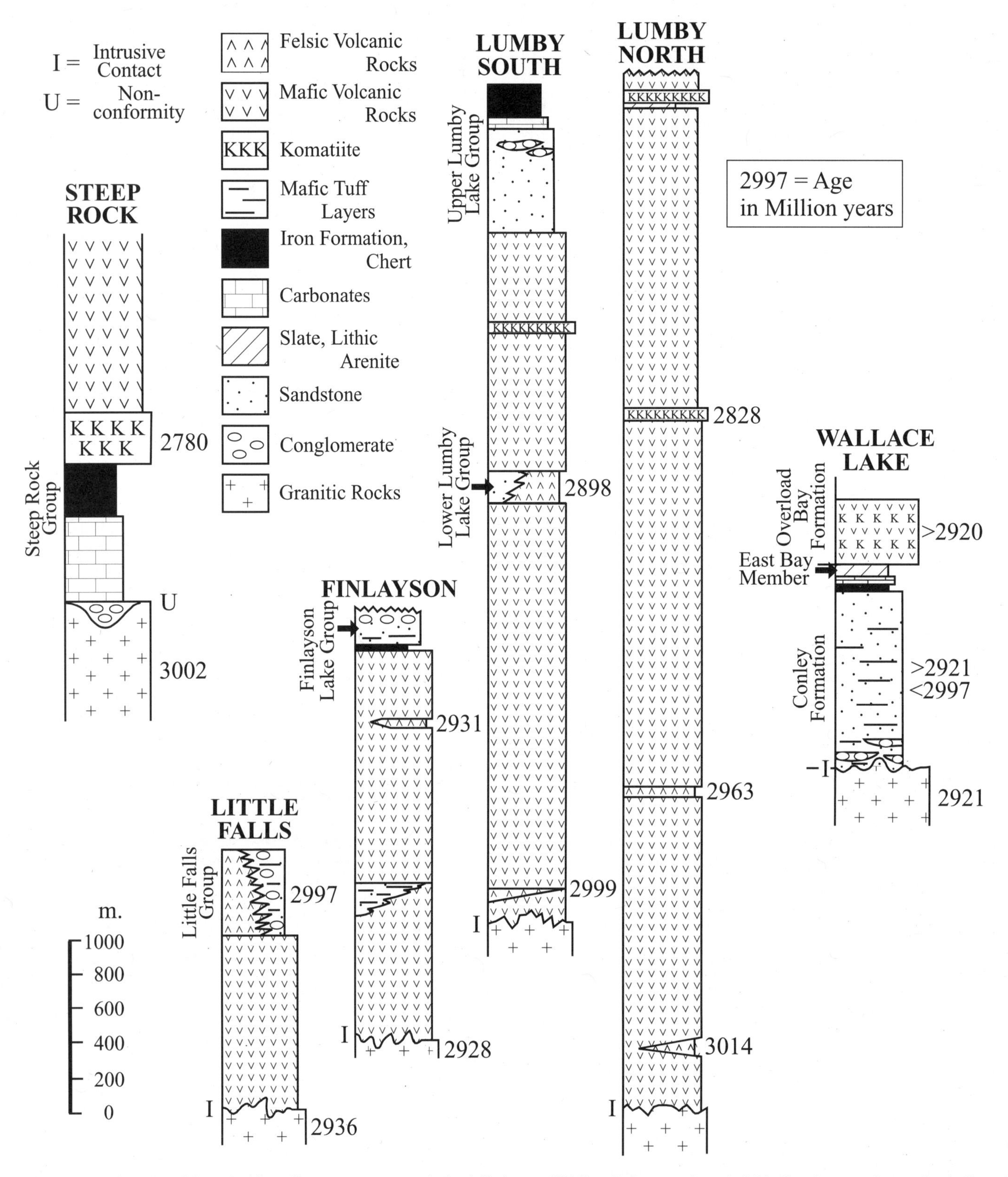


Figure 2. Stratigraphy of the volcanic-sedimentary sequences in the Atikokan and Wallace Lake areas (ages in Ma). Consistency of younging indicators throughout the stratigraphic successions is supported by U-Pb zircon age determinations and appropriate stratigraphic position of correlative units in adjacent areas. Units have been tectonically flattened. Age determinations are from: Steep Rock, 3001.6 ± 1.7 Ma (Tomlinson et al., 2003), 2780.4 ± 1.4 Ma (Tomlinson et al., 2003); Little Falls, 2936 Ma (analysis by D. Davis, cited in Stone et al., 1992), 2996.9 ± 0.8 Ma (analysis by D. Davis, new data); Finlayson, 2928 Ma (analysis by D. Davis, cited in Stone et al., 1992), 2931 ± 2 Ma (Tomlinson et al., 2003); South Lumby, 2999 Ma (Davis and Jackson, 1988), 2897.6 ± 2.1 Ma (Tomlinson et al., 2003); North Lumby, 3013.6 ± 1.3 Ma, 2963.3 ± 4.5 Ma, and 2828 ± 2 Ma (Tomlinson et al., 2003); and Wallace Lake, tonalite is 2921 ± 1 Ma (Tomlinson et al., 2001), dike cutting mafic volcanic rocks is 2921 ± 2 Ma (Tomlinson et al., 2001), dike cutting sediments is 2921 Ma (Davis, 1994), detrital zircons in sediment are 2997–3000 Ma (Davis, 1994).

intruded by 2936 Ma tonalite to granodiorite (age determination by D. Davis, reported in Stone et al., 1992).

The sedimentary rocks of the Little Falls Group consist of medium- and coarse-grained arkosic sandstone, monomictic conglomerate, and slate. The sandstones dominate and range from coarse-grained, quartz and feldspar granule-rich layers averaging 75 cm in thickness (up to 600 cm thick) to medium-grained sandstones averaging 25 cm in thickness (up to 100 cm thick). All have low amounts of clay-rich matrix. The coarse-grained sandstones show grading to medium sand only in their uppermost few centimeters and are commonly separated by slate partings less than 1 cm thick. The granules in these sandstones are angular to subangular. Most beds are massive with only rare, small- to medium-scale, trough cross-stratification highlighted by thin mud flasers. The medium-grained sandstones are massive and exhibit more pronounced grading. Both types of sandstone layers fill shallow scours in underlying beds and contain rare mud rip-up clasts. Less common, but distinctive, layers of dark-colored, chlorite-rich sandstone are interbedded with the other lithologies. They are similar to the other sandstones except for the higher concentration of mafic detritus.

Clast- to matrix-supported conglomerates are interlayered with the sandstones. Clast sizes are dominated by pebbles and cobbles with rare small boulders. All clasts are fragments of felsic volcanic rock similar to the felsic volcanic pile they grade into to the west. The conglomerates are generally poorly sorted, although thicker units exhibit some internal organization into clast-rich and clast-poor zones, or zones with differing clast sizes. These zones, as well as sandstone lenses within conglomerates, suggest that the thick conglomerate units are composed of smaller, 50–100-cm-thick lenses. The boundaries between the lenses are indistinct and are commonly gradational over a few centimeters. Conglomeratic beds range from 10 to 700 cm in thickness. Contacts between the sandstones and conglomerates are generally sharp. Chloritic mafic tuffaceous layers and mafic, matrix-supported lapilli tuffs with felsic clasts are interlayered with the sedimentary rocks.

**Finlayson Lake**

Two areas of sedimentary rocks are present in the Finlayson Lake greenstone belt. The lower succession is the laterally continuous extension of the Little Falls Group, and it occurs within the mafic volcanic assemblage, which dominates the Finlayson area (Figs. 1 and 2). The upper sediment package (Finlayson Lake Group) forms a coarsening-upward succession that caps the volcanic assemblage (Fig. 3). A felsic pyroclastic unit near the top of the volcanic pile has a U-Pb zircon age of 2931.4 ± 2.0 Ma (Tomlinson et al., 2003), and the base of the volcanic-sedimentary sequence is intruded by 2928 Ma tonalite (age determination by D. Davis reported in Stone et al., 1992) (Fig. 2).

Isolated pockets of fine- to medium-grained sandstones of the lower sedimentary unit are contained within the mafic volcanic rocks. These sedimentary rocks have bedding thicknesses ranging from 10 to 25 cm, and they are similar to the medium-grained sandstones found in the Little Falls area. They laterally correlate with, and are included within, the Little Falls Group.

The 300 m of sedimentary rocks capping the volcanic succession (Finlayson Lake Group) occur in the center of the main syncline and are truncated at their top by a vertical fault running down the synclinal axis that cuts stratigraphically downward to the north (Fralick and King, 1996). Oxide-facies iron formation, chert, and fine-grained clastic rocks conformably overlie the volcanic sequence (Fig. 3A). The lower siliciclastic rocks are composed of thinly bedded, fine-grained, chlorite-rich sandstone or siltstone couplets with silty slate tops (Fig. 3B). Thick successions of these units are gradational upwards into 15–20-cm-thick, graded, medium- and coarse-grained sandstones (Fig. 3C). They are massive except for abundant parallel lamination in their upper portions. Rare thicker beds have scattered quartz and tonalitic and felsic volcanic granules and pebbles at their base. This lithofacies is succeeded upward by similar beds with abundant internal cut-and-fill structures (Fig. 3D). These graded sandstones have basal zones composed of a number of lenses of coarser sand separated by the medium sand of the bed. Thin mud flasers may be present in the medium sand. The coarsening-upward sedimentary succession is capped by polymictic, pebble-cobble conglomerates. Clasts are dominantly rounded to subrounded and composed of granitoid rock fragments (50%) with lesser amounts of felsic (26%) and mafic (13%) volcanic detritus and quartz (11%). Beds are nongraded, clast sorting is moderate to poor, and a medium-grained sand matrix is present. Clast support dominates, but there is a complete gradation to pebbly sandstones. Layers are up to 4 m in thickness and are internally composed of conglomeratic, and rarer sandstone, lenses averaging 20 cm in thickness. The conglomerate-dominated succession is sharply overlain by sandstones similar to the underlying scoured beds, which are, in turn, upwardly gradational into more conglomerate. Six detrital zircons from an upper sandstone layer have given U-Pb ages of: 2997.0 ± 2.5 Ma, 2999.1 ± 1.2 Ma, 2999.7 ± 1.5 Ma, 3001.4 ± 1.2 Ma, 3001.6 ± 0.8 Ma, and 3002.3 ± 0.9 Ma (age determined by D. Davis, reported in Fralick and King, 1996).

A somewhat different laterally correlative succession is locally present in the northern portion of the Finlayson greenstone belt. Instead of basal chemical sediment overlying mafic volcanic flows, in one restricted area, a 120-m-thick dacitic epiclastic conglomerate floors the sedimentary succession. This assemblage overlies an extensive assemblage of massive and pillowed basalt flows with interflow units of graphitic slate, chert, and sulfide-facies iron formation. The epiclastic conglomerate is well layered and forms a fining- and thinning-upward succession. Clast-supported cobble layers near the base are up to 12 m thick, whereas matrix-supported pebble layers near the top do not exceed 4 m. The epiclastic conglomerate is overlain by 6.5 m of graphitic slate, chert, and sulfide-facies iron formation, which in turn is overlain by fine-grained siliciclastic rocks at the base of the coarsening-upward succession. Conglomerates higher in the assemblage are dominated by igneous extrusive

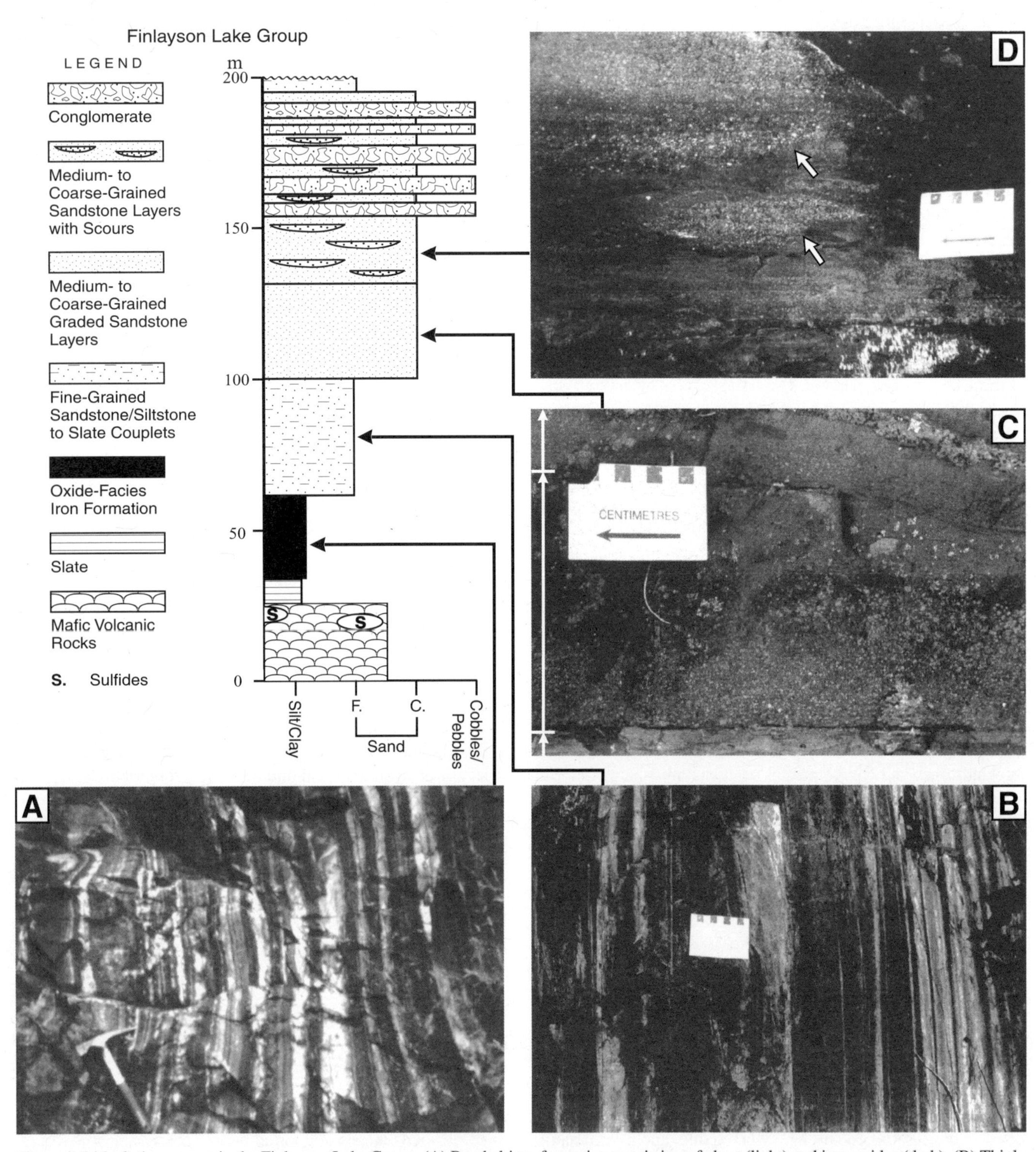


Figure 3. Lithofacies present in the Finlayson Lake Group. (A) Banded-iron formation consisting of chert (light) and iron oxides (dark). (B) Thinly layered succession of beds graded from very fine-grained sandstone to slate. (C) Thicker beds grading from coarse- to fine-grained sandstone with slate drapes. (D) Broad lenses of coarse-grained sandstone in medium-grained sandstone (lower arrow) and coarse-grained sandstone beds with erosive bases.

rocks with: 52% rhyolite and rhyodacite, 38% mafic and intermediate volcanic, 5% iron formation, and 5% chert clasts.

**Lumby Lake**

The Lumby Lake greenstone belt is the fault-disrupted extension of the Finlayson greenstone belt (Fralick and King, 1996; King, 1998; Wyman and Hollings, 1998; Fig. 2). As in the Finlayson belt, beds are subvertical and form a trough-shaped syncline that is disrupted along its axis by a fault, where the sedimentary succession forms the top of the southern limb but is missing from the fault-truncated top of the northern limb. The volcanic-sedimentary assemblage in both limbs is intruded at its base by tonalites. The northern limb contains a felsic volcanic unit near its base that has a U-Pb zircon age of 3014 ± 1 Ma (Tomlinson et al., 2003) (Fig. 2). Zircons recovered from a thin felsic tuff interbedded with komatiites near the top of the succession have given an age of 2828 ± 2 Ma (Tomlinson et al., 2003). The southern limb contains a laterally extensive felsic volcanic unit near its base that has a U-Pb zircon age of 2999 Ma (Davis and Jackson, 1988) to 3001 ± 1 Ma (Tomlinson et al., 2003), making it the lateral chronostratigraphic equivalent of the felsic volcanic-sedimentary succession at Little Falls and the lower sedimentary zone at Finlayson Lake (Little Falls Group). Another zone of felsic volcanic rocks (2897.6 ± 2.1 Ma; Tomlinson et al., 2003) occurs higher in this succession and is laterally transitional into sandstones, carbonate, and chert to the west (Lower Lumby Lake Group) (Fig. 2). The Upper Lumby Lake Group caps the volcanic pile. The siliciclastic rocks appear to be overlain by carbonate, which is in turn overlain by cherty iron formation, although intense isoclinal folding and poor exposure complicate interpreting the stratigraphy here. The sedimentary succession is probably ~500–800 m thick at its maximum.

The 2898 Ma Lower Lumby Lake Group is dominated by medium- to fine-grained sandstones. The beds are laterally continuous, commonly nongraded, and massive. The medium-grained sandstones are 5–30 cm thick, whereas the fine-grained sandstones are 0.5–3 cm thick. Very thin slate laminae commonly separate the sandstone layers. An 8-m-thick, low-Fe ankeritic carbonate overlies the sandstones and is overlain by slate with fine-grained sandstone interbeds.

The Upper Lumby Lake Group varies from west to east. In the west, fine- to medium-grained, massive, commonly indistinctively graded sandstones overlie the mafic volcanic rocks. This ~300-m-thick succession is overlain by 50 m of polymictic, pebble-cobble conglomerate with a chloritic, fine-grained sand matrix. Mafic (53%) and felsic (44%) volcanic rock fragments dominate the clast lithologies, with minor amounts of gabbro, chert, and iron formation. The conglomerate is clast supported, poorly sorted, nongraded, and massive. An 80-m-thick marble overlies the conglomerate. It consists of interlayered calcite (layers 0.2–4 cm thick) and chert (layers 0.1–1 cm thick) arranged into decimeter-scale packages where they alternately dominate. A 200-m-thick chert-magnetite unit overlies the marble. This unit is composed of interlayered chert and magnetite-rich chert and also contains Fe-carbonate laminations plus thin and thick (up to tens of meters) successions of carbonaceous (maximum 15 wt% C), pyritiferous slate.

The conglomerates and marble wedge out to the east and the basal sandstones decrease in bed thickness and total unit thickness in this direction. In the central area of the Lumby Lake greenstone belt, the cherts and siliceous iron formation directly overlie thin-bedded, fine-grained sandstone. In the eastern Lumby Lake belt, the cherts and iron formation lie directly on the mafic volcanic assemblage without intervening clastic material (Fig. 1).

**Wallace Lake**

Siliciclastic rocks of the 1-km-thick Conley Formation are the oldest supercrustal rocks in the Wallace Lake area. They are predominantly composed of sandstone and conglomerate capped by dolostone, iron formation, and argillites. The formation contains detrital zircons with U-Pb ages of 2997–3000 Ma and is cut by a 2921 felsic dike (Davis, 1994). Nearby 3003 Ma tonalites (Turek and Weber, 1994) of the Wanipigow plutonic complex are the probable source of the detrital zircons. The siliciclastic succession is overlain by 50 m of carbonate, which is itself overlain by chlorite-rich sandstones and slates with interbedded iron formation of the East Bay member. An assemblage of komatiitic and mafic volcanic units with associated iron formation and mafic intrusives, the Overload Bay Formation (Tomlinson et al., 2001), conformably overlies the Conley Formation (Sasseville et al., 2006). These are cut by a 2921 ± 2 Ma dacitic dike and intruded by a 2921 ± 1 Ma tonalite batholith (Tomlinson et al., 2001; Sasseville et al., 2006).

The lower portion of the Conley Formation is composed of medium- and coarse-grained sandstones, pebbly sandstones, and conglomerates organized into a stacked succession of lenses. The lenses have erosive basal contacts that commonly undulate and are irregular (Fig. 4A). Where larger channel-ways are visible, the elongate lenses fill the channels as irregular sheets with onlapping ends. The lenses are commonly less than 30 cm thick and are filled with either single or multiple trough cross-stratified units. Small, irregular lenses appear as patches of differing grain size, some of which represent sand shadows next to pebble-cobble mounds. The sandstones contain pebbles, cobbles, and small boulders as stringers and conglomerate lenses. Clasts are dominated by granitoids and felsic volcanic fragments. Fine-grained, chloritic tuffs are common interbedded with the succession. They are irregularly scoured by the sandstones and conglomerates, and some sandstones have acquired a darker color through incorporation of this material. The basal lithofacies association is gradational upward into, and in some areas alternates with, trough cross-stratified sandstones with mud flasers, ripple laminated tops, and slate drapes (Fig. 4B) interbedded with meter-thick assemblages of laterally continuous, thinly bedded, fine-grained sandstones. They are graded

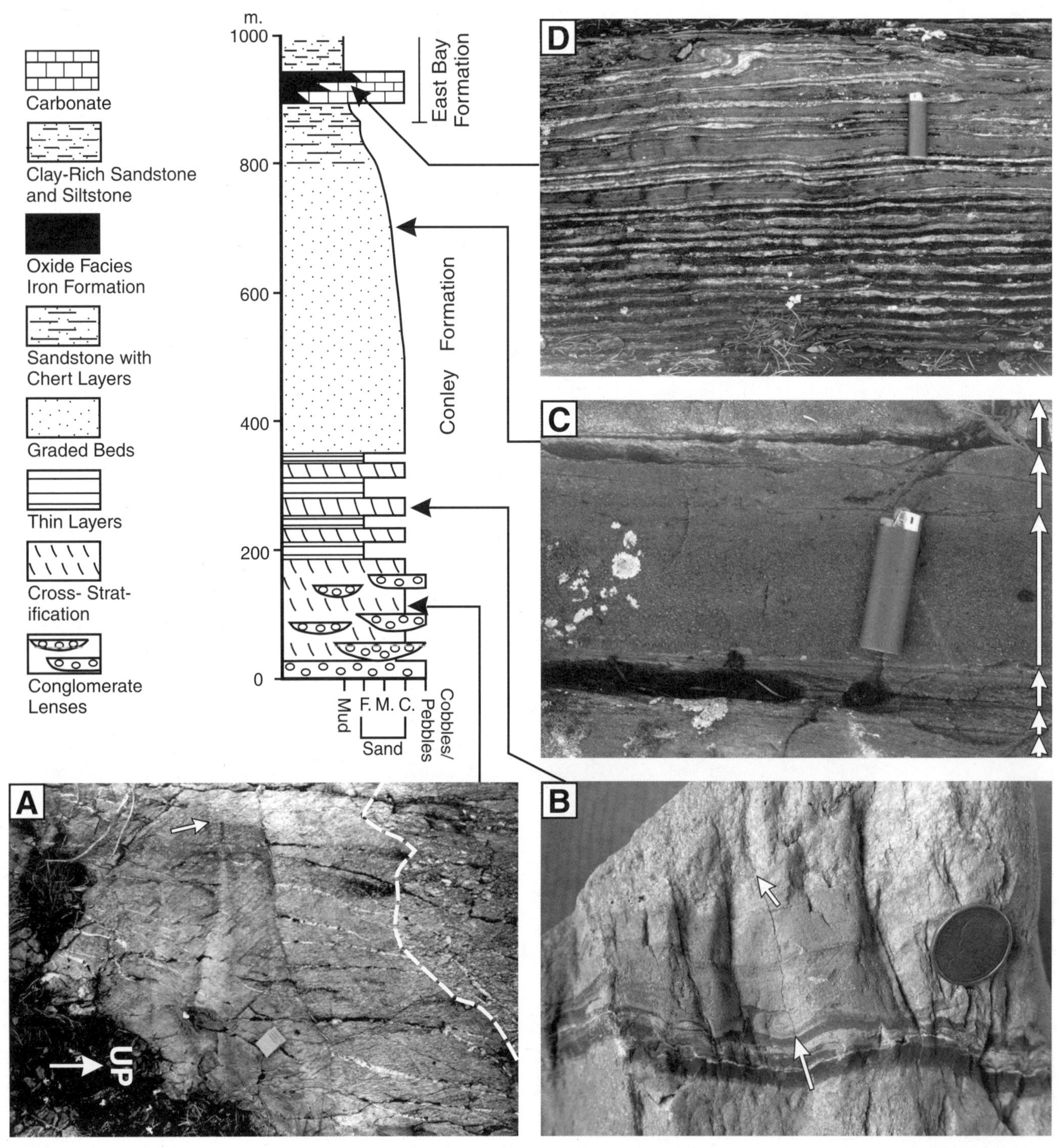


Figure 4. Lithofacies present in the Conley Formation. (A) The base of a large conglomeratic lens irregularly cutting down into a coarse-grained, pebbly sandstone. A sandstone lens is present immediately above the matches (for scale), and the arrow shows the position of a small conglomerate lens. (B) Cross-stratified, coarse-grained sandstone (upper arrow showing dipping cross-set) interlayered with fine-grained sandstones, siltstones, and slates. The lower arrow shows a fine-grained sandstone loading into a silty slate. (C) Graded coarse- to fine-grained sandstone with thin slate drapes forms the majority of the formation. (D) Oxide-facies banded-iron formation. The light layers are chert; the medium gray layers are chloritic, very fine-grained sandstone and siltstone; and the dark gray to black layers are iron oxides.

and have a varve-like appearance. Slightly higher in the succession, these meter-thick assemblages interbed with successions of amalgamated, coarse-tail-graded, pebbly sandstones; individual beds average 1 m thick. The assemblages of thin beds become less prominent upward through the Conley Formation. At ~200 m above where they first appeared, they are rare, and the succession is dominated by graded, coarse- to fine-grained sandstones organized into meter-scale, thicker-bedded (30 cm) and thinner-bedded (10 cm) packages (Fig. 4C). Medium- to coarse-grained pebbly sandstones and matrix-supported conglomerates are sporadically interbedded in this succession. Some are graded with pebbles concentrated near their base; others are disorganized. Chloritic tuffs present here are laterally continuous. Some outcrops contain stratigraphically limited areas of trough cross-stratified and rippled sandstone lenses with mud drapes and flasers similar to the underlying lithofacies association. The laterally continuous graded and nongraded sandstones comprise the majority of the formation and persist up to ~50 m below the overlying chemical sedimentary rocks. The uppermost siliciclastic rocks are dominated by thin-bedded, graded, fine-grained, chlorite-rich sandstone and siltstone interlayered with 0.1–1-cm-thick chert laminae. Over ~1 m, the sandstones become even more chlorite-rich and develop thin to, less commonly, decimeter-thick slate layers separating them. This marks the base of the East Bay member. Meters to tens of meters above this contact, the chloritic sandstones develop thin magnetite interlayers and are transitional to iron formation (Fig. 4D) or, in other areas, are overlain by dolomitic carbonate. The carbonate is recrystallized and massive, although where silicification has preserved original millimeter-scale, wavy layering, domal stromatolites are visible. Chloritic argillites overlie the iron formation and carbonate.

## GEOCHEMISTRY

Seventy-two samples were collected from the study areas. Of these, 62 were fine or medium-grained sandstones, and 10 were clasts from conglomerates and samples of mafic tuff that were obtained for use in constructing mixing models. Major, minor, and trace elements, except $SiO_2$, were analyzed using inductively coupled plasma–atomic-emission spectrometry (ICP-AES) techniques (Church, 1981). $SiO_2$ was determined using standard X-ray fluorescence (XRF) glass bead techniques. Accuracy and precision were evaluated using a suite of international and in-house standards and duplicated analyses for all samples. Accuracy for major and minor elements has an average error of 2.6%, varying from 0.7% for $K_2O$ to 4.3% for $Al_2O_3$. Accuracy for trace elements has an average error of 4.5%, varying from 1.1% for Zr to 10.6% for Ni. Precision for major and minor elements has an average error of 3.1%, and prevision for trace elements is 1.6%. All elemental concentrations were well above detection limits. Element mobility was assessed using techniques described in Fralick and Kronberg (1997) and Fralick (2003). Data for the 62 sandstones analyzed are given in Tables 1 and 2.

Figure 5 depicts the medium-grained sandstone samples from the southern study areas on a felsic-mafic versus alkalinity plot that was first derived for volcanic rocks (Winchester and Floyd, 1976) but is also useful for sedimentary rocks composed of igneous debris (Fralick, 2003). The 20 samples from the Little Falls Group, its extension in the Finlayson area, and the Lower Lumby Lake Group either have compositions similar to the Marmion Batholith or plot along a well-defined trend from Marmion compositions toward a more felsic region of Figure 5. As the Little Falls sandstones have Marmion age zircons, and are laterally traceable into, and derived from, felsic pyroclastic rocks, it is likely that their detritus is the eruptive equivalent of the nearby Marmion Batholith, and the linear arrangement of data points represents its differentiation trend. The lower Lumby sedimentary rocks are younger (2898 Ma as opposed to 2997 Ma) and derived from a different felsic volcanic event. Even so, they exhibit a very similar composition (Fig. 5).

The 16 samples from the Finlayson Lake Group and the Upper Lumby Lake Group plot in a separate, but partially overlapping, area on Figure 5. They form a mixing trend from average Marmion compositions into the alkaline basalt field. The implied mixed intermediate to mafic source is consistent with intermediate to mafic clast compositions of interstratified conglomerates. The mafic contribution to the sediment results in elevated concentrations of V, Ni, and Cr compared to the tonalite (Fig. 6). Figure 5 also highlights the alkaline nature of the mafic detritus. The type of alkaline mafic material was further investigated using $Ni/Al_2O_3$ and V/Zr ratios (Fig. 7). This separated the upper Finlayson and Lumby samples into two groups; the upper Finlayson samples lie in an array between Marmion tonalite compositions and the compositions of Finlayson and Lumby mafic volcanic rocks. The majority of Lumby samples lie on a mixing line between Marmion tonalite compositions and Lumby Al-depleted komatiites. This agrees with field observations, since no komatiites have been found in the mafic volcanic sequence in the Finlayson greenstone belt, but the Lumby belt has pyroclastic Al-depleted komatiites near its top underlying the sedimentary rocks (Tomlinson et al., 1999; Hollings and Wyman, 1999).

The 26 medium-grained sandstone samples from the Conley Formation of the Wallace Lake Group exhibit an assortment of geochemical trends (Fig. 8). Tonalite clasts from the conglomerates have rhyodacitic compositions, whereas a chloritic tuff interlayered with the sedimentary succession is similar in composition to komatiitic basalt ($SiO_2$ = 43.7%, $Al_2O_3$ = 8.24%, MgO = 12.3%, Cr = 1088 ppm, Ni = 272 ppm). Samples from the lower Conley Formation generally plot along a tie line between these two compositions (Fig. 8). Middle Conley sandstones define another mixing line between lower Conley compositions and an alkaline mafic composition. The remaining Conley samples (excluding the East Bay member) lie close to a tie line between the alkaline mafic and rhyodacitic compositions. The East Bay member forms another trend; its points are concentrated in the subalkaline basalt field (Fig. 8). The major source area for these chloritic sandstones was apparently composed of basalt.

TABLE 1. GEOCHEMISTRY OF SANDSTONES FROM THE LITTLE FALLS, FINLAYSON, AND LUMBY LAKE AREAS

| Sample # | 2 | 7 | 8 | 9 | 10 | 11 | 16 | 20 | 23 | 27 | 29 | 30 | 141 | 142 | 143 | 144 | 145 | 148 |
|---|---|---|---|---|---|---|---|---|---|---|---|---|---|---|---|---|---|---|
| Location | Little Falls | Little Falls | Little Falls | Little Falls | Little Falls | Little Falls | Little Falls | Little Falls | Little Falls | Little Falls | Little Falls | Little Falls | Lower Lumby | Lower Lumby | Lower Lumby | Lower Lumby | Lower Lumby | Lower Lumby |
| Lithology | m-c sst | m sst | m-c sst | m-c sst | m-c sst | m-c sst | m-c sst | c sst | c sst | m-c sst | c sst | c sst | m sst | m sst | m-f sst | m-f sst | m-c sst | m-f sst |
| wt% | | | | | | | | | | | | | | | | | | |
| $SiO_2$ | 67.35 | 68.67 | 65.70 | 75.69 | 66.64 | 74.22 | 66.67 | 65.31 | 67.21 | 72.19 | 66.54 | 64.88 | 65.85 | 78.05 | 46.82 | 81.43 | 74.40 | 47.91 |
| $TiO_2$ | 0.31 | 0.40 | 0.41 | 0.25 | 0.56 | 0.20 | 0.25 | 0.28 | 0.33 | 0.22 | 0.43 | 0.39 | 0.30 | 0.41 | 0.45 | 0.14 | 0.23 | 0.72 |
| $Al_2O_3$ | 12.16 | 13.53 | 14.78 | 12.15 | 16.58 | 11.72 | 14.20 | 14.57 | 11.75 | 12.80 | 15.01 | 14.78 | 14.64 | 8.52 | 9.57 | 5.33 | 6.88 | 13.98 |
| $Fe_2O_3t$ | 4.05 | 3.98 | 4.09 | 3.08 | 5.30 | 2.75 | 4.28 | 4.25 | 3.90 | 3.05 | 4.67 | 3.90 | 2.30 | 6.87 | 7.94 | 7.45 | 6.89 | 6.56 |
| MnO | 0.02 | 0.05 | 0.05 | 0.03 | 0.05 | 0.03 | 0.05 | 0.05 | 0.03 | 0.04 | 0.06 | 0.05 | 0.03 | 0.07 | 0.15 | b.d. | 0.11 | 0.16 |
| MgO | 1.59 | 2.24 | 2.15 | 1.01 | 2.11 | 0.41 | 2.21 | 1.63 | 1.37 | 1.20 | 0.49 | 0.32 | b.d. | b.d. | 9.76 | b.d. | b.d. | 2.43 |
| CaO | 0.25 | 3.01 | 3.30 | 1.74 | 1.41 | 1.15 | 2.48 | 3.34 | 1.84 | 1.73 | 1.88 | 2.89 | 2.73 | 0.11 | 7.13 | b.d. | 2.14 | 7.86 |
| $Na_2O$ | 3.12 | 2.81 | 3.95 | 2.81 | 4.43 | 4.20 | 4.72 | 2.58 | 4.66 | 3.40 | 5.44 | 6.06 | 4.10 | 0.50 | 1.54 | 0.38 | 0.60 | 2.53 |
| $K_2O$ | 2.50 | 2.26 | 1.40 | 1.88 | 2.27 | 2.03 | 1.19 | 2.69 | 0.56 | 2.12 | 1.01 | 0.59 | 2.25 | 1.00 | 0.04 | 1.08 | 1.18 | 1.58 |
| $P_2O_5$ | 0.07 | 0.07 | 0.08 | 0.05 | 0.10 | 0.02 | 0.07 | 0.07 | 0.08 | 0.05 | 0.07 | 0.07 | 0.12 | 0.04 | 0.22 | 0.01 | 0.03 | 0.67 |
| $H_2O$ | 1.98 | 1.62 | 1.53 | 1.53 | 2.07 | 0.63 | 1.80 | 2.16 | 1.35 | 1.35 | 1.26 | 1.44 | 1.26 | 2.34 | 3.87 | 0.72 | 1.17 | 2.16 |
| $CO_2$ | 0.07 | 2.24 | 0.81 | 1.36 | 0.77 | 0.84 | 2.05 | 2.60 | 1.54 | 1.47 | 1.06 | 2.35 | 3.59 | 1.03 | 11.92 | 0.37 | 4.80 | 11.81 |
| ppm | | | | | | | | | | | | | | | | | | |
| Cr | 16 | 58 | 38 | b.d. | 4 | 1 | 33 | 10 | 15 | 6 | 54 | 39 | 13 | 50 | 656 | 25 | 46 | 19 |
| Ni | 26 | 48 | 34 | 9 | 15 | b.d. | 40 | 22 | 25 | 7 | 37 | 32 | 13 | 54 | 380 | 32 | 30 | 68 |
| V | 82 | 65 | 76 | 4 | 92 | 2 | 70 | 73 | 79 | 26 | 68 | 58 | 25 | 50 | 111 | 11 | 23 | 88 |
| Ba | 554 | 545 | 542 | 207 | 27 | 402 | 190 | 550 | 70 | 580 | 515 | 200 | 920 | 109 | 51 | 241 | 222 | 710 |
| Sr | 101 | 142 | 248 | 82 | 109 | 70 | 142 | 88 | 80 | 44 | 152 | 237 | 391 | 30 | 316 | 26 | 61 | 720 |
| Nb | 3.0 | 3.9 | 3.9 | 5.1 | 6.5 | 5.0 | 2.6 | 3.3 | 3.7 | 3.2 | 3.8 | 3.6 | 2.4 | 4.1 | 3.5 | 1.4 | 2.5 | 7.8 |
| Zr | 90 | 107 | 36 | 226 | 154 | 218 | 97 | 120 | 132 | 113 | 90 | 84 | 97 | 118 | 82 | 91 | 107 | 190 |
| Y | 5.9 | 7.9 | 7.4 | 18 | 16 | 20 | 6.5 | 8.9 | 9.5 | 10 | 7.9 | 7.6 | 4.4 | 10 | 7.8 | 4.6 | 7.4 | 19 |
| Sample # | 78 | 79 | 62 | 64 | 73 | 63a | 63b | 38 | 39 | 40 | 41 | 46 | 47 | 161 | 162 | 163 | 92 | 96 |
| Location | Lower Finlay-son | Lower Finlay-son | Finlay-son | Finlay-son | Finlay-son | Finlay-son | Finlay-son | Finlay-son | Finlay-son | Finlay-son | Finlay-son | Finlay-son | Finlay-son | Upper Lumby | Upper Lumby | Upper Lumby | Upper Lumby | Upper Lumby |
| Lithology | m sst | m sst | m sst | m sst | m-c sst | m sst | m sst | m-c sst | m-c sst | m-c sst | m-c sst | c sst | c sst | m-c sst | m-c sst | m-c sst | m sst | m sst |
| wt% | | | | | | | | | | | | | | | | | | |
| $SiO_2$ | 72.25 | 74.12 | 69.35 | 81.15 | 65.96 | 65.44 | 66.86 | 78.81 | 60.93 | 59.86 | 69.41 | 70.24 | 72.38 | 50.75 | 64.41 | 55.28 | 55.91 | 55.89 |
| $TiO_2$ | 0.37 | 0.36 | 0.35 | 0.43 | 0.60 | 0.65 | 0.58 | 0.39 | 0.65 | 0.34 | 0.40 | 0.26 | 0.54 | 0.87 | 0.31 | 0.64 | 0.52 | 0.61 |
| $Al_2O_3$ | 13.86 | 13.68 | 10.70 | 8.52 | 12.34 | 13.39 | 11.50 | 8.05 | 14.10 | 10.78 | 12.33 | 8.95 | 11.70 | 13.84 | 16.62 | 15.02 | 12.75 | 9.71 |
| $Fe_2O_3t$ | 3.20 | 3.21 | 5.31 | 6.67 | 6.87 | 7.51 | 7.38 | 4.85 | 8.53 | 6.82 | 6.46 | 11.09 | 8.52 | 7.67 | 3.24 | 7.65 | 7.74 | 7.68 |
| MnO | 0.03 | 0.04 | 0.20 | 0.10 | 0.13 | 0.09 | 0.07 | 0.09 | 0.07 | 0.20 | 0.10 | 0.09 | 0.08 | 0.12 | 0.13 | 0.15 | 0.13 | 0.12 |
| MgO | b.d. | b.d. | 0.42 | 2.42 | 1.26 | 3.25 | 2.47 | 1.46 | 1.83 | 3.54 | 2.07 | 2.06 | 1.92 | 2.15 | b.d. | 5.35 | 8.02 | 5.05 |
| CaO | 2.52 | 2.04 | 6.55 | 0.85 | 2.76 | 2.23 | 1.39 | 0.49 | 0.59 | 6.57 | 1.27 | 1.30 | 0.75 | 7.37 | 2.31 | 7.35 | 8.22 | 5.24 |
| $Na_2O$ | 4.17 | 4.41 | 2.20 | 1.46 | 2.25 | 1.88 | 1.46 | 1.69 | 1.05 | 1.57 | 1.10 | 1.47 | 2.25 | 5.52 | 1.48 | 3.70 | 2.88 | 2.09 |
| $K_2O$ | 1.21 | 1.63 | 1.63 | 0.57 | 1.11 | 2.40 | 2.00 | 0.90 | 2.45 | 1.67 | 2.12 | 0.22 | 0.96 | 2.61 | 4.39 | 0.74 | 1.74 | 1.04 |
| $P_2O_5$ | 0.06 | 0.06 | 0.05 | 0.03 | 0.10 | 0.09 | 0.08 | 0.05 | 0.09 | 0.04 | 0.04 | 0.05 | 0.05 | 0.47 | 0.10 | 0.25 | 0.19 | 0.20 |
| $H_2O$ | 0.63 | 0.45 | 2.07 | 2.97 | 2.70 | 2.79 | 2.70 | 0.90 | 3.33 | 1.17 | 2.16 | 3.15 | 2.70 | 1.26 | 1.98 | 1.35 | 0.99 | 2.16 |
| $CO_2$ | 0.22 | 0.37 | 5.10 | 1.17 | 2.68 | 3.08 | 1.43 | 0.37 | 0.84 | 11.18 | 3.92 | 0.95 | 0.51 | 6.23 | 1.76 | 1.10 | 0.59 | 1.83 |
| ppm | | | | | | | | | | | | | | | | | | |
| Cr | 11 | 44 | 67 | 73 | 193 | 153 | 115 | 64 | 133 | 67 | 86 | 65 | 93 | 98 | 29 | 371 | 408 | 394 |
| Ni | 7 | 17 | 34 | 58 | 86 | 78 | 68 | 29 | 66 | 41 | 48 | 57 | 50 | 35 | 6 | 170 | 305 | 224 |
| V | 19 | 18 | 108 | 94 | 105 | 165 | 146 | 83 | 203 | 125 | 142 | 99 | 131 | 120 | 10 | 142 | 119 | 130 |
| Ba | 290 | 577 | 175 | 91 | 220 | 297 | 241 | 147 | 109 | 100 | 201 | 20 | 127 | 720 | 554 | 171 | 259 | 120 |
| Sr | 219 | 171 | 103 | 31 | 80 | 60 | 39 | 44 | 40 | 164 | 80 | 38 | 53 | 1547 | 163 | 205 | 346 | 190 |
| Nb | 5.1 | 5.4 | 3.0 | 3.5 | 5.1 | 4.9 | 4.2 | 3.3 | 4.4 | 2.8 | 3.0 | 1.9 | 4.2 | 7.9 | 2.6 | 6.3 | 5.1 | 5.1 |
| Zr | 90 | 102 | 72 | 61 | 83 | 95 | 89 | 55 | 102 | 74 | 74 | 74 | 88 | 118 | 72 | 134 | 90 | 102 |
| Y | 15 | 17 | 6.1 | 5.0 | 9.5 | 6.1 | 4.8 | 5.2 | 3.1 | 5.9 | 4.3 | 4.0 | 5.4 | 15 | 3.9 | 14 | 11 | 9.2 |

*Note:* b.d.—below detection; m-c sst—medium- to coarse-grained sandstone; m sst—medium-grained sandstone; c sst—coarse-grained sandstone; m-f sst—medium- to fine-grained sandstone; sst—sandstone.

None of the sedimentary rocks analyzed is compositionally supermature. Average $SiO_2$ contents for medium-grained sandstones of the various units range from just 52 wt% for the East Bay member to 82 wt% for the middle Conley Formation (Fig. 9). For comparison, moderately mature, Paleoproterozoic fluvial sandstones of the Matinenda Formation, Huronian Supergroup, average 86 wt% $SiO_2$, and supermature Mesoproterozoic shelf sands of the Baraboo Formation average 98 wt% $SiO_2$. Chemical index of alteration (CIA) values for sandstones of the study area and the Huronian and Baraboo produce similar results (Fig. 9). Most sandstones in the study area are composed of only slightly more chemically weathered debris than fresh tonalite, and the Upper Lumby Lake Group and East Bay member have indexes lower than tonalite due to the significant amount of ultramafic and mafic detritus they, respectively, contain.

## DISCUSSION

### Depositional Environments

Based on the lateral relationships previously described, the sandstones and conglomerates of the 2997 Ma Little Falls Group formed as a clastic apron and derived sediment from the adjacent felsic volcanic center. Sediment transport was via suspension clouds, where both dispersive pressure and turbulence acted to keep the sand in the water column (i.e., Lowe, 1982). This resulted in a succession dominated by grain flows and deposits of high-density turbidity currents. Current reworking of some sediment-gravity deposits, indicated by trough cross-stratification at the tops of some coarse-grained sandstone beds, possibly denotes that water depths were shallow enough to allow wave-

TABLE 2. GEOCHEMISTRY OF SANDSTONES FROM THE WALLACE LAKE AREA

| Sample # | W-9 | W-10 | W-11 | W-12 | W-45 | W-46 | W-48 | W-49 | W-30 | W-31 | W-32 | W-33 | W-34 |
|---|---|---|---|---|---|---|---|---|---|---|---|---|---|
| Location | Lower Conley | Lower Conley | Lower Conley | Lower Conley | Lower Conley | Lower Conley | Lower Conley | Lower Conley | Middle Conley | Middle Conley | Middle Conley | Middle Conley | Middle Conley |
| Lithology | sst + tuff | m-c sst | m-c sst | m-c sst | m-c sst | m-c sst | m-c sst | m-c sst | m-c sst | m-c sst | m-c sst | sst + tuff | m-c sst |
| wt% | | | | | | | | | | | | | |
| $SiO_2$ | 59.39 | 73.77 | 79.20 | 70.21 | 83.46 | 73.96 | 72.00 | 72.48 | 81.39 | 79.12 | 88.14 | 53.06 | 87.83 |
| $TiO_2$ | 0.23 | 0.25 | 0.34 | 0.33 | 0.30 | 0.46 | 0.28 | 0.37 | 0.26 | 0.41 | 0.19 | 1.30 | 0.19 |
| $Al_2O_3$ | 9.46 | 13.61 | 10.05 | 12.61 | 7.06 | 12.07 | 14.75 | 13.83 | 6.95 | 8.51 | 5.84 | 12.90 | 4.55 |
| $Fe_2O_3t$ | 7.62 | 2.10 | 2.64 | 5.26 | 2.58 | 3.88 | 2.38 | 2.84 | 2.65 | 2.71 | 1.28 | 13.13 | 2.16 |
| MnO | 0.13 | 0.02 | 0.04 | 0.09 | 0.03 | 0.03 | 0.03 | 0.05 | 0.05 | 0.05 | 0.02 | 0.18 | 0.03 |
| MgO | 11.78 | 0.71 | 0.74 | 1.54 | 1.10 | 1.88 | 1.37 | 1.31 | 1.47 | 1.66 | 0.74 | 5.58 | 0.91 |
| CaO | 5.90 | 2.03 | 1.63 | 4.33 | 2.36 | 1.80 | 1.21 | 2.33 | 2.12 | 1.83 | 0.30 | 5.43 | 1.26 |
| $Na_2O$ | 2.92 | 5.66 | 3.69 | 2.91 | 1.70 | 2.95 | 4.75 | 4.62 | 1.76 | 1.94 | 0.37 | 1.95 | 0.50 |
| $K_2O$ | 0.21 | 1.01 | 1.02 | 1.09 | 0.12 | 1.28 | 1.79 | 1.17 | 0.86 | 1.58 | 1.63 | 0.23 | 0.92 |
| $P_2O_5$ | 0.02 | 0.08 | 0.07 | 0.08 | 0.03 | 0.06 | 0.08 | 0.07 | 0.02 | 0.02 | 0.02 | 0.10 | 0.02 |
| $H_2O$ | 2.34 | 0.36 | 0.45 | 1.17 | 0.63 | 1.26 | 1.08 | 0.63 | 1.08 | 1.17 | 0.99 | 4.32 | 0.81 |
| $CO_2$ | b.d. | 0.40 | 0.11 | 0.40 | 0.62 | 0.37 | 0.29 | 0.29 | 1.39 | 1.02 | 0.44 | 1.61 | 0.81 |
| ppm | | | | | | | | | | | | | |
| Cr | 984 | 5.5 | 12 | 10 | 12 | 14 | 5.3 | 12 | 6.1 | 11 | 8.7 | 122 | 5.8 |
| Ni | 300 | 1.0 | 8.3 | 17 | 13 | 15 | 3.1 | 12 | 11 | 16 | 12 | 92 | 12 |
| V | 166 | 21 | 34 | 37 | 29 | 47 | 22 | 44 | 24 | 37 | 23 | 359 | 18 |
| Ba | 36 | 204 | 172 | 223 | 11 | 257 | 245 | 189 | 136 | 230 | 142 | 110 | 120 |
| Sr | 90 | 230 | 95 | 137 | 74 | 118 | 138 | 122 | 77 | 48 | 20 | 276 | 28 |
| Nb | 6.6 | 4.5 | 5.5 | 6.7 | 4.5 | 5.2 | 3.9 | 4.5 | 7.2 | 8.5 | 6.3 | 14 | 7.2 |
| Zr | 31 | 76 | 61 | 64 | 44 | 68 | 85 | 67 | 39 | 52 | 37 | 74 | 30 |
| Y | 7.1 | 3.7 | 7.0 | 12 | 5.0 | 6.3 | 3.9 | 6.6 | 4.4 | 4.5 | 2.0 | 20 | 2.0 |
| Sample # | W-36 | W-43 | W-44 | W-40 | W-41 | W-42 | W-56 | W-1 | W-2 | W-6 | W-52 | W-54 | W-55 |
| Location | Middle Conley | Middle Conley | Middle Conley | Upper Conley | Upper Conley | Upper Conley | Upper Conley | East Bay | East Bay | East Bay | East Bay | East Bay | East Bay |
| Lithology | m-c sst | m-c sst | m-c sst | m-c sst | m-c sst | m-c sst | m-c sst | f-m sst | f-m sst | f-m sst | f-m sst | f-m sst | f-m sst |
| wt% | | | | | | | | | | | | | |
| $SiO_2$ | 88.41 | 70.45 | 69.76 | 91.15 | 68.20 | 71.55 | 89.69 | 46.34 | 44.14 | 46.78 | 59.51 | 56.78 | 50.72 |
| $TiO_2$ | 0.23 | 0.30 | 0.35 | 0.13 | 0.25 | 0.25 | 0.20 | 1.09 | 0.73 | 0.88 | 0.71 | 1.73 | 1.60 |
| $Al_2O_3$ | 4.59 | 13.09 | 13.74 | 3.33 | 14.08 | 13.75 | 4.22 | 13.9 | 11.32 | 13.22 | 15.96 | 13.53 | 15.82 |
| $Fe_2O_3t$ | 1.85 | 2.80 | 3.03 | 1.40 | 2.69 | 2.71 | 1.79 | 14.74 | 11.69 | 13.18 | 6.20 | 13.89 | 14.22 |
| MnO | 0.02 | 0.05 | 0.04 | 0.02 | 0.03 | 0.03 | 0.02 | 0.22 | 0.22 | 0.19 | 0.10 | 0.21 | 0.22 |
| MgO | 1.03 | 0.93 | 1.00 | 0.93 | 1.02 | 1.01 | 0.71 | 5.88 | 8.52 | 7.27 | 5.51 | 2.96 | 5.22 |
| CaO | 0.93 | 2.66 | 2.52 | 0.56 | 4.45 | 2.36 | 0.71 | 7.45 | 9.92 | 5.94 | 5.20 | 5.32 | 1.62 |
| $Na_2O$ | 0.59 | 2.55 | 2.98 | 0.92 | 2.69 | 3.86 | 0.18 | 2.41 | 1.36 | 1.05 | 1.20 | 2.47 | 3.72 |
| $K_2O$ | 1.05 | 3.43 | 3.22 | 0.74 | 2.67 | 1.79 | 0.77 | 0.69 | 0.20 | 2.42 | 1.96 | 1.43 | 0.56 |
| $P_2O_5$ | 0.01 | 0.09 | 0.12 | 0.02 | 0.12 | 0.11 | 0.02 | 0.10 | 0.40 | 0.07 | 0.06 | 0.17 | 0.13 |
| $H_2O$ | 0.72 | 1.53 | 1.44 | 0.81 | 0.72 | 1.26 | 0.99 | 4.32 | 4.86 | 5.04 | 2.70 | 1.26 | 5.49 |
| $CO_2$ | 0.55 | 2.13 | 1.80 | 0.81 | 3.08 | 1.32 | 0.70 | 2.86 | 6.64 | 3.96 | 0.88 | 0.22 | 0.66 |
| ppm | | | | | | | | | | | | | |
| Cr | 6.0 | 3.5 | 7.7 | 6.2 | 6.3 | 5.8 | 7.6 | 107 | 656 | 148 | 335 | 6.4 | 152 |
| Ni | 9.7 | 4.1 | 6.5 | 11 | 5.2 | 6.1 | 10 | 117 | 150 | 116 | 106 | 49 | 170 |
| V | 20 | 29 | 40 | 16 | 41 | 37 | 19 | 336 | 201 | 288 | 201 | 428 | 361 |
| Ba | 128 | 370 | 363 | 83 | 279 | 343 | 68 | 163 | 25 | 66 | 270 | 266 | 83 |
| Sr | 21 | 131 | 147 | 13 | 248 | 246 | 24 | 138 | 106 | 22 | 46 | 168 | 33 |
| Nb | 4.1 | 6.2 | 5.6 | 3.0 | 6.3 | 5.6 | 1.8 | 12 | 11 | 9.2 | 6.1 | 12 | 8.7 |
| Zr | 39 | 110 | 52 | 38 | 65 | 60 | 48 | 36 | 91 | 40 | 40 | 110 | 45 |
| Y | 4.1 | 6.7 | 4.9 | 2.0 | 4.6 | 4.6 | 2.9 | 23 | 18 | 17 | 13 | 30 | 16 |

*Note:* b.d.—below detection; sst—sandstone; m-c sst—medium- to coarse-grained sandstone; f-m sst—fine- to medium-grained sandstone.

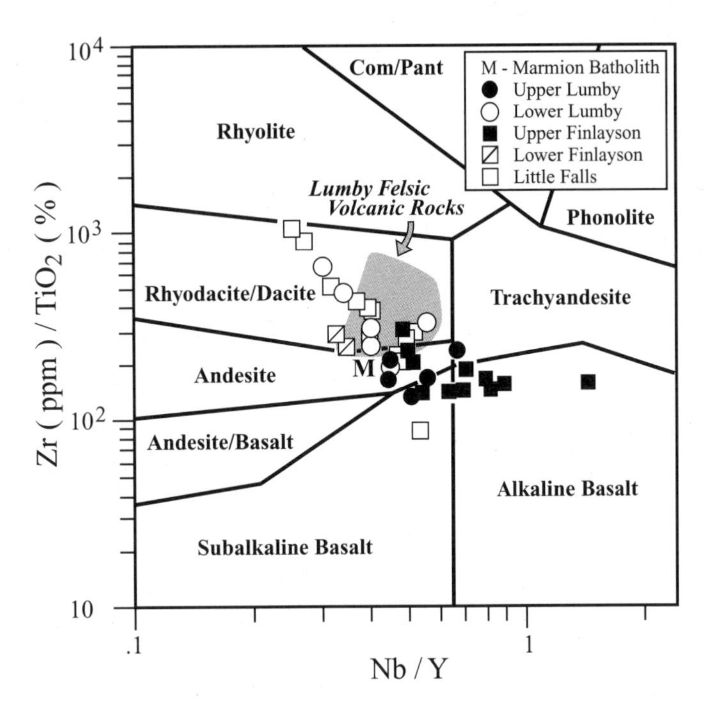


Figure 5. Ratio plot (Winchester and Floyd, 1976) depicting mafic-felsic trends ($Zr/TiO_2$) versus alkalinity (Nb/Y). One end-member composition of both the Lower Lumby Lake Group–Little Falls Group and the Upper Lumby Lake Group–Finlayson Group is represented by the geochemistry of the Marmion Batholith (M on the diagram, average analyses of 3.0 Ga tonalite; Stone et al., 1992). The former groups are composed of pyroclastic-sourced sandstones and outline a differentiation trend extending from the Marmion composition to more felsic geochemistries. The latter Groups (Finlayson–Upper Lumby) appear to represent mixtures of Marmion composition and an alkaline basalt component. The presence of significant amounts of both tonalite and basaltic clasts in the conglomerates supports this interpretation. The field for Lumby felsic volcanic rocks is from Tomlinson (2004, personal commun.).

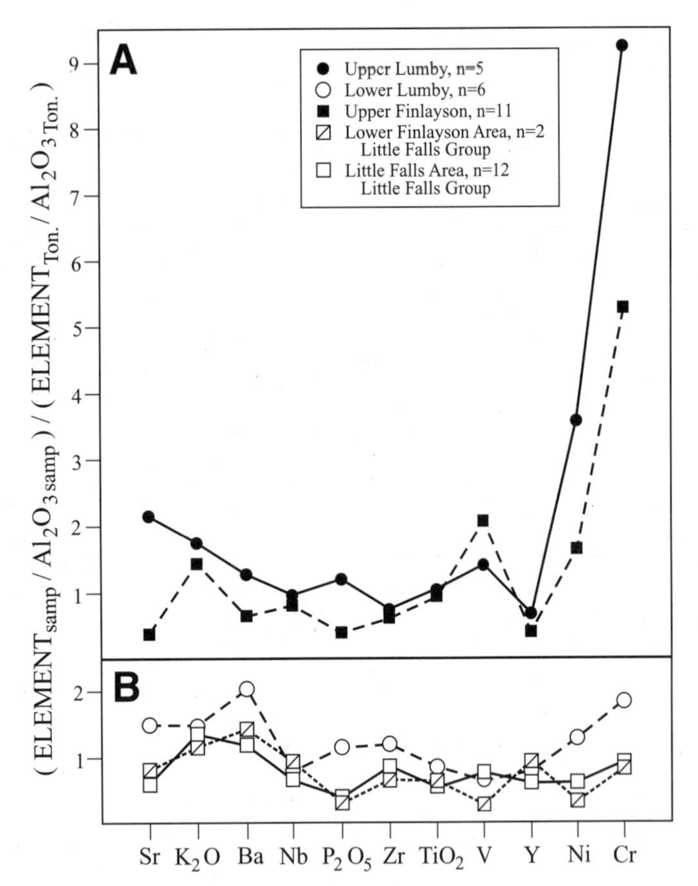


Figure 6. Normalized extended-ratio plots comparing the geochemistries of the sedimentary units in the Atikokan area with tonalites of the Marmion Batholith (Ton.). Division by $Al_2O_3$ corrects for quartz enrichment in the sandstones, depressing values for the elements plotted (for a discussion of this technique, see Fralick, 2003). The element/$Al_2O_3$ ratio for a group of sandstone samples will be similar to the average, weighted composition of the source area given chemical immobility of the element and a similar hydrodynamic behavior of its major mineral phases and those of Al (Fralick and Kronberg, 1997; Fralick, 2003). A ratio of 1 for an element results from the same element/$Al_2O_3$ ratios for both the sandstone samples and the tonalite. Ratios >1 represent enrichment of that element in the sandstones relative to the tonalite, and those <1 reflect depletion. The geochemistries of the Lower Lumby Lake and Little Falls Groups are similar to the tonalite. The geochemistries of the Upper Lumby Lake and Finlayson Lake Groups are similar to the other three sets but with enrichment in V, Ni, and Cr. This is consistent with a mixed tonalite–mafic to ultramafic source for these sedimentary units.

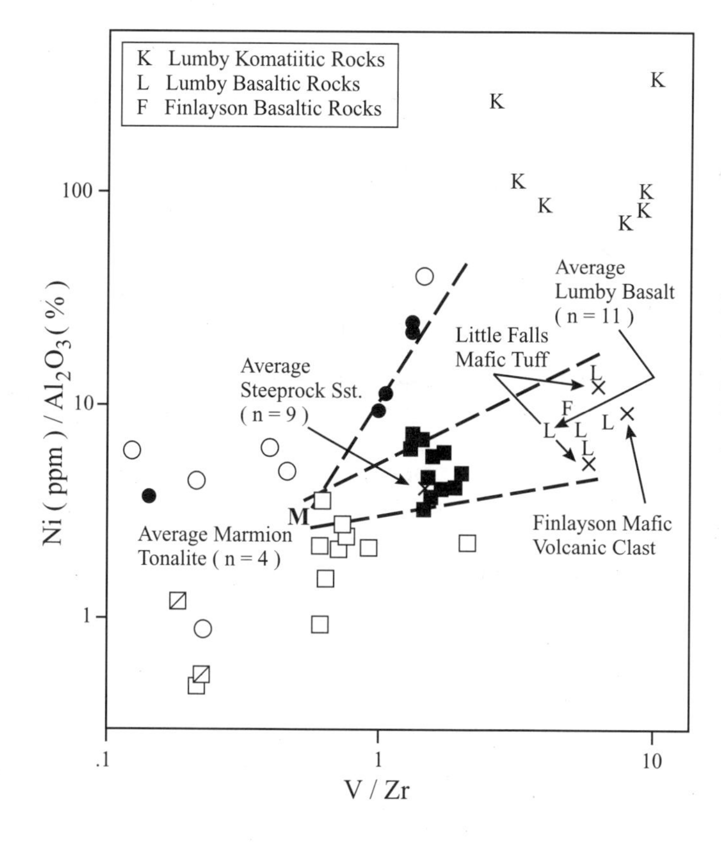


Figure 7. $Ni/Al_2O_3$ and V/Zr ratios are useful for separating mafic and ultramafic components in the sediment. The symbols are the same as in Figure 5. The Upper Finlayson samples fall in an array between Marmion (M) compositions and Finlayson and Lumby mafic volcanic samples. The majority of upper Lumby sandstone samples fall on a mixing line between Marmion compositions and Lumby Al-depleted komatiites (data from Hollings and Wyman, 1999). Individual mafic volcanic rock analyses are from Tomlinson et al. (1999), and the average Lumby basalt is from Hollings and Wyman (1999).

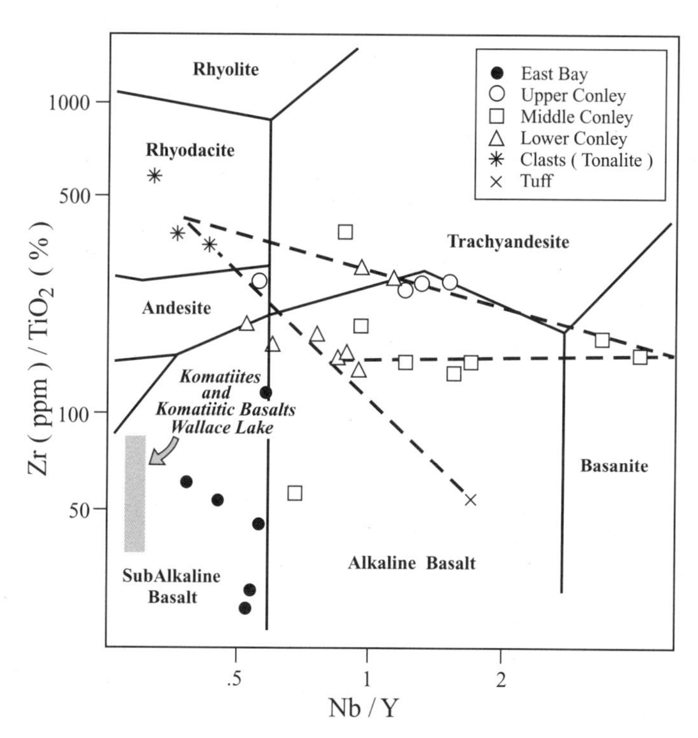


Figure 8. Plot showing trends of samples from the Wallace Lake area, similar to Figure 4. The middle Conley sample, which plots low in the alkaline basalt field, near the East Bay samples, had abundant chloritic tuff mixed with the sand. The diagram is discussed in the text.

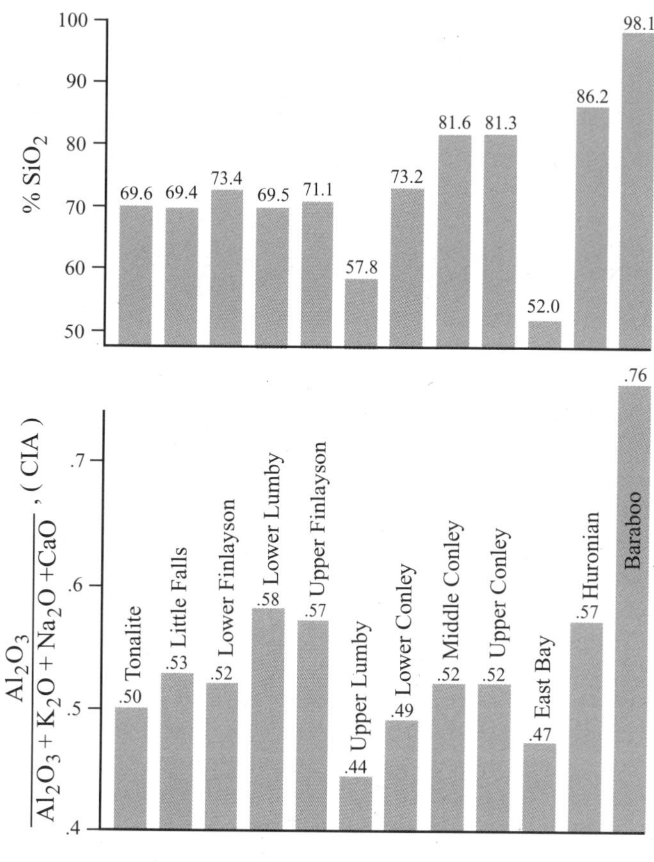


Figure 9. Average wt% $SiO_2$ and chemical index of alteration (CIA) for medium-grained sandstones from the sedimentary units in the study areas and the Paleoproterozoic lower Huronian Supergroup and Baraboo Quartzite of the southern Canadian Shield. Comparison of the study area units with the moderately mature Huronian and mature Baraboo highlights their mostly immature nature. This contradicts previous assertions on the maturity of Mesoarchean, Superior Province sandstones (Thurston and Chivers, 1990). The use of sandstones rather than shales suppresses all the CIA values but does not interfere with comparisons to samples of other units with similar grain sizes.

and/or tide-generated flow to rework the bottom in some locations. Eruptions of mafic ash occurred during this time period and blanketed the bottom with this debris. Substantial incorporation of this mafic material into some sediment flows denotes that it was accumulating in areas that were actively slumping.

The Lower Lumby Lake Group, though almost 100 m.y. younger, is similar to the Little Falls Group. It forms a clastic apron that is laterally transitional to pyroclastic rocks, which form a felsic center. Clastic transport was via subaqueous, sand-rich dispersion clouds that deposited turbidites. As deposition waned, the upper portion of the succession became dominated by finer-grained siliciclastic and carbonate units.

As its lowest unit, the capping sedimentary assemblage in the Finlayson greenstone belt has a starved succession of chemical sediments deposited after cessation of mafic volcanism. The first clastics delivered to the area were clays and silts transported by low-density turbidity flows. These probably represent prodelta debris (e.g., Coleman and Prior, 1982), although sediment transport via storm currents is possible (Aigner and Reineck, 1982; Allen, 1982). Shoreline progradation caused the coarsening-upward succession (Fig. 3) to form, culminating in density current deposition at depths affected by wave-induced scour (similar to deposits described by Pudsey, 1984; Hwang and Chough, 1990; Martinsen, 1990). Delta-top sands and gravels cap the sediment pile and show the effects of successive episodes of outbuilding and abandonment and subsequent subsidence-induced flooding (e.g., Marzo and Anadon, 1988). The sedimentary succession that caps the volcanic rocks in the northern Finlayson Lake greenstone belt begins with a localized thick assemblage of epiclastic conglomerates deposited by cohesive debris flows (i.e., Nemec and Steel, 1984) in an otherwise sediment-starved, subaqueous setting. This is overlain by carbon-rich muds sedimented by rainout from the water column and low-density turbidity current deposits. Polymictic conglomerates patchily overlying the finer-grained deposits were laid down by subaqueous mass flows and are buried by a coarsening-upward assemblage similar to the one present in southern Finlayson Lake.

The sedimentary succession overlying the komatiitic and mafic volcanic rocks in the Lumby Lake greenstone belt begins with a thin unit of iron formation and carbonaceous slate. In the west, this is overlain by low-density turbidity current deposits interlayered with subaqueous debris flows. Here, the succession is capped by a thick carbonate unit that recrystallization and poor outcrop has rendered indecipherable. The overlying assemblage of various types of iron formation and chert interlayered with very carbonaceous, sulfitic slates represents a possibly extended period of sediment starvation in a deep-water setting. The conglomerates and thick carbonate are only present in the western portion of the belt, reflecting the existence of a more source-proximal, possibly shallower, environment here.

The Conley Formation is interpreted to reflect a braid delta depositional system in a transgressive setting. The lower conglomeratic sediments were deposited in a braided fluvial channel network (e.g., Miall, 1977). The lack of macroforms (i.e., Miall, 1985, 1994) and the small-scale, wispy nature of lenses imply shallow flow in very high width-to-depth ratio channels. This sequence is gradationally succeeded upward by distributary mouth bars with higher mud contents than the fluvial system and abundant mud flasers and drapes (Fig. 4B) (e.g., Vos, 1981; Fralick and Miall, 1989). The distal bar environment is dominated by the presence of thin-bedded, graded sandstone and clay-rich sandstone couplets that have a varve-like appearance and may denote seasonal flow. The lack of traction-produced bed forms indicates flow separation from the bottom (i.e., Wright, 1977). Turbidites and grain flows dominate the basin floor offshore from the bar complexes. These were probably initiated by delta-front slumping, similar to many well-documented modern examples (Coleman et al., 1974; Coleman and Prior, 1982). Interbedding of distributary mouth bar deposits with the turbidites indicates that sediment deposition at times was able to reverse the overall transgressive nature of the system. Reworking of these bars into shoals may explain the increased maturity of some sandstone layers (Percival, 1992). The overlying thinly bedded sandstones and cherts of the Conley Formation imply a reduction in sediment supply. This trend is continued in the uppermost Conley with the stromatolitic carbonates and the East Bay member. Though the photosynthesizing capabilities of the microorganism that built these mounds are not known, it is certainly a distinct possibility that the carbonate was limited to areas of good light penetration to bottom, and the iron formation accumulated in deeper zones. Chlorite-rich sandstones overlying the chemical sedimentary rocks denote a return to clastic deposition via low-density turbidity currents.

## Evolution of the Depositional Systems

The combination of interpretation of depositional environments with sediment geochemistry and precise U-Pb geochronology of volcanic units enables reconstruction of paleogeography for this area spanning a 230 m.y. period from the Mesoarchean to Neoarchean. A thin unit of felsic volcanic rocks in the dominantly tholeiitic basalt sequence at Lumby Lake is the oldest dated unit (3014 Ma) in the study areas. Thick sequences (>800 m) of mafic tholeiitic volcanic flow rocks older than 2997 Ma are also present in the Little Falls belt. This implies that a subaqueous lava plateau existed in the southern area by 3.0 Ga (Fig. 10). At 3000 Ma, tonalitic magmas intruded the tholeiitic lava plateau of the southern study area. In both the Lumby Lake and Little Falls areas, Marmion magmas erupted to the surface, resulting in felsic volcanism and deposition of clastic aprons surrounding the volcanic centers (Little Falls Group; Fig. 11). Ongoing mafic volcanism throughout this time interval resulted in ash layers from this source interlayered with the felsic volcanic debris. From 3000 to 2920 Ma, mafic volcanism continued to thicken the volcanic pile in the southern area. Periodic bursts of felsic volcanism linked to tonalitic intrusions also occurred (Fig. 12). This was the time of major sedimentation in the northern area (Wallace Lake), where a fluvial-deltaic system fed off a hinterland, with rocks of rhyodacitic and alkaline mafic composition, and built extensive clastic

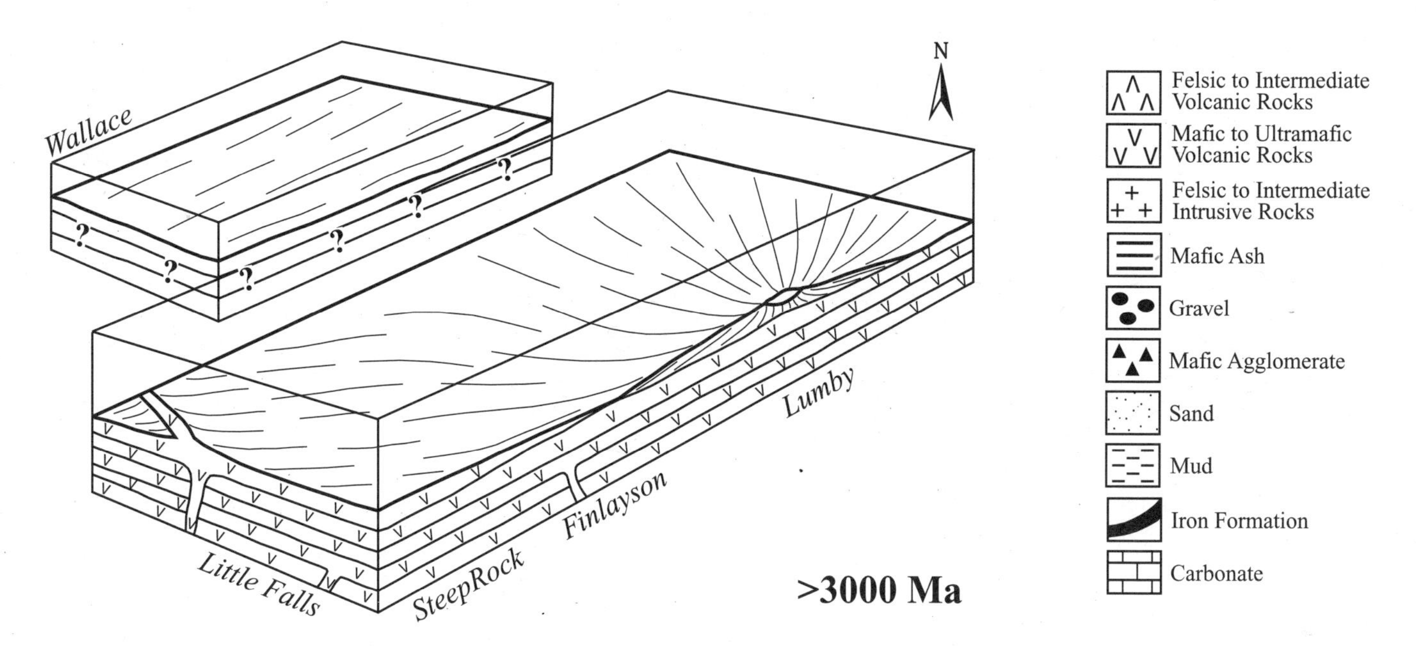


Figure 10. Mafic volcanism in the Steep Rock–Lumby Lake area prior to 3.0 Ga. No record is known to exist for the Wallace Lake area prior to 3.0 Ga.

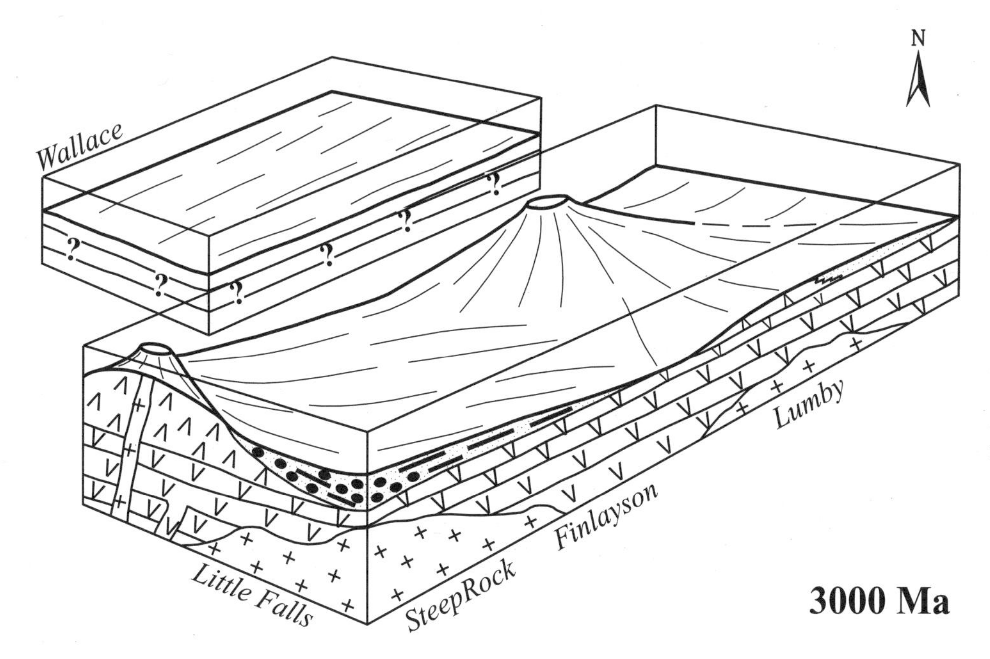


Figure 11. At 3000 Ma, tonalitic magmas intruded into the Steep Rock–Lumby Lake ocean plateau, and resultant surface volcanism built positive relief features from which sediment aprons composed of eroded felsic volcanic debris spread out over the plateau. Synchronous intrusives are also present in the Wallace Lake area, although subsequent tectonic activity has obscured original relationships.

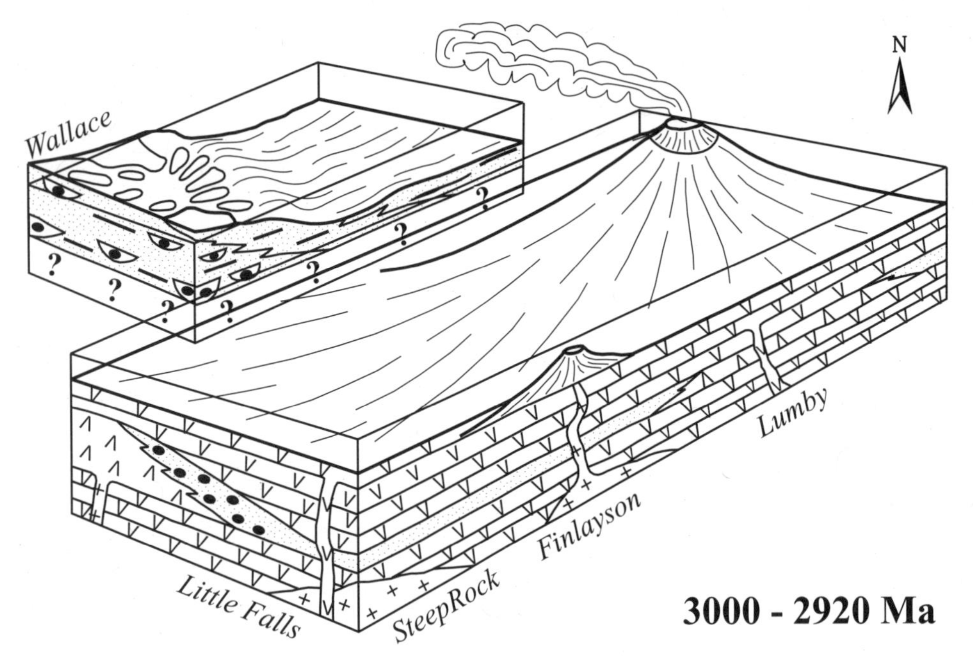


Figure 12. The Steep Rock–Lumby Lake oceanic plateau continued to grow during 3000–2920 Ma. Sporadic tonalitic intrusion fed minor intermediate to felsic volcanic events, but the dominant volcanism erupted tholeiitic basalt. Clastic deposition began in the Wallace Lake area. Braided streams delivered sediment to delta complexes from where it slumped into deeper water. Clast compositions, detrital zircon U-Pb geochronology, and sediment geochemistry indicate the sediment was derived from 3.0 Ga tonalites and rhyolites, mafic to ultramafic ashes, and an alkaline mafic rock type. An upward subareal to subaqueous transition reflects the inability of sedimentation to keep pace with relative sea-level rise. This culminated in carbonate and iron formation accumulation and the deposition of thin turbidites entirely sourced by mafic volcanism, which then commenced in this area.

deposits (Fig. 12). Mafic volcanism was active somewhere in the northern region leading to abundant ash layers in the sandstones. However, the area was subsiding relative to sea level more rapidly than sedimentation could keep up, and the transgression eventually led to sediment starvation, possibly due to submergence of the local source area. This resulted in replacement of clastic deposition by carbonates, chert, iron formation, and eventually a new, deep-water clastic influx of sediment derived from mafic volcanism. This was the harbinger of the commencement of mafic to ultramafic volcanism in the Wallace Lake area.

Throughout the next 100 m.y. mafic volcanism sporadically continued in the southern area with felsic volcanic pulses, such as the one that erupted the 2898 Ma rhyodacites in the Lumby Lake belt and supplied the clastic apron forming the Lower Lumby Lake Group. At ca. 2828 Ma, komatiitic volcanism commenced in the Lumby Lake belt. With the cessation of volcanism throughout the southern area, minor block faulting, possibly triggered by adjustment to thermal decay in the region, created localized depocenters for epiclastic sediments in the northern Finlayson area and possibly upraised the tonalites of the Steep Rock area (Fig. 13). These tonalities and mafic volcanic country rocks underwent erosion, and sediment was channeled through paleovalleys to the shoreline where deltaic deposits accumulated in the shallow areas and mass flows off the deltas dominated the deeper regions. Subsidence of the area throughout this time period led to submergence of abandoned portions of the delta-top. As subsidence and erosion lowered the source area to base level, the bedrock channel system backfilled, forming the basal Steep Rock clastics of the Wagita Formation (Wilks and Nisbet, 1988), and starved the now largely submerged Finlayson deltaic complexes of sediment. Continued subsidence led to the development of carbonates in the shallower areas, which was succeeded by iron formation as the area continued to deepen (Fig. 14). Sediments in the Lumby Lake area had a different source than the Steep Rock and southern Finlayson deposits. Upraised volcanics, including komatiites, provided the sediment to the Upper Lumby Lake Group via sediment-gravity flows. This clastic wedge rapidly terminates to the east, highlighting the limited nature of the source area and depositional system. This sub-basin also became starved of clastic detritus, and carbonates were deposited in the shallower areas, with iron formation in the deeper regions (Fig. 14). Continuing subsidence caused carbonate deposition to be succeeded by accumulation of iron formation throughout the entire region.

## IMPLICATIONS

Previously, the sedimentary rocks in the Steep Rock–Little Falls–Finlayson–Lumby Lake areas have been ascribed to a rift setting (Wilks and Nisbet, 1988; Thurston and Chivers, 1990; Tomlinson et al., 1996). The major problem with this interpretation is that some of the volcanic units and associated sedimentary aprons are as old as the supposedly rifted tonalitic basement. This still allows the younger, upper sedimentary units to be rift-related. However, recent investigations have indicated that the mafic to komatiitic volcanic pile associated with the sedimentary rocks was erupted in an oceanic-plateau setting (Wyman and Hollings, 1998; Hollings et al., 1999; Hollings and Wyman, 1999).

Modern oceanic plateaus consist of thick piles of mafic, plume-derived, volcanic rocks overlain by sedimentary sequences that may exceed 1 km in thickness, i.e., the Ontong Java Plateau (Furumoto et al., 1976) and the Kerguelen Plateau (Fritsch et

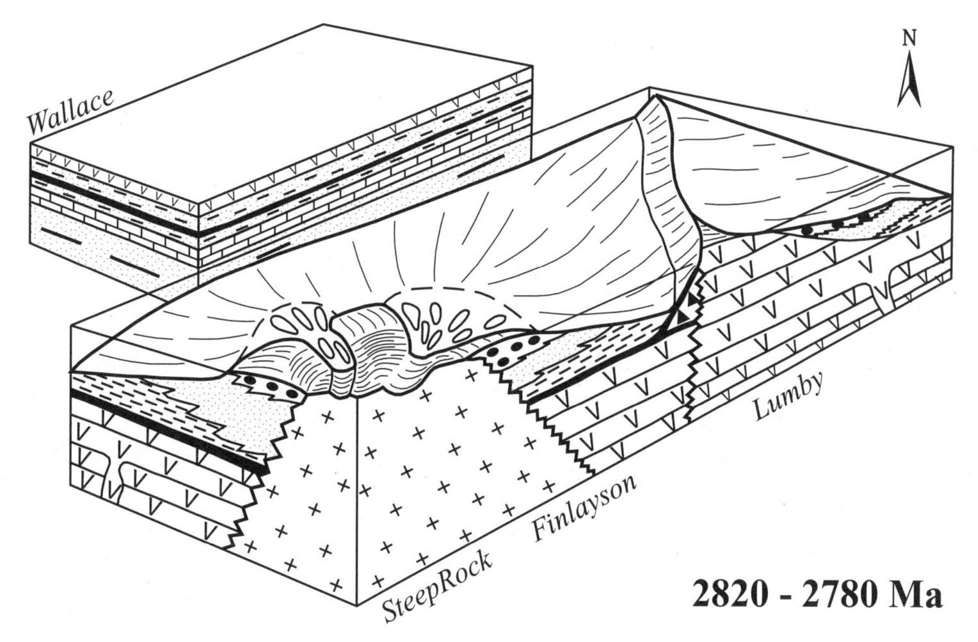


Figure 13. Volcanism ceases on the Steep Rock–Lumby Lake oceanic plateau between 2820 and 2780 Ma, and the major sedimentary assemblage is deposited. Localized relief produces slope failures and resultant agglomerates in the north Finlayson area. These are overlain by a deltaic clastic wedge prograding from the incised channels of the Steep Rock area. Sediments in the Lumby Lake area are mainly derived from mass flows of lithologies in the upper volcanic pile.

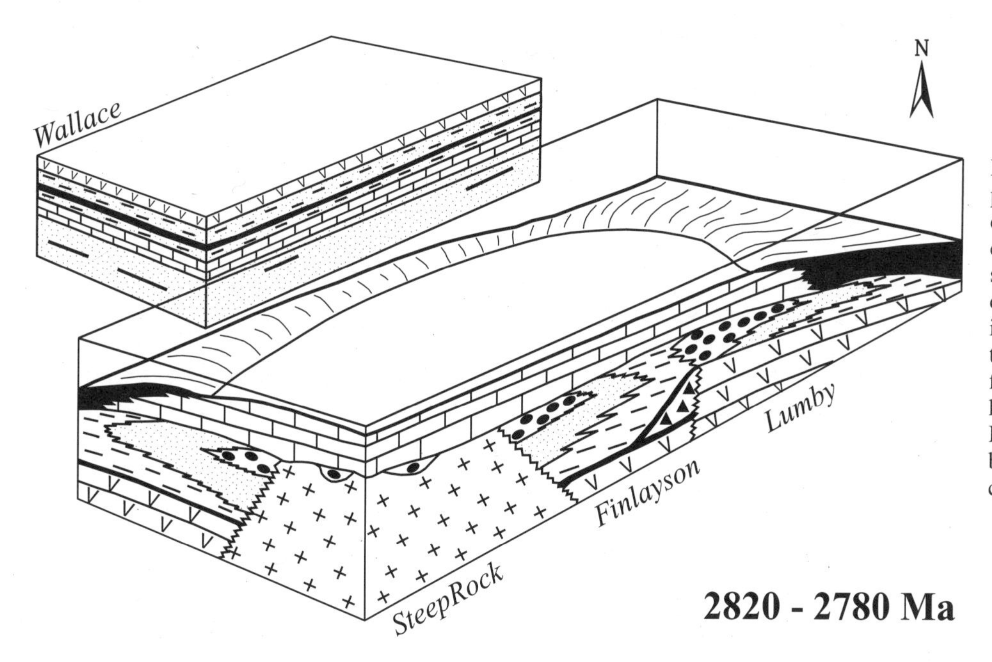


Figure 14. The siliciclastic system depicted in Fig. 13 is succeeded by chemical/biochemical deposits as transgression causes starvation. With flooding of the source terrain, stromatolitic carbonates developed in the shallower areas, and iron formation, which had dominated the deeper regions to the north, migrated first over the adjacent clastics, then blanketed the carbonate assemblage as well. Iron formation accumulation was ended by a pulse of komatiitic volcanism at ca. 2780 Ma.

al., 1992). After volcanism ceases, the sediment begins to accumulate, commonly in subareal to shallow-water environments (Schlich et al., 1989; Coffin, 1992), which are gradually drowned as thermal decay leads to contraction of the crust (Coffin, 1992). Sediment sources are local, as the oceanic plateau is isolated from exotic areas.

The Kerguelen Plateau represents a modern analogue to the sedimentary successions studied here. This plateau was built off Antarctica by extensive volcanic outpourings at ca. 120–110 Ma (Schlich et al., 1989), and it underwent at least three extensional phases with localized graben development (Fritsch et al., 1992; Munschy et al., 1992a), similar to the graben phase at the termination of volcanism in the Steep Rock–Lumby area. A core recovered from drilling indicates that up to 500,000 km$^2$ of the southern plateau was subaerially exposed after cessation of volcanism, and some of this probably remained above sea level for 40 m.y., with a resultant 400–8000 m of erosion (Coffin et al., 1990; Coffin, 1992). Similar erosive stripping is necessary for the exposure of the Marmion Batholith in the Steep Rock area. The basal fluvial to coastal sediments of the Kerguelen Plateau were derived by erosion of basalt (Holmes, 1992), in particular, tholeiitic and alkaline basalts (Schlich et al., 1989). Similarly, basalts and mafic alkaline rocks formed a major sediment source for the fluvial-deltaic deposits in the Finlayson and Wallace Lake areas. At some sites on the Kerguelen Plateau, additional Cr-rich sediment was supplied by subareal erosion of ultramafic rocks and transport by mass-flow processes (Bitschene et al., 1992), analogous to the Cr- and Ni-rich mass-flow sandstones derived from erosion of komatiites in the Lumby Lake area. In addition to glauconitic sediment, the Kerguelen Plateau contains algal limestones. These appear to have been common in the shallow-marine environments (Barron et al., 1989), which developed as subareal sedimentation failed to keep pace with subsidence (Coffin, 1992). The clastic successions at Steep Rock, Lumby Lake, and Wallace Lake are all capped by stromatolitic limestones deposited after base-level rise flooded source areas. As load- and thermal decay–induced subsidence continued to outpace sedimentation on the Kerguelen Plateau, the calcareous and glauconitic shallow-water sequences were succeeded by marine chalk and marl. Deep-water cherts and iron formation record the same processes on the Mesoarchean Steep Rock–Lumby Lake and Uchi (Wallace Lake) Plateaus. The western Kerguelen Plateau remained upraised much longer than the eastern part (Watkins et al., 1992). The east-west lithofacies transitions in the Lumby Lake greenstone belt reflect a similar water depth gradient. The most recent development of the Kerguelen Plateau system was 43 Ma (Munschy et al., 1992b) rejuvenation of the northern portion by a heating event accompanied by renewed ocean island volcanism (Storey et al., 1988), which built a subareal platform. Pulses of plume events have been inferred for the Steep Rock–Lumby area (Tomlinson et al., 1999).

The striking analogies between the upper units at Finlayson and Lumby Lake and the sediments of the Kerguelen Plateau reinforce interpretations based on igneous petrochemistry (Wyman and Hollings, 1998; Hollings and Wyman, 1999; Hollings et al., 1999) that these old assemblages are oceanic plateaus. It is unlikely that the Wallace Lake and Steep Rock–Lumby Lake areas formed one plateau as their geologic histories are quite different (Tomlinson et al., 2003). Both were intruded by tonalite at 3000 Ma, but the Wallace Lake area experienced early subsidence-induced sediment accumulation and komatiitic volcanism, whereas the Steep Rock–Lumby Lake area had its komatiitic eruptions and major sedimentation phase ~150 m.y. later. The two areas may have originated as one plateau that rifted and separated at ca. 2950 Ma, in a manner similar to the extensional events of the Kerguelen Plateau (Fritsch et al., 1992; Munschy et al., 1992a, 1992b), or the two areas may have evolved separately as concluded by Hollings et al. (1999) and Tomlinson et al. (2001, 2003).

There is a major difference between the Mesoarchean plateaus and modern plateaus; the old assemblages have long histories of intrusive and extrusive intermediate to felsic igneous activity synchronous with mafic volcanism. This has been ascribed to plume-arc interaction by Wyman and Hollings (1998), who argued that a mantle plume component was carried along a spreading center to impart the plume signature into the belt, and the late Al-depleted komatiite signature represents the impact of the plume itself. Although this model is appealing, it has several problems. Plume-arc interaction should not last for hundreds of millions of years, although modern hotspots such as Tristan da Cunha, Galapagos, and Heard Island preserve evidence of volcanism over 80–130 m.y. (Storey, 1995; Bercovici and Mahoney, 1994; Class et al., 1996). With higher heat flow in the Archean, the duration would be expected to be even shorter than today. Abbott (1996), who originally speculated on plume-arc interaction as a source of greenstone belts, outlined a history of continental-arc development, tonalite-trondhjemite-granodiorite (TTG) intrusion in the accretionary wedge with melt generation from thickened oceanic crust, and subsequent eruption of mid-ocean-ridge basalt (MORB) and oceanic-plateau basalt (OPB) in localized rifts in the forearc area. This is not consistent with the geology of the two areas studied, where there is no evidence, other than the presence of TTGs, of Mesoarchean subduction, such as an accretionary complex or orogenic activity. An alternative explanation for the juxtaposition of tonalites and plume-related basalts can be developed by applying Smithies (2000) contention that Mesoarchean TTG series developed by melting of hydrous basaltic rocks deep in thickened portions of the crust (also see the review of TTG production by basal, Archean plateau melting in Zegers, 2004). This process of tonalitic melt generation has occurred in the recent past (Atherton and Petford, 1993) and is consistent with interpretations by Tarney and Jones (1994), who contended that Archean TTGs originated by partial melting of plume-related basalt, not MORB. If modern examples are indicative of the Mesoarchean, the root zones of ocean plateaus would have provided these overthickened crustal sites (the crust formed by the Ontong Java Plateau is 35–42 km thick; Furumoto et al., 1976; Hussong et al., 1979) where partial melting could occur. This model makes the generation of tonalites an integral phase in the development of thick Mesoarchean plateaus and does not necessitate the synchronous eruption of plume- and arc-derived magmas.

The long-lasting nature of the Mesoarchean plume-generated volcanism creates problems for models of early Earth that promote abundant ridges and/or fast spreading to dissipate Earth's greater heat flow. Obviously, if these fast-spreading or multiple-ridge models are accurate, the life span of any Mesoarchean ocean plateau would be short. The existence of the Steep Rock–Lumby Lake Plateau for over 200 m.y. without showing any evidence of collision-induced tectonism during this period may indicate that substantial heat loss was accomplished through Mesoarchean plumes. It is not known whether the pulses of plume magmatism were the result of the plate moving over different plumes or a linked lithosphere-asthenosphere system affected by a pulsing single plume. More detailed work on Mesoarchean sequences, including ample use of precise zircon geochronology, is necessary before some of the fundamental questions on this portion of the history of Earth can be answered.

## ACKNOWLEDGMENTS

An early draft of this paper benefited from constructive reviews by K.Y. Tomlinson, Phil Thurston, and Chris Fedo. Word processing was conducted by K. Carey, and figures were drafted by S. Spivak. This research was supported by the Natural Science and Engineering Research Council of Canada and Lithoprobe Western Superior Transect.

## REFERENCES CITED

Abbott, D.H., 1996, Plumes and hotspots as sources of greenstone belts: Lithos, v. 37, p. 113–127, doi: 10.1016/0024-4937(95)00032-1.

Aigner, T., and Reineck, H.-E., 1982, Proximity trends in modern storm sands from the Helegoland Bight (North Sea) and their implications for basin analysis: Senckenbergiana Maritime, v. 14, p. 183–215.

Allen, J.R.L., 1982, Sedimentary Structures: Their Character and Physical Basis: Elsevier, Amsterdam, Developments in Sedimentology, v. 30A and 30B, 663 p.

Atherton, M.P., and Petford, N., 1993, Generation of sodium-rich magmas from newly underplated basaltic crust: Nature, v. 362, p. 144–146, doi: 10.1038/362144a0.

Barron, J., Larsen, B., Baldauf, J.G., Alibert, C., Berkowitz, S.P., Caulet, J.-P., Chambers, S., Cooper, A., Cranston, R., Dorn, W., Ehrmann, W., Fox, R., Fryxell, G., Hambrey, M., Huber, B., Jenkins, C., Kang., S.-H., Keating, B., Mehl, K., Noh, I., Ollier, G., Pittenger, A., Sakai, H., Schroder, C., Solheim, A., Stockwell, D., Thierstein, H., Tocher, B., Turner, B., and Wei, W., 1989, Proceedings of the Ocean Drilling Program, Initial Reports, Volume 119: College Station, Texas, Ocean Drilling Program, 942 p.

Bercovici, D., and Mahoney, J., 1994, Double flood basalts and plume head separation at the 660 kilometer discontinuity: Science, v. 266, p. 1367–1369, doi: 10.1126/science.266.5189.1367.

Bitschene, P.R., Holmes, M.A., and Breza, J.R., 1992, Composition and origin of Cr-rich glauconitic sediments from the southern Kerguelen Plateau (site 748), *in* Wise, S.W., Jr., and Schlich, R., et al., Proceedings of the Ocean Drilling Program, Scientific Results, Volume 120: College Station, Texas, Ocean Drilling Program, p. 113–121.

Busby, C.J., and Ingersoll, R.V., 1995, Tectonics of Sedimentary Basins: Cambridge, Massachusetts, U.S.A., Blackwell Science, 579 p.

Church, S.E., 1981, Multielement analysis of fifty-four geochemical reference samples using inductively coupled plasma–atomic emission spectrometry: Geostandards and Geoanalytical Research, v. 5, p. 133–160, doi: 10.1111/j.1751-908X.1981.tb00320.x.

Class, C., Goldstein, S.L., and Galer, S.J.G., 1996, Temporal evolution of the Kerguelen plume: Geochemical evidence from approximately 38 to 82 Ma lavas forming the Ninetyeast Ridge—Discussion: Contributions to Mineralogy and Petrology, v. 124, p. 98–103, doi: 10.1007/s004100050177.

Coffin, M.F., 1992, Subsidence of the Kerguelen Plateau: The Atlantis concept, *in* Wise, S.W., Jr., and Schlich, R., et al., Proceedings of the Ocean Drilling Program, Scientific Results, Volume 120: College Station, Texas, Ocean Drilling Program, p. 945–949.

Coffin, M.F., Munschy, M., Colwell, J.B., Schlich, R., Davies, H.L., and Li, Z.G., 1990, Seismic stratigraphy of the Aggatt Basin, southern Kerguelen Plateau: Tectonic and paleoceanographic implications: Geological Society of America Bulletin, v. 102, p. 563–579, doi: 10.1130/0016-7606(1990)102<0563:SSOTRB>2.3.CO;2.

Coleman, J.M., and Prior, D.B., 1982, Deltaic Sand Bodies: American Association of Petroleum Geologists Short Course 15, 171 p.

Coleman, J.M., Suhayda, J.N., Whelan, T., and Wright, L.D., 1974, Mass movement of Mississippi River delta sediments: Transactions of the Gulf Coast Association of Geological Societies, v. 24, p. 49–68.

Davis, D.W., 1993, Report on U-Pb Geochronology in the Atikokan Area, Wabigoon Subprovince: Toronto, Ontario, Canada, Royal Ontario Museum Report, 9 p.

Davis, D.W., 1994, Report on the Geochronology of Rocks from the Rice Lake Belt, Manitoba: Toronto, Ontario, Canada, Royal Ontario Museum Report, 11 p.

Davis, D.W., and Jackson, M.C., 1988, Geochronology of the Lumby Lake greenstone belt: A 3 Ga complex within the Wabigoon subprovince, northwest Ontario: Geological Society of America Bulletin, v. 100, p. 818–824, doi: 10.1130/0016-7606(1988)100<0818:GOTLLG>2.3.CO;2.

Eriksson, K.A., Krapez, B., and Fralick, P.W., 1994, Sedimentology of Archean greenstone belts: Signatures of tectonic evolution: Earth-Science Reviews, v. 37, p. 1–88, doi: 10.1016/0012-8252(94)90025-6.

Eriksson, K.A., Krapez, B., and Fralick, P.W., 1997, Sedimentological aspects, *in* de Wit, M.J., and Ashwal, L.D., eds., Greenstone Belts: Oxford, UK, Oxford University Press, p. 33–54.

Fralick, P.W., 2003, Geochemistry of clastic sedimentary rocks: Ratio techniques, *in* Lentz, D.R., ed., Geochemistry of Sediments and Sedimentary Rocks: Toronto, Ontario, Canada, Geological Association of Canada, Geotext 4, p. 85–104.

Fralick, P.W., and King, D., 1996, Mesoarchean evolution of western Superior Province: Evidence from metasedimentary sequences near Atikokan, *in* Harrap, R.M., and Helmstaedt, H., eds., Lithoprobe Report 53: Vancouver, British Columbia, Canada, Lithoprobe Secretariat, University of British Columbia, p. 29–35.

Fralick, P.W., and Kronberg, B.I., 1997, Geochemical discrimination of clastic sedimentary rock sources: Sedimentary Geology, v. 113, p. 111–124, doi: 10.1016/S0037-0738(97)00049-3.

Fralick, P.W., and Miall, A.D., 1989, Sedimentology of the lower Huronian Supergroup (Early Proterozoic), Elliot Lake area, Ontario, Canada: Sedimentary Geology, v. 63, p. 127–153, doi: 10.1016/0037-0738(89)90075-4.

Fritsch, B., Schlich, R., Munschy, M., Fezga, F., and Coffin, M.F., 1992, Evolution of the southern Kerguelen Plateau deduced from seismic stratigraphic studies and drilling at sites 748 and 750, *in* Wise, S.W., Jr., and Schlich, R., et al., Proceedings of the Ocean Drilling Program, Scientific Results, Volume 120: College Station, Texas, Ocean Drilling Program, p. 895–905.

Furumoto, A.S., Webb, J.P., Odegard, M.E., and Hussong, D.M., 1976, Seismic studies on the Ontong Java Plateau, 1970: Tectonophysics, v. 34, p. 71–90, doi: 10.1016/0040-1951(76)90177-3.

Hollings, P., and Wyman, D., 1999, Trace element and Sm-Nd systematics of volcanic and intrusive rocks from the 3 Ga Lumby Lake greenstone belt, Superior Province: Evidence for Archean plume-arc interaction: Lithos, v. 46, p. 189–213, doi: 10.1016/S0024-4937(98)00062-0.

Hollings, P., Wyman, D., and Kerrich, R., 1999, Komatiite-basalt-rhyolite volcanic associations in northern Superior Province greenstone belts: Significance of plume-arc interaction in the generation of the proto-continental Superior Province: Lithos, v. 46, p. 137–161, doi: 10.1016/S0024-4937(98)00058-9.

Holmes, M.A., 1992, Cretaceous subtropical weathering followed by cooling at 60S latitude: The mineral composition of southern Kerguelen Plateau sediment, Leg 120, *in* Wise, S.W., Jr., and Schlich, R., et al., Proceedings of the Ocean Drilling Program, Scientific Results, Volume 120: College Station, Texas, Ocean Drilling Program, p. 99–109.

Hussong, D.M., Wipperman, L.K., and Kroenke, L.W., 1979, The crustal structure of the Ontong-Java and Manihiki oceanic plateaus: Journal of Geophysical Research, v. 84, p. 6003–6010.

Hwang, I.G., and Chough, S.K., 1990, The Miocene Chunbuk Formation, southeastern Korea: Marine Gilbert-type fan-delta system, *in* Colella, A., and Prior, D.B., eds., Coarse-Grained Deltas: International Association of Sedimentologists Special Publication 10, p. 235–254.

King, D., 1998, Depositional environments of the 3.0 Ga Finlayson and Lumby Lake greenstone belts, Superior Province, Canada [M.Sc. thesis]: Thunder Bay, Ontario, Canada, Lakehead University, 200 p.

Langford, F.F., and Morin, J.A., 1976, The development of the Superior Province of northwestern Ontario by merging island arcs: American Journal of Science, v. 276, p. 1023–1034.

Lowe, D.R., 1982, Sediment gravity flows: II. Depositional modes with special reference to the deposits of high-density turbidity currents: Journal of Sedimentary Petrology, v. 52, p. 279–297.

Martinsen, O.J., 1990, Fluvial, inertia-dominated deltaic deposition in the Namurian (Carboniferous) of northern England: Sedimentology, v. 37, p. 1099–1113, doi: 10.1111/j.1365-3091.1990.tb01848.x.

Marzo, M., and Anadon, P., 1988, Anatomy of a conglomeratic fan-delta complex: The Eocene Montserrat Conglomerate, Elbro Basin, northeastern Spain, *in* Nemec, W., and Steele, R.J., Fan Deltas: Sedimentology and Tectonic Settings: London, Blackie, p. 318–340.

Miall, A.D., 1977, A review of the braided-river depositional environment: Earth-Science Reviews, v. 13, p. 1–62, doi: 10.1016/0012-8252(77)90055-1.

Miall, A.D., 1985, Architectural-element analysis: A new method of facies analysis applied to fluvial deposits: Earth-Science Reviews, v. 22, p. 261–308, doi: 10.1016/0012-8252(85)90001-7.

Miall, A.D., 1994, Reconstructing fluvial macroform architecture from two-dimensional outcrops: Examples from the Castlegate Sandstone, Book Cliffs, Utah: Journal of Sedimentary Research, v. B64, p. 146–158.

Munschy, M., Fritsch, B., Schlich, R., Fezga, F., Rotstein, Y., and Coffin, M.F., 1992a, Structure and evolution of the central Kerguelen Plateau deduced from seismic stratigraphic studies and drilling at site 747, *in* Wise, S.W., Jr., and Schlich, R., et al., Proceedings of the Ocean Drilling Program, Scientific Results, Volume 120: College Station, Texas, Ocean Drilling Program, p. 881–893.

Munschy, M., Dyment, J., Boulanger, M.O., Boulanger, D., Tissot, J.D., Schlich, R., Rotstein, Y., and Coffin, M.F., 1992b, Breakup and seafloor spreading between the Kerguelen Plateau-Labuan Basin and the Broken Ridge–Diamantina Zone, *in* Wise, S.W., Jr., and Schlich, R., et al., Proceedings of the Ocean Drilling Program, Scientific Results, Volume 120: College Station, Texas, Ocean Drilling Program, p. 931–942.

Nemec, W., and Steel, R.J., 1984, Alluvial and coastal conglomerates: Their significant features and some comments on gravelly mass-flow deposits, *in* Koster, E.H., and Steel, R.J., eds., Sedimentology of Gravels and Conglomerates: Canadian Society of Petroleum Geologists Memoir 10, p. 1–31.

Percival, C.J., 1992, The Harthorpe Gannister–transgressive barrier island to shallow marine sand-ridge from the Namurian of northern England: Journal of Sedimentary Petrology, v. 62, p. 442–454.

Pudsey, N.P., 1984, Fluvial to marine transition in the Ordovician of Ireland—A humid-region fan-delta?: Geological Journal, v. 19, p. 143–172.

Sasseville, C., Tomlinson, K.Y., Hynes, A., and McNicoll, V., 2006, Stratigraphy, structure and geochronology of the 3.0–2.7 Ga Wallace Lake greenstone belt, western Superior Province, southeast Manitoba, Canada: Canadian Journal of Earth Sciences, v. 43, p. 929–945, doi: 10.1139/E06-041.

Schaefer, S.J., and Morton, P., 1991, Two komatiitic pyroclastic units, Superior Province, northwestern Ontario: Their geology, petrography, and correlation: Canadian Journal of Earth Sciences, v. 28, p. 1455–1470.

Schlich, R., Wise, S.W., Jr., Palmer, J.A.A., Aubry, M.P., Berggren, W.A., Bitschene, P.R., Blackburn, N.A., Breza, J., Coffin, M.F., Harwood, D.M., Heider, F., Holmes, M.A., Howard, W.R., Inokuchi, H., Kelts, K., Lazarus, D.B., Mackensen, A., Maruyama, T., Munschy, M., Pratson, E., Quilty, P.G., Rack, F., Salters, V.J.M., Sevigny, J.H., Storey, M., Takemura, A., Watkins, D.K., Whitechurch, H., Zachos, J., Hos, D.P.C., and Mohr, B., 1989, Principal results and summary, *in* Schlich, R., Wise, S.W., Jr., Palmer, J.A.A., et al., Proceedings of the Ocean Drilling Program, Initial Reports, Volume 120: College Station, Texas, Ocean Drilling Program, p. 73–85.

Smithies, R.H., 2000, The Archean tonalite-trondhjemite-granodiorite (TTG) series is not an analogue of Cenozoic adakite: Earth and Planetary Science Letters, v. 182, p. 115–125, doi: 10.1016/S0012-821X(00)00236-3.

Stone, D., Kamineni, D.C., and Jackson, M.C., 1992, Precambrian Geology of the Atikokan Area, Northwestern Ontario: Geological Survey of Canada Bulletin 405, 106 p.

Stone, D., Percival, J.A., Davis, D.W., Fralick, P.W., Halle, J., Pufahl, P.K., and Tomlinson, K., 2001, South central Wabigoon, *in* Western Superior NATMAP Compilation Series: Toronto, Ontario, Canada, Ontario Geological Survey, Preliminary Map, scale 1:250,000.

Storey, B.C., 1995, The role of mantle plumes in continental breakup: Case histories from Gondwanaland: Nature, v. 377, p. 301–308, doi: 10.1038/377301a0.

Storey, M., Saunders, A.D., Tarney, J., Leat, P., Thirlwall, M.F., Thompson, R.N., Menzies, M.A., and Marriner, G.E., 1988, Geochemical evidence for plume-mantle interactions beneath Kerguelen and Heard Islands, Indian Ocean: Nature, v. 336, p. 371–374, doi: 10.1038/336371a0.

Stott, G.M., 1997. The Superior Province, Canada, *in* deWit, M.J., and Ashwal, L.D., eds., Greenstone Belts: Oxford, UK, Oxford University Press, p. 480–507.

Tarney, J., and Jones, C.E., 1994, Trace element geochemistry of orogenic igneous rocks and crustal growth models: Journal of the Geological Society of London, v. 151, p. 855–868.

Thurston, P.G., and Chivers, K.M., 1990, Secular variations in greenstone sequence development emphasizing Superior Province, Canada: Precambrian Research, v. 46, p. 21–58, doi: 10.1016/0301-9268(90)90065-X.

Thurston, P.C., Williams, H.R., Sutcliffe, R.H., and Stott, G.M., 1991, Geology of Ontario: Ontario Geological Survey Special Volume 4, Part 1, 711 p.

Tomlinson, K.Y., 2000, Neodymium isotopic data from the central Wabigoon subprovince, Ontario: Implications for crustal recycling in 3.1 to 2.7 Ga sequences, *in* Radiogenic Age and Isotopic Studies: Geological Survey of Canada, Current Research 2000-F8, Report 13, p. 10.

Tomlinson, K.Y., Thurston, P.C., Hughes, D.J., and Keays, R.R., 1996, The central Wabigoon region: Petrogenesis of mafic-ultramafic rocks in the Steep Rock, Lumby Lake, and Obonga Lake greenstone belts (continental rifting and drifting in the Archean), *in* Harrap, R.M., and Helmstaedt, H., eds., Western Superior Lithoprobe Transect: Vancouver, British Columbia, Canada, Lithoprobe Sectretariat, University of British Columbia, Report 53, p. 65–73.

Tomlinson, K.Y., Hughes, D.J., Thurston, P.C., and Hall, R.P., 1999, Plume magmatism and crustal growth at 2.9 to 3.0 Ga in the Steep Rock and Lumby Lake area, western Superior Province: Lithos, v. 46, p. 103–136, doi: 10.1016/S0024-4937(98)00057-7.

Tomlinson, K.Y., Davis, D.W., and Scott, G.M., 2000, Nd isotopes in the central and eastern Wabigoon subprovince: Implications for crustal recycling and regional correlations, *in* Harrap, R.M., and Helmstaedt, H.H., eds., Western Superior Lithoprobe Transect, $6^{th}$ Annual Workshop, 3–4 February 2000: Ottawa, Canada, Lithoprobe Secretariat, University of British Columbia, Lithoprobe Report 77, p. 119–126.

Tomlinson, K.Y., Sasseville, C., and McNicoll, V., 2001, New U-Pb geochronology and structural interpretations from the Wallace Lake greenstone belt (North Caribou terrane): Implications for new regional correlations, *in* Harrap, R.M., and Helmstaedt, H., eds., Western Superior Lithoprobe Transect: Vancouver, British Columbia, Canada, Lithoprobe Secretariat, University of British Columbia, Report 78.

Tomlinson, K.Y., Davis, D.W., Stone, D., and Hart, T.R., 2003, U-Pb age and Nd isotopic evidence for Archean terrane development and crustal recycling in the south-central Wabigoon subprovince, Canada: Contributions to Mineralogy and Petrology, v. 144, p. 684–702.

Turek, A., and Weber, W., 1994, The 3 Ga Basement of the Rice Lake Supercrustal Rocks, Southeast Manitoba: Winnipeg, Manitoba, Canada, Manitoba Energy and Mines Mineral Division Report of Activities, p. 167–169.

Vos, R.G., 1981, Sedimentology of an Ordovician fan delta complex, western Libya: Sedimentary Geology, v. 29, p. 153–170, doi: 10.1016/0037-0738(81)90005-1.

Watkins, D.K., Quilty, P.G., Mohr, B.A.R., Mao, S., Francis, J.E., Gee, C.T., and Coffin, M.F.R., 1992, Paleontology of the Cretaceous of the central Kerguelen Plateau, *in* Wise, S.W., Jr., and Schlich, R., et al., Proceedings of the Ocean Drilling Program, Scientific Results, Volume 120: College Station, Texas, Ocean Drilling Program, p. 951–959.

Wilks, M.E., 1986, The geology of the Steep Rock Group, northwestern Ontario: A major Archean unconformity and Archean stromatolites [M.Sc. thesis]: Saskatoon, Saskatchewan, Canada, University of Saskatchewan, 206 p.

Wilks, M.E., and Nisbet, E.G., 1988, Stratigraphy of the Steep Rock Group, northwest Ontario: A major Archaean unconformity and Archaean stromatolites: Canadian Journal of Earth Sciences, v. 25, p. 370–391.

Winchester, J.A., and Floyd, P.A., 1976, Geochemical discrimination of different magma series and their differentiation products using immobile elements: Chemical Geology, v. 39, p. 81–92.

Wright, L.D., 1977, Sediment transport and deposition at river mouths: A synthesis: Geological Society of America Bulletin, v. 88, p. 857–868, doi: 10.1130/0016-7606(1977)88<857:STADAR>2.0.CO;2.

Wyman, D., and Hollings, P., 1998, Long-lived mantle-plume influence on an Archean protocontinent: Geochemical evidence from the 3 Ga Lumby Lake greenstone belt, Ontario, Canada: Geology, v. 26, p. 719–722, doi: 10.1130/0091-7613(1998)026<0719:LLMPIO>2.3.CO;2.

Zegers, T.E., 2004, Granite formation and emplacement as indicators of Archean tectonic processes, *in* Eriksson, P.G., Altermann, W., Nelson, D.R., Mueller, W.U., and Catuneanu, O., eds., The Precambrian Earth: Tempos and Events: New York, Elsevier, Developments in Precambrian Geology, v. 12, p. 103–118.


MANUSCRIPT ACCEPTED BY THE SOCIETY 14 AUGUST 2007

Printed in the USA

The Geological Society of America
Special Paper 440
2008

# *Characteristic thermal regimes of plate tectonics and their metamorphic imprint throughout Earth history: When did Earth first adopt a plate tectonics mode of behavior?*

**Michael Brown***
*Laboratory for Crustal Petrology, Department of Geology, University of Maryland, College Park, Maryland 20742-4211, USA*

## ABSTRACT

**Where plates converge, one-sided subduction generates two contrasting thermal environments in the subduction zone (low *dT/dP*) and in the arc and subduction zone backarc or orogenic hinterland (high *dT/dP*). This duality of thermal regimes is the hallmark of modern plate tectonics, which is imprinted in the ancient rock record as penecontemporaneous metamorphic belts of two contrasting types, one characterized by higher-pressure–lower-temperature metamorphism and the other characterized by higher-temperature–lower-pressure metamorphism.**

**Granulite facies ultrahigh-temperature metamorphism (G-UHTM) is documented in the rock record predominantly from the Neoarchean to the Cambrian, although it may be inferred at depth in some younger Phanerozoic orogenic systems. Medium-temperature eclogite–high-pressure granulite metamorphism (E-HPGM) also is first recognized in the Neoarchean, although well-characterized examples are rare in the Neoarchean-to-Paleoproterozoic transition, and occurs at intervals throughout the Proterozoic and Paleozoic rock record. The first appearance of E-HPGM belts in the rock record registers a change in geodynamics that generated sites of lower heat flow than previously seen, inferred to be associated with subduction-to-collision orogenesis. The appearance of coeval G-UHTM belts in the rock record registers contemporary sites of high heat flow, inferred to be similar to modern arcs, abd backarcs, or orogenic hinterlands, where more extreme temperatures were imposed on crustal rocks than previously recorded. Blueschists first became evident in the Neoproterozoic rock record, and lawsonite blueschists, low-temperature eclogites (high-pressure metamorphism, HPM), and ultrahigh-pressure metamorphism (UHPM) characterized by coesite or diamond are predominantly Phanerozoic phenomena. HPM-UHPM registers low to intermediate apparent thermal gradients typically associated with modern subduction zones and the eduction of deeply subducted lithosphere, including the eduction of continental crust subducted during the early stage of the collision process in subduction-to-collision orogenesis. During the Phanerozoic, most UHPM belts have developed by closure of relatively short-lived ocean basins that opened due to rearrangement of the continental lithosphere within a continent-dominated hemisphere as Eurasia was formed from Rodinian orphans and joined with Gondwana in Pangea,**

*mbrown@umd.edu

Brown, M., 2008, Characteristic thermal regimes of plate tectonics and their metamorphic imprint throughout Earth history: When did Earth first adopt a plate tectonics mode of behavior?, *in* Condie, K.C., and Pease, V., eds., When Did Plate Tectonics Begin on Planet Earth?: Geological Society of America Special Paper 440, p. 97–128, doi: 10.1130/2008.2440(05). For permission to copy, contact editing@geosociety.org. 

**and then due to successive closure of the Paleo-Tethys and Neo-Tethys Oceans as the East Gondwanan sector of Pangea began to fragment and disperse.**

**The occurrence of both G-UHTM and E-HPGM belts since the Neoarchean manifests the onset of a "Proterozoic plate tectonics regime," which evolved during a Neoproterozoic transition to the "modern plate tectonics regime" characterized by HPM-UHPM. The "Proterozoic plate tectonics regime" may have begun locally during the Mesoarchean to Neoarchean and may only have become global during the Neoarchean-to-Paleoproterozoic transition. The age distribution of metamorphic belts that record extreme conditions of metamorphism is not uniform. Extreme metamorphism occurs at times of amalgamation of continental lithosphere into supercratons (Mesoarchean to Neoarchean) and supercontinents (Paleoproterozoic to Phanerozoic), and along sutures due to the internal rearrangement of continental lithosphere within a continent-dominated hemisphere during the life of a supercontinent.**

**Keywords:** blueschist metamorphism, eclogite–high-pressure granulite metamorphism, paired metamorphic belts, ultrahigh-pressure metamorphism, ultrahigh-temperature metamorphism.

## INTRODUCTION

"There is something fascinating about science. One gets such a wholesale return of conjecture out of such a trifling investment of fact." Mark Twain (1884, *Life on the Mississippi*)

In the late 1960s the plate tectonics revolution introduced a robust paradigm within which to understand the Cenozoic and Mesozoic tectonics of the lithosphere—the strong outer layer of Earth above the softer asthenosphere (Isacks et al., 1968). Since that time there has been much debate about when Earth might have adopted a plate tectonics mode of behavior. A consensus may be emerging that Earth had partially or completely adopted a plate tectonics mode of behavior by sometime in the Mesoarchean to Neoarchean eras (Brown, 2006; Cawood et al., 2006; Dewey, 2007). However, there are those who argue for (some form of) plate tectonics as early as the Hadean (Harrison et al., 2006; Davies, 2006, 2007; Shirey et al., this volume)—consistent with the null hypothesis that plate tectonics was the mode of convection throughout recorded Earth history—and those who argue against worldwide modern-style subduction before the Neoproterozoic (Stern, 2005, this volume)—requiring an alternative hypothesis to plate tectonics for the Hadean to Mesoproterozoic interval. There is also a view that the plate tectonics era began ca. 2.0 Ga (Hamilton, 2003). Another challenge is to assess whether once a plate tectonics mode of behavior was established it was maintained to the present day or whether geodynamics on Earth alternated between plate tectonics and some other mode, particularly in the Mesoarchean to Mesoproterozoic interval (Sleep, 2000; Silver and Behn, 2008; Stern, this volume).

### Historical Perspective

The concept of plates was introduced by Wilson (1965) as follows: "Many geologists have maintained that movements of the Earth's crust are concentrated in mobile belts, which may take the form of mountains, mid-ocean ridges or major faults with large horizontal movements.... This article suggests that these features are not isolated ... but that they are connected into a continuous network of mobile belts about the Earth which divide the surface into several large rigid plates...." This elegant concept is well illustrated by the map in Figure 1, which shows a model strain rate field for Earth determined from geodetic velocities worldwide (Kreemer et al., 2003).

Currently we identify a mosaic of 15 or 16 lithosphere plates on Earth—seven major and eight or nine minor (by my count, Wilson [1965] identified seven major and five [poorly defined] minor plates in his Fig. 1). The plates are separated by three fundamentally different types of boundary, two of which record opposite displacements—spreading and upwelling at ocean ridges and net convergence and subduction at ocean trenches—whereas the third—transforms, which were introduced as a new type of fault (Wilson, 1965)—connects these first two, making a coherent kinematic system (Fig. 1). A number of implications of the Wilson hypothesis were developed in a series of papers during 1967–1968 (McKenzie and Parker, 1967; Morgan, 1968; LePichon, 1968; Isacks et al., 1968), after which the paradigm explaining how the lithosphere behaves on modern Earth became known as plate tectonics.

Plate tectonics is a kinematic description of how the plates interact, based on the notion that motions of torsionally rigid plates may be understood as rotations on a sphere. These motions are identified from data sets such as age maps of the ocean floors, the worldwide location of major earthquakes and first-motion studies of these earthquakes, and geodetic velocities. The geodetic measurements are current, the earthquake data are historical, and the oldest ocean floor is Jurassic, so these data sets do not help us understand how far back in time Earth has operated in a plate tectonics mode. Instead, we must turn our attention to the imprints that plate tectonics processes leave in the rock record.

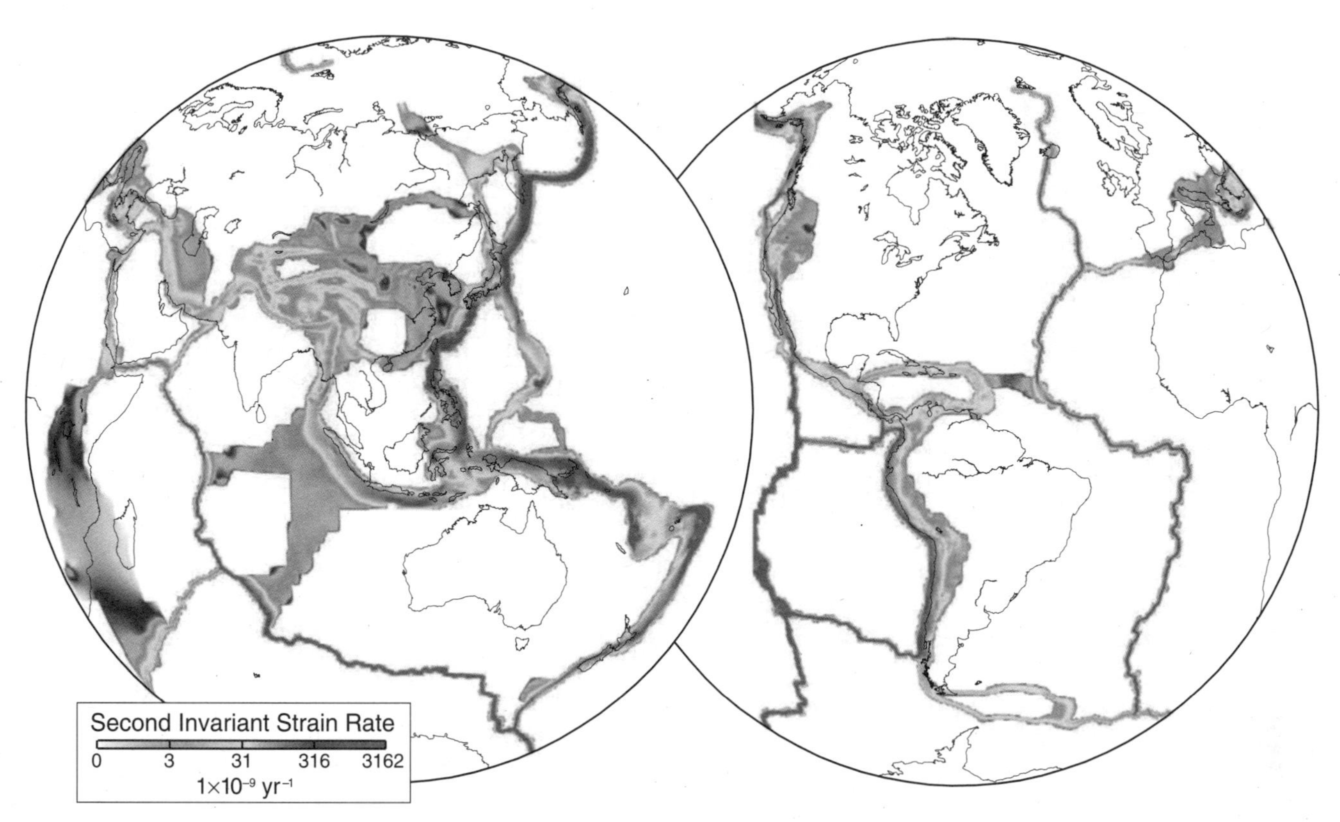


Figure 1. Contours of the second invariant of a model strain rate field determined using 3000 geodetic velocities worldwide, and Quaternary slip rates in Asia. All areas in white are assumed to behave rigidly. The contour scale is quasi-exponential. (Reproduced from Kreemer et al., 2003, with permission.)

It had become clear by the early 1970s that the concept of torsionally rigid plates separated by narrow zones of high strain could not be applied to the continents. Plate boundary zones at continental margins (e.g., western Americas) and within continents (e.g., central Asia) are broader than in the ocean basins (Fig. 1), and deformation in these broad zones tends to be partitioned into bands of higher and lower strain (e.g., central Asia). Additionally, plate mosaics evolve continuously, and convergent plate boundary zones are dynamic entities. As explained by Dewey (1975), dynamic interactions along convergent plate boundary zones lead to global and local episodicity of events, generating complex sequences of orogenic evolution globally and complicated geological histories locally. According to Dewey (1975), inversion of the geological histories retrieved from ancient orogenic belts—in which composite terranes accreted by subduction have been smeared into linear suture zones during (commonly oblique) terminal collision—to make rigorous plate tectonics analyses may not be possible.

## Phanerozoic Earth

Currently, there are two main zones of subduction into the mantle—the circum-Pacific subduction system and the Alpine–Himalayan–Indonesian subduction system—and two major zones of upwelling or superswells (McNutt, 1998, 1999; Davaille, 1999; Zhao, 2001)—under Africa and under the South Pacific—which define a simple pattern of long-wavelength mantle convection. The Phanerozoic has been distinguished by rearrangement of the continental lithosphere during the final steps in supercontinent assembly—formation of Eurasia and its linking to Gondwana forming Pangea—followed by the early stage of supercontinent breakup and dispersal. The formation of Eurasia, its linking to Gondwana, and the fragmentation and dispersal of Pangea involved successive subduction-to-collision orogenic systems within a single (Pangean) convection cell—the Appalachian/Caledonian–Variscide–Altaid and the Cimmerian-Himalayan-Alpine orogenic systems—whereas a complementary (Panthalassan) convection cell, centered on the Pacific superswell, is composed of ocean lithosphere and defined at its outer edge by the circum-Pacific accretionary orogenic systems (Collins, 2003). The Panthalassan convection cell became established toward the end of the Neoproterozoic, reached a maximum around the Devonian-Carboniferous boundary, and has been in decline since as the Pangean cell began expanding in the Jurassic. Amaesia is a possible future supercontinent that could be formed by merger of Asia and the Americas along a Siberian suture if the Atlantic Ocean continues to expand and the Eurasian continent continues to rotate clockwise, and by closure of the South Pacific Ocean if Australia continues to be driven northeastward (Bleeker, 2003).

## Modes of Convective Style

For our purpose in this chapter, we may consider plate tectonics as a mode of convection by which the mantle is recycled by creation and destruction of ocean lithosphere; this is the mechanism that currently controls major heat loss from Earth's interior and drives the supercontinent cycle. On modern Earth, deep mantle plumes play a secondary role in heat removal, primarily removing heat from the core through the mantle to the surface, but plumes may have had a greater role in transferring heat from Earth's interior to space earlier in Earth history. From a modeling perspective (e.g., Lenardic et al., 2004; Hansen et al., 2006; Loddoch et al., 2006), plate tectonics is referred to as a "mobile-lid regime."

As identified above, an important question concerns when a plate tectonics mode began on Earth. A hotter Earth may have produced thicker or thinner ocean crust than modern Earth, although in either case the thermal boundary layer most likely was thinner (Davies, 2006, 2007; van Hunen and van den Berg, 2007). As a result, the early lithosphere may have been more buoyant but weaker than modern ocean lithosphere; consequently early plate tectonics may or may not have been viable (Davies, 2007; van Hunen et al., 2004; van Hunen and van den Berg, 2007) and an unstable mode of subduction rather than a stable mode might have dominated (Toussaint et al., 2004; Burov and Watts, 2006).

Early Earth may have been characterized by a different style of mantle convection in which a rigid and largely immobile layer developed at the surface of the mantle (Lenardic et al., 2004). In this tectonic mode, heat loss by vigorous convection beneath the lid may have been configured as plume-like (Hansen et al., 2006; Loddoch et al., 2006). From a modeling perspective, this type of convection is referred to as a "stagnant-lid regime." In a stagnant lid-regime, horizontal motions and convergence may occur due to nonrigidity, and this may generate symmetric subduction (in contrast to the asymmetric subduction of the plate tectonics paradigm) of the lower (mantle) part of ocean lithosphere as postulated by Davies (1992, subcrustal subduction; see also Burov and Watts, 2006). However, a stagnant-lid regime is not an efficient cooling mechanism, which is likely why a variety of alternative tectonic regimes have been suggested for early Earth (van Hunen et al., this volume).

A third behavior is an "episodic regime" in which the mode of convection may alternate between a stagnant-lid and a mobile-lid regime, or there may be a transition from one to the other as the dominant mode via an episodic regime (e.g., Hansen et al., 2006; Loddoch et al., 2006). A scenario of changing from a stagnant-lid to a mobile-lid mode via an episodic regime is one possible evolutionary path for Earth. Indeed, Davies (1995) has suggested a punctuated tectonic evolution for Earth in which the Archean might be characterized by periods of plate tectonics alternating with mantle overturn events; Davies hypothesized that the early rock record might register mainly the effects of the overturn events.

## The Hallmark of Plate Tectonics

From a geological perspective, we may break down into several components the question of when plate tectonics began on Earth. For example, we may ask when did the lithosphere first behave as a mosaic of plates—that is, a mosaic of largely torsionally rigid lithosphere elements bounded by zones of generation, destruction, or transform displacement—and how far back in Earth history may we identify independent horizontal motions from different cratons (e.g., Evans and Pisarevsky, this volume)? Also, we may ask when in the rock record do we first identify zones of convergence and subduction of ocean lithosphere (e.g., Pease et al., this volume). However, care is required—what imprint in the rock record do we take to answer each of these questions (Cawood et al., 2006; Condie and Kröner, this volume)?

The first record of blueschist facies series metamorphism occurs in the Neoproterozoic, and this is commonly taken as one indicator of the beginning of the modern style of subduction (Stern, 2005). However, due to secular cooling, Brown (2006, 2007a) asked whether plate tectonics on a warmer Earth might have left a different imprint in the ancient rock record. To answer this question, it is necessary to establish the distinctive geological characteristics of plate tectonics that might be preserved or imprinted in the ancient rock record. Brown (2006, 2007a) argued that the definitive hallmark of plate tectonics is a duality of thermal environments at convergent plate margins (the subduction zone, and the arc and subduction zone backarc or orogenic hinterland), and that the imprint of this hallmark will be found in the rock record as the contemporary occurrence of spatially distinct contrasting types of metamorphism. In effect, the rock record of metamorphism will tell us when plate tectonics had become established on Earth.

Early plate-like behavior may have allowed crust to float to form thick stacks above zones of boundary-layer downwelling (Davies, 1992), or, for a slightly cooler Earth, thick stacks above zones of "sublithospheric" subduction (van Hunen et al., this volume), but these behaviors are unlikely to have created dual thermal environments at the site of tectonic stacking, although thickening (with or without delamination of eclogite sinkers) might have induced melting deep in the pile to generate tonalite-trondhjemite-granodiorite (TTG) magmas (Foley, this volume). Therefore, in principle, the metamorphic imprint imposed by either of these behaviors during the early part of Earth history should be distinguishable from that imposed by a plate tectonics mode of convection.

Metamorphic rocks record evidence of the change in the pressure ($P$) and temperature ($T$) of the crust with time ($t$), which is commonly expressed as a $P$-$T$-$t$ path, recording burial and exhumation in the particular thermal environment in which the mineral assemblages crystallized and/or equilibrated (e.g., Brown, 1993, 2001). In effect, the $P$-$T$-$t$ path records the changing spectrum of transient metamorphic geotherms characteristic of a particular thermal environment. Different thermal environments are characterized by different ranges of $dT/dP$ and register different metamorphic imprints in the geologic record.

## Orogenic Systems

The circum-Pacific and Alpine–Himalayan–Cimmerian orogenic systems define two orthogonal great-circle distributions of the continents, which may reflect a simple pattern of mantle convection. Along each great-circle distribution the convergent plate margins are characterized by a different type of orogenic system, and this has been the situation during at least the Cenozoic and Mesozoic eras (Zwart, 1967; Maruyama, 1997; Ernst, 2001; Liou et al., 2004). These two types are (1) accretionary orogenic systems (variants have been called "Pacific-type" [e.g., Matsuda and Uyeda, 1971] or "Cordilleran-type" [e.g., Coney et al., 1980]), which form during ongoing plate convergence, as exemplified by the Phanerozoic evolution of the Pacific Ocean rim, and (2) collisional orogenic systems (sometimes called "Himalayan-type" [Liou et al., 2004] or "Turkic-type" [Şengör and Natal'in, 1996]), in which an ocean is closed and arcs and/or allochthonous terranes and/or continents collide, as exemplified by the Tethysides.

Accretionary orogenic systems may be grouped into several different types. "Extensional-contractional accretionary orogenic systems" (e.g., the Tasmanides) and "contractional-erosional orogenic systems" (e.g., the South American Cordillera) are distinguished according to the velocity field and whether the overriding plate is retreating or advancing (Doglioni et al., 2006; Husson et al., 2007), and "terrane accretion orogenic systems" (e.g., the North American Cordillera) are characterized by suturing of suspect terranes, some of which may be far-traveled (cf. Coney et al., 1980; Johnston, 2001; English and Johnston, 2005; Johnston and Borel, 2007; Colpron et al., 2007). Although Şengör and Natal'in (1996) argued that the precollision history of one, or both, of the colliding margins might involve the growth of very large accretionary orogenic systems, with significant juvenile arc magmatism, a distinction remains useful because some accretionary orogenic systems exist for hundreds of millions of years without disruption by collision, whereas others have the continuity of subduction interrupted—but not necessarily terminated if a new trench is formed outboard of the collider—by subduction of ocean floor debris or an oceanic plateau or by nonterminal collision and suturing of allochthonous terranes or arcs (e.g., Cloos, 1993; Collins, 2002; Koizumi and Ishiwatari, 2006; Windley et al., 2007).

The major mountain belts of the circum-Pacific orogenic systems are located in subduction zone backarcs, which are characterized by high heat flow, >70 mW $m^{-2}$ for continental crust with average radiogenic heat production, and uniformly thin and weak lithosphere over considerable widths (Hyndman et al., 2005a; Currie and Hyndman, 2006). Subduction zone backarcs are hot perhaps due to shallow convection in the mantle wedge asthenosphere, where convection is inferred to result from a reduction in viscosity induced by water from the underlying subducting plate. According to Hyndman et al. (2005a; see also Currie and Hyndman, 2006), Moho temperatures in subduction zone backarcs are 800–900 °C and lithosphere thicknesses are 50–60 km, compared to 400–500 °C and 200–300 km for cratons; the difference results in backarc lithosphere being at least an order of magnitude weaker than cratons.

Parameters such as the contemporaneous surface heat flow cannot be measured directly in ancient orogens, and once again we must look to the metamorphic record for information. Also, thermal expansion due to the high lithosphere temperatures is argued to account for ~2500 m of mountain belt elevation in subduction zone backarcs without significant crustal thickening (Hyndman et al., 2005b)—which may be recorded in an ancient orogen as variation in the pressure field. Most of these circum-Pacific mountain belts are broad zones of long-lived tectonic activity because they are sufficiently weak to be deformed by the forces developed at plate boundaries, and complex histories of deformation in these orogenic systems most likely relate to changing dynamics along convergent plate boundaries (cf. Dewey, 1975). In terrane accretion orogenic systems, former backarcs remain a locus of deformation during terrane suturing because they continue to be weaker than the hinterland.

Shortening and crustal thickening in large hot subduction-to-collision orogenic systems, such as the Himalayan orogenic system, and in large hot contractional-erosional orogenic systems, such as the South American Cordillera, appear to be accommodated mostly in the weak lower crust (Beaumont et al., 2004, 2006; Sobolev and Babeyko, 2005). The upper crust may be uplifted as a plateau, which may be underthrust by the adjacent stable craton, and the upper crust may remain largely undeformed internally with deformation localized into high-strain belts (e.g., Kreemer et al., 2003).

## The Metamorphic Realm

The metamorphic realm traditionally has been divided into facies, each represented by a group of mineral assemblages associated in space and time that are inferred to register equilibration within a limited range of *P-T* conditions. Some rocks characteristic of the granulite facies record temperatures >1000 °C at pressures around 1 GPa (ultrahigh-temperature metamorphism; e.g., Harley, 1998; Brown, 2007a), whereas some eclogite facies rocks record pressures >5 GPa at temperatures of 600–1000 °C, and in one case mineralogical evidence suggests pressures of at least 10 GPa (ultrahigh-pressure metamorphism; e.g., Chopin, 2003; Liu et al., 2007; Brown, 2007a). We recognize a transition between these two facies, referred to as medium-temperature eclogite–high-pressure granulite metamorphism (E-HPGM; e.g., O'Brien and Rötzler, 2003; Brown, 2007a). The *P-T* regimes for metamorphic facies are shown in Figure 2. On modern Earth, the different types of metamorphic facies series leading to granulite facies ultrahigh-temperature metamorphism (G-UHTM), medium-temperature eclogite–high-pressure granulite metamorphism (E-HPGM), and high-pressure metamorphism–ultrahigh-pressure metamorphism (HPM-UHPM) are generated in different tectonic settings with contrasting thermal regimes at convergent plate boundary zones (Brown, 2006, 2007a; summarized later).

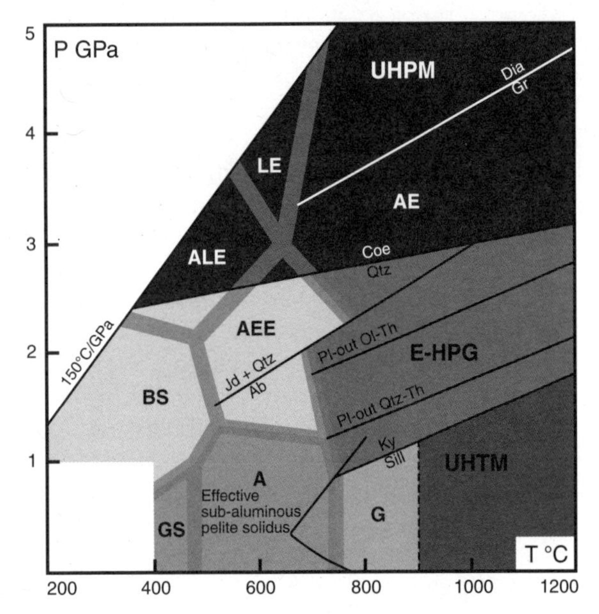


Figure 2. *P-T* diagram to show the location of the principal metamorphic facies in *P-T* space and the *P-T* ranges of different types of extreme metamorphism. HPM-UHPM includes the following: BS—blueschist; AEE—amphibole-epidote eclogite facies; ALE—amphibole-lawsonite eclogite facies; LE—lawsonite eclogite facies; AE—amphibole eclogite facies; UHPM—ultrahigh-pressure metamorphism. Other abbreviations: GS—greenschist facies; A—amphibolite facies; E-HPG—medium-temperature eclogite–high-pressure granulite metamorphism; G—granulite facies; UHTM—the ultrahigh-temperature metamorphic part of the granulite facies.

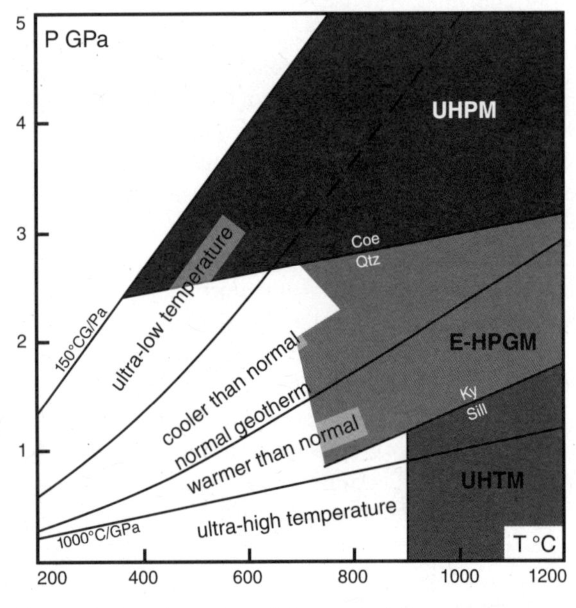


Figure 3. An alternative division of *P-T* space based on whether thermal conditions implied by peak metamorphic mineral assemblages in orogenic crust were warmer or cooler than a normal (conductive) continental geotherm (cf. Stüwe, 2007). On this diagram we may distinguish *P-T* fields that are reached as a function of different tectonic processes. See discussion in text.

Recently, Stüwe (2007) has introduced an alternative division of *P-T* space based on whether thermal conditions implied by peak metamorphic mineral assemblages in orogenic crust were warmer or cooler than a normal (conductive) continental geotherm (Fig. 3). On this diagram we may distinguish *P-T* fields that are reached as a function of different tectonic processes. For thermal conditions warmer than a normal continental geotherm, a thermal gradient of ~1000 °C/GPa is a practical limit for a conductive response; metamorphic belts that record apparent thermal conditions hotter than this limit (I introduce the term "ultrahigh temperature" for this *P-T* field) require a component of intracrustal advection driven by heating, perhaps in an arc or due to thinning of the subcrustal lithospheric mantle (e.g., Sandiford and Powell, 1991), consistent with the subduction zone backarc model discussed above (Hyndman et al., 2005a, 2005b; Currie and Hyndman, 2006). For thermal conditions cooler than a normal continental geotherm, Stüwe (2007) suggests a twofold division into a "cooler than normal" (note, Stüwe used "colder") and an "ultralow temperature" *P-T* field. The "cooler than normal" *P-T* field is a thermal regime that may be reached by whole lithospheric thickening for an acceptable range of thermal parameters, and these *P-T* conditions do not necessarily reflect thermal regimes attendant with subduction or that require unusually rapid exhumation (cf. Sandiford and Dymoke, 1991). In contrast, the "ultralow temperature" *P-T* field may only be reached by processes other than normal crustal thickening; subduction is one possible process by which rocks may enter this thermal regime (Stüwe, 2007).

## Metamorphism at Extreme *P-T* Conditions—Definitions

For our purpose in this chapter, I follow the definitions from Brown (2007a).

Granulite facies ultrahigh-temperature metamorphism (G-UHTM) is *metamorphism of crustal rocks in which peak temperature exceeds 900 °C, recognized either by robust thermobarometry or by the presence of a diagnostic mineral assemblage in an appropriate bulk composition and oxidation state, such as assemblages with orthopyroxene + sillimanite + quartz, sapphirine + quartz or spinel + quartz, generally at pressure conditions of sillimanite stability in metapelites*. The current state-of-the-art procedure in robust thermobarometry for granulite metamorphism in general for a variety of bulk compositions is either the method proposed by Pattison et al. (2003) or the average *P-T* mode in THERMOCALC (Powell et al., 1998) with the internally consistent thermodynamic data set of Holland and Powell (1998, most recent update).

Eclogite facies ultrahigh-pressure metamorphism (UHPM) is *metamorphism of crustal rocks under conditions of pressure that exceed the stability field of quartz, recognized either by the presence of coesite or diamond, or by equivalent P-T conditions determined using robust thermobarometry*. Examples of robust thermobarometry currently in use, as recently evaluated by Hacker (2006), include the method of Ravna and Terry (2004) and the average *P-T* mode in THERMOCALC (Powell et al., 1998) with the internally consistent thermodynamic data set of Holland and Powell (1998, most recent update).

In between the *P-T* regimes for these two types of extreme metamorphism, there is a transition in pressure from common granulite facies conditions to high-pressure granulite facies conditions that is not well defined, but which may be taken to correspond approximately to conditions of the change from sillimanite to kyanite stability in metapelitic rocks (Fig. 2). The field of high-pressure granulite facies metamorphism overlaps the transition from amphibolite/granulite to eclogite in basaltic rocks, which is a transition over more than 1 GPa according to chemical composition of the protolith (O'Brien and Rötzler, 2003). For these reasons I refer to this *P-T* field as medium-temperature eclogite–high-pressure granulite metamorphism (E-HPGM).

## Some Cautionary Notes

We piece together the changes in pressure (from which we infer changes in depth) and temperature (from which we infer heating and cooling) with time (from which we obtain rates) from the mineralogical, chemical, and microstructural evidence preserved in the rock. However, unless the effects of overprinting of multiple thermal events (polymetamorphism) where present or suspected are distinguished (e.g., Hensen and Zhou, 1995; Hensen et al., 1995), any evaluation of the imprint of metamorphism in the rock record will be potentially flawed. As a community during the past decade we have an increased confidence in our ability as forensic scientists correctly to interpret the evidence recovered from eroded orogens (e.g., O'Brien, 1997, 1999; Brown, 2001, 2002a; White et al., 2002; O'Brien and Rötzler, 2003; Johnson and Brown, 2004; Baldwin et al., 2005; Powell et al., 2005); this credence was lacking in the early days of studies of metamorphism under extreme *P-T* conditions (e.g., Green, 2005).

Our surety is in part due to improvements in our ability to interrogate rocks and to recognize the effects of overprinting and "see through" them due to better technical capabilities for rapid chemical mapping of mineral grains and high-spatial-resolution in situ geochronology. The wider availability of high-precision ages on accessory phases linked to specific microstructural sites and metamorphic *P-T* conditions—commonly through the identification of mineral assemblages preserved as microinclusions in the core, mantle, or rim of zircon grains—has increased our confidence that we know the true "age" of peak metamorphism in many metamorphic belts, even though we also know that zircon may grow continuously along the part of the *P-T* evolution around the metamorphic peak in UHPM rocks (e.g., McLelland et al., 2006; Mattinson et al., 2006; Leech et al., 2007; Rubatto and Hermann, 2007) and along the prograde and/or retrograde segments of the *P-T* evolution in UHTM rocks, but generally not at the metamorphic peak (e.g., Harley et al., 2007; Baldwin and Brown, 2008). Measured timescales for both thermal and burial-exhumation events have become shorter as precision in dating has increased, and we now find acceptable rates for tectonic processes that would have been considered too fast only a few years ago (e.g., Hermann, 2003; Camacho et al., 2005). As a result of this progress we may now assess with some confidence the imprint of metamorphism through Earth history.

The *P-T-t* path and *P-t* and *T-t* data provide information about the dynamic evolution of the metamorphic (transient) geotherms, which relates to tectonic processes such as rates of burial and heating, and the map distribution of peak *P-T* conditions and *P-t* and *T-t* data provides information about the spatial variation in thermal regime and rates of burial and heating. When this information is combined with structural data to constrain the kinematics, the data set may allow the particular tectonic setting in which the metamorphism occurred to be identified.

The *P-T-t* paths recovered by this forensic science on rocks are generally described as clockwise or counterclockwise (in *P-T* space), although the *P-T* loops may vary from hairpin to open. Clockwise paths may include a segment with significant isentropic decompression at high *T*, counterclockwise paths may exhibit approximately isobaric cooling at moderate *P*, and portions of either type may be stepped. Although these features alone may not be diagnostic of the tectonic setting, there is a relationship between the retrograde metamorphic evolution and changes in the cooling rate that is characteristic of the underlying tectonic process (e.g., Ehlers et al., 1994). Also, for regional-scale metamorphism—such as that found in large accretionary orogenic systems or subduction-to-collision orogenic systems—the relationship between metamorphic grade and the timing of peak metamorphism, and the magnitude of the departure from a stable geotherm, are different for metamorphism driven by conductive heating versus metamorphism that requires a component of advected heat and for metamorphism terminated by erosion versus that terminated by extension (e.g., Stüwe et al., 1993a, 1993b; Stüwe, 1998a, 1998b).

Orogenic events in accretionary orogenic systems are driven by variations in the coupling across the convergent plate boundary, which may be due to changes in plate kinematics (Dewey, 1975) or caused by features on the subducting plate (Cloos, 1993; Collins, 2002). Accretionary orogenic systems are progressively terminated by subduction-to-collision orogenesis. Therefore, as a generalization, we may expect that the age of metamorphism in the subduction zone environment is older than, or may date, the transition from subduction to collision at any particular locality—although it may be diachronous along the orogenic system—whereas the age of metamorphism in the arc and subduction zone backarc or orogenic hinterland is likely to be close to or younger than the age of the transition from subduction to collision.

As high-precision geochronology becomes more widely available from a larger number of orogenic belts of all ages, it should be possible to evaluate whether the high-pressure metamorphism for a particular orogenic belt is older than the associated high-temperature metamorphism. In the case of the European Variscides, this sequence—high-pressure metamorphism older than high-temperature metamorphism—is, indeed, the case (O'Brien and Rötzler, 2003). However, a word of caution is necessary since orogen-parallel translation may juxtapose metamorphic belts formed at slightly different times along a common major plate boundary system, as was the case for the Mesozoic metamorphic belts in southwest Japan, where the outboard HP-LT Sanbagawa metamorphism records younger ages of peak metamorphism than the inboard HT-LP Ryoke metamorphism (Brown, 1998a, 1998b, 2002b).

The relationship between plate tectonics and metamorphism was addressed in the early days by Miyashiro (1972) and Brothers and Blake (1973). In this chapter, I review sites of contemporary metamorphism (<200 Ma) and consider metamorphism in relation to different orogenic systems during the earlier part of the Phanerozoic rock record (>200 Ma) before extending the analysis back throughout Earth's history using a published data set compiled in 2005 (Brown, 2007a) and updated as necessary in the discussion below. I discuss the implications of these data in relation to dual thermal environments as the hallmark of plate tectonics and identify when the imprint of this distinctive feature first appears in the rock record (cf. Brown, 2006). Finally, I evaluate the age distribution of extreme metamorphism in the Neoarchean to Cenozoic rock record and consider how this relates to the supercontinent cycle (cf. Brown, 2007b).

## METAMORPHISM ON EARTH DURING THE PAST 600 MILLION YEARS

### Contemporary Metamorphism (<200 Ma)

Metamorphism during the Cenozoic and Mesozoic eras has taken place at (1) divergent plate boundaries, where newly generated ocean crust is metamorphosed following hydrothermal circulation of seawater, (2) convergent plate boundaries, where subduction takes crustal rocks deep into the mantle before partial exhumation and accretion, and arcs and subduction zone backarcs may be shortened and thickened during subduction-to-collision orogenesis, and (3) plate boundaries that involve mainly lateral displacement, although these also may involve transpression or transtension.

Rocks of the blueschist to eclogite facies series (higher-pressure–lower-temperature [HP-LT] metamorphism) are created in the subduction zone. We know this from samples of glaucophane schist entrained in serpentine mud volcanoes in the Mariana forearc (e.g., Salisbury et al., 2002). Incipient blueschists in the Mariana forearc with the assemblage lawsonite-pumpellyite-hematite yield an estimated $T$ of 150–250 °C at $P$ of 0.5–0.6 GPa (Maekawa et al., 1993) and $dT/dP$ of 300–400 °C/GPa (approximately equivalent to 9–11 °C/km). These results are consistent with thermal models for warm subduction zones (e.g., Hacker et al., 2003) and thermal gradients derived from Eocene low-grade blueschist facies series rocks (e.g., Potel et al., 2006).

Ultrahigh-pressure metamorphic (UHPM) rocks—where the minimum $P$ must exceed ~2.7 GPa at moderate $T$ for coesite to occur—register deep subduction and exhumation of continental lithosphere during the early stage of collision. We know this from samples of coesite-bearing eclogites of Eocene age at Tso Morari in the frontal Himalayas (Sachan et al., 2004; Leech et al., 2005, 2007), and from Pliocene age coesite-bearing eclogites in eastern Papua New Guinea (Baldwin et al., 2004; Monteleone et al., 2007) that have been exposed exceptionally rapidly by rifting. Furthermore, Searle et al. (2001) have argued that deep earthquakes of the Hindu Kush seismic zone represent a tracer of contemporary coesite and diamond formation in eclogite facies ultrahigh-pressure metamorphism occurring at the present day.

Lawsonite blueschists and lawsonite eclogites (HPM-UHPM) record cold subduction. We know this from thermodynamic modeling of Eocene lawsonite blueschists and lawsonite eclogites exhumed to Earth's surface in the Pam Peninsula, New Caledonia (Clarke et al., 1997, 2006), and from studies of xenoliths of lawsonite eclogite of Eocene age, brought up to the surface in Tertiary kimberlite pipes at Garnet Ridge in the Colorado Plateau, United States, which are inferred to record conditions in excess of 3.5 GPa (Usui et al., 2003, 2006).

Rocks of the granulite facies series (higher-temperature–lower-pressure [HT-LP] metamorphism) are scavenged from the deeper parts of Mesozoic oceanic plateaus (e.g., Gregoire et al., 1994; Shafer et al., 2005) and occur in exposed middle to lower crust of young continental arcs (e.g., Lucassen and Franz, 1996). Although the interpretation is contentious, metapelitic xenoliths retrieved from Neogene volcanoes in central Mexico (Hayob et al., 1989, 1990) and at El Joyazo in southeast Spain (Cesare and Gomez-Pugnaire, 2001) have been argued to record evidence of Cenozoic G-UHTM, and evidence of melt-related processes in lower crustal garnet granulite xenoliths from Kilbourne Hole, Rio Grande Rift, suggests contemporary G-UHTM in the lower crust of rifts (Scherer et al., 1997).

The mechanism that provides the enhanced thermal flux necessary to drive granulite facies series metamorphism remains elusive in most circumstances. Radiogenic heating is important and sometimes is sufficient (Andreoli et al., 2006; McLaren et al., 2006). However, commonly the apparent thermal gradients retrieved from granulite terranes exceed the conductive limit, and significant advection of heat within the crust is required (e.g., Sandiford and Powell, 1991; Stüwe, 2007). Underplating by basaltic magma commonly is implicated to provide heat (e.g., Dewey et al., 2006), but the temporal relations between extension, crustal melting, and basalt emplacement are rarely simple (e.g., Barboza et al., 1999; Peressini et al., 2007). Heating by multiple intraplating of sills may be effective at melting the lower crust in continental arcs (e.g., Jackson et al., 2003; Dufek and Bergantz, 2005; Annen et al., 2006).

Hyndman et al. (2005a, 2005b; see also Currie and Hyndman, 2006) have argued that mountain belts and orogenic plateaus occur in former subduction zone backarcs, which are characterized by high surface heat flow and thin weak lithosphere (cf. Sandiford and Powell, 1991). Currie and Hyndman (2006) report observations that indicate uniformly high temperatures in the shallow mantle and a thin lithosphere (1200 °C at ~60 km depth) over subduction zone backarc widths of 250 to >900 km. Similar high temperatures are inferred for extensional backarcs of the western Pacific and southern Europe, but the thermal structures are complicated by extension and spreading. Following termination of subduction by collision, the high temperatures decay over a timescale of ~300 m.y. (Curie and Hyndman, 2006). During terrane accretion and/or subduction-to-collision orogenesis, subduction zone backarcs represent a particularly suitable tectonic setting for the generation of granulite metamorphism.

It is likely that granulites are being generated today under orogenic plateaus, such as Tibet and the Altiplano, based on an inference of the presence of melt, derived from mica breakdown melting, in the interpretation of multiple geophysical data sets (e.g., Nelson et al., 1996; Schilling and Partzsch, 2001; Unsworth et al., 2005). Also, there is evidence from crustal xenoliths of high-pressure–high-temperature metamorphism under Tibet (Hacker et al., 2000, 2005b; Ding et al., 2007; Chan et al., 2007). Modeling of metamorphic *P-T-t* paths in large convergent orogens produces peak *P* followed by peak *T* (clockwise) *P-T-t* paths with maximum *P-T* conditions that may reach those typical of E-HPGM (Jamieson et al., 2002, 2006), although temperatures in the lower crust beneath the plateau in recent models approach 1000 °C (Beaumont et al., 2006). Modeling by McKenzie and Priestly (2008) suggests that temperatures in the middle crust of thickened lithosphere similar to that of Tibet may exceed 1000 °C, sufficient to generate widespread granulites and ultrahigh-temperature metamorphic rocks at moderate pressures. However, a cautionary note is necessary, since melting is an endothermic process that buffers heating (e.g., Stüwe, 1995).

Heating by viscous dissipation may be important in some subduction-to-collision orogenic systems, particularly in the early stages of subduction. Although viscous dissipation is a mechanism potentially to generate heat (e.g., Kincaid and Silver, 1996; Stüwe, 1998a; Leloup et al., 1999; Burg and Gerya, 2005), it requires initially strong lithosphere (e.g., a differential stress of 100–300 MPa); England and Houseman (1988) suggest that a differential stress of 100–200 MPa is necessary to generate a plateau, consistent with this requirement. The strong positive correlation between overall intensity of viscous heating in crustal rocks and the instantaneous convergence rate suggests that a significant contribution (>0.1 $\mu W\ m^{-3}$) of viscous heating into the crustal heat balance may be expected when the convergence rate exceeds 1 cm $yr^{-1}$, particularly if the lower crust is strong (Burg and Gerya, 2005). Therefore, heating by viscous dissipation may become a dominant heat source in the early stage of subduction-to-collision orogenesis if convergence rates are rapid (e.g., the Himalayas), and viscous dissipation may be generally significant in the thermal energy budget of orogens (e.g., Stüwe, 2007).

We may conclude that contemporary metamorphism on Earth occurs in two very different thermal environments. The first environment is characterized by low *dT/dP*, mirrored by higher-pressure–lower-temperature metamorphism, corresponding to the subduction zone environment; this may be modified by and/or terminated by terrane accretion and/or continental subduction and terminal collisional suturing. The second environment is characterized by high *dT/dP*, registered by higher-temperature–lower-pressure metamorphism, corresponding to the arc and a subduction zone backarc, which will be inverted during collision; a thick hot orogen may become modified by erosion and/or subsequent collapse and/or extension.

## Metamorphism in the Earlier Phanerozoic Record (>200 Ma)

The ocean lithosphere is older than 200 Ma. A plethora of papers since the 1968 plate tectonics revolution suggest that we may reasonably extrapolate back through one or two ocean lithosphere cycles. Thus, we may expect similar tectonics since the last stages of Gondwana amalgamation at the dawn of the Phanerozoic eon. Prior to this time the extrapolation becomes more contentious, particularly for rocks older than Neoproterozoic (Stern, 2005).

Metamorphism in extensional-contractional accretionary orogenic systems associated with west-directed subduction (Doglioni et al., 2006; Husson et al., 2007), such as the Tasmanides of eastern Australia or the Acadian of northeastern North America, is dominated by HT-LP metamorphism, with near-isobaric clockwise hairpin-like heating/cooling *P-T-t* paths or counterclockwise *P-T-t* paths in which deformation and metamorphism proceed contemporaneously. G-UHTM, E-HPGM, and UHPM rocks are absent at outcrop, although blueschists may occur sporadically early in the orogenic cycle. Short-lived phases of orogenesis probably relate to interruptions in the continuity of subduction by topographic features on the ocean plate, particularly oceanic plateaus (Cloos, 1993; Collins, 2002; Koizumi and Ishiwatari, 2006).

Metamorphism in contractional-erosional orogenic systems associated with east-directed subduction (Doglioni et al., 2006; Husson et al., 2007), such as the late Paleozoic accretionary prism of the Coastal Cordillera in Chile (Willner, 2005) and the Cretaceous Diego de Almagro Metamorphic Complex in Chilean Patagonia (Willner et al., 2004), is characterized by low-temperature (low-*dT/dP*) metamorphism, including blueschists, registering a thermal environment with an apparent thermal gradient around 350 °C/GPa. Such conditions may be more typical of the west-facing advancing (Cordilleran) convergent plate margin in comparison with an east-facing retreating (extensional-contractional) convergent plate margin.

In terrane accretion orogenic systems, allochthonous elements were accreted to continental margins in convergent systems that commonly involved oblique relative plate motion vectors. Paired metamorphic belts sensu Miyashiro (1961) are characteristic due to orogen-parallel terrane migration and juxtaposition by accretion of contemporary belts of contrasting type (Brown, 1998a, 1998b, 2002b), with a HP-LT metamorphic belt outboard and a HT-LP metamorphic belt inboard, exemplified by the Mesozoic metamorphic belts of Japan (Miyashiro, 1961) and the Paleozoic-Mesozoic metamorphic belts of the western United States (Patrick and Day, 1995). In some systems an important additional feature was ridge subduction, which is reflected in the pattern of both HT-LP metamorphism and the associated magmatism (Brown, 1998b). Granulites may occur at the highest grade of metamorphism in the HT-LP belt, but UHPM rocks generally appear to be absent in the outboard HP-LT belt.

In subduction-to-collision orogenic systems, subduction and initial collision may generate HPM-UHPM of continental crust as it is being subducted and until the subduction zone is choked. Deformation of the hinterland may generate clockwise (or less commonly counterclockwise) metamorphic *P-T-t* paths in granulite facies series rocks according to the mechanisms and relative rates of thickening and thinning versus heat transfer by conduction and advection. In many subduction-to-collision orogenic systems, radiogenic heat production and heat transfer by conduction rather than advection are dominant, and penetrative deformation largely precedes the peak of metamorphism because rates of continental deformation typically are about an order of magnitude faster than rates of thermal equilibration over the length scale of the crust (Stüwe, 2007). The European Variscides represent a classic example of a subduction-to-collision orogenic system, in which early HPM-UHPM is followed by E-HPGM (e.g., O'Brien and Rötzler, 2003) and, possibly, G-UHTM, although this point is contentious (Rötzler and Romer, 2001; Štípská and Powell, 2005). At a larger scale, the variation in metamorphic style of subduction-to-collision orogenic systems is well illustrated by the variation along the Appalachian/Caledonian–Variscide–Altaid and the Cimmerian–Himalayan–Alpine chains.

Each of these older Phanerozoic orogenic systems preserves metamorphic belts with contrasting metamorphic facies series that may be inferred to record contrasting thermal regimes. These examples are consistent with the premise that plate tectonics may be extrapolated back through at least two ocean lithosphere cycles (to ca. 600 Ma); they also highlight the variations in imprint in the rock record generated by the complex interactions and diversity of processes along plate boundary zones.

## Paired Metamorphic Belts Revisited

Orogens may be composed of belts with contrasting types of metamorphism that record different apparent thermal gradients. Classic paired metamorphic belts—those composed of an inboard higher-temperature–lower-pressure (HT-LP) metamorphic belt, commonly with associated penecontemporaneous granites, juxtaposed along a tectonic contact against an outboard higher-pressure–lower-temperature (HP-LT) metamorphic belt—are found in accretionary orogens of the circum-Pacific (Miyashiro, 1961).

In accretionary orogens, HP-LT blueschists and low-temperature eclogites are generated in the subduction zone. A majority of known localities of lawsonite eclogite are associated with the HP-LT terranes of the circum-Pacific accretionary orogens (Tsujimori et al., 2006). Complementary HT-LP metamorphism may be generated in an arc and subduction zone backarc or orogenic hinterland. Events that may occur at the trench include ridge subduction, entry into the subduction channel of an ocean floor topographic high, and evolution of the plate kinematics, leading to a change from a retreating hinge to an advancing arc. These events may affect the thermal structure of the overriding plate, may lead to a change in the displacement vector across the axis of plate divergence or across an associated transform as a ridge is subducted, and may change the tectonics in the orogen from extension to shortening (e.g., Brown, 1998b, 2002b; Collins, 2002). HP-LT blueschists and low-temperature eclogites represented educted and accreted materials that have been translated along a convergent margin as a forearc terrane—due to changes in plate kinematics—to juxtapose metamorphic belts of contrasting type during a single orogenic cycle along a common convergent margin (e.g., Brown, 1998a, 1998b, 2002b; Tagami and Hasebe, 1999).

In Miyashiro's original classification of types of metamorphism (Miyashiro, 1961), an intermediate-*P/T* type of metamorphism was included for unpaired belts such as those in the Scottish Highlands and the northern Appalachians, although in both cases the medium-*P/T* metamorphic belt (Barrovian type) is juxtaposed against an intermediate lower-*P/T* metamorphic belt (Buchan type), which is also the case in the eastern Himalayas in Nepal (Goscombe and Hand, 2000). Miyashiro (1973) subsequently suggested that "paired and unpaired (single) metamorphic belts form by the same mechanism, and an unpaired belt represents paired belts in which the contrast between the two belts is obscure, or in which one of the two belts is undeveloped or lost." One issue to consider is whether to extend the concept of "paired metamorphic belts" more widely than accretionary orogens, outside the original usage by Miyashiro (1961), to subduction-to-collision orogenic systems, as perhaps was implied by Miyashiro in his 1973 publication.

The modern plate tectonics regime is characterized by a duality of thermal environments in which two principal types of regional-scale metamorphic belts are being formed contemporaneously. Brown (2006) considers this duality to be the hallmark of plate tectonics, and the characteristic imprint of plate tectonics in the ancient rock record to be the broadly contemporaneous occurrence of two contrasting types of metamorphism reflecting the duality of thermal environments. On this basis, I propose the following: *Penecontemporaneous belts of contrasting type of metamorphism that record different apparent thermal gradients, one warmer and the other colder, commonly juxtaposed by plate tectonics processes, may be called "paired metamorphic belts."*

Thus, the combination of penecontemporaneous G-UHTM with E-HPGM in adjacent terranes may be described as a paired metamorphic belt. This extends the original concept of Miyashiro (1961) beyond the simple pairing of HT-LP and HP-LT metamorphic belts in circum-Pacific accretionary orogenic systems, and makes it more useful in the context of our better understanding of the relationship between thermal regimes and tectonic setting. This is particularly useful in subduction-to-collision orogenic systems, where an accretionary phase is overprinted by a collision phase that will be registered in the rock record by the imprint of penecontemporaneous higher-pressure–lower-temperature metamorphism at the suture and higher-temperature–lower-pressure metamorphism in the arc and subduction zone backarc or orogenic hinterland (Brown, 2006).

## METAMORPHISM FROM THE MESOARCHEAN TO THE CENOZOIC

Metamorphic belts that exhibit extreme metamorphism are classified into three types according to the characteristic metamorphic facies series registered by the belt (Fig. 4), as follows: (1) HPM-UHPM, characterized by lawsonite blueschist to lawsonite eclogite facies series rocks and blueschist to eclogite to ultrahigh-pressure facies series rocks, where *T* plotted in Figure 4 is that registered at maximum *P*; (2) E-HPGM, characterized by facies series that reach peak *P-T* in the high-pressure granulite facies (O'Brien and Rötzler, 2003), where maximum *P* and *T* generally are achieved approximately contemporaneously (Fig. 4); and (3) G-UHTM, characterized by granulite facies series rocks that may reach ultrahigh-temperature metamorphic conditions, where *P* plotted in Figure 4 is that registered at maximum *T*.

The *P-T* value for each terrane shown in Figure 5 records a point on a metamorphic (transient) geotherm, and different apparent thermal gradients are implied by each type of metamorphism. These apparent thermal gradients are inferred to reflect different tectonic settings. HPM-UHPM is characterized by apparent thermal gradients of 150–350 °C/GPa (approximately equivalent to 4–10 °C/km), and plots across the boundary between the "cooler than normal" and "ultralow temperature" fields. About half of these terranes require a process other than simple thickening to achieve such cold gradients. We know from the global context that all of these terranes were associated with subduction, so it is likely that subduction was the process that created the ultralow-temperature environment. E-HPGM is characterized by apparent thermal gradients of 350–750 °C/GPa (10–20 °C/km), and plots across the normal continental geotherm but mostly in the field where heating is a conductive response to thickening. G-UHTM is characterized by apparent thermal gradients >>750 °C/GPa (>>20 °C/km), and mostly plots across the boundary into the field that requires an advective component of heating.

Figure 6 illustrates *P* of metamorphism, which is taken as a proxy for depth of burial to the metamorphic peak, plotted against age of peak metamorphism. There is a marked change in *P* of metamorphism and implied depth of burial beginning in the late Neoproterozoic, which records the first appearance in the orogenic record of HPM-UHPM rocks. The period from the Neoarchean to the Neoproterozoic is characterized by *P* of metamorphism from 0.5 GPa to 2.5 GPa, which yields an implied maximum depth of burial up to 90 km. This does not necessarily require subduction. Figure 7 illustrates apparent thermal gradient, which is inferred to relate to tectonic setting, plotted against age of peak metamorphism. Each type of metamorphism has a distinct range of apparent thermal gradient, as expected from Figure 5, and HPM-UHPM is restricted to the late Neoproterozoic and Phanerozoic, as expected from Figure 6. However, what is now clear is the dual

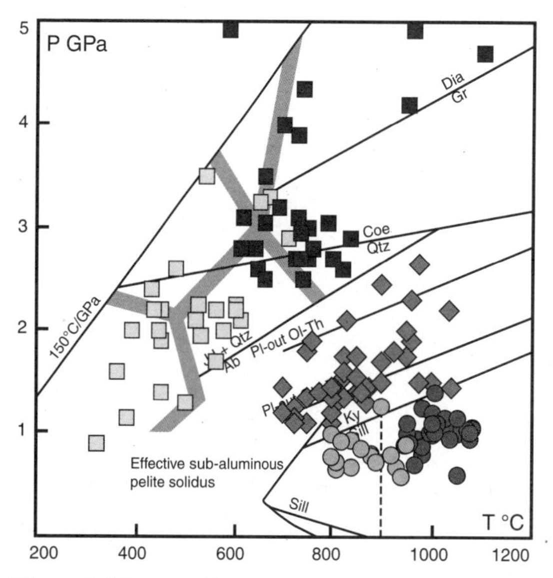


Figure 4. Metamorphic patterns based on representative "peak" metamorphic *P-T* conditions of metamorphic belts in relation to the metamorphic facies in Figure 2. Common granulite belts and granulite facies ultrahigh-temperature metamorphism—G-UHTM belts (*P* at maximum *T*; light circles are common granulite belts and dark circles are G-UHTM belts; data from Tables 1 and 2 of Brown, 2007a); medium-temperature eclogite–high-pressure granulite (E-HPGM) belts (peak *P-T*; diamonds; data from Table 3 of Brown, 2007a); lawsonite blueschist–lawsonite eclogite and ultrahigh-pressure metamorphic (HPM-UHPM) belts (*T* at maximum *P*; light squares are lawsonite-bearing rocks and dark squares are ultrahigh-pressure rocks; data from Tables 4 and 5 of Brown, 2007a).

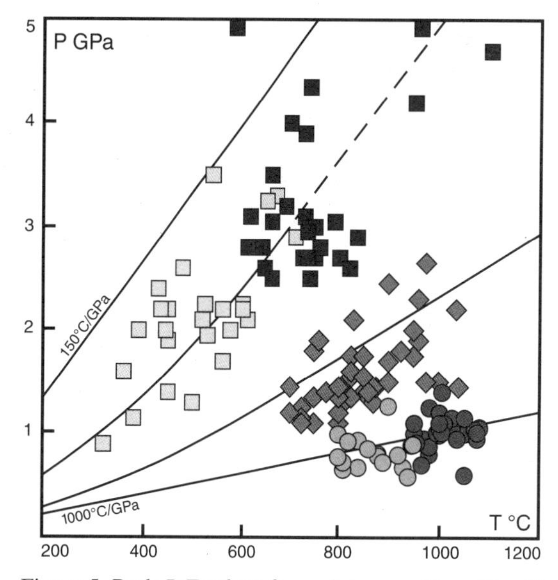


Figure 5. Peak *P-T* values for each terrane (from Fig. 4) in relation to a normal (conductive) continental geotherm. Each of these data records a point on a metamorphic (transient) geotherm, and the different apparent thermal gradients implied by each type of metamorphism are inferred to reflect different tectonic settings and thermal regimes.

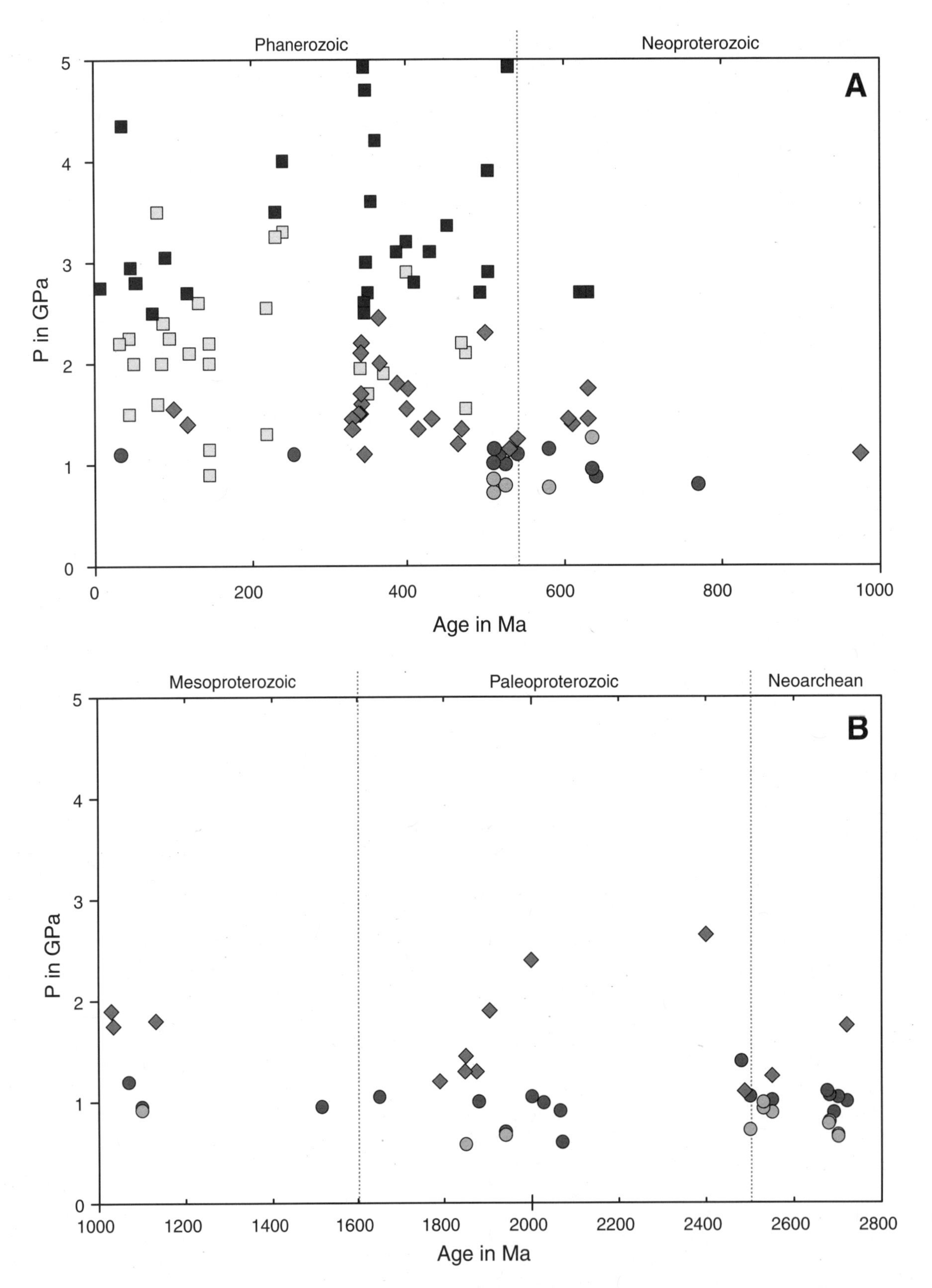


Figure 6. *P* of metamorphism in GPa plotted against age of metamorphism in Ma (data from Tables 1–5 of Brown, 2007a) for the three main types of extreme metamorphic belt—G-UHTM (circles), E-HPGM (diamonds), and HPM-UHPM (squares)—for two time intervals. Shading as in Figure 4. (A) Phanerozoic to Neoproterozoic. (B) Mesoproterozoic to Neoarchean.

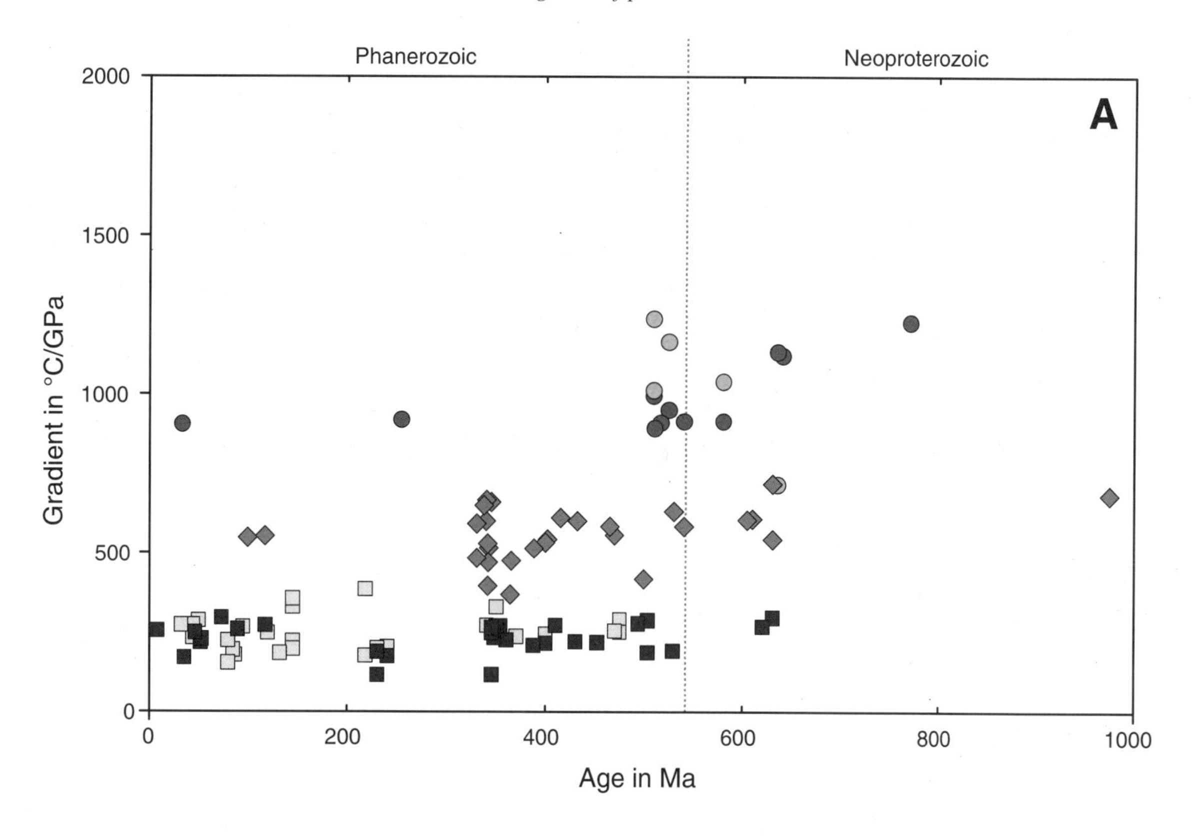


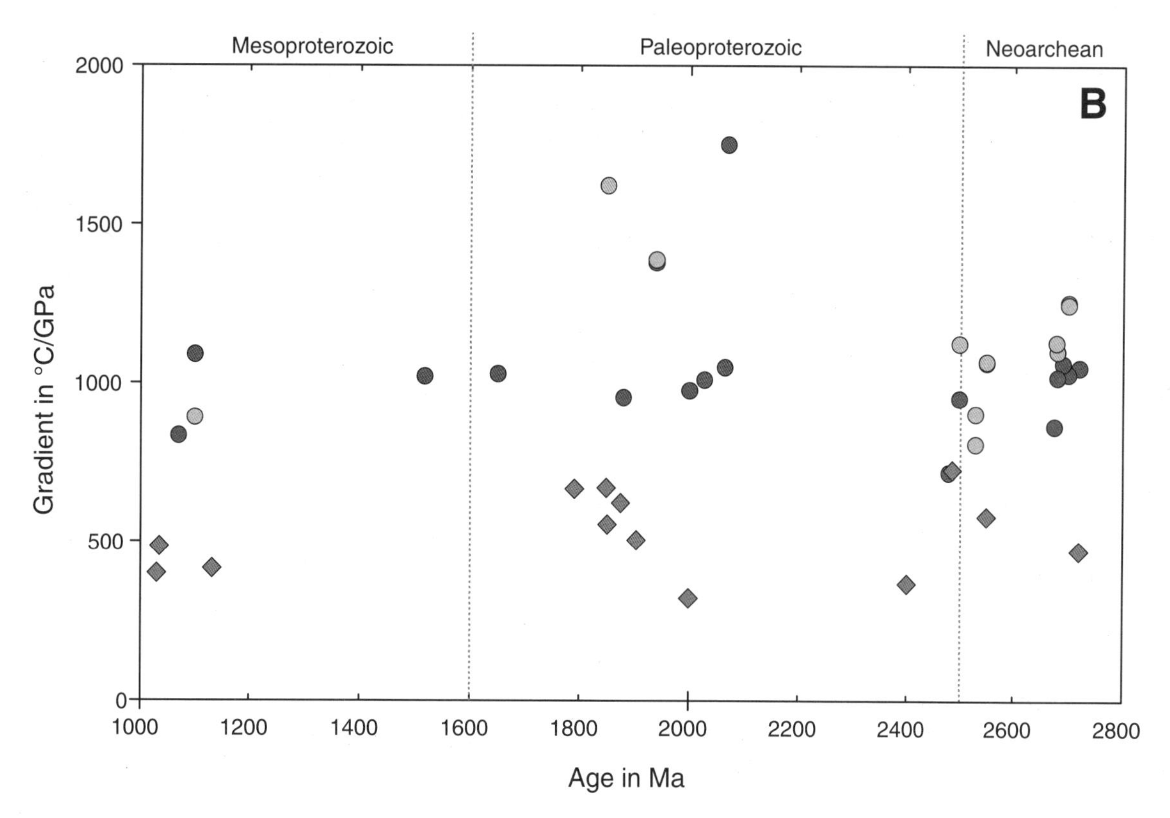


Figure 7. Apparent thermal gradient in °C/GPa plotted against age of peak metamorphism in Ma (data from Tables 1–5 of Brown, 2007a) for the three main types of extreme metamorphic belt—G-UHTM (circles), E-HPGM (diamonds), and HPM-UHPM (squares)—for two time intervals. Shading as in Figure 4. (A) Phanerozoic to Neoproterozoic. (B) Mesoproterozoic to Neoarchean.

nature of the thermal regimes represented in the metamorphic record since the Neoarchean. The Neoarchean to Neoproterozoic period is characterized by G-UHTM and E-HPGM, whereas the late Neoproterozoic and the Phanerozoic are characterized by HPM-UHPM and E-HPGM together with G-UHTM, although the latter occurs only sporadically post-Cambrian.

There are several caveats about possible bias in this record. It is commonly argued that going back through time increases loss of information by erosion of the older record. However, the data in Figures 5 and 6 plot in particular periods, and there is a clear distinction between the pre-Neoproterozoic, where HPM-UHPM does not occur, and Neoproterozoic and younger belts, where HPM-UHPM is common. These observations are inconsistent with a progressively degraded record with increasing age. Extrapolation back in time also raises questions about partial to complete overprinting by younger events, which is a concern in any metamorphic study. However, our ability to recognize the effects of overprinting and to "see through" them has improved significantly, and overprinting has been avoided in compiling the data set used for this analysis (Brown, 2007a). Finally, as discussed earlier, it is likely that some Phanerozoic G-UHTM rocks have not yet been exposed at Earth's surface, leading to bias in the younger part of the record.

Although patterns based on first occurrences may be challenged by new discoveries, the analysis that follows provides a set of compelling first-order observations from which to argue that the modern era of ultralow-temperature subduction began in the Neoproterozoic, as registered by the occurrence of HPM-UHPM, but that ultralow-temperature subduction alone is not the hallmark of plate tectonics. In contrast, G-UHTM and E-HPGM are present in the exposed rock record back to the Neoarchean, registering a duality of thermal regimes, which has been argued to represent the hallmark of plate tectonics (Brown, 2006). Based on this observation, plate tectonics processes likely were operating in the Neoarchean as recorded by the imprints of dual types of metamorphism in the rock record, and this may manifest the beginning of a global plate tectonics mode on Earth.

The distribution of these types of metamorphism throughout Earth history, based on the age of metamorphism as defined above, is displayed in Figure 8 together with periods of super-

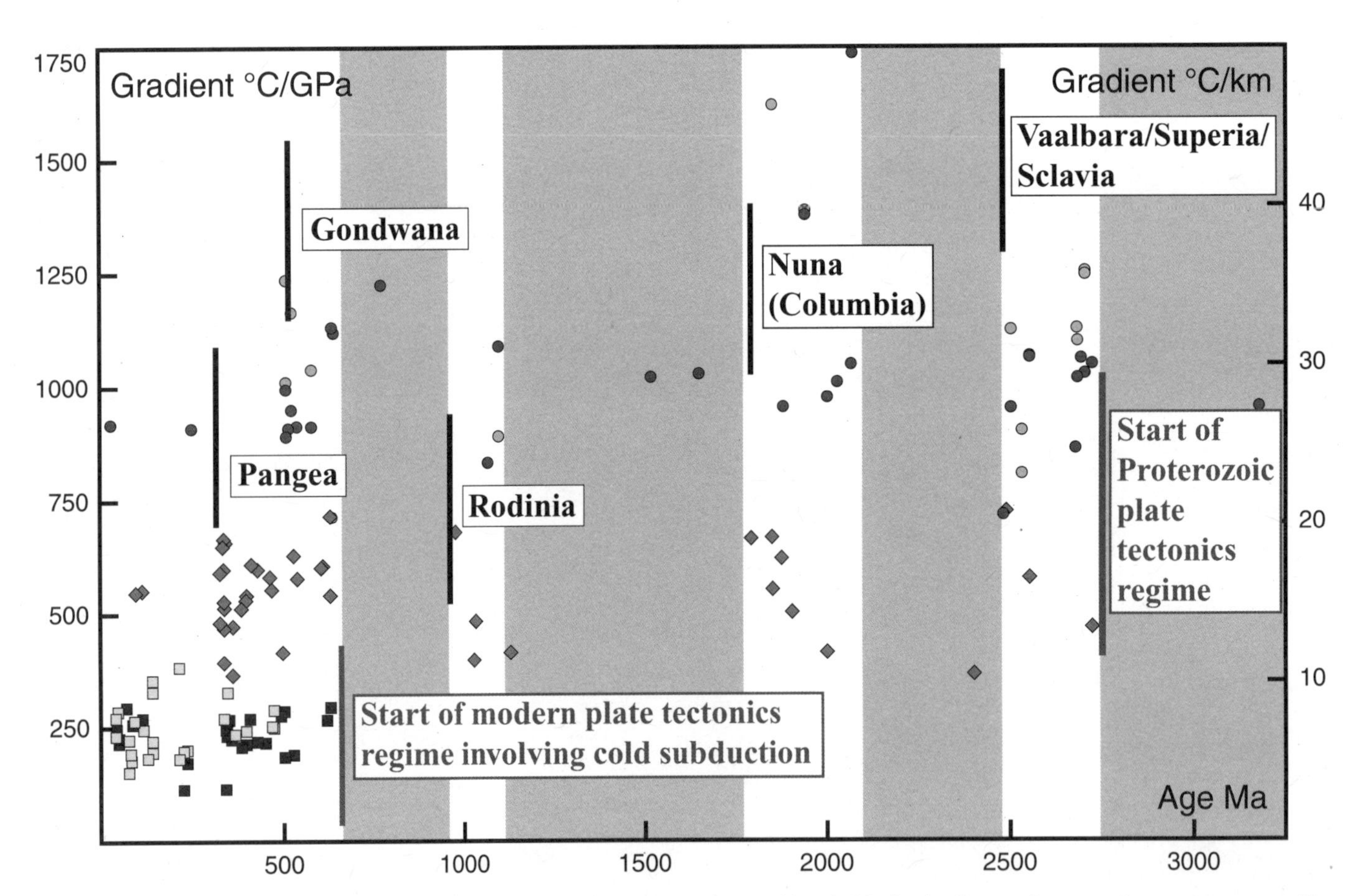


Figure 8. Plot of apparent thermal gradient in °C/GPa versus age of peak metamorphism in Ma for the three main types of extreme metamorphic belt—G-UHTM (circles), E-HPGM (diamonds), and HPM-UHPM (squares). Shading as in Figure 4. An approximate conversion to °C/km is also shown (data from Tables 1–5 of Brown, 2007a). Also shown are the timing of supercraton amalgamation (Vaalbara, Superia, and Sclavia) and supercontinent amalgamation (Nuna [Columbia], Rodinia, Gondwana as a step to Pangea), and the start of the Proterozoic plate tectonics and modern plate tectonics regimes in relation to thermal gradients for the three main types of extreme metamorphic belt.

continent amalgamation. Changes in the metamorphic record broadly coincide with the transitions from the Archean to Proterozoic and Proterozoic to Phanerozoic eons, and imply a different style of tectonics in the Archean eon in comparison with the Proterozoic eon and in the Proterozoic eon in comparison with the Phanerozoic eon. Overall, the restricted time span of different types of metamorphism through Earth history and the periods of metamorphic quiescence during the Proterozoic eon suggest a link with the supercontinent cycle and major events in the mantle. This issue is discussed in more detail later in this chapter.

### Granulite Facies Ultrahigh-Temperature Metamorphism (G-UHTM)

G-UHTM indicates regional-scale deep crustal metamorphism at temperatures generally >750 °C, and in extreme cases >1100 °C, well above the solidus. At these temperatures it is challenging to determine accurately intensive variables such as the peak temperature, and it is also challenging to determine accurately the age of the metamorphic peak and the rate of prograde and retrograde evolution around peak temperatures. Zircon is the preferred chronometer for G-UHTM, especially in garnet-bearing granulites where some of the controversy concerning the timing of zircon growth, recrystallization, and reequilibration with respect to peak temperatures may be reduced via an assessment of REE patterns (Harley et al., 2007). In spite of these issues, compilation of the best available data does reveal a global pattern of G-UHTM that I interpret to have geodynamic significance.

G-UHTM is dominantly a Neoarchean to Cambrian phenomenon documented during four distinct periods in Earth history (Figs. 6 and 7), with Neoproterozoic-Cambrian G-UHTM being as common in the geological record as Neoarchean examples. The four periods are (1) the Neoarchean (e.g., Kaapvaal Craton and the Southern Marginal Zone of the Limpopo belt [southern Africa] and the Badcallian event in the Lewisian Complex of the Assynt terrane [Scotland] at ca. 2.72–2.69 Ga; the Napier Complex [East Antarctica] and the Andriamena Unit [Madagascar] at ca. 2.56–2.46 Ga), (2) the Orosirian (e.g., the Hoggar [northern Africa] and the Lewisian Complex on South Harris [Scotland] at ca. 2.0–1.88 Ga), (3) the Ectasian to Stenian (e.g., the Eastern Ghats belt [India] and the Mollendo-Camana Block [Peru] at ca. 1.4–1.0 Ga), and (4) the Ediacaran to Lower Cambrian (e.g., the Brasiliano and Pan-African belts of the southern continents at ca. 630–510 Ma). The four main periods of G-UHTM show a remarkable coincidence with the first formation of supercratons (Vaalbara, Superior, and Sclavia) and the supercontinent cycle (Nuna [Columbia], Rodinia, and Gondwana).

An accurate age for the oldest high-grade metamorphism in these terranes commonly has been difficult to retrieve, and the age of the oldest G-UHTM for many of these examples is disputed. Although the area of the Lewisian Complex in Scotland is small on a world scale, it represents one of the most intensively studied pieces of Archean crust reworked in the Paleoproterozoic, yet the age of the G-UHTM (Badcallian event) is unresolved. It may be as old as ca. 2.7 Ga (Crowley et al., 2006; Corfu, 2007) or as young as ca. 2.5 Ga (Friend et al., 2007). In the Limpopo belt of southern Africa, a Neoarchean age for G-UHTM in the Southern Marginal Zone is well established (Kreissig et al., 2000, 2001). However, controversy is associated with the age of G-UHTM in the Central Zone, where there is clearly a younger G-UHTM event at ca. 2.03 Ga (Buick et al., 2003, 2006), but where there is accumulating evidence for an event at ca. 2.56 Ga (Kröner et al., 1999; Boshoff et al., 2006).

There is controversy about the age of G-UHTM in the Napier Complex, Antarctica, which is perhaps the hottest regionally developed metamorphic terrane on Earth. An early Paleoproterozoic age commonly has been suggested, but more recent work using well-characterized zircons suggests a late Neoarchean age to be more likely as the true age for the G-UHTM event (Kelly and Harley, 2005). U-Pb zircon ages of ca. 2.49 Ga from garnet-bearing orthogneiss provide an absolute minimum age for the foliation associated with the G-UHTM event, whereas in paragneiss, internal zircon zoning relationships and estimated zircon-garnet rare earth element partitioning data suggest an absolute minimum age of ca. 2.51 Ga for this event. Furthermore, Harley (2006) suggests that the real age for the G-UHTM event is more likely to be older than 2.56 Ga, based on higher Ti concentration in zircons in localized leucosomes formed on the crystallization of late-stage dry melts, and argues that the many published U-Pb zircon ages of 2.51–2.45 Ga reflect late, postpeak zircon growth associated with near-isobaric cooling and fluid-melt-rock interactions.

There is only one confirmed example of G-UHTM >>2.5 Ga in age, which is the Voronezh Crystalline Massif of Sarmatia (ca. 3.2 Ga; Fonarev et al., 2006). G-UHTM is rare in the exposed post-Cambrian rock record but may be inferred for the lower crust of large Cenozoic orogenic systems as discussed above. In the Phanerozoic eon, parts of the European Variscides evolve from E-HPGM to G-UHTM, but there is only one confirmed example of G-UHTM younger than Paleozoic exposed at Earth's surface, which is the Gruf Unit in the central Alps (ca. 33 Ma; Liati and Gebauer, 2003).

### Medium-Temperature Eclogite–High-Pressure Granulite Metamorphism (E-HPGM)

There are about 50 documented terranes of this type, with approximately equal numbers of Proterozoic and Phanerozoic age, although the Phanerozoic examples are dominated by various segments of the European Variscides. About one-third of recognized high-pressure granulite belts have evolved from medium-temperature eclogite facies assemblages, and about one-quarter of the remainder evolve into granulite facies ultrahigh-temperature metamorphic conditions, both apparently without distinction between Proterozoic and Phanerozoic examples. Compositional zoning of minerals, and reaction coronas, symplectites, and kelyphites commonly develop as multivariant reactions are crossed during a clockwise *P-T* evolution.

Periods of E-HPGM coincide with the Nuna, Rodinia, and Gondwana supercontinent cycles; E-HPGM also is common in the lower Paleozoic Caledonides and Variscides, and rare examples occur in the exposed post-Carboniferous rock record. There are sporadic occurrences of E-HPGM during the Neoarchean-to-Paleoproterozoic transition, although the age of several of these is poorly constrained or in dispute. This group of occurrences is critical to establishing the early record of E-HPGM, and so it is worth discussing them in more detail.

The oldest occurrence is represented by eclogite blocks within mélange in the Gridino Zone of the Eastern Domain of the Belomorian province (Slabunov et al., 2006). The eclogite facies metamorphism appears to have been reliably dated at 2.72 Ga (Bibikova et al., 2004), and the *P-T* data of 1.40–1.75 GPa and 740–865 °C are well characterized (Volodichev et al., 2004; Slabunov et al., 2006). Nonetheless, this occurrence is one of the earliest records of E-HPGM within a suture zone (see also the discussion about the Inyoni shear zone below, under "Early Archean Metamorphism")—a critical piece of evidence in evaluating the start of plate tectonics from the continental record of metamorphism and early occurrences of low apparent thermal gradients—and it will be worthwhile to replicate these results, although the fact that these Neoarchean ages come from blocks in a mélange cannot be avoided.

The age of granulite facies metamorphism in the North China Craton is a matter of dispute (Zhao and Kröner, 2007; Polat et al., 2007), although the balance of evidence favors an age of ca. 1.85 Ga for the regionally developed E-HPGM in the Trans–North China orogen that cuts north-south through the center of the craton (Kröner et al., 2006; Zhang et al., 2006). The age of 2.49 Ga quoted in Brown (2007a) is for the Jianping Complex, in western Liaoning province, which is part of the Eastern Block of the North China Craton (Kröner et al., 1998). The age was derived from four metamorphic zircon grains from BIF sample Ji 7 that were analyzed by SHRIMP (2487 ± 2 Ma) and an additional two metamorphic zircon grains that were evaporated (2487.2 ± 1.1 Ma), and is interpreted to date an older high-pressure granulite metamorphism (1.1 GPa, 800 °C; Wei et al., 2001), which is widely regarded to be ca. 2.5 Ga in both the Western and Eastern Blocks of the North China Craton (e.g., Zhao et al., 2006a). There is no disagreement about the existence of an event at 1.85 Ga, or that this event produces E-HPGM mineral assemblages, but it is unclear whether there is sufficient evidence to support an earlier regionally developed high-pressure granulite event at ca. 2.5 Ga.

Metamorphism of the Sare Sang series in the South Badakhshan Block of the western Hindu Kush in Afghanistan has not been directly dated and is assumed to be the same age as imprecisely dated rocks from the Pamirs (Faryad, 1999; Shanin et al., 1979), which is clearly unsatisfactory. However, in the absence of modern U-Pb analysis, these old ages may bear no relationship to the high-grade metamorphism in the Pamirs—let alone the Hindu Kush to the west—but might record disequilibrium. In the $^{40}Ar/^{39}Ar$ geochronology of Hubbard et al. (1999), there is no evidence for a Precambrian event in the Pamirs, which also call into doubt the inferred Precambrian age for the Sare Sang rocks. A modern U-Pb study of metamorphic zircon or monazite from the Sare Sang rocks is necessary to resolve the age of E-HPGM.

Finally, there are two episodes of Neoarchean to Paleoproterozoic granulite facies metamorphism in the Sharizhalgai salient of the southern Siberian Craton, one at ca. 2.6 Ga and another at 1.88–1.86 Ga (Poller et al., 2005). Following the interpretation of Ota et al. (2004), metamorphism in the garnet-websterites from the tectonically intercalated peridotites of the Saramta Massif is inferred to record the older event, and the peridotites are inferred to represent part of a subarc mantle associated with Neoarchean subduction. However, the garnet-websterite has not been directly dated, which does leave in question the true age of the E-HPGM.

Overall, eclogites from the Gridino Zone of the Eastern Domain of the Belomorian province appears to provide the most robust evidence for subduction associated with Neoarchean supercraton formation by craton amalgamation. The next-younger well-dated example is the eclogite from the Usagaran orogen, Tanzania, at ca. 2.00 Ga (Möller et al., 1995; Collins et al., 2004).

### Blueschist and High-Pressure to Ultrahigh-Pressure Metamorphism (HPM-UHPM)

Blueschist belts and HPM-UHPM appear for the first time in the Neoproterozoic era. The oldest blueschists are found in Asia, where blueschists older than 900 Ma occur in the Jiangnan belt of southern China (Shu and Charvet, 1996), and blueschists of the Aksu Group in western China are overlain by sediments of Ediacaran age (Maruyama et al., 1996). Blueschist-bearing mélange of Ediacaran age occurs in the Bou Azzer inlier of the Anti-Atlas belt in central Morocco (Hefferan et al., 2002), which may be part of a continuous plate boundary zone with the Trans-Saharan belt to the southeast where well-characterized UHPM in Mali is dated at ca. 620 Ma by Jahn et al. (2001).

HPM-UHPM has been recognized during multiple periods in Earth history: the Ediacaran (e.g., Mali, western Africa); the Cambrian (e.g., Kokchetav Massif, northern Kazakhstan; southwest Altyn Tagh, western China); the Ordovician-Silurian (e.g., Tromsø Nappe, Norway; Dulan belt, North Qaidam, western China); the Devonian (e.g., western Gneiss Region, Norway, northeastern Greenland; Maksyutov Complex, southern Urals; Saxonian Erzgebirge, Germany; Bohemian Massif, central Europe; the French Massif Central); the Permo-Triassic (e.g., Dabie-Sulu belt, east-central China; western Tianshan (?), western China); the Lower Cretaceous (e.g., , SW Sulawesi, central Indonesia); through the Cenozoic era (e.g., Dora Maira Massif, Italy; Zermatt-Saas Zone, western Alps; Tso Morari Complex, India; Kaghan Valley, Pakistan); and finally, the Miocene-Pliocene, Earth's youngest UHPM rocks (Fergusson Island, eastern Papua New Guinea). All known examples of UHPM are located within the Pangean convection cell of Collins (2003), and none are to be found associated with long-lived subduction around the edges of the Panthalassan convection cell. In contrast, a majority of occurrences of lawsonite eclogite occur in association with Pacific subduction, the principal exceptions being Spitsbergen, Corsica, Turkey, and Sulawesi (Tsujimori et al., 2006).

Many UHPM terranes record peak temperatures that are warmer than those calculated for steady-state subduction of old ocean lithosphere, which is consistent with the common absence of a coeval volcano-plutonic arc. These two features suggest that UHPM may not be a product of long-lived, steady-state subduction (Liou et al., 2000). *P-T-t* paths from UHPM terranes vary from hairpins, in which the exhumation path is close to the burial path in terms of *dT/dP* and is in the range 150–350 °C/GPa (4–10 °C/km), implying synsubduction exhumation, to clockwise paths with isentropic decompression segments down to Barrovian-type thermal regimes of 350–750 °C/GPa (10–20 °C/km), consistent with thermal relaxation after initial fast exhumation to Moho depths.

## Early Archean Metamorphism

The Eoarchean through Mesoarchean rock record generally records *P-T* conditions characteristic of low- to moderate-*P*, moderate- to high-*T* metamorphic facies. In greenstone belts, metamorphic grade varies from prehnite-pumpellyite facies through greenschist facies to amphibolite facies and, rarely, into the granulite facies; ocean floor metamorphism of the protoliths is common. In high-grade terranes, ordinary granulite facies metamorphism and multiple episodes of anatexis are the norm.

In southern West Greenland, in the Isua supracrustal belt the metamorphism is polyphase—Eoarchean and Neoarchean—and an age of ca. 3.7 Ga has been argued for the early metamorphism (summarized in Rollinson, 2002). *P-T* conditions of 0.5–0.7 GPa and 500–550 °C (Hayashi et al., 2000) or up to 600 °C (Rollinson, 2002) have been retrieved from the Isua supracrustal belt. These data yield warm apparent thermal gradients of 800–1000 °C/GPa (~23–28 °C/km), which is just within the range for a purely conductive response to thickening (Fig. 3), but may reflect thinner lithosphere with higher heat flow than later in Earth history. Also in southern West Greenland, in mafic rocks on the eastern side of Innersuartuut Island in the Itsaq Gneiss Complex of the Færingehavn terrane, where the gneisses record several events in the interval 3.67 Ga to 3.50 Ga (Friend and Nutman, 2005b), early orthopyroxene + plagioclase assemblages record poorly defined but rather ordinary granulite facies conditions, whereas overprinting garnet + clinopyroxene + quartz assemblages, thought by Griffin et al. (1980) to be Eoarchean, probably record the widespread Neoarchean high-pressure granulite facies metamorphism associated with final terrane assembly in the interval 2.715–2.650 Ga (Nutman and Friend, 2007). This is generally consistent with a Mesoarchean to Neoarchean evolution involving a postulated island arc complex in the eastern Akia terrane of southern West Greenland that was built on a volcanic substrate in the interval 3.07–3.00 Ga and subsequently was thickened and metamorphosed in the interval 3.00–2.97 Ga (Garde, 2007), with the concept of progressive terrane assembly during the interval 2.95–2.65 Ga as espoused by Friend and Nutman (2005a), and with the view that plate tectonics processes were operating on Earth by the Mesoarchean to Neoarchean (Brown, 2006).

It has been suggested that some Archean TTG suites and eclogite xenoliths scavenged by kimberlites from deeper levels in Archean cratons are respectively melt and complementary residue from a basaltic source, such as the low-MgO eclogites from the Koidu kimberlite within the TTG crust of the Man Shield in western Africa (Rollinson, 1997). Imprecise Paleoarchean formation ages for the eclogites and TTG crust overlap in time, which is permissive for crustal growth by partial melting of the protolith of the eclogite xenoliths (Barth et al., 2002). Values of $\delta^{18}O$ for the eclogites outside the mantle range suggest that the protolith may represent ocean floor, which in turn permits Archean crustal growth in the Man Shield to have occurred in a convergent margin setting involving subduction of ocean lithosphere. High-MgO eclogites in the Man Shield and elsewhere are inferred to correspond to metamorphosed cumulates from hydrous basalt magmas emplaced beneath thick continental arcs, also supporting a role for arc magmatism in the formation of cratonic roots in the Mesoarchean to Neoarchean (Horodyskyj et al., 2007).

In South Africa, recent work on the ca. 3.23 Ga metamorphism of the Barberton and related greenstone belts has yielded the following *P-T* data: in the Onverwacht Group greenstone remnants, *P-T* conditions of 0.8–1.1 GPa and 650–700 °C (Dziggel et al., 2002); in the southern Barberton greenstone belt, *P-T* conditions of 0.9–1.2 GPa and 650–700 °C (Diener et al., 2005, 2006); in amphibolite-dominated blocks of supracrustal rocks in tectonic mélange from the Inyoni shear zone, *P-T* conditions of 1.2–1.5 GPa and 600–650 °C (Moyen et al., 2006). Apparent thermal gradients for these belts are in the range 450–700 °C/GPa (~13–20 °C/km) and are within the limit for thickening alone (Fig. 3). Consequently, the conclusion drawn by Moyen et al. (2006) for the Inyoni shear zone samples with the lowest apparent thermal gradients that "these high-pressure, low-temperature conditions represent metamorphic evidence for … subduction-driven tectonic processes during the evolution of the early Earth" may be premature. The lowest of these apparent thermal gradients is close to the highest apparent thermal gradients retrieved from Phanerozoic HPM-UHPM terranes inferred to have been associated with subduction. The Inyoni shear zone may represent a suture recording a former site of subduction, but the apparent thermal gradients derived from the *P-T* conditions retrieved from the high-pressure rocks do not provide unambiguous evidence of subduction.

Blueschists are not documented in the Archean eon, and there is no metamorphic imprint of subduction of continental crust to mantle conditions and return to crustal depths (although this could be interpreted to mean that the subducted continental crust was not returned). However, the chemistry of Mesoarchean to Neoarchean eclogite xenoliths and the chemistry and paragenesis of Neoarchean diamonds in kimberlites within cratons suggest that some process associated with supercraton formation at convergent margins was operating to take basalt and other supracrustal material into the mantle by the Neoarchean era (e.g., Jacob, 2004; Shirey et al., 2004). This may have been some form of subduction, but the hotter ocean lithosphere may have been weaker due to lower viscosity leading to more frequent slab breakoff and slab breakup, and preventing exhumation (van Hunen and van den Berg, 2007). Another way of addressing this issue is to ask "When did the heat flow through the ocean basins permit the transition from a 'crème brûlée' lithosphere rheology to a 'jelly sandwich' lithosphere rheology to permit integrity of a subducted slab?" (Figs. 9 and 10) (Burov and Watts, 2006).

Worldwide, the oldest surviving Paleoarchean crustal remnants generally are composed of juvenile tonalite-trondhjemite-granodiorite rock suites (TTGs) formed prior to a period of polyphase granulite facies metamorphism in the interval 3.65–3.60 Ga (Friend and Nutman, 2005b). Extreme thermal conditions typical of G-UHTM are not generally registered before the Neoarchean. This appears to be counterintuitive, since we might expect that the higher abundance of heat-producing elements would lead to higher crustal heat production and hotter orogens. However, it is possible that contemporary heat loss through oceans and continents was higher and lithosphere rheology was generally weaker on early Earth. This is permissive for a "crème brûlée" lithosphere rheology structure (Figs. 9 and 10); also, it is consistent with (1) modeling by Toussaint et al. (2004), who suggested dominance of unstable subduction for plate collision regimes with very hot geotherms in which convergence is accommodated by pure shear thickening and development of gravitational (Rayleigh-Taylor) instabilities, (2) modeling by van Hunen et al. (2004), who ruled out the commonly proposed flat subduction model for early Earth tectonics, (3) the proposition that early plate-like behavior may have allowed crust to float to form thick stacks above zones of boundary-layer downwelling (Davies, 1992), or, for a slightly cooler Earth, thick stacks above zones of "sublithospheric" subduction (van Hunen et al., this volume), and (4) the absence of E-HPGM and HPM-UHPM in the early Earth rock record, inferred herein to be more typical of plate collisions in the Proterozoic and Phanerozoic.

Bailey (1999) offered the interesting alternative of continental overflow in the Archean whereby continental-style crust would

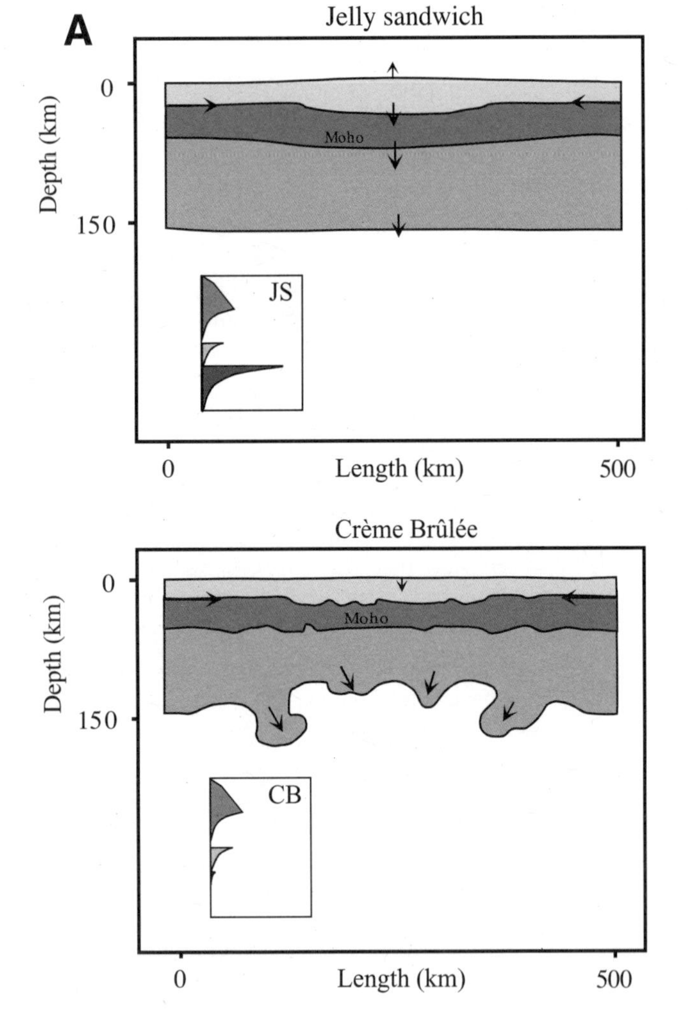


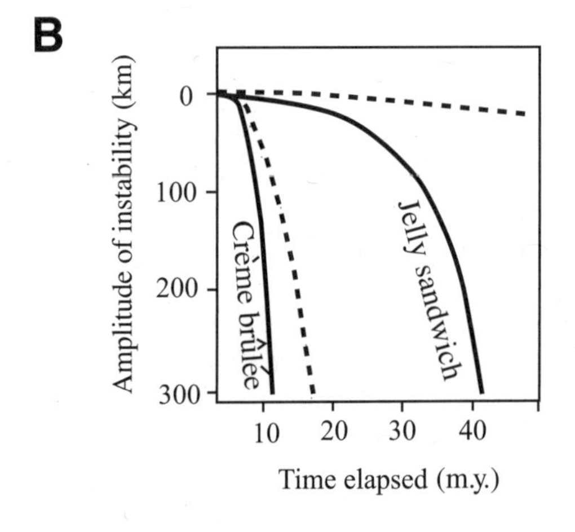


Figure 9. Numerical models to test the stability of a mountain range using the "jelly sandwich" lithosphere rheology structure (JS) and "crème brûlée" lithosphere rheology structure (CB) models. The model assumes a free upper surface and a hydrostatic boundary condition at the lower surface. The stability test is based on a mountain range height of 3 km and width of 200 km that is initially in isostatic equilibrium with a zero-elevation 36-km-thick crust. The thermal structure is equivalent to that of a 150-Ma plate. The isostatic balance was disturbed by applying a horizontal compression to the edges of the lithosphere at a rate of 5 mm $yr^{-1}$. The displacements of both the surface topography and Moho were then tracked through time. (A) Crustal and mantle structure after 10 m.y. have elapsed. (B) The amplitude of the mantle root instability as a function of time. The figure shows the evolution of a marker that was initially positioned at the base of the mechanical lithosphere (the depth where the strength is 10 MPa). This initial position is assumed to be at 0 km on the vertical axis. The solid and dashed lines show the instability for a weak, young (thermotectonic age = 150 Ma) and strong, old (thermotectonic age = 500 Ma) plate, respectively. (Reproduced from Burov and Watts, 2006, with permission.)

override oceanic-style lithosphere to facilitate partial melting of the underlying plate and formation of TTGs in a manner similar to the model of Foley et al. (2003). A modern analogue might be the recent model proposed by Copley and McKenzie (2007) for southward gravity-driven flow of southern Tibet over the underlying Indian lithosphere. Although Bailey (1999) proposes that continental overflow may lead to TTG formation without subduction, such a process might have created an environment favorable for subduction and might have been the initial step required to initiate subduction and plate tectonics (van Hunen et al., this volume). Continental overflow might lead to shallow subduction, as preferred by Foley et al. (2003), but this requires that the mantle viscosity was large enough to resist rapid sinking of the oceanic-type lithosphere plate, which may not have been the case for the Archean mantle (van Hunen et al., 2004).

An unusual feature of some Archean high-grade gneiss terranes is retrograde replacement of granulite facies orthopyroxene by "blebby" hornblende ± biotite ± quartz aggregates in leucosome pods and sometimes more widely in the host rocks that define a mappable, but retrograde and not prograde, granulite facies–amphibolite facies isograd (e.g., McGregor and Friend, 1997; Friend and Nutman, 2005b). Widespread retrogression requires the availability of an $H_2O$-rich fluid phase (volatile phase and/or crystallizing melt that may exsolve a volatile phase) during the waning stages of the metamorphic evolution. Together with the proposal that low-temperature water-saturated melting may be common early in Earth history (Watson and Harrison, 2005), this widely developed retrograde feature of Eoarchean to Mesoarchean orthopyroxene-bearing rocks may suggest a tectonic setting and/or process particular to this eon (although such a unique interpretation of the data in Watson and Harrison has been questioned by Coogan and Hinton, 2006).

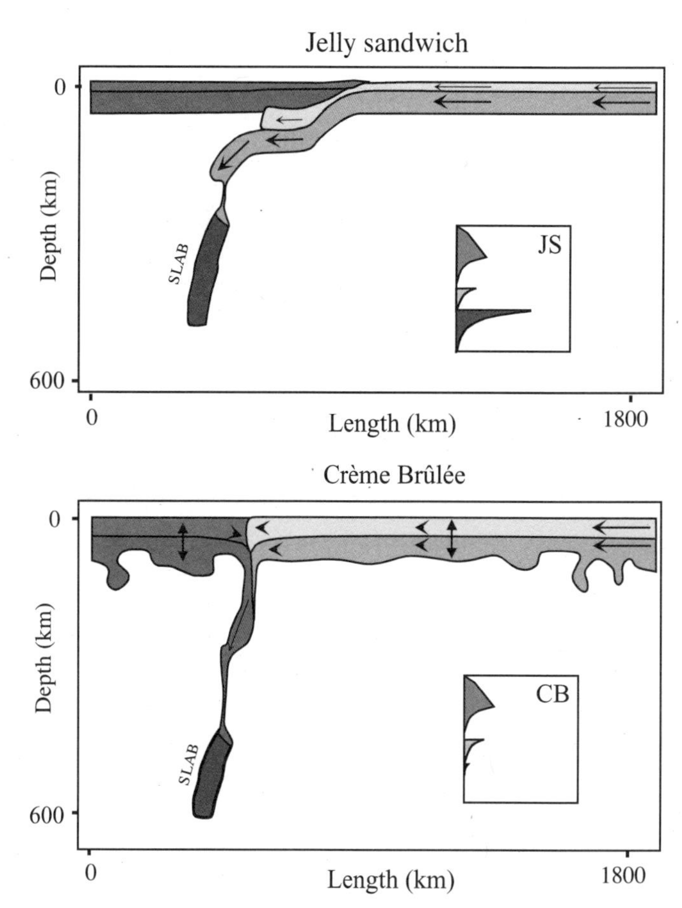


Figure 10. Numerical models to test the stability of a continental collision system using the "jelly sandwich" lithosphere rheology and "crème brûlée" lithosphere rheology structures. The model assumes a free upper surface and a hydrostatic boundary condition at the lower surface. The collision test is based on a continent-continent collision initiated by subduction of a dense, downgoing ocean plate using the failure envelopes associated with the "jelly sandwich" lithosphere rheology structure (JS) and "crème brûlée" lithosphere rheology structure (CB) models. A normal ocean crust thickness of 7 km, a total convergence rate of 60 mm $yr^{-1}$, and a serpentinized subducted ocean crust were assumed. The elastic thickness, Te, and Moho temperature are ~20 km and 600 °C, respectively, for both models. The figure shows a snapshot at 5 m.y. of the tectonic styles that develop after 300 km of shortening. (Reproduced from Burov and Watts, 2006, with permission.)

## INFERENCES ABOUT PLATE TECTONICS REGIMES FROM THE METAMORPHIC RECORD

### Thermal Regimes and Tectonic Settings

The maximum pressures and thermal regime of granulite facies ultrahigh-temperature metamorphism (1.0–1.2 GPa, >>750 °C/GPa [>>20 °C/km]) place it across the boundary between the "warmer than normal" and "ultrahigh temperature" fields in Figure 5, which suggests that many of these terranes require an advective component of heating. Given the *P-T* conditions and the residual compositions of such terranes, anatectic melt fluxing through these rocks advecting heat must have occurred. The concentration of G-UHTM belts in the interval Neoarchean to Cambrian (Figs. 6 and 7), and their broad correlation in time with amalgamation of continental crust into supercratons and supercontinents (Fig. 8), suggests that ultrahigh-temperature metamorphism, which commonly records tight clockwise *P-T* paths, may record closure and thickening of sedimentary basins and underlying lithosphere in a convergent margin tectonic setting characterized by high heat flow and thin lithosphere. On modern Earth, high heat flow and thin lithosphere are characteristic of subduction zone backarcs, which are developed around the margins of large ocean basins such as the modern Pacific Ocean basin (Currie and Hyndman, 2006; Hyndman et al., 2005a, 2005b).

The maximum pressures and thermal regime of medium-temperature eclogite–high-pressure granulite metamorphism (up to 2.5 GPa, 350–750 °C/GPa [~10–20 °C/km]) place it across the boundary between the "cooler than normal" and "warmer than normal" fields in Figure 5, which suggests that lithosphere thickening may be sufficient for this type of metamorphism.

The higher pressures common in E-HPGM terranes suggest that crustal rocks have been taken to mantle depths. On modern Earth, this is accomplished during subduction-to-collision orogenesis, although for E-HPGM terranes the pressures could simply reflect collision and doubling in thickness by crustal stacking following subduction of ocean plates (cf. Štípská et al., 2004).

The maximum pressures and thermal regime of high-pressure metamorphism to ultrahigh-pressure metamorphism (up to 5.0 GPa, 150–350 °C/GPa [approximately equivalent to 4–10 °C/km]) place it across the boundary between the "cooler than normal" and "ultralow temperature" fields in Figure 5 and suggest that some tectonic process other than simple thickening of the lithosphere, such as subduction, must control this type of metamorphism. The repeated occurrence of HPM-UHPM (with the inference of subducted continental crust) since the Ediacaran period suggests a common process in convergent tectonic settings, and the *P-T* conditions coupled with high-precision ages indicate fast rates of subduction to generate the low thermal gradient (150–350 °C/GPa) and rates of exhumation similar to modern plate velocities (1–2 cm $yr^{-1}$). It is pertinent to note that the ocean basins that have been subducted during production of HPM-UHPM belts generally appear to have been smaller and shorter-lived than, for example, the modern Pacific Ocean basin, based on recent plate tectonics models for the Phanerozoic that integrate geologic, geodynamic, and geophysical data (Stampfli and Borel, 2002, 2004; Collins, 2003; Hafkenscheid et al., 2006). This issue is discussed further below.

## Secular Change and Plate Tectonics

The rock record indicates a dramatic change in the thermal environments of crustal metamorphism through the Neoarchean to the Archean-to-Proterozoic transition, which may register the switch to a "Proterozoic plate tectonics regime" (the protoplate tectonics of Dewey, 2007). This premise is consistent with aggregation of the crust into progressively larger units to form supercratons (Bleeker, 2003), perhaps indicating a change in the pattern of mantle convection during the transition to the Proterozoic eon. The emergence of plate tectonics requires forces sufficient to initiate and drive subduction, and lithosphere with sufficient strength to subduct coherently. These requirements likely were met as basalt became able to transform to eclogite. Secular change in thermal regimes to allow this transformation appears to have been gradual, occurring regionally first during the Mesoarchean to Neoarchean—leading to the successive formation of the supercratons Vaalbara, Superia, and Sclavia—and worldwide during the Paleoproterozoic—evidenced in orogenic belts that suture Nuna (Columbia)—unless this distribution is an artifact of (lack of) preservation or thorough overprinting of eclogite in the exposed Archean cratons.

The transition to a Proterozoic plate tectonics regime resulted in stabilized lithosphere in which cratons formed the cores of continents that subsequently grew dominantly by marginal accretion. Furthermore, the transition coincides with the first occurrence of ophiolite-like complexes in Proterozoic suture zones (Moores, 2002) and with the increase in $\delta^{18}O$ of magmas through the Paleoproterozoic, which Valley et al. (2005) argue may reflect maturation of the crust, the beginning of recycling of supracrustal rocks, and their increasing involvement in magma genesis via subduction. Although the style of Proterozoic subduction remains cryptic, the change in tectonic regime whereby interactions between discrete lithospheric plates generated tectonic settings with contrasting thermal regimes was a landmark event in Earth history.

The transition to the "modern plate tectonics regime" through the Neoproterozoic is one in which ocean crust became thinner while the lithosphere became thicker (e.g., Moores, 2002), and the pattern of the Brasiliano and Pan-African orogens that sutured the disparate continental fragments of Gondwana was replaced by semicircumferential large-scale plate-margin orogenic systems (e.g., the Terra Australis, the Appalachian/Caledonian–Variscide–Altaid, and the Cimmerian–Himalayan–Alpine orogenic systems). One of the most dramatic features in global mantle shear-wave attenuation models is a very low-Q anomaly at the top of the lower mantle beneath eastern Asia (Lawrence and Wysession, 2006). Lawrence and Wysession (2006) believe this to be due to water that has been taken into the lower mantle through the long history of subduction during the Phanerozoic in the western Pacific. These features may indicate a modification of the pattern of mantle convection driven by deeper subduction that facilitated transport of water deeper into the mantle and enabled subduction and eduction of continental crust. Many authors have noted the confluence of other changes to the Earth system during the Neoproterozoic transition to the Phanerozoic (e.g., Hoffman, 1991; Maruyama and Liou, 1998, 2005), which may have been related to the change to deeper subduction.

## Thermal Structure of Subduction Zones

Intraslab earthquakes, arc volcanism, the viscosity of the mantle above the slab, and the development of backarcs are intimately linked to transfer of water into the mantle wedge as a result of metamorphic dehydration reactions in the subducting slab. Several studies have argued that the thermal structure of the subduction zone, which is a function of incoming plate age and plate velocity, controls the depth of dehydration events and gaps between dehydration events in both the crust and the mantle of the subducting slab (e.g., Peacock and Wang, 1999; van Keken et al., 2002; Hacker et al., 2003, 2005a; Omori et al., 2004). Modeling predicts that the thermal structure of a subduction zone may be several hundred degrees cooler at any particular depth where older ocean lithosphere is subducted (e.g., northeast Japan) compared to subduction of younger ocean lithosphere (e.g., southwest Japan or Cascadia).

If the apparent thermal gradient is very low (lower than ~250 °C/GPa), water may be carried deep into the mantle wedge by lawsonite eclogite and hydrated harzburgite, consistent with cooler subduction zones where arc volcanism is common and intraslab earthquakes reach >200 km depth. In contrast, if the apparent thermal gradient is not so extreme (higher than ~250 °C/GPa), at least the ocean crust segment of the slab may dehydrate at shallow depths and less water may be transported into the sub-lithospheric mantle, consistent with observations from warmer subduction zones where arc volcanism is sparse and intraslab earthquakes extend only to 60–70 km depth.

The higher heat flow measured in subduction zone backarcs may be due to fast upward convective transfer of heat beneath the thin lithosphere (Hyndman et al., 2005a; Currie and Hyndman, 2006), and the uniform high heat flow across backarcs might be achieved by small-scale vigorous convection with flow rates faster than relative plate velocities (Currie et al., 2004). On modern Earth, shallow vigorous convection is promoted by a decrease in the mantle viscosity inferred to be due to incorporation of water expelled from the subducting slab (Hyndman et al., 2005a; Currie and Hyndman, 2006). The backarc convection system is poorly understood, but Hyndman et al. (2005a; see also Currie et al., 2004; Currie and Hyndman, 2006) suggest that vigorous convection may mix the water throughout the whole mantle wedge.

Apparent thermal gradients for the Proterozoic—derived from E-HPGM terranes—are warmer than modern subduction zone thermal gradients—based on numerical modeling of intra-oceanic subduction (Maresch and Gerya, 2005), where the apparent thermal gradient along the slab is ~350 °C/GPa—consistent with an expectation of secular cooling. Assuming a warmer mantle wedge and warmer ocean lithosphere in the Proterozoic than in the Phanerozoic, the viscosity of the mantle wedge may have been sufficiently low to allow induced convective transfer of heat beneath backarcs regardless of the shallower dehydration of the slab and the lower amount of water likely transported to sublithosphere depths by subduction.

## THE SUPERCONTINENT CYCLE

The continental lithosphere has been amalgamated into a supercontinent at several intervals during Earth history with dramatic effects on both surface and deep Earth processes. Aggregated continental lithosphere influences mantle heat loss by acting as an insulator to the convecting mantle and by imposing its own wavelength on mantle convection by impeding downwelling (Gurnis, 1988; Guillou and Jaupart, 1995; Cooper et al., 2006; Coltice et al., 2007). Therefore, during periods of aggregated continental lithosphere, we may anticipate longer-wavelength convection and global variations in mantle heat loss and mantle potential temperature (Grigné and Labrosse, 2001; Grigné et al., 2005; Lenardic et al., 2005; Lenardic, 2006; Coltice et al., 2007).

### Correlations

The temporal relationship between metamorphism at extreme *P-T* conditions, plate tectonics, and the supercontinent cycle is shown in Figure 8. If we accept the formation of Gondwana by ca. 500 Ma as an intermediate step between Rodinia and Pangea, then there are four principal periods of crustal amalgamation since the Neoarchean, as follows: formation of the supercratons Vaalbara, Superia, and Sclavia from older superterranes and juvenile crustal additions during the Mesoarchean to Neoarchean interval (Vaalbara slightly older and Sclavia slightly younger), and formation of the supercontinents Nuna (Columbia), Rodinia, and Pangea by ca. 1.8 Ga, ca. 1.0 Ga, and ca. 250 Ma, respectively. The supercratons were fragmented during the early Paleoproterozoic. Nuna may have begun breaking up as early as ca. 1.6 Ga, and Rodinia was fragmenting by ca. 850 Ma. The period between the beginning of breakup and amalgamation into a new supercontinent is ~600 m.y. There is a correspondence between the timing of extreme metamorphism in the rock record and suturing of continental lithosphere by subduction-to-collision orogenesis into supercratons and supercontinents (Fig. 8).

I speculate that it is the stability of the South Pacific superswell and its precursors during the late Neoproterozoic to early Phanerozoic that forced the transformation of Rodinia to Pangea by breakout of Laurentia from the Rodinia supercontinent, its separation from Baltica and Siberia by creation of internally generated oceans, and closure of these oceans along the Appalachian/Caledonian–Variscide–Altaid orogenic system. Subsequently, the Mesozoic-Cenozoic Tethysides were built by accretion to the northern European "core" of terranes calved off from the southern (Gondwana) side of Pangea. These terranes were transferred across short-lived internally generated oceans that closed by subduction either at an intraoceanic arc system or at the leading edge of a ribbon-continent terrane as it accreted to the European core. The opening of a new ocean basin along the trailing edge of a terrane as it was transferred from the south to the north side of the former Pangea generated the next short-lived ocean to be closed by subduction.

### Wilson-Type Cycles and Hoffman-Type Breakups

There are suggested to be two end-member models for the formation of supercontinents (Murphy and Nance, 2003), based on different concepts introduced by Wilson (1966, closing and reopening an ocean) and Hoffman (1991, turning a supercontinent inside out), respectively. In one model, continental lithosphere simply fragments and reassembles along the same (internal) contacts or introverts (the "Wilson cycle" sensu stricto; Wilson, 1966)—which I regard as a process of rearrangement of the continental lithosphere within a continent-dominated hemisphere rather than a mechanism for supercontinent fragmentation, dispersal, and reassembly—whereas in the alternative model a supercontinent fragments, disperses, and reassembles (or extroverts) by closure of the complementary superocean (Hoffman, 1991).

Wilson-type cycles operated during the Phanerozoic, when the continental lithosphere was restricted to one hemisphere. In contrast, a Hoffman-type breakup was the process by which the Gondwanan elements of Rodinia were dispersed and reassembled into Gondwana (the type example) and most likely was the process by which Nuna (Columbia) was reconfigured into Rodinia, although in this case the continental fragments were fewer and larger (Condie, 2002). Whether any Wilson-type cycles occurred within Rodinia or Nuna remains to be determined, but there is no a priori reason to suppose that Wilson-type cycles did not occur during the Proterozoic.

### Intrasupercontinent Rearrangements by Wilson-Type Cycles (Introversion)

During introversion, "expansion" of the continent-dominated hemisphere—the hemisphere that includes the supercontinent—may be limited by the persistence of an opposed ocean-dominated hemisphere, as has been the case for the Pacific Ocean through the Phanerozoic. In this case, modification of the way the continental lithosphere is aggregated requires splitting the continental lithosphere to generate an internal ocean that separates a terrane from the parent continent, followed by transport across a closing ocean and suturing at a new location (commonly referred to as "terrane export"). Initiation of subduction within the internal ocean generated by the initial splitting leads to calving of a second terrane by renewed splitting inboard from the margin of the parent continent, which, in turn, is exported and sutured to the new location as this second internal ocean also closes, and so on. It is the successive closure of these newly generated but relatively short-lived oceans by subduction and terrane export from one side of a supercontinent to another that leads to a new arrangement of the continental lithosphere within the continental hemisphere.

The metamorphic style associated with Wilson-type cycles during the Phanerozoic is blueschist-HPM-UHPM (Liou et al., 2004), which may be linked with or transitional to E-HPGM. In contrast, in the complementary oceanic hemisphere, ongoing circumferential subduction creates accretionary orogenic systems that may develop paired metamorphic belts sensu Miyashiro (1961) with an outboard HP-LT terrane and an inboard HT-LP belt.

The limited extent of ocean lithosphere subduction and choking of subduction by continental lithosphere during Phanerozoic subduction-to-collision orogenesis may inhibit transport of water into the mantle wedge, and as a result the development of small-scale convection and thinner backarc lithosphere is retarded, and arc volcanism is sparse. Ernst (2005) previously noted that calc-alkaline volcanic-plutonic rocks are rare in what he called "Alpine-type" orogenic belts, a feature that he related to the difficulty of dehydrating subducted continental lithosphere during closure of short-lived oceans that typify many intracratonic suture zones.

Numerical modeling of the minimum subduction necessary to generate thermal conditions suitable for blueschist metamorphism is consistent with limited subduction of short-lived ocean basins (Maresch and Gerya, 2005). Because of the geometric interplay between the nose of the subducting ocean lithosphere and the evolving array of isotherms, subduction of younger and hotter ocean lithosphere may lead to earlier formation of blueschists than subduction of older and cooler ocean lithosphere. Maresch and Gerya (2005) identify an optimum age of 40–60 m.y. for subducted ocean lithosphere to generate blueschist conditions. The thermal structure also explains the rarity of arc volcanics in these orogens, because the necessary flux of water into the mantle wedge to drive melting will be weak or even absent.

### Phanerozoic Examples of Wilson-Type Cycles and Intrasupercontinent Orogenesis

Consider the Paleozoic Appalachian/Caledonian–Variscide–Altaid orogenic system, which records Wilson-type cycles of orogenesis. The transformation of Rodinia to Pangea involved the breakout of Laurentia from the Rodinia supercontinent by the end of the Neoproterozoic, which was achieved by rifting the eastern Gondwanan continental lithosphere fragments from western Laurentia, followed by rifting the western Gondwanan continental lithosphere fragments from eastern Laurentia (Hoffman, 1991). Baltica was left behind and separated from Laurentia by creation of the Iapetus Ocean; Laurentia and Baltica were separated from Gondwana by creation of the Rheic Ocean.

The Iapetus Ocean was closed along the Appalachian/Caledonian orogenic system, and the intervening internally generated lithosphere was consumed during reassembly of Laurentia, Avalonia, and Baltica in the Early Devonian, forming Laurussia (Murphy and Nance, 2005; see also Collins, 2003). UHPM rocks have not yet been identified in the southern Appalachians—although HPM rocks are present (Dilts et al., 2006) and are a characteristic feature of the final (Alleghanian) phase of orogenesis—but UHPM rocks are a common feature of the Norwegian Caledonides (e.g., Carswell and Compagnoni, 2003; Liou et al., 2004).

The Rheic Ocean was closed along the Variscide–Altaid orogenic system. In the Variscide sector, the intervening internally generated lithosphere was consumed by subduction during clockwise rotation of Gondwana and collision with Laurasia (Laurussia and Siberia sutured along the Uralides) during final assembly of Pangea in the Carboniferous. UHPM rocks are a common feature of the European Variscides (e.g., Carswell and Compagnoni, 2003; Liou et al., 2004).

The Altaid Central Asian orogenic belt is something of an enigma in this context, for it includes some of the oldest-formed and most extreme UHPM rocks known (at Kotchatev; Kaneko et al., 2000) but also ridge-trench interactions, expressed as HT-LP metamorphism, and arc collisions (Windley et al., 2007; see also Collins, 2003, and Filippova et al., 2001). Furthermore, isotopic tracer studies over the past decade have shown that central Asia has been the dominant place on Earth for production of juvenile crust in the Phanerozoic eon (Jahn et al., 2000a, 2000b, 2000c, 2004). In the Central Asian orogenic belt many granites sensu lato include a component derived from older continental crust, evident in the isotope chemistry, perhaps indicating melting of young ocean lithosphere and contamination of the magma during soft collision between cratonic continental lithosphere to the north and accreting terranes derived from the south. The greater variety in the Altaid sector most likely reflects the widening of the Rheic Ocean to the east.

Within the Mesozoic-Cenozoic Tethysides, the central orogenic belt of China (the Tianshan–Qinling–Dabie-Sulu orogenic system; e.g., Yang et al., 2005) and the Alpides (the Alpine–Zagros–Himalayan orogenic system; e.g., Şengör, 1987; Hafkenscheid et al., 2006) are products of multiple accretion events or collisions. In each case, the orogenic event involved the destruction of a relatively short-lived ocean, and the sutures commonly are decorated with occurrences of HPM and/or UHPM rocks.

Intrasupercontinent orogenesis is widely recognized during the Phanerozoic because we have available large data sets that have not been overprinted by subsequent supercontinent cycles, as well as age maps of the ocean floors since the Mesozoic. It is not yet clear whether this style of orogenesis was less common in earlier supercontinent cycles or simply has not yet been recognized to be as common.

### Supercontinent Fragmentation, Dispersal, and Reassembly by Hoffman-Type Breakups (Extroversion)

In extroversion, continental lithosphere tends to amalgamate over cold downwellings to form a supercontinent, which inhibits subduction and mantle cooling (Gurnis, 1988; Collins, 2003). Stagnant slabs may avalanche into the lower mantle, forming a slab "graveyard." The mantle beneath the supercontinent may eventually overheat and become the site of a new upwelling that fragments the overlying continental lithosphere under tension to dissipate the thermal anomaly. The next supercontinent amalgamates above areas of mantle downwelling by subduction-to-collision orogenesis in a cycle with a period of ~500 m.y. This is the basic principle that leads to breakup of one supercontinent and formation of another. As a result of Hoffman-type breakup, the internal rifted margins of the old supercontinent become the external margins of the new supercontinent, and the external accretionary margins of the old supercontinent become deformed and smeared out along the internal sutures of the new supercontinent.

### The Transformation of Rodinia to Gondwana by Hoffman-Type Breakup

Consider the transformation of Rodinia to Gondwana, which is the original example of a Hoffman-type breakup (Hoffman, 1991). This transformation involved the fragmentation, dispersal, and reassembly of the continental lithosphere by subduction-to-collision orogenesis to form the network of Brasiliano and Pan-African belts (e.g., Cordani et al., 2003; Collins and Pisarevsky, 2005), leaving the orphaned Laurasian continental fragments to combine with each other and then Gondwana to form Pangea at a later time.

The internal geometry of Rodinia changed considerably during its few hundred million years of existence. Geologic and paleomagnetic data suggest that the supercontinent consolidated at 1100–1000 Ma and most likely disintegration began between 850 and 800 Ma (e.g., Torsvik, 2003; Cordani et al., 2003). Reassembly to form Gondwana occurred by destruction of parts of the complementary superocean as Rodinia progressively disintegrated, although the exact configuration and global location of the different continental lithosphere fragments in Rodinia at the time of breakup is uncertain (e.g., Murphy et al., 2004).

Although G-UHTM and E-HPGM are associated with the Brasiliano and Pan-African orogenic belts, the Trans-Saharan segment of the Pan-African orogenic system also records the first coesite-bearing eclogites, and sutures within the Anti-Atlas belt and the South China Block record the first blueschists. These features point to a transition due to secular cooling and the changeover to the "modern plate tectonics regime" characterized by colder subduction.

### Distinguishing Wilson-Type Cycles from Hoffman-Type Breakups

Murphy and Nance (2003, 2005) have suggested that Sm-Nd isotope data may be used to distinguish internally generated ocean lithosphere formed during Wilson-type cycles (characterized by younger mantle extraction ages than the age of an intrasupercontinent rifting event) from older external ocean lithosphere associated with Hoffman-type breakups (characterized by older mantle extraction ages than the age of a supercontinent fragmentation event). For Gondwana, mafic complexes accreted during amalgamation yield mantle extraction ages older than the breakup of Rodinia and so represent fragments of an exterior superocean. In contrast, mafic complexes in the Appalachian/Caledonian–Variscan belt yield mantle extraction ages consistent with subduction of an interior ocean in which the lithosphere was formed after the intrasupercontinent rifting event.

### Pre-Neoproterozoic Supercontinents

The balance between consumption of an external ocean after supercontinent fragmentation and consumption of an internally generated ocean formed during a period of intrasupercontinent rearrangement remains to be determined for Proterozoic supercontinent cycles. The transformation of Nuna (Columbia) into Rodinia is thought to have occurred by splitting the supercontinent into a small number of large fragments that were rearranged by consumption of an external ocean and amalgamated by suturing along Grenville-age orogens (Condie, 2002). This model is supported by limited Sm-Nd isotope data that suggest the Grenville belt records closure of a exterior ocean (Murphy and Nance, 2005).

Farther back in Earth history the situation is more speculative. The formation of Nuna (Columbia) occurred in the interval 2.1–1.8 Ga (Zhao et al., 2002, 2004, 2006b). Murphy and Nance (2005) discuss a limited Sm-Nd isotope data set that suggests closure of an internal ocean for part of the Trans-Hudson orogen. Accepting the correlations among cratons proposed by Bleeker

(2003, his Fig. 6), it is possible that the Laurentian cratons—Slave, Superior, Sask, Hearne, Rae, and Wyoming—were part of at least two different supercratons at the end of the Neoarchean. If this is correct, it is probable that at least some of the Paleoproterozoic orogens that suture Nuna, and maybe the majority, formed by consumption of an exterior ocean. However, this hypothesis remains to be tested.

## WIDER CORRELATIONS

Assuming a hotter early Earth, Davies (2006, 2007) and van Hunen and van den Berg (2007) have used numerical models of convection to address the viability of plate tectonics. Modeling by Davies (2006, 2007) takes into account early depletion of the upper mantle, which leads him to suggest that the ocean crust might have been of similar thickness to modern Earth but that the thermal boundary layer might have been thinner overall. On Earth, this might have limited the ability of the ocean lithosphere to subduct and might have impacted the viability of early plate tectonics. However, the numerical modeling suggests that the spatial and temporal variability in crustal thicknesses that applied to Earth might have allowed some plates to subduct while other plates might have been blocked to form protocontinental crust.

In an alternative scenario without early depletion of the upper mantle, modeling by van Hunen and van den Berg (2007) suggests that the lower viscosity and higher degree of melting for a hotter fertile mantle might have led to both a thicker crust and a thicker depleted harzburgite layer forming the ocean lithosphere. Compositional buoyancy resulting from the thicker crust and harzburgite layers might be a serious limitation for subduction initiation, and a different mode of downwelling (Davies, 1992) or "sublithospheric" subduction (van Hunen et al., this volume) might have characterized early Earth. An additional problem for deep subduction is posed by the lower viscosity, which van Hunen and van den Berg (2007) suggest might lead to more frequent slab breakoff and possible breakup of the crust and harzburgite layers of the subducting ocean lithosphere. In this case the lower viscosity of the slab components might have been the principal limitation inhibiting plate tectonics on early Earth.

The absence of HPM-UHPM terranes before the Ediacaran may relate to weakness of the subducting lithosphere, which might have been strong enough to allow shallow subduction, but the rheology may have been too weak to allow deep subduction of coherent slabs and/or provide a mechanism for eduction of continental crust if, indeed, it was ever subductable before the Ediacaran. This does not conflict with Late Archean and Proterozoic subduction of ocean lithosphere, since early slab breakoff and slab breakup still might have transported materials with a surface chemical signature deep into the mantle as recorded in some diamonds (e.g., Farquhar et al., 2002).

Tonalite-trondhjemite-granodiorite suites (TTGs) dominate the Archean rock record, but their origin has been controversial. In a series of papers that integrate results from experimental petrology and major and trace element geochemistry, Foley (this volume; Foley et al., 2002, 2003) has argued that most TTGs formed by melting of garnet amphibolite of broadly basaltic composition hydrated by interaction with seawater. Further, he argues that melting of eclogite increased in importance through the Mesoarchean to Neoarchean (cf. van Hunen et al., this volume), as shown by an increase in Nb/Ta in TTGs. Melting of garnet amphibolite may be achieved in the lower part of thickened basaltic crust or by subduction on a warmer Earth. Conceptually, the first alternative is consistent with early plate-like behavior that allowed crust to float to form thick stacks above zones of boundary-layer downwelling (Davies, 1992), or, for a slightly cooler Earth, thick stacks above zones of "sublithospheric" subduction (van Hunen et al., this volume), and this indeed may have been the convective mode of the Eoarchean to Paleoarchean. By the Mesoarchean to Neoarchean, and possibly as early as the Paleoarchean, subduction was operating in some regions of Earth (e.g., Pease et al., this volume); this subduction was likely characterized by warmer geotherms that enabled melting of subducting garnet amphibolite and, with secular cooling, a change to melting of subducting eclogite.

Calculations by Hynes (this volume) for thinner Archean ocean lithosphere suggest that plate motions in the Archean might have been faster than on modern Earth. Faster plate velocities may explain the scarcity of accretionary prisms in the Archean rock record since high convergence rates will favor subduction erosion over subduction accretion. Another consequence of a hotter Earth is the reduced subsidence of lithosphere in extension, which will limit the accommodation space for passive-margin sediments and contribute to the scarcity of passive-margin sequences in the Archean rock record (Hynes, this volume). The formation of diamonds in the Mesoarchean to Neoarchean requires geotherms similar to those of modern Earth, which in turn probably reflects the presence of cool mantle roots beneath the continents. Stretching of continents underlain by cool mantle roots would yield passive margins similar to those on modern Earth. Thus, the development of recognizable passive margins might only have occurred after the emergence of cratons in the Mesoarchean to Neoarchean.

Bradley's (2007) analysis of the geological record indicates that passive margins have existed on Earth since at least 2.69 Ga, consistent with the prediction by Hynes (this volume). The abundance of passive margins has varied with the supercontinent cycle. Passive margins were abundant in the interval 2.25–1.75 Ga, terminating with subduction-to-collision orogenesis and the assembly of Nuna (Columbia). Passive margins are rare in the interval 1.65–1.00 Ga, which is consistent with the suggestion that the transition from Nuna to Rodinia did not involve complete breakup and re-formation of the supercontinent (Condie, 2002). A maximum in passive-margin activity during the Ediacaran-Cambrian corresponds to dispersal of Laurentia, Baltica, and Siberia following staged breakup of Rodinia during the Neoproterozoic and formation of Gondwana during the Ediacaran-Cambrian. A minimum in the interval 350–250 Ma corresponds to the existence of Pangea, whereas the present-day maximum in number and aggregate length of passive margins corresponds to a time of continental dispersal following fragmentation of Pangea. Furthermore, Bradley's (2007) analysis demonstrates that passive margins in the Mesoproterozoic-Neoproterozoic were long-lived compared with those in the Phanerozoic, which may reflect the preponderance of Hoffman-type breakups of supercontinents in the Proterozoic compared with the dominance of shorter-lived Wilson-type cycles during the Phanerozoic.

The earliest supercraton (Vaalbara) and the Nuna supercontinent correlate broadly with peaks in the age distribution of major additions of juvenile material to the continental crust at ca. 3.3 Ga and ca. 1.9 Ga, based on Hf model ages of magmatic zircons with a variety of U-Pb crystallization ages (Hawkesworth and Kemp, 2006). A closer correlation has been argued between the younger supercratons (Superia and Sclavia) and the supercontinents Nuna and Rodinia and peaks in the distribution of U-Pb crystallization ages of zircons from isotopically juvenile igneous rocks at ca. 2.7 Ga, ca. 1.9 Ga, and ca. 1.2 Ga (McCulloch and Bennett, 1994; Condie, 1998). However one interprets these data, the observations suggest a relationship between aggregation of the continental lithosphere into supercontinents and large melting events in the mantle and/or melting of recently underplated juvenile material of the lower crust, at least during the Neoarchean to Mesoproterozoic interval.

## CONCLUSIONS

The hallmark of plate tectonics is the development of two contrasting thermal regimes where plates converge; on modern Earth these are (1) the subduction zone and (2) the arc and subduction zone backarc or orogenic hinterland. The first imprint of this hallmark in the rock record is the appearance of contemporaneous contrasting types of metamorphism, i.e., E-HPGM and G-UHTM, in Neoarchean belts. For the period as far back as the Neoarchean, I posit a common genesis for E-HPGM and G-UHTM in former subduction zones choked by collision, and in arcs and subduction zone backarcs or orogenic hinterlands thickened by collision. Further, I propose that we extend the use of the term "paired metamorphic belts" to include metamorphic belts with penecontemporaneous higher-pressure–lower-temperature and higher-temperature–lower-pressure metamorphism produced during subduction-to-collision orogenesis as well as HP-LT and HT-LP metamorphic belts formed in circum-ocean accretionary orogenic systems.

Plate tectonics has operated on Earth since the Neoarchean. However, the "Proterozoic plate tectonics regime" may have begun locally during the Mesoarchean to Neoarchean and may only have become global during the Neoarchean-to-Paleoproterozoic transition. A pre-Neoarchean tectonic regime may have involved plates, but if so these were probably thin with an ultramafic crust; if contrasting thermal regimes were generated, they do not appear to be preserved in the record of metamorphism. This view of when Earth adopted a plate tectonics mode of convection may be changed with new findings and acquisition of new data, and there are inevitably local details of relative age and spatial location that are obscured by generalizing at a global scale. The "modern plate tectonics regime" is one in which it became possible for ocean lithosphere to be deeply subducted and transfer water into the top of the lower mantle, and for continental lithosphere to be subducted and educted as HPM-UHPM terranes that mark sutures in Phanerozoic subduction-to-collision orogenic belts.

There is a close relationship between the age of metamorphic belts characterized by extreme metamorphism and the supercontinent cycle, which reflects the fact that two contrasting thermal regimes are generated where plates converge and are registered in the rock record by subduction-to-collision orogenesis during amalgamation of the continental lithosphere into supercontinents. The fragmentation of a supercontinent occurs by Hoffman-type breakups. Wilson-type cycles are the mechanism whereby the aggregated continental lithosphere within a continent-dominated hemisphere may be rearranged.

## ACKNOWLEDGMENTS

I thank Barry Reno for help in drafting the figures, Alan Collins and Reiner Klemd for provocative reviews, and many friends for discussion. My research in high-grade metamorphism is supported by the University of Maryland and the U.S. National Science Foundation (grant EAR-0227553).

## REFERENCES CITED

Andreoli, M.A., Hart, R.J., Ashwal, L.D., and Coetzee, H., 2006, Correlations between U, Th content and metamorphic grade in the western Namaqualand Belt, South Africa, with implications for radioactive heating of the crust: Journal of Petrology, v. 47, p. 1095–1118, doi: 10.1093/petrology/egl004.

Annen, C., Blundy, J.D., and Sparks, R.S.J., 2006, The genesis of intermediate and silicic magmas in deep crustal hot zones: Journal of Petrology, v. 47, p. 505–539, doi: 10.1093/petrology/egi084.

Bailey, R.C., 1999, Gravity-driven continental overflow and Archaean tectonics: Nature, v. 398, p. 413–415, doi: 10.1038/18866.

Baldwin, J.A., and Brown, M., 2008, Age and duration of ultrahigh-temperature metamorphism in the Anápolis-Itauçu Complex, Southern Brasília Belt, central Brazil—constraints from U-Pb geochronology, mineral rare-earth element chemistry and trace element thermometry: Journal of Metamorphic Geology, v. 26, p. 213–233, doi:10.1111/j.1525-1314.2007.00759.x.

Baldwin, J.A., Powell, R., Brown, M., Moraes, R., and Fuck, R.A., 2005, Mineral equilibria modeling of ultrahigh-temperature metamorphism: An example from the Anápolis-Itauçu Complex, central Brazil: Journal of Metamorphic Geology, v. 23, p. 511–531, doi: 10.1111/j.1525-1314.2005.00591.x.

Baldwin, S.L., Monteleone, B.D., Webb, L.E., Fitzgerald, P.G., Grove, M., and Hill, E.J., 2004, Pliocene eclogite exhumation at plate tectonic rates in eastern New Guinea: Nature, v. 431, p. 263–267, doi: 10.1038/nature02846.

Barboza, S.A., Bergantz, G.W., and Brown, M., 1999, Regional granulite facies metamorphism in the Ivrea zone: Is the Mafic Complex the smoking gun or a red herring?: Geology, v. 27, p. 447–450, doi: 10.1130/0091-7613(1999)027<0447:RGFMIT>2.3.CO;2.

Barth, M.G., Rudnick, R.L., Carlson, R.W., Horn, I., and McDonough, W.F., 2002, Re-Os and U-Pb geochronological constraints on the eclogite-tonalite connection in the Archean Man Shield, West Africa: Precambrian Research, v. 118, p. 267–283, doi: 10.1016/S0301-9268(02)00111-0.

Beaumont, C., Jamieson, R.A., Nguyen, M.H., and Medvedev, S., 2004, Crustal channel flows, 1: Numerical models with applications to the tectonics of the Himalayan-Tibetan orogen: Journal of Geophysical Research, v. 109, B06406, doi: 10.1029/2003JB002809.

Beaumont, C., Nguyen, M.H., Jamieson, R.A., and Ellis, S., 2006, Crustal flow modes in large hot orogens, *in* Law, R.D., et al., eds., Channel flow, ductile extrusion and exhumation in continental collision zones: Geological Society [London] Special Publication 268, p. 91–145.

Bibikova, E.V., Bogdanova, S.V., Glebovitsky, V.A., Claesson, S., and Skiold, T., 2004, Evolution of the Belomorian Belt: NORDSIM U-Pb zircon dating of the Chupa paragneisses, magmatism, and metamorphic stages: Petrology, v. 12, p. 195–210.

Bleeker, W., 2003, The Late Archean record: A puzzle in ca. 35 pieces: Lithos, v. 71, p. 99–134, doi: 10.1016/j.lithos.2003.07.003.

Boshoff, R., Van Reenen, D.D., Smit, C.A., Perchuk, L.L., Kramers, J.D., and Armstrong, R., 2006, Geologic history of the Central Zone of the Limpopo Complex: The West Alldays area: Journal of Geology, v. 114, p. 699–716, doi: 10.1086/507615.

Bradley, D.C., 2007, Age distribution of passive margins through Earth history and tectonic implications: Eos (Transactions, American Geophysical Union), v. 88, Joint Assembly Supplement, abstract U44A-03.

Brothers, R.N., and Blake, M.C., 1973, Tertiary plate tectonics and high-pressure metamorphism in New Caledonia: Tectonophysics, v. 17, p. 337–358, doi: 10.1016/0040-1951(73)90046-2.

Brown, M., 1993, *P-T-t* evolution of orogenic belts and the causes of regional metamorphism: Geological Society [London] Journal, v. 150, p. 227–241.

Brown, M., 1998a, Unpairing metamorphic belts: *P-T* paths and a tectonic model for the Ryoke Belt, southwest Japan: Journal of Metamorphic Geology, v. 16, p. 3–22, doi: 10.1111/j.1525-1314.1998.00061.x.

Brown, M., 1998b, Ridge-trench interactions and high-*T*–low-*P* metamorphism, with particular reference to the Cretaceous evolution of the Japanese Islands, *in* Treloar, P.J., and O'Brien, P.J., eds., What drives metamorphism and metamorphic reactions?: Geological Society [London] Special Publication 138, p. 131–163.

Brown, M., 2001, From microscope to mountain belt: 150 years of petrology and its contribution to understanding geodynamics, particularly the tectonics of orogens: Journal of Geodynamics, v. 32, p. 115–164, doi: 10.1016/S0264-3707(01)00018-7.

Brown, M., 2002a, Prograde and retrograde processes in migmatites revisited: Journal of Metamorphic Geology, v. 20, p. 25–40, doi: 10.1046/j.0263-4929.2001.00362.x.

Brown, M., 2002b, Plate margin processes and "paired" metamorphic belts in Japan: Comment: Earth and Planetary Science Letters, v. 199, p. 483–492, doi: 10.1016/S0012-821X(02)00582-4.

Brown, M., 2006, A duality of thermal regimes is the distinctive characteristic of plate tectonics since the Neoarchean: Geology, v. 34, p. 961–964, doi: 10.1130/G22853A.1.

Brown, M., 2007a, Metamorphic conditions in orogenic belts: A record of secular change: International Geology Review, v. 49, p. 193–234.

Brown, M., 2007b, Metamorphism, plate tectonics and the supercontinent cycle: Earth Science Frontiers, v. 14, p. 1–18, doi: 10.1016/S1872-5791(07)60001-3.

Buick, I.S., Williams, I.S., Gibson, R.L., Cartwright, I., and Miller, J.A., 2003, Carbon and U-Pb evidence for a Palaeoproterozoic crustal component in the Central Zone of the Limpopo Belt, South Africa: Geological Society [London] Journal, v. 160, p. 601–612.

Buick, I.S., Hermann, J., Williams, I.S., Gibson, R.L., and Rubatto, D., 2006, A SHRIMP U-Pb and LA-ICP-MS trace element study of the petrogenesis of garnet-cordierite-orthoamphibole gneisses from the Central Zone of the Limpopo Belt, South Africa: Lithos, v. 88, p. 150–172, doi: 10.1016/j.lithos.2005.09.001.

Burg, J.-P., and Gerya, T.V., 2005, The role of viscous heating in Barrovian metamorphism of collisional orogens: Thermomechanical models and application to the Lepontine Dome in the central Alps: Journal of Metamorphic Geology, v. 23, p. 75–95, doi: 10.1111/j.1525-1314.2005.00563.x.

Burov, E.B., and Watts, A.B., 2006, The long-term strength of continental lithosphere: "Jelly sandwich" or "crème brûlée"?: GSA Today, v. 16, no. 1, p. 4–10, doi: 10.1130/1052-5173(2006)016<4:TLTSOC>2.0.CO;2.

Camacho, A., Lee, J.K.W., Hensen, B.J., and Braun, J., 2005, Short-lived orogenic cycles and the eclogitization of cold crust by spasmodic hot fluids: Nature, v. 435, p. 1191–1196, doi: 10.1038/nature03643.

Carswell, D.A., and Compagnoni, R., 2003, Ultrahigh-pressure metamorphism: Budapest, Eötvös University Press, 508 p.

Cawood, P.A., Kröner, A., and Pisarevsky, S., 2006, Precambrian plate tectonics: Criteria and evidence: GSA Today, v. 16, no. 7, p. 4–11, doi: 10.1130/GSAT01607.1.

Cesare, B., and Gomez-Pugnaire, M.T., 2001, Crustal melting in the Alboran domain: Constraints from xenoliths of the Neogene volcanic province: Physics and Chemistry of the Earth, Part A: Solid Earth and Geodesy, v. 26, p. 255–260, doi: 10.1016/S1464-1895(01)00053-9.

Chan, G.H.-N., Waters, D., Aitchison, J., Searle, M., Horstwood, M., Lo, C.-H., and Crowley, Q., 2007, What lies beneath southern Tibet? Clues from crustal xenoliths hosted in a Miocene ultrapotassic dyke: Metamorphic Studies Group Research in Progress and Annual General Meeting, 4 April 2007, University of Cambridge, UK, Abstracts and Program, p. 12–13.

Chopin, C., 2003, Ultrahigh-pressure metamorphism: Tracing continental crust into the mantle: Earth and Planetary Science Letters, v. 212, p. 1–14, doi: 10.1016/S0012-821X(03)00261-9.

Clarke, G.L., Aitchison, J.C., and Cluzel, D., 1997, Eclogites and blueschists of the Pam Peninsula, NE New Caledonia: A reappraisal: Journal of Petrology, v. 38, p. 843–876, doi: 10.1093/petrology/38.7.843.

Clarke, G.L., Powell, R., and Fitzherbert, J.A., 2006, The lawsonite paradox: A comparison of field evidence and mineral equilibria modelling: Journal of Metamorphic Geology, v. 24, p. 715–725, doi: 10.1111/j.1525-1314.2006.00664.x.

Cloos, M., 1993, Lithospheric buoyancy and collisional orogenesis: Subduction of oceanic plateaus, continental margins, island arcs, spreading ridges, and seamounts: Geological Society of America Bulletin, v. 105, p. 715–737, doi: 10.1130/0016-7606(1993)105<0715:LBACOS>2.3.CO;2.

Collins, A.S., and Pisarevsky, S.A., 2005, Amalgamating eastern Gondwana: The evolution of the Circum-Indian orogens: Earth-Science Reviews, v. 71, p. 229–270, doi: 10.1016/j.earscirev.2005.02.004.

Collins, A.S., Reddy, S.M., Buchan, C., and Mruma, A., 2004, Temporal constraints on Paleoproterozoic eclogite formation and exhumation (Usagaran orogen, Tanzania): Earth and Planetary Science Letters, v. 224, p. 175–192, doi: 10.1016/j.epsl.2004.04.027.

Collins, W.J., 2002, Nature of extensional accretionary orogens: Tectonics, v. 21, 1024, doi: 10.1029/2000TC001272.

Collins, W.J., 2003, Slab pull, mantle convection, and Pangaean assembly and dispersal: Earth and Planetary Science Letters, v. 205, p. 225–237, doi: 10.1016/S0012-821X(02)01043-9.

Colpron, M., Nelson, J.L., and Murphy, D.C., 2007, Northern Cordilleran terranes and their interactions through time: GSA Today, v. 17, no. 4/5, p. 4–10, doi: 10.1130/GSAT01704-5A.1.

Coltice, N., Phillips, B.R., Bertrand, H., Ricard, Y., and Rey, P., 2007, Global warming of the mantle at the origin of flood basalts over supercontinents: Geology, v. 35, p. 391–394, doi: 10.1130/G23240A.1.

Condie, K.C., 1998, Episodic continental growth and supercontinents: A mantle avalanche connection?: Earth and Planetary Science Letters, v. 163, p. 97–108, doi: 10.1016/S0012-821X(98)00178-2.

Condie, K.C., 2002, Breakup of a Paleoproterozoic supercontinent: Gondwana Research, v. 5, p. 41–43, doi: 10.1016/S1342-937X(05)70886-8.

Condie, K.C., and Kröner, A., 2008, this volume, When did plate tectonics begin? Evidence from the geologic record, *in* Condie, K.C., and Pease, V., eds., When did plate tectonics begin on planet Earth?: Geological Society of America Special Paper 440, doi: 10.1130/2008.2440(14).

Coney, P.J., Jones, D.L., and Monger, J.W.H., 1980, Cordilleran suspect terranes: Nature, v. 288, p. 329–333, doi: 10.1038/288329a0.

Coogan, L.A., and Hinton, R.W., 2006, Do the trace element compositions of detrital zircons require Hadean continental crust?: Geology, v. 34, p. 633–636, doi: 10.1130/G22737.1.

Cooper, C.M., Lenardic, A., and Moresi, L., 2006, Effects of continental insulation and the partitioning of heat producing elements on the Earth's heat loss: Geophysical Research Letters, v. 33, L13313, doi: 10.1029/2006GL026291.

Copley, A., and McKenzie, D., 2007, Models of crustal flow in the India-Asia collision zone: Geophysical Journal International, v. 169, p. 683–698, doi: 10.1111/j.1365-246X.2007.03343.x.

Cordani, U.G., D'Agrella-Filho, M.S., Brito-Neves, B.B., and Trindade, R.I.F., 2003, Tearing up Rodinia: The Neoproterozoic palaeogeography of South American cratonic fragments: Terra Nova, v. 15, p. 350–359, doi: 10.1046/j.1365-3121.2003.00506.x.

Corfu, F., 2007, Comment to paper: Timing of magmatism and metamorphism in the Gruinard Bay area of the Lewisian gneiss complex: Comparison with the Assynt Terrane and implications for terrane accretion by G.J. Love, P.D. Kinny, and C.R.L. Friend: Contributions to Mineralogy and Petrology, 2004, v. 146, p. 620 636: Contributions to Mineralogy and Petrology, v. 153, p. 483–488, doi: 10.1007/s00410-006-0157-5.

Crowley, Q.G., Noble, S.R., and Key, R., 2006, U-Pb Geochronology and Hf-isotope constrains on formation of Archaean crust from the Lewisian of NW Scotland, Great Britain: Eos (Transactions, American Geophysical Union), v. 87, Fall Meeting Supplement, abstract V11D-0608.
Currie, C.A., and Hyndman, R.D., 2006, The thermal structure of subduction zone back arcs: Journal of Geophysical Research, v. 111, B08404, doi: 10.1029/2005JB004024.
Currie, C.A., Wang, K., Hyndman, R.D., and He, J., 2004, The thermal effects of slab-driven mantle flow above a subducting plate: The Cascadia subduction zone and backarc: Earth and Planetary Science Letters, v. 223, p. 35–48, doi: 10.1016/j.epsl.2004.04.020.
Davaille, A., 1999, Simultaneous generation of hotspots and superswells by convection in a heterogeneous planetary mantle: Nature, v. 402, p. 756–760, doi: 10.1038/45461.
Davies, G.F., 1992, On the emergence of plate tectonics: Geology, v. 20, p. 963–966, doi: 10.1130/0091-7613(1992)020<0963:OTEOPT>2.3.CO;2.
Davies, G.F., 1995, Punctuated tectonic evolution of the Earth: Earth and Planetary Science Letters, v. 136, p. 363–379, doi: 10.1016/0012-821X(95)-0167-B.
Davies, G.F., 2006, Gravitational depletion of the early Earth's upper mantle and the viability of early plate tectonics: Earth and Planetary Science Letters, v. 243, p. 376–382, doi: 10.1016/j.epsl.2006.01.053.
Davies, G.F., 2007, Controls on density stratification in the early mantle: Geochemistry Geophysics Geosystems, v. 8, doi:10.1029/2006GC001414.
Dewey, J.F., 1975, Finite plate implications: Some implications for evolution of rock masses at plate margins: American Journal of Science, v. A275, p. 260–284.
Dewey, J.F., 2007, The secular evolution of plate tectonics and the continental crust: An outline, *in* Hatcher, R.D., Jr., Carlson, M.P., McBride, J.H., and Martínez Catalán, J.R., eds., 4-D Framework of Continental Crust: Geological Society of America Memoir 200, p. 1–7, doi: 10.1130/2007.1200(01).
Dewey, J.F., Robb, L., and Van Schalkwyk, L., 2006, Did Bushmanland extensionally unroof Namaqualand?: Precambrian Research, v. 150, p. 173–182, doi: 10.1016/j.precamres.2006.07.007.
Diener, J.F.A., Stevens, G., Kisters, A.F.M., and Poujol, M., 2005, Metamorphism and exhumation of the basal parts of the Barberton greenstone belt, South Africa: Constraining the rates of Mesoarchean tectonism: Precambrian Research, v. 143, p. 87–112, doi: 10.1016/j.precamres.2005.10.001.
Diener, J., Stevens, G., and Kisters, A., 2006, High-pressure intermediate-temperature metamorphism in the southern Barberton granitoid-greenstone terrain, *in* Benn, K., et al., eds., Archean geodynamics and environments: American Geophysical Union Geophysical Monograph 164, p. 239–254.
Dilts, S.L., Stewart, K.G., and Loewy, S.L., 2006, Analysis of mineral inclusions in zircon from the eclogite-bearing Ashe Metamorphic Suite, North Carolina: Implications for exhumation history: Eos (Transactions, American Geophysical Union), v. 87, Joint Assembly Supplement, abstract T41A-04.
Ding, L., Kapp, P., Yue, Y., and Lai, Q., 2007, Postcollisional calc-alkaline lavas and xenoliths from the southern Qiangtang terrane, central Tibet: Earth and Planetary Science Letters, v. 254, p. 28–38, doi: 10.1016/j.epsl.2006.11.019.
Doglioni, C., Carminati, E., and Cuffaro, M., 2006, Simple kinematics of subduction zones: International Geology Review, v. 48, p. 479–493.
Dufek, J., and Bergantz, G.W., 2005, Lower crustal magma genesis and preservation: A stochastic framework for the evaluation of basalt–crust interaction: Journal of Petrology, v. 46, p. 2167–2195, doi: 10.1093/petrology/egi049.
Dziggel, A., Stevens, G., Poujol, M., Anhaeusser, C.R., and Armstrong, R.A., 2002, Metamorphism of the granite-greenstone terrane south of the Barberton greenstone belt, South Africa: An insight into the tectono-thermal evolution of the "lower" portions of the Onverwacht Group: Precambrian Research, v. 114, p. 221–247, doi: 10.1016/S0301-9268(01)00225-X.
Ehlers, K., Stüwe, K., Powell, R., Sandiford, M., and Frank, W., 1994, Thermometrically inferred cooling rates from the Plattengneis, Koralm region, eastern Alps: Earth and Planetary Science Letters, v. 125, p. 307–321, doi: 10.1016/0012-821X(94)90223-2.
England, P.C., and Houseman, G.A., 1988, The mechanics of the Tibetan Plateau: Royal Society of London Philosophical Transactions, ser. A, v. 326, p. 301–320.
English, J.M., and Johnston, S.T., 2005, Collisional orogenesis in the northern Canadian Cordillera: Implications for Cordilleran crustal structure, ophiolite emplacement, continental growth, and the terrane hypothesis: Earth and Planetary Science Letters, v. 232, p. 333–344, doi: 10.1016/j.epsl.2005.01.025.
Ernst, W.G., 2001, Subduction, ultrahigh-pressure metamorphism, and regurgitation of buoyant crustal slices: Implications for arcs and continental growth: Physics of the Earth and Planetary Interiors, v. 127, p. 253–275, doi: 10.1016/S0031-9201(01)00231-X.
Ernst, W.G., 2005, Alpine and Pacific styles of Phanerozoic mountain building: Subduction-zone petrogenesis of continental crust: Terra Nova, v. 17, p. 165–188, doi: 10.1111/j.1365-3121.2005.00604.x.
Evans, D.A.D., and Pisarevsky, S.A., 2008, this volume, Plate tectonics on early Earth? Weighing the paleomagnetic evidence, *in* Condie, K.C., and Pease, V., eds., When did plate tectonics begin on planet Earth?: Geological Society of America Special Paper 440, doi: 10.1130/2008.2440(12).
Farquhar, J., Wing, B.A., McKeegan, K.D., Harris, J.W., Cartigny, P., and Thiemens, M.H., 2002, Mass-independent sulfur of inclusions in diamond and sulfur recycling on early Earth: Science, v. 298, p. 2369–2372, doi: 10.1126/science.1078617.
Faryad, S.W., 1999, Metamorphic evolution of the Precambrian South Badakhshan block, based on mineral reactions in metapelites and metabasites associated with whiteschists from Sare Sang (Western Hindu Kush, Afghanistan): Precambrian Research, v. 98, p. 223–241, doi: 10.1016/S0301-9268(99)00051-0.
Filippova, I.B., Bush, V.A., and Didenko, A.N., 2001, Middle Paleozoic subduction belts: The leading factor in the formation of the central Asian fold-and-thrust belt: Russian Journal of Earth Sciences, v. 3, p. 405–426.
Foley, S., 2008, this volume, A trace element perspective on Archean crust formation and on the presence or absence of Archean subduction, *in* Condie, K.C., and Pease, V., eds., When did plate tectonics begin on planet Earth?: Geological Society of America Special Paper 440, doi: 10.1130/2008.2440(02).
Foley, S.F., Tiepolo, M., and Vannucci, R., 2002, Growth of early continental crust controlled by melting of amphibolite in subduction zones: Nature, v. 417, p. 837–840, doi: 10.1038/nature00799.
Foley, S.F., Buhre, S., and Jacob, D.E., 2003, Evolution of the Archaean crust by delamination and shallow subduction: Nature, v. 421, p. 249–252.
Fonarev, V.I., Pilugin, S.M., Savko, K.A., and Novikova, M.A., 2006, Exsolution textures of orthopyroxene and clinopyroxene in high-grade BIF of the Voronezh Crystalline Massif: Evidence of ultrahigh-temperature metamorphism: Journal of Metamorphic Geology, v. 24, p. 135–151, doi: 10.1111/j.1525-1314.2006.00630.x.
Friend, C.R.L., and Nutman, A.P., 2005a, New pieces to the Archaean terrane jigsaw puzzle in the Nuuk region, southern West Greenland: Steps in transforming a simple insight into a complex regional tectonothermal model: Geological Society [London] Journal, v. 162, p. 147–162, doi: 10.1144/0016-764903-161.
Friend, C.R.L., and Nutman, A.P., 2005b, Complex 3670–3500 Ma orogenic episodes superimposed on juvenile crust accreted between 3850 and 3690 Ma, Itsaq Gneiss Complex, southern West Greenland: Journal of Geology, v. 113, p. 375–397, doi: 10.1086/430239.
Friend, C.R.L., Kinny, P.D., and Love, G.J., 2007, Timing of magmatism and metamorphism in the Gruinard Bay area of the Lewisian Gneiss Complex: Comparison with the Assynt Terrane and implications for terrane accretion: Reply: Contributions to Mineralogy and Petrology, v. 153, p. 489–492, doi: 10.1007/s00410-006-0156-6.
Garde, A.A., 2007, A mid-Archaean island arc complex in the eastern Akia terrane, Godthåbsfjord, southern West Greenland: Geological Society [London] Journal, v. 164, p. 565–579, doi: 10.1144/0016-76492005-107.
Goscombe, B., and Hand, M., 2000, Contrasting *P-T* paths in the Eastern Himalaya, Nepal: Inverted isograds in a paired metamorphic mountain belt: Journal of Petrology, v. 41, p. 1673–1719, doi: 10.1093/petrology/41.12.1673.
Green, H.W., 2005, Psychology of a changing paradigm: 40+ years of high-pressure metamorphism: International Geology Review, v. 47, p. 439–456.
Gregoire, M., Mattielli, N., Nicollet, C., Cottin, J.Y., Leyrit, H., Weis, D., Shimizu, N., and Giret, A., 1994, Oceanic mafic granulite xenoliths from the Kerguelen archipelago: Nature, v. 367, p. 360–363, doi: 10.1038/367360a0.
Griffin, W.L., McGregor, V.R., Nutman, A., Taylor, P.N., and Bridgwater, D., 1980, Early Archean granulite-facies metamorphism south of Ameralik, West Greenland: Earth and Planetary Science Letters, v. 50, p. 59–74, doi: 10.1016/0012-821X(80)90119-3.
Grigné, C., and Labrosse, S., 2001, Effects of continents on Earth cooling: Thermal blanketing and depletion in radioactive elements: Geophysical Research Letters, v. 28, p. 2707–2710, doi: 10.1029/2000GL012475.
Grigné, C., Labrosse, S., and Tackley, P.J., 2005, Convective heat transfer as a function of wavelength: Implications for the cooling of the Earth: Journal of Geophysical Research, v. 110, B03409, doi: 10.1029/2004JB003376.

Guillou, L., and Jaupart, C., 1995, On the effect of continents on mantle convection: Journal of Geophysical Research, v. 100, p. 24,217–24,238, doi: 10.1029/95JB02518.

Gurnis, M., 1988, Large-scale mantle convection and the aggregation and dispersal of supercontinents: Nature, v. 332, p. 695–699, doi: 10.1038/332695a0.

Hacker, B.R., 2006, Pressures and temperatures of ultrahigh-pressure metamorphism: Implications for UHP tectonics and $H_2O$ in subducting slabs: International Geology Review, v. 48, p. 1053–1066.

Hacker, B.R., Abers, G.A., and Peacock, S.M., 2003, Subduction factory, 1: Theoretical mineralogy, densities, seismic wave speeds, and $H_2O$ contents: Journal of Geophysical Research, B, Solid Earth and Planets, v. 108, article 2029, doi: 10.1029/2001JB001127.

Hacker, B.R., Gnos, E., Ratschbacher, L., Grove, M., McWilliams, M., Sobolev, S.V., Wan, J., and Wu, Z.H., 2000, Hot and dry deep crustal xenoliths from Tibet: Science, v. 287, p. 2463–2466, doi: 10.1126/science.287.5462.2463.

Hacker, B.R., Abers, G.A., Peacock, S.M., and Johnson, S., 2005a, Subduction factory, 1: Theoretical mineralogy, densities, seismic wave speeds and $H_2O$ contents: Reply: Journal of Geophysical Research, B, Solid Earth and Planets, v. 110, B02207, doi: 10.1029/2004JB003490.

Hacker, B., Luffi, P., Lutkov, V., Minaev, V., Ratschbacher, L., Plank, T., Ducea, M., Patino-Douce, A., McWilliams, M., and Metcalf, J., 2005b, Near-ultrahigh pressure processing of continental crust: Miocene crustal xenoliths from the Pamir: Journal of Petrology, v. 46, p. 1661–1687, doi: 10.1093/petrology/egi030.

Hafkenscheid, E., Wortel, M.J.R., and Spakman, W., 2006, Subduction history of the Tethyan region derived from seismic tomography and tectonic reconstructions: Journal of Geophysical Research, v. 111, B08401, doi: 10.1029/2005JB003791.

Hamilton, W.B., 2003, An alternative Earth: GSA Today, v. 13, no. 11, p. 4–12, doi: 10.1130/1052-5173(2003)013<0004:AAE>2.0.CO;2.

Hansen, U., Harder, H., Stein, C., and Loddoch, A., 2006, Numerical modelling of the internal dynamics and surface tectonics of the terrestrial planets, *in* Münster, G., Wolf, D., and Kremer, M., eds., Proceedings, NIC Symposium 2006, NIC Series Volume 32: Jülich, Germany, John von Neumann Institut für Computing (NIC), p. 289–296.

Harley, S.L., 1998, Ultrahigh temperature granulite metamorphism (1050 °C, 12 kbar) and decompression in garnet (Mg 70)–orthopyroxene–sillimanite gneisses from the Rauer Group, East Antarctica: Journal of Metamorphic Geology, v. 16, p. 541–562, doi: 10.1111/j.1525-1314.1998.00155.x.

Harley, S.L., 2006, The hottest crust: Geochimica et Cosmochimica Acta, v. 70, p. A230.

Harley, S.L., Kelly, N.M., and Moller, A., 2007, Zircon behaviour and the thermal histories of mountain chains: Elements, v. 3, p. 25–30.

Harrison, T.M., Blichert-Toft, J., Müller, W., Albarède, F., Holden, P., and Mojzsis, S.J., 2006, Heterogeneous Hadean hafnium: Evidence of continental crust at 4.4 to 4.5 Ga: Reply: Science, v. 312, p. 1139, doi: 10.1126/science.1125408.

Hawkesworth, C.J., and Kemp, A.I.S., 2006, Evolution of the continental crust: Nature, v. 443, p. 811–817, doi: 10.1038/nature05191.

Hayashi, M., Komiya, T., Nakamura, Y., and Maruyama, S., 2000, Archean regional metamorphism of the Isua supracrustal belt, southern West Greenland: Implications for a driving force for Archean plate tectonics: International Geology Review, v. 42, p. 1055–1115.

Hayob, J.L., Essene, E.J., Ruiz, J., Ortega-Gutiérrez, F., and Aranda-Gómez, J.J., 1989, Young high-temperature granulites from the base of the crust in central Mexico: Nature, v. 342, p. 265–268, doi: 10.1038/342265a0.

Hayob, J.L., Essene, E.J., and Ruiz, J., 1990, Reply: Nature, v. 347, p. 133–134, doi: 10.1038/347133b0.

Hefferan, K.P., Admou, H., Hilal, R., Karson, J.A., Saquaque, A., Juteau, T., Bohn, M.M., Samson, S.D., and Kornprobst, J.M., 2002, Proterozoic blueschist-bearing mélange in the Anti-Atlas Mountains, Morocco: Precambrian Research, v. 118, p. 179–194, doi: 10.1016/S0301-9268(02)00109-2.

Hensen, B.J., and Zhou, B., 1995, Retention of isotopic memory in garnets partially broken down during an overprinting granulite-facies metamorphism: Implications for the Sm-Nd closure temperature: Geology, v. 23, p. 225–228, doi: 10.1130/0091-7613(1995)023<0225:ROIMIG>2.3.CO;2.

Hensen, B.J., Zhou, B., and Thost, D.E., 1995, Are reaction textures reliable guides to metamorphic histories? Timing constraints from garnet Sm-Nd chronology for "decompression" textures in granulites from Sostrene Island, Prydz Bay, Antarctica: Geological Journal, v. 30, p. 261–271, doi: 10.1002/gj.3350300306.

Hermann, J., 2003, Experimental evidence for diamond-facies metamorphism in the Dora-Maira Massif: Lithos, v. 70, p. 163–182, doi: 10.1016/S0024-4937(03)00097-5.

Hoffman, P.F., 1991, Did the breakout of Laurentia turn Gondwanaland inside out?: Science, v. 252, p. 1409–1412, doi: 10.1126/science.252.5011.1409.

Holland, T.J.B., and Powell, R., 1998, An internally consistent thermodynamic data set for phases of petrological interest: Journal of Metamorphic Geology, v. 16, p. 309–343, doi: 10.1111/j.1525-1314.1998.00140.x.

Horodyskyj, U.N., Lee, C.-T.A., and Ducea, M.N., 2007, Similarities between Archean high MgO eclogites and Phanerozoic arc-eclogite cumulates and the role of arcs in Archean continent formation: Earth and Planetary Science Letters, v. 256, p. 510–520, doi: 10.1016/j.epsl.2007.02.006.

Hubbard, M.S., Grew, E.S., Hodges, K.V., Yates, M.G., and Pertsev, N.N., 1999, Neogene cooling and exhumation of upper-amphibolite-facies "whiteschists" in the southwest Pamir Mountains, Tajikistan: Precambrian Research, v. 305, p. 325–337.

Husson, L., Faccenna, C., and Conrad, C.P., 2007, Westward drift of the Pacific plates, trenches and upper mantle: Geophysical Research Abstracts, v. 9, no. 04169.

Hyndman, R.D., Currie, C.A., and Mazzotti, S.P., 2005a, Subduction zone backarcs, mobile belts, and orogenic heat: GSA Today, v. 15, no. 2, p. 4–10.

Hyndman, R.D., Currie, C., and Mazzotti, S., 2005b, The origin of global mountain belts: Hot subduction zone backarcs: Eos (Transactions, American Geophysical Union), v. 86, abstract T11B-0367.

Hynes, A., 2008, this volume, Effects of a warmer mantle on the characteristics of Archean passive margins, *in* Condie, K.C., and Pease, V., eds., When did plate tectonics begin on planet Earth?: Geological Society of America Special Paper 440, doi: 10.1130/2008.2440(07).

Isacks, B., Sykes, L.R., and Oliver, J., 1968, Seismology and the new global tectonics: Journal of Geophysical Research, v. 73, p. 5855–5899.

Jackson, M.D., Cheadle, M.J., and Atherton, M.P., 2003, Quantitative modeling of granitic melt generation and segregation in the continental crust: Journal of Geophysical Research, B, Solid Earth and Planets, v. 108, article 2332.

Jacob, D.E., 2004, Nature and origin of eclogite xenoliths from kimberlites: Lithos, v. 77, p. 295–316, doi: 10.1016/j.lithos.2004.03.038.

Jahn, B.-M., Griffin, W.L., and Windley, B.F., editors, 2000a, Continental growth in the Phanerozoic: Evidence from Central Asia: Tectonophysics, v. 328, p. vii–x, doi: 10.1016/S0040-1951(00)00174-8.

Jahn, B.-M., Wu, F.Y., and Chen, B., 2000b, Granitoids of the Central Asian Orogenic Belt and continental growth in the Phanerozoic: Transactions of the Royal Society of Edinburgh, Earth Sciences, v. 91, p. 181–193.

Jahn, B.-M., Wu, F.Y., and Chen, B., 2000c, Massive granitoid generation in Central Asia: Nd isotope evidence and implication for continental growth in the Phanerozoic: Episodes, v. 23, p. 82–92.

Jahn, B.-M, Caby, R., and Monie, P., 2001, The oldest UHP eclogites of the World: Age of UHP metamorphism, nature of protoliths and tectonic implications: Chemical Geology, v. 178, p. 143–158, doi: 10.1016/S0009-2541(01)00264-9.

Jahn, B.-M., Windley, B., Natal'in, B., and Dobretsov, N., editors, 2004, Phanerozoic continental growth in central Asia: Journal of Asian Earth Sciences, v. 23, p. 599–603.

Jamieson, R.A., Beaumont, C., Nguyen, M.H., and Lee, B., 2002, Interaction of metamorphism, deformation and exhumation in large convergent orogens: Journal of Metamorphic Geology, v. 20, p. 9–24, doi: 10.1046/j.0263-4929.2001.00357.x.

Jamieson, R.A., Beaumont, C., Nguyen, M.H., and Grujic, D., 2006, Prominence of the Greater Himalayan Sequence and associated rocks: Predictions of channel flow models, *in* Law, R.D., et al., eds., Channel flow, ductile extrusion and exhumation in continental collision zones: Geological Society [London] Special Publication 268, p. 165–182.

Johnson, T., and Brown, M., 2004, Quantitative constraints on metamorphism in the Variscides of southern Brittany—A complementary pseudosection approach: Journal of Petrology, v. 38, p. 1237–1259.

Johnston, S.T., 2001, The great Alaskan terrane wreck: Reconciliation of paleomagnetic and geological data in the northern Cordillera: Earth and Planetary Science Letters, v. 193, p. 259–272, doi: 10.1016/S0012-821X(01)00516-7.

Johnston, S.T., and Borel, G.D., 2007, The odyssey of the Cache Creek terrane, Canadian Cordillera: Implications for accretionary orogens, tectonic setting of Panthalassa, the Pacific superswell, and break-up of Pangea: Earth and Planetary Science Letters, v. 253, p. 415–428, doi: 10.1016/j.epsl.2006.11.002.

Kaneko, Y., Maruyama, S., Terabayashi, M., Yamamoto, H., Ishikawa, M., Anma, R., Parkinson, C.D., Ota, T., Nakajima, Y., Katayama, I., Yamamoto, J., and Yamauchi, K., 2000, Geology of the Kokchetav UHP-HP metamorphic belt, northern Kazakhstan: Island Arc, v. 9, p. 264–283, doi: 10.1046/j.1440-1738.2000.00278.x.

Kelly, N.M., and Harley, S.L., 2005, An integrated microtextural and chemical approach to zircon geochronology: Refining the Archaean history of the Napier Complex, East Antarctica: Contributions to Mineralogy and Petrology, v. 149, p. 57–84, doi: 10.1007/s00410-004-0635-6.

Kincaid, C., and Silver, P., 1996, The role of viscous dissipation in the orogenic process: Earth and Planetary Science Letters, v. 142, p. 271–288, doi: 10.1016/0012-821X(96)00116-1.

Koizumi, K., and Ishiwatari, A., 2006, Oceanic plateau accretion inferred from late Paleozoic greenstones in the Jurassic Tamba accretionary complex, southwest Japan: Island Arc, v. 15, p. 58–83, doi: 10.1111/j.1440-1738.2006.00518.x.

Kreemer, C., Holt, W.E., and Haines, A.J., 2003, An integrated global model of present-day plate motions and plate boundary deformation: Geophysical Journal International, v. 154, p. 8–34, doi: 10.1046/j.1365-246X.2003.01917.x.

Kreissig, K., Nagler, T.F., Kramers, J.D., van Reenen, D.D., and Smit, C.A., 2000, An isotopic and geochemical study of the northern Kaapvaal Craton and the Southern Marginal Zone of the Limpopo Belt: Are they juxtaposed terranes?: Lithos, v. 50, p. 1–25, doi: 10.1016/S0024-4937(99)00037-7.

Kreissig, K., Holzer, L., Frei, R., Villa, I.M., Kramers, J.D., Kröner, A., Smit, C.A., and van Reenen, D.D., 2001, Geochronology of the Hout River shear zone and the metamorphism in the southern marginal zone of the Limpopo Belt, southern Africa: Precambrian Research, v. 109, p. 145–173, doi: 10.1016/S0301-9268(01)00147-4.

Kröner, A., Cui, W.Y., Wang, S.Q., Wang, C.Q., and Nemchin, A.A., 1998, Single zircon ages from high-grade rocks of the Jianping Complex, Liaoning province, NE China: Journal of Asian Earth Sciences, v. 16, p. 519–532, doi: 10.1016/S0743-9547(98)00033-6.

Kröner, A., Jaeckel, P., Brandl, G., Nemchin, A.A., and Pidgeon, R.T., 1999, Single zircon ages for granitoid gneisses in the Central Zone of the Limpopo Belt, southern Africa, and geodynamic significance: Precambrian Research, v. 93, p. 299–337, doi: 10.1016/S0301-9268(98)00102-8.

Kröner, A., Wilde, S.A., Zhao, G.C., O'Brien, P.J., Sun, M., Liu, D.Y., Wan, Y.S., Liu, S.W., and Guo, J.H., 2006, Zircon geochronology and metamorphic evolution of mafic dykes in the Hengshan Complex of northern China: Evidence for late Palaeoproterozoic extension and subsequent high-pressure metamorphism in the North China Craton: Precambrian Research, v. 146, p. 45–67, doi: 10.1016/j.precamres.2006.01.008.

Lawrence, J.F., and Wysession, M.E., 2006, Seismic evidence for subduction-transported water in the lower mantle, *in* Jacobsen, S.D., and van der Lee, S., eds., Earth's deep-water cycle: American Geophysical Union Geophysical Monograph 168, p. 251–261.

Leech, M.L., Singh, S., Jain, A.K., Klemperer, S.L., and Manickavasagam, R.M., 2005, The onset of India-Asia continental collision: Early, steep subduction required by the timing of UHP metamorphism in the western Himalaya: Earth and Planetary Science Letters, v. 234, p. 83–97, doi: 10.1016/j.epsl.2005.02.038.

Leech, M.L., Singh, S., and Jain, A.K., 2007, Continuous metamorphic zircon growth and interpretation of U-Pb SHRIMP dating: An example from the Western Himalaya: International Geology Review, v. 49, p. 313–328.

Leloup, Ph.D., Ricard, Y., Battaglia, J., and Lacassin, R., 1999, Shear heating in continental strike-slip shear zones: Model and field examples: Geophysical Journal International, v. 136, p. 19–40, doi: 10.1046/j.1365-246X.1999.00683.x.

Lenardic, A., 2006, Continental growth and the Archean paradox, *in* Condie, K.C., et al., eds., Archean geodynamics and environments: American Geophysical Union Geophysical Monograph 164, p. 33–45.

Lenardic, A., Nimmo, F., and Moresi, L., 2004, Growth of the hemispheric dichotomy and the cessation of plate tectonics on Mars: Journal of Geophysical Research, v. 109, E02003, doi: 10.1029/2003JE002172.

Lenardic, A., Moresi, L.-N., Jellinek, M., and Manga, M., 2005, Continental insulation, mantle cooling, and the surface area of oceans and continents: Earth and Planetary Science Letters, v. 234, p. 317–333, doi: 10.1016/j.epsl.2005.01.038.

LePichon, X., 1968, Seafloor spreading and continental drift: Journal of Geophysical Research, v. 73, p. 3661–3697.

Liati, A., and Gebauer, D., 2003, Geochronological constraints of the time of metamorphism in the Gruf Complex (central Alps) and implications for the Adula-Cima Lunga Nappe system: Schweizerische Mineralogische und Petrographische Mitteilungen, v. 83, p. 159–172.

Liou, J.G., Hacker, B.R., and Zhang, R.Y., 2000, Into the forbidden zone: Science, v. 287, p. 1215–1216, doi: 10.1126/science.287.5456.1215.

Liou, J.G., Tsujimori, T., Zhang, R.Y., Katayama, I., and Maruyama, S., 2004, Global UHP metamorphism and continental subduction/collision: The Himalayan model: International Geology Review, v. 46, p. 1–27.

Liu, L., Zhang, J., Green, H.W., Jin, Z., and Bozhilov, K.N., 2007, Evidence of former stishovite in metamorphosed sediments, implying subduction to >350 km: Earth and Planetary Science Letters, v. 263, p. 180–191.

Loddoch, A., Stein, A., and Hansen, U., 2006, Temporal variations in the convective style of planetary mantles: Earth and Planetary Science Letters, v. 251, p. 79–89, doi: 10.1016/j.epsl.2006.08.026.

Lucassen, F., and Franz, G., 1996, Magmatic arc metamorphism: Petrology and temperature history of metabasic rocks in the Coastal Cordillera of northern Chile: Journal of Metamorphic Geology, v. 14, p. 249–265, doi: 10.1046/j.1525-1314.1996.59011.x.

Maekawa, H., Shozui, M., Ishii, T., Fryer, P., and Pearce, J.A., 1993, Blueschist metamorphism in an active subduction zone: Nature, v. 364, p. 520–523, doi: 10.1038/364520a0.

Maresch, W.V., and Gerya, T.V., 2005, Blueschists and blue amphiboles: How much subduction do they need?: International Geology Review, v. 47, p. 688–702.

Maruyama, S., 1997, Pacific-type orogeny revisited: Miyashiro-type orogeny proposed: Island Arc, v. 6, p. 91–120, doi: 10.1111/j.1440-1738.1997.tb00042.x.

Maruyama, S., and Liou, J.G., 1998, Initiation of ultrahigh-pressure metamorphism and its significance on the Proterozoic-Phanerozoic boundary: Island Arc, v. 7, p. 6–35, doi: 10.1046/j.1440-1738.1998.00181.x.

Maruyama, S., and Liou, J.G., 2005, From snowball to Phanerozoic Earth: International Geology Review, v. 47, p. 775–791.

Maruyama, S., Liou, J.G., and Terabayashi, M., 1996, Blueschists and eclogites of the world and their exhumation: International Geology Review, v. 38, p. 485–594.

Matsuda, T., and Uyeda, S., 1971, On the Pacific-type orogeny and its model—Extension of the paired belts concept and possible origin of marginal seas: Tectonophysics, v. 11, p. 5–27, doi: 10.1016/0040-1951(71)90076-X.

Mattinson, C.G., Wooden, J.L., Liou, J.G., Bird, D.D., and Wu, C.L., 2006, Age and duration of eclogite-facies metamorphism, North Qaidam HP/UHP terrane, western China: American Journal of Science, v. 306, p. 683–711, doi: 10.2475/09.2006.01.

McCulloch, M.T., and Bennett, V.C., 1994, Progressive growth of the Earth's continental crust and depleted mantle: Geochemical constraints: Geochimica et Cosmochimica Acta, v. 58, p. 4717–4738, doi: 10.1016/0016-7037(94)90203-8.

McGregor, V.R., and Friend, C.R.L., 1997, Field recognition of rocks totally retrogressed from granulite facies: An example from Archaean rocks in the Paamiut region, South-West Greenland: Precambrian Research, v. 86, p. 59–70, doi: 10.1016/S0301-9268(97)00041-7.

McKenzie, D.P., and Parker, R.L., 1967, The North Pacific: An example of tectonics on a sphere: Nature, v. 216, p. 1276–1280, doi: 10.1038/2161276a0.

McKenzie, D., and Priestley, K., 2008, The influence of lithospheric thickness variations on continental evolution: Lithos, v. 102, p. 1–11, doi: 10.1016/j.lithos.2007.05.005.

McLaren, S., Sandiford, M., Powell, R., Neumann, N., and Woodhead, J., 2006, Palaeozoic intraplate crustal anatexis in the Mount Painter province, South Australia: Timing, thermal budgets and the role of crustal heat production: Journal of Petrology, v. 47, p. 2281–2302, doi: 10.1093/petrology/egl044.

McLelland, W.C., and Power, S.E., Gilotti, J.A., Mazdab, F.K., and Wopenka, B., 2006, U-Pb SHRIMP geochronology and trace-element geochemistry of coesite-bearing zircons, North-East Greenland Caledonides, *in* Hacker, B.R., et al., eds., Ultrahigh-pressure metamorphism: Deep continental subduction: Geological Society of America Special Paper 403, p. 23–43, doi: 10.1130/2006.2403(02).

McNutt, M.K., 1998, Superswells: Reviews of Geophysics, v. 36, p. 211–244, doi: 10.1029/98RG00255.

McNutt, M.K., 1999, Earth science: The mantle's lava lamp: Nature, v. 402, p. 739–740, doi: 10.1038/45416.

Miyashiro, A., 1961, Evolution of metamorphic belts: Journal of Petrology, v. 2, p. 277–311.

Miyashiro, A., 1972, Metamorphism and related magmatism in plate tectonics: American Journal of Science, v. 272, p. 629–656.

Miyashiro, A., 1973, Paired and unpaired metamorphic belts: Tectonophysics, v. 17, p. 241–254, doi: 10.1016/0040-1951(73)90005-X.

Möller, A., Appel, P., Mezger, K., and Schenk, V., 1995, Evidence for a 2 Ga subduction zone: Eclogites in the Usagaran belt of Tanzania: Geology, v. 23, p. 1067–1070, doi: 10.1130/0091-7613(1995)023<1067:EFAGSZ>2.3.CO;2.

Monteleone, B.D., Baldwin, S.L., Webb, L.E., Fitzgerald, P.E., Grove, M., and Schmitt, A.K., 2007, Late Miocene–Pliocene eclogite facies metamorphism, D'Entrecasteaux Islands, SE Papua New Guinea: Journal of Metamorphic Geology, v. 25, p. 245–265, doi: 10.1111/j.1525-1314.2006.00685.x.

Moores, E.M., 2002, Pre–1 Ga (pre-Rodinian) ophiolites: Their tectonic and environmental implications: Geological Society of America Bulletin, v. 114, p. 80–95, doi: 10.1130/0016-7606(2002)114<0080:PGPROT>2.0.CO;2.

Morgan, W.J., 1968, Rises, trenches, great faults, and crustal blocks: Journal of Geophysical Research, v. 73, p. 1959–1982.

Moyen, J.-F., Stevens, G., and Kisters, A., 2006, Record of mid-Archaean subduction from metamorphism in the Barberton terrain of South Africa: Nature, v. 442, p. 559–562, doi: 10.1038/nature04972.

Murphy, J.B., and Nance, R.D., 2003, Do supercontinents introvert or extrovert? Sm-Nd isotope evidence: Geology, v. 31, p. 873–876, doi: 10.1130/G19668.1.

Murphy, J.B., and Nance, R.D., 2005, Do supercontinents turn inside-in or inside-out?: International Geology Review, v. 47, p. 591–619.

Murphy, J.B., Pisarevsky, S.A., Nance, R.D., and Keppie, J.D., 2004, Neoproterozoic–early Paleozoic evolution of peri-Gondwanan terranes: Implications for Laurentia-Gondwana connections: International Journal of Earth Sciences, v. 93, p. 659–682, doi: 10.1007/s00531-004-0412-9.

Nelson, K.D., Zhao, W.J., Brown, L.D., Kuo, J., Che, J.K., Liu, X.W., Klemperer, S.L., Makovsky, Y., Meissner, R., Mechie, J., Kind, R., Wenzel, F., Ni, J., Nabelek, J., Chen, L.S., Tan, H.D., Wei, W.B., Jones, A.G., Booker, J., Unsworth, M., Kidd, W.S.F., Hauck, M., Alsdorf, D., Ross, A., Cogan, M., Wu, C.D., Sandvol, E., and Edwards, M., 1996, Partially molten middle crust beneath southern Tibet: Synthesis of project INDEPTH results: Science, v. 274, p. 1684–1688, doi: 10.1126/science.274.5293.1684.

Nutman, A., and Friend, C.R.L., 2007, Adjacent terranes with c. 2715 and 2650 Ma high-pressure metamorphic assemblages in the Nuuk region of the North Atlantic Craton, southern West Greenland: Complexities of Neoarchean collisional orogeny: Precambrian Research, v. 155, p. 159–203, doi: 10.1016/j.precamres.2006.12.009.

O'Brien, P.J., 1997, Garnet zoning and reaction textures in overprinted eclogites, Bohemian Massif, European Variscides: A record of their thermal history during exhumation: Lithos, v. 41, p. 119–133, doi: 10.1016/S0024-4937(97)82008-7.

O'Brien, P.J., 1999, Asymmetric zoning profiles in garnet from HP-HT granulite and implications for volume and grain boundary diffusion: Mineralogical Magazine, v. 63, p. 227–238, doi: 10.1180/002646199548457.

O'Brien, P.J., and Rötzler, J., 2003, High-pressure granulites: Formation, recovery of peak conditions and implications for tectonics: Journal of Metamorphic Geology, v. 21, p. 3–20, doi: 10.1046/j.1525-1314.2003.00420.x.

Omori, S., Komabayashi, T., and Maruyama, S., 2004, Dehydration and earthquakes in the subducting slab: Empirical link in intermediate and deep seismic zones: Physics of the Earth and Planetary Interiors, v. 146, p. 297–311, doi: 10.1016/j.pepi.2003.08.014.

Ota, T., Gladkochubb, D.P., Sklyarov, E.V., Mazukabzov, A.M., and Watanabe, T., 2004, *P-T* history of garnet-websterites in the Sharyzhalgai complex, southwestern margin of Siberian craton: Evidence for Paleoproterozoic high-pressure metamorphism: Precambrian Research, v. 132, p. 327–348, doi: 10.1016/j.precamres.2004.03.009.

Patrick, B.E., and Day, H.W., 1995, Cordilleran high-pressure metamorphic terranes: Progress and problems: Journal of Metamorphic Geology, v. 13, p. 1–8, doi: 10.1111/j.1525-1314.1995.tb00201.x.

Pattison, D.R.M., Chacko, T., Farquhar, J., and McFarlane, C.R.M., 2003, Temperatures of granulite-facies metamorphism: Constraints from experimental phase equilibria and thermobarometry corrected for retrograde exchange: Journal of Petrology, v. 44, p. 867–900, doi: 10.1093/petrology/44.5.867.

Peacock, S.M., and Wang, K., 1999, Seismic consequences of warm versus cool subduction metamorphism: Examples from southwest and northeast Japan: Science, v. 286, p. 937–939, doi: 10.1126/science.286.5441.937.

Pease, V., Percival, J., Smithies, H., Stevens, G., and Van Kranendonk, M., 2008, this volume, When did plate tectonics begin? Evidence from the orogenic record, *in* Condie, K.C., and Pease, V., eds., When did plate tectonics begin on planet Earth?: Geological Society of America Special Paper 440, doi: 10.1130/2008.2440(10).

Peressini, G., Quick, J.E., Sinigoi, S., Hofmann, A.W., and Fanning, M., 2007, Duration of a large mafic intrusion and heat transfer in the lower crust: A SHRIMP U-Pb zircon study in the Ivrea-Verbano Zone (western Alps, Italy): Journal of Petrology, v. 48, p. 1185–1218, doi:10.1093/petrology/egm014.

Polat, A., Kusky, T., and Li, J., 2007, Geochemistry of Neoarchean (ca. 2.55–2.50 Ga) volcanic and ophiolitic rocks in the Wutaishan greenstone belt, central orogenic belt, North China craton: Implications for geodynamic setting and continental growth: Reply: Geological Society of America Bulletin, v. 119, p. 490–492, doi: 10.1130/B26163.1.

Poller, U., Gladkochub, D., Donskaya, T., Mazukabzov, A., Sklyarov, E., and Todt, W., 2005, Multistage magmatic and metamorphic evolution in the Southern Siberian Craton: Archean and Palaeoproterozoic zircon ages revealed by SHRIMP and TIMS: Precambrian Research, v. 136, p. 353–368, doi: 10.1016/j.precamres.2004.12.003.

Potel, S., Ferreiro Mählmann, R., Stern, W.B., Mullis, J., and Frey, M., 2006, Very low-grade metamorphic evolution of pelitic rocks under high-pressure/low-temperature conditions, NW New Caledonia (SW Pacific): Journal of Petrology, v. 47, p. 991–1015, doi: 10.1093/petrology/egl001.

Powell, R., Holland, T., and Worley, B., 1998, Calculating phase diagrams involving solid solutions via nonlinear equations, with examples using THERMOCALC: Journal of Metamorphic Geology, v. 16, p. 577–588.

Powell, R., Guiraud, M., and White, R.W., 2005, Truth and beauty in metamorphic phase equilibria: Conjugate variables and phase diagrams: Canadian Mineralogist, v. 43, p. 21–33.

Ravna, E.J.K., and Terry, M.P., 2004, Geothermobarometry of UHP and HP eclogites and schists: An evaluation of equilibria among garnet-clinopyroxene-kyanite-phengite-coesite/quartz: Journal of Metamorphic Geology, v. 22, p. 570–592.

Rollinson, H.R., 1997, Eclogite xenoliths in west African kimberlites as residues from Archaean granitoid crust formation: Nature, v. 389, p. 173–176, doi: 10.1038/38266.

Rollinson, H.R., 2002, The metamorphic history of the Isua Greenstone Belt, West Greenland, *in* Fowler, C.M.R., et al., eds., The early Earth: Geological Society [London] Special Publication 199, p. 329–350.

Rötzler, J., and Romer, R.L., 2001, *P-T-t* evolution of ultrahigh-temperature granulites from the Saxon Granulite Massif, Germany, part I: Petrology: Journal of Petrology, v. 42, p. 1995–2013.

Rubatto, D., and Hermann, J., 2007, Zircon behavior in deeply subducted rocks: Elements, v. 3, p. 31–35.

Sachan, H.K., Mukherjee, B.K., Ogasawara, Y., Maruyama, S., Ishida, H., Muko, A., and Yoshioka, N., 2004, Discovery of coesite from Indus suture zone (ISZ), Ladakh, India: Evidence for deep subduction: European Journal of Mineralogy, v. 16, p. 235–240, doi: 10.1127/0935-1221/2004/0016-0235.

Sandiford, M., and Dymoke, P., 1991, Some remarks on the stability of blueschists and related high *P*-low *T* assemblages in continental orogens: Earth and Planetary Science Letters, v. 102, p. 14–23, doi: 10.1016/0012-821X(91)90014-9.

Sandiford, M., and Powell, R., 1991, Some remarks on high-temperature–low-pressure metamorphism in convergent orogens: Journal of Metamorphic Geology, v. 9, p. 333–340, doi: 10.1111/j.1525-1314.1991.tb00527.x.

Scherer, E.E., Cameron, K.L., Johnson, C.M., Beard, B.L., Barovich, K.M., and Collerson, K.D., 1997, Lu-Hf geochronology applied to dating Cenozoic events affecting lower crustal xenoliths from Kilbourne Hole, New Mexico: Chemical Geology, v. 142, p. 63–78, doi: 10.1016/S0009-2541(97)00076-4.

Schilling, F.R., and Partzsch, G.M., 2001, Quantifying partial melt fraction in the crust beneath the central Andes and the Tibetan Plateau: Physics and Chemistry of the Earth, Part A: Solid Earth and Geodesy, v. 26, p. 239–246, doi: 10.1016/S1464-1895(01)00051-5.

Searle, M., Hacker, B.R., and Bilham, R., 2001, The Hindu Kush seismic zone as a paradigm for the creation of ultrahigh-pressure diamond- and coesite-bearing continental rocks: Journal of Geology, v. 109, p. 143–153, doi: 10.1086/319244.

Şengör, A.M.C., 1987, Tectonics of the Tethysides: Orogenic collage development in a collisional setting: Annual Review of Earth and Planetary Sciences, v. 15, p. 213–244, doi: 10.1146/annurev.ea.15.050187.001241.

Şengör, A.M.C., and Natal'in, B.A., 1996, Turkic-type orogeny and its role in the making of the continental crust: Annual Review of Earth and Planetary Sciences, v. 24, p. 263–337, doi: 10.1146/annurev.earth.24.1.263.

Shafer, J.T., Neal, C.R., and Mahoney, J.J., 2005, Crustal xenoliths from Malaita, Solomon Islands: A window to the lower crust of the Ontong Java Plateau: Eos (Transactions, American Geophysical Union), v. 86, abstract T11C-0397.

Shanin, L.L., Volkov, V.N., Litsarev, M.A., Arakelyants, M.M., Gol'tsman, Yu.V., Ivanenko, V.V., and Bairova, E.D., 1979, Criteria for reliability of radiological dating methods (in Russian): Moscow, Nauka, 208 p.

Salisbury, M.H., Shinohara, M., Richter, C., Araki, E., Barr, S.R., D'Antonio, M., Dean, S.M., Diekmann, B., Edwards, K.M., Fryer, P.B., Gaillot, P.J., Hammon, W.S., Hart, D.J., Januszczak, N., Komor, S.C., Kristensen, M.B., Lockwood, J.P., Mottl, M.J., Moyer, C.L., Nakahigashi, K., Savov, I.P., Xin, S., Wei, K.Y., and Yamada, T., 2002, Leg 195 summary, *in* Salisbury, M.H., et al., eds., Proceedings of the Ocean Drilling Program, Initial Reports, Volume 195: College Station, Texas, Ocean Drilling Program, p. 1–63, http://www-odp.tamu.edu/publications/195_IR/VOLUME/CHAPTERS/IR195_01.PDF.

Shirey, S.B., Richardson, S.H., and Harris, J.W., 2004, Integrated models of diamond formation and craton evolution: Lithos, v. 77, p. 923–944, doi: 10.1016/j.lithos.2004.04.018.

Shirey, S.B., Kamber, B.S., Whitehouse, M.J., Mueller, P.A., and Basu, A.R., 2008, this volume, A review of the isotopic and trace element evidence for mantle and crustal processes in the Hadean and Archean: Implications for the onset of plate tectonic subduction, *in* Condie, K.C., and Pease, V., eds., When did plate tectonics begin on planet Earth?: Geological Society of America Special Paper 440, doi: 10.1130/2008.2440(01).

Shu, L., and Charvet, J., 1996, Kinematics and geochronology of the Proterozoic Dongxiang-Shexian ductile shear zone: With HP metamorphism and ophiolitic mélange (Jiangnan region, South China): Tectonophysics, v. 267, p. 291–302, doi: 10.1016/S0040-1951(96)00104-7.

Silver, P.G., and Behn, M.D., 2008, Intermittent plate tectonics: Science, v. 319, p. 85–88.

Slabunov, A.I., Lobach-Zhuchenko, S.B., Bibikova, E.V., Balagansky, V.V., Sorjonen-Ward, P., Volodichev, O.I., Shchipansky, A.A., Svetov, S.A., Chekulaev, V.P., Arestova, N.A., and Stepanov, V.S., 2006, The Archean of the Baltic Shield: Geology, geochronology, and geodynamic settings: Geotectonics, v. 40, p. 409–433, doi: 10.1134/S001685210606001X.

Sleep, N.H., 2000, Evolution of the mode of convection within terrestrial planets: Journal of Geophysical Research, v. 105, p. 17,563–17,578, doi: 10.1029/2000JE001240.

Sobolev, S.V., and Babeyko, A.Y., 2005, What drives orogeny in the Andes?: Geology, v. 33, p. 617–620, doi: 10.1130/G21557.1.

Stampfli, G.M., and Borel, G.D., 2002, A plate tectonic model for the Paleozoic and Mesozoic constrained by dynamic plate boundaries and restored synthetic oceanic isochrons: Earth and Planetary Science Letters, v. 196, p. 17–33, doi: 10.1016/S0012-821X(01)00588-X.

Stampfli, G.M., and Borel, G.D., 2004, The TRANSMED transects in space and time: Constraints on the paleotectonic evolution of the Mediterranean domain, *in* Cavazza, W., et al., eds., The TRANSMED atlas: The Mediterranean region from crust to mantle: New York, Springer, p. 53–80.

Stern, R.J., 2005, Evidence from ophiolites, blueschists, and ultrahigh-pressure metamorphic terranes that the modern episode of subduction tectonics began in Neoproterozoic time: Geology, v. 33, p. 557–560, doi: 10.1130/G21365.1.

Stern, R.J., 2008, this volume, Modern-style plate tectonics began in Neoproterozoic time: An alternative interpretation of Earth's tectonic history, *in* Condie, K.C., and Pease, V., eds., When did plate tectonics begin on planet Earth?: Geological Society of America Special Paper 440, doi: 10.1130/2008.2440(13).

Štípská, P., and Powell, R., 2005, Does ternary feldspar constrain the metamorphic conditions of high-grade meta-igneous rocks? Evidence from orthopyroxene granulites, Bohemian Massif: Journal of Metamorphic Geology, v. 23, p. 627–647, doi: 10.1111/j.1525-1314.2005.00600.x.

Štípská, P., Schulmann, K., and Kröner, A., 2004, Vertical extrusion and middle crustal spreading of omphacite granulite: A model of syn-convergent exhumation (Bohemian Massif, Czech Republic): Journal of Metamorphic Geology, v. 22, p. 179–198, doi: 10.1111/j.1525-1314.2004.00508.x.

Stüwe, K., 1995, Thermal buffering effects at the solidus: Implications for the equilibration of partially melted metamorphic rocks: Tectonophysics, v. 248, p. 39–51, doi: 10.1016/0040-1951(94)00282-E.

Stüwe, K., 1998a, Heat sources of Cretaceous metamorphism in the eastern Alps: A discussion: Tectonophysics, v. 287, p. 251–269, doi: 10.1016/S0040-1951(98)80072-3.

Stüwe, K., 1998b, Tectonic constraints on the timing relationships of metamorphism, fluid production and gold-bearing quartz vein emplacement: Ore Geology Reviews, v. 13, p. 219–228, doi: 10.1016/S0169-1368(97)00019-X.

Stüwe, K., 2007, Geodynamics of the lithosphere: New York, Springer, 504 p.

Stüwe, K., Will, T.M., and Zhou, S.H., 1993a, On the timing relationship between fluid production and metamorphism in metamorphic piles: Some implications for the origin of post-metamorphic gold mineralization: Earth and Planetary Science Letters, v. 114, p. 417–430, doi: 10.1016/0012-821X(93)90073-I.

Stüwe, K., Sandiford, M., and Powell, R., 1993b, Episodic metamorphism and deformation in low-pressure, high-temperature terranes: Geology, v. 21, p. 829–832, doi: 10.1130/0091-7613(1993)021<0829:EMADIL>2.3.CO;2.

Tagami, T., and Hasebe, N., 1999, Cordilleran-type orogeny and episodic growth of continents: Insights from the circum-Pacific continental margins: Island Arc, v. 8, p. 206–217, doi: 10.1046/j.1440-1738.1999.00232.x.

Torsvik, T.H., 2003, The Rodinia jigsaw puzzle: Science, v. 300, p. 1379–1381, doi: 10.1126/science.1083469.

Toussaint, G., Burov, E., and Jolivet, L., 2004, Continental plate collision: Unstable vs. stable slab dynamics: Geology, v. 32, p. 33–36, doi: 10.1130/G19883.1.

Tsujimori, T., Sisson, V.B., Liou, J.G., Harlow, G.E., and Sorensen, S.S., 2006, Very-low-temperature record of the subduction process: A review of worldwide lawsonite eclogites: Lithos, v. 92, p. 609–624, doi: 10.1016/j.lithos.2006.03.054.

Unsworth, M.J., Jones, A.G., Wei, W., Marquis, G., Gokarn, S.G., and Spratt, J.E., 2005, Crustal rheology of the Himalaya and southern Tibet inferred from magnetotelluric data: Nature, v. 438, p. 78–81, doi: 10.1038/nature04154.

Usui, T., Nakamuri, E., Kobayashi, K., Maruyama, S., and Helmstaedt, H., 2003, Fate of the subducted Farallon plate inferred from eclogite xenoliths in the Colorado Plateau: Geology, v. 31, p. 589–592, doi: 10.1130/0091-7613(2003)031<0589:FOTSFP>2.0.CO;2.

Usui, T., Nakamura, E., and Helmstaedt, H., 2006, Petrology and geochemistry of eclogite xenoliths from the Colorado Plateau: Implications for the evolution of subducted oceanic crust: Journal of Petrology, v. 47, p. 929–964, doi: 10.1093/petrology/egi101.

Valley, J.W., Lackey, J.S., Cavosie, A.J., Clechenko, C.C., Spicuzza, M.J., Basei, M.A.S., Bindeman, I.N., Ferreira, V.P., Sial, A.N., King, E.M., Peck, W.H., Sinha, A.K., and Wei, C.S., 2005, 4.4 billion years of crustal maturation: Oxygen isotope ratios of magmatic zircon: Contributions to Mineralogy and Petrology, v. 150, p. 561–580, doi: 10.1007/s00410-005-0025-8.

van Hunen, J., and van den Berg, A.P., 2007, Plate tectonics on the early Earth: Limitations imposed by strength and buoyancy of subducted lithosphere: Lithos, doi: 10.1016/j.lithos.2007.09.016 (in press).

van Hunen, J., van den Berg, A.P., and Vlaar, N.J., 2004, Various mechanisms to induce present-day shallow flat subduction and implications for the younger Earth: A numerical parameter study: Physics of the Earth and Planetary Interiors, v. 146, p. 179–194, doi: 10.1016/j.pepi.2003.07.027.

van Hunen, J., van Keken, P.E., Hynes, A., and Davies, G.F., 2008, this volume, Tectonics of early Earth: Some geodynamic considerations, *in* Condie, K.C., and Pease, V., eds., When did plate tectonics begin on planet Earth?: Geological Society of America Special Paper 440, doi: 10.1130/2008.2440(08).

van Keken, P.E., Kiefer, B., and Peacock, S.M., 2002, High-resolution models of subduction zones: Implications for mineral dehydration reactions and the transport of water into the deep mantle: Geochemistry, Geophysics, Geosystems, v. 3, article 1056.

Volodichev, O.I., Slabunov, A.I., Bibikova, E.V., Konilov, A.N., and Kuzenko, T.I., 2004, Archean eclogites in the Belomorian Mobile Belt, Baltic Shield: Petrology, v. 12, p. 540–560.

Watson, E.B., and Harrison, T.M., 2005, Zircon thermometer reveals minimum melting conditions on earliest Earth: Science, v. 308, p. 841–844, doi: 10.1126/science.1110873.

Wei, C.J., Zhang, C.G., Zhang, A.L., Wu, T.H., and Li, J.H., 2001, Metamorphic *P-T* conditions and geological significance of high-pressure granulite from the Jianping complex, western Liaoning province: Acta Petrologica Sinica, v. 17, p. 269–282.

White, R.W., Powell, R., and Clarke, G.L., 2002, The interpretation of reaction textures in Fe-rich metapelitic granulites of the Musgrave Block, central Australia: Constraints from mineral equilibria calculations in the system $K_2O$-FeO-MgO-$Al_2O_3$-$SiO_2$-$H_2O$-$TiO_2$-$Fe_2O_3$: Journal of Metamorphic Geology, v. 20, p. 41–55, doi: 10.1046/j.0263-4929.2001.00349.x.

Willner, A.P., 2005, Pressure-temperature evolution of a late Palaeozoic paired metamorphic belt in north-central Chile (34°00′–35°30′ S): Journal of Petrology, v. 46, p. 1805–1833, doi: 10.1093/petrology/egi035.

Willner, A.P., Hervé, F., Thomson, S.N., and Massonne, H.J., 2004, Converging *P-T* paths of Mesozoic HP-LT metamorphic units (Diego de Almagro Island, southern Chile): Evidence for juxtaposition during late shortening of an active continental margin: Mineralogy and Petrology, v. 81, p. 43–84, doi: 10.1007/s00710-004-0033-9.

Wilson, J.T., 1965, A new class of faults and their bearing on continental drift: Nature, v. 207, p. 343–347, doi: 10.1038/207343a0.

Wilson, J.T., 1966, Did the Atlantic close and then re-open?: Nature, v. 211, p. 676–681, doi: 10.1038/211676a0.

Windley, B.F., Alexeiev, D., Xiao, W., Kröner, A., and Badarch, G., 2007, Tectonic models for accretion of the Central Asian Orogenic Belt: Geological Society [London] Journal, v. 164, p. 31–47, doi: 10.1144/0016-76492006-022.

Yang, J., Liu, F., Wu, C., Xu, Z., Shi, R., Chen, S., Deloule, E., and Wooden, J.L., 2005, Two ultrahigh-pressure metamorphic events recognized in the central orogenic belt of China: Evidence from the U-Pb dating of coesite-bearing zircons: International Geology Review, v. 47, p. 327–343.

Zhang, J., Zhao, G.C., Sun, M., Wilde, S.A., Li, S.Z., and Liu, S.W., 2006, High-pressure mafic granulites in the Trans–North China orogen: Tectonic significance and age: Gondwana Research, v. 9, p. 349–362, doi: 10.1016/j.gr.2005.10.005.

Zhao, D., 2001, Seismic structure and origin of hotspots and mantle plumes: Earth and Planetary Science Letters, v. 192, p. 251–265, doi: 10.1016/S0012-821X(01)00465-4.

Zhao, G., and Kröner, A., 2007, Geochemistry of Neoarchean (ca. 2.55–2.50 Ga) volcanic and ophiolitic rocks in the Wutaishan greenstone belt, central orogenic belt, North China craton: Implications for geodynamic setting and continental growth: Discussion: Geological Society of America Bulletin, v. 119, p. 487–489, doi: 10.1130/B26022.1.

Zhao, G.C., Cawood, P.A., Wilde, S.A., and Sun, M., 2002, Review of global 2.1–1.8 Ga orogens: Implications for a pre-Rodinia supercontinent: Earth-Science Reviews, v. 59, p. 125–162, doi: 10.1016/S0012-8252(02)00073-9.

Zhao, G.C., Sun, M., Wilde, S.A., and Li, S., 2004, A Paleo-Mesoproterozoic supercontinent: Assembly, growth and breakup: Earth-Science Reviews, v. 67, p. 91–123, doi: 10.1016/j.earscirev.2004.02.003.

Zhao, G.C., Sun, M., Wilde, S.A., Li, S.Z., Liu, S.W., and Zhang, H., 2006a, Composite nature of the North China granulite-facies belt: Tectonothermal and geochronological constraints: Gondwana Research, v. 9, p. 337–348, doi: 10.1016/j.gr.2005.10.004.

Zhao, G.C., Sun, M., Wilde, S.A., Li, S.Z., and Zhang, J., 2006b, Some key issues in reconstructions of Proterozoic supercontinents: Journal of Asian Earth Sciences, v. 28, p. 3–19, doi: 10.1016/j.jseaes.2004.06.010.

Zwart, H.J., 1967, The duality of orogenic belts: Geologie en Mijnbouw, v. 46, p. 283–309.

Manuscript Accepted by the Society 14 August 2007

Printed in the USA

The Geological Society of America
Special Paper 440
2008

# *Evidence for modern-style subduction to 3.1 Ga: A plateau–adakite–gold (diamond) association*

**D.A. Wyman**
*School of Geosciences, University of Sydney, Sydney, NSW 2006, Australia*

**C. O'Neill**
*ARC National Key Centre for Geochemical Evolution and Metallogeny of Continents (GEMOC), Department of Earth & Planetary Science, Macquarie University, Sydney, NSW 2109, Australia*

**J.A. Ayer**
*Precambrian Geoscience Section, Ontario Geological Survey, 933 Ramsey Lake Road, Sudbury, Ontario P3E 6B5, Canada*

## ABSTRACT

**The distribution of lamprophyre-hosted Neoarchean diamond deposits in the southern Superior Province, Canada, corresponds to that of coeval giant (>100 t Au) examples of "orogenic" gold deposits, which are typically associated with quartz-carbonate veining. This common association in the southern Superior Province, and at Yellowknife in the Slave craton, suggests that the occurrence of both diamonds and gold was promoted by the same geodynamic factors. A previously proposed subduction diamond model invokes flat subduction of buoyant oceanic plateau crust as the only means of entraining diamonds in shoshonitic lamprophyres, which are generally derived from relatively shallow mantle depths. Computer modeling of the thermal evolution and dehydration processes associated with this tectonic scenario clarifies observations made in present-day flat subduction settings and suggests many factors that should enhance the hydrothermal mineralization systems responsible for Archean and post-Archean orogenic gold deposits. The case for links among mantle plume–derived oceanic plateaus, crustal growth, and anomalously large gold deposits is strengthened by the newly recognized association of oxidized granites, lamprophyres, and diamonds at 3.1 Ga in the Kaapvaal craton and by evidence for similar recurring gold-diamond ± lamprophyre associations throughout the geologic record, including the Mother Lode deposits of California.**

**Keywords:** diamond, lamprophyre, Archean, subduction.

## INTRODUCTION

Despite a broad consensus in support of some type of Neoarchean plate tectonics, controversies remain over specific aspects of Archean geodynamics (e.g., Witze, 2006). Study of the mass-transfer processes associated with certain types of mineralization can help to resolve these issues, given that they require complex relationships between diverse sets of geological parameters (Fig. 1). For example, small "orogenic gold" occurrences are typically associated with quartz-carbonate veining and are widespread on most cratons. Giant examples of these hydrothermal deposits (>100 t Au), however, exhibit a far more restricted distribution (Fig. 2), which is inferred to result from a confluence of favorable factors extending from deep sources to the structural geometry of the

Wyman, D.A., O'Neill, C., and Ayer, J.A., 2008, Evidence for modern-style subduction to 3.1 Ga: A plateau–adakite–gold (diamond) association, *in* Condie, K.C., and Pease, V., eds., When Did Plate Tectonics Begin on Planet Earth?: Geological Society of America Special Paper 440, p. 129–148, doi: 10.1130/2008.2440(06).

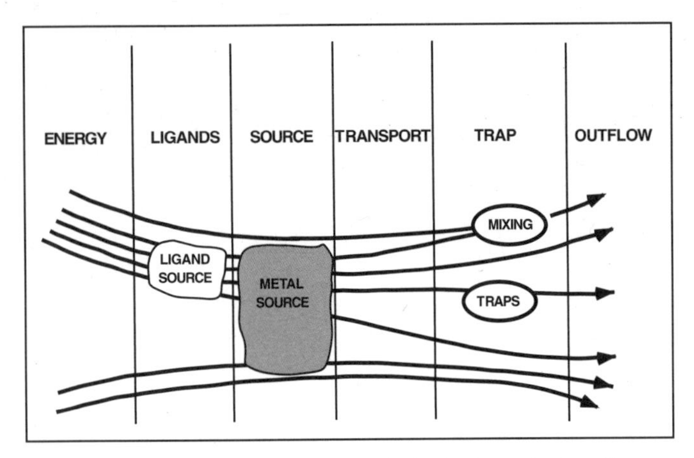


Figure 1. A schematic representation of a hydrothermal mineralization system illustrating the factors that contribute to ore formation (after Wyborn et al., 1994). Archean and post-Archean orogenic gold deposits share a wide array of characteristics, which indicate that they are derived from analogous systems (McCuaig and Kerrich, 1998).

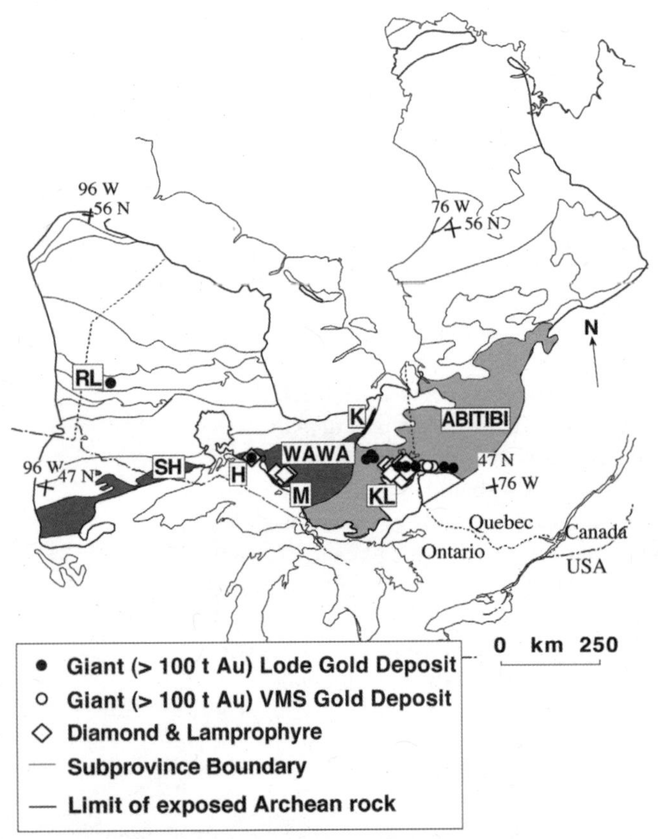


Figure 2. A summary map of the Superior Province craton illustrating the location of Neoarchean giant gold deposits and diamond-bearing lamprophyre localities. Note that the cluster of localities in eastern Ontario reflects the results of a concerted survey of lamprophyres in that area by the Ontario Geological Survey (Grabowski and Wilson, 2005). Giant-class gold deposits are indicated; most are orogenic gold types, but several volcanogenic massive sulfide (VMS) examples are known (Robert and Poulsen, 1997; Robert et al., 2005). RL—Red Lake deposit; SH—Shebandowan belt, Wawa subprovince; H—Hemlo giant orogenic gold deposit; M—Michipicoten belt, Wawa subprovince; K—Kapuskasing structural zone; KL—Kirkland Lake.

immediate host rocks (Phillips et al., 1996; Wyman et al., 1999b; Bierlein et al., 2006). In the case of the southern Superior Province, a trend of giant orogenic gold deposits overlaps the distribution of contemporaneous in situ Neoarchean diamond deposits hosted by lamprophyric dikes and breccias. Given the spatial and temporal overlap between the diamond and gold deposits, and their mutual association with shoshonitic lamprophyres, it is likely that factors that favor one form of deposit also favor the other. It is therefore notable that the only other occurrence of diamonds in Archean shoshonitic lamprophyres has been reported from the Slave Province in the vicinity of the Con Deposit (Armstrong and Barnett, 2003), one of only two Archean giant orogenic gold deposits in Canada outside of the Superior Province (Robert and Poulsen, 1997).

Wyman et al. (2006) and O'Neill and Wyman (2006) argued that the relatively shallow mantle sources of the shoshonitic host magmas can only be reconciled with the pressure-temperature (*P-T*) requirements of diamond stability in a shallow subduction setting. If diamonds are not a universal feature of Neoarchean granite-greenstone terranes, then it is possible that their occurrence correlates with other distinctive features of granite-greenstone host terranes that are also tied to shallow subduction. In addition to host lamprophyres, the model developed for diamonds highlighted two apparently unrelated rock types: (1) komatiites, which are widely interpreted to be volcanic components of thick mantle plume–generated oceanic-plateau crust (Polat et al., 1998; Condie, 2004), and (2) adakitic members of the high-Al tonalite-trondhjemite-granodiorite (TTG) suite. The adakite igneous suite is dominated by rocks that have $SiO_2 > 56$ wt%, $Al_2O_3 > 15$ wt%, and $Na_2O > 3.5$ wt%, along with low Y and heavy rare earth element abundance (HREE) plus high Sr/Y and La/Yb ratios (Defant and Kepezhinskas, 2001). One mechanism for generating adakitic rocks involves subduction of buoyant crust, such as oceanic plateaus, which produces a shallow subduction angle and allows slab heating to occur at comparatively shallow depths (Gutscher et al., 2000; Defant and Kepezhinskas, 2001).

In this paper, we review the distribution of komatiites, adakitic suites, and lamprophyres along with the known distribution of Archean diamond and giant gold deposits in the Superior Province in order to evaluate whether they support an indirect link between the diamonds and optimal conditions for orogenic gold mineralization. Two-dimensional geodynamic simulations are used to characterize the least accessible parts of orogenic gold hydrothermal systems (i.e., the left side of Fig. 1) within the shallow subduction scenario. Given the novel aspects of this approach and its results, we present a preliminary overview of evidence that the geodynamic model developed for the southern Superior Province may be applicable to other major gold mineralization events since the Mesoarchean (ca. 3.1 Ga).

## OROGENIC GOLD MINERALIZATION

A number of detailed recent reviews have summarized features of a class of gold deposit formed in "late- to post-tectonic" or orogenic environments since the Archean. Based on their tectonic setting, Groves et al. (1998) classified the type as "orogenic gold" deposits, although they are also known as mesothermal, lode, or low-S gold deposits. Archean examples share a wide range of attributes with post-Archean deposits and are considered to be metallogenically equivalent (McCuaig and Kerrich, 1998; Kerrich et al., 2005). This style of mineralization is recognized back to the Mesoarchean but became common in the late Neoarchean. The deposits are regarded as the product of fluid advection from deep sources such as the lower crust or accumulated subducted material (e.g., Wyman and Kerrich, 1988; Groves et al., 2005). Thermal reequilibration during orogeny rapidly heated this material, which had previously been subjected to the cold-finger effect of subduction. Therefore, the potential for this type of mineralization within an orogen is influenced by prior subduction histories, which largely determine the magnitude of the available gold reservoir and the extent of the subarc thermal anomaly relative to typical presubduction or postorogenic mantle. Low- to moderate-salinity $CO_2$-rich aqueous fluids are driven upward out of this reservoir and transport gold, which has been liberated from metamorphic mineral assemblages, along major structures to sites in the shallow crust. In the case of late Cenozoic gold mineralization in the European Alps, de Boorder et al. (1998) used mantle tomography to demonstrate that mineralization was associated with the emplacement of hot asthenosphere following slab detachment.

McCuaig and Kerrich (1998) reviewed orogenic gold deposit characteristics and noted that the ore systems responsible for Archean and post-Archean counterparts replicate the following features: (1) low fluid salinities that have generally <3 wt% NaCl equivalent, and $CO_2$ ($\pm CH_4$) concentrations of 10–24 wt%. combined with temperatures between 200 °C and 400 °C; (2) metal inventories (enrichments of Au, Ag, ± As, Sb, Te, W, Mo, Bi, B); (3) gangue mineral associations (quartz, carbonate, albite, tourmaline, etc); (4) syn- to post–peak metamorphic timing in orogenic settings; (5) a preferred siting in second-order or higher splays off of large-scale structures; and (6) deposit architecture (e.g., lode systems with vertical extents to 2 km) and vein types indicating fluid pressures that fluctuate between supralithostatic and sublithostatic. In addition, Archean and younger deposits commonly display a spatial and temporal association with distinctive orogenic rock types, particularly shoshonitic lamprophyres (e.g., mica-bearing minettes and hornblende-bearing spessartites). For example, lamprophyres were emplaced into the giant-class Golden Mile deposit synchronously with mineralization in the Yilgarn craton (McNaughton et al., 2005), and mutually crosscutting relationships between lamprophyres and gold mineralization have long been recognized in the Abitibi belt (McNeil and Kerrich, 1986). Given the shared attributes of the Neoarchean and younger deposits, Wyman and Kerrich (1988) proposed that their widespread occurrence during Neoarchean cratonization reflected the growing dominance of modern-style plate tectonics at this time. Various lines of geochemical evidence, particularly the change in Th/Nb ratio of the upper mantle during the Archean, have also been used to establish the occurrence of subduction tectonics in the Archean (Collerson and Kamber, 1999). More recently, Smithies et al. (2005) reviewed lithostratigraphic, geochemical, isotopic, and other evidence from the Pilbara craton indicating that modern-style subduction had emerged at least locally by ca. 3.1 Ga.

Although many alternative Archean geodynamic models have been proposed over the last two decades, they have not attempted to explain why Archean orogenic gold mineralization could so comprehensively mimic post-Archean occurrences. According to recent reviews, the general features of orogenic gold deposits outlined here are applicable to all orogenic gold deposits no matter their size. Optimal structural and geochemical geologic features present around major deposits clearly play a role in the specific location of anomalously large ore bodies (Phillips et al., 1996), but other factors must be involved in order to define the large-scale distribution of major deposits within the Superior Province as a whole (Fig. 2). These factors are likely to be distinctive geodynamic processes that are operative at scales approaching that of orogenic belts.

## SETTINGS

### Giant Gold Deposits in the Southern Superior Province

The Abitibi subprovince and the eastern part of the Wawa subprovince host numerous gold and base-metal deposits and have been the subject of a disproportionately large number of studies relative to the Superior Province as a whole. Rock types in the two areas consist mainly of (1) mantle plume–related komatiites and tholeiite basalt sequences; (2) ca. 2700 Ma arc-like tholeiitic and calc-alkaline volcanic suites of basalts, andesites, and more evolved volcanic rock types; (3) turbiditic sedimentary rocks; and (4) 2700–2690 Ma tonalite-trondhjemite-granodiorite (TTG) batholiths (Corfu, 1993; Wyman et al., 2002; Ayer et al., 2002). Porphyry dike swarms share adakitic compositions with the contemporaneous, and relatively young (ca. 2687 Ma), volcanic rocks in the Abitibi (Krist Formation: MacDonald et al., 2005). Late-tectonic intrusions include a shoshonitic suite (ca. 2685–2674 Ma) of monzonites, syenites, and trachytes, which respectively crosscut or unconformably overlie mantle plume– and arc-like volcanic sequences in the Superior Province (Wyman and Kerrich, 1989, 1993; Feng and Kerrich, 1992; Ayer et al., 2002). Based on field relations, compositional similarities, and absolute ages, minettes, spessartites, and other Neoarchean lamprophyres are part of this shoshonite suite and are now recognized on several Archean cratons, and they commonly display a preferred association with major structures (Wyman and Kerrich, 1988; Perring et al., 1989). Regional thrusting in Abitibi and Wawa greenstone assemblages during orogeny (2696 Ma to 2690 Ma in the Abitibi) resulted in east-west–oriented anticlines and synclines and was followed by localized crustal extension along the major structures between 2680 Ma and 2670 Ma, which corresponds to the time of lamprophyre emplacement (Ayer et al., 2002).

The Abitibi and Wawa subprovinces define a continuous east-west granite-greenstone trend interrupted by the Kapuskasing tectonic zone, which exposes midcrustal lithologies as a result of

Early Proterozoic thrusting. The early relationship of the Wawa and Abitibi subprovinces remains unresolved, and, as a result, the Kapuskasing structural zone may be the site of an earlier suture, if the two subprovinces evolved independently through most of their development (Williams et al., 1991). In any case, the eastern Wawa subprovince includes assemblages ranging in age between 2.97 Ga and 2.7 Ga, whereas pre–2.75 Ga crust has not been identified in the southern Abitibi (Ayer et al., 2002).

It is noteworthy that komatiite-tholeiite associations, generally interpreted as oceanic-plateau sequences and attributed to mantle plumes (e.g., Ernst et al., 2005), are not uniformly common in ca. 2.7 Ga greenstone terranes of the Superior Province. In fact, they are most abundant in the areas associated with the larger gold deposits of the southern Abitibi subprovince and southeastern margin of the Wawa subprovince rather than in the Wabigoon or the western Wawa subprovinces. Relationships between the 2.7 Ga komatiite-tholeiite associations and arc-like volcanic associations differ slightly between the eastern Wawa and Abitibi subprovinces. In the eastern Wawa, the association is a structural one and is interpreted to reflect accretionary tectonic processes (Polat et al., 1998), whereas in the southern Abitibi, there is evidence for direct interaction between a mantle plume and an existing arc (Dostal and Mueller, 1997; Wyman et al., 1999a, 1999b; Ayer et al., 2002; Scott et al., 2002). In contrast to the southeastern margin of the Wawa subprovince, the Shebandowan area of the western Wawa subprovince (Fig. 2) contains no ca. 2.7 Ga komatiitic units. A shoshonitic suite, including lamprophyres, is prominent in the Shebandowan area, yet the region is poorly endowed with orogenic gold deposits. By comparison, the Kirkland Lake area of the Abitibi contains a similar association of shoshonitic syenites, trachytes, and lamprophyres and hosts the second largest orogenic gold deposit in Canada (Robert and Poulsen, 1997).

**Neoarchean Diamond Deposits**

Over the last decade, in situ diamond deposits were discovered in late Neoarchean lamprophyric dikes and diatreme breccias of the eastern Wawa and Abitibi subprovinces (Sage, 2000; Stott et al., 2002). More recently, a significant diamond deposit has also been found in conglomeratic sediments in the Michipicoten belt of the eastern Wawa subprovince. Bulk sampling has revealed diamonds up to 1.5 carats (cts) from the conglomerate unit, which has an average thickness of 200 m (Dianor, 2007). Based on an associated heavy mineral suite and the presence of native gold in caustic fusion residues of bulk samples, this deposit has been interpreted as a paleoplacer by Dianor Resources, which is currently developing the property.

Recent studies have demonstrated that diamonds in the lamprophyric rocks display highly variable characteristics, and occurrences separated by only a few kilometers can be "entirely different" in terms of morphology, nitrogen characteristics, and carbon isotope compositions (Stachel et al., 2006). De Stefano et al. (2006) described the unusual mixed affinities of the eastern Wawa occurrences, where a single diamond may contain primary inclusions of both peridotitic and eclogitic paragenesis. In the eastern Wawa subprovince, tens of thousands of microdiamonds and hundreds of macrodiamonds have been recovered from lamprophyric host rocks, where the largest in situ diamond reported to date is 0.95 cts. Based on several lines of evidence, the eastern Wawa lamprophyric deposits are considered to be distinct from the diamondiferous conglomeratic deposits rather than resulting from lamprophyre contamination. They are geographically removed from the conglomeratic occurrences or possible subsurface extensions of the conglomerate; marked variation in diamond characteristics between nearby lamprophyric deposits is inconsistent with sampling of a placer sedimentary source, which is likely to contain a more homogeneous suite of diamonds than that found in the eastern Wawa subprovince; lamprophyre diamond content increases with the presence of ultramafic xenoliths, which have been shown to contain diamonds when analyzed as separated samples; and diamondiferous lamprophyres are now known from a large area of the southern Abitibi subprovince as well as from the Yellowknife area of the Slave craton (Armstrong and Barnett, 2003).

Given that the more recently discovered conglomeratic deposits are not yet well characterized, we focus in this paper on the lamprophyre-hosted diamond deposits. A late Archean age for these host rocks is confirmed by tectonic fabrics related to cratonization and by a U-Pb isotopic age of 2674 ± 8 Ma for titanate obtained from one of the Michipicoten greenstone belt (eastern Wawa) diamond-bearing dikes (Stott et al., 2002). This age corresponds to a U-Pb titanite age of 2674 ± 2 Ma previously determined for an apparently nondiamondiferous Abitibi minette (Wyman and Kerrich, 1993). Some fragmental diamond host rocks in the eastern Wawa subprovince, such as those of the Festival locality in the Michipicoten belt, have been interpreted as volcaniclastic, based in part on a U-Pb zircon age of 2744 ± 44 Ma (Stachel et al., 2006). Vaillancourt et al. (2004), however, obtained a range of zircon ages from these rocks including 2687 ± 2 Ma, 2683 ± 2 Ma, 2681 ± 2 Ma, and a pair of precisely overlapping 2679.2 ± 2.1 Ma ages. All of these ages are distinctly younger than the 2701.4 ± 2.1 Ma age of the surrounding felsic volcanic unit and indicate that this fragmental diamond host rock is an intrusive breccia, which is consistent with the fact that its whole rock and matrix compositions trend toward those of nearby lamprophyres (Vaillancourt et al., 2004; Williams, 2002). The lamprophyre emplacement ages emphasize that diamond entrainment occurred at the same time as the formation of large hydrothermal gold deposits, such as the giant-class Hollinger-McIntyre Deposit Timmins (~1000 t Au), which has been dated at 2672 ± 7 Ma on the basis of a Re-Os age for gold-associated molybdenite (Robert and Poulsen, 1997; Ayer et al., 2005).

The diamond hosts compositionally overlap other Archean lamprophyres of the Superior Province but have primitive compositions indicting little or no magma evolution in shallow chambers, as required to preserve diamonds (Williams, 2002; Wyman et al., 2006). They also contain abundant ultramafic xenoliths. A reconnaissance study of 14 of the xenoliths selected from Michipicoten occurrences by Wyman et al. (2006) found 11 to have websteritic compositions similar to the low-Al websterites commonly found at the base of the crust in post-Archean arcs (e.g., Parlak et al., 2002). Variable trace-element abundances in the xenoliths indicate metasomatism by a light rare earth element–enriched fluid. The fluid must have contained an aged component, given that it displays $\varepsilon_{NdT} \approx 8.5$, versus typical mantle values of ~2.4–4.0 in the southern Superior Province at 2.7 Ga (Wyman et al., 2006; Machado et al., 1986).

## THE SUBDUCTION DIAMOND MODEL

The diversity in diamond morphology, N characteristics, and C isotopic compositions, along with the mixed eclogitic and peridotitic affinity of mineral inclusions are difficult to account for in existing models of cratonic or orogenic diamonds. Based on preliminary (i.e., uncalibrated) depth estimates of 250 km to 300 km derived from majoritic components in Cr-rich garnet inclusions, Stachel et al. (2006) postulated the existence of preexisting deep cratonic roots prior to orogeny. They were, however, unable to develop a plausible exhumation mechanism to deliver the diamonds to the shallow depths typically associated with shoshonitic magmas during orogeny.

Apart from the shallow mantle depths associated with the lamprophyric magmas, there are several other factors that do not favor a deep cratonic root scenario for the Archean diamonds. As summarized by Wyman and Kerrich (2002), the southern Abitibi was unlikely to have progressively developed a thick mantle root during the main episode of greenstone belt volcanism between ca. 2750 Ma and ca. 2700 Ma. For example, rifted-arc–style mafic tholeiitic to calc-alkaline volcanism, derived from shallow mantle sources, continued to erupt in the center of the belt until 2698 ± 1 Ma (e.g., Wyman and Hollings, 2006; Mortensen, 1993). Widespread plutonism occurred after 2700 Ma and included ca. 2698 Ma adakitic phases of the Round Lake Batholith. Shoshonitic syenites and monzonites, derived at least in part from subduction-modified asthenosphere, were emplaced between 2690 Ma and ca. 2674 Ma (Corfu, 1993). Adakitic volcanism manifest in the Krist Formation in the Timmins area and coeval albitic porphyries suggest continued slab melting until 2687 Ma (Wyman et al., 2002; Ayer et al., 2002, 2005). Abitibi-Pontiac subprovince convergence until ca. 2685 Ma, as evidenced by regional-scale deformation and the late occurrence of detrital zircons from older (>2800 Ma), more distal terranes (Daigneault et al., 2002; Ayer et al., 2002), also requires the subduction of intervening oceanic crust into the region where a thick cratonic root would occur.

Studies of diamond placer deposits in Phanerozoic southeastern Australia have accounted for a similar absence of a deep cratonic root through variants of a "subduction diamond" model (e.g., Griffin et al., 2000; Barron et al., 1996). Diamonds exhibit a variety of oxidation states (McCammon and Kopylova, 2003), and they even exhibit oxidized species such as $CO_2$, carbonate, and water in fluid inclusions. Thus, there is potential for diamond formation in numerous tectonic environments under highly variable redox conditions, providing suitable *P-T* conditions for diamond stability and a fluid source exist. A common feature of the Australian models is the perturbation of the geotherm by subduction in order to stabilize diamonds at relatively shallow depths. Griffin et al. (2000) proposed that diamonds formed near the top of a subducted slab were entrained by lamproitic magmas that passed through the slab. Barron et al. (2005) invoked partial exhumation of the diamond-bearing slab from ultrahigh-pressure (UHP) conditions followed by diamond capture in shallow mantle magmas (e.g., basanite found in proximity to some of the diamond occurrences).

The discovery of diamonds in exhumed continental UHP terranes, such as the Norwegian Caledonides, and diamond pseudomorphs in fossil mantle wedge peridotites, such as the Ronda Massif, demonstrates that diamonds may be generated and sometimes preserved in the oxidized environments associated with subduction (Brueckner et al., 2002; Davies et al., 1993; Pearson et al., 1993). Given the tectonic setting of shoshonitic lamprophyres, and the depths associated with their parental magmas (50–160 km), Wyman et al. (2006) argued that a subduction diamond model could apply to the Neoarchean diamonds but only if subduction had been shallow prior to orogeny. Based on the thermal model of Gutscher et al. (2000), flat subduction initially results in a transient episode of slab melting and adakitic magmatism. Continued flat subduction, however, rapidly produces a shallow cold finger effect in the mantle that generates a wide range of temperatures in the top of the slab and the overlying mantle wedge. As shown by the geodynamic modeling of O'Neill and Wyman (2006), these perturbations can produce a diamond stability window at depths shallow enough to be sampled by typical shoshonitic magmas, provided that the shallow subduction is not the "wedge-less" variety postulated by Smithies et al. (2003) and instead resembles the narrow wedge configuration identified under parts of the Andes (Wagner et al., 2006; Gilbert et al., 2006).

Figure 3 illustrates an idealized depiction of the flat subduction model and highlights the fact that more than one diamond stability window may be accessible to lamprophyric magmas in the flat subduction setting. In principle, the flat subduction scenario

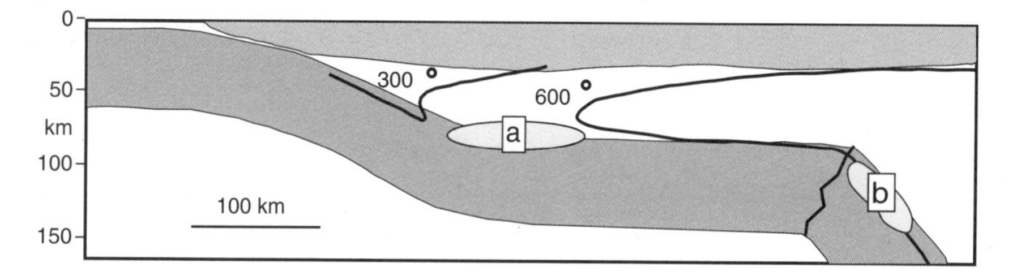


Figure 3. Conceptual subduction diamond model based on Andean idealized flat subduction scenario of Gutscher et al. (2000). In this subduction configuration, the top of the slab intersects the diamond stability field twice. A shallow diamond stability "window" occurs at "a" and the top of another window occurs at "b" (from Wyman et al., 2006).

provides a shallow mantle window into the diamond stability field as well as a deeper stability window analogous to that proposed by Barron et al. (1996) and Griffin et al. (2000) for normal subduction (Fig. 3). It is evident from this model that diamonds may be carried (in a metastable state) by the subducting slab from a shallow stability window through a region of diamond instability and then into another, deeper zone of diamond stability. Minor changes in subduction parameters may produce continuous diamond stability along the upper part of the flat slab segment. Irrespective of these details, the asthenospheric wedge overlying the slab is eventually frozen, and fragments of diamond-bearing asthenosphere may also be dragged along the tectonized slab-mantle interface toward the deeper regions of diamond stability. The process could ultimately generate a complex assemblage of diamonds with mixed peridotitic and eclogitic affinities and varied *P-T* histories as illustrated by previous geodynamic modeling (Wyman et al., 2006; O'Neill and Wyman, 2006).

Arguments against the subduction diamond model include the claim that it is not viable because carbonate minerals in the slab are stable until partial melting (Stachel et al., 2006). Irrespective of the experimental basis for this argument, it does not address the fact that thermal models of flat subduction imply early formation of slab-melt adakites (Gutscher et al., 2000) or the potential of $H_2O$ infiltration to induce decarbonation in subducted basalts, as required to account for the high $CO_2$ content of arc magmas (Kerrick and Connolly, 2001), orogenic lamprophyres, and orogenic gold systems associated with quartz-carbonate veining. Another argument against subduction diamonds suggests that activation energies for the graphite-to-diamond transition could only be overcome at shallow depths with the aid of subduction-related shearing and then will only result in microscopic crystallites (Stachel et al., 2006). This claim may not be relevant, given that diamonds in continental subduction settings, such as the Western Gneiss Region of Norway, are derived from slab-derived C-O-H fluids rather than graphite. As noted by van Roermund et al. (2002), diamonds in this setting were only preserved in strong-pressure vessels such as in spinels that were themselves enclosed in garnet. This requirement selectively preserves microdiamonds but does not preclude the formation of larger diamonds that were destroyed during the relatively slow return of the ultrahigh-pressure terrane to shallow crustal levels.

## IMPLICATIONS FOR OROGENIC GOLD MINERALIZATION AND NEOARCHEAN TECTONICS

### A Komatiite (Tholeiite)–Adakite–Shoshonite (±Diamond) Association

The flat subduction scenario can account for the entrainment of diamonds in shoshonitic lamprophyres and appears to resolve the "enigmatic" features of the southern Superior Province diamonds and their retrieval from the mantle. It would therefore constitute a robust model of Neoarchean tectonics prior to orogeny in the eastern Wawa and Abitibi subprovinces if the scenario also provided an explanation for the formation of closely associated giant-class orogenic gold deposits.

The role of komatiite-tholeiite sequences in the model presented by Wyman et al. (2006) for the southern Superior Province corresponds to that of present-day buoyant oceanic-plateau sequences or hotspot chains, which are commonly associated with modern flat subduction (Gutscher et al., 2000; Reich et al., 2003). Some Neoarchean komatiite-tholeiite sequences preserved in other terranes overlie much older sialic crust and therefore may not correspond to oceanic plateau. Nonetheless, many of these occurrences are closely associated with slightly younger subduction-style volcanic sequences that erupted a few tens of millions of years prior to crustal shortening and orogeny. This relationship leaves open the possibility that subduction of oceanic counterparts of the preserved "cratonic" plume sequences took place, as is inferred for the Abitibi belt, despite the fact that some komatiites erupted through the existing arc (Dostal and Mueller, 1997; Wyman et al., 1999a, 1999b). The Yellowknife belt, the site of a giant Con deposit and diamondiferous Archean lamprophyre, in the southwest part of Slave craton is one example. It contains the Kam Group, which is a thick sequence of ca. 2730–2700 Ma tholeiitic basalts with minor komatiite intercalations that is interpreted as a flood basalt sequence "approaching LIP [large igneous province] proportions" (Bleeker et al., 2001). The younger Banting Group (2660 Ma) extends from tholeiitic mafic volcanics to adakitic evolved rock types that could have been derived from either the melting of underplated mafic crust or slab melting, according to Cousens et al. (2002). Subduction-related geodynamic scenarios have previously been suggested for the area, however (Davis and Bleeker, 1999, and references therein), and they are consistent with the presence of shoshonitic magmas such as the diamondiferous lamprophyre found in the area. Similarly, komatiites overlying older crust in the Eastern Goldfields superterrane (EGST) of the Yilgarn craton are generally accounted for in scenarios that incorporate the close association of mantle plumes and subduction zones (Nelson, 1998; Smithies and Champion, 1999; Barley et al., 2007). Key geologic features of the Eastern Goldfields superterrane include the presence of the 2705 Ma komatiite–tholeiite assemblage and two younger (2690–2680 Ma) suites of evolved rock types. One of these is characterized by low $(La/Yb)_{cn}$ ratios between 4 and 9 and is most plausibly derived from intracrustal melting, while the other is an adakitic dacite-rhyolite suite considered to be derived from subducted oceanic crust (Morris and Witt, 1997). Shoshonites are represented by a series of late intrusions and lamprophyres that extends from Kambalda in the south of the Kalgoorlie terrane of the Eastern Goldfields superterrane to Kalgoorlie where a synmineralization lamprophyre, dated at 2642 ± 6 Ma, occurs in the giant (1500 t Au) Golden Mile deposit (Perring et al., 1989; McNaughton et al., 2005).

### Varieties of Neoarchean Subduction Zones

Despite the common association of slab melting with Archean subduction, there is abundant evidence for a diversity of subduction styles by 2.7 Ga. For example, many mafic magmas of the southern Superior Province display pronounced Zr-Hf anomalies that are more plausibly attributed to selective mobilization of REE in hydrous fluids rather than to slab melts. Examples include the youngest volcanic sequences of the Blake River Group in the southern Abitibi (Wyman et al., 2002), parts of volcanic belts in the central Wawa subprovince, (e.g., the Winston Lake belt), and the northeastern margin of the Wawa province (e.g., the Manitouwadge belt), where there is no primary or tectonic association with komatiite-tholeiite suites (Polat and Kerrich, 2001). Adakites and high-Mg andesites also occur in these belts but display trends from typical examples that lack negative Zr-Hf anomalies (e.g., Martin et al., 2005) to occurrences with pronounced anomalies, possibly reflecting competing metasomatic mechanisms similar to those observed in the later stages of Abitibi volcanism.

Although flat subduction may be associated with many Pliocene-Quaternary adakites, as argued by Gutscher et al. (2000) and others, there are a variety of ways to generate adakitic magmas, including (1) the subduction of unusually young and hot crust, (2) the heating of normal subducted crust along slab tears or where remnant slabs are warmed to typical mantle temperatures following subduction, and (3) melting of eclogitic material at the base of thickened subcontinental lithosphere (Defant and Kepezhinskas, 2001; Petford and Atherton, 1996; Wang et al., 2007). Kelemen et al. (2003) and Prouteau et al. (2001) have suggested that slab melts are actually very common, although rarely observed in their purest form; the Aleutians present a rare case where the effect of slab melting can be observed in oceanic high-Mg andesitic magmas. Given the variety of ways adakites may be generated, it is evident that, by themselves, they are an imprecise monitor of slab melting or a particular style of Archean subduction.

One approach to establishing the presence of slab melting in a subduction zone is to look for its effects in the compositions of mafic volcanic rocks that were derived from subduction-modified mantle sources. Wyman (2003) and Wyman and Hollings (2006) have shown that the high field strength element (HFSE) systematics of mafic rocks (MgO > 5 wt% and $SiO_2$ < 55 wt%) in the Winston Lake and Manitouwadge belts of the Wawa subprovince and of the 2.7 Ga Confederation assemblage in the central Superior Province are distinct from those of postplume volcanic sequences in the southern Abitibi belt, as summarized in Figure 4. The Abitibi suites display a crosscutting trend on the plots consistent with slab-melt metasomatism of a mantle wedge, whereas the mafic rocks from the other areas conform to the trends defined by mafic rocks from normal (nonadakitic) modern arcs. The Nb/Y versus Zr/Y plot illustrates that the southern Abitibi volcanic sequences overlap with the Aleutian high-Mg andesites that Kelemen et al. (2003) characterized as displaying a strong slab-melt component. The distinction between the trend of "normal" arcs and adakitic arcs may be one of degree, as suggested by the observations Kelemen et al. (2003) and the presence of at least minor occurrences of adakite rock types along with other felsic rock types in many Neoarchean greenstone belts, including the

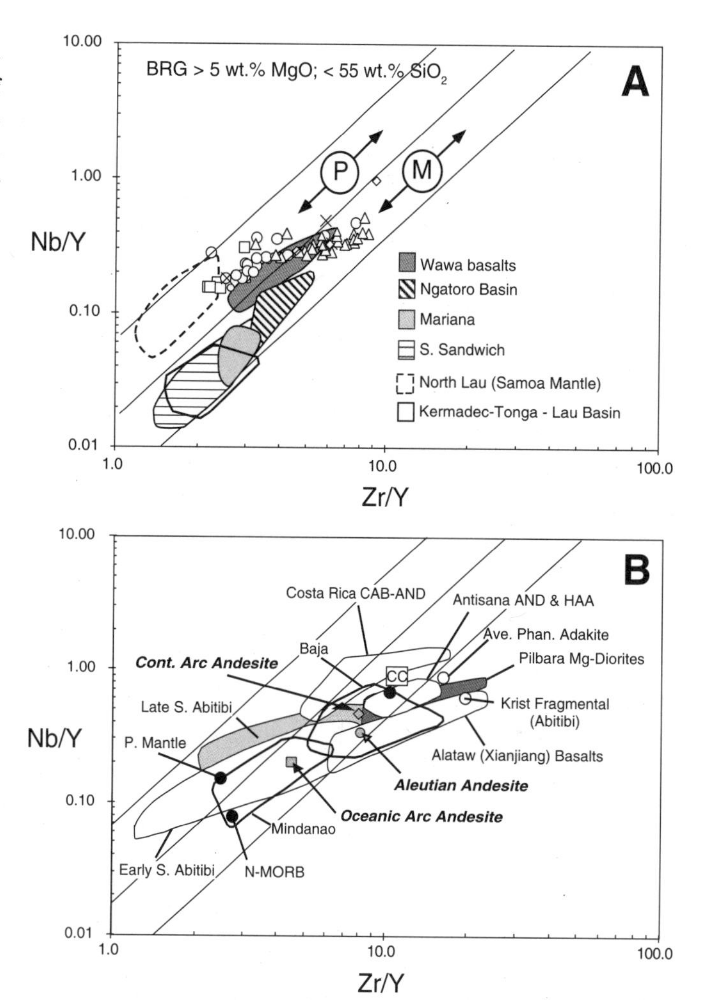


Figure 4. (A) Plots of Nb/Y versus Zr/Y showing the trends for modern plume (P) and mid-ocean-ridge basalt (MORB) + arc (M) magmas (Fitton et al., 1997). The slopes of the trends for most arc basalts are comparable to MORB because hydrous fluids transport little Zr or Nb. Data for primitive Abitibi rifted-arc samples from the Blake River Group (BRG) superimposed on the field for similar mafic rocks of the Manitouwadge and Winston Lake greenstone belts (Wawa) and Phanerozoic arc fields, which are not limited to samples with less than 5 wt% MgO. (B) A trend for slab-melt metasomatized mantle from the Archean to the Phanerozoic. Fields for Phanerozoic mafic and mafic-intermediate magmas from settings inferred to involve slab melting are shown. Also shown are trends for the early and late (rift) assemblages of the southern Abitibi belt (Wyman, 2003). CAB—calc-alkaline basalts; AND—andesites; HAA—high-Al andesites. Phan. Adakite—average Phanerozoic adakite of Drummond et al. (1996). Krist Abitibi—adakites. Continental, oceanic, and Aleutian andesite data are from Kelemen et al. (2003). See Wyman and Hollings (2006) for additional references.

Confederation assemblage and Manitouwadge belt (Wyman and Hollings, 2006; Hart et al., 2004; Polat and Kerrich, 2001).

In any case, the significance of these observations in the present context is that (1) the HFSE systematics of southern Abitibi volcanic sequences correspond to those found in Archean and post-Archean settings that have specifically been associated with slab melting by previous workers; and (2) this is not a universal feature of mafic Neoarchean arc-style magmas. Crustal contamination could theoretically replicate similar trends (Condie, 2005), but it is not consistent with overall chemical trends in the Abitibi data (Wyman, 2003). Although it may be difficult in some cases to rule out crustal contamination as a cause of the "slab-melt" trends where extensive older crust is present, it is notable that in the Confederation assemblage, where contamination might be likely to occur, mafic volcanic sequences still define a trend comparable to that of normal modern arcs (see Fig. 8 in Wyman and Hollings, 2006). In summary, the Abitibi mafic volcanic rocks are not only distinctive in terms of their association with giant gold deposits and the diamondiferous lamprophyres they host, they are also unusual in the extent to which a slab-melt effect is evident in their HFSE systematics.

## A Role for Slab Melting in the Genesis of Giant Orogenic Gold Deposits?

An obvious and important question arising from the association of diamonds and major gold deposits is whether or not flat subduction could play any role in enhancing the size of later orogenic gold deposits. Several potential mechanisms for optimizing later orogenic gold hydrothermal systems relate to increasing amounts of gold in a subarc reservoir.

Mungall (2002) has shown that oxidized adakite magmas (or oxidized supercritical fluids, depending on lithosphere thickness) associated with the early hot stages of a flat subduction zone are unique in their ability to break down gold-bearing mantle sulfide minerals, allowing transport of the gold to form porphyry-style deposits. Similarly oxidized magmas have existed since the Neoarchean. Ishihara (2004) and Ishihara et al. (2006) reported that transitional ilmenite-magnetite granitoid suites developed on the Kaapvaal craton at ca. 3230 Ma and that younger (3105 Ma) batholiths on that craton were magnetite series. By the late Archean, oxidized granitoids were common and included magnetite series suites in the Eastern Goldfields of the Yilgarn. Although a genetic link between adakitic ("albitite") porphyries and hydrothermal gold deposits was once commonly suggested because of their spatial association, the mineralization is now known to postdate the emplacement of porphyry suites in the Abitibi and other Neoarchean terranes by millions of years (MacDonald et al., 2005).

Despite the time gap between adakites and orogenic gold mineralization, the flat subduction scenario does provide a possible role for the slab melts in the development of major orogenic gold deposits, given that slab melts would have increased gold transport to the base of the crust and (or) remained as magmatic veins in the frozen and thinned asthenospheric wedge. Either case would represent an enhancement of the orogenic gold reservoir. Another distinctive feature of the southern Abitibi, relative to Superior Province as a whole, is the presence of several giant gold-rich volcanogenic massive sulfide deposits, containing 120–345 t Au. These deposits were formed during a short-lived episode of arc extension (2701–2697 Ma) but occur along the same trend defined by major orogenic gold deposits and diamond-bearing lamprophyres (Fig. 2; Robert and Poulsen, 1997; Robert et al., 2005). Slab-melt metasomatism of the sub-Abitibi mantle, which is evident in the trace-element signature of transitional tholeiite–calc-alkaline host rocks of the massive sulfides (Fig. 4; Wyman, 2003), may account for the otherwise unusually coincidental localization of giant pre-orogenic (massive sulfide) and orogenic gold deposit types. The massive sulfide deposits demonstrate the capacity of flat subduction settings to promote oxidation of gold-bearing mantle sulfide minerals, whether the overlying crust is thick enough to host porphyry deposits or has been thinned in a rifted-arc environment.

## Flat Subduction and Slab Dehydration

Mungall (2002) indicated that, apart from melts and supercritical fluids, the only other viable mechanism for oxidation of the wedge and the liberation of gold from mantle sulfide minerals is via hydrous fluids derived from the slab, although their oxidation capacity is only 1/400th (by weight) the capacity of adakitic magmas. The impact of aqueous fluids could nonetheless still have been considerable in a flat subduction scenario because convection in the wedge would initially be restricted by the presence of the shallow slab and later by dropping temperatures. As a result, large cumulative volumes of fluids may have percolated through this nonconvecting asthenosphere rather than being added to a convecting mantle wedge where asthenosphere was continuously replenished and replaced. Not only would fluids driven off of altered oceanic slab and subducted sediments carry gold from these sources, they would contribute significantly to the oxidation of the overlying trapped asthenosphere.

Indirect observations relating to the process of slab dehydration during flat subduction have been made on the south-central Chilean subduction zone via three-dimensional seismic tomography studies. Wagner et al. (2005, 2006) found low P- and S-wave velocities and low $V_p/V_s$ that indicate a lack of melt or hydrated minerals (e.g., serpentine). They inferred that the results were most consistent with large proportions of orthopyroxene in the mantle wedge. The orthopyroxenite apparently formed during transient fluxing by silica-rich fluids derived from sediments, which left little free water in the mantle. Although much of the wedge was relatively anhydrous, a pocket of high-$V_s$, high-$V_p$ material near the trench was considered to represent serpentinized material formed from pooled water that may have been driven off during prior episodes of normal subduction (Wagner et al., 2005). Gilbert et al. (2006) employed the same seismometer array to study the structure of the lithosphere and upper mantle of central Chile and Argentina. Their observations indicate that the lower crust in the area of flat subduction was tectonically thickened, despite the relatively low-lying nature of surface topography. Based on the density requirements needed to account for these observations, they inferred that the lower crust was partially eclogitized, probably by transient episodes of channelized water flow (e.g., Miller et al., 2003). These observations highlight several factors that may promote formation of anomalously large hydrothermal gold systems. The sediment-derived aqueous siliceous fluids that modified the wedge may have scavenged gold from seafloor hydrothermal sediments and would also have promoted oxidation of mantle sulfides. Tectonically thickened and eclogitized arc lower-crust represents an enhanced (larger) gold and fluid source, if flat subduction is terminated by orogeny. This crust also provides a type of additional stored energy for future hydrothermal systems, given that it has the potential for greater than normal isostatic uplift and the enhanced development of transcrustal faults, if slab break off results in the heating of eclogite.

Key features of the Andean flat subduction zone are consistent with the tectonic model proposed for the southern Superior Province. For example, the conclusion that the narrow wedge associated with flat subduction contains abundant orthopyroxene is consistent with the limited data presently available for xenoliths derived from the shoshonitic lamprophyres of the southern Superior Province (Wyman et al., 2006). Highly variable trace-element and isotope systematics between the Archean xenoliths also point to channelized fluid flow. Frozen asthenosphere identified in the mantle wedge along the Chilean convergent margin is consistent with the models of Gutscher et al. (2000) and O'Neill and Wyman (2006), although, in this instance, a large proportion appears to have been displaced to the south of the flat slab as subduction shallowed (Wagner et al., 2006).

## DIAMOND STABILITY AND SLAB DEHYDRATION DURING FLAT SUBDUCTION—ELLIPSIS MODELS

### Methodology

We have modeled the thermal evolution and dehydration processes associated with general examples of flat subduction zones in order to further assess the metallogenic potential of flat subduction for hydrothermal ore deposits. Specifically, we used a numerical code to simulate flat subduction to clarify two factors in the synchronous diamond and gold deposits: (1) the distribution of slab-derived fluid flux, which controls where later large ion lithophile element–enriched shoshonitic magmas are likely to form in the wedge, and which also provides a major fluid source for gold deposits; and (2) the regions in a flat subduction zone where diamonds are likely to be stable.

Ellipsis is a particle-in-cell finite-element code designed for mantle convection studies (Moresi et al., 2003). It solves the standard convective equations in a two-dimensional (2-D) Cartesian geometry, and, in our examples, with constant temperature top and bottom boundary conditions, reflective side conditions, and free-slip bottom conditions. The velocity condition on the top is fixed over the "continent" or "craton" region, and a constant convergence velocity is imposed in the "oceanic" domain. We explore time-dependant convergence rates further in the following sections. An additional velocity condition is imposed in the mantle in the vicinity of the subducting slab to force it to "flatten-out." We used a temperature-dependent viscosity (Frank-Kamenetski approximation, see O'Neill and Wyman 2006), which varies over five orders of magnitude, and a brittle-failure criterion to allow the highly viscous plates to fail.

### Steady-State Flat Subduction

Figure 5 shows a typical temperature field for steady-state flat subduction, with a convergence velocity of 10 cm/yr and a slab age of 55 m.y. These two parameters are fundamental in determining the thermal structure of the subducting slab, and they are highly variable between different subduction zones. Given the difficulties in determining these parameters for the Archean, we have adopted these representative values because they are both within reasonable bounds for today (if slightly faster and younger than the present averages, respectively) but are still plausible for the late Archean thermal regime, which has been suggested to have had faster plate velocities with plates subducting at younger ages (Davies, 1992).

Figure 5A shows the dewatering curve, and Figure 5B shows the geotherms for our flat subduction scenario (Figs. 5C and 5D), based on the computed phase equilibria and *P-T* plots of Rupke et al. (2004). The steep subduction profiles are from Rupke et al. (2004) for both 40 Ma and 120 Ma slabs. The water content is based on the stability of serpentinized mantle at different *P-T* conditions (see Rupke et al. [2004] for details on the calculations). In these calculations, we have neglected water release from the dehydration of sediments or mafic oceanic crust, which is most important in the forearc mantle (Hyndman and Peacock, 2003). Following Rupke et al. (2004), we expect serpentinized mantle to be the main water carrier into the deeper mantle. Sediments and crust, in general, dehydrate at shallower depths, although crustal material may contribute at the depths investigated here. The main effect of crustal dehydration, in addition to surplus water, is a smearing of the water release pattern due to the complicated dehydration reactions of oceanic crust (Rupke et al., 2004). Highly serpentinized mantle contains a number of water-bearing phases, including brucite, chlorite, and antigorite. Completely serpentinized mantle has the capacity to store up to 13 wt% water. Obviously, this is an extremely high estimate for the bulk oceanic lithosphere, however, it is sensible to use this value as a starting value for dehydration curves because while serpentinization is likely to be highly heterogeneous (around fracture zones, etc.), the parts of the lithosphere that are serpentinized are likely to be highly serpentinized (e.g., Li and Lee, 2006). The *P-T* path for flat subduction shows a simplified dehydration reaction sequence compared to steep subduction. First, brucite breaks down, releasing a small amount of free water. The main water carrier, antigorite, breaks down at ~600–630 °C, releasing the majority of water carried by the slab. The subsequent breakdown of chlorite completes the dehydration sequence in this simple example. It should be noted, however, that this is a simple closed-system reaction, and variations in initial water content and time-varying water loss (i.e., for an open system) will alter the phase relations somewhat. Given the

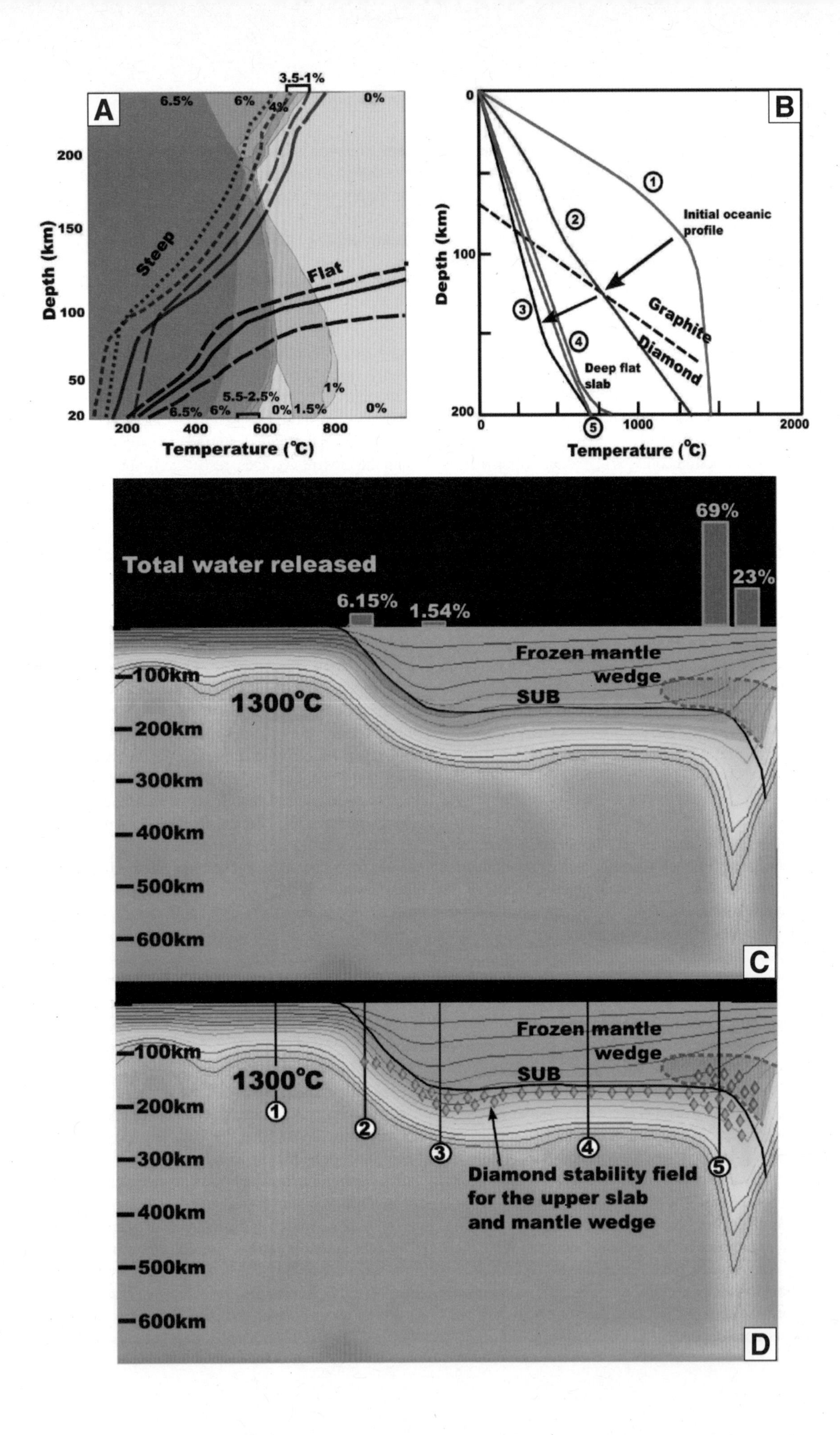
A
3.5-1%
6.5%
6%
4%
0%
Steep
Flat
5.5-2.5%
1%
6.5%
6%
0%
1.5%
0%
Depth (km)
Temperature (℃)
B
Initial oceanic profile
Graphite
Diamond
Deep flat slab
Depth (km)
Temperature (℃)
Total water released
6.15%
1.54%
69%
23%
Frozen mantle wedge
SUB
1300℃
100km
200km
300km
400km
500km
600km
C
Frozen mantle wedge
SUB
1300℃
Diamond stability field for the upper slab and mantle wedge
100km
200km
300km
400km
500km
600km
D

Figure 5. (A) Temperature-depth profiles of steep and shallow subducting slabs superposed over the deserpentinization reactions in a highly altered slab (from Rupke et al., 2004). The different shades represent different stable mineral assemblages, with the total percentage of bound water in each case labeled (for details see Rupke et al., 2004). The steep slab profiles are also from Rupke et al. (2004); the flat slab is from our models. The flat slab loses most of its bound water at ~630 °C when antigorite breaks down, becoming completely dehydrated at ~800 °C. Steep subduction profiles (in order of left to right at the top) are 7 km depth/40 Ma slab age, 17 km/40 Ma, 7 km/120 Ma, and 17 km/120 Ma. The flat slab profiles (top to bottom) are 7 km/55 Ma, 17 km/55 Ma, and 7 km/90 Ma. (B) Evolution of geotherms through the flat subducting slab shown in D. Positions labeled. Diamond-graphite transition is also shown. (C) Total water released in a steady-state flat subduction zone with slab age of 55 Ma and convergence rate of 10 cm/yr. Isotherms are plotted every 100 °C until 1300 °C. Hydrous phases are stable in the cooler portions of the slab below 800 °C, shown as transparent blue. The breakdown of brucite and release of bound water account for the minor early water released. The majority of bound water is released in the breakdown of antigorite (69%), which is closely followed by chlorite breakdown in this example (23%). Most of this water is released into the chilled mantle wedge above the subducting slab (red, dashed transparent region). There is no vertical exaggeration in these figures. (D) Thermal structure of the flat slab showing the regions of diamond stability in the downgoing slab and in the chilled mantle wedge. Diamond stability in the continent or bulk mantle is not shown. Numbers correspond to profiles shown in B.

uncertainties in degree and extent of serpentinization and initial composition of the oceanic lithospheric mantle, this approach is adequate for our illustrative purposes here.

The initial subduction phase results in the breakdown of brucite and a small amount of water released from the slab (Fig. 5C). The stability of the isotherms in the flat subducting slab retards further reactions until farther inland. The result is similar to the observations reported by Gilbert et al. (2006) and suggests that the unexplained region of serpentinized asthenosphere near the Chilean trench may be the product of brucite breakdown in the slab. When the upper portions of the slab pass the 600–630 °C isotherm, antigorite breaks down, releasing a large proportion of the slab's stored water (~69%). This is closely followed by the late breakdown of chlorite in this example, liberating the remaining water in the slab. The main result here is that the flux of water from a subducting slab is dominated by the breakdown of antigorite, and this result is likely to be insensitive to the specifics of the computed phase equilibria. Antigorite breakdown generally occurs around 600–630 °C and will generally provide the greatest mantle-derived water flux from the subducting slab. In our model, this also correlates with the position of the frozen mantle wedge, and thus this region has the potential to become a huge fluid reservoir.

Figure 5B shows the evolution of geotherms at the positions labeled in Figure 5D. The initial oceanic temperature profile (1) does not pass through the diamond transition within the depth of the lithosphere. During subduction, however, the thrusting of the cool oceanic lithosphere to great depths results in the depression of these geotherms, and they pass through the diamond transition. Although diamonds are also stable at greater depths, we have only shown diamonds for the upper portions of the subducting slab and mantle wedge in Figure 5D because these are regions where fluids are likely to be mobilized and diamonds are accessible to later shoshonitic magmas. The diamond stability field also intersects the cratonic lithosphere (e.g., O'Neill and Wyman, 2006) but is not included in this example. The main result for steady-state subduction is that the conditions for diamond occur in much of the eclogitic "flattened slab" and in the peridotitic frozen mantle wedge.

## Time-Dependent Subduction

Flat subduction is inherently time-dependent in that the factors that give rise to shallow subduction—such as buoyant crust due to oceanic plateaus, or a fast convergence rate of the continent with respect to the subduction zone—are themselves transient. In Figure 6, we have explored the thermal effects of a flattening slab, in particular, the evolution of fluid flux and temperatures in the mantle wedge between the slab and the cratonic lithosphere.

The model starts from an initially steep subduction configuration—the input from a previous simulation with steep subduction, where the modeling domain has reflective side boundary conditions. This boundary condition restricts lateral mantle flow, and so there is no net motion of the mantle reference frame compared to the stationary craton. At the onset of Figure 6, this boundary condition is set to periodic, and basal velocities are set to half that of the oceanic plate, allowing a net eastward drift of the mantle (or, equivalently, a net westward drift of the craton with respect to the mantle). Due to the motion of the craton compared to the mantle, the slab flattens in response to the mantle "wind." Although we do not explicitly include a buoyant plateau in this example, the thermal effects of slab flattening are equivalent.

Diamond stability conditions are prevalent in the flat slab for all times. After the flat slab has reached a stable configuration (~40 m.y. in this model), the main effect is a chilling of the trapped mantle wedge due to cool slab-surface temperatures beneath it. This can be seen in the relative position of the 630 °C isotherm between 40.4 m.y. and 50.1 m.y.; the mantle wedge develops the *P-T* conditions for stability of antigorite, an important hydrous phase, implying that such a cooled wedge could potentially be an important reservoir of fluids. The superposition of the cool wedge retards the dehydration reactions within the flat slab, however, and most water will be carried ~500 km further inland, where it will hydrate large portions of the chilled wedge and overlying lithosphere. Thus, two important criteria for gold-carrying fluids are met: delivery of significant quantities of fluids over large distances (100–500 km), and a reservoir that traps these fluids.

The time series in Figure 7 continues from Figure 6, but at the onset of the simulation, the bottom velocity boundary condition was set in the opposite direction to the subducting plate, which eliminates the net drift between the craton and the mantle and forces subduction into its previous steep configuration. This

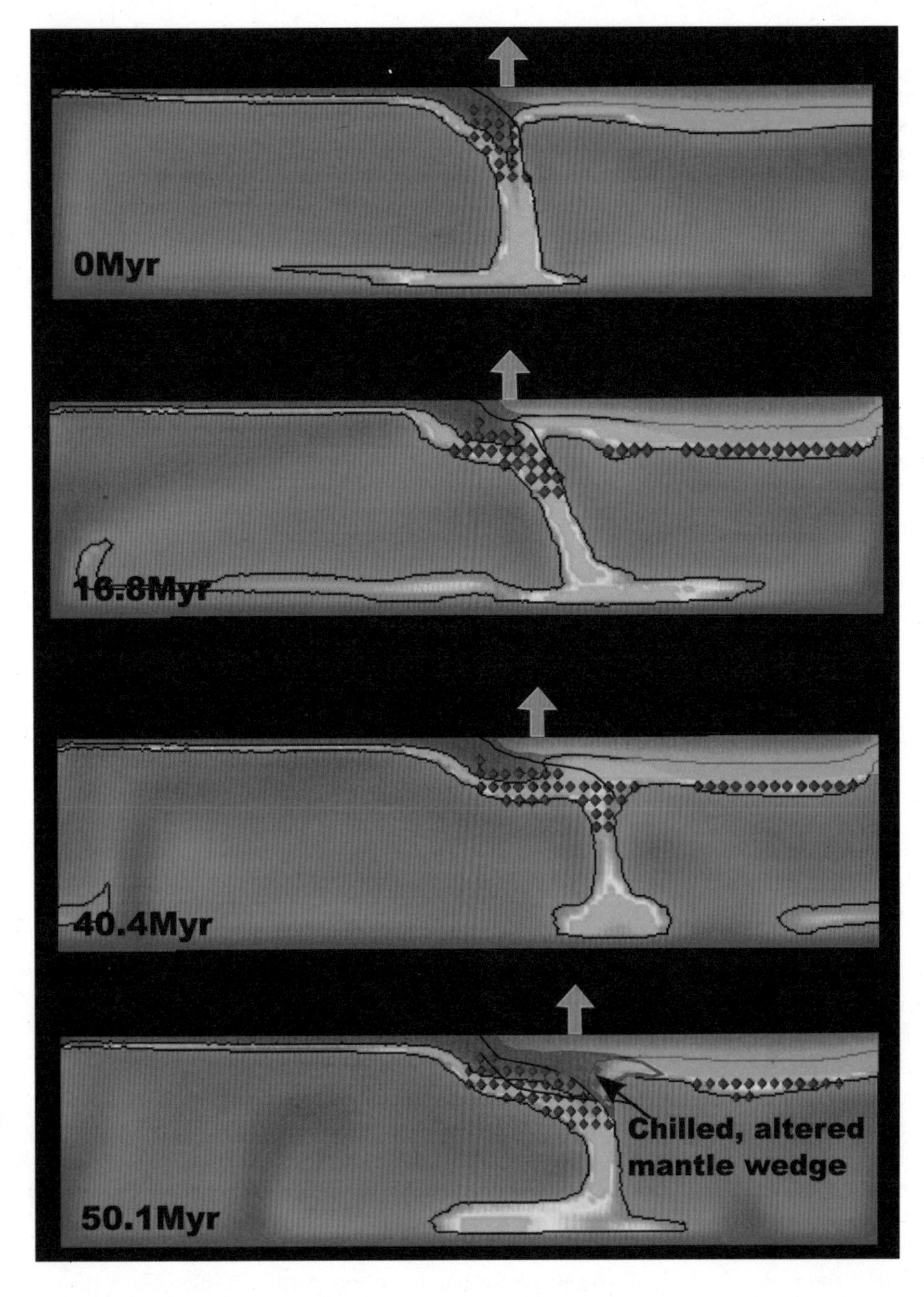


Figure 6. Time steps in the transition from steep to flat subduction. Convergence rate is 10 km/yr, and slab age at the subduction zone is 55 Ma. The initial steep subduction zone is obtained by using reflective side boundary conditions and restricting net motion between the stationary continent and the mantle reference frame. At the onset of this simulation, that condition was changed to periodic, so that there is a net eastward drift of the mantle (equivalent to a westward drift of the continent with respect to the mantle). Craton is shown in transparent green, and two isotherms (630 °C and 1300 °C) are shown as black lines. Thick black line from subduction zone represents slab upper boundary. Diamond stability in the shallow portions of the slab, wedge, and lithosphere are shown as small diamonds. The slab takes ~50 m.y. to become a stable flat subduction zone. Approximate position of antigorite breakdown in the subducting slab is shown by arrow.

is equivalent to a cessation, or reversal, of motion of the craton, as a result of collision and orogenesis, for example.

As previously shown by O'Neill and Wyman (2006), orogeny following flat subduction provides a limited time window during which diamond emplacement may occur. The slab rolls back, and hot asthenosphere upwells into the previously chilled, fertile wedge, which is heated and melted by the intruding asthenosphere. The upwelling asthenosphere itself may also melt as it intrudes into the old wedge. For a few million years, diamonds coexist with these wedge-derived melts, but the thermal pulse of the asthenospheric intrusions and the end of subduction will destroy both the conditions for diamond stability and the stability of altered mantle phases like antigorite. This would result in the mobilization of large amounts of fluids stored in the cool wedge and provide a correspondingly large pulse of fluids for gold mineralization in the overlying shallow crust.

## PRE- AND POST-NEOARCHEAN EXAMPLES OF FLAT SUBDUCTION GIANT GOLD DEPOSITS?

### Mantle Plumes and Giant Gold Deposits

If anomalously large Neoarchean orogenic gold deposits are the product of modern-style flat subduction, then did this geody-

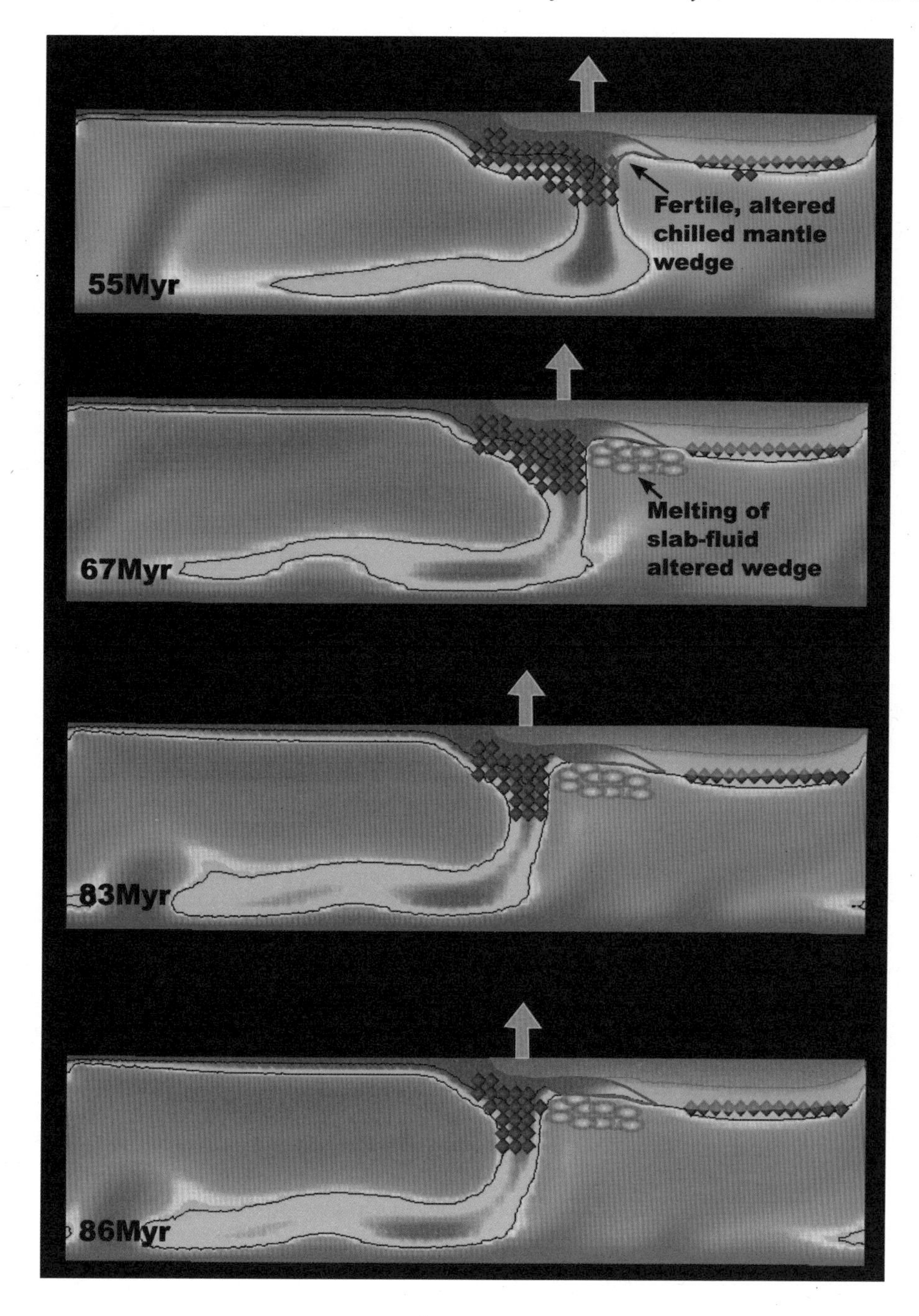


Figure 7. Time steps in the transition from flat to steep subduction. Initial conditions from the end of Figure 6; time continues from Figure 6. The bottom boundary condition is changed at the start of this simulation to be in the opposite direction to the incoming slab, at half its speed, which eliminates the net motion between the craton and the mantle, equivalent to a collision or orogeny documented in the Wawa belt. As the slab rolls back to a steep configuration, upwelling asthenospheric material instigates melting (shaded ovals) in the fertile old mantle wedge. This not only provides the source for shoshonitic lamprophyres in the Wawa, but explains the coexistence of the source of these magmas and diamonds in the chilled mantle wedge. Furthermore, the thermal pulse produced by the intruding asthenosphere destroys the stability conditions for hydrous mantle phases in the chilled wedge, resulting in the mobilization of large amounts of free water, which provides a source for the fluid pulse associated with major gold mineralization contemporaneous with the shoshonitic lamprophyres.

namic scenario produce other large concentrations of gold prior to, or following, late Archean cratonization? Strictly speaking, the observations made here do not require that flat subduction must be associated with mantle plumes in order to generate giant gold deposits. Nonetheless, some correlation might be expected between major plume episodes and larger examples of this type of mineralization, as is observed when superplume events, crustal growth, and orogenic gold mineralization rates through time are compared (Condie, 2004; Goldfarb et al., 2001). In this section, we summarize evidence that the plateau-related flat subduction scenario may have contributed to major gold events throughout a large part of the geologic record. It is stressed that "subduction diamonds" may not have been preserved in all such gold districts, given that appropriate primitive lamprophyric magmas must travel rapidly through the crust in order to preserve them.

## The Mesoarchean Kaapvaal Craton

As previously noted, economically significant examples of the orogenic gold deposit type are known to extend back to the Mesoarchean. Significantly, the Barberton gold vein deposits of the Kaapvaal craton formed at ca. 3126–3084 Ma (de Ronde et al., 1991); this is directly comparable to the age of pyrite from the Witwatersrand deposits, which has been Re-Os dated at 2990

± 110 Ma (Kirk et al., 2002). If orogenic gold occurrences contemporary with those in the Barberton belt were the source for the extremely large Witwatersrand placer gold deposits (40,000t Au), then they were likely to have included one or more giant orogenic gold deposits (Goldfarb et al., 2001).

Evidence in support of a Kaapvaal flat subduction model is indirect but nonetheless significant. Konstantinovskii (2003) reported that hundreds of diamonds have been recovered from at least two of six Witwatersrand gold-bearing paleofans, and they are typically 0.1–1.5 carats in size (maximum 8 carats). Modern gold-mining techniques do not favor diamond recovery, and therefore the actual diamond contents of the Witwatersrand reefs can only be more significant than these numbers indicate. Other components of the southern Superior Province flat subduction geodynamic scenario are also present in the consolidating Kaapvaal craton, starting with accretion of komatiite-bearing oceanic plateau during the ca. 3200 Ma assembly of the Barberton greenstone belt (Poujol et al., 2003, and references therein). The craton is also host to the oldest known oxidized (magnetite-series) granitoids that date back to the 3230 Ma adakitic Kaap Valley tonalite (e.g., 0.45 wt% $TiO_2$; 576 ppm Sr; 7 ppm Y; Cr/Ni = 4.5; Ishihara et al., 2006; cf. Martin et al., 2005). The oldest known shoshonitic lamprophyres (ca. 3120 Ma) have also recently been identified in the Kaapvaal craton (Anhaeusser, 1999, Prevac et al., 2004). Collectively, the evidence from across the Kaapvaal craton provides strong indications that modern-style subduction extended back to at least ca. 3100 Ma, although it may not have been the dominant tectonic style at that time.

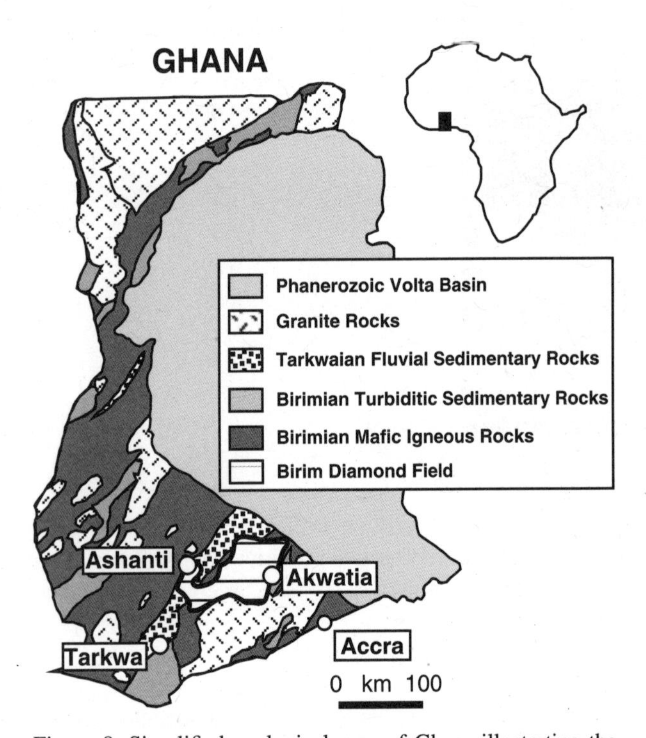


Figure 8. Simplified geological map of Ghana illustrating the association of the Birim diamond field with the Ashanti orogenic gold district (adapted from Allibone et al., 2002; Stachel and Harris, 1997). Diamonds have been identified in breccias in the Birimian basement in faults located to the west of, and parallel to, the Tarkwa basin (Konstantinovskii, 2003).

## Paleoproterozoic West Africa

It is notable that a global plume peak or superplume event at ca. 2450 Ma is not associated with significant orogenic gold mineralization. Condie (2004) characterized this event as a "shielding" superplume event linked to supercontinent breakup rather than the "catastrophic" style of superplume event that is triggered by other mechanisms (slab avalanche). The latter corresponds with and perhaps contributes to periods of enhanced crustal growth. Following the Archean, the next major orogenic gold event appears to have occurred at ca. 2100 Ma, based on several mineralization episodes in West Africa, where vein and placer deposits such as the Obuasi (Ashanti) have collectively produced more than 1500 t of gold over many centuries (Oberthür et al., 1998).

Condie (2004) and Abbott and Isley (2002) inferred some level of superplume activity at ca. 2100 Ma, but it is far from being a peak period, based on their analysis of plume proxies. Despite this categorization, Poidevin (1994) described a primary association of komatiites, tholeiites, and boninites in the Central African Republic that is probably ca. 2100 Ma in age (Toteu et al., 2001) and is reminiscent of the southern Abitibi subprovince at ca. 2700 Ma. For example, Poidevin (1994) inferred that trace-element systematics of the boninitic rocks display the involvement of trondhjemite magmas, and alluvial diamonds are also found in the Central African Republic.

Further to the west in Ghana and Burkina Faso, construction of the West African Man Shield is considered to have involved the Birimian accretion of ocean-plateau sequences in a main sequence of growth that lasted ~50 m.y. (Boher et al., 1992). The event was associated with both vein and Paleoproterozoic placer gold deposits that occur in the Tarkwanian sediments of the Ashanti belt, Ghana (250 t Au; Fig. 8). In past decades, diamonds have been extracted from conglomeratic sediments as a by-product of gold mining in this area and from pebble deposits in two tributaries of the Birim River (7.5 million carats; 50 million carat diamond reserves: Konstantinovskii, 2003). Based on in situ finds, the highly weathered mafic-ultramafic diamond source rocks have been identified as "peculiar greenstone-altered breccias" that "make up chains of lenticular bodies elongated along ancient faults" (Konstantinovskii, 2003, p. 535). Although information is limited, the breccias are most plausibly similar to the Archean lamprophyric breccias of the eastern Wawa subprovince, which have whole-rock MgO contents between 9 and 13 wt% and matrix MgO contents near 20.5 wt% MgO (Wyman et al., 2006). The Ghanian and other large Paleoproterozoic gold districts are often described as being similar to Archean greenstone belt deposits but with an increased importance of supracrustal metasedimentary rocks (e.g., Goldfarb et al., 2001). The features summarized here extend that comparison and imply that common geodynamic features linked to flat subduction were responsible for the magnitude of orogenic gold mineralization and related placer deposits.

## Paleoproterozoic Guyana Shield

Gold deposits of the Guyana Shield represent a substantial but slightly younger occurrence of Paleoproterozoic orogenic-style mineralization (Goldfarb et al., 2001). Diamond-bearing mafic-ultramafic rocks (19–16 wt% MgO), mainly altered to albite-carbonate-chlorite-talc schist, have also been reported from the Guyana Shield at the Dachine locality (Bailey, 1999; Capdevila et al., 1999). The unit is poorly exposed, but soil sampling has been used to infer variable widths between 350 m and 1000 m and strike length of 5 km (Bailey, 1999), which is similar to the 100 m × 4 km dimensions of the diamondiferous Wawa breccia described by Lefebvre et al. (2005). Features such as locally preserved olivine pseudomorphs in least altered samples indicate that the high-Mg contents are a primary feature. The host rocks were initially considered to be meta-kimberlite or meta-lamproites, but later workers interpreted them as komatiitic (Capdevila et al., 1999; Bailey, 1999). As with the Wawa lamprophyric breccias, the Dachine rocks contain ellipsoidal monolithic ultramafic fragments. The diamond-bearing fragmental rocks from both localities display variable major-element compositions that may in part reflect the incorporation of the ultramafic xenoliths (Wyman et al., 2006). Given their high $CO_2$ content and strong metamorphic overprint, it is likely that local loss of moderately immobile elements such as the LREE has occurred. A comparison of three Dachine samples containing the highest LREE contents and breccia matrix from the Wawa Cristal locality is given in Table 1 and highlights significant compositional overlap. Mantle-normalized plots of the three Dachine samples (Fig. 9) also indicate immobile trace-element contents that are similar to the Wawa breccia. Further study of the Dachine samples is required, but the available data suggest that the occurrence is likely to represent another example of the shoshonitic lamprophyre–diamond association and that the gold mineralization of the Guyana Shield was enhanced by a process analogous to that proposed for the southern Superior Province.

TABLE 1. SUMMARY DATA FOR DACHINE DIAMOND HOST ROCKS AND WAWA BRECCIA MATRIX

| | Dachine 1 | Dachine 2 | Dachine 3 | 02DS89 |
|---|---|---|---|---|
| | Major-element oxides (wt%) | | | |
| $SiO_2$ | 44.2 | 48.2 | 48.3 | 47.2 |
| $TiO_2$ | 0.94 | 0.81 | 0.84 | 0.75 |
| $Al_2O_3$ | 9.6 | 8.0 | 8.8 | 9.0 |
| $Fe_2O_3$ | 13.7 | 13.0 | 13.2 | 10.8 |
| MgO | 25.2 | 25.0 | 23.0 | 21.6 |
| | Trace elements (ppm) | | | |
| Ni | 976.0 | 920.0 | 761.0 | 821.0 |
| Nb | 7.9 | 6.5 | 7.5 | 4.2 |
| Zr | 67.0 | 61.0 | 61.0 | 91.0 |
| Th | 1.2 | 0.9 | 1.0 | 3.2 |
| Y | 14.8 | 12.7 | 15.1 | 14.3 |
| La | 14.4 | 11.3 | 13.5 | 15.2 |
| Ce | 32.8 | 27.0 | 32.3 | 35.5 |
| Nd | 17.8 | 14.8 | 16.2 | 19.4 |
| Sm | 3.6 | 3.1 | 3.8 | 4.3 |
| Eu | 0.9 | 0.77 | 0.8 | 1.25 |
| Er | 1.4 | 1.2 | 1.4 | 1.4 |
| Yb | 1.4 | 1.1 | 1.3 | 1.3 |
| Lu | 0.22 | 0.2 | 0.19 | 0.18 |

*Note*: Data are from Capdevilla et al. (1999) and Wyman et al. (2006).

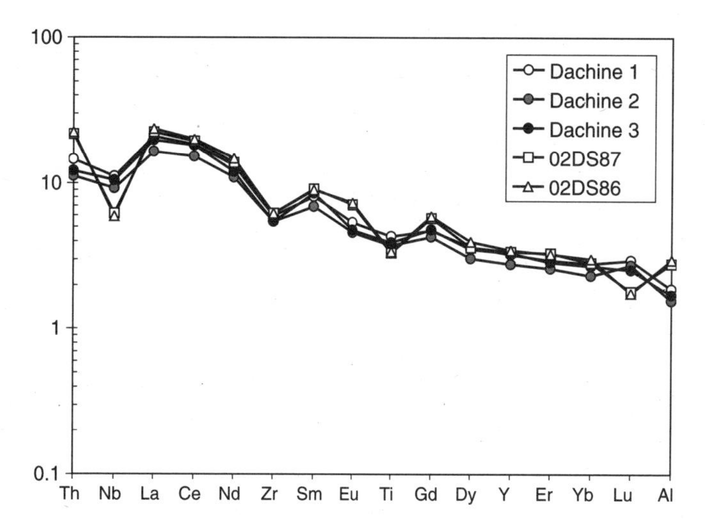


Figure 9. Primitive mantle–normalized multi-element plots for least-altered diamondiferous Dachine host rocks and Wawa breccia matrix. Normalizing factors are from Sun and McDonough (1989); data are from Capdevila et al. (1999) and Wyman et al. (2006).

## Post-Paleoproterozoic Flat Subduction

Although komatiites are much less common following the Paleoproterozoic, major plume events have occurred since this time. They coincide with episodes of crustal growth and orogenic gold mineralization that define major peaks at ca. 1.8 Ga and between ca. 500 Ma and 50 Ma (Condie, 2004; Goldfarb et al. 2001). The Quaternary-Pliocene occurrences of flat subduction listed by Gutscher et al. (2000) are not associated with orogeny and therefore are not associated with orogenic-style gold deposits. Any orogens of similar age would also be too young to expose orogenic gold mineralization in overlying terranes, which generally require at least 50 m.y. of erosion in order to be exposed, based on the known occurrences of these deposits worldwide (Goldfarb et al., 2001). Mesozoic examples of flat subduction, however, may have been associated with recognized major episodes of orogenic gold mineralization.

High-grade metamorphic blocks in the Franciscan mélange indicate that subducted material initially incorporated unusually young and warm proto-arc sequences, whereas later subduction involved a combination of mid-ocean-ridge–style basalts and ocean-island–style basalts (OIB) (Saha et al., 2005). The presence of OIB-style basalt protoliths in lower-temperature tectonic blocks suggests that a strong cooling effect may have involved buoyant ridge or hotspot crust. Prevailing orogenic gold models would characterize the preserved blueschist terranes of the Franciscan mélange as an untapped fluid (±Au) source that avoided reheating during or after orogeny. As expected, giant orogenic gold deposits are not found in the mélange, but the prodigious Mother Lode trend (~1000 t Au; Goldfarb et al., 2001) occurs to the east and has an age that is slightly younger than the main episode of cooling in the Franciscan mélange (Bohlke and Kistler, 1986). A transition from proto-arc to plume-related lithologies in the Franciscan mélange is strongly reminiscent of models for the evolution of the southern Abitibi subprovince, and the analogy may extend further. As summarized by Hausel (1998), placer diamonds are relatively common in the Klamath and northern Mother Lode districts of California and were recovered in the Feather River area near Alleghany during hydraulic mining of placer gold deposits between the 1850s and 1913 (Fig. 10).

## SUMMARY AND CONCLUSIONS

A merely coincidental relationship between the Superior Province diamond and giant gold deposits is shown to be unlikely by the documented association of diamondiferous lamprophyre and giant gold deposit in the Yellowknife belt of the Slave craton and by the more circumstantial evidence outlined herein from any or all of the several major gold districts listed. Given the strong correlation between in situ diamond deposits and giant gold deposits of the southern Superior Province (Fig. 2), any model intended to account for one of these Neoarchean features must also be consistent with the occurrence of the other. In fact, the flat subduction geodynamic scenario first proposed to account only for the novel occurrence of Neoarchean diamonds has obvious implications for key features of prevailing orogenic gold mineralization models.

Figure 10. Distribution of diamond recovery sites in the southwestern United states (diamond symbols; no occurrences have been recorded for Nevada) according to Hausel (1998) and the location of the main districts for orogenic and related placer gold deposit localities (shaded) in California (Young, 1998).

The existence of the Barberton orogenic gold deposits and the recent identification of oxidized granites and shoshonitic lamprophyres of similar age in the Kaapvaal craton represent firm evidence in support of recycling of fluids through modern-style subduction zone processes at 3.1 Ga. Here, we considered features of current orogenic gold genetic models, observations from the present-day Chilean margin, the association of some major gold districts with subduction diamonds, and our own modeling to develop a plausible scenario that accounts for the specific subduction processes responsible for the sporadic and localized development of giant hydrothermal gold systems since that time. Rather than acting on a single feature of the mineralizing system (e.g., Fig. 1), it is likely that the geodynamic events in the southern Superior Province enhanced multiple features in order to generate a trend of large-scale deposits that far surpass typical gold deposits across the Superior Province. The flat subduction scenario suggests numerous ways by which this may have occurred.

Plateaus represent anomalously thick accumulations of hydrothermally altered (ultra) mafic volcanic flows, their differentiates, and melt residues. Compared to average oceanic crust and lithosphere, they are unusually large sources of fluids and partially liberated gold in hydrothermal sediments and altered igneous rocks. Independently of these system-enhancing characteristics, the positive buoyancy of plateaus or hotspot tracks modifies the dip of the subducting slab in a way that initially promotes slab melting to form oxidized magmas that are particularly efficient at breaking down mantle wedge sulfides and transporting the associated gold into the subcontinental lithosphere. Compared to normal subduction, which is associated with dehydration over a wide depth range into convecting asthenosphere, continued flat subduction promotes the physical isolation and freezing of the mantle wedge while also liberating a large proportion of contained fluids over a narrow subduction interval. The dehydration occurs at relatively shallow depths into the isolated and/or nonconvecting asthenosphere and therefore contributes significantly to oxidation of the wedge and the liberation of gold mantle sulfide minerals. More importantly, the fluids also have the potential to transport large amounts of gold from hydrated subducting oceanic-plateau crust, sediments, and serpentinized lithosphere to the base of the arc. This transfer process may be enhanced by channelized flow once the wedge begins to freeze.

If lithosphere thickening and eclogitization of thickened arc crust above flat slabs is common (Gilbert et al., 2006), then it represents another means of enhancing orogenic gold systems. As a transient isostatic feature, the anomaly would be eliminated by thermal reequilibration during orogeny. Differential uplift of the terrane, in addition to transpression associated with oblique plate convergence (Kerrich and Wyman, 1990), would promote development of the major transcrustal fault systems that channel gold upward. Flat subduction also appears to generate at least two extra fluid (±Au) reservoirs compared to normal subduction. Fluids stored in thickened, eclogitized, lower arc crust and the metasomatized frozen asthenosphere represent additional sources compared to those available in normal orogens (stalled slab, subcreted crust, and sediments). The frozen asthenospheric wedge is reheated during orogeny as the slab is "peeled" back and sinks into the mantle. Fluids from all of the reservoirs, would be driven upward and trenchward in response to this advancing thermal pulse. Finally, the fluids would be channeled via major structures to midcrustal depths. Emplacement of mantle-derived volatile-rich lamprophyric magmas along the same structures must have coincided with this period of fluid advection, based on field evidence and the stability requirements of diamonds (O'Neill and Wyman, 2006).

The formation of volatile-rich lamprophyres on Archean cratons, particularly primitive examples that are strongly enriched in large ion lithophile elements, is difficult to account for in non–plate tectonic scenarios, as are the numerous features shared by Archean and post-Archean orogenic gold deposits (Wyman and Kerrich, 1988, 1993). We have shown that the association of diamonds with these lamprophyres can be extended from the Archean Superior Province and Slave cratons to compositionally and texturally equivalent Paleoproterozoic host rocks of the Guyana Shield. The Paleoproterozoic association of orogenic gold, diamonds, and ultramafic breccias in Ghana presents strong circumstantial evidence for another example of this type of occurrence. The recent identification of Mesoarchean shoshonitic lamprophyres, oxidized granites, and the presence of diamonds in the Witswatersrand gold placers do not rise to the level of proof for the flat subduction model we have developed here. This association does, however, imply that the model must be seriously considered and may provide the key to understanding the origin of this anomalously large placer deposit. The distribution of Archean-aged diamond and gold deposits combined with the results of modeling suggest that a modern style of flat subduction involving a thin asthenospheric wedge developed sporadically in response to the subduction of oceanic plateau or hotspot tracks at least as early as 3.1 Ga.

## ACKNOWLEDGMENTS

We thank Kent Condie for the opportunity to contribute this paper. The many constructive comments of Richard Goldfarb, Ken Collerson, and an anonymous reviewer greatly improved upon the original manuscript. O'Neill acknowledges support from a Macquarie University Research Fellowship (ARC National Key Centre for Geochemical Evolution and Metallogeny of Continents (GEMOC) Publication no. 508). This paper was published with the permission of the Senior Manager of the Precambrian Geoscience Section of the Ontario Geological Survey.

## REFERENCES CITED

Abbott, D.H., and Isley, A.E., 2002, The intensity, occurrence and duration of superplume events and eras over geological time: Journal of Geodynamics, v. 34, p. 265–307, doi: 10.1016/S0264-3707(02)00024-8.

Allibone, A., Teasdeale, J., Cameron, G., Etheridge, M., Uttley, P., Soboh, A., Appiah-Kubi, J., Adanu, A., Arthur, R., Mamphey, J., Odoom, B., Zuta, J., Tsikata, A., Pataye, F., and Famiyeh, S., 2002, Timing and structural controls on gold mineralization at the Bogoso Gold Mine, West Africa: Economic Geology and the Bulletin of the Society of Economic Geologists, v. 97, p. 949–969.

Anhaeusser, C.R., 1999, Archaean crustal evolution of the central Kaapvaal craton, South Africa: Evidence from the Johannesburg Dome: South African Journal of Geology, v. 102, p. 303–322.

Armstrong, J.P., and Barnett, R.L., 2003, The association of Zn-chromite with diamondiferous lamprophyres and diamonds: Unique compositions as a guide to the diamond potential of non-traditional diamond host rocks, *in* 8th International Kimberlite Conference Extended Abstracts FLA_0230: Victoria, Canada, Venue West Conference Services, Ltd., 3 pages on CD.

Ayer, J., Amelin, Y., Corfu, F., Kamo, S., Ketchum, J., Kwok, K., and Trowell, N., 2002, Evolution of the southern Abitibi greenstone belt based on U-Pb geochronology: Autochthonous volcanic construction followed by plutonism, regional deformation and sedimentation: Precambrian Research, v. 115, p. 63–95, doi: 10.1016/S0301-9268(02)00006-2.

Ayer, J.A., Thurston, P.C., Bateman, R., Dubé, B., Gibson, H.L., Hamilton, M.A., Hathway, B., Hocker, S.M., Houlé, M.G., Hudak, G., Ispolatov, V.O., Lafrance, B., Lesher, C.M., MacDonald, P.J., Péloquin, A.S., Piercey, S.J., Reed, L.E., and Thompson, P.H., 2005, Overview of Results from the Greenstone Architecture Project: Discover Abitibi Initiative: Ontario Geological Survey Open-File Report 6154, 146 p.

Bailey, L.M., 1999, An Unusual Diamond-Bearing Talc Schist from the Dachine Area of French Guiana [M.S. thesis]: Kingston, Ontario, Canada, Queen's University, 161 p.

Barley, M.E., Brown, S.J.A., Krapez, B., and Kositcin, N., 2007, Physical volcanology and geochemistry of a late Archean volcanic arc: Kurnalpi and Gindalbie terranes, Eastern Goldfields Superterrane, Western Australia: Precambrian Research, v. 161, p. 53–76, doi:10.1016/j.precamres.2007.06.019.

Barron, J.J., Barron, L.M., and Duncan, G., 2005, Eclogitic and ultrahigh-pressure crustal garnets and their relationship to Phanerozoic subduction diamonds, Bingra area, New England fold belt, Eastern Australia: Economic Geology and the Bulletin of the Society of Economic Geologists, v. 100, p. 1565–1582.

Barron, L.M., Lishmund, S.R., Oakes, G.M., Barron, B.J., and Sutherland, F.L., 1996, Subduction model for the origin of some diamonds in the Phanerozoic of eastern New South Wales: Australian Journal of Earth Sciences, v. 43, p. 257–267, doi: 10.1080/08120099608728253.

Bierlein, F.P., Groves, D.I., Goldfarb, R.J., and Dube, B., 2006, Lithospheric controls on the formation of provinces hosting giant orogenic gold deposits: Mineralium Deposita, v. 40, p. 874–886, doi: 10.1007/s00126-005-0046-2.

Bleeker, W., Davis, W.J., Ketchum, J.W., Sircombe, K., and Stern, R.A., 2001, Tectonic evolution of the Slave craton, Canada, *in* Cassidy, K.F., Dunphy, J.M., and Van Kranendonk, M.J., eds., 4th International Archean Symposium Extended Abstracts: AGSO Geoscience Australia Record 2001/37, p. 288–290.

Boher, M., Abouchami, W., Michard, A., Alberade, F., and Arndt, N.T., 1992, Crustal growth in West Africa at 2.1 Ga: Journal of Geophysical Research, v. 97, no. B1, p. 345–369.

Bohlke, J.K., and Kistler, R.W., 1986, Rb-Sr, K-Ar, and stable isotope evidence for ages and sources of fluid components of gold-bearing quartz veins in the northern Sierra Nevada foothills metamorphic belt, California: Economic Geology and the Bulletin of the Society of Economic Geologists, v. 81, p. 296–322.

Brueckner, H.K., Carswell, D.A., and Griffin, W.L., 2002, Paleozoic diamonds within a Precambrian peridotite lens in UHP gneisses of the Norwegian Caledonides: Earth and Planetary Science Letters, v. 203, p. 805–816, doi: 10.1016/S0012-821X(02)00919-6.

Capdevila, R., Arndt, N., Letendre, J., and Sauvage, J.-F., 1999, Diamonds in volcaniclastic komatiite from French Guiana: Nature, v. 399, p. 456–458, doi: 10.1038/20911.

Collerson, K.D., and Kamber, B.S., 1999, Evolution of the continents and the atmosphere inferred from Th-U-Nb systematics of the depleted mantle: Science, v. 283, p. 1519–1522, doi: 10.1126/science.283.5407.1519.

Condie, K.C., 2004, Supercontinents and superplume events: Distinguishing between signals in the geologic record: Physics of the Earth and Planetary Interiors, v. 146, p. 319–332, doi: 10.1016/j.pepi.2003.04.002.

Condie, K.C., 2005, High field strength element ratios in Archean basalts: A window to evolving sources of mantle plumes?: Lithos, v. 79, p. 491–504, doi: 10.1016/j.lithos.2004.09.014.

Corfu, F., 1993, The evolution of the southern Abitibi greenstone belt in light of precise U-Pb geochronology: Economic Geology and the Bulletin of the Society of Economic Geologists, v. 88, p. 1323–1340.

Cousens, B., Facey, K., and Falck, H., 2002, Geochemistry of the late Archean Bangting Group, Yellowknife greenstone belt, Slave Province, Canada: Simultaneous melting of the upper mantle and juvenile mafic crust: Canadian Journal of Earth Sciences, v. 39, p. 1635–1656, doi: 10.1139/e02-070.

Daigneault, R., Mueller, W.U., and Chown, E.H., 2002, Oblique Archean subduction: Accretion and exhumation of an oceanic arc during dextral transpression, Southern volcanic zone, Abitibi subprovince, Canada: Precambrian Research, v. 115, p. 261–290, doi: 10.1016/S0301-9268(02)00012-8.

Davies, G.R., 1992, On the emergence of plate tectonics: Geology, v. 20, p. 963–966, doi: 10.1130/0091-7613(1992)020<0963:OTEOPT>2.3.CO;2.

Davies, G.R., Nixon, P.H., Pearson, D.G., and Obata, M., 1993, Tectonic implications of graphitized diamonds from the Ronda peridotite massif, southern Spain: Geology, v. 21, p. 471–474, doi: 10.1130/0091-7613(1993)021<0471:TIOGDF>2.3.CO;2.

Davis, W.J., and Bleeker, W., 1999, Timing of plutonism, deformation, and metamorphism in the Yellowknife domain, Slave Province, Canada: Canadian Journal of Earth Sciences, v. 36, p. 1169–1187, doi: 10.1139/cjes-36-7-1169.

de Boorder, H., Spakman, W., White, S.H., and Wortel, M.J.R., 1998, Late Cenozoic mineralization, orogenic collapse and slab detachment in the European Alpine belt: Earth and Planetary Science Letters, v. 164, p. 569–575, doi: 10.1016/S0012-821X(98)00247-7.

Defant, M.J., Kepezhinskas, P., 2001, Evidence suggests slab melting in arc magmas: Eos (Transactions, American Geophysical Union), v. 82, p. 65 and 68–69.

de Ronde, C.E.J., Kamo, S.L., Davis, D.W., de Wit, M.J., and Spooner, E.T.C., 1991, Field, geochemical, and U-Pb isotopic constraints from hypabyssal felsic intrusion within the Barberton greenstone belt, South Africa: Implications for tectonics and the timing of gold mineralization: Precambrian Research, v. 49, p. 261–280, doi: 10.1016/0301-9268(91)90037-B.

De Stefano, A., Lefebvre, N., and Kopylova, M., 2006, Enigmatic diamonds in Archean calc-alkaline lamprophyres of Wawa, southern Ontario, Canada: Contributions to Mineralogy and Petrology, v. 151, p. 158–173, doi: 10.1007/s00410-005-0052-5.

Dianor Resources Inc. (Dianor), 2007, Leadbetter Diamond Property, Northwestern Ontario, Canada: A new deposit model for diamond exploration in Canada, *in* Prospectors and Developers Association of Canada 2007 Convention, Technical Session Abstract: http://www.pdac.ca/pdac/conv/2007/pdf/ts-ryder.pdf (January 2008).

Dostal, J., and Mueller, W., 1997, Komatiite flooding of a rifted Archean rhyolitic arc complex: Geochemical signature and tectonic significance of the Stoughton-Roquemaure Group, Abitibi greenstone belt, Canada: The Journal of Geology, v. 105, p. 545–563.

Drummond, M.S., Defant, M.J., and Kepezhinsksas, P.K., 1996, Petrogenesis of slab-derived trondhjemite-tonalite-dacite/adakite magmas: Transactions of the Royal Society of Edinburgh, v. 87, p. 205–215.

Ernst, R.E., Buchan, K.L., and Campbell, I.H., 2005, Frontiers in large igneous province research: Lithos, v. 79, p. 271–297, doi: 10.1016/j.lithos.2004.09.004.

Feng, R., and Kerrich, R., 1992, Geochemical evolution of granitoids from the Archean Abitibi Southern volcanic zone and the Pontiac subprovince, Superior Province, Canada: Implications for tectonic history and source regions: Chemical Geology, v. 98, p. 23–70, doi: 10.1016/0009-2541(92)90090-R.

Fitton, J.G., Saunders, A.D., Norry, M.J., Hardarson, B.S., and Taylor, R.N., 1997, Thermal and chemical structure of the Iceland plume: Earth and Planetary Science Letters, v. 153, p. 197–208, doi: 10.1016/S0012-821X(97)00170-2.

Gilbert, H., Beck, S., and Zandt, G., 2006, Lithospheric and upper mantle structure of central Chile and Argentina: Geophysical Journal International, v. 165, p. 383–398, doi: 10.1111/j.1365-246X.2006.02867.x.

Goldfarb, R.J., Groves, D.I., and Gardoll, S., 2001, Orogenic gold and geologic time: A global synthesis: Ore Geology Reviews, v. 18, p. 1–75, doi: 10.1016/S0169-1368(01)00016-6.

Grabowski, G.P.B., and Wilson, A.C., 2005, Sampling Lamprophyre Dikes for Diamonds: Discover Abitibi Initiative: Ontario Geological Survey Open-File Report 6170, 262 p.

Griffin, W.L., O'Reilly, S.Y., and Davies, R.M., 2000, Subduction-related diamond deposits? Constraints, possibilities and new data from eastern Australia: Reviews in Economic Geology, v. 11, p. 291–310.

Groves, D.I., Goldfarb, R.J., Gebre-Mariam, H., Hagemann, S.G., and Robert, F., 1998, Orogenic gold deposits—A proposed classification in the context of their crustal distribution and relationship to other gold deposit types: Ore Geology Reviews, v. 13, p. 7–27, doi: 10.1016/S0169-1368(97)00012-7.

Groves, D.I., Condie, K., Goldfarb, R.J., Hronsky, J.M.A., and Vielreicher, R.M., 2005, 100th Anniversary Special Paper: Secular changes in global tectonic processes and their influences on the temporal distribution of gold-bearing mineral deposits: Economic Geology and the Bulletin of the Society of Economic Geologists, v. 100, p. 203–224.

Gutscher, M.A., Maury, R., Eissen, J.P., and Bourdon, E., 2000, Can slab melting be caused by flat subduction?: Geology, v. 28, p. 535–538, doi: 10.1130/0091-7613(2000)28<535:CSMBCB>2.0.CO;2.

Hart, T.R., Gibson, H.L., and Lesher, C.M., 2004, Trace element geochemistry and petrogenesis of felsic volcanic rocks associated with volcanogenic massive Cu-Zn-Pb sulfide deposits: Economic Geology and the Bulletin of the Society of Economic Geologists, v. 99, p. 1003–1013.

Hausel, W.D., 1998, Diamonds and Mantle Source Rocks in the Wyoming Craton, with a Discussion of Other U.S. Occurrences: Wyoming State Geological Survey Report of Investigations 53, 93 p.

Hyndman, R.D., and Peacock, S.M., 2003, Serpentinization of the forearc mantle: Earth and Planetary Science Letters, v. 212, p. 417–432, doi: 10.1016/S0012-821X(03)00263-2.

Ishihara, S., 2004, The redox state of granitoids relative to tectonic setting and Earth history: The magnetite-ilmenite series 30 years later: Transactions of the Royal Society of Edinburgh, Earth Sciences, v. 95, p. 23–33.

Ishihara, S., Ohmoto, H., Anhaeusser, C.R., Imai, A., and Robb, L.J., 2006, Discovery of the oldest oxic granitoids in the Kaapvaal Craton and its implications for the redox evolution of early Earth, *in* Kesler, S.E., and Ohmoto, H., eds., Evolution of Early Earth's Atmosphere, Hydrosphere, and Biosphere—Constraints from Ore Deposits: Geological Society of America Memoir 198, p. 67–80.

Kelemen, P.B., Yogodzinski, G.M., and Scholl, D.W., 2003, Along-strike variation in lavas of the Aleutian Island arc: Genesis of high Mg# andesite and implications for continental crust, *in* Eiler, J., ed., Inside the Subduction Factory: American Geophysical Union Geophysical Monograph 138, p. 223–276.

Kerrich, R., and Wyman, D., 1990, Geodynamic setting of mesothermal gold deposits: An association with accretionary tectonic regimes: Geology, v. 18, p. 882–885, doi: 10.1130/0091-7613(1990)018<0882:GSOMGD>2.3.CO;2.

Kerrich, R., Goldfarb, R.J., Richards, J.P., 2005, Metallogenic provinces in an evolving geodynamic framework: Economic Geology 100th Anniversary Volume: Economic Geology and the Bulletin of the Society of Economic Geologists, v. 100, p. 1097–1136.

Kerrick, D.M., and Connolly, J.A.D., 2001, Metamorphic devolatilization of subducted oceanic metabasalts: Implications for seismicity, arc magmatism and volatile recycling: Earth and Planetary Science Letters, v. 189, p. 19–29, doi: 10.1016/S0012-821X(01)00347-8.

Kirk, J., Ruiz, J., Chesley, J., Walshe, J., and England, G., 2002, A major Archean, gold- and crust-forming event in the Kaapvaal craton, South Africa: Science, v. 297, p. 1856–1858, doi: 10.1126/science.1075270.

Konstantinovskii, A.A., 2003, Epochs of diamond placer formation in the Precambrian and Phanerozoic: Lithology and Mineral Resources, v. 38, p. 530–546, doi: 10.1023/A:1027316611376.

Lefebvre, N., Kopylova, M., and Kivi, K., 2005, Archean calc-alkaline lamprophyres of Wawa, Ontario, Canada: Unconventional diamondiferous volcaniclastic rocks: Precambrian Research, v. 138, p. 57–87, doi: 10.1016/j.precamres.2005.04.005.

Li, Z.-X.A., and Lee, C.-T.A., 2006, Geochemical investigation of serpentinized oceanic lithospheric mantle in the Feather River ophiolite, California: Implications for the recycling rate of water by subduction: Chemical Geology, v. 235, p. 161–185, doi: 10.1016/j.chemgeo.2006.06.011.

MacDonald, P.J., Piercey, S.J., and Hamilton, M.A., 2005, An Integrated Study of Intrusive Rocks Spatially Associated with Gold and Base Metal Mineralization in the Abitibi Greenstone Belt, Timmins Area and Clifford Township: Discover Abitibi Initiative: Ontario Geological Survey Open-File Report 6160, 190 p.

Machado, N., Brooks, C., and Hart, S.R., 1986, Determination of initial $^{87}Sr/^{86}Sr$ and $^{143}Nd/^{144}Nd$ in primary minerals from mafic and ultramafic rocks: Experimental procedure and implications for the isotopic characteristics of the Archean mantle under the Abitibi greenstone belt, Canada: Geochimica et Cosmochimica Acta, v. 50, p. 2335–2348, doi: 10.1016/0016-7037(86)90086-4.

Martin, H., Smithies, R.H., Rapp, R., Moyen, J.-F., and Champion, D., 2005, An overview of adakite, tonalite-trondhjemite-granodiorite (TTG) and sanukitoid: Relationships and some implications for crustal evolution: Lithos, v. 79, p. 1–24, doi: 10.1016/j.lithos.2004.04.048.

McCammon, C.A., and Kopylova, M.G., 2003, Mantle oxygen fugacity and diamond formation, *in* 8th International Kimberlite Conference: Victoria, BC, Canada, Venue West Conference Services Ltd., 5 pages on CD.

McCuaig, T.C., and Kerrich, R., 1998, *P-T-t*-deformation-fluid characteristics of lode gold deposits: Evidence from alteration systematics: Ore Geology Reviews, v. 12, p. 381–453.

McNaughton, N.J., Mueller, A.G., and Groves, D.I., 2005, The age of the Giant Golden Mile deposit, Kalgoorlie, Western Australia: Ion-microprobe zircon and monazite U-Pb geochronology of a synmineralization lamprophyre dike: Economic Geology and the Bulletin of the Society of Economic Geologists, v. 100, p. 1427–1440.

McNeil, A.M., and Kerrich, R., 1986, Archean lamprophyric dykes and gold mineralisation, Matheson, Ontario: The conjunction of LIL-enriched mafic magmas, deep crustal structures and Au concentration: Canadian Journal of Earth Sciences, v. 23, p. 324–342.

Miller, S.A., van der Zee, W., Olgaard, D.L., and Connolly, J.A.D., 2003, A fluid-pressure feedback model of dehydration reactions: Experiments, modelling, and applications to subduction zones: Tectonophysics, v. 370, p. 241–251, doi: 10.1016/S0040-1951(03)00189-6.

Moresi, L., Dufour, F., and Muhlhaus, H.-B., 2003, A Lagrangian integration point finite element method for large deformation modeling of viscoelastic geomaterials: Journal of Computational Physics, v. 184, p. 476–497, doi: 10.1016/S0021-9991(02)00031-1.

Morris, P.A., and Witt, W.K., 1997, Geochemistry and tectonic setting of two contrasting Archaean felsic volcanic association in the Eastern Goldfields, Western Australia: Precambrian Research, v. 83, p. 83–107, doi: 10.1016/S0301-9268(97)00006-5.

Mortensen, J.K., 1993, U-Pb geochronology of the eastern Abitibi subprovince. Part 2: Noranda-Kirkland Lake area: Canadian Journal of Earth Sciences, v. 30, p. 29–41.

Mungall, J.E., 2002, Roasting the mantle: Slab melting and the genesis of major Au and Au-rich Cu deposits: Geology, v. 30, p. 915–918, doi: 10.1130/0091-7613(2002)030<0915:RTMSMA>2.0.CO;2.

Nelson, D.R., 1998, Granite-greenstone crust formation on the Archaean Earth; a consequence of two superimposed processes: Earth and Planetary Science Letters, v. 158, p. 109–119, doi: 10.1016/S0012-821X(98)00049-1.

Oberthür, T., Vetter, U., Davis, D.W., and Amanor, J.A., 1998, Age constraints on gold mineralization and Paleoproterozoic crustal evolution in the Ashanti belt of southern Ghana: Precambrian Research, v. 89, p. 129–143, doi: 10.1016/S0301-9268(97)00075-2.

O'Neill, C., and Wyman, D.A., 2006, Geodynamic modeling of late Archean subduction: *P-T* constraints from greenstone belt diamond deposits, *in* Benn, K., Mareschal, J.-C., and Condie, K., eds., Archean Geodynamics and Environments: American Geophysical Union, Geophysical Union Monograph 164, p. 177–188.

Parlak, O., Höck, V., and Delaloye, M., 2002, The supra-subduction zone Pozanti–Karsanti ophiolite, southern Turkey: Evidence for high-pressure crystal fractionation of ultramafic xenoliths: Lithos, v. 65, p. 205–224, doi: 10.1016/S0024-4937(02)00166-4.

Pearson, D.G., Davies, G.R., Nixon, P.H., Greenwood, P.B., and Mattey, D.P., 1991, Oxygen isotope evidence for the origin of pyroxenites in the Beni Bousera peridotite massif, North Morocco: Derivation from subducted oceanic lithosphere: Earth and Planetary Science Letters, v. 102, 289–301.

Perring, C.S., Rock, N.M.S., Golding, S.D., and Roberts, D.E., 1989, Criteria for the recognition of metamorphosed lamprophyres: A case study from the Archaean of Kambalda, Western Australia: Precambrian Research, v. 43, p. 215–237, doi: 10.1016/0301-9268(89)90057-0.

Petford, N., and Atherton, M., 1996, Na-rich partial melts from newly underplated basaltic crust: The Cordillera Blanca Batholith, Peru: Journal of Petrology, v. 37, p. 1491–1521, doi: 10.1093/petrology/37.6.1491.

Phillips, G.N., Groves, D.I., and Kerrich, R., 1996, Factors in the control of the giant Kalgoorlie gold deposit: Ore Geology Reviews, v. 10, p. 295–317, doi: 10.1016/0169-1368(95)00028-3.

Poidevin, J.-L., 1994, Boninite-like rocks from the Palaeoproterozoic greenstone belt of Bogoin, Central African Republic: Geochemistry and petrogenesis: Precambrian Research, v. 68, p. 97–113, doi: 10.1016/0301-9268(94)90067-1.

Polat, A., and Kerrich, R., 2001, Magnesian andesites, Nb-enriched basalt-andesites, and adakites from late Archean 2.7 Ga Wawa greenstone belts, Superior Province, Canada: Implications for late Archean subduction zone petrogenetic processes: Contributions to Mineralogy and Petrology, v. 141, p. 36–52.

Polat, A., Kerrich, R., and Wyman, D.A., 1998, The late Archean Screiber–Hemlo and White River–Dayohessarah greenstone belts, Superior Province: Collages of oceanic plateaus, oceanic arcs, and subduction-accretion complexes: Tectonophysics, v. 289, p. 295–326, doi: 10.1016/S0040-1951(98)00002-X.

Poujol, M., Robb, L.J., Anhaeusser, C.R., and Gricke, B., 2003, A review of the geochronological constraints on the evolution of the Kaapvaal craton, South Africa: Precambrian Research, v. 127, p. 181–213, doi: 10.1016/S0301-9268(03)00187-6.

Prevac, S.A., Anhaeusser, C.R., and Poujol, M., 2004, Evidence for Archean lamprophyre from the Kaapvaal craton, South Africa: South African Journal of Science, v. 100, p. 549–555.

Prouteau, G., Scaillet, B., Pichavant, M., and Maury, R., 2001, Evidence for mantle metasomatism by hydrous silicic melts derived from subducted oceanic crust: Nature, v. 410, p. 197–200, doi: 10.1038/35065583.

Reich, M., Parada, M.A., Palacios, C., Diedtrich, A., Schults, F., and Lehmann, B., 2003, Adakite-like signature of late Miocene intrusions at the Los Pelambres giant porphyry copper deposit in the Andes of central Chile: Metallogenic implications: Mineralium Deposita, v. 38, p. 876–885, doi: 10.1007/s00126-003-0369-9.

Robert, F., and Poulsen, K.H., 1997, World-class Archean gold deposits in Canada: An overview: Australian Journal of Earth Sciences, v. 44, p. 329–351, doi: 10.1080/08120099708728316.

Robert, F., Poulsen, K.H., Cassidy, K.F., and Hidgson, C.J., 2005, Gold metallogeny of the Superior and Yilgarn cratons: Economic Geology 100th Anniversary Volume: Economic Geology and the Bulletin of the Society of Economic Geologists, v. 100, p. 1001–1033.

Rupke, L.H., Morgan, J.P., Hort, M., and Connolly, J.A.D., 2004, Serpentine and the subduction zone water cycle: Earth and Planetary Science Letters, v. 223, p. 17–34, doi: 10.1016/j.epsl.2004.04.018.

Sage, R.P., 2000, The "Sandor" Diamond Occurrence, Michipicoten Greenstone Belt, Wawa, Ontario: A Preliminary Study: Ontario Geological Survey Open-File Report 6016, 49 p.

Saha, A., Basu, A.R., Wakabayashi, J., and Wortman, G.L., 2005, Geochemical evidence for a subducted infant arc in Franciscan high-grade-metamorphic tectonic blocks: Geological Society of America Bulletin, v. 117, p. 1318–1335, doi: 10.1130/B25593.1.

Scott, C.R., Mueller, W.U., and Pilote, P., 2002, Physical volcanology, stratigraphy, and lithogeochemistry of an Archean volcanic arc: Evolution from plume-related volcanism to arc rifting of SE Abitibi greenstone belt, Val d'Or, Canada: Precambrian Research, v. 115, p. 223–260, doi: 10.1016/S0301-9268(02)00011-6.

Smithies, R.H., and Champion, D.C., 1999, Late Archean felsic alkaline rocks in the Eastern Goldfields, Yilgarn craton, Western Australia: A result of lower crustal delamination?: Journal of the Geological Society of London, v. 156, p. 561–576.

Smithies, R.H., Champion, D.C., and Cassidy, K.F., 2003, Formation of Earth's early Archaean continental crust: Precambrian Research, v. 127, p. 89–101, doi: 10.1016/S0301-9268(03)00182-7.

Smithies, R.H., Champion, D.C., Van Kranendonk, M.J., Howard, H.M., and Hickman, A.H., 2005, Modern-style subduction processes in the Mesoarchean: Geochemical evidence from the 3.12 Ga Whundo intra-oceanic arc: Earth and Planetary Science Letters, v. 231, p. 221–237, doi: 10.1016/j.epsl.2004.12.026.

Stachel, T., and Harris, J.W., 1997, Syngenetic inclusions in diamond from the Barim field (Ghana)—A deep peridotitic profile with a history of depletion and re-enrichment: Contributions to Mineralogy and Petrology, v. 127, p. 336–352, doi: 10.1007/s004100050284.

Stachel, T., Banas, A., Muehlenbachs, K., Kurszlaukis, S., and Walker, E.C., 2006, Archean diamonds from Wawa (Canada): Samples from deep cratonic roots predating cratonization of the Superior Province: Contributions to Mineralogy and Petrology, v. 151, p. 737–750.

Stott, G.M., Ayer, J.A., Wilson, A.C., and Grabowski, G.P.B., 2002, Are the Neoarchean diamond-bearing breccias in the Wawa area related to late-orogenic alkalic and "sanukitoid" intrusions?, *in* Summary of Field Work and Other Activities 2002: Ontario Geological Survey Open-File Report 6100, p. 9-1–9-10.

Sun, S.-S., and McDonough, W.F., 1989, Chemical and isotopic systematics of oceanic basalts: Implications for mantle composition and processes, *in*

Saunders, A.D., and Norry, M.J., eds., Magmatism in the Ocean Basins: Geological Society of London Special Publication 42, p. 313–345.

Toteu, S.F., Van Schmus, W.R., Penaye, J., and Michard, A., 2001, New U-Pb and Sm-Nd data from north-central Cameroon and its bearing on the pre–Pan African history of central Africa: Precambrian Research, v. 108, p. 45–73, doi: 10.1016/S0301-9268(00)00149-2.

Vaillancourt, C., Ayer, J.A., Zubowski, S.M, and Kamo, S.L., 2004, Synthesis and timing of Archean geology and diamond-bearing rocks in the Michipicoten greenstone belt: Menzies and Musquash Townships, *in* Summary of Field Work and Other Activities 2004: Ontario Geological Survey Open-File Report 6145, p. 6-1–6-9.

van Roermund, H.L.M., Carswell, D.A., Drury, M.R., and Heijboer, T.C., 2002, Microdiamonds in a megacrystic garnet websterite pod from Bardane on the island of Fjortoft, western Norway: Evidence for diamond formation in mantle rocks during deep continental subduction: Geology, v. 30, p. 959–962, doi: 10.1130/0091-7613(2002)030<0959:MIAMGW>2.0.CO;2.

Wagner, L.S., Beck, S., and Zandt, G., 2005, Upper mantle structure in the south central Chilean subduction zone (30° to 36°S): Journal of Geophysical Research, v. 110, B01308, doi: 10.1029/2004JB003238.

Wagner, L.S., Beck, S., Zandt, G., and Ducea, M.N., 2006, Depleted lithosphere, cold, trapped asthenosphere, and frozen melt puddles above the flat slab in central Chile and Argentina: Earth and Planetary Science Letters, v. 245, p. 289–301, doi: 10.1016/j.epsl.2006.02.014.

Wang, Q., Wyman, D.A., Xu, J.F., Zhao, Z.H., Jian, P., and Zi, F., 2007, Partial melting of thickened or delaminated lower crust in the middle of eastern China: Implications for Cu-Au mineralization: The Journal of Geology, v. 115, p. 149–161, doi: 10.1086/510643.

Williams, F., 2002, Diamonds in Late Archean Calc-Alkaline Lamprophyres Ontario, Canada: Origins and Implications [Honor's thesis]: Sydney, University of Sydney, 82 p.

Williams, H.R., Stott, G.M., Heather, K.B., Muir, T.L., and Sage, R.P., 1991, Wawa subprovince, *in* Thurston, P.C., Williams, H.R., Sutcliffe, R.H., and Stott, G.M., eds., Geology of Ontario: Ontario Geological Survey Special Volume 4, Part 1, p. 485–539.

Witze, A., 2006, The start of the world as we knew it—News Feature: Nature, v. 442, p. 128–131, doi: 10.1038/442128a.

Wyborn, L.A.I., Gallagher, R., and Jagodzinski, E.A., 1994, A conceptual approach to metallogenic modelling using GIS: Examples from the Pine Creek Inlier: Glenside, South Australia, Australian Research on Ore Genesis Symposium, Australian Mineral Foundation, Poster Paper Abstracts, p. 15.1–15.5.

Wyman, D.A., 2003, Upper mantle processes beneath the 2.7 Ga Abitibi belt, Canada: A trace element perspective: Precambrian Research, v. 127, p. 143–165, doi: 10.1016/S0301-9268(03)00185-2.

Wyman, D.A., and Hollings, P., 2006, Late-Archean convergent margin volcanism in the Superior Province: A comparison of the Blake River Group and Confederation assemblage, *in* Benn, K., Mareschal, J.-C., and Condie, K., eds., Archean Geodynamics and Environments: American Geophysical Union, Geophysical Union Monograph 164, p. 215–237.

Wyman, D.A., and Kerrich, R., 1988, Alkaline magmatism, major structures, and gold deposits: Implications for greenstone belt gold metallogeny: Economic Geology and the Bulletin of the Society of Economic Geologists, v. 83, p. 454–461.

Wyman, D.A., and Kerrich, R., 1989, Archean shoshonitic lamprophyres of the Superior Province, Canada: Distribution, petrology and geochemical characteristics: Journal of Geophysical Research, v. 94B, p. 4667–4696.

Wyman, D.A., and Kerrich, R., 1993, Archean shoshonitic lamprophyres of the Abitibi subprovince, Canada: Petrogenesis, age, and tectonic setting: Journal of Petrology, v. 34, p. 1067–1109.

Wyman, D.A., and Kerrich, R., 2002, Formation of Archean Continental Lithospheric Roots: The role of mantle plumes: Geology, v. 30, p. 543–546.

Wyman, D.A., Bleeker, W., and Kerrich, R., 1999a, A 2.7 Ga plume, proto-arc, to arc transition and the geodynamic setting of the Kidd Creek deposit: Evidence from precise ICP MS trace element data, *in* Hannington, M.D., and Barrie, C.T., eds., Economic Geology Monograph 10: The Giant Kidd Creek Massive Sulfide Deposit, Western Abitibi Subprovince, Canada, p. 511–528.

Wyman, D.A., Kerrich, R., and Groves, D.I., 1999b, Lode gold deposits and Archean mantle plume–island arc interaction, Abitibi subprovince, Canada: The Journal of Geology, v. 107, p. 715–725, doi: 10.1086/314376.

Wyman, D.A., Kerrich, R., and Polat, A., 2002, Assembly of Archean cratonic mantle lithosphere and crust: Plume-arc interaction in the Abitibi-Wawa subduction-accretion complex: Precambrian Research, v. 115, p. 37–62, doi: 10.1016/S0301-9268(02)00005-0.

Wyman, D.A., Ayer, J.A., Conceição, R.V., and Sage, R.P., 2006, Mantle processes in an Archean orogen: Evidence from 2.67 Ga diamond-bearing lamprophyres and xenoliths: Lithos, v. 89, p. 300–328, doi: 10.1016/j.lithos.2005.12.005.

Young, L., 1998, Map of California Historic Gold Mines: Sacramento, California, California Department of Conservation, Division of Mines and Geology, scale 1:1,500,000.

Manuscript Accepted by the Society 14 August 2007

Printed in the USA

The Geological Society of America
Special Paper 440
2008

# *Effects of a warmer mantle on the characteristics of Archean passive margins*

**Andrew Hynes***
*Department of Earth and Planetary Sciences, McGill University, 3450 University Street, Montreal, Quebec H3A 2A7, Canada*

## ABSTRACT

**The Archean mantle was probably warmer than the modern one. Continental plates underlain by such a warmer mantle would have experienced less subsidence than modern ones following extension because extension would have led to widespread melting of the underlying mantle and the generation of large volumes of mafic rock. A 200 °C increase in mantle temperature leads to the production of nearly 12 km of melt beneath a continental plate extended by a factor of 2, and the resulting thinned plate rides with its upper surface little below sea level. The thick, submarine, mafic-to-ultramafic volcanic successions on continental crust that characterize many Archean regions could therefore have resulted from extension of continental plates above warm mantle.**

**Long-term subsidence of passive margins is driven by thermal relaxation of the stretched continental plate (cf. McKenzie). With a warmer mantle, the relaxation is smaller. For a continental plate stretched by a factor of 2, underlain by a 200 °C warmer mantle than at present, the cooling-driven subsidence drops from 2.3 km to 1.1 km. The combined initial and thermal subsidence declines by more than 40%, and by even more than this if initial continental crustal thicknesses were lower. The greatly reduced subsidence results in a concomitant decline in accommodation space for passive-margin sediments and may explain the scarcity of passive-margin sequences in the Archean record.**

**The formation of diamonds in the Archean requires geotherms similar to modern ones, which in turn probably reflect the presence of cool mantle roots beneath the continents. Stretching of continents underlain by cool mantle roots would yield passive margins similar to modern ones. Thus, development of significant passive margins may have occurred only through rifting of continents underlain by cool mantle roots. Furthermore, the widespread subcontinental melting associated with rifting of continents devoid of roots may have been a significant contributor to development of the roots themselves.**

**Keywords:** Archean, passive margins, volcanism, subsidence.

*andrew.hynes@mcgill.ca

Hynes, A., 2008, Effects of a warmer mantle on the characteristics of Archean passive margins, *in* Condie, K.C., and Pease, V., eds., When Did Plate Tectonics Begin on Planet Earth?: Geological Society of America Special Paper 440, p. 149–156, doi: 10.1130/2008.2440(07). For permission to copy, contact editing@geosociety.org. 

## INTRODUCTION

Although questions remain (e.g., Stern, 2005), it is widely believed that the surface tectonics of Earth have been governed predominantly by the motions of lithospheric plates since at least the Early Proterozoic. The operation of plate tectonics in the Archean is a more controversial topic. Plate-tectonic scenarios have been sketched for many Archean terrains (e.g., de Wit, 1991, 1998; Calvert et al., 1995; Percival et al., 2004). The geological characteristics of Archean terrains differ, however, in significant ways from those of younger terrains, and as a result, plate-tectonic models for their evolution are still called into question by some scientists (e.g., Hamilton, 1998; Bleeker, 2002) and alternative models, involving primarily vertical tectonic processes, have been advanced for a number of Archean terrains (e.g., Chardon et al., 1996; Collins et al., 1998; Bédard et al., 2003; Van Kranendonk et al., 2004), although vertical tectonic processes do not preclude the coeval operation of plate tectonics (e.g., Chardon et al., 2002; Lin, 2005). The most severe criticisms of plate-tectonic models for Archean processes have involved the scarcity or absence of some features that might be expected in such a regime. Notable among these features are passive-margin sedimentary successions. In this paper, the operation of plate tectonics in the Archean is adopted as a working hypothesis. What is pursued is the question of the effect of higher mantle temperatures in the Archean on the development and nature of passive-margin sequences.

## TEMPERATURE OF THE ARCHEAN MANTLE

It is generally thought that the temperature of the mantle was higher in the Archean than it is today. First-order evidence for this comes from the presence of komatiites in Archean supracrustal successions. Komatiites require high degrees of partial melting of mantle parents. Archean komatiitic melts are estimated to have been as hot as 1580 °C, considerably warmer than the 1400 °C estimated for the most magnesian melts known from Tertiary or later time (Nisbet et al., 1993). Estimates of komatiitic melt temperatures have in turn been used to estimate mantle potential temperatures of 1900, 1800, and 1600 °C for their source regions at 3.45 Ga, 2.7 Ga, and 0.16 Ga respectively (Nisbet et al., 1993). These figures do not, however, provide reliable estimates of ambient mantle potential temperatures at these ages, both because komatiites may well not have formed in typical Archean oceanic crust (Bickle et al., 1994) and because the temperatures of komatiitic melts may have been overestimated if they were rich in water (Parman et al., 1997; Grove and Parman, 2004). They do, however, indicate that mantle temperatures were probably 200 °C or more higher in the Archean than today. There is at present no other direct evidence for the temperature of the Archean mantle; support for its higher temperature is argued on the basis of the progressive decline with time of the amount of heat produced in Earth by the decay of unstable isotopes of K, U, and Th and on the basis of secular cooling models for Earth.

Uncertainties concerning the relative amounts of K and U + Th in Earth (Wasserburg et al., 1964) dictate that the bulk half-life for radioactive decay in Earth may be anywhere between 1.4 and 2.1 b.y., so that the heat generation due to radioactivity at 3.0 Ga may have been anywhere between 2.4 and 3.5 times as large as it is today. The effect of this difference on heat loss at the surface is not well constrained, both because a proportion of the heat being lost from Earth is probably due to growth of the inner core, and because the efficiency with which heat generated in Earth is transported to the surface is unclear. It is commonly assumed that the heat loss from Earth was considerably greater at 2.7 Ga than it is today and that there is a positive relationship between the temperature of the mantle and the heat loss from Earth (see, for example, the reviews by Korenaga, 2006; van Hunen et al., this volume), thus leading to higher temperatures in the mantle in the past. Although it has recently been argued that heat loss from Earth was not significantly higher in the past than at present, this argument is itself dependent on higher temperatures in the mantle because plate motions may have been slower (Korenaga, 2003, 2006). Thus, the suggestion that mantle temperatures in the Archean were higher than they are at present is not controversial. In this paper, I assume mantle temperatures were up to 200 °C higher than at present and investigate the effects of such higher temperatures on the characteristics of passive margins.

## INITIAL ELEVATION CHANGE

Passive-margin development was modeled by McKenzie (McKenzie, 1978) as a two-step procedure, with a first phase of instantaneous lithospheric stretching and a second phase of long-term cooling. Both phases were associated with change in elevation to maintain isostatic equilibrium of the stretched lithosphere. The model was based on a simple set of assumptions, including an estimate of the thickness of the lithosphere affected by the stretching (125 km) and uniform stretching of the crustal and mantle part of that lithosphere. Although data from present passive margins make it clear that stretching of the lithosphere is not necessarily uniform (e.g., Royden and Keen, 1980), I first adopt a uniform-stretching model, and comment later on the effect of nonuniformity.

Algebraic expressions for the initial elevation due to stretching used in this paper are, following McKenzie (1978) and Royden and Keen (1980), based on an assumed linear geothermal gradient in the plate being stretched, a constant temperature in the underlying mantle, and a linear density gradient within the crustal and mantle part of the plate. The thickness of the plate and the temperature at its base are important parameters affecting the resulting subsidence. The characteristics of the geotherm beneath a continent were estimated using the method of McKenzie and Bickle (McKenzie and Bickle, 1988, Appendix B therein). In this method, a "mechanical boundary layer," a cool surficial layer not involved in convection, is underlain by a "thermal boundary layer" in which heat transfer is both conductive and convective. The thermal boundary layer is underlain by a region in which heat transfer is purely convective. Geothermal gradients are steep and linear in the mechanical boundary layer and shallow and linear (adiabatic) in the purely convective upper mantle (e.g., Fig. 1). The base of the "plate" for calculations of initial subsidence lies in the thermal boundary layer, where the geothermal gradient shallows markedly. If the mantle potential temperature ($T_p$) and the thickness of the mechanical boundary layer are known, a steady-state geotherm can be calculated once a value has been assigned for the viscosity of the convecting upper mantle.

For the geotherms depicted on Figure 1, the viscosity of the upper mantle was estimated using the mantle-viscosity relationships of Karato and Wu (1993), with applied stress of 1 MPa and grain size of 1 mm. Viscosities were calculated at 1 km intervals from the base of the thermal boundary layer to depths of 350 km. At each depth, the lesser of the effective viscosities due to dislocation and diffusion creep was selected, and the effective viscosity assigned was the geometric mean of the viscosities for wet and dry olivine. The viscosity calculated for the convecting upper mantle was the geometric mean of the viscosities calculated every kilometer from the base of the thermal boundary layer to a depth of 350 km. For a $T_p$ of 1300 °C, comparable with the present day, and a mechanical boundary layer thickness of 112 km, the calculated upper-mantle viscosity is $1.4 \times 10^{20}$ Pa s. For a 200 °C higher $T_p$, and a mechanical boundary layer thickness of 52 km, the viscosity of the convecting upper mantle declines to $5.6 \times 10^{17}$ Pa s.

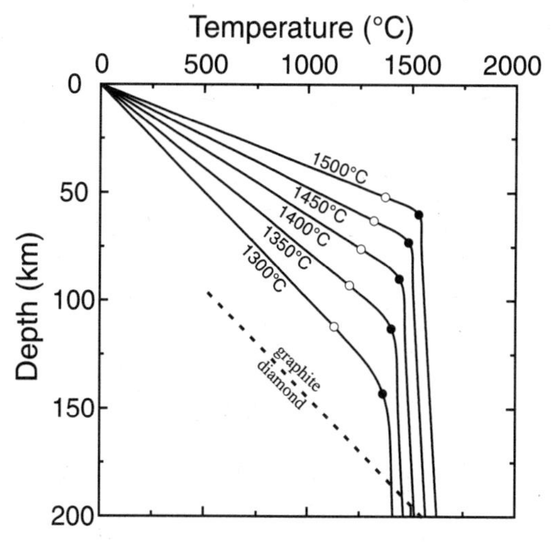


Figure 1. Geothermal gradients for different $T_p$, calculated using the methods of McKenzie and Bickle (McKenzie and Bickle, 1988, Appendix B therein). Open circles—base of mechanical boundary layer; filled circles—base of "plate" for initial subsidence calculations. Diamond stability field after Liu (2002).

The temperature and depth used for the base of the plate in calculation of the initial subsidence were derived from the calculated geotherm. In the treatment of McKenzie and Bickle, depth within the thermal boundary layer was specified by the dimensionless variable

$$\bar{z} = Ra^{0.281}\left(\frac{z}{d}\right), \quad (1)$$

where $Ra$ is the Rayleigh number, $z$ is the actual depth in the thermal boundary layer, and $d$ is the thickness of the convecting upper mantle. I specified the position in the thermal boundary layer for which $\bar{z}$ was 3.0 as the base of the plate for purposes of calculation of initial subsidence. (Figure 1 indicates where these depths sit on the calculated geotherms.)

The base of the plate in my calculations is governed primarily by the depth at which small-scale convection is initiated beneath the plate. The procedures used by McKenzie and Bickle were based on the analytical treatment of Parsons and McKenzie (Parsons and McKenzie, 1978). Numerical methods have provided better controls on the depths of initiation of such small-scale subduction (e.g., Korenaga and Jordan, 2002). The treatment of McKenzie and Bickle is retained here because it supplies an analytical expression that lends itself easily to characterization of the effects of initial rifting. The mechanical boundary layer thicknesses used in the calculations have, however, been adjusted to yield thermal boundary layer thicknesses comparable with those obtained by Korenaga (2003) (Fig. 1).

The major result of an increase in $T_p$ is a profound decrease in the thickness of the thermal boundary layer, and in consequence in the thickness of the plate (Fig. 1). A seminal feature of Korenaga's arguments concerning Archean tectonics is that rising mantle temperatures may in fact have resulted in thicker rather than thinner plates, because greater degrees of melting of the mantle beneath oceanic spreading ridges may have extended partial-melting-induced hardening to greater depths (Korenaga, 2003, 2006). There is, however, no reason to argue that this feature should have applied to the mantle beneath the continents, which is the mantle under discussion here.

Initial subsidence due to stretching was calculated from

$$S_i = h_T\left[\frac{h_c}{h_T}\left(\rho_c\left(1-\frac{\alpha T_T}{2}\frac{h_c}{h_T}\right)-\rho_m\right)\left(1-\frac{1}{\delta}\right)+\right.$$

$$\left.\rho_m\frac{\alpha T_T}{2}\left\{\left(1+\left(\frac{h_c}{h_T}\right)^2\right)\left(1-\frac{1}{\gamma}\right)+2\frac{h_c}{h_T}\left(\frac{1}{\gamma}-\frac{1}{\delta}\right)\right\}\right]$$

$$\div\left[\rho_w-\rho_m\left(1-\alpha T_T\right)\right], \quad (2)$$

where $h_T$ is the depth selected as the base of the plate within the thermal boundary layer, $h_c$ is the thickness of the crust, $\rho_c$ is the STP density of the crust, $\alpha$ is the coefficient of thermal expansion, $T_T$ is the temperature at depth $h_T$, $\rho_m$ is the STP density of the mantle, $\gamma$ is the stretching factor in the mantle, $\delta$ is the stretching factor in the crust, and $\rho_w$ is the density of water. This expression is based only on the assumption of isostatic equilibrium after

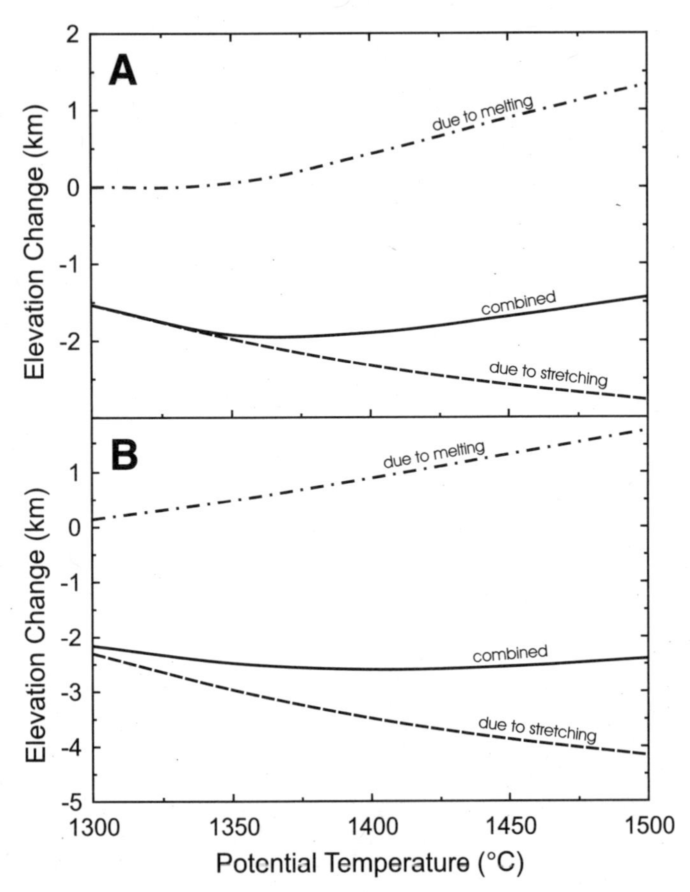


Figure 2. Initial change in elevation as a function of $T_P$. Dashed lines—due purely to stretching; dash-dot lines—due to production of melt; solid lines—combination of the two. (A) For stretching factors (δ and γ) of 2. (B) For δ = γ = 4.

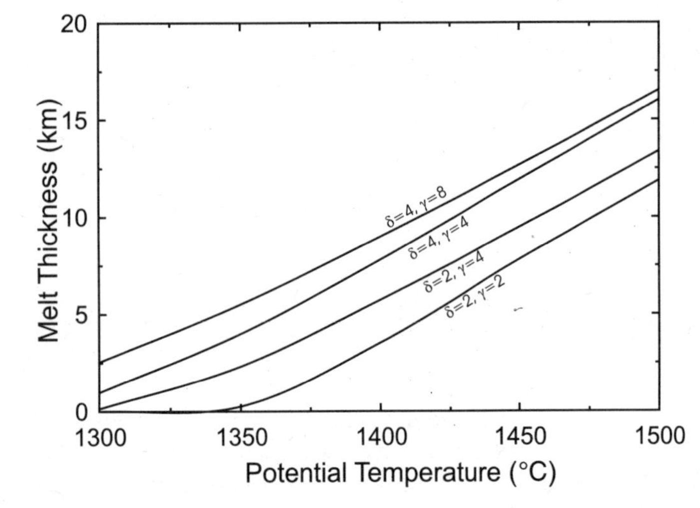


Figure 3. Melt production during continental thinning as a function of $T_P$ for various crustal and mantle stretching factors (δ and γ).

stretching, and that the subsided plate is overlain by water. If there is rise following stretching, $\rho_w$ is omitted from the denominator. In calculating initial subsidence, $\rho_m$, $\rho_c$, and $\rho_w$ were assigned values of 3330, 2900, and 1000 kg m$^{-3}$, and α was set at 3.2 × 10$^{-5}$ °C$^{-1}$. The values of $h_T$ and $T_T$ were derived from the calculated geotherms, as discussed above.

The initial response of continental plates to extension is strongly dependent on the relative thicknesses of the crust and mantle in the plate. With decreased plate thickness and constant initial crustal thicknesses, the lower ratio of mantle to crust in the plate leads to markedly increased subsidence due to extension (Fig. 2, dashed lines; subsidence values calculated for a crustal thickness of 40 km). For a 200 °C warmer mantle, subsidence increases by more than 1 km for stretching factors (δ and γ) of 2 (Fig. 2A), and more than 2 km for stretching factors of 4 (Fig. 2B). Thus, it would appear that conditions in the Archean should have resulted in much greater initial subsidence at continental margins than at present. This conclusion, however, ignores the effect of melt production during continental extension, and melt production would have been substantially greater with warmer mantle material rising beneath the stretched lithosphere.

I estimate the production of melt during continental extension using the methods of McKenzie and Bickle (McKenzie and Bickle, 1988), in which melt is derived from all parts of the column of rising mantle material that overstep the mantle solidus. The amount of melt generated is proportional to the temperature difference between the rising peridotite and the solidus. Increased temperatures in the rising asthenosphere beneath the stretched plate result in the production of large amounts of melt (Fig. 3).

One effect of the generation of melt is to reduce the intrinsic density of the part of the mantle undergoing melting, both because the broadly gabbroic rock derived from the melt has lower density than its parent peridotite and because depleted peridotite has lower density than fertile peridotite. Thus, production of melt in the rising mantle raises the isostatic position of the resulting column of depleted mantle, generated melt, and stretched continental plate. To estimate the magnitude of this rise, I assign a density of 3000 kg m$^{-3}$ to rocks solidified from the melt and a 10 kg m$^{-3}$ drop in the density of the mantle in the depleted column. The calculated rises are substantial (Fig. 2, dash-dot curves). For a 200 °C increase in $T_P$, the rise essentially cancels the effect of increased subsidence due to the smaller mantle/crust ratio in the extending plate. Thus, the net initial subsidence associated with stretching of the continental lithosphere in the Archean may have been similar to that today, provided crustal thicknesses were similar and provided the crust and the mantle stretched by similar amounts. A significant difference would have been, however, that there were much larger volumes of melt produced, which would have been represented by lava flows erupted onto the stretched continent, underplates of the continental crust, or a combination of the two.

In the calculations represented on Figure 2, I have adopted a conservative approach to differences between the modern and Archean situations, permitting changes only in $T_P$. I have not considered other possible differences between the Archean and today, such as the degree to which the mantle stretched more than the crust and the initial thickness of the continental crust.

Modeling of the subsidence of present-day continental margins (e.g., Royden and Keen, 1980) indicates that the mantle may stretch twice as much as the crust during continental rifting. This is not surprising given the substantially higher temperatures of the mantle compared with the crust. Since the temperature difference between Archean geotherms and the modern one is greater at mantle depths than at crustal depths (Fig. 2), it is likely that the difference between the stretching in the mantle and the stretching in the crust was even greater in the Archean than it is today. The effect of this may be illustrated by comparing the results of stretching the mantle and crust uniformly with those of stretching the mantle twice as much as the crust (Fig. 4). Increasing the stretching factors for the mantle reduces the amount of initial subsidence due to stretching alone. Indeed, it was precisely the need to account for lower observed initial subsidence than that predicted from uniform-stretching models that led Royden and Keen to suggest that stretching on the North American margin had increased with increasing depth. There is also an extra elevation due to the production of more melt, so that the net effect of increasing the stretching in the mantle by a factor of 2 is to decrease the total initial subsidence, with greater melt production.

The amount of continental crust present in the Archean, and its thickness, remain the subject of considerable debate, with suggestions ranging all the way from more continental crust than at present (Fyfe, 1978) through constant amounts of crust through geological time (Armstrong, 1991) to models of progressive growth of the continents (e.g., Reymer and Schubert, 1987). The latter has been the prevailing wisdom and appears to be most compatible with isotopic (e.g., Collerson and Kamber, 1999) and freeboard (Hynes, 2001) arguments. Whether the presence of less continental crust on Earth in the past would have been reflected in thinner continental crust depends on the age distribution of the oceanic crust, and on the total volume of oceanic water. Only if the amount of oceanic water in the past was significantly less than it is today is it likely that the continental crust would have been thinner than at present, even if there was less of it on Earth (Hynes, 2001). Under these circumstances, however, stable Archean crust would have become submerged subsequently. Thus, it appears unlikely that Archean continental crust was markedly thinner than modern continental crust, and it may indeed have been slightly thicker, in which case it could have kept pace with the freeboard requirements through subsequent erosion. Again, the geological record limits the extent of this erosion, so that continental crustal thickness cannot have been significantly greater than that today. The effect of increase in continental crustal thickness is to increase the initial subsidence due to stretching (Fig. 5). There would be no appreciable effect on the subsidence due to melt generation. The net effect of decreasing continental crustal thicknesses would have been to produce ~150 m of decreased initial subsidence for each 1 km decrease in the thickness of the continental crust.

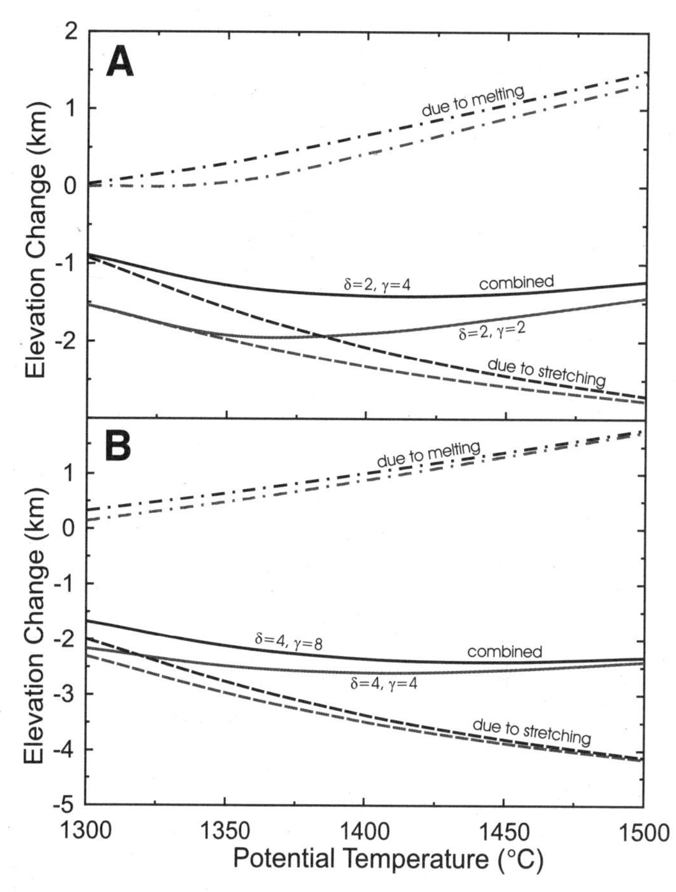


Figure 4. Initial change in elevation as a function of $T_p$; effect of increasing γ. Curves are shown for equal values for δ and γ (gray), and for values of γ twice those of δ (black). (A) For δ = 2. (B) For δ = 4. Line ornament as for Figure 2.

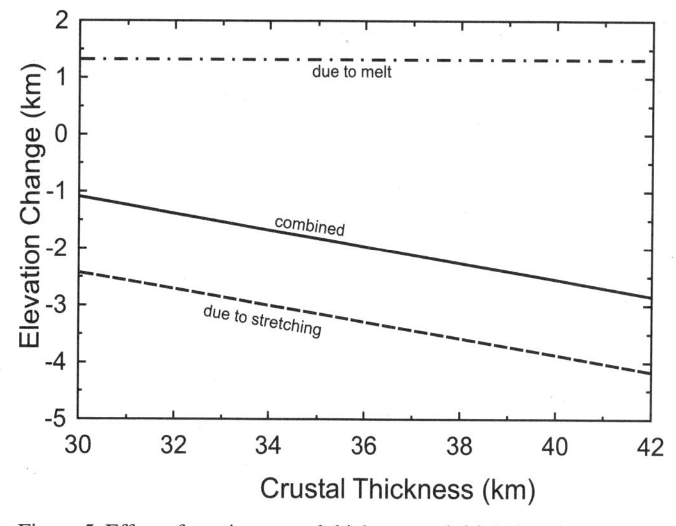


Figure 5. Effect of varying crustal thickness on initial elevation change, for δ = γ = 4 and $T_p$ = 1450 °C. Line ornament as for Figure 2.

In summary, increased temperature of the mantle leads to a predicted decrease in the thicknesses of continental plates in the Archean, which decreases the amount of initial subsidence during continental extension. The modeling indicates, however, that this effect is almost completely offset by the generation of melt in the rising mantle beneath the plate. If the ratio of stretching in the crust to stretching in the mantle is assumed to have been larger in the Archean than today, or the continental crust is assumed to have been thinner, more melt is predicted, and its effect more than offsets the subsidence due to stretching. Given that the modeling of subsidence at some modern continental margins indicates little or no subsidence at the onset of stretching (e.g., Royden and Keen, 1980), the likelihood is that the stretching of continental crust associated with passive-margin formation in the Archean was associated with widespread uplift of the margins above sea level, much of which was achieved by the eruption and/or intrusion of large volumes of mafic igneous rock.

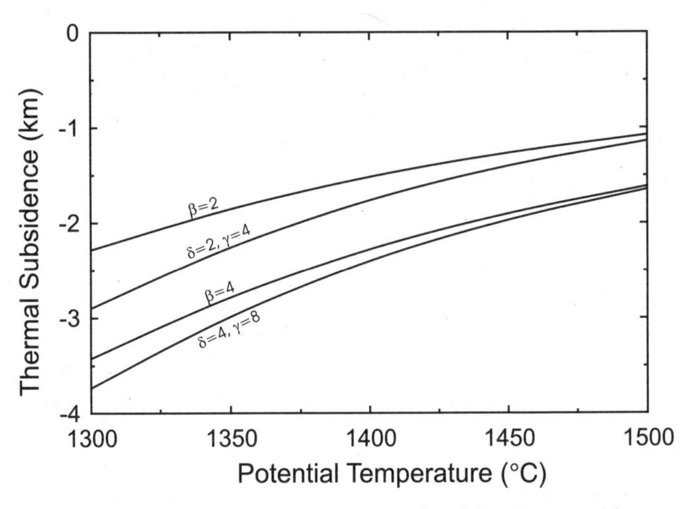


Figure 6. Long-term subsidence due to cooling following stretching, as a function of $T_p$, for uniform stretching ($\beta = 2, 4$) and for twice as much stretching in the mantle as in the crust.

## THERMAL SUBSIDENCE

Long-term subsidence of the continental margins following extension results from relaxation of the geotherm, perturbed from its equilibrium form by the extension, back into equilibrium (McKenzie, 1978). For uniform extension, such subsidence may be estimated from

$$S_t = \frac{h_T \rho_0 \alpha T_T}{(\rho_0 - \rho_w)} \frac{4}{\pi^2} \sum_{n=0}^{\infty} \left\{ \frac{1}{(2n+1)^2} \left[ \frac{\beta}{(2n+1)\pi} \sin\frac{(2n+1)\pi}{\beta} \right] \exp\left(-(2n+1)^2\right) \right\}, \qquad (3)$$

where $\rho_0$ is the STP density of the plate, and $\beta$ is the uniform stretching factor for the whole plate (adapted from McKenzie, 1978, equation 8 therein).

The calculated subsidence is markedly reduced if the mantle is warmer (Fig. 6). Uniform stretching of a continental plate by a factor of 2 with, for example, a mantle 200 °C warmer than that today results in subsidence of only 1.1 km, compared with 2.3 km today, and the differences are even more marked for greater degrees of extension. These differences are due to the thinner nature of the mechanical and thermal boundary layers (Fig. 1), which provide lower volumes in which thermal relaxation can take place. Similar reductions may be expected in the total thermal subsidence exhibited by oceanic plates as they cool passing away from spreading ridges, because the greater amount of heat being delivered to the thermal boundary layer by the convecting mantle limits the amount of cooling the plates experience before thermal equilibrium is obtained.

Thermal subsidence would increase slightly if the mantle stretched more than the crust. Indeed, the effects would essentially cancel the decreases in initial subsidence (due to stretching alone, not due to melt generation) calculated for increased mantle stretching. For twice as much stretching in the mantle as in the crust, these increases are as much as 0.6 km with $\delta = 2$ and $T_p =$ 1300 °C, but decrease with increasing mantle temperatures and increased stretching (Fig. 6). Thus, the long-term subsidence of stretched continental margins is predicted to have been substantially lower than at the present day, even if there was substantially more stretching of the mantle than the crust.

A gauge of the importance of this effect is provided by some numbers. For a $T_p$ of 1300 °C, approximating modern conditions, the thermal subsidence predicted with uniform stretching of the plate by a factor of 4 is 3.43 km (Fig. 6). At 200 °C higher values for $T_p$, the thermal subsidence is predicted to have been only 1.62 km, rising to 1.65 km if the mantle stretched twice as much as the crust under these hotter conditions. This thermal subsidence is predicted to have been imposed on an initial subsidence of 2.40 km, compared with modern initial subsidence of 2.16 km (Fig. 2b). With twice the stretching in the mantle in the past, this initial subsidence declines to 2.32 km (Fig. 4b), and if the initial crustal thickness was 35 km instead of 40 km, it declines to 1.57 km (Fig. 5). Thus, the overall effect of the hotter mantle is potentially to reduce the accommodation space due to subsidence from 5.59 km (3.43 + 2.16) to 4.02 km (1.62 + 2.40), or 3.97 km (1.65 + 2.32) if there was twice as much stretching in the mantle, and even less with thinner initial crust. Accommodation space is reduced by more than 40%, and the proportional reduction in accommodation space is even higher for lower stretching values (Fig. 7). The space available for the development of passive-margin sedimentary sequences was therefore probably substantially smaller than it is today.

## DISCUSSION AND CONCLUSIONS

The calculations presented in this paper show that it would not be at all surprising if the Archean geological record

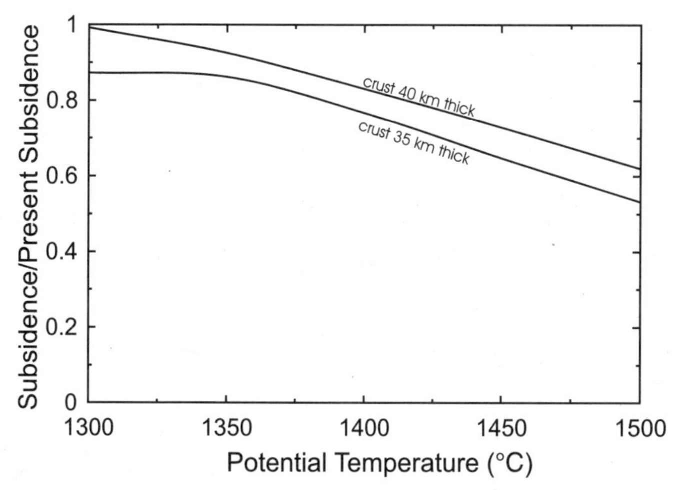


Figure 7. Proportional reduction in accommodation space with increased $T_p$, compared with uniform stretching under modern conditions. Curves are for $\beta = 2$. Modern crust is taken to be 40 km thick. The lower curve shows the effect of reducing continental crustal thicknesses by 5 km.

contained a relative paucity of well-preserved, thick, passive-margin successions, even if Archean geological history had been characterized by the stretching and rifting of stable continental plates. Furthermore, the stretching and ultimate rifting apart of continental plates should have been accompanied by voluminous mantle-derived magma. If this magma reached the surface, it would have been erupted as mafic to ultramafic submarine to shallow-water lava flows, onto continental basement. The frequency of this association has already been identified as an important characteristic of Archean and Early Proterozoic terrains (Arndt, 1999). Another noteworthy feature of the modeling addressed here is that thicknesses of melt produced due to thinning of the continental plates are of the same order as the thinning of the continental crust due to stretching. The net thickness of the continental crust therefore may have changed little during passive-margin formation. Following cooling, such crust may have been less susceptible than that of modern passive margins to significant entrainment in subsequent orogeny. This may have contributed to the widespread preservation of low-grade supracrustal successions in Archean terrains.

My calculations of the response of the continental lithosphere to stretching at passive margins are based on geotherms that place the base of plates in the Archean at depths much shallower than the field of diamond stability (Fig. 1). With geotherms such as this, regions of the mantle within the diamond stability field would be entrained in convective flow, and it would be difficult to preserve diamonds of Archean age beneath Archean cratons, contrary to what is observed (e.g., Richardson et al., 1984; Shirey et al., 2004). The preservation of Archean diamonds suggests that at least locally the geotherms beneath Archean continents must have been similar to those of today, and under these circumstances well-developed passive-margin sequences might be expected. In this context, it is noteworthy that the Kaapvaal craton has a well-developed 3.0–2.9 Ga passive-margin succession (de Wit et al., 1992). The production of thick passive-margin sedimentary successions in the Archean may, therefore, be associated largely with cratonization of the continental regions through the development of deep mantle roots. In contrast, with secular cooling of Earth over time, such margins could develop even in the absence of such cratonic roots.

## ACKNOWLEDGMENTS

This work was stimulated particularly by discussions with Tom Skulski. Discussions with him and Herb Helmstaedt were particularly helpful in this work.

## REFERENCES CITED

Armstrong, R.L., 1991, The persistent myth of crustal growth: Australian Journal of Earth Sciences, v. 38, p. 613–630, doi: 10.1080/08120099108727995.

Arndt, N., 1999, Why was flood volcanism on submerged platforms so common in the Precambrian?: Precambrian Research, v. 97, p. 155–164, doi: 10.1016/S0301-9268(99)00030-3.

Bédard, J.H., Brouillette, P., Madore, L., and Berclaz, A., 2003, Archean cratonization and deformation in the northern Superior province, Canada: An evaluation of plate tectonic versus vertical tectonic models: Precambrian Research, v. 127, p. 61–87, doi: 10.1016/S0301-9268(03)00181-5.

Bickle, M.J., Nisbet, E.G., and Martin, A., 1994, Archean greenstone belts are not oceanic crust: Journal of Geology, v. 102, p. 121–138.

Bleeker, W., 2002, Archean tectonics: A review with illustrations from the Slave craton: Geological Society [London] Special Publication 199, p. 151–181.

Calvert, A.J., Sawyer, E.W., Davis, W.J., and Ludden, J.N., 1995, Archaean subduction inferred from seismic images of a mantle suture in the Superior province: Nature, v. 375, p. 670–674, doi: 10.1038/375670a0.

Chardon, D., Choukroune, P., and Jayanand, M., 1996, Strain patterns, décollement and incipient sagducted greenstone terrains in the Archaean Dharwar craton (South India): Journal of Structural Geology, v. 18, p. 991–1004, doi: 10.1016/0191-8141(96)00031-4.

Chardon, D., Peucat, J.-J., Jayananda, M., Choukroune, P., and Fanning, C.M., 2002, Archean granite-greenstone tectonics at Kolar (South India): Interplay of diapirism and bulk inhomogeneous contraction during juvenile magmatic accretion: Tectonics, v. 21, 1016, doi: 10.1029/2001TC901032.

Collerson, K.D., and Kamber, B.S., 1999, Evolution of the continents and the atmosphere inferred from Th-U-Nb systematics of the depleted mantle: Science, v. 283, p. 1519–1523, doi: 10.1126/science.283.5407.1519.

Collins, W.J., Van Kranendonk, M.J., and Teyssier, C., 1998, Partial convective overturn of Archaean crust in the east Pilbara Craton, Western Australia: Driving mechanisms and tectonic implications: Journal of Structural Geology, v. 20, p. 1405–1424, doi: 10.1016/S0191-8141(98)00073-X.

de Wit, M.J., 1991, Archaean greenstone belt tectonism and basin development: Some insights from the Barberton and Pietersburg greenstone belts, Kaapvaal Craton, South Africa: Journal of African Earth Sciences, v. 13, p. 45–63, doi: 10.1016/0899-5362(91)90043-X.

de Wit, M.J., 1998, On Archean granites, greenstones, cratons and tectonics: Does the evidence demand a verdict?: Precambrian Research, v. 91, p. 181–226, doi: 10.1016/S0301-9268(98)00043-6.

de Wit, M.J., Roering, C., Hart, R.J., Armstrong, R.A., de Ronde, C.E.J., Green, R.W.E., Tredoux, M., and Hart, R.A., 1992, Formation of an Archaean continent: Nature, v. 357, p. 553–562, doi: 10.1038/357553a0.

Fyfe, W.S., 1978, The evolution of the Earth's crust: Modern plate tectonics to ancient hot spot tectonics?: Chemical Geology, v. 23, p. 89–114, doi: 10.1016/0009-2541(78)90068-2.

Grove, T.L., and Parman, S.W., 2004, Thermal evolution of the Earth as recorded by komatiites: Earth and Planetary Science Letters, v. 219, p. 173–187, doi: 10.1016/S0012-821X(04)00002-0.

Hamilton, W.B., 1998, Archean magmatism and deformation were not products of plate tectonics: Precambrian Research, v. 91, p. 143–179, doi: 10.1016/S0301-9268(98)00042-4.

Hynes, A., 2001, Freeboard revisited: Continental growth, crustal thickness change and Earth's thermal efficiency: Earth and Planetary Science Letters, v. 185, p. 161–172, doi: 10.1016/S0012-821X(00)00368-X.

Karato, S.-I., and Wu, P., 1993, Rheology of the upper mantle: A synthesis: Science, v. 260, p. 771–778, doi: 10.1126/science.260.5109.771.

Korenaga, J., 2003, Energetics of mantle convection and the fate of fossil heat: Geophysical Research Letters, v. 30, p. 1437, doi: 10.1029/2003GL016982.

Korenaga, J., 2006, Archean geodynamics and the thermal evolution of Earth: American Geophysical Union Geophysical Monograph 164, p. 7–32.

Korenaga, J., and Jordan, T.H., 2002, Onset of convection with temperature- and depth-dependent viscosity: Geophysical Research Letters, v. 29, p. 1923, doi: 10.1029/2002GL015672.

Lin, S., 2005, Synchronous vertical and horizontal tectonism in the Neoarchean: Kinematic evidence from a synclinal keel in the northwestern Superior craton, Canada: Precambrian Research, v. 139, p. 181–194, doi: 10.1016/j.precamres.2005.07.001.

Liu, L.-G., 2002, Critique of stability limits of the UHPM index minerals diamond and coesite: International Geology Review, v. 44, p. 770–778.

McKenzie, D.P., 1978, Some remarks on the development of sedimentary basins: Earth and Planetary Science Letters, v. 40, p. 25–32, doi: 10.1016/0012-821X(78)90071-7.

McKenzie, D.P., and Bickle, M.J., 1988, The volume and composition of melt generated by extension of the lithosphere: Journal of Petrology, v. 29, p. 625–679.

Nisbet, E.G., Cheadle, M.J., Arndt, N.T., and Bickle, M.J., 1993, Constraining the potential temperature of the Archaean mantle: A review of the evidence from komatiites: Lithos, v. 30, p. 291–307, doi: 10.1016/0024-4937(93)90042-B.

Parman, S., Dann, J., Grove, T.L., and de Wit, M.J., 1997, Emplacement conditions of komatiite magmas from the 3.49 Ga Komati Formation, Barberton Greenstone Belt, South Africa: Earth and Planetary Science Letters, v. 150, p. 303–323, doi: 10.1016/S0012-821X(97)00104-0.

Parsons, B., and McKenzie, D., 1978, Mantle convection and the thermal structure of the plates: Journal of Geophysical Research, v. 83, p. 4485–4496.

Percival, J.A., McNicoll, V., Brown, J.L., and Whalen, J.B., 2004, Convergent margin tectonics, central Wabigoon subprovince, Superior province, Canada: Precambrian Research, v. 132, p. 213–244, doi: 10.1016/j.precamres.2003.12.016.

Reymer, A.P.S., and Schubert, G., 1987, Phanerozoic and Precambrian crustal growth, *in* Kröner, A., ed., Proterozoic lithospheric evolution, volume 130: International Lithosphere Program: Washington, D.C., American Geophysical Union, p. 1–8.

Richardson, S.H., Gurney, J.J., Erlank, A.J., and Harris, J.W., 1984, Origin of diamonds in old enriched mantle: Nature, v. 310, p. 198–202, doi: 10.1038/310198a0.

Royden, L., and Keen, C.E., 1980, Rifting process and thermal evolution of the continental margin of eastern Canada determined from subsidence curves: Earth and Planetary Science Letters, v. 51, p. 343–361, doi: 10.1016/0012-821X(80)90216-2.

Shirey, S.B., Richardson, S.H., and Harris, J.W., 2004, Integrated models of diamond formation and craton evolution: Lithos, v. 77, p. 923–944, doi: 10.1016/j.lithos.2004.04.018.

Stern, R.J., 2005, Evidence from ophiolites, blueschists, and ultrahigh-pressure metamorphic terranes that the modern episode of subduction tectonics began in Neoproterozoic time: Geology, v. 33, p. 557–560, doi: 10.1130/G21365.1.

van Hunen, J., van Keken, P.E., Hynes, A., and Davies, G.F., 2008, this volume, Tectonics of early Earth: Some geodynamic considerations, *in* Condie, K.C., and Pease, V., eds., When did plate tectonics begin on planet Earth?: Geological Society of America Special Paper 440, doi: 10.1130/2008.2440(08).

Van Kranendonk, M.J., Collins, W.J., Hickman, A., and Pawley, M.J., 2004, Critical tests of vertical vs horizontal tectonic models for the Archean East Pilbara granite-greenstone terrane, Pilbara craton, Western Australia: Precambrian Research, v. 131, p. 173–211, doi: 10.1016/j.precamres.2003.12.015.

Wasserburg, G.J., MacDonald, G.J.F., Hoyle, F., and Fowler, W.A., 1964, Relative contributions of uranium, thorium and potassium to heat production in the Earth: Science, v. 143, p. 465–467, doi: 10.1126/science.143.3605.465.

Manuscript Accepted by the Society 14 August 2007

Printed in the USA

The Geological Society of America
Special Paper 440
2008

# *Tectonics of early Earth: Some geodynamic considerations*

**Jeroen van Hunen***
*Durham University, Department of Earth Sciences, Science Site, Durham DH1 3LE, UK*

**Peter E. van Keken**
*University of Michigan, Department of Geological Sciences, 2534 CC Little Building, Ann Arbor, Michigan 48109-1005, USA*

**Andrew Hynes**
*McGill University, Department of Earth and Planetary Sciences, 3450 University Street, Montreal, Quebec H3A 2A7, Canada*

**Geoffrey F. Davies**
*Australian National University, Research School of Earth Sciences, Canberra ACT 0200, Australia*

## ABSTRACT

**Today, plate tectonics is the dominant tectonic style on Earth, but in a hotter Earth tectonics may have looked different due to the presence of more melting and associated compositional buoyancy as well as the presence of a weaker mantle and lithosphere. Here we review the geodynamic constraints on plate tectonics and proposed alternatives throughout Earth's history. Observations suggest a 100–300 °C mantle potential temperature decrease since the Archean. The use of this range by theoretical studies, parameterized convection studies, and numerical simulations puts a number of constraints on the viability of the different tectonic styles. The ability to sufficiently cool early Earth with its high radiogenic heat production forms one of the major constraints on the success of any type of tectonics. The viability of plate tectonics is mainly limited by the availability of sufficient driving forces and lithospheric strength. Proposed alternative mechanisms include local or global magma oceans, diapirism, independent dynamics of crust and underlying mantle, and large-scale mantle overturns. Transformation of basaltic crust into dense eclogite is an important driving mechanism, regardless of the governing tectonic style.**

**Keywords:** plate tectonics, Archean, geodynamics, mantle temperature.

## 1. INTRODUCTION

Plate tectonics provides a well-established framework for the present-day dynamics of solid Earth, in which semirigid plates are driven primarily by the negative buoyancy of cold and dense oceanic lithosphere (Forsyth and Uyeda, 1975). This buoyancy force, which overcomes the viscous resistance of the underlying mantle, is commonly termed "slab pull." It is unclear, however, how long the present-day form of plate tectonics has been operating and whether Earth in earlier times had a modified or dramatically different method for cooling its interior. At present, there is no oceanic lithosphere older than 200 Ma, and observational evidence supporting earlier plate tectonics needs to be found on the continents. Geologic remnants of oceanic lithosphere in the form of ophiolites mostly date back to ca. 800–900 Ma. A few have been reported as far

*jeroen.van-hunen@durham.ac.uk

van Hunen, J., van Keken, P.E., Hynes, A., and Davies, G.F., 2008, Tectonics of early Earth: Some geodynamic considerations, *in* Condie, K.C., and Pease, V., eds., When Did Plate Tectonics Begin on Planet Earth?: Geological Society of America Special Paper 440, p. 157–171, doi: 10.1130/2008.2440(08). For permission to copy, contact editing@geosociety.org. 

back as 3.8 Ga (Kontinen, 1987; Scott et al., 1992; Furnes et al., 2007), but reports of these older ophiolites remain controversial (Kusky and Li, 2002; Zhai et al., 2002). A comparison with Venus (a dry planet on which plate tectonics is absent) suggests that the presence of liquid water may play an important role in the decoupling and cooling of plates. The onset of plate tectonics has been suggested to have been anywhere between as recent as 1.0–2.0 Ga (marked by the earliest observations of ultrahigh-pressure terranes, blueschists, and ophiolites; Hamilton, 1998; Stern, 2005; Condie et al., 2006; Brown, this volume) and >4.0 Ga (shortly after freezing of an early magma ocean; e.g., de Wit, 1998). The earliest estimates for the start of plate tectonics are based on the evidence from zircons for the presence of liquid water and generation of continental crust early in Earth's history (Peck et al., 2001; Harrison et al., 2005). If plate tectonics was not a viable mechanism for Hadean and Archean tectonics, then other mechanisms were required to ensure sufficient surface heat flow of Earth through time.

The main objective of this paper is to summarize and discuss different geodynamic arguments for and against plate tectonics in the Archean. Some of the presented models and constraints are new, but many are well established in the literature. By compiling them we hope to give a clarifying overview.

This paper has the following organization. We first discuss the thermal constraints that apply to any tectonic regime, such as its ability to provide a sufficient cooling agent. Then we discuss the circumstances under which the present-day style of subduction and therefore plate tectonics is possible, and we give a description of the various proposed alternative models and their geodynamic viability. We end with suggestions for research areas in which improved data or techniques would be most crucial for answering the question "When did plate tectonics start?"

## 2. THERMAL CONSTRAINTS

The evolution of the temperature of Earth's (upper) mantle is a fundamental constraint that needs to be taken into account in our discussion of the viability of different tectonic regimes in the past. It is essential to evaluate whether a particular tectonic regime was able to operate under the supposed thermal regime, and, conversely, whether this tectonic regime could explain the observations of secular changes in mantle temperature. Since mantle temperature influences the range of viable tectonic mechanisms but is in itself controlled by those mechanisms, the problem is nonlinear, and more than one solution may be possible, even if we could perfectly understand the relationship between tectonic regime and thermal evolution. However, by bracketing the mantle temperature through time, we should eventually be able to solve this puzzle.

### 2.1. Observations of Mantle Temperature in the Archean

Direct measurement of mantle temperature is not possible, but several indirect methods have been developed using the compositions of igneous rocks to infer mantle potential temperatures at their time of formation.

Pressure-temperature data from Archean high-grade terrains yield bounds on continental heat fluxes that suggest an Archean continental geotherm similar to the present-day one (England and Bickle, 1984). However, this may reflect relative constancy of continental lithosphere thickness rather than the temperature of the underlying mantle (Davies, 1979). The oceanic environment appears to have been hotter in the Archean. Inferred liquidus temperatures of MORBs from ophiolites and greenstone belts from the past 3.7 b.y. record a gradual 200 °C drop in the extrusion temperature of basalts over this period (Fig. 1A). This data set explicitly excluded komatiites under the assumptions that these are plume-derived and therefore not representative of the average upper mantle temperature. Liquidus temperatures can be derived from the MgO content of lavas due to the well-established Mg-Fe partitioning upon melting (e.g., Langmuir et al., 1992). Inclusion of high-MgO komatiitic data also indicates a strong cooling trend (Fig. 1B), with possibly a stepwise decrease at the end of the Archean, potentially suggesting a change in convective mechanism.

It is convenient to discuss the upper-mantle temperature in terms of its potential temperature, which is the ambient temperature extrapolated to the surface along the adiabat. Extrusion temperatures are always lower than the potential temperature of the mantle from which the melt is formed due to the consumption of latent heat upon melting, but one can derive the mantle potential temperature from the extrusion temperatures using models of decompression melting (e.g., McKenzie and Bickle, 1988). It is common to assume that basalts are generated by pressure-release melting, and since melting starts deeper in a hotter Earth, the amount of latent heat consumption and percentage of partial melting increases. An extrapolation of the (McKenzie and Bickle, 1988) melt formalism to higher potential temperature (e.g., Vlaar et al., 1994) shows that the extrusion temperature is linearly related to the potential temperature and that a 200 °C drop in extrusion temperature since the Archean corresponds to a 300 °C drop in mantle potential temperature (Fig. 2A). This modeling approach provides some insights into the potential changes in dynamic regime: Since the melting starts deeper, the amount of melt generated is larger, leading to a thicker crust as well as a thicker depleted harzburgite layer (Fig. 2B). This leads to a compositionally stable stratification that is increasingly resistant to thermal cooling (Fig. 2C). We shall return to the consequences of this deeper mantle melting in a hotter Earth in section 3.

A high-end estimate of mantle potential temperature can be made by inclusion of the komatiitic data. Nisbet et al. (1993) derived an upper limit of Archean mantle potential temperature of 1800 °C, which, under the assumption that mantle plumes are at least 200 °C hotter than ambient mantle, would suggest an average mantle potential temperature of perhaps 1600 °C. However, Parman et al. (1997) suggest that the high MgO contents of those komatiites are due to alteration during greenschist metamorphism and should be treated with care. Grove and Parman

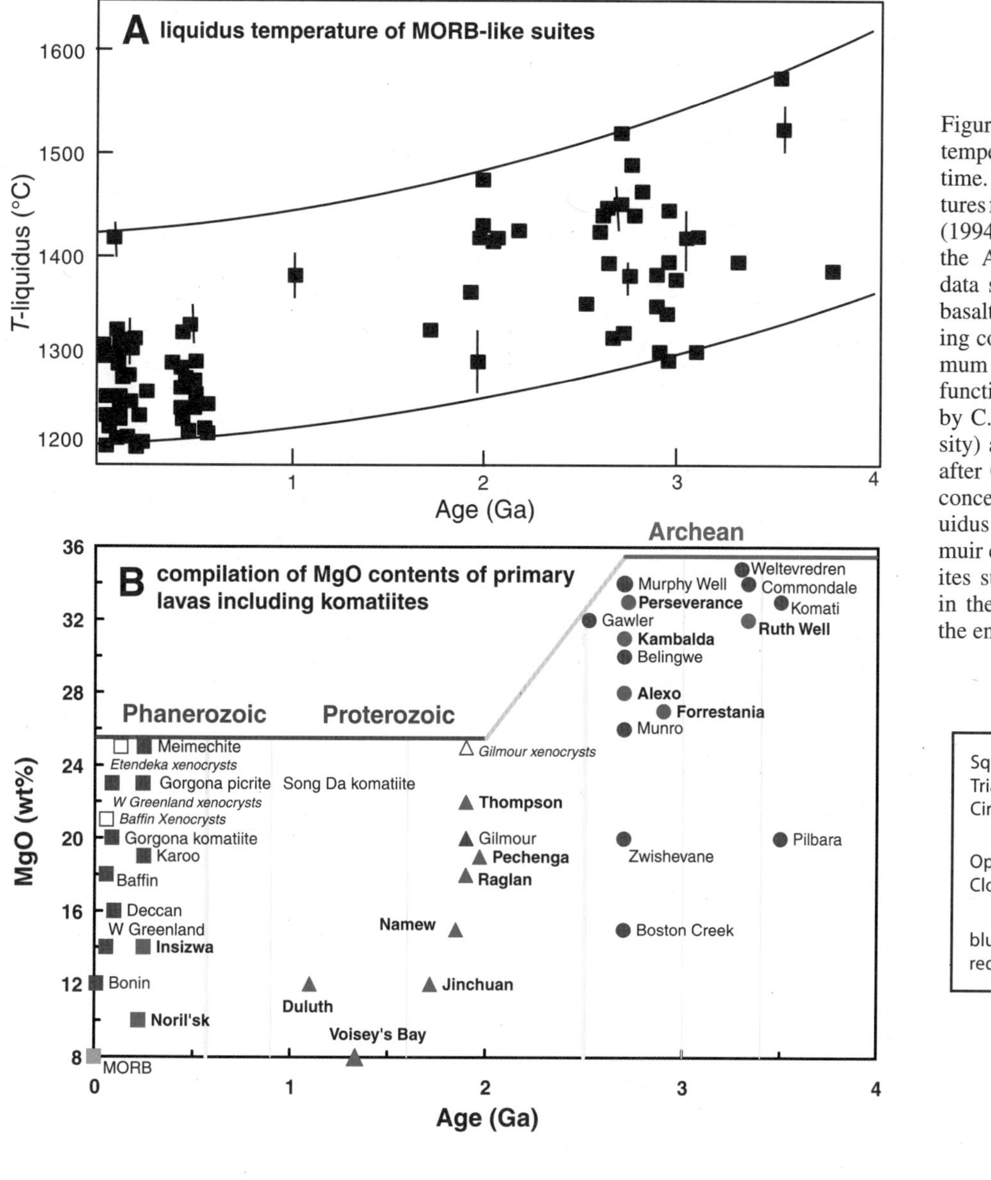


Figure 1. Constraints on mantle potential temperature from basaltic lavas through time. (A) Compilation of liquidus temperatures for MORB-like suites from Abbott et al. (1994a) (reproduced with permission from the American Geophysical Union). This data set excludes komatiitic lavas and arc basalts, and suggests a gradual and continuing cooling of the upper mantle. (B) Maximum MgO content of basaltic melts as a function of extrusion age. Figure provided by C. Michael Lesher (Laurentian University) and Nick Arndt (Grenoble), modified after Campbell and Griffiths (1992). MgO concentrations are directly linked to the liquidus temperature of the lavas (e.g., Langmuir et al., 1992). The inclusion of komatiites suggests a possible stepwise decrease in the temperature of the upper mantle at the end of the Archean.

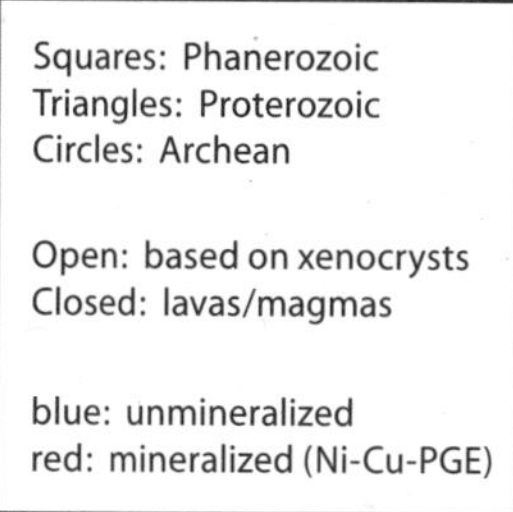


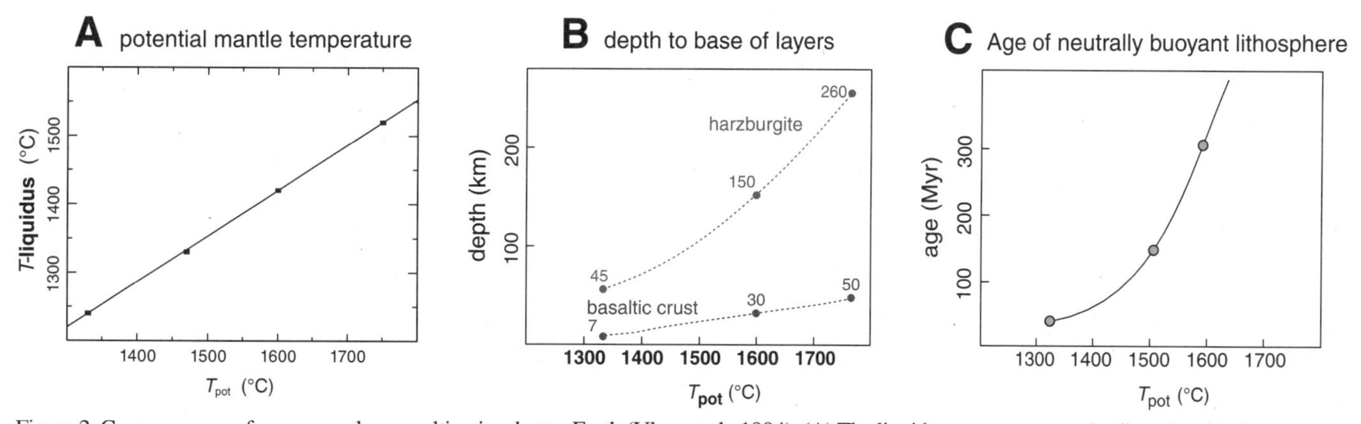


Figure 2. Consequences of pressure-release melting in a hotter Earth (Vlaar et al., 1994). (A) The liquidus temperature scales linearly with the potential temperature. The 200 °C decrease in liquidus temperature over 4 b.y. suggested in Fig. 1A corresponds to a 300 °C increase in mantle potential temperature. (B) Decompression melting starts deeper in a hotter Earth, causing a thicker layering of basaltic crust and depleted harzburgite. (C) The increased compositional buoyancy causes the lithosphere to remain stable against thermal cooling for significantly longer periods of time in a hotter Earth.

(2004) argue that komatiites could also have been formed in a wet rather than a hot environment by hydrous melting at shallow depths in a subduction environment, in which case the Archean mantle only needed to be ~100 °C hotter than at present. A recent overview for the various constraints on the thermal evolution of the mantle is given in Jaupart et al. (2007).

For the period prior to 4 Ga, very few observations are available, with virtually no constraints on mantle potential temperature. This implies that anything between rapid cooling and even warming up of the Hadean mantle cannot be ruled out on the basis of observations. Based on planetary evidence, however, it is most likely that Earth got a hot start, likely with a magma ocean, due to its rapid and highly energetic accretion.

## 2.2. A Simple Heat Balance

Independent estimates of the evolution of Earth's mantle temperature can be obtained by theoretical considerations using the first law of thermodynamics. Changes in mantle temperature are caused by the imbalance of heat conducted into and out of the mantle and any internally generated heat. Earth loses heat to space through its surface. The main internal heat sources are primordial heat from early accretion and differentiation, and radioactive heating through time from $^{238}U$, $^{235}U$, $^{232}Th$, and $^{40}K$. Some proportion of the primordial heat and heat from past radiogenic heating is still stored in Earth in the form of thermal energy. The gradual secular cooling of the mantle can be seen as a present-day heat source, so that the present-day energy balance is of the form "heat out at surface = heat in from core + present-day radiogenic heating + present-day secular cooling of the mantle."

A simple, first-order heat balance gives a good idea of the effects of heating and cooling on the mantle temperature through time (Sleep, 2000). Today, the globally averaged surface heat flow $q_s$ is ~80 mW/m$^2$ (Pollack et al., 1993). For the convective heat loss it is perhaps more relevant to use the oceanic average (100 mW/m$^2$). Approximately half of the present-day heat loss (44 TW) is estimated to come from cooling, the other half from the present-day rate of radioactive heating of ~20 TW (McKenzie and Richter, 1981; Pollack et al., 1993; McDonough and Sun, 1995; Jaupart et al., 2007). Radioactive heating was about three to four times higher in the Archean and Hadean (e.g., Turcotte and Schubert, 2002). If we now make the crude but convenient assumption that Earth's surface heat flow was always roughly what it is today (because, for example, plate tectonics always operated in the same way and with the same vigor), we must conclude that in the past more heat was produced by radioactivity than escaped through the surface, and Earth was heating up instead of cooling down. We could, instead, make another crude assumption that the cooling rate (i.e., the part of the surface heat flow that results in secular cooling) was always 40 mW/m$^2$ (but the total $q_s$ was larger due to the larger radioactive component), which is equivalent to a constant (average) cooling rate for Earth of ~100 °C/b.y. (Sleep, 2000).

Although the observations (discussed in the former section) seem to favor the latter case of constant cooling (although not necessarily as fast as 100 °C/b.y.), the scenario in the former case (of an early Earth that warmed up instead of cooled down) cannot be excluded. Obviously, to proceed we would like to avoid these crude assumptions and provide theoretical arguments related to the efficiency of convection.

## 2.3. Relation Between Convection and Thermal Evolution of the Mantle

Tectonic deformation can be regarded as the surface expression of convection cells with a rather complicated rheology. The vigor of mantle convection strongly depends on the mantle viscosity, which in turn is largely dependent on mantle temperature. If we only take the temperature dependence of viscosity into account, it seems reasonable to assume that in an early, hotter Earth, mantle convection was more vigorous, surface heat flow greater, and therefore cooling more efficient. In the presence of a dynamic lithosphere (such as with plate tectonics), the story is likely to be more complicated.

### *2.3.1. Parameterized Convection*

For simple, isoviscous convection, the relationship between total surface heat flow and vigor of convection is described by

$$Nu = cRa^{\beta}, \tag{1}$$

with *Nu* the Nusselt number, a nondimensional measure of the surface heat flux, and *Ra* the Rayleigh number, a measure of convective vigor. The nondimensional parameters $c$ and $\beta$ can be determined from boundary layer theory or fluid dynamical experiments. The use of this simple relationship as a proxy for full convection calculations is usually referred to as "parameterized convection." For isoviscous flow, $\beta \approx 1/3$ (e.g., Turcotte and Schubert, 2002). For temperature-dependent rheology, $\beta$ can drop below 0.1 (Christensen, 1985; van Keken et al., 2001). However, Gurnis (1989) argued that such low values apply only within the transition from the mobile thermal boundary layer to the so-called "stagnant-lid regime," in which the cool boundary layer is so stiff that much of it does not take part in the convection (Davaille and Jaupart, 1993; Solomatov, 1995). He further showed that a plate tectonics type of convection, featuring stiff but mobile plates, would again lead to $\beta \approx 1/3$. The correct value of $\beta$ for Earth's mantle is still under debate, and most present estimates range from 0 to ~1/3 (see an overview in McNamara and van Keken, 2000), but even negative values have been proposed (Korenaga, 2003, 2006). $\beta$ can be regarded as a parameter that describes how much the surface heat flow is dependent on mantle temperature: A smaller $\beta$ leads to a smaller dependence, and for $\beta = 0$, surface heat flow is totally independent of mantle temperature. Once the heat flux versus vigor of convection relationship is decided, the thermal evolution of the mantle can be calculated with a simple balance (Davies, 1999):

$$C\frac{dT}{dt} = H + \frac{Q_{in} - Q_{out}}{M}, \qquad (2)$$

or in words, the net rate of mantle heat loss (cooling rate $dT/dt$ times average mantle heat capacity $C$) equals heat input (by radioactive heating $H$ and core influx $Q_{in}$) less heat output (from surface heat $Q_{out}$) for a mantle mass $M$.

Although elegant, the simple *Nu-Ra* relationship can yield problematic results. One possible implication of the parameterization is what has been termed "thermal catastrophe": Starting from today's cooling rate, a higher mantle temperature in earlier times gives higher heat flux, a faster temperature drop, and therefore an even larger mantle temperature to begin with. Using this parameterization to extrapolate mantle temperature back in time could lead to unacceptably high temperatures in the Early Archean for certain parameters. Of course, such potential catastrophes may be avoided by integrating forward in time, but the practice of reversing the arrow of time and using the present-day mantle temperature as an initial condition rather than something to shoot for is remarkably prevalent in the literature. Davies (1980) showed that the thermal catastrophe can be avoided if the present-day heat generation rate is more than ~65% of the heat loss rate, a value that is contradicted by geochemical models.

The problem with the *Nu-Ra* relationship, however, is more fundamental: It implies that only three parameters ($Ra$, $\beta$, and $c$) govern the cooling history of the mantle. In reality, even for today's plate tectonics situation, the relationship between plate age and speed is not easily described in terms of physical parameters (Jarrard, 1986; Lallemand et al., 2005; Labrosse and Jaupart, 2007). This indicates that also the resultant mantle cooling should be described by more complicated relationships. Korenaga (2006) suggests a plate tectonic cooling parameterization with a more complicated relationship between Ra and Nu: A hotter mantle gives more melting and related dehydration stiffening of the upper mantle. This, in turn, leads to more sluggish rather than faster plate motions in the past, which leads to significantly less or no secular cooling during the early evolution, and consequently an almost constant potential temperature for Earth throughout its early history. He shows that such a parameterization fits the inferred mantle temperatures through time, though his parameterization depends sensitively on some parameters that are not very well constrained, especially the bending radius as a function of plate thickness.

Other shortcomings of the standard Nu-Ra parameterization are illustrated in (Grigné et al., 2005). Their numerical modeling results of mantle convection with incorporated large lithospheric plates show that (1) cooling of such a wide system is much slower than in square convection cells with an aspect ratio of one, and (2) the parameterized convection approach does not account for significant heat flow variations on an ~400 m.y. timescale (the duration of a Wilson cycle). Labrosse and Jaupart (2007) provide a thermal history based on (1) the observation that, at present, oceanic lithosphere of all ages can subduct with equal probability, and (2) the assumption of constant maximum seafloor age throughout Earth's history. They suggest a modest mantle temperature decrease of only ~150 °C over the last 3 b.y. Their model, however, requires a significant extrapolation from the 200 m.y. of observations on oceanic lithosphere to the 3000 m.y. of model span.

#### *2.3.2. Layered and Whole-Mantle Convection*

Geochemical differences between mid-oceanic-ridge basalts (MORBs) and ocean-island basalts (OIBs) favor a layered mantle and associated layered mantle convection throughout large periods of Earth's history as a possibility to maintain separate large-scale geochemical reservoirs (e.g., Hofmann, 1997; van Keken et al., 2002). Seismic tomography of the mantle (e.g., van der Hilst, 1995; Bijwaard et al., 1998; Grand, 2002) indeed show slabs that linger at 660 km depth. At this depth, the endothermic phase transition from spinel to perovskite forms a barrier for descending cold material and ascending hot material, and therefore forms a good candidate for layered convection (e.g., Christensen and Yuen, 1985). Seismic tomography also convincingly shows, however, that subducting slabs penetrate into the lower mantle, and the (at least local) mass exchange between upper and lower mantle makes present-day completely layered mantle convection unlikely. McKenzie and Richter (1981) were the first to discuss the influence of layered mantle on the thermal history of Earth. On the basis of parameterized convection, Davies (1995) showed that mantle convection may well have been layered in the early history, and later on evolved into whole-mantle convection through more and more frequent overturns of the mantle. This would suggest a rather complicated pulsating temperature variation in the upper mantle, and might in turn also have influenced the viable range of tectonic mechanisms. Davies (1995) links these mantle overturns to the observed peaks in crust formation (McCulloch and Bennett, 1994), although it should be noted that a pure plate tectonic origin has also recently been suggested for these peaks (O'Neill et al., 2007).

Complete present-day mantle layering at 660 km has by now convincingly been ruled out from tomography results (e.g., van der Hilst, 1995), but deeper, chemical layering has been proposed as well (e.g., Davaille, 1999; Kellogg et al., 1999). Following Spohn and Schubert (1982), McNamara and van Keken (2000) applied the simple Nu-Ra relationship to layered mantle convection. They compared their secular cooling results for layered and whole-mantle convection, and concluded that it is very difficult or perhaps impossible to create a bottom layer that is below its solidus, while simultaneously having a top layer that is not too cool, and they suggest that whole-mantle convection was more likely to have dominated throughout Earth's history. This suggests that deep chemical layering such as proposed by Kellogg et al. (1999) may not have existed throughout Earth's history, but some form of material exchange should have been present at times, possibly in the form of a "doming" regime (Davaille, 1999).

#### *2.3.3. The Influence of a Tectonic Regime on the Thermal History*

An important test of proposed tectonic models is whether they provide sufficient cooling of Earth. Different tectonic mechanisms have different cooling efficiencies. In an Earth with no active tectonics, mantle cooling occurs purely by diffusion, which is far less efficient than by advection of heat in a convective system. Van Thienen et al. (2005) show that, for the hotter early Earth scenario, plate tectonics can remove the heat from a steadily cooling Earth, provided that plate tectonics operated at least at a present-day rate or slightly more slowly. Their model calculations suggest that in a hotter Earth, plate tectonics was a more efficient cooling mechanism, because plates were thinner. They also report that a mechanism of simple extrusion of basalt onto the surface would require about one to two orders of magnitude more eruption than the rate of Phanerozoic flood basalt eruption to provide a similar cooling.

The fact that these tectonic mechanisms could provide sufficient cooling doesn't necessarily mean they did. It still has to be shown that these mechanisms were able to operate (and fast enough) in a hotter Earth. It is possible (although not indicated by mantle temperature estimates) that no tectonic mechanism existed for some period of time in the past that could achieve cooling of Earth at the same rate as radiogenic heating. Earth would have warmed through time throughout this period and would not have cooled (as is assumed to be the case at present). The viability of several of the proposed tectonic regimes in a hotter Earth is the topic of the next sections.

## 3. IS THE PRESENT-DAY STYLE OF SUBDUCTION POSSIBLE IN EARLY EARTH?

The main driving mechanism of plate tectonics is found at subduction zones. Here, the relatively high density of the subducting slab in comparison with the ambient mantle material causes the lithosphere to founder and pull the surface plate in its wake. This slab pull is widely accepted to be the dominant driving force for plate tectonics (e.g., Forsyth and Uyeda, 1975; Conrad and Lithgow-Bertelloni, 2002). The viability of present-day plate tectonics therefore hinges primarily on the presence of subduction zones. However, a few more constraints apply: Mid-ocean ridges must not be too strong to resist spreading (Sleep, 2000), and lithospheric plates must be strong enough to resist internal deformation, while weak enough to bend into the subduction zone (Conrad and Hager, 1999). In addition, their edges must be weak enough to decouple neighboring plates from each other.

### 3.1. Buoyancy Arguments

A major challenge for the viability of subduction, and therefore for plate tectonics, in an earlier, hotter Earth is the sustainability of the slab pull driving force. Decompression melting in a hotter mantle results in more melting and therefore a thicker crust (which in a plate tectonic regime would be oceanic crust) and underlying harzburgitic mantle (Fig. 2B) (Vlaar and van den Berg, 1991; Davies, 1992; van Thienen et al., 2004b). The combined effects upon subduction are likely to be similar to Phanerozoic oceanic large igneous provinces such as Ontong Java, which are unsubductable. The Ontong Java Plateau has indeed caused a reversal of the orientation of subduction in the Solomon trench (Fitton and Godard, 2004). This is caused by the stronger compositional buoyancy, compared with normal oceanic crust. In general, the oceanic lithosphere has positive compositional buoyancy since (1) the crust itself is compositionally less dense than the peridotitic mantle material from which it originated and (2) the harzburgitic melting residue is also less dense than mantle material. For the present day, the compositional buoyancy of the ~7 km thick oceanic crust and underlying residual harzburgite is compensated by the negative thermal buoyancy (increased density due to cooling) after the lithosphere has cooled on average for 20–30 m.y. (Fig. 2C) (e.g., Vlaar et al., 1994; Turcotte and Schubert, 2002). The age at which this neutral buoyancy is reached increases dramatically with higher potential mantle temperature (Fig. 2C). It is interesting to note that the thermal state of oceanic lithosphere reaches a steady state at ca. 70 Ma (possibly due to convective erosion of the bottom of the lithosphere—usually referred to as "small-scale convection") (Parsons and McKenzie, 1978; Stein and Stein, 1992; Ritzwoller et al., 2004). In a hotter, weaker early Earth mantle, this equilibration could possibly have happened even earlier (Huang et al., 2003; van Hunen et al., 2005). It is therefore not clear that the oceanic lithosphere in a hotter Earth can reach neutral buoyancy, let alone obtain sufficient negative buoyancy to founder into the mantle. This may render subduction inoperative in a hotter Earth. It has been suggested that the crossover point from dense subducting plates (that can actually drive subduction and therefore plate tectonics) to buoyant ones (that resist subduction) may have occurred when Earth was only 50 °C hotter than today (Davies, 1992).

At least two mechanisms have been suggested to compensate for the buoyancy problem: eclogitization of the basaltic crust, and intense depletion of the early upper mantle.

The transformation of buoyant basalt into dense eclogite (e.g., Cloos, 1993) at ~40 km depth would make slabs still dense below this depth, and slab pull could still be viable, although it is not clear how one could initiate subduction to achieve the eclogite transition in the first place. Another complication in this might be that this transformation has been suggested to be kinetically slow (Ahrens and Schubert, 1975; Rubie, 1990; Hacker, 1996; Austrheim, 1998). Numerical models of shallow flat subduction of oceanic plateaus (see section 3.4) seem to suggest that basalt probably remains metastable for a geologically significant time (van Hunen et al., 2002b). On the other hand, it has been argued that this basalt metastability might not apply in a wet environment such as that of a subduction zone (Rubie, 1990). The effect of these phase change kinetics on hot, buoyant subduction was examined by van Hunen and van den Berg (2007), who concluded that kinetic delay times up to 1 m.y. may not significantly hamper subduction, but delay times as large as 5 m.y. do so. Note that this basalt-to-eclogite transition does not influence the difficulty of initiating subduction (see section 3.3), when no part of the plate/slab is yet in the eclogite stability field.

Davies (2006) has argued for the possibility of quick and intense depletion of the early upper mantle, because subducting crust, once in the eclogite assemblage, would tend to settle out of the upper mantle. Settling would be enhanced in the early mantle by the mantle's lower viscosity due to its higher temperature. Also, the present-day upper mantle is ~30 times less viscous than the lower mantle, and this may have been the case in the Archean as well. Thus both factors would enhance gravitational segregation of heavier eclogitic material from lighter depleted residue in the upper mantle, leaving it quite strongly depleted. This mantle depletion would lead to less melting, and therefore to a thinner crust and underlying depleted residue, similar to today's situation.

### 3.2. Lithospheric Rheology Arguments

In addition to buoyancy we should consider the rheology of slabs. If plates become weaker, they may bend more easily into the mantle at subduction zones and less bending energy may be required (Conrad and Hager, 1999), which favors subduction. On the other hand, if a slab is too weak to support its own weight, it may break off, in which case a significant proportion of the slab pull will be lost (van Hunen and van den Berg, 2007). The latter process can be compared with present-day slab detachment as suggested for the Alps (Davies and von Blanckenburg, 1995), the Mediterranean-Carpathian region (Wortel and Spakman, 2000), or Kamchatka (Levin et al., 2002). The main arguments for slabs being weaker in an early, hotter Earth are (1) slabs were hotter and, on average, probably younger at the trench, and therefore rheologically thinner, and (2) their crust was thicker, and basalt/eclogite is intrinsically weaker than cold parts of the mantle lithosphere beneath (Shelton and Tullis, 1981; Karato and Wu, 1993; Kohlstedt et al., 1995; Stöckhert and Renner, 1998). The efficiency of plate tectonics may therefore decrease significantly for a 200 °C higher ambient mantle temperature, mainly as a result of early and frequent slab breakoff and the inability to form long, subduction-enhancing slabs (van Hunen and van den Berg, 2007) (Fig. 3). However, depletion due to partial melting dehydrates the mantle and could make it up to two orders of magnitude stronger (Hirth and Kohlstedt, 1996). This would make the rheologically strong part of the mantle (the literal definition of "lithosphere") thicker and might not allow for any slab breakoff (van Hunen and van den Berg, 2007). Thicker lithosphere (in a rheological sense) could slow down plate tectonics, because more energy was necessary to bend the oceanic plate into the mantle. The net strength of the lithosphere is difficult to estimate a priori, because of the competing effects of temperature weakening and dehydration strengthening on the ambient viscosity. As a rule of thumb, mantle material weakens by an order of magnitude for every 100 °C temperature increase. Korenaga (2006) argues that if plates were indeed stronger in the past, this would also slow down plate tectonics, which partly removes the buoyancy arguments against plate tectonics in a hotter Earth, as discussed above. Van Hunen and van den Berg (2007) found that slab breakoff is indeed suppressed when dehydration strengthening is applied, but that subduction speed still increases with increasing mantle temperature.

Slab suction (the transmission of stress from the subducting slab to the plate at the surface through the mantle surrounding the lithosphere) has been identified as an important component of slab pull (Conrad and Lithgow-Bertelloni, 2002). In a hotter and thus weaker mantle, the contribution of slab suction to driving plate motion might have been substantially smaller.

Lawsonite blueschists and lawsonite eclogites are formed by high- to ultrahigh-pressure metamorphism (UHPM) and are recording cold subduction (Brown, this volume). A characteristic feature of the Precambrian is the absence of UHPM (e.g., Maruyama and Liou, 1998; Brown, this volume). This absence is sometimes regarded as an indication for either absence of plate tectonics (Stern, this volume) or a significant change in plate tectonics. In addition, temporal changes in *P-T* conditions in subduction could lead to different *P-T* paths in which no or different UHPM is formed (Maruyama and Liou, 1998). Biased preservation of Archean rocks may be another explanation (Möller et al., 1995). Yet another possibility might be related to the secular changes in the rheological structure of the subducting slab (van Hunen and van den Berg, 2007). During closure of oceanic basins, subducting oceanic lithosphere is followed by an attempt at continental subduction. At present, the high integrated strength of subducting lithosphere enables the dragging of buoyant continental lithosphere to some depth into the subduction zone. Breakoff then results in rebound of the buoyant continental rocks, and consequently brings UHPM rocks toward the surface. Weaker lithosphere in the past may have favored early detachment of the subducted oceanic slab from its continental counterpart, thereby preventing the development of the high-pressure rocks and the possibilities for substantial rebound.

### 3.3. Initiation of Subduction

Our present understanding of the initiation of subduction is rather poor, and this renders an extrapolation to different geodynamic conditions in the past difficult. Various mechanisms for subduction initiation have been suggested. Cloetingh et al. (1982) and Regenauer-Lieb et al. (2001) investigated sediment loading of passive margins to create subduction initiation. Hall et al. (2003) proposed conversion of a fracture zone into a self-sustaining subduction zone. Niu et al. (2003) use the principle of lateral compositional density variation within the lithosphere to create the required stress field to initiate subduction. Solomatov (2004) suggested convective thinning (or small-scale convective erosion) of the lithosphere, which creates a preferred location for subduction initiation. Whether these mechanisms could have been active in early Earth has to be evaluated for each mechanism individually. In a weaker mantle, small-scale convection probably keeps lithosphere thinner than today, which favors Solomatov's model for early Earth. On the other hand, stress transmission might have been more difficult when mantle and lithosphere were weaker, which makes localization of stress more difficult and could imply that Hall's mechanism applies more to the modern lithosphere.

### 3.4. Shallow Flat Subduction

Today, shallow flat subduction occurs in southern Peru and central Chile and has been at least partially associated with the presence of buoyant plateaus (caused by large igneous provinces, aseismic ridges, or seamount chains), which have thicker-than-average oceanic crust (Cross and Pilger, 1978; McGeary et al., 1985; Gutscher et al., 2000). Such shallow subduction needs either (1) thick crust ("buoyant plateau") or (2) an actively seaward-advancing overriding plate (van Hunen et al., 2002a, 2002b) or (3) a low-pressure mantle wedge above the subducting slab, which causes rise of the slab (Jischke, 1975; Stevenson and Turner, 1977; Tovish et al., 1978), and possibly a combination of these mechanisms (van Hunen et al., 2004). It has been suggested that a similar type of subduction might have been widespread in the past (Vlaar, 1986; Abbott et al., 1994b; O'Neill and Wyman, 2006), because the thick oceanic crust suggested for an early, hotter Earth might have had much in common with buoyant plateaus. In terms of observational evidence for this, it has been suggested that the common tonalite-trondhjemite-granodiorite (TTG) suites of the Archean originated from shallow flat subduction in the early history of Earth. TTGs were suggested as Archean analogues of Phanerozoic adakites (Drummond and Defant, 1990), which are products of slab crustal melting and occur in association with subduction of young oceanic lithosphere in South America. Smithies (2000) showed that slab crustal melting could generate TTGs, but only if there was no chemical interaction of melt with the mantle wedge. Shallow flat subduction is therefore a suitable setting.

However, the driving force for the shallow flat subduction observed today is provided by other, steeper parts of same subducting slab (van Hunen et al., 2004). If large parts of the oceanic plate are negatively buoyant, and subduct steeply, these can drag down smaller parts of the same oceanic plate that are positively buoyant. Once subducted, these buoyant plateaus resist being dragged down into the deep mantle, thereby causing temporary and local shallow flat subduction. In a hotter mantle, where the total oceanic plate was buoyant, shallow flat subduction would not have operated, because no driving force would have been available to subduct the buoyant plate. Instead, subduction would simply have ceased, or would never have started. One possible mechanism that could have provided shallow flat subduction of the total slab would have been active seaward motion of the overriding plate, but it is unclear why this should generally have been the case in the Archean. Other processes may need to be invoked, such as obduction by "continental overflow" (Bailey, 1999), although this would probably no longer be regarded as plate tectonics. This obduction would only lead to flat subducting slabs if the mantle viscosity was large enough to resist rapid sinking of the subducting slab. For the hotter Archean mantle this was probably not the case (van Hunen et al., 2004) (Fig. 4).

### 3.5. Plate Velocities and Plate Ages in the Past

The above arguments can be combined to estimate plate speeds and the residence time of oceanic plates in the past. This could in principle be related to surface heat flow values, and combined with observed plate velocities throughout Earth's history to assess the question of whether plate tectonics was viable in the past. It is often suggested that plate motions were faster than today based on a rheological argument that the viscosity of the mantle is lower in a hotter Earth. However, the overall scenario is much more complicated since we can expect major changes with temperature in the subduction-favoring ingredients (weaker mantle, easier plate bending) and the inhibiting effects (plate compositional buoyancy, weak slab breakoff, dehydration strengthening). Van Hunen and van den Berg (2007) estimate the turnover point to net inhibition to have been at around 100–200 °C hotter than today. Davies (1995) furthermore argues that plate tectonics may not have been continuously active throughout Earth's history, but might instead have been periodically switched off, when mantle overturns suddenly increased the upper-mantle temperature and increased the average thickness of the oceanic crust significantly, thereby blocking the subduction process. O'Neill et al. (2007) suggested episodic rather than continuous plate tectonics in an early, weaker Earth with lower stresses.

The evaluation of tectonic mechanisms would be greatly aided by reliable observations that constrain the motion of oceanic plates. Unfortunately, direct information is unavailable for ages older than 200 Ma. Observations on Precambrian plate motion are extremely sparse (and so may not represent global features) and indirect (and so involve several assumptions) and should be handled with care. The most quantitative method available to estimate Precambrian plate motion is paleomagnetism. Details on the methodology, results, and limitations can be found in Evans and Pisarevsky (this volume).

## 4. ALTERNATIVE TECTONIC MODELS

Today, plate tectonics dominates the large-scale dynamics of Earth's surface and governs the associated surface heat flow. It could be that other mechanisms cannot operate because plate tectonics is the most efficient and therefore the currently dominant mechanism. As shown in the previous section, however, this plate tectonics monopoly was not necessarily the case throughout Earth's history; plate tectonics might have had a harder time to run efficiently in the past, which might have left room for competitors to operate. The two most obvious dynamical arguments against Archean plate tectonics are the gravitational stability and the weakness of the oceanic lithosphere. These arguments have been the main reason for proposing alternative Archean tectonic regimes. Several regimes have been suggested in the last few

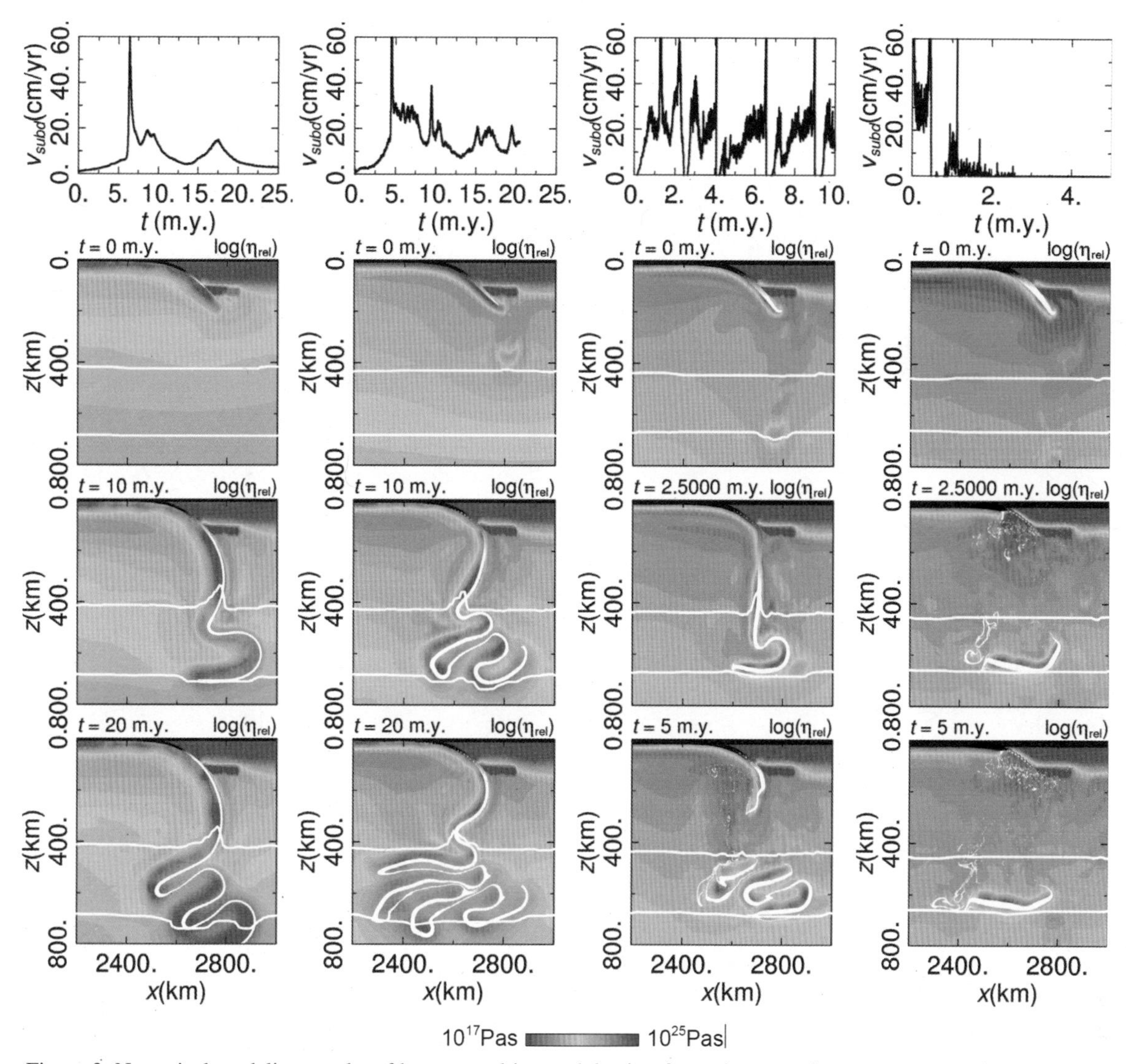


Figure 3. Numerical modeling results of buoyancy-driven subduction for various mantle temperatures and corresponding crustal thicknesses. Columns represent mantle potential temperatures of (from left to right) 0, 100, 200, and 300 °C hotter than today. The top row shows subduction velocity, and the lower panels show snapshots of the effective viscosity (colors) and presence of crustal material: basalt (black) transforming into eclogite (white). Horizontal lines indicate the 400-km and 660-km mantle phase transitions. Model results indicate that for a 100–200 °C hotter mantle, subduction remains qualitatively similar to today's situation. But an even hotter Earth shows frequent slab breakoff that may frustrate the subduction process (from van Hunen and van den Berg, 2007).

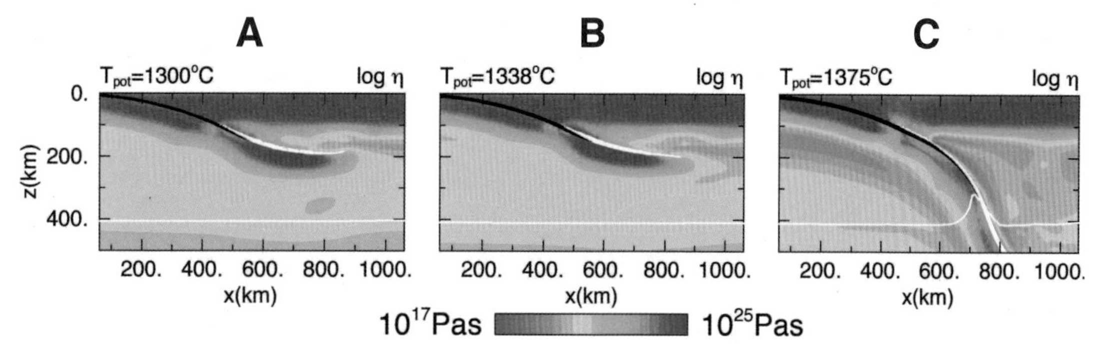


Figure 4. Numerical simulations of subduction with an overriding plate for different mantle potential temperatures $T_{pot}$: (A) $T_{pot}$ = 1350 °C (present-day situation); (B) $T_{pot}$ = 1338 °C; (C) $T_{pot}$ = 1375 °C. Results show that shallow flat subduction is not a viable scenario if the mantle temperature becomes 75 °C or more hotter than at present (van Hunen et al., 2004).

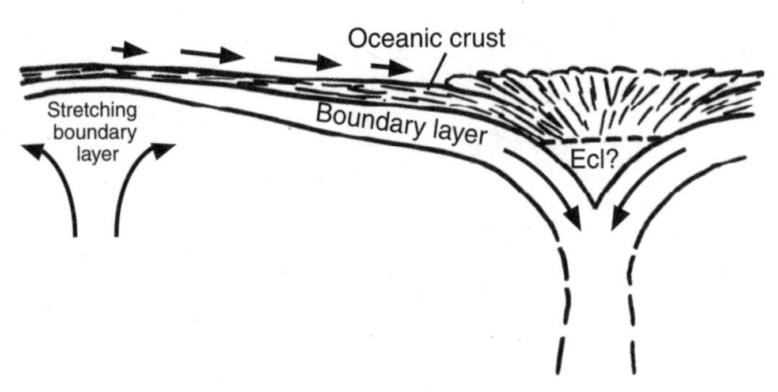


Figure 5. Sketch of a possible pre-plate tectonics dynamic regime. Buoyant crust and denser mantle part of the lithosphere decouple. The crust remains floating and forms thick stacks, of which the bottom may transform to eclogite. Eclogitized crust and mantle part of the lithosphere could "drip" down in the mantle (from Davies, 1992).

decades, and discussion of their mechanism, viability, and uncertainties may aid an appreciation of the long-term debate on "the timing of initiation of plate tectonics." Most mechanisms have been suggested to provide an alternative cooling mechanism for plate tectonics that is more efficient than a stagnant-lid regime.

### 4.1. Drip Tectonics

Hoffman and Ranalli (1988) estimated Archean oceanic crustal production to be 130–195 km$^3$/yr. Their estimate was based on extrapolation of the present-day rate of crust formation at mid-ocean ridges, under the assumption that the higher mantle temperature leads to increased spreading rates or ridge lengths. They concluded that only one-fourth of this produced Archean crust is preserved today. A nonuniformitarian method for crustal production is by diapirism and subsequent eruption of basaltic melt (Vlaar et al., 1994; van Thienen et al., 2004b). For this diapirism to provide adequate cooling of Archean Earth, van Thienen et al. (2005) concluded that eruption rates needed to be one or two orders of magnitude higher than the amount of Phanerozoic flood basalt magmatism, which again suggests lack of preservation of the majority of the Archean oceanic crust. Most of this buoyant crust must, then, somehow have been recycled into the mantle. A commonly proposed mechanism for recycling is eclogitization of part of the oceanic crust. Model-dependent estimates of the thickness of the Archean basaltic crust are 30–40 km (Fig. 2B). This brings the bottom of the crust into the eclogite stability field, and with a modest amount of water present, the basalt will transform geologically fast into eclogite (Rubie, 1990). Eclogite is denser than peridotite and may, under the right circumstances, sink into the mantle to be recycled. If the crustal thickness exceeds the thermal boundary layer, the eclogitic bottom of the crust might form drip-like instabilities (Fig. 5) (Campbell and Griffiths, 1992; Davies, 1992; Vlaar et al., 1994; van Thienen et al., 2004a). Zegers and van Keken (2001) propose that this mechanism could have formed the earliest continental crust and the TTG suites recorded in Middle Archean cratons (Fig. 6). A similar mechanism, although with incorporation of a mantle upwelling, is proposed by Bédard (2006) to explain the near-coeval

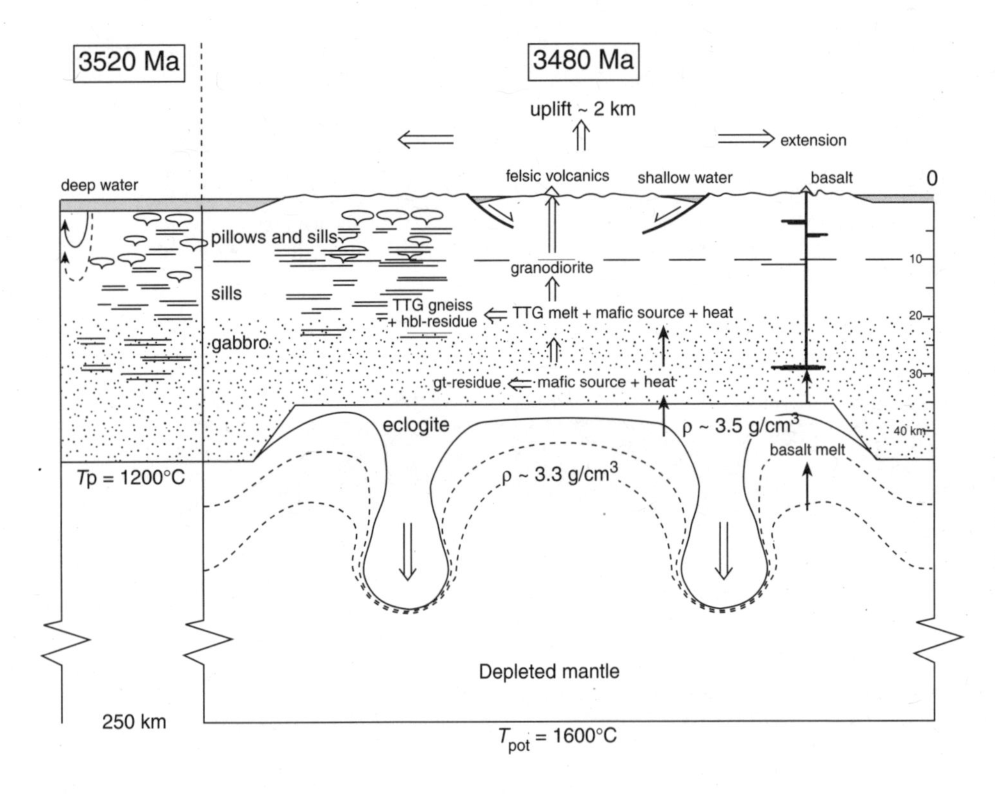


Figure 6. Model for delamination of thick oceanic crust causing both cooling of Earth and Early and Middle Archean TTG formation. Partial melting in a mantle with high potential temperature leads to a thickened crust and thick depleted mantle. Oceanic crust is hydrothermally altered by the circulation of seawater. Delamination of the eclogitized lower crust causes upwelling of depleted melting with renewed melt production. The intrusion of those melts into the protocrust results in the formation of TTG melts. After Zegers and van Keken (2001).

formation of Archean crust and mantle lithosphere of the Minto Block in the northeastern Superior province.

### 4.2. Sandwich Rheology

The thicker Archean oceanic crust leads to a weaker lithosphere, because diabase is intrinsically weaker than peridotite under the same pressure and temperature conditions (e.g., Kohlstedt et al., 1995). Hoffman and Ranalli (1988) pointed out that the thick Late Archean crust in combination with a reasonably thick thermal boundary layer may have led to a sandwich rheology, similar to that proposed for the present-day continental crust: A weak lower part of the crust is positioned between the stronger upper part of the crust and the mantle lithosphere below. This mechanical decoupling of crust and mantle lithosphere may have given rise to a variety of tectonic regimes in which the upper crust and mantle part of the lithosphere operated more or less independently. The mantle lithosphere may have delaminated and been recycled into the mantle while the crust may have been (partly) preserved. Archean flake tectonics (Hoffman and Ranalli, 1988) has been proposed to illustrate the independent dynamics of crust and mantle. Davies (1992) proposed subcrustal subduction as a possible mechanism (Fig. 7). While the mantle part of the lithosphere subducts, the crustal part of the lithosphere may stay at the surface and stack to form the greenstone belts that are characteristic of Archean terrains. Archean continental overflow (Bailey, 1999) exploits the same type of sandwich rheology to enable juxtaposition of continental-style crust with oceanic-style crust, which would then lead to the partial melting of the latter and formation of Archean TTG rock suites. Although Bailey suggests that this leads to TTG formation without actual subduction, such a process would probably create an environment favorable for subduction, and might have been the initial step toward full subduction and plate tectonics. Archean continental overflow would have ceased once the mean continental geothermal gradient dropped to values below 25–30 °C/km, when the ductile lower part of the crust would have become too strong. A problematic aspect of these sandwich-rheology styles of tectonics is that these may not act as a sufficient cooling agent for Earth. The "subduction" of the mantle part of the lithosphere is probably a sluggish process, because only a small part of the total thermal boundary layer is available to provide the necessary energy to drive this mantle convection (Davies, 1999). Eclogitization of the lower crust, similar to drip tectonics, can potentially provide the majority of the cooling.

### 4.3. Large-Scale Convective Overturns and Resurfacing Events

The endothermic mantle phase transition at 660 km may have, under certain circumstances, given rise to episodic large-scale overturns every several hundred million years during the early history of Earth (e.g., Machetel and Weber, 1991; Tackley, 1993; Solheim and Peltier, 1994). Davies (1995) shows that the

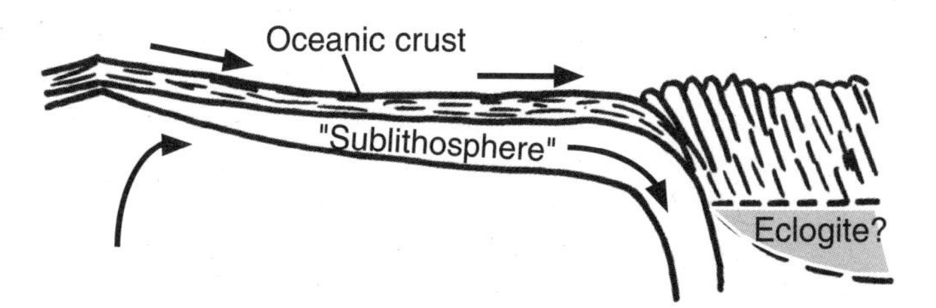


Figure 7. Sketch of a possible pre-plate tectonics dynamic regime. As Figure 5, but for a slightly cooler Earth and, consequently, a more rigid mantle part of the lithosphere. While eclogitized crust may still "drip" down in the mantle, the mantle part of the lithosphere may form "sublithospheric" subduction (from Davies, 1992).

continuing overturns can eventually provide a more continuous mass flux between upper and lower mantle such as we observe today (e.g., van der Hilst, 1997). The effects of such overturns would have been rather dramatic: Sudden increase in melting due to the suddenly 200–300 °C hotter upper mantle would have produced much thicker oceanic crust and influenced the viability of the discussed tectonic regimes. Plate tectonics might have ceased until the upper mantle had sufficiently cooled down again. These overturns might have triggered pulses of crustal formation, and might explain the various observed degrees of (upper and lower) mantle depletion.

A possibly equally dramatic scenario is presented by van Thienen et al. (2004a) with their self-consistent modeling of large-scale (on the order of 1000 km) resurfacing of Archean oceanic crust. This process provides an alternative explanation for the formation of TTGs: After triggering a resurfacing event of gravitationally unstable eclogitic crust, greenstone-producing melting events occur in the hot upwelling counterflow; remelting of the bottom part of this greenstone formation gives TTG plutons.

### 4.4. Magma Ocean

A totally different tectonic environment would have been provided early in Earth's history if a large-scale magma ocean was present. This requires a very different thermal and rheological regime than that representative of the present-day solid-state convection. Earth probably accreted hot enough for an initial magma ocean to develop and to extend perhaps down to the lower mantle (Abe, 1997). Also, giant impacts or continuous smaller impacts may have led to temporary or more permanent magma oceans. Magma oceans are, however, extremely efficient in cooling the mantle if Earth's surface is relatively cold, which would lead to a rapid freezing of those oceans (Davies, 1990, 1999). Such magma oceans are unlikely to have been maintained for longer than 20–200 m.y. after Earth's accretion or the major impact period (Abe, 1997; Sleep, 2000), and probably did not exist after 4 Ga because Earth would have cooled down sufficiently by then. The preservation of continental fragments since 4.4 Ga (Harrison et al., 2005) also suggests that any global magma ocean stage would have been short-lived. The presence of an early magma ocean is important in our discussion because (1) it provides bounds on the thermal regime at the time the magma ocean freezes and (2) it provides estimates of the initial differentiation, and therefore the mantle composition, of the very young Earth (Abe, 1997). Early massive differentiation is suggested by Hf isotope data that suggest a depleted mantle that is at least 4.08 Ga (Amelin et al., 2000; van Thienen et al., 2004a) or even older, 4.4–4.5 Ga (Harrison et al., 2005), although other mechanisms have been proposed (de Smet et al., 2000; Davies, 2006).

## 5. DISCUSSION AND CONCLUSIONS

We have provided a broad overview of proposals for the dynamics of early Earth and its transition into present-day plate tectonics. Observations suggest a 100–300 °C drop in mantle temperature since the Archean (although not necessarily a monotonic temperature drop), and many of the proposed geodynamic arguments against the viability of a certain tectonic regime are based on its failure to match this observation, because the surface heat flux in those regimes would be too low to provide the necessary cooling. The viability of plate tectonics in a hotter mantle may have been restricted by lack of driving forces or lithospheric strength required for subduction. However different the proposed tectonic regimes, they seem to have one aspect in common: In order to recycle buoyant crust, transformation to eclogite needs to be efficient.

Observational evidence for Archean and Proterozoic tectonics is limited to the older cratons within the continents. Even though plate tectonic activity is usually restricted to oceanic lithosphere, occasionally evidence for its operation is preserved in the continents. It is these (relatively rare) occasions that will give us better observational insight into the operation of plate tectonics, and possibly other tectonic regimes, in the past. Therefore, it is useful to understand the dynamics of continents. Zircon ages suggest continental growth as early as 4.5–4.4 Ga (Harrison et al., 2005). It has been proposed that this provides evidence for the operation of plate tectonics. More precisely, however, this provides evidence for liquid water and remelting in the presence of liquid water, which does not exclude the operation of an alternative tectonic mechanism. Numerical modeling of continent formation and growth (e.g., Lenardic et al., 2003; Lenardic, 2006) provides helpful insight into the geodynamic role of continents as part of a convecting Earth, and their preservation throughout Earth's history forms another geodynamic constraint. Most of the tectonic regimes presented here would probably be able to grow continents somehow. Being able to distinguish continental growth procedures under different tectonic regimes would be a significant step forward in our search for the tectonic regimes of early Earth. Lee (2006), for example, suggests a plate tectonic origin for the formation of cratonic continents on the basis of geochemical and petrologic methods. Whether the equivalent of oceanic lithosphere (i.e., the short-lived counterpart of stable continents) was possible under alternative tectonic regimes remains to be seen.

A prominent mantle seismic reflector extending to depths of 60 km occurs beneath the central Superior province of North America and projects to the surface at the junction between two different Neoarchean (ca. 2.7 Ga) terranes (Calvert et al., 1995). It has been interpreted as a fossil subduction zone (Calvert and Ludden, 1999; Ludden and Hynes, 2000). Its present dip of ~40° provides little support for a shallow-subduction scenario. Farther west in the Superior province, north-dipping imbrications on the Moho and a high-velocity, shallowly dipping unit at the base of the crust have been interpreted as subduction scars and underplated oceanic crust, respectively (White et al., 2003). Although they do not extend sufficiently deep into the mantle to provide any indication of the dips of the associated subduction zones, a magnetotelluric study of the same region indicates the presence of steeply dipping resistivity boundaries to significant depths in the mantle (Ferguson et al., 2005). Seismic reflection data in the Svecofennian region indicate that the crust is composed of accreted terrains, which suggest continental collision and plate tectonics (Korja and Heikkinen, this volume).

The sometimes large uncertainties in the predictions from the discussed geodynamic models are largely due to poorly constrained input parameters. Most importantly, we do not understand the full dynamics of present-day plate tectonics in large part because of the uncertainty of the interaction between the viscous hot mantle and the brittle-elastic-plastic lithosphere. In addition, rheological measurements are often significantly complicated by the presence of volatiles or melt (e.g., Rubie, 1990; Kohlstedt et al., 1995; Hirth and Kohlstedt, 1996). Also, phase change kinetics, in particular for the basalt-to-eclogite transition, is poorly constrained, again mostly due to the unknown volatile contents (Hacker, 1996). More accurate estimates of the volatile and melt contents of the materials involved in tectonic processes would therefore significantly enhance the reliability of geodynamic models. The modeling tools also have room for improvement: Self-consistent modeling of plate tectonics without the use of ad hoc rheologies has only started to appear (e.g., Tackley, 1998; Bercovici, 2003). Without these modeling tools, extrapolation to a hotter Earth regime will clearly retain its uncertainties. Due to the fast technological advances in computational resources and accurate measurements (such as zircon dating and rheology measurements), insight into the geodynamical aspects of "When did plate tectonics start?" is rapidly improving.

## ACKNOWLEDGMENTS

We thank Peter van Thienen and Stephane Labrosse for detailed and constructive reviews. We also thank the organizing committee and participants of the Penrose Conference "When Did Plate Tectonics Begin on Earth?," which featured frank and insightful discussions about Archean geodynamics that helped formulate and refine the ideas presented here.

## REFERENCES CITED


Abbott, D., Burgess, L., Longhi, J., and Smith, W.H.F., 1994a, An empirical thermal history of the Earth's upper mantle: Journal of Geophysical Research, B, Solid Earth and Planets, v. 99, p. 13,835–13,850, doi: 10.1029/94JB00112.

Abbott, D., Drury, R., and Smith, W.H.F., 1994b, Flat to steep transition in subduction style: Geology, v. 22, p. 937–940, doi: 10.1130/0091-7613(1994)022<0937:FTSTIS>2.3.CO;2.

Abe, Y., 1997, Thermal and chemical evolution of the terrestrial magma ocean: Physics of the Earth and Planetary Interiors, v. 100, p. 27–39, doi: 10.1016/S0031-9201(96)03229-3.

Ahrens, T.J., and Schubert, G., 1975, Gabbro-eclogite reaction rate and its geophysical significance: Reviews of Geophysics and Space Physics, v. 13, p. 383–400.

Amelin, Y., Lee, D.C., and Halliday, A.N., 2000, Early-middle Archaean crustal evolution deduced from Lu-Hf and U-Pb isotopic studies of single zircon grains: Geochimica et Cosmochimica Acta, v. 64, p. 4205–4225, doi: 10.1016/S0016-7037(00)00493-2.

Austrheim, H., 1998, Influence of fluid and deformation on metamorphism of the deep crust and consequences for the geodynamics of collision zones, *in* Hacker, B., and Liou, J., When continents collide: Geodynamics and geochemistry of ultrahigh-pressure rocks: Kluwer, p. 297–323.

Bailey, R.C., 1999, Gravity-driven continental overflow and Archaean tectonics: Nature, v. 398, p. 413–415, doi: 10.1038/18866.

Bédard, J.H., 2006, A catalytic delamination-driven model for coupled genesis of Archaean crust and sub-continental lithospheric mantle: Geochimica et Cosmochimica Acta, v. 70, p. 1188–1214, doi: 10.1016/j.gca.2005.11.008.

Bercovici, D., 2003, The generation of plate tectonics from mantle convection: Earth and Planetary Science Letters, v. 205, p. 107–121, doi: 10.1016/S0012-821X(02)01009-9.

Bijwaard, H., Spakman, W., and Engdahl, E.R., 1998, Closing the gap between regional and global travel time tomography: Journal of Geophysical Research, v. 103, p. 30,055–30,078, doi: 10.1029/98JB02467.

Brown, M., 2008, this volume, Characteristic thermal regimes of plate tectonics and their metamorphic imprint throughout Earth history: When did Earth first adopt a plate tectonics mode of behavior?, *in* Condie, K.C., and Pease, V., eds., When did plate tectonics begin on planet Earth?: Geological Society of America Special Paper 440, doi: 10.1130/2008.2440(05).

Calvert, A.J., and Ludden, J.N., 1999, Archean continental assembly in the southeastern Superior province of Canada: Tectonics, v. 18, p. 412–429, doi: 10.1029/1999TC900006.

Calvert, A.J., Sawyer, E.W., Davis, W.J., and Ludden, J.N., 1995, Archaean subduction inferred from seismic images of a mantle suture in the Superior province: Nature, v. 375, p. 670–674, doi: 10.1038/375670a0.

Campbell, I.H., and Griffiths, R.W., 1992, The changing nature of mantle hotspots through time: Implications for the chemical evolution of the mantle: Journal of Geology, v. 100, p. 497–523.

Christensen, U.R., 1985, Thermal evolution models for the Earth: Journal of Geophysical Research, B, Solid Earth and Planets, v. 90, p. 2995–3007.

Christensen, U.R., and Yuen, D.A., 1985, Layered convection induced by phase transitions: Journal of Geophysical Research, v. 99, p. 10,291–10,300.

Cloetingh, S., Wortel, M.J.R., and Vlaar, N.J., 1982, Evolution of passive continental margins and initiation of subduction zones: Nature, v. 297, p. 139–142, doi: 10.1038/297139a0.

Cloos, M., 1993, Lithospheric buoyancy and collisional orogenesis: Subduction of oceanic plateaus, continental margins, island arcs, spreading ridges, and seamounts: Geological Society of America Bulletin, v. 105, p. 715–737, doi: 10.1130/0016-7606(1993)105<0715:LBACOS>2.3.CO;2.

Condie, K.C., Kroener, A., and Stern, R.J., 2006, Penrose Conference Report: When did plate tectonics start?: GSA Today, v. 16, no. 10, p. 40–41, doi: 10.1130/1052-5173(2006)16[40:PCRWDP]2.0.CO;2.

Conrad, C.P., and Hager, B.H., 1999, Effects of plate bending and fault strength at subduction zones on plate dynamics: Journal of Geophysical Research, v. 104, p. 17,551–17,571, doi: 10.1029/1999JB900149.

Conrad, C.P., and Lithgow-Bertelloni, C., 2002, How mantle slabs drive plate tectonics: Science, v. 298, p. 207–209, doi: 10.1126/science.1074161.

Cross, T.A., and Pilger, R.H., Jr., 1978, Tectonic controls of Late Cretaceous sedimentation, western interior, USA: Nature, v. 274, p. 653–657, doi: 10.1038/274653a0.

Davaille, A., 1999, Simultaneous generation of hotspots and superswells by convection in a heterogeneous planetary mantle: Nature, v. 402, p. 756–760, doi: 10.1038/45461.

Davaille, A., and Jaupart, C., 1993, Transient high-Rayleigh-number thermal convection with large viscosity variations: Journal of Fluid Mechanics, v. 253, p. 141–166, doi: 10.1017/S0022112093001740.

Davies, G.F., 1979, Thickness and thermal history of continental crust and root zones: Earth and Planetary Science Letters, v. 44, p. 231–238, doi: 10.1016/0012-821X(79)90171-7.

Davies, G.F., 1980, Thermal histories of convective Earth models and constraints on radiogenic heat-production in the Earth: Journal of Geophysical Research, v. 85, p. 2517–2530.

Davies, G.F., 1990, Heat and mass transport in the early Earth, *in* Newsom, H.E., and Jones, J.H., eds., Origin of the Earth: New York, Houston, Oxford University Press, Lunar and Planetary Institute, p. 175–194.

Davies, G.F., 1992, On the emergence of plate tectonics: Geology, v. 20, p. 963–966, doi: 10.1130/0091-7613(1992)020<0963:OTEOPT>2.3.CO;2.

Davies, G.F., 1995, Penetration of plates and plumes through the lower mantle transition zone: Earth and Planetary Science Letters, v. 133, p. 507–516, doi: 10.1016/0012-821X(95)00039-F.

Davies, G.F., 1999, Dynamic Earth: Plates, plumes, and mantle convection: Cambridge, UK, Cambridge University Press, 458 p.

Davies, G.F., 2006, Gravitational depletion of the early Earth's upper mantle and the viability of early plate tectonics: Earth and Planetary Science Letters, v. 243, p. 376–382, doi: 10.1016/j.epsl.2006.01.053.

Davies, J.H., and von Blanckenburg, F., 1995, Slab breakoff: A model of lithospheric detachment and its test in the magmatism and deformation of collisional orogens: Earth and Planetary Science Letters, v. 129, p. 85–102, doi: 10.1016/0012-821X(94)00237-S.

de Smet, J.H., van den Berg, A.P., and Vlaar, N.J., 2000, Early formation and long-term stability of continents resulting from decompression melting in a convecting mantle: Tectonophysics, v. 322, p. 19–33, doi: 10.1016/S0040-1951(00)00055-X.

de Wit, M.J., 1998, On Archean granites, greenstones, cratons and tectonics: Does the evidence demand a verdict?: Precambrian Research, v. 91, p. 181–226, doi: 10.1016/S0301-9268(98)00043-6.

Drummond, M.S., and Defant, M.J., 1990, A model for trondhjemite-tonalite-dacite genesis and crustal growth via slab melting: Archean to modern comparisons: Journal of Geophysical Research, v. 95, p. 21,503–21,521.

England, P., and Bickle, M., 1984, Continental thermal and tectonic regimes during the Archaean: Journal of Geology, v. 92, p. 353–367.

Evans, D.A.D., and Pisarevsky, S.A., 2008, this volume, Plate tectonics on early Earth? Weighing the paleomagnetic evidence, *in* Condie, K.C., and Pease, V., eds., When did plate tectonics begin on planet Earth?: Geological Society of America Special Paper 440, doi: 10.1130/2008.2440(12).

Ferguson, I.J., Craven, J.A., Kurtz, R.D., Boerner, D.E., Bailey, R.C., Wu, X., Orellana, M.R., Spratt, J., Wennberg, G., and Norton, M., 2005, Geoelectric response of Archean lithosphere in the western Superior province, central Canada: Physics of the Earth and Planetary Interiors, v. 150, p. 123–143, doi: 10.1016/j.pepi.2004.08.025.

Fitton, J.G., and Godard, M., 2004, Origin and evolution of magmas on the Ontong Java Plateau, *in* Fitton, J.G., et al., eds., Origin and evolution of the Ontong Java Plateau: Geological Society [London] Special Publication 229, p. 151–178.

Forsyth, D., and Uyeda, S., 1975, On the relative importance of the driving forces of plate motion: Geophysical Journal of the Royal Astronomical Society, v. 43, p. 163–200.

Furnes, H., de Wit, M., Staudigel, H., Rosing, M., and Muehlenbachs, K., 2007, A vestige of Earth's oldest ophiolite: Science, v. 315, p. 1704–1707, doi: 10.1126/science.1139170.

Grand, S.P., 2002, Mantle shear-wave tomography and the fate of subducted slabs: Royal Society of London Philosophical Transactions, ser. A, v. 360, p. 2475–2491.

Grigné, C., Labrosse, S., and Tackley, P.J., 2005, Convective heat transfer as a function of wavelength: Implications for the cooling of the Earth: Journal of Geophysical Research, v. 110, B03409, doi: 10.1029/2004JB003376.

Grove, T.L., and Parman, S.W., 2004, Thermal evolution of the Earth as recorded by komatiites: Earth and Planetary Science Letters, v. 219, p. 173–187, doi: 10.1016/S0012-821X(04)00002-0.

Gurnis, M., 1989, A reassessment of the heat-transport by variable viscosity convection with plates and lids: Geophysical Research Letters, v. 16, p. 179–182.

Gutscher, M.-A., Spakman, W., Bijwaard, H., and Engdahl, E.R., 2000, Geodynamics of flat subduction: Seismicity and tomographic constraints from the Andean margin: Tectonics, v. 19, p. 814–833, doi: 10.1029/1999TC001152.

Hacker, B.R., 1996, Eclogite formation and the rheology, buoyancy, seismicity and $H_2O$ content of oceanic crust, *in* Bebout, G.E., Scholl, D., Kirby, S., and Platt, J.P., Subduction: Top to Bottom, AGU Monograph Ser. 96: Washington, DC, American Geophysical Union, p. 337–346.

Hall, C.E., Gurnis, M., Sdrolias, L.L., and Muller, R.D., 2003, Catastrophic initiation of subduction following forced convergence at transform boundaries: Earth and Planetary Science Letters, v. 212, p. 15–30, doi: 10.1016/S0012-821X(03)00242-5.

Hamilton, W.B., 1998, Archean magmatism and deformation were not products of plate tectonics: Precambrian Research, v. 91, p. 143–179, doi: 10.1016/S0301-9268(98)00042-4.

Harrison, T.M., Blichert-Toft, J., Muller, W., Albarede, F., Holden, P., and Mojzsis, S.J., 2005, Heterogeneous Hadean hafnium: Evidence of continental crust at 4.4 to 4.5 Ga: Science, v. 310, p. 1947–1950, doi: 10.1126/science.1117926.

Hirth, G., and Kohlstedt, D.L., 1996, Water in the oceanic upper mantle: Implications for rheology, melt extraction and the evolution of the lithosphere: Earth and Planetary Science Letters, v. 144, p. 93–108, doi: 10.1016/0012-821X(96)00154-9.

Hofmann, A.W., 1997, Mantle geochemistry: The message from oceanic volcanism: Nature, v. 385, p. 219–229, doi: 10.1038/385219a0.

Hoffman, P.F., and Ranalli, G., 1988, Archean oceanic flake tectonics: Geophysical Research Letters, v. 15, p. 1077–1080.

Huang, J., Zhong, S., and van Hunen, J., 2003, Controls on sub-lithospheric small-scale convection: Journal of Geophysical Research, v. 108, no. B8, p. 2405, doi: 10.1029/2003JB002456.

Jarrard, R.D., 1986, Relations among subduction parameters: Reviews of Geophysics, v. 24, p. 217–284.

Jaupart, C., Labrosse, S., and Mareschal, J.-C., 2007, Temperatures, heat, and energy in the mantle of the Earth, *in* Bercovici, D., and Schubert, G., eds., Treatise on Geophysics, Volume 7: Mantle dynamics: Elsevier, p. 253–303.

Jischke, M.C., 1975, On the dynamics of descending lithospheric plates and slip zones: Journal of Geophysical Research, v. 80, p. 4809–4813.

Karato, S., and Wu, P., 1993, Rheology of the upper mantle: A synthesis: Science, v. 260, p. 771–778, doi: 10.1126/science.260.5109.771.

Kellogg, L.H., Hager, B., and van der Hilst, R., 1999, Compositional stratification in the deep mantle: Science, v. 99, p. 276–289.

Kohlstedt, D.L., Evans, B., and Mackwell, S.J., 1995, Strength of the lithosphere: Constraints imposed by laboratory experiments: Journal of Geophysical Research, v. 100, p. 17,587–17,602, doi: 10.1029/95JB01460.

Kontinen, A., 1987, An Early Proterozoic ophiolite: The Jormua mafic-ultramafic complex, northeastern Finland: Precambrian Research, v. 35, p. 313–341, doi: 10.1016/0301-9268(87)90061-1.

Korenaga, J., 2003, Energetics of mantle convection and the fate of fossil heat: Geophysical Research Letters, v. 30, p. 1437, doi: 10.1029/2003GL016982.

Korenaga, J., 2006, Archean geodynamics and the thermal evolution of Earth, *in* Benn, K., et al., eds., Archean geodynamics and environments: American Geophysical Union Geophysical Monograph 164, p. 7–32.

Korja, A., and Heikkinen, P., 2008, this volume, Seismic images of Paleoproterozoic microplate boundaries in the Fennoscandian Shield, *in* Condie, K.C., and Pease, V., eds., When did plate tectonics begin on planet Earth?: Geological Society of America Special Paper 440, doi: 10.1130/2008.2440(11).

Kusky, T.M., and Li, J.H., 2002, Is the Dongwanzi complex an Archean ophiolite? Response: Science, v. 295, p. 923, doi: 10.1126/science.295.5557.923a.

Labrosse, S., and Jaupart, C., 2007, The thermal evolution of the Earth: Long term and fluctuations: Earth and Planetary Science Letters, v. 260, p. 465–481.

Lallemand, S., Heuret, A., and Boutelier, D., 2005, On the relationships between slab dip, back-arc stress, upper plate absolute motion, and crustal nature in subduction zones: Geochemistry, Geophysics, Geosystems, v. 6, article Q09006.

Langmuir, C.H., Klein, E.M., and Plank, T., 1992, Petrological constraints on mid-oceanic ridge basalts: Constraints on melt generation beneath ocean ridges, *in* Phipps Morgan, J., et al., eds., Mantle flow and melt generation at mid-oceanic ridges: American Geophysical Union Geophysical Monograph 71, p. 183–280.

Lee, C.-T.A., 2006, Geochemical/petrologic constraints on the origin of cratonic mantle, *in* Benn, K., et al., eds., Archean geodynamics and environments: Washington, D.C., American Geophysical Union, p. 89–114.

Lenardic, A., 2006, Continental growth and the Archean paradox, *in* Benn, K., et al., eds., Archean geodynamics and environments: Washington, D.C., American Geophysical Union, p. 33–46.

Lenardic, A., Moresi, L.N., and Muhlhaus, H., 2003, Longevity and stability of cratonic lithosphere: Insights from numerical simulations of coupled mantle convection and continental tectonics: Journal of Geophysical Research, B, Solid Earth and Planets, v. 108, no. 2303, doi: 10.1029/2002JB001859.

Levin, V., Shapiro, N., Park, J., and Ritzwoller, M., 2002, Seismic evidence for catastrophic slab loss beneath Kamchatka: Nature, v. 418, p. 763–767, doi: 10.1038/nature00973.

Ludden, J., and Hynes, A., 2000, The Abitibi-Grenville Lithoprobe project: Two billion years of crust formation and recycling in the Precambrian Shield of Canada: Canadian Journal of Earth Sciences, v. 37, p. 459–476, doi: 10.1139/cjes-37-2-3-459.

Machetel, P., and Weber, P., 1991, Intermittent layered convection in a model mantle with an endothermic phase change at 670 km: Nature, v. 350, p. 55–57, doi: 10.1038/350055a0.

Maruyama, S., and Liou, J.G., 1998, Initiation of ultrahigh-pressure metamorphism and its significance on the Proterozoic-Phanerozoic boundary: Island Arc, v. 7, p. 6–35, doi: 10.1046/j.1440-1738.1998.00181.x.

McCulloch, M.T., and Bennett, V.C., 1994, Progressive growth of the Earth's continental crust and depleted mantle: Geochemical constraints: Geochimica et Cosmochimica Acta, v. 58, p. 4717–4738, doi: 10.1016/0016-7037(94)90203-8.

McDonough, W.F., and Sun, S.-S., 1995, The composition of the Earth: Chemical Geology, v. 120, p. 223–253, doi: 10.1016/0009-2541(94)00140-4.

McGeary, S., Nur, A., and Ben-Avraham, Z., 1985, Spatial gaps in arc volcanism: The effect of collision or subduction of oceanic plateaus: Tectonophysics, v. 119, p. 195–221, doi: 10.1016/0040-1951(85)90039-3.

McKenzie, D.P., and Bickle, M.J., 1988, The volume and composition of melt generated by extension of the lithosphere: Journal of Petrology, v. 29, p. 625–679.

McKenzie, D.P., and Richter, F.M., 1981, Parameterized thermal convection in a layered region and the thermal history of the Earth: Journal of Geophysical Research, v. 86, p. 1667–1680.

McNamara, A.K., and van Keken, P.E., 2000, Cooling the Earth: A parameterized convection study of whole versus layered convection: Geochemistry, Geophysics, Geosystems, v. 1, doi: 10.1029/2000GC000045.

Möller, A., Appel, P., Mezger, K., and Schenk, V., 1995, Evidence for a 2 Ga subduction zone: Eclogites in the Usagaran Belt of Tanzania: Geology, v. 23, p. 1067–1070, doi: 10.1130/0091-7613(1995)023<1067:EFAGSZ>2.3.CO;2.

Nisbet, E.G., Cheadle, M.J., Arndt, N.T., and Bickle, M.J., 1993, Constraining the potential temperature of the Archean mantle: A review of the evidence from komatiites: Lithos, v. 30, p. 291–307, doi: 10.1016/0024-4937(93)90042-B.

Niu, Y.L., O'Hara, M.J., and Pearce, J.A., 2003, Initiation of subduction zones as a consequence of lateral compositional buoyancy contrast within the lithosphere: A petrological perspective: Journal of Petrology, v. 44, p. 851–866, doi: 10.1093/petrology/44.5.851.

O'Neill, C., and Wyman, D., 2006, Geodynamic modeling of Late Archean subduction: Pressure-temperature constraints from greenstone belt diamond deposits, *in* Benn, K., et al., eds., Archean geodynamics and environments: Washington, D.C., American Geophysical Union, p. 177–188.

O'Neill, C., Lenardic, A., Moresi, L., Torsvik, T.H., and Lee, C.-T.A., 2007, Episodic Precambrian subduction: Earth and Planetary Science Letters, v. 262, p. 552–562.

Parman, S.W., Dann, J.C., Grove, T.L., and de Wit, M.J., 1997, Emplacement conditions of komatiite magmas from the 3.49 Ga Komati Formation, Barberton Greenstone Belt, South Africa: Earth and Planetary Science Letters, v. 150, p. 303–323, doi: 10.1016/S0012-821X(97)00104-0.

Parsons, B., and McKenzie, D., 1978, Mantle convection and thermal structure of the plates: Journal of Geophysical Research, v. 83, p. 4485–4496.

Peck, W.H., Valley, J.W., Wilde, S.A., and Graham, C.M., 2001, Oxygen isotope ratios and rare earth elements in 3.3 to 4.4 Ga zircons: Ion microprobe evidence for high $\delta^{18}O$ continental crust and oceans in the Early Archean: Geochimica et Cosmochimica Acta, v. 65, p. 4215–4229, doi: 10.1016/S0016-7037(01)00711-6.

Pollack, H.N., Hurter, S.J., and Johnson, J.R., 1993, Heat-flow from the Earth's interior: Analysis of the global data set: Reviews of Geophysics, v. 31, p. 267–280, doi: 10.1029/93RG01249.

Regenauer-Lieb, K., Yuen, D.A., and Branlund, J., 2001, The initiation of subduction: Criticality by addition of water?: Science, v. 294, p. 578–580, doi: 10.1126/science.1063891.

Ritzwoller, M.H., Shapiro, N.M., and Zhong, S., 2004, Cooling history of the Pacific lithosphere: Earth and Planetary Science Letters, v. 226, p. 69–84, doi: 10.1016/j.epsl.2004.07.032.

Rubie, D.C., 1990, Role of kinetics in the formation and preservation of eclogites, *in* Carswell, D.A., ed., Eclogite facies rocks: Glasgow, Blackie, p. 111–140.

Scott, D.J., Helmstaedt, H., and Bickle, M.J., 1992, Purtuniq ophiolite, Cape Smith Belt, northern Quebec, Canada: A reconstructed section of Early Proterozoic oceanic crust: Geology, v. 20, p. 173–176, doi: 10.1130/0091-7613(1992)020<0173:POCSBN>2.3.CO;2.

Shelton, G., and Tullis, J., 1981, Experimental flow laws for crustal rocks: Eos (Transactions, American Geophysical Union), v. 62, p. 396.

Sleep, N.H., 2000, Evolution of the mode of convection within terrestrial planets: Journal of Geophysical Research, v. 105, p. 17,563–17,578, doi: 10.1029/2000JE001240.

Smithies, R.H., 2000, The Archean tonalite-trondhjemite-granodiorite (TTG) series is not an analogue of Cenozoic adakite: Earth and Planetary Science Letters, v. 182, p. 115–125, doi: 10.1016/S0012-821X(00)00236-3.

Solheim, L.P., and Peltier, W.R., 1994, Avalanche effects in phase transition modulated thermal convection: A model of Earth's mantle: Journal of Geophysical Research, v. 99, p. 6997–7018, doi: 10.1029/93JB02168.

Solomatov, V.S., 1995, Scaling of temperature- and stress-dependent viscosity convection: Physics of Fluids, v. 7, p. 266–274, doi: 10.1063/1.868624.

Solomatov, V.S., 2004, Initiation of subduction by small-scale convection: Journal of Geophysical Research, B, Solid Earth and Planets, v. 109, B01412, doi: 10.1029/2003JB002628.

Spohn, T., and Schubert, G., 1982, Modes of mantle convection and the removal of heat from the Earth's interior: Journal of Geophysical Research, v. 87, p. 4682–4696.

Stein, C.A., and Stein, S., 1992, A model for the global variation in oceanic depth and heat flow with lithospheric age: Nature, v. 359, p. 123–129, doi: 10.1038/359123a0.

Stern, R.J., 2005, Evidence from ophiolites, blueschists, and ultrahigh-pressure metamorphic terranes that the modern episode of subduction tectonics began in Neoproterozoic time: Geology, v. 33, p. 557–560, doi: 10.1130/G21365.1.

Stern, R.J., 2008, this volume, Modern-style plate tectonics began in Neoproterozoic time: An alternative interpretation of Earth's tectonic history, *in* Condie, K.C., and Pease, V., eds., When did plate tectonics begin on planet Earth?: Geological Society of America Special Paper 440, doi: 10.1130/2008.2440(13).

Stevenson, D.J., and Turner, S.J., 1977, Angle of subduction: Nature, v. 270, p. 334–336, doi: 10.1038/270334a0.

Stöckhert, B., and Renner, J., 1998, Rheology of crustal rocks at ultrahigh pressure, *in* Hacker, B., and Liou, J., eds., When continents collide: Geodynamics and geochemistry of ultrahigh-pressure rocks: Kluwer Academic Publishers, p. 57–95.

Tackley, P.J., 1993, Effects of strongly temperature-dependent viscosity on time-dependent, three-dimensional models of mantle convection: Geophysical Research Letters, v. 20, p. 2187–2190.

Tackley, P.J., 1998, Self-consistent generation of tectonic plates in three-dimensional mantle convection: Earth and Planetary Science Letters, v. 157, p. 9–22, doi: 10.1016/S0012-821X(98)00029-6.

Tovish, A., Schubert, G., and Luyendyk, B.P., 1978, Mantle flow pressure and the angle of subduction: Non-Newtonian corner flows: Journal of Geophysical Research, v. 83, p. 5892–5898.

Turcotte, D.L., and Schubert, G., 2002, Geodynamics: Applications of continuum physics to geological problems (2nd edition): Cambridge, UK, Cambridge University Press, 456 p.

van der Hilst, R., 1995, Complex morphology of subducted lithosphere in the mantle beneath the Tonga trench: Nature, v. 374, p. 154–157, doi: 10.1038/374154a0.

van der Hilst, R., 1997, Evidence for deep mantle circulation from global tomography: Nature, v. 386, p. 578–584, doi: 10.1038/386578a0.

van Hunen, J., and van den Berg, A.P., 2007, Plate tectonics on the early Earth: Limitations imposed by strength and buoyancy of subducted lithosphere: Lithos, doi:10.1016/j.lithos.2007.09.016.

van Hunen, J., van den Berg, A.P., and Vlaar, N.J., 2002a, The impact of the South American plate motion and the Nazca Ridge subduction on the flat subduction below South Peru: Geophysical Research Letters, v. 29, no. 14, 1690, doi: 10.1029/2001GL014004.

van Hunen, J., van den Berg, A.P., and Vlaar, N.J., 2002b, On the role of subducting oceanic plateaus in the development of shallow flat subduction: Tectonophysics, v. 352, p. 317–333, doi: 10.1016/S0040-1951(02)00263-9.

van Hunen, J., van den Berg, A.P., and Vlaar, N.J., 2004, Various mechanisms to induce present-day shallow flat subduction and implications for the younger Earth: A numerical parameter study: Physics of the Earth and Planetary Interiors, v. 146, p. 179–194, doi: 10.1016/j.pepi.2003.07.027.

van Hunen, J., Zhong, S., Shapiro, N.M., and Ritzwoller, M.H., 2005, New evidence for dislocation creep from 3-D geodynamic modeling of the Pacific upper mantle structure: Earth and Planetary Science Letters, v. 238, p. 146–155, doi: 10.1016/j.epsl.2005.07.006.

van Keken, P.E., Ballentine, C.J., and Porcelli, D., 2001, A dynamical investigation of the heat and helium imbalance: Earth and Planetary Science Letters, v. 188, p. 421–434, doi: 10.1016/S0012-821X(01)00343-0.

van Keken, P.E., Hauri, E.H., and Ballentine, C.J., 2002, Mantle mixing: The generation, preservation, and destruction of chemical heterogeneity: Annual Review of Earth and Planetary Sciences, v. 30, p. 493–525, doi: 10.1146/annurev.earth.30.091201.141236.

van Thienen, P., van den Berg, A.P., and Vlaar, N.J., 2004a, Production and recycling of oceanic crust in the early Earth: Tectonophysics, v. 386, p. 41–65, doi: 10.1016/j.tecto.2004.04.027.

van Thienen, P., Vlaar, N.J., and van den Berg, A.P., 2004b, Plate tectonics on the terrestrial planets: Physics of the Earth and Planetary Interiors, v. 142, p. 61–74, doi: 10.1016/j.pepi.2003.12.008.

van Thienen, P., Vlaar, N.J., and van den Berg, A.P., 2005, Assessment of the cooling capacity of plate tectonics and flood volcanism in the evolution of Earth, Mars, and Venus: Physics of the Earth and Planetary Interiors, v. 150, p. 287–315, doi: 10.1016/j.pepi.2004.11.010.

Vlaar, N.J., 1986, Archaean global dynamics: Geologie en Mijnbouw, v. 65, p. 91–101.

Vlaar, N.J., and van den Berg, A.P., 1991, Continental evolution and archeao-sea-levels, *in* Sabadini, R., Lambeck, K., and Boschi, E., eds., Glacial isostasy, sea-level and mantle rheology: Kluwer.

Vlaar, N.J., van Keken, P.E., and van den Berg, A.P., 1994, Cooling of the Earth in the Archaean: Consequences of pressure-release melting in a hotter mantle: Earth and Planetary Science Letters, v. 121, p. 1–18, doi: 10.1016/0012-821X(94)90028-0.

White, D.J., Musacchio, G., Helmstaedt, H.H., Harrap, R.M., Thurston, P.C., van der Velden, A., and Hall, K., 2003, Images of a lower-crustal oceanic slab: Direct evidence for tectonic accretion in the Archean western Superior province: Geology, v. 31, p. 997–1000, doi: 10.1130/G20014.1.

Wortel, M.J.R., and Spakman, W., 2000, Subduction and slab detachment in the Mediterranean-Carpathian region: Science, v. 290, p. 1910–1917, doi: 10.1126/science.290.5498.1910.

Zegers, T.E., and van Keken, P.E., 2001, Middle Archean continent formation by crustal delamination: Geology, v. 29, p. 1083–1086, doi: 10.1130/0091-7613(2001)029<1083:MACFBC>2.0.CO;2.

Zhai, M.G., Zhao, G.C., and Zhang, Q., 2002, Is the Dongwanzi complex an Archean ophiolite?: Science, v. 295, p. 923, doi: 10.1126/science.295.5557.923a.

Manuscript Accepted by the Society 14 August 2007

Printed in the USA

The Geological Society of America
Special Paper 440
2008

# *The Late Archean Abitibi-Opatica terrane, Superior Province: A modified oceanic plateau*

**Keith Benn***
*Ottawa-Carleton Geoscience Centre and Department of Earth Sciences, University of Ottawa, Ottawa, Ontario K1N 6N5, Canada*

**Jean-François Moyen***
*Department of Geology, Stellenbosch University, Private Bag X1, Matieland 7602, Western Cape, South Africa*

## ABSTRACT

**The Abitibi-Opatica terrane is defined to include the Abitibi granite-greenstone Subprovince and the Opatica granite-gneiss domain in southeastern Superior Province, Canada. We combine the geological, structural, geochronological, and geochemical knowledge base for the region, with new geochemical data for suites of granitic rocks, in order to establish a testable model for the geodynamic setting and the tectonomagmatic evolution of the Late Archean crust. The geochemistry of TTG orthogneiss and plutons are correlated, petrogenetically and temporally, with the published data and interpretations for the volcanic stratigraphy. The geochemistry of later granodiorite plutons is correlated with the crustal melting signatures of the youngest volcanic assemblage. Putting the geochemical data and interpretations into a framework with data on crustal structure, crustal thickness, and geochronology allows us to define the precollisional tectonomagmatic history of the Abitibi-Opatica terrane. A geodynamic-tectonic model is proposed, involving subduction of an ocean basin beneath an existing, magmatically active and partially differentiated oceanic plateau. The geochemical signature of "plume-arc interaction" is attributed to subduction that was initiated under the magmatically active oceanic plateau, in the presence of a still-active plume. The proposed plate tectonic model explains the presence of plume-type and subduction-type signatures in the volcanic stratigraphy, and in the TTG gneiss and plutons, and requires a single period of plate convergence and subduction that lasted for ~35 million years, ending in a tectonic collision event, ca. 2700 Ma. We propose that the interstratification of plume-type and subduction-type lavas, and the concomitant emplacement of TTG plutons with slab-melting characteristics, might be explained by the formation of a slab window in the subplateau subduction zone.**

**Keywords:** Greenstone belt, TTG, oceanic plateau, slab window.

*Benn, corresponding author: kbenn@uottawa.ca; Moyen: moyen@sun.ac.za

Benn, K., and Moyen, J.-F., 2008, The Late Archean Abitibi-Opatica terrane, Superior Province: A modified oceanic plateau, *in* Condie, K.C., and Pease, V., eds., When Did Plate Tectonics Begin on Planet Earth?: Geological Society of America Special Paper 440, p. 173–197, doi: 10.1130/2008.2440(09). For permission to copy, contact editing@geosociety.org. 

## INTRODUCTION

The Superior Province of the Canadian Shield (Fig. 1) is the largest known Archean craton and forms the core of the North American continent. It has classically been divided into a number of *subprovinces* based on differences in structural trends, metamorphic grades, lithological makeup, and geochronology (Card and Poulsen, 1998). Other regions of the Superior Province are divided into *domains* (Card and Poulsen, 1998) that also have distinctive geological characteristics from surrounding subprovinces and domains, but that remain poorly understood. The division of the Superior Province into distinct subprovinces and domains has proven to be a very useful step on the way to identifying and defining tectonic terranes, terrane boundaries, and the tectonic processes involved in the assembly of the terranes. This study is mainly concerned with the southeastern part of the Superior Province, which is made up of greenstones and plutonic rocks of the Abitibi Subprovince, the metaplutonic Opatica domain, and the metasedimentary Pontiac Subprovince (Fig. 1).

Until recently, it has been generally accepted that the Abitibi Subprovince and the Opatica domain represented two distinct tectonic terranes that would have collided ca. 2700 Ma. That interpretation was supported by the presence of an apparent subduction zone scar within the sub-Opatica domain lithosphere, interpreted from Lithoprobe seismic reflection profiles (Calvert et al., 1995). Other potential terrane sutures, corresponding to high-strain lineaments, were also identified within the Abitibi Subprovince (Ludden et al., 1986; Kerrich and Feng, 1992; Mueller et al., 1996; Daigneault et al., 2002, 2004). As a result, tectonic models for the southeastern portion of the Superior Province needed to explain a collage of allochthonous arc and oceanic plateau terranes, assembled during multiple collisions (e.g., Kimura et al., 1993).

A recent interpretation of the crustal structure of the Abitibi Subprovince and the Opatica domain removed at least two previously interpreted tectonic terrane boundaries and led to the proposal that the Abitibi Subprovince and the Opatica domain would represent a single tectonic terrane, the Abitibi-Opatica terrane (Fig. 1) (Benn, 2006), rather than a collage of allochthonous crustal fragments. The interpretation that the Abitibi Subprovince and the Opatica domain would form a single tectonic terrane is based on a reevaluation of Lithoprobe seismic profiles and a synthesis of geochronological and geological data (Benn, 2006), and on new interpretations of high-strain lineaments within the Abitibi Subprovince (Benn and Peschler, 2005; Peschler et al., 2006).

It is important to consider the ramifications of the proposed Abitibi-Opatica terrane for our understanding of the southeastern part of the Superior Province, ramifications that stem from the size of the proposed terrane, its composition, and its tectonomagmatic

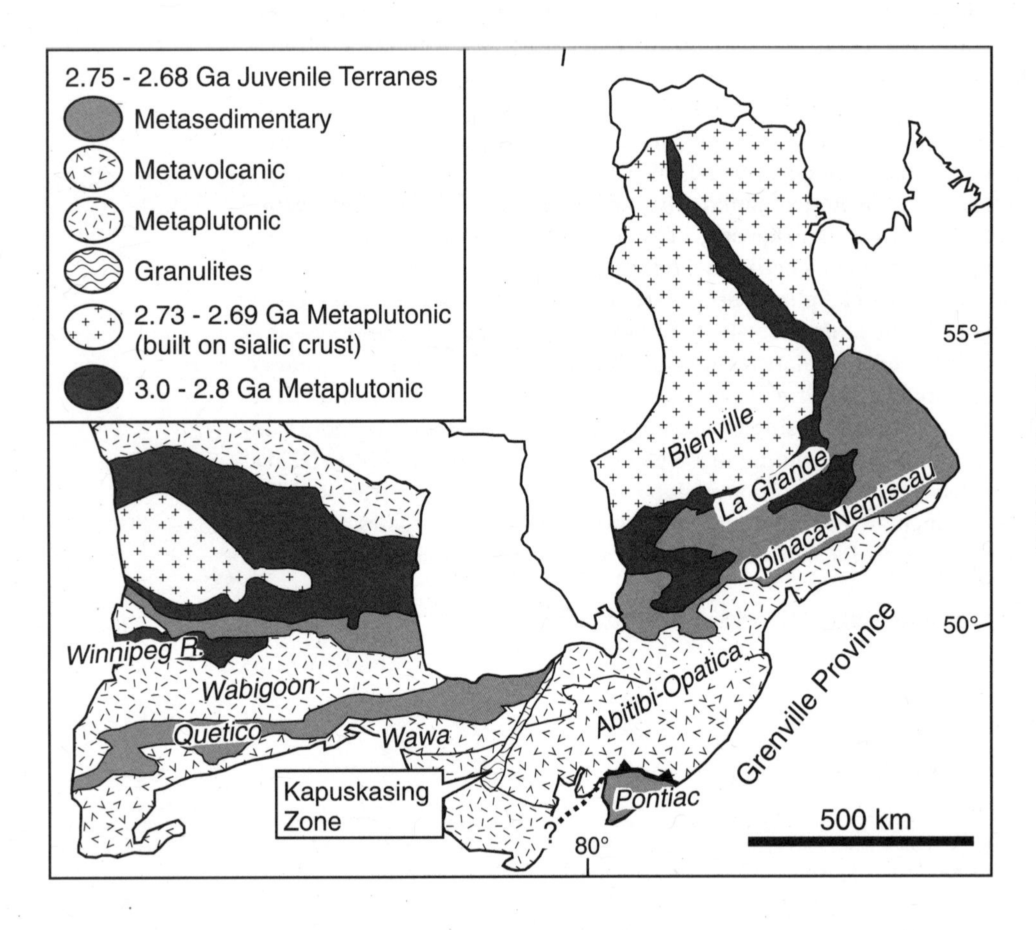


Figure 1. Thematic geological map of the Superior Province, in the Canadian Shield. The names indicated on the map pertain to geological subprovinces, geological domains, and the Abitibi-Opatica terrane. Modified from Percival et al. (1994).

history. A refined understanding of the geodynamic origin and the tectonomagmatic history of the Abitibi-Opatica terrane will certainly have consequences for our comprehension of Archean plate tectonic processes and better constrain the application of plate tectonic models to the formation of Late Archean crust and lithosphere (Condie and Benn, 2006).

It is the purpose of this paper to use published, and new, data on the Abitibi Subprovince, the Opatica domain, and adjacent geological regions, in order to propose a new model for the geodynamic origins of the Abitibi-Opatica terrane, and to clarify the shared tectonomagmatic evolution of the greenstones and the plutonic rocks. We synthesize available data and we incorporate new geochemical data from the Kenogamissi complex, a large crystalline complex that crops out within the Abitibi Subprovince. The Kenogamissi complex includes plutonic units whose ages span much of the regional magmatic history. The new data make it possible to better understand the magmatic history that is recorded in the volcanic stratigraphy and in the plutonic units.

We show that the magmatic history of the Abitibi-Opatica terrane began with the formation and differentiation of a thick mafic crust in response to a mantle plume, followed by magmatic modification of the crust during lower crustal melting and later subduction-related magmatism. The result is a model for a magmatically modified Late Archean oceanic plateau. The proposed model explains the apparent plume-subduction interaction that is recorded in the geochemistry of a portion of the volcanic stratigraphy, as well as in the geochemical signatures of the plutonic rocks, as the result of a slab window formed beneath the Abitibi Subprovince.

The model indicates that the formation and evolution of the Late Archean Abitibi-Opatica terrane can be explained by a plate tectonic model involving plume-related magmatism and subduction of an ocean basin beneath a magmatically active oceanic plateau.

## SOUTHEASTERN SUPERIOR PROVINCE

A tectonic context is provided for the Abitibi-Opatica terrane by reviewing the geology and geochronology of those regions that lie to the north and the south of it. In the subsequent section we will discuss the better-known geology and the more abundant geochronology of the Abitibi-Opatica terrane within the established context. Geochronological data for the southeastern Superior Province are compiled in Table 1 with the reported errors, along with references to the literature sources of the data.

The western boundary of the Abitibi-Opatica terrane is the Kapuskasing structural zone, where Late Archean granulite- and amphibolite-grade rocks were uplifted in Proterozoic time (Percival, 1988; Halls and Zhang, 1998). In the east, the boundary is marked by the Grenville Front that marks the foreland of the Proterozoic Grenville orogeny.

In the north, the granitic rocks and orthogneisses of the Opatica domain are bounded by the poorly known Opinaca-Nemiscau domain (Fig. 1), which is composed predominantly of metasedimentary rocks with lesser amounts of metavolcanic rocks, intruded by biotite granites. There are no published geochronological data for the Opinaca-Nemiscau domain, nor do we have published descriptions of the regional metamorphism. Personal observations (K.B.) during reconnaissance fieldwork near the border of the Opinaca-Nemiscau domain with the Opatica domain indicate the presence of metapsammites and metapelites, metamorphosed to the greenschist grade, in the southernmost Opinaca-Nemiscau domain.

To the north of the Opinaca-Nemiscau domain lies the La Grande domain, the northern part of which is described as a complex assemblage of "basement gneisses," migmatites, diorites, and tonalites. Geochronological data reveal that those rocks range in age from 2880 Ma to 2760 Ma, and that some of the older basement gneiss contains inherited zircons dated ca. 3037 Ma (data reported in Isnard and Gariépy, 2004). Isotopic study of the gneisses revealed a mantle extraction age of 3000 Ma (Isnard and Gariépy, 2004). It is likely that tectonometamorphic reworking of pre–3000 Ma crust gave rise to the younger components of the basement gneisses, dated by the 2880 Ma zircons.

In the northern part of the La Grande domain, metavolcanic rocks are dated 2820 Ma (Isnard and Gariépy, 2004). The same authors compiled data that indicate dates of 2728 Ma to 2701 Ma for "syn- to post-tectonic" tonalites and granodiorites in the La Grande domain, falling within the range of dates from tonalite and granodiorite plutons within the Abitibi-Opatica terrane. Detrital zircons from conglomerates and arenites that overlie the basement gneisses are dated 3800 Ma to 3100 Ma (data reported in Isnard and Gariépy, 2004). The autochthonous, or allochthonous, origin of the metasedimentary rocks is unknown.

To the south of the Abitibi Subprovince, the supracrustal rocks of the Pontiac Subprovince (Fig. 1) are predominantly metapsammites and metapelites of the Pontiac Group, with lesser amounts of metavolcanic rocks that crop out mainly in the southwestern part of the region where they represent the Belleterre belt (Fig. 2). The Belleterre belt has yielded dates of 2690 Ma through 2682 Ma (Mortensen and Card, 1993). The metamorphic grade of the Pontiac Group metasedimentary rocks ranges from greenschist, around the margins of the subprovince, to amphibolite grade in its central parts, where kyanite facies rocks are intruded by anatectic granite (Benn et al., 1994). Regional metamorphism in the Pontiac Subprovince is attributed to burial under the overthrust southern Abitibi Subprovince (Benn et al., 1994; Benn, 2006). Detrital zircons from the Pontiac Group metasedimentary rocks have provided dates ranging from ca. 3042 Ma to as young as 2683 Ma (Feng and Kerrich, 1991; Mortensen and Card, 1993; Davis, 2002).

The metasedimentary and metavolcanic rocks of the Pontiac Subprovince were intruded by plutons of tonalite, granodiorite, quartz syenite, biotite granite, and muscovite-biotite granite, the last containing magmatic sillimanite. An orthogneiss of tonalite composition was dated 2695 Ma (Mortensen and Card, 1993).

TABLE 1. GEOCHRONOLOGY OF THE ABITIBI-OPATICA TERRANE AND THE LA GRANDE AND PONTIAC SUBPROVINCES

| Date (Ma) | Rock type | Geological unit | Notes |
|---|---|---|---|
| La Grande domain | | | |
| 3037 ± 14 | Tonalite | Langelier complex | Parent (1998) reported in Isnard and Gariépy (2004); inherited zircon in tonalitic enclave |
| 2881 ± 2 | Tonalite | Langelier complex | Parent (1998) reported in Isnard and Gariépy (2004) |
| 2832 ± 5 | Tonalite enclave | Langelier complex | Parent (1998) reported in Isnard and Gariépy (2004) |
| 2794 ± 2 | Tonalite | Langelier complex | David (1996) reported in Isnard and Gariépy (2004) |
| 2788 ± 4 | Hb tonalite | Langelier complex | David (1996) reported in Isnard and Gariépy (2004) |
| ~2751 | Tonalite | Duncan Suite | Parent (1998) reported in Isnard and Gariépy (2004) |
| 2728 ± 4 | Tonalite | Eastmain Suite | Parent (1998) reported in Isnard and Gariépy (2004) |
| 2716 ± 3 | Tonalite | Amisach Wat pluton (Duncan Suite) | Isnard and Gariépy (2004) |
| 2709 ± 2 | Granodiorite | Duxbury pluton | Gauthier (1982) reported in Isnard and Gariépy (2004) |
| 2701 ± 8 | Granodiorite | Duxbury pluton | Parent (1998) reported in Isnard and Gariépy (2004) |
| 2699 ± 4 | Bt monzogranite | Lac Taylor pluton | Isnard and Gariépy (2004) |
| 2657 ± 4 | Bt granite | Vieux comptoir | David and Parent (1997) reported in Isnard and Gariépy (2004) |
| 2618 ± 3 | Pegmatite | Vieux comptoir | David and Parent (1997) reported in Isnard and Gariépy (2004) |
| Abitibi-Opatica terrane | | | |
| Plutonism | | | |
| 2825 ± 3 | Lac Rodayer pluton | Opatica domain | Davis et al. (1995) |
| 2807 ± 13 | Tonalite orthogneiss | Opatica domain | Davis et al. (1995); age determined by regression of three zircon fractions |
| 2773 ± 23 | Tonalite orthogneiss | Opatica domain | Davis et al. (1995); age determined by regression of three zircon fractions |
| 2761 +42 / −15 | Tonalite orthogneiss | Opatica domain | Davis et al. (1995); age determined by regression of three zircon fractions |
| ≥2753 | Tonalite | Mistaouac pluton | Davis et al. (2000); inherited zircons |
| 2744 ± 0.9 | Tonalite | Component of Round Lake batholith | Ayer et al. (2004) |
| 2742 ± 2 | Hb-Bt tonalite | Rice Lake batholith (Kenogamissi complex) | Ayer et al. (2004) |
| 2740 +5 / −3 | Tonalite | Unnamed pluton (Opatica domain) | Davis et al. (1995) |
| 2740 ± 2 | Trondhjemite, tonalite | Chester complex | Heather and Shore (1999) |
| 2735 | Tonalite | Gogama orthogneiss | Heather and Shore (1999); inherited zircons |
| 2727 +11 / −9 | Bt tonalite | Arbutus pluton | Heather and Shore (1999) |
| 2727 ± 1.3 | Qz-bearing ferric pyroxenite | Lac Doré gabbro-anorthosite-granophyre complex | Mortensen (1993a) |
| 2726 ± 2 | Tonalite | Mistaouac pluton | Davis et al. (2000) |
| 2725 +3 / −2 | Granophyre | Bell River gabbro-anorthosite-granophyre complex | Mortensen (1993a) |
| 2723 ± 3 | Hb-Bt tonalite | Gogama tonalite | Heather and Shore (1999) |
| 2719 +3 / −0.6 | No rock type reported | La Dauversière pluton | Mortensen (1993a) |
| 2713 +2 / −3 | Hb-Bt tonalite | Regan pluton (Kenogamissi complex) | Heather and Shore (1999) |
| 2713 ± 1 | Hb-Bt tonalite | Component of Round Lake batholith | Ayer et al. (2004) |
| 2713 ± 3–2711 ± 1 | Tonalite | Lapparent complex | Mortensen (1993c) |
| 2713 ± 2 | Tonalite | Boivin pluton | Davis et al. (2000) |
| 2703 +3 / −2 | Tonalite | Rousseau pluton | Davis et al. (2000) |
| 2702 ± 3 | Tonalite | Unnamed tonalite pluton (Opatica domain) | Davis et al. (1995) |
| 2701 ± 1 | Bt granodiorite | Rice Lake batholith (Kenogamissi complex) | Ayer et al. (2004) |
| 2700 ± 2 | No rock type reported | Renaud pluton (Lapparent complex) | Mortensen (1993a) |

(*continued*)

TABLE 1. GEOCHRONOLOGY OF THE ABITIBI-OPATICA TERRANE AND THE LA GRANDE AND PONTIAC SUBPROVINCES (*continued*)

| Date (Ma) | Rock type | Geological unit | Notes |
|---|---|---|---|
| 2699 ± 3 | Qz diorite | Indian Chute stock (Watabeag batholith) | Frarey and Krogh (1986) |
| 2698 +5 / –4 | Tonalite | Round Lake batholith | Mortensen (1993b) |
| 2697 ± 2 | Granodiorite | Indian Chute stock (Round Lake batholith) | Mortensen (1993b) |
| 2696 | Monzodiorite | Barlow pluton | Davis et al. (1995) |
| 2696 ± 3 | Bt granodiorite | Northrup pluton (Kenogamissi complex) | Heather and Shore (1999) |
| 2696 +3 / –2 | Granodiorite | Colombourg pluton | Mortensen (1993b) |
| 2696 ± 2 | Tonalite | La Reine pluton | Davis et al. (2000) |
| 2695 ± 3 | Tonalite | Waswanipi pluton | Davis et al. (2000) |
| 2695 ± 2 | Tonalite | Côté stock | Ayer et al. (2004) |
| 2695 ± 2 | Tonalite | Île Nepawa pluton (Lake Abitibi batholith) | Mortensen (1993b) |
| 2694 +2 / –4 | Tonalite | Nat River complex | Ayer et al. (2004) |
| 2693 +3 / –2 | Tonalite | Canet pluton (Opatica domain) | Davis et al. (1995) |
| 2693 ± 1 | Tonalite | Lac Ouescapis (Opatica domain) | Davis et al. (1995) |
| 2691 +3 / –5 | Bt granodiorite | Isaiah Creek stock | Heather and Shore (1999) |
| 2690 ± 1 | Tonalite | Lake Abitibi batholith | Mortensen (1987) |
| 2690 ± 2 | Granodiorite | Lac Dufault pluton | Mortensen (1993b) |
| 2690 ± 1 | Granodiorite | Lake Abitibi batholith | Mortensen (1993b) |
| 2689 ± 1 | Qz diorite | Claris Lake stock | Corfu and Noble (1992) |
| 2686 ± 2 | Hb diorite | Muskego diorite (Nat River complex) | Heather and Shore (1999) |
| 2686 ± 3 | Granodiorite | Adams stock | Frarey and Krogh (1986) |
| 2686 ± 2 | Granodiorite | Paradis pluton | Davis et al. (2000) |
| 2684 ± 3 | Bt granodiorite | Hoodoo pluton | Frarey and Krogh (1986) |
| 2682 ± 3 | Syenite | Cléricy syenite | Mortensen (1993b) |
| 2682 ± 3 | Hb granodiorite | Neville pluton | Heather and Shore (1999) |
| 2681 ± 3 | Hb granodiorite | Hillary pluton | Heather and Shore (1999) |
| 2681 ± 3 | Hb monzonite | Kukatush pluton | Heather and Shore (1999) |
| 2681 ± 2 | Qz monzonite | Watabeag batholith | Frarey and Krogh (1986) |
| 2680 ± 3 | Hb diorite | Ivanhoe pluton | Percival and Krogh (1983) |
| 2679 ± 4 | Granodiorite | Garrison stock | Frarey and Krogh (1986) |
| 2679 ± 1 | Syenite | Otto stock | Corfu and Noble (1992) |
| 2678 ± 2–2690 ± 2 | Bt granite | Late granite in Opatica domain | Davis et al. (1995) |
| 2677 ± 2 | Granite | Winnie Lake stock | Frarey and Krogh (1986) |
| 2676 ± 2 | Qz monzonite | Watabeag batholith | Frarey and Krogh (1986) |
| 2676 ± 2 | Bt granite | Lac Case pluton | Davis et al. (2000) |
| 2676 ± 2 | Syenite | Cairo stock | Ayer et al. (2004) |
| 2676 +6 / –5 | Syenite | Douay pluton | Davis et al. (2000) |
| 2673 ± 2 | Syenite | Lebel stock | Ayer et al. (2004) |
| 2672 ± 2 | Syenite | Murdock Creek stock | Ayer et al. (2004) |
| 2662 ± 4 | Bt granite | Somme pluton | Heather and Shore (1999) |
| 2660 ± 3 | Mu granite pegmatite | Lac Case pluton | Davis et al. (2000) |
| Volcanism in Quebec Abitibi—compiled ages of volcanic units in Ontario indicated in Figure 4 | | | |
| Frotet-Evans belt | | | |
| 2793–2755 | | Greenstones | Reported in Boily and Dion (2002) |
| Northern Volcanic Zone—Quebec | | | |
| 2766–2802 | Qz-Fs porphyry block in magmatic breccia | Obatagamau Formation | Mortensen (1993a); xenocrystic zircons |
| 2759 ± 2 | Rhyodacite | Obatagamau Formation | Mortensen (1993a) |
| 2730 ± 2–2728 ± 2 | Rhyolite | Waconichi Formation | Mortensen (1993a) |
| 2724 ± 2–2723 ± 1 | Rhyolite | Lac Watson Formation | Mortensen (1993a) |
| 2728 ± 1–2721 ± 1 | Several rock types | Joutel volcanic complex | Legault et al. (2002) |
| Southern Volcanic Zone—Quebec | | | |
| 2747 ± 1 | Felsic tuff | Pacaud Assemblage | Mortensen (1993b) |

(*continued*)

TABLE 1. GEOCHRONOLOGY OF THE ABITIBI-OPATICA TERRANE AND THE LA GRANDE AND PONTIAC SUBPROVINCES (*continued*)

| Date (Ma) | Rock type | Geological unit | Notes |
|---|---|---|---|
| 2730 ± 2 | Rhyolite | Hunter Mine Group | Mortensen (1993b) |
| 2715–2698 | | Various volcanic units, including Blake River Assemblage | Mortensen (1993b) |
| Angular unconformity | | | |
| Siliciclastic sedimentation | | | |
| 3017 ± 4 | Graywacke | Lac Caste Group | Davis (2002); one zircon |
| 2696–2690 | Graywacke, sandstone | Porcupine Assemblage | Compiled in Ayer et al. (2002) |
| 2697 ± 3 | Sandstone | Duparquet Group | Davis (2002); youngest detrital zircon |
| 2694 ± 3 | Graywacke | Lac Caste Group | Davis (2002); youngest detrital zircon |
| 2687 ± 3 | Graywacke | Cadillac Group | Davis (2002); youngest detrital zircon |
| 2686 ± 4 | Graywacke | Kewagama Group | Davis (2002); youngest detrital zircon |
| Angular unconformity—alluvial-fluvial sedimentary deposits, sandstone, conglomerate, interbedded volcaniclastites | | | |
| 2687–2675 | Graywacke, conglomerate | Timiskaming Assemblage | Compiled in Ayer et al. (2002) |
| Pontiac subprovince | | | |
| Plutonism | | | |
| 2695 ± 2 | Tonalite | Lac des Quinze orthogneiss | Mortensen and Card (1993) |
| 2685 ± 3 | Qz syenite | Lac Frechette pluton | Mortensen and Card (1993) |
| 2682 ± 1 | Qz diorite | Lac Fournière pluton | Davis (2002); cuts Pontiac Group |
| 2681 ± 2 | Qz syenite | Lac Frechette pluton | Mortensen and Card (1993) |
| 2680 ± 1 | Qz syenite | Lac Rémigny pluton | Mortensen and Card (1993) |
| 2675 ± 8–2671 ± 4 | Hb granodiorite | Lac Rémigny pluton | Feng and Kerrich (1991) |
| 2672 ± 2 | Syenite porphyry | Unnamed plutons | Davis (2002); cuts Timiskaming Assemblage |
| 2668 ± 1 | Mu pegmatite sill | Decelles batholith | Mortensen and Card (1993); monazite from sill in Pontiac Group |
| 2663 ± 1 | Bt pegmatite | Decelles batholith | Mortensen and Card (1993); monazite from dike in Pontiac Group |
| 2655 ± 8–2643 ± 4 | Bt-Mu granite | Decelles batholith | Feng and Kerrich (1991) |
| Volcanism | | | |
| 2690 ± 2 | Qz-phyric felsic volcanic breccia | Belleterre belt | Mortensen and Card (1993) |
| 2685 ± 1 | Synvolcanic Qz-Fs porphyry | Belleterre belt | Mortensen and Card (1993) |
| 2682 ± 1 | Rhyodacite | Belleterre belt | Mortensen and Card (1993) |
| Siliciclastic sedimentation | | | |
| 3177 +132 / −102 | Graywacke | Pontiac Group | Gariépy et al. (1984) |
| 3028 ± 4; 3021 ± 4 | Graywacke | Pontiac Group | Davis (2002); two zircons |
| 2925 +99 / −85 | Graywacke | Pontiac Group | Gariépy et al. (1984) |
| 2850–2750 | Graywacke | Pontiac Group | Compilation in Davis (2002) |
| 2685 ± 3 | Graywacke | Pontiac Group | Davis (2002); youngest detrital zircon |
| 2683 ± 1 | Graywacke | Pontiac Group | Mortensen and Card (1993); youngest detrital zircon |
| 3042 ± 6–2691 ± 8 | Mica schist and amphibolite | Lacorne Block | Feng and Kerrich (1991) |
| Angular unconformity | | | |
| 2678 ± 4 | Graywacke | Timiskaming Assemblage | Davis (2002); youngest detrital zircon |
| 2674 ± 1 | Conglomerate | Timiskaming Assemblage | Davis (2002); youngest detrital zircon |
| 2673 ± 3 | Interbedded volcaniclastite | Timiskaming Assemblage | Davis (2002); dates deposition |

*Note:* Hb—hornblende; Bt—biotite; Mu—muscovite; Qz—quartz; Fs—feldspar.

Several dates for quartz syenite and syenitic porphyry intrusions range from 2685 Ma through 2672 Ma (Mortensen and Card, 1993; Davis, 2002). Muscovite-bearing and biotite-bearing granites and pegmatites yielded dates between 2668 Ma and 2643 Ma (Feng and Kerrich, 1991; Mortensen and Card, 1993).

The lead isotopic compositions of K-feldspars, determined for late felsic plutonic rocks in the Pontiac Subprovince, indicate they were derived from crustal melts that included components older than 3000 Ma, providing compelling evidence for the presence of pre–3000 Ma basement below the Pontiac Subprovince that is not present below the Abitibi Subprovince (Carignan et al., 1993). The older crustal rocks below the exposed Pontiac Subprovince were tectonically emplaced during the Late Archean orogenic episode that deformed the whole of the Pontiac Subprovince and the Abitibi-Opatica terrane (Benn, 2006). Geochemical and isotopic investigations of the Pontiac Group metasedimentary rocks indicate that some of the detritus received by the basin within which they were deposited included components 3000 Ma to 3500 Ma in age, possibly derived from the older terrane that is now wedged beneath the Pontiac Subprovince (Feng et al., 1993).

## THE ABITIBI-OPATICA TERRANE

Figure 2 is a map of the Abitibi-Opatica terrane and the Pontiac Subprovince, with the metavolcanic units of the Abitibi-Opatica terrane grouped according to ages. The metavolcanic units are grouped in that way because stratigraphic interpretations and nomenclature are not uniform across the Ontario-Quebec provincial boundary. In grouping the metavolcanic units, we chose temporal boundaries that can be traced with confidence across the provincial border and that allow us to discuss the petrogenetic evolution, in time, of volcanism in the study area. The relevant stratigraphic assemblage names, as used in Ontario, are indicated along with ages for the units in Figure 2.

The crustal structure is shown in Figure 3, which is an interpretation of Lithoprobe deep seismic reflection profiles that cross the entire region (Benn, 2006). The interpretation in Figure 3 differs from an earlier-published one that placed a paleosubduction zone beneath the southern Opatica domain (Calvert et al., 1995); we interpret that structure as a thrust that offsets the lower crust. Deformation of the Abitibi-Opatica terrane resulted from the wedging of an older crustal block, buried below the metasedimentary rocks of the Pontiac Subprovince, into the middle crust of the southern Abitibi Subprovince, resulting in the tectonic delamination of the lower crust (Fig. 3). A fuller justification for the interpretation in Figure 3 is provided in Benn (2006).

Figure 3 shows the difference in deformation style between the lower crust of the Abitibi-Opatica terrane and its middle and upper crust. The relatively stiff lower crust was tectonically delaminated in the southern Abitibi Subprovince, and deformed by thrust faulting below the Abitibi-Opatica boundary (Fig. 3) (Benn, 2006). Modeling of the rheological response to collision of the Abitibi crust suggests the stiffer lower crust was composed of mafic granulites, which may be represented in the uplifted, granulite-grade metamorphic rocks of the Kapuskasing structural zone (Fig. 1) (Percival, 1988).

The middle and upper crust deformed in a more ductile manner, resulting in the formation of the upright to slightly overturned regional folds that are characteristic of the entire Abitibi-Opatica terrane (Benn et al., 1992; Sawyer and Benn, 1993; Daigneault et al., 1990; Benn and Peschler, 2005; Peschler et al., 2006). In Figure 3, it can be seen that mapped folds can, in many instances, be correlated with folds that are interpreted from the seismic profiles, in the middle crust. Composite batholiths that are made up of several generations of granitic rocks tend to crop out within, or to be centered on, the cores of anticlines (Peschler et al., 2006). Some batholith complexes are themselves folded along with the greenstones (Benn, 2004).

Inspection of Figure 2 reveals that the ages of volcanic units that crop out in the Abitibi Subprovince tend to decrease from north to south. In the northern part of the Abitibi Subprovince, a region referred to as the Northern Volcanic Zone (Ludden et al., 1986), the volcanic assemblages are dated 2760 Ma to 2723 Ma (Mortensen, 1993a; Legault et al., 2002). In the southern part of the Abitibi Subprovince, referred to as the Southern Volcanic Zone (Ludden et al., 1986), the youngest volcanic units are dated 2698 Ma (Ayer et al., 2002). In the Southern Volcanic Zone, the older assemblages (Pacaud and Deloro Assemblages) crop out in the cores of anticlines and near the margins of batholiths, many of which are themselves hosted by anticlines (Peschler et al., 2004, 2006).

The base of the Abitibi greenstones is shallower in the north of the belt than in the south (Fig. 3), consistent with the outcropping of older volcanic stratigraphy in the northern Abitibi Subprovince. The shallowing of the base of the greenstones in the north may be due, in part, to a northward vergence of the regional folds. It is also due, in part, to the presence of deeply rooted thrusts that are linked to the Casa Berardi fault zone (Fig. 3). Exposure of deeper (older) levels of the volcanic stratigraphy in the northern Abitibi Subprovince is consistent with the interpretation that the orthogneisses of the Opatica domain would represent yet deeper levels of the Abitibi Subprovince.

### Volcanic History

Several studies have documented the geochemistry of the volcanic stratigraphy in the greenstones of the Abitibi Subprovince, and interpreted the geodynamical significance of the volcanic rock compositions. Here we synthesize those results and interpretations with emphasis on the Abitibi Subprovince in Ontario where we can relate the changes in petrogenesis of the volcanic rocks to the well-defined stratigraphy. We also refer to correlative volcanic units in Quebec, and to possibly correlative units in the small Frotet-Evans greenstone belt that lies within the Opatica domain (Fig. 2).

Volcanic rocks from the Abitibi-Opatica terrane belong to two broad associations: (1) an association dominated by medium- to high-$TiO_2$, high-Mg, high-Fe tholeiitic basalts with minor komatiites and komatiitic basalts, and some tholeiitic felsic

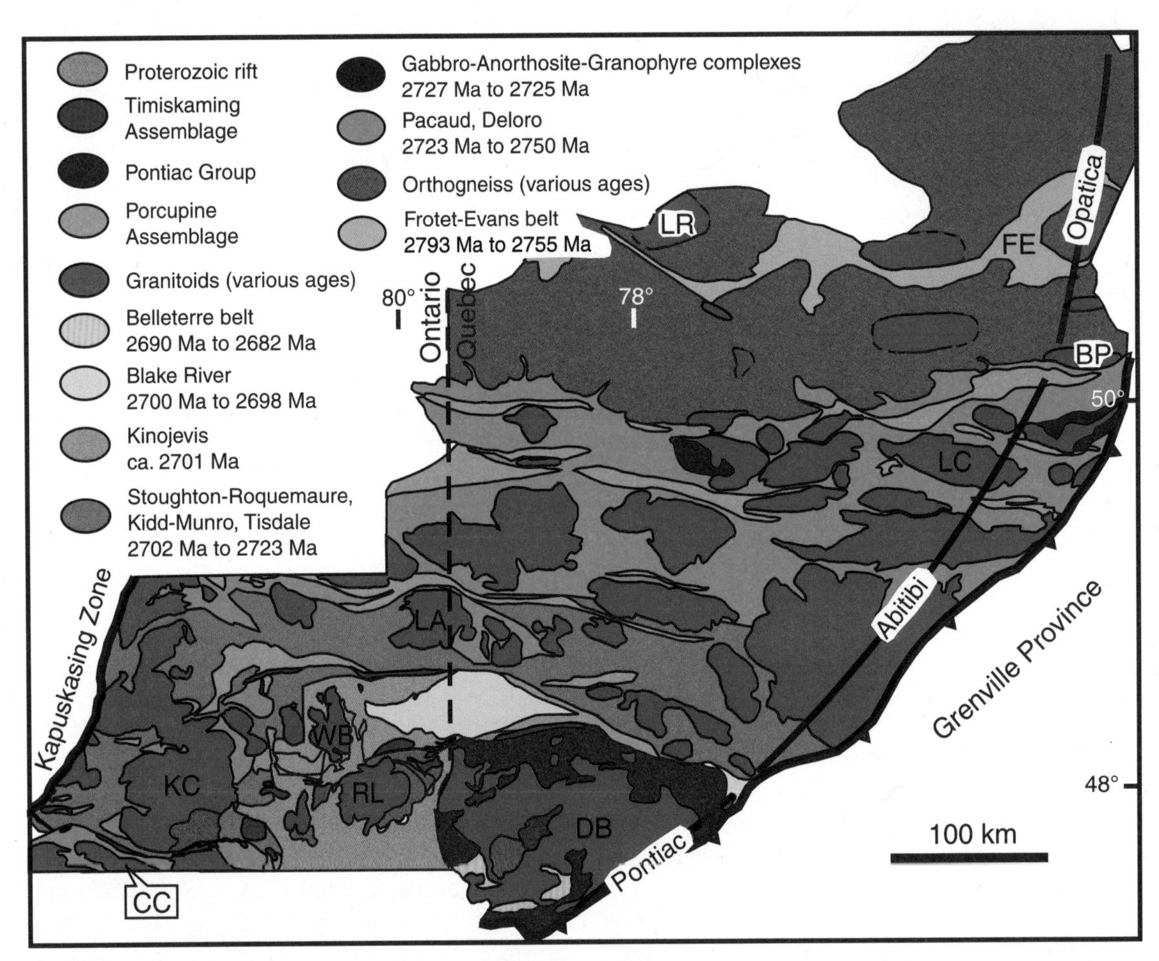


Figure 2. Geological compilation map of the Abitibi-Opatica terrane. CC—Chester complex; LR—Lac Rodayer pluton; RL—Round Lake batholith; KC—Kenogamissi complex; WB—Watabeag batholith; LA—Lake Abitibi batholith; LC—Lapparent complex; BP—Barlow pluton; FE—Frotet-Evans greenstone belt.

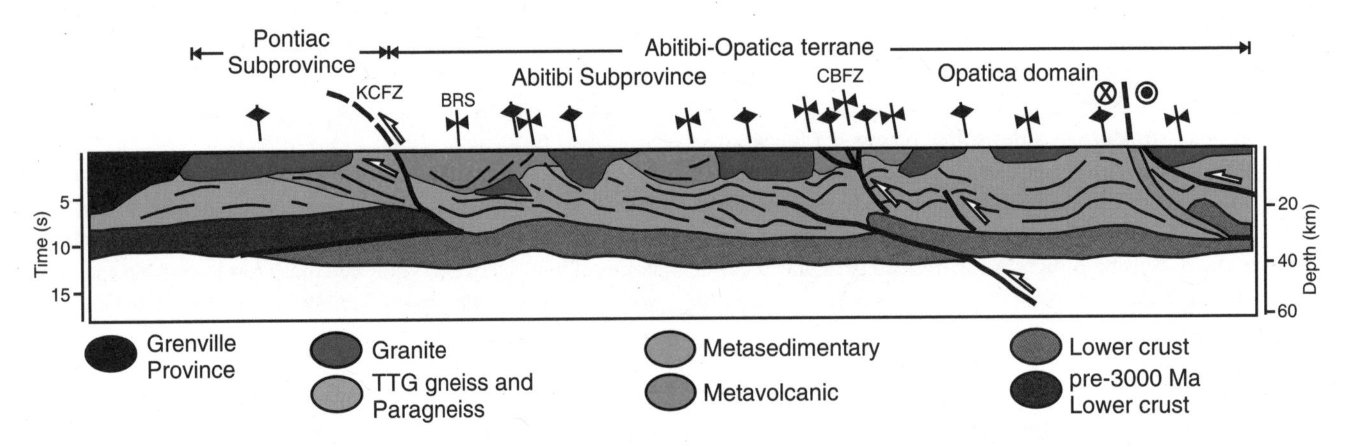


Figure 3. Crustal structure profile of the Abitibi-Opatica terrane, interpreted from surface geology and Lithoprobe seismic reflection profiles. Modified from Benn (2006). BRS—Blake River syncline; CBFZ—Casa Berardi fault zone; KCFZ—Kirkland-Cadillac fault zone.

rocks, that is interpreted as reflecting an intraplate plume-related setting (Kerrich et al., 1999; Sproule et al., 2002), and (2) a more diverse association, composed of both tholeiitic and calc-alkaline lavas and including minor andesites and dacites, boninites, and adakites (Kerrich et al., 1998; Wyman, 1999a; Wyman et al., 2002). Collectively, this assemblage is suggestive of subduction-related magmatism; the presence of adakites is interpreted as reflecting a "hot subduction" in which the magmas include a slab-melt component (Polat and Kerrich, 2006). Wyman (2003) subdivided this group into three assemblages of depleted tholeiites and boninites, arc tholeiites and calc-alkaline rocks, and adakites and high-Mg andesites. For simplicity, in this paper we refer to the two magmatic associations as "plume-related" and "subduction-related," bearing in mind that the direct petrogenetic links between the geochemical signatures of volcanic rocks and geodynamical settings are not everywhere unambiguous.

We summarize the volcanic history of the Abitibi-Opatica terrane and the geochemical signatures of the volcanic rocks with reference to Figure 4, which provides a summary of the tectonomagmatic history.

***The Frotet-Evans Greenstone Belt***

The Frotet-Evans greenstone belt lies to the north of the Abitibi greenstones and is surrounded by the metaplutonic rocks of the Opatica domain (Fig. 2). In the Frotet-Evans greenstone belt, a package of magnesian- and ferrotholeiites may be interstratified with volcanic rocks of reportedly boninitic affinity and are either interlayered with or overlying calc-alkaline volcanic units (Boily and Dion, 2002). Reported dates for the Frotet-Evans belt (Boily and Dion, 2002) indicate it includes volcanic units with ages between 2793 Ma, which is 30 m.y. older than any known volcanic units in the Abitibi Subprovince, and 2755 Ma, which is time-correlative to some of the oldest volcanic stratigraphy in the Abitibi Subprovince (Fig. 4A). Here we speculate that the youngest units in the Frotet-Evans belt may be stratigraphically correlated with the lower parts of the volcanic stratigraphy in the Abitibi Subprovince. If that correlation is correct, it is likely that the ca. 2755 Ma units in the Frotet-Evans belt would rest upon an unconformity, structurally above the much older volcanic units.

***Initial Plume-Related Volcanism (2760–2735 Ma)***

In Ontario, the Pacaud Assemblage represents a 15 m.y. period of volcanism, 2750–2735 Ma (Fig. 4B). Volcanic units as old as 2760 Ma, in the Northern Volcanic Zone in Quebec, may be correlative with the Pacaud Assemblage or may include older units. The Pacaud Assemblage includes a relative volume of <1% komatiitic rocks (komatiites, komatiitic basalts) that are reported to be Ti-depleted and that are interpreted to have been generated by partial melting of a deep, garnet-rich mantle source, strongly suggesting an origin in a mantle plume (Sproule et al., 2002). A study of the lithogeochemistry of the Pacaud Assemblage, in the western Abitibi Subprovince, shows the remainder is predominantly composed of Mg- and Fe-rich tholeiites that can also be attributed to melting within a mantle plume (Oliver et al., 2000). This stage of plume-related magmatism was followed by a 5 m.y. period of apparent volcanic quiescence, documented in the stratigraphy of the Ontario portion of the Abitibi Subprovince (Fig. 4B) (Ayer et al., 2002).

***Subduction-Related Volcanism (2730–2724 Ma)***

The Deloro Assemblage represents a period of eruption of magmas with compositions that suggest a suprasubduction zone environment. It has a large component of intermediate to felsic rocks that have calc-alkaline geochemical signatures (Oliver et al., 2000). The Deloro Assemblage is similar in age to, and has been stratigraphically correlated with, the volcanic Hunter-Mine Group in the Southern Volcanic Zone in Quebec (Ayer et al., 2002). Dikes of dacitic and rhyolitic composition in the Hunter Mine Group also have arc-like compositions (Dostal and Mueller, 1996). Units of the Joutel volcanic complex, in the Northern Volcanic Zone, are also in the same age range and also have calc-alkaline geochemistry (Legault et al., 2002). Significantly, komatiitic rocks are absent from the Deloro Assemblage and from time-correlative units in Quebec. The Deloro period of volcanism was also a period during which gabbro-anorthosite complexes were emplaced within the Northern Volcanic Zone (Figs. 2 and 4B).

***Renewed Komatiitic-Tholeiitic Volcanism (2723–2720 Ma) and Concomitant Subduction-Related Volcanism (2719–2701 Ma)***

The Stoughton-Roquemaure Assemblage marks a shift in the petrogenesis of volcanic rocks, suggesting renewed plume-related volcanism following the deposition of the Deloro Assemblage that represents subduction-related volcanism (Dostal and Mueller, 1997). Komatiites represent ~2% of the Stoughton-Roquemaure Assemblage (Sproule et al., 2002), the remainder being tholeiitic basalts and minor amounts of felsic volcanic rocks. In Quebec, the Stoughton-Roquemaure Assemblage overlies the Hunter Mine Group, which is correlated to the Deloro Assemblage. The komatiites of the Stoughton-Roquemaure Assemblage have geochemical signatures that suggest melting of a more depleted harzburgitic mantle source, and at shallower depths, compared to the komatiites of the Pacaud Assemblage, which has been interpreted to indicate melting of the head of a mantle plume (Sproule et al., 2002).

Subduction-related volcanic products are interstratified with plume-related volcanic rocks in the Kidd-Munro and Tisdale Assemblages (Fig. 4B), which represent ~16 m.y. of the Abitibi volcanic history (Ayer et al., 2002). Komatiitic rocks represent ~5% of the volume of each of the Kidd-Munro and Tisdale Assemblages, and they have geochemical signatures similar to the komatiites of the Stoughton-Roquemaure Assemblage (Sproule et al., 2002), suggesting melting of a shallow, depleted mantle source.

The geochemical compositions of subduction-related rocks suggest interactions of slab-derived fluids and/or melts with a depleted mantle wedge. In the Kidd-Munro Assemblage and

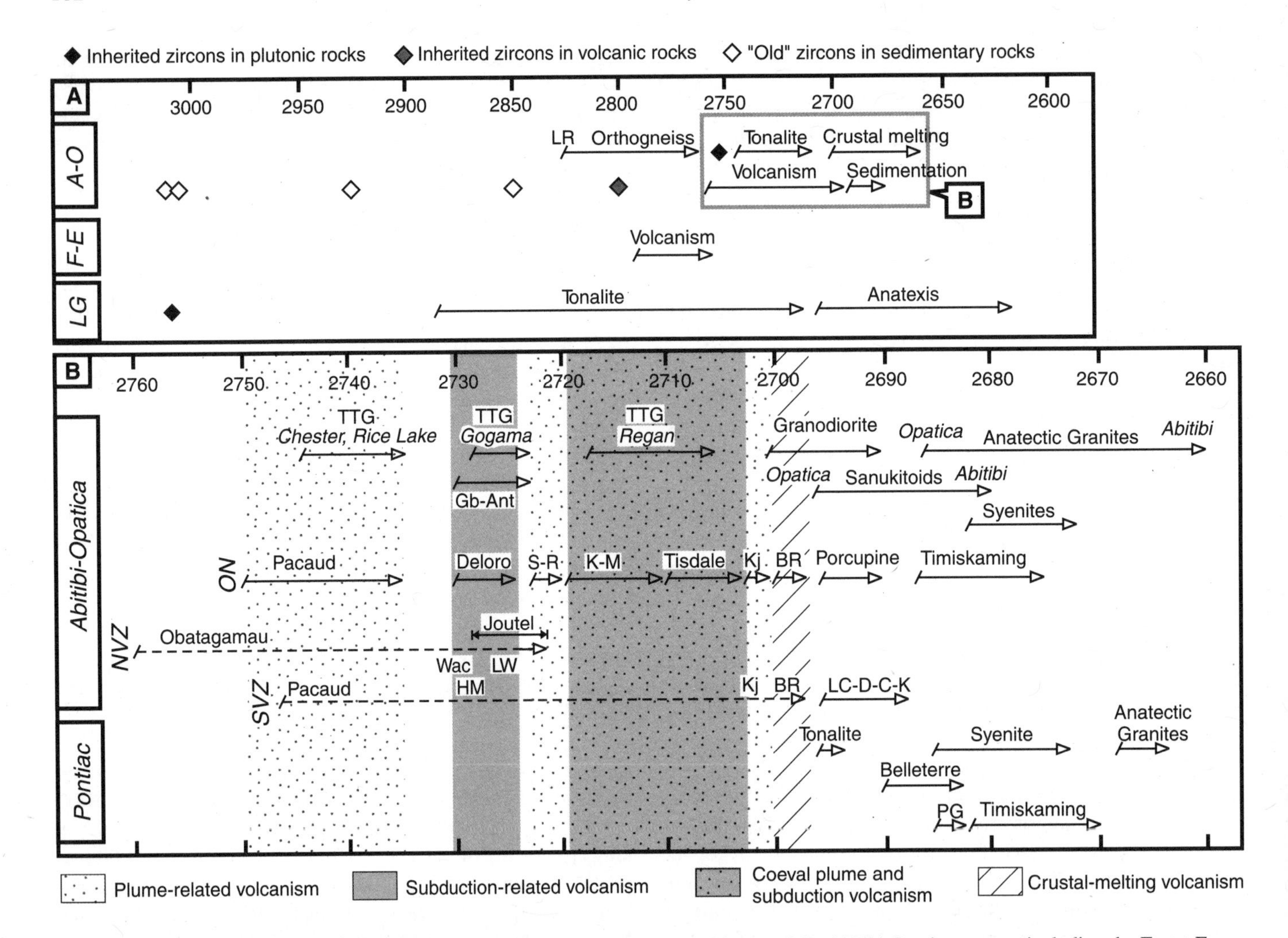


Figure 4. Compilation of geochronological data, depicting the tectonomagmatic history of the Abitibi-Opatica terrane, including the Frotet-Evans greenstone belt, and of the La Grande and Pontiac Subprovinces. Ages given in Ma. (A) Reported occurrences of inherited zircons in plutonic and volcanic rocks are indicated, as are occurrences of detrital zircons in clastic sedimentary rocks, which are thought to be much older than the rocks themselves. A-O—Abitibi-Opatica terrane; F-E—Frotet-Evans greenstone belt; LG—La Grande domain; LR—Lac Rodayer pluton. (B) Shaded areas indicate different magmatic periods, as interpreted by us, based on the published interpretations of the geochemistry of volcanic units in the Abitibi Subprovince greenstones. ON—Ontario; NVZ—Northern Volcanic Zone (in Quebec); SVZ—Southern Volcanic Zone (in Quebec); Gb-Ant—gabbro-anorthosite complexes; S-R—Stoughton-Roquemaure Assemblage; K-M—Kidd-Munro Assemblage; Kj—Kinojevis Assemblage; BR—Blake River Assemblage; Wac—Waconichi Formation; LW—Lac Watson Formation; HM—Hunter Mine Group; LC-D-C-K—Lac Caste, Duparquet, Cadillac, and Kewagama Groups, in the Southern Volcanic Zone, Quebec; PG—Pontiac Group.

in the Malartic Block, in the Southern Volcanic Zone in Quebec, depleted tholeiites are interpreted to be of boninitic affinity (Wyman, 1999a) indicative of wedge melting induced by slab-derived fluids or melts (Wyman et al., 2002). The presence of high-Mg andesites, Nb-enriched basalts, and adakites are also indicative of interactions of slab melts with a mantle wedge, suggesting they formed in a "hot subduction" environment (Kerrich and Polat, 2006; Polat and Kerrich, 2006).

The Kinojevis Assemblage (2702–2701 Ma) is composed of high-Mg and high-Fe tholeiitic basalts with a stratigraphic evolution to Fe enrichment, and to tholeiitic andesite, dacite, and rhyolite toward the top of the assemblage. Rare komatiitic flows are also reported in the Kinojevis Assemblage (Lafleche et al., 1992). The geochemistry of the Kinojevis Assemblage is suggestive of plume-type magmatism (Fowler and Jensen, 1989).

***Terminal Volcanism: Crustal Melting (2700–2698 Ma)***

The uppermost volcanic stratigraphy in the Abitibi Subprovince is the Blake River Assemblage, which represents a 2 m.y. period of volcanism with a significant component derived from crustal melting. The Blake River Assemblage is composed of interstratified mafic-tholeiitic and mafic to rhyolitic calc-alkaline volcanic rocks. Geochemical modeling shows that the calc-alkaline components of the Blake River Assemblage could have been generated by partial melting of pyroxene-garnet granulite or garnet eclogite (Lafleche et al., 1992).

## History of Granitic Plutonism

The Opatica domain represents the northern part of the Abitibi-Opatica terrane and is composed of tonalite orthogneiss and younger tonalite, granodiorite, and monzodiorite plutons, all crosscut by late biotite granites (Benn et al., 1992; Sawyer and Benn, 1993).

Three tonalite orthogneiss samples from the northern part of the Opatica domain provided dates between 2810 Ma and 2760 Ma. The three dates were calculated based on regression of several zircon fractions, possibly explaining the 10 Ma to 40 Ma analytical errors associated with the dates (Table 1) (Davis et al., 1995). They overlap, in time, with zircon dates from the tonalite basement gneisses of the La Grande domain. The Lac Rodayer pluton, a tonalite-diorite intrusion situated near the boundary of the Opatica domain with the Opinaca-Nemiscau domain (Fig. 2), yielded a more precise date, 2825 ± 3 Ma (Davis et al., 1995), also falling within the range of ages determined for the older tonalites within the northern part of the La Grande domain. Therefore, the geochronological data suggest that the northern Opatica domain, as mapped in Figure 2, may contain remnants of a margin that was once part of the La Grande domain.

The shared plutonic history of the Abitibi Subprovince and the Opatica domain began with the emplacement of tonalite plutons, dated between 2744 Ma and 2740 Ma. Tonalite plutons of those ages are documented in the Opatica domain (Davis et al., 1995), and in the Round Lake batholith, the Kenogamissi complex, and the Chester complex (Table 1), which crop out within the greenstones of the Abitibi Subprovince (Fig. 2). The ca. 2740 Ma tonalite plutonism postdates the oldest known volcanic stratigraphy within the Abitibi Subprovince by ~20 m.y.

Inspection of Table 1 reveals that tonalite plutons as young as 2690 Ma are reported from the Abitibi Subprovince (for example, a component of the Lake Abitibi batholith; Table 1). However, our experience shows that in some cases nonporphyritic granodiorites in the Abitibi Subprovince have been reported as tonalites. Based on our analysis of the Kenogamissi complex (next section), we suggest that tonalite plutonism probably ended ca. 2700 Ma, though other workers have suggested that tonalite plutonism may have continued until ca. 2690 Ma (Feng and Kerrich, 1992a, 1992b). Alternatively, it is possible that the transition from tonalitic to granodioritic magmatism was diachronous, occurring at ca. 2700 Ma in the west (Kenogamissi complex) but only at 2695 or even 2690 Ma farther east (Round Lake and Lake Abitibi batholiths).

The emplacement of biotite-bearing, porphyritic and nonporphyritic granodiorite plutons occurred between 2700 Ma and 2685 Ma (Fig. 4B). A recognizably different suite of rocks, consisting of hornblende granodiorite, hornblende monzodiorite, hornblende monzonite, and hornblende diorite intrusions overlapped in time with the biotite granodiorites, occurring at 2696 Ma in the Opatica domain (Barlow pluton, Fig. 2) and between 2686 Ma and 2680 Ma in the Abitibi Subprovince (Fig. 4B). Our analysis of the geochemistry of the Kenogamissi complex (next section) leads us to interpret those intrusions as representing sanukitoids. The last plutonic event, the emplacement of anatectic granites, began ~26 m.y. earlier in the Opatica domain (2686 Ma) than in the Abitibi Subprovince (2660 Ma) (Table 1; Fig. 4B).

## The Kenogamissi Complex

The Kenogamissi complex is located in the southwestern part of the Abitibi Subprovince (Fig. 2). It is a large crystalline complex composed of multiple plutons with distinct petrogenetic signatures, spanning the range of plutonic types in the Abitibi-Opatica terrane. The western part of the Kenogamissi complex was mapped in detail by Heather and Shore (1999). We have mapped the remainder of the complex in order to establish the spatial distribution of the principal plutonic units (Fig. 5).

### *Tonalite*

The oldest dated plutonic unit is the 2742 Ma hornblende-biotite tonalite (Ayer et al., 2004) that forms a unit in the southwestern lobe of the Kenogamissi complex, referred to as the Rice Lake batholith (Becker and Benn, 2003).

A second tonalite unit, the Gogama orthogneiss, is composed of several phases of hornblende-biotite tonalite (to quartz diorite) and biotite tonalite (to trondhjemite) with numerous enclaves of amphibolite-grade diorite to quartz diorite. A date of 2723 Ma is reported for a hornblende-biotite tonalite component of the Gogama orthogneiss (Heather and Shore, 1999). The same rock yielded a date of 2735 Ma that was interpreted to represent inherited zircons (Heather and Shore, 1999), and that is interpreted by us to be derived from the ca. 2740 Ma tonalite unit.

A third tonalite unit, the Regan tonalite, contains xenoliths of the Rice Lake and Gogama units. The Regan tonalite has yielded a 2713 Ma date. It forms a very large body in the north-central part of the Kenogamissi complex (Fig. 5), and rocks of that unit also form outcrops near the southern and eastern margins of the Kenogamissi complex.

### *Biotite Granodiorite*

Two suites of biotite granodiorites are identified. The oldest is a nonporphyritic white granodiorite, dated 2701 Ma, that makes up part of the Rice Lake batholith (Fig. 5) where it intruded the 2742 Ma tonalite (Becker and Benn, 2003). The McOwen pluton (Fig. 5) has a very similar appearance in the field, and we interpret that it is the equivalent to the Rice Lake granodiorite. A typically K-feldspar porphyritic, white to gray granodiorite, dated 2696 Ma, is named the Roblin pluton (Fig. 5).

### *Hornblende Granodiorite*

The hornblende-bearing, K-feldspar porphyritic granodiorite unit is dated 2682 Ma. It is easily distinguished in the field by the presence of hornblende and of centimeter- to decimeter-scale,

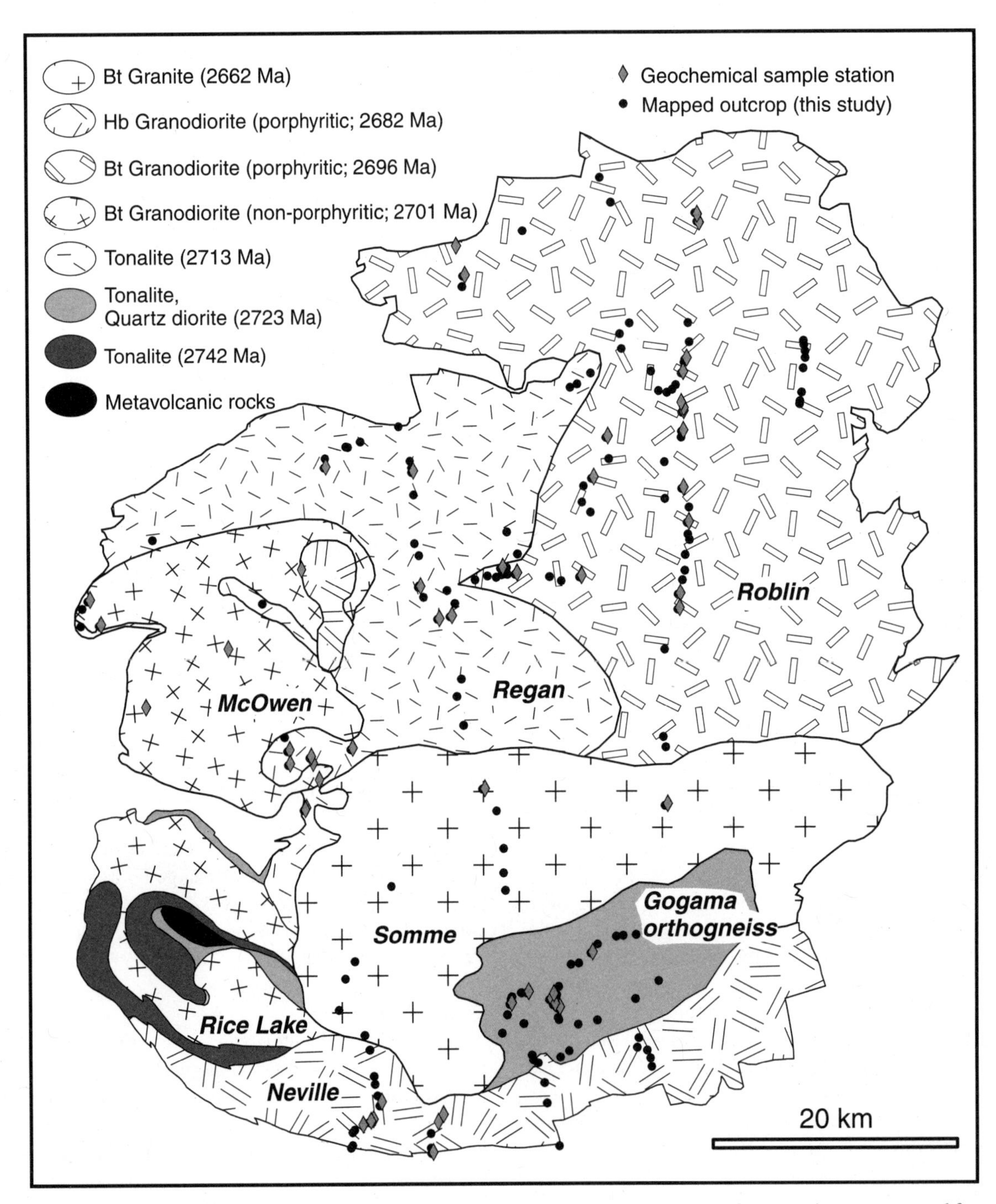


Figure 5. Geological map of the Kenogamissi complex, which can be located in Figure 2. Outcrops that were mapped for this study, and outcrops that were sampled by us for geochemical analyses are indicated. Geological contacts in the western part of the Kenogamissi complex are from Heather and Shore (1999).

angular to subrounded amphibolite enclaves. The largest mapped body of this unit is the Neville pluton (Fig. 5).

### *Granite*

The tonalite and granodiorite units of the Kenogamissi complex are intruded by pink (to gray) biotite granites and associated pegmatites, referred to as the Somme granite (Fig. 5). The Somme granite was dated 2662 Ma. It is most abundant in the south-central part of the Kenogamissi complex, where dikes are present in virtually every outcrop of the Gogama orthogneiss, the Neville granodiorite, and the southern part of the Roblin pluton. The granite is, in general, undeformed, though deformed and foliated dikes of the Somme granite are observed within the Neville pluton close to the southern margin of the Kenogamissi complex.

## Geochemistry of the Kenogamissi Complex

We collected 86 samples from the different plutonic units in the Kenogamissi complex (Fig. 5), and also seven samples from the 2740 Ma (Heather and Shore, 1999) Chester trondhjemite-tonalite complex, to the south of the Kenogamissi complex (Fig. 2) that is age-correlative to the 2742 Ma Rice Lake tonalite. Part of the sample set was analyzed for major elements using XRF at Stellenbosch University, and for trace elements using LA-ICP-MS at the University of Cape Town. The analytical procedure is based on the method described by Eggins (2003). The rest of the sample set was analyzed by ICP-AES and ICP-MS at the Geoscience Laboratories, Mines and Minerals Division of the Ontario Ministry of Northern Development and Mines, in Sudbury. No systematic differences appear between the data sets. In addition, 36 analyses of plutonic rocks in the Kenogamissi complex were incorporated from an Abitibi geological database, compiled and marketed by the Ontario Geological Survey (Ayer et al., 2004).

The geochemistry of the 2740–2660 Ma rocks record a progressive change toward more and more potassic rocks, from tonalites to granodiorites and eventually granites (Fig. 6). Two important changes occurred in the geochemical signatures of the plutonic rocks during the magmatic history of the study area. A first change, ca. 2723–2713 Ma, is revealed in the petrogenetic signatures of tonalites, and a second one, ca. 2710–2700 Ma, corresponds to the transition from sodic plutons (tonalites) to potassic ones (granodiorites and granites).

### *Tonalites (2740–2700 Ma)*

The three generations of tonalite broadly belong to the TTG (tonalite-trondhjemite-granodiorite) series. There is reasonable agreement on the fact that TTGs are generated by partial melting of metabasite rocks (amphibolites or eclogites) within the garnet stability field, although the details of the melting are still debated (see review in Moyen and Stevens, 2006). On the other hand, the geodynamical environment of TTG generation is hotly debated. Two main geodynamic settings have been proposed: One is slab melting in a hot subduction zone (Arth and Hanson, 1975; Moorbath, 1975; Barker, 1976; Condie, 1986;

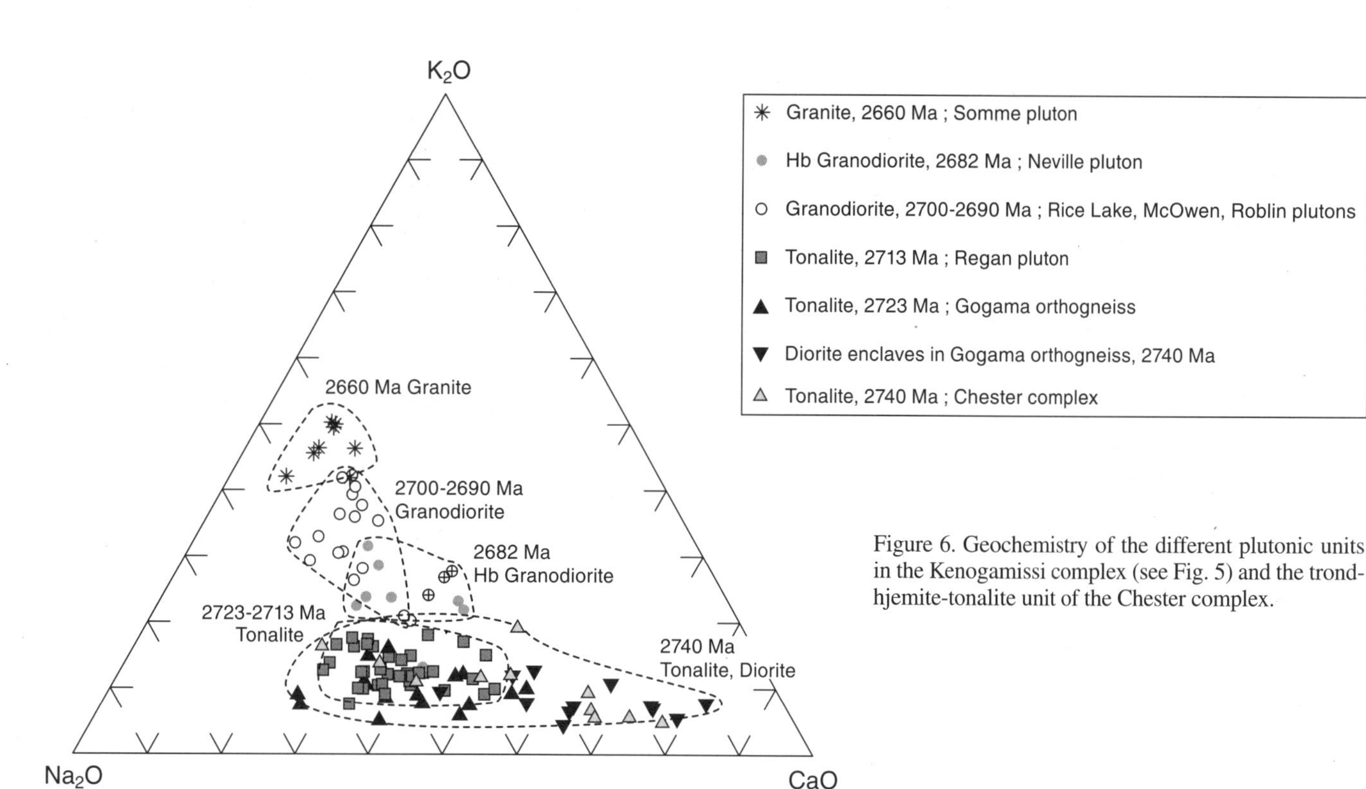


Figure 6. Geochemistry of the different plutonic units in the Kenogamissi complex (see Fig. 5) and the trondhjemite-tonalite unit of the Chester complex.

Jahn et al., 1981; Rapp et al., 1991; Martin, 1986, 1994, 1999; Rapp and Watson, 1995; Foley et al., 2002; Martin et al., 2005), the other is melting of the lower parts of a thick, mafic crust in an intraplate setting (Maaløe, 1982; Kay and Kay, 1991; Collins et al., 1998; White et al., 1999; Zegers and Van Keken, 2001; Van Kranendonk et al., 2004; Bédard, 2006).

The TTGs generated in each of the two contrasting geodynamic contexts should be distinguishable because the pressure-temperature conditions of melting are significantly different in the two cases (Moyen and Stevens, 2006; Moyen et al., 2007). Intraplate situations are expected to correspond to relatively low pressures (12–15 kbar), for temperatures of 1000 °C to 1100 °C, whereas subducting slabs (even when relatively hot) would undergo melting at higher pressures (15–25 kbar) but somewhat lower temperature (900–1000 °C). Melting of a subducted slab would correspond to lower melt fractions, coexisting with a garnet-rich assemblage (30%–50% of the residuum), and yielding high-$SiO_2$, high-Na/Ca tonalites, with high Sr/Y (Moyen et al., 2007), fairly similar to modern, subduction-related "high-silica adakites" (Martin et al., 2005). In contrast, intraplate melting would correspond to higher melt fractions, coexisting with garnet-poorer assemblages (although still containing an appreciable amount of this mineral, around 10%), resulting in relatively low-$SiO_2$ (60–65 wt%), low-Na/Ca tonalites with moderate Sr/Y ratios. Interestingly, the Aruba tonalitic pluton (White et al., 1999), which probably formed by remelting of the Caribbean plateau, shows similar geochemical features.

Of course, the chemistry of TTG melts is not simply a function of melting processes. Later processes, such as fractional crystallization and/or AFC (assimilation–fractional crystallization) (Kamber et al., 2002; Kleinhanns et al., 2003; Bédard, 2006), or interactions with the mantle wedge (Smithies and Champion, 2000; Martin and Moyen, 2002; Martin et al., 2005), can potentially affect the primitive melt's compositions. Fractional crystallization (especially at high pressure, involving epidote and garnet; Schmidt, 1993; Schmidt and Thompson, 1996; Bédard, 2006) can produce high-silica, high-Sr/Y differentiated melts from tonalites. However, closer examination of the trends and modeling of both major and trace element behavior during crystallization (Moyen et al., 2007) reveals that low- and high-$SiO_2$ TTGs cannot be related to each other by fractionation of any realistic mineral assemblage.

On the other hand, interactions with the mantle, by assimilation of, or reaction with, mantle peridotites, would to some degree blur the geochemical identity of the "deep," high-silica melts and make them more similar to lower-*P*, higher melt-fraction liquids (Smithies and Champion, 2000; Rapp et al., 1999; Moyen et al., 2003; Rapp, 2003). However, the existence of $SiO_2$-rich leucocratic melts with high Sr/Y is a strong indication that such interactions, for some reason, did not occur to a significant degree in the formation of at least some of the TTGs, and points to deep, low melt-fraction melting, probably in a subduction setting. Indeed, melts generated in subduction zones rise very quickly through the mantle, as demonstrated by both geochronological (Sigmarsson et al., 2002) and physical (Spiegelman et al., 2001) considerations, typically in less than $10^4$ yr. Since, on the other hand, typical diffusion rates between melts and crystals are slow (1–10 mm/m.y.; Fourcade et al., 1992; Spear, 1993), e.g., $10^{-2}$ to $10^{-1}$ mm during the time it takes to transfer magmas from the subducted slab to the surface, it is very possible to preserve the pristine geochemical signature of melts generated in subduction zones even if they have to cross a relatively large mantle wedge.

The tonalite units dated ca. 2740 Ma, including the 2742 Ma Rice Lake tonalite (Fig. 5) and the 2740 Ma Chester complex, as well as the dioritic enclaves hosted by the Gogama tonalites, typically belong to the low-$SiO_2$ group (Fig. 7A). This suggests high melt fractions during melting, coexisting with a relatively garnet-poor assemblage, the situation attributed to a relatively high geothermal gradient in an intraplate situation, e.g., in an active oceanic plateau above a hot mantle plume, or in a continental rift above an active plume. In contrast, most samples of the younger 2723 Ma Gogama tonalites and the 2713 Ma Regan tonalite correspond to higher-$SiO_2$ rocks (Fig. 7A), probably generated under a lower geothermal gradient, at higher pressure; they would therefore more likely correspond to melting of a subducting slab.

The dichotomy in tonalite compositions is also seen in Figure 7B where the older, ca. 2740 Ma tonalites, and the dioritic enclaves in the 2723 Gogama tonalites, have very low Sr/Y ratios, whereas the younger tonalites of the Gogama orthogneiss and the Regan pluton, trend to higher Sr/Y ratios. Although Sr mobility due to alteration or metamorphism has been suspected in Archean plutonic rocks (mostly on the basis of Rb-Sr isochrons with wrong apparent ages), the differences observed in Y and Sr/Y ratios are first-order features. Massive gain or loss of Sr (50%–100% or even more) would be required to affect the compositional difference between the two groups. It is very unlikely that modifications of this magnitude could occur without also significantly modifying the mineralogy or major element compositions of the rocks.

Interestingly, published data (Feng and Kerrich, 1992a) reveal the existence of the same two TTG "subseries" farther east in the Abitibi Subprovince (Round Lake and Lake Abitibi batholiths) within their "TTGM" group, a relatively low-$SiO_2$ group defining a low-Sr, moderate-Sr/Y trend, and a higher-$SiO_2$ group with higher Sr and Sr/Y values. This suggests that our observation is fairly robust at the scale of the Abitibi Subprovince.

### *Granodiorites and Granites (2700–2660 Ma)*

In contrast to the TTGs, the granodiorites and granites show a potassic composition ($K_2O$ > 2 wt%, whereas it is always <2 wt% for both TTG groups). They are typically higher in $SiO_2$ than the TTGs, at >70 wt% (except for the Neville pluton, a hornblende granodiorite that has $SiO_2$ around 65 wt% for the same $K_2O$ values). They are otherwise fairly similar to the TTGs, showing the same high Sr/Y values, fractionated REE patterns,

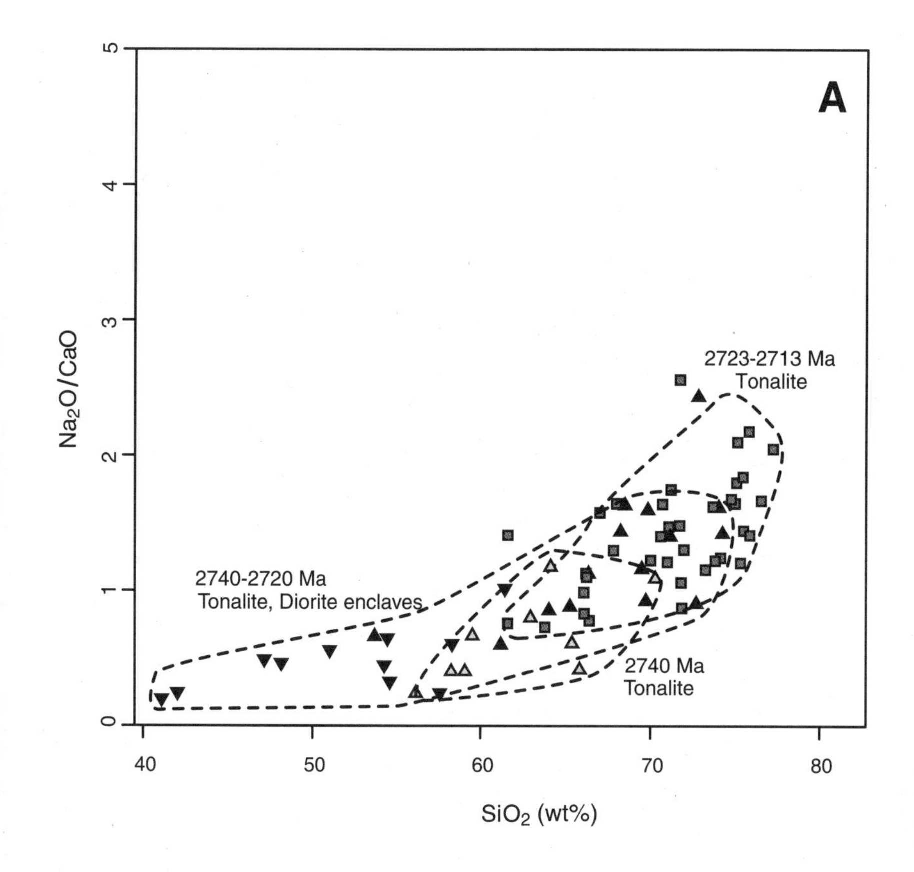


- ■ Tonalite, 2713 Ma ; Regan pluton
- ▲ Tonalite, 2723 Ma ; Gogama orthogneiss
- ▼ Diorite enclaves in Gogama orthogneiss, 2740 Ma
- △ Tonalite, 2740 Ma ; Chester complex

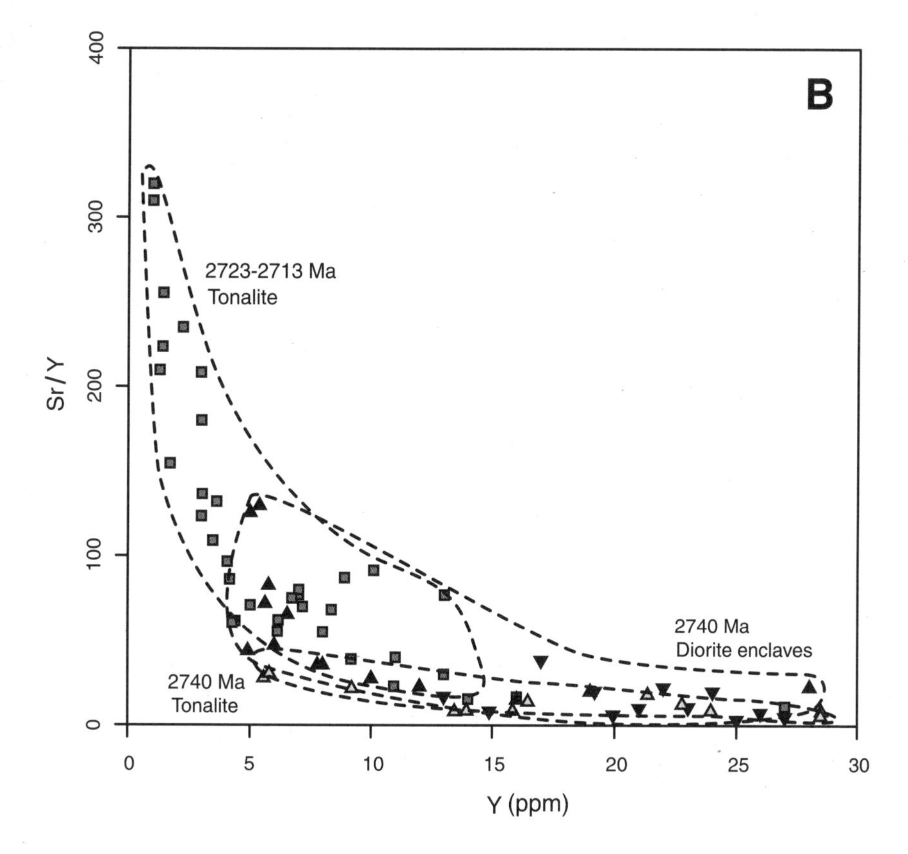


Figure 7. Geochemistry of the TTG units and their dioritic enclaves in the Kenogamissi complex (see Fig. 5) and the trondhjemite-tonalite unit of the Chester complex.

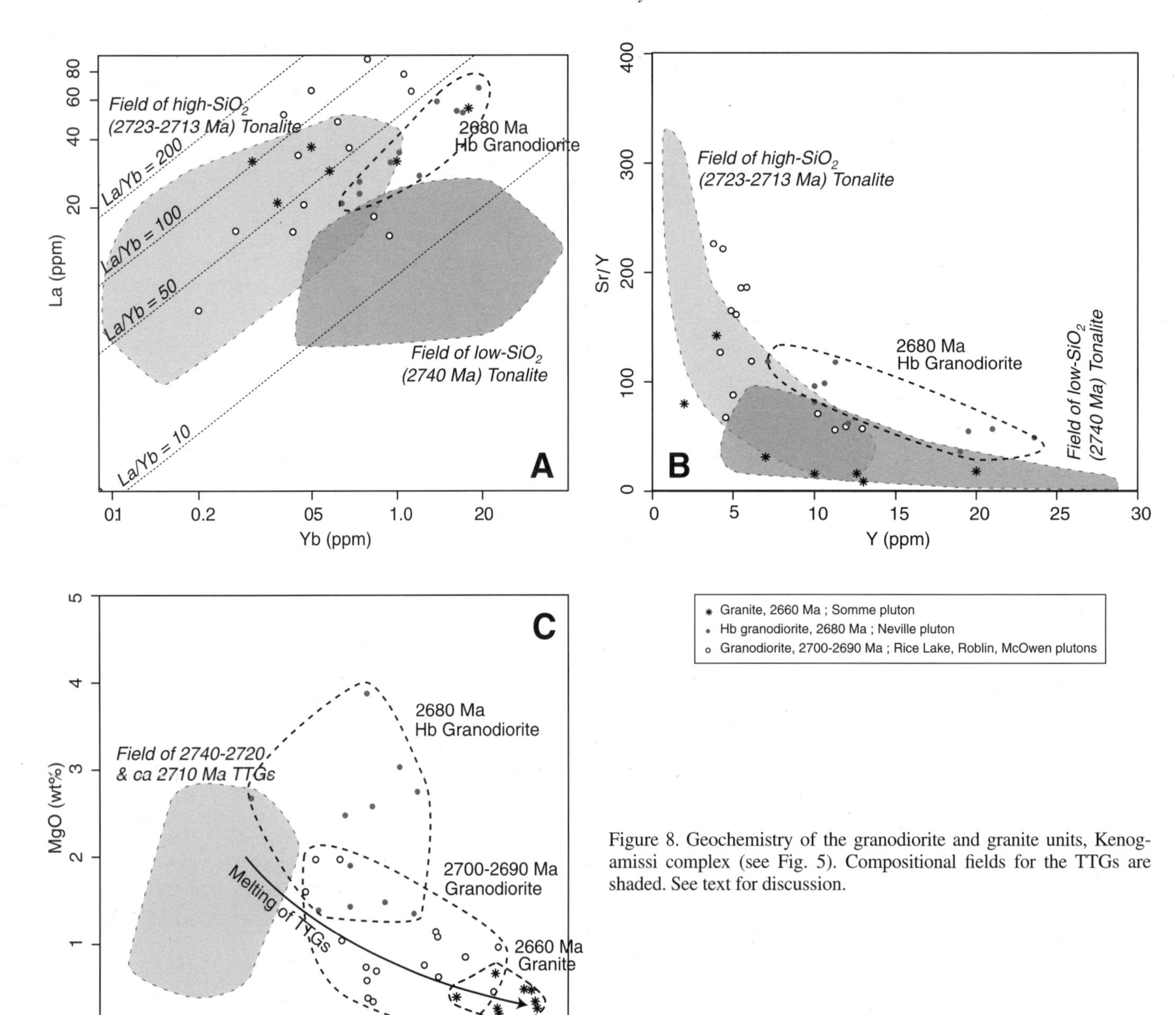


Figure 8. Geochemistry of the granodiorite and granite units, Kenogamissi complex (see Fig. 5). Compositional fields for the TTGs are shaded. See text for discussion.

and high La/Yb (Figs. 8A and 8B); this probably explains why the published literature commonly describes these granodiorites as belonging to the TTG series.

On the basis of both geochemical and experimental studies, comparable rocks are classically interpreted to be the product of remelting of preexisting TTGs (Gardien et al., 1995; Patiño-Douce and Beard, 1995; Champion and Smithies, 1998; Patiño-Douce, 2005). In the Kenogamissi complex, that interpretation is supported by the presence of leucocratic dikes that crosscut the tonalite units that underlie the granodiorite plutons. The dikes are interpreted to have been formed during folding of the partly molten tonalites, leading to the mobilization of melt into dikes that formed on fold limbs (Benn, 2004).

Some differences are recognized between the geochemical signatures of the biotite granodiorite, the hornblende granodiorite, and the granite. The oldest of the three units, the ca. 2700–2690 Ma biotite granodiorites, are moderately potassic (2–4 wt% $K_2O$) with high $SiO_2$ (68–72 wt%) (Fig. 8C). Forming them from TTGs (~1 wt% $K_2O$) requires relatively large melt fractions, from 25% to 50%, which is consistent with the significant volume of the granodiorite plutons (Fig. 5). This points to widespread melting of the crust, ca. 2700–2690 Ma.

The slightly younger Neville pluton (2680 Ma) is volumetrically less important than the biotite granodiorites, forming a sheet-like body inside the southern contact of the Kenogamissi complex, and it is geochemically distinctive. It features lower $SiO_2$ (60–68 wt%) and somewhat higher MgO for similar $K_2O$ contents (2–4 wt%) (Fig. 8C), consistent with its hornblende-bearing nature. It also tends to be enriched in REEs and LILEs compared to the other rocks. These characteristics are evocative of the so-called sanukitoid suite, well known in the Superior Province (Shirey and Hanson, 1984; Stern et al., 1989; Sutcliffe et al., 1990; Stern and Hanson, 1991; Henry et al., 1998; King et al., 1998; Stevenson et al., 1999; Beakhouse and Davis, 2005). The sanukitoids are a high-$K_2O$, high-MgO group of Late Archean plutonic rocks, with trace element signatures similar to the TTGs. They are formed by interaction between slab melts (melts from garnet-bearing metabasites) and mantle peridotites (Shirey and Hanson, 1984; Stern et al., 1989; Smithies and Champion, 1999; Moyen et al., 2003; Rapp, 2003; Martin et al., 2005).

The geodynamic significance of the slab melts–mantle peridotites interaction is a subject of debate. It may point to a "hot" subduction, in which the TTG melts generated in the slab interact with the overlying mantle wedge (Rapp et al., 1999; Moyen et al., 2003), or it may occur due to remelting of a mantle that was premetasomatized by slab melts, typically in a postcollision situation (Shirey and Hanson, 1984; Stern et al., 1989; Stern and Hanson, 1991; Smithies and Champion, 2000; Moyen et al., 2003). The sanukitoid affinity of the Neville pluton is not unambiguously demonstrated but is likely. If the hornblende granodiorites are indeed sanukitoids, then they would have been generated during collisional to postcollisional melting of the premetasomatized mantle, since they postdate all tonalite plutonism and all volcanism in the Abitibi Subprovince.

The ca. 2660 Ma biotite granites correspond to relatively small volumes (although they are geographically widespread, they appear to be thin sheets), and are commonly associated with aplites and pegmatites. They plot along the same trends as the 2700–2690 Ma granodiorites, but show higher $SiO_2$ (>72 wt%) and $K_2O$ (4–5 wt%) (Figs. 6 and 8C). This suggests late, possibly fluid-present, low-melt-fraction melting of preexisting tonalites or granodiorites. In the Opatica domain, diatexites in tonalitic rocks contain leucosomes that may be correlated with late biotite granite plutons that were emplaced at higher structural levels (Sawyer, 1998).

In terms of geodynamic history, both the 2700–2690 Ma granodiorites and the 2660 Ma granites are exclusively products of melting of felsic crustal lithologies (TTGs, or the first generation of granodiorite), and as such are not very diagnostic of a given geodynamical context. They are, however, consistent with a collisional context, having been generated after the cessation of volcanism and emplaced during regional shortening. The transition from tonalitic to granodioritic (and sanukitoid) magmatism is observed at the scale of the whole Abitibi Subprovince (and also the Wawa Subprovince, farther to the west, Fig. 1) (Sutcliffe et al., 1990; Feng and Kerrich, 1992a, 1992b; Chown et al., 2002; Beakhouse and Davis, 2005). The transition occurs between 2700 Ma and 2685 Ma throughout the Abitibi Subprovince.

## DISCUSSION

The identification of the large tectonic terrane that includes the Abitibi Subprovince and the Opatica granite-gneiss domain, referred to as the Abitibi-Opatica terrane, is a fundamentally important development in our understanding of the southern Superior Province. It represents a shift in our thinking of Late Archean tectonics in the region from the accretion of multiple, relatively small allochthonous crustal fragments, to the building of a large volcano-plutonic province during a period of some 80–100 m.y. We have synthesized the data on the region and we have introduced an important new set of data, the geochemistry of the plutonic suites, in order to provide the framework for the geodynamic origin and the tectonomagmatic history of the region.

In this discussion we focus on the salient features of the Abitibi-Opatica terrane that are important to understand how it formed, and the plate interactions that preceded its tectonic accretion. We also point out some of the limits of our knowledge. A model for the geodynamic origin and the tectonomagmatic evolution of the Abitibi-Opatica terrane is presented that satisfies the first-order constraints imposed by the data.

### Terrane Boundaries

The northern boundary of the Abitibi-Opatica terrane is not precisely located, nor has it been shown whether the boundary is of a depositional nature, or rather a tectonic terrane suture. The geological, metamorphic, and structural history of the Opinaca-Nemiscau metasedimentary domain that borders the Opatica domain on its northern side is virtually unknown, and there are no geochronological data available for that region. However, based on available geochronological data for neighboring regions that are discussed above and that are compiled in Table 1 and summarized in Figure 4, we suggest that the original northern boundary of the Abitibi-Opatica terrane was most likely of a depositional nature.

It has been argued that some greenstone belts in the southern Superior Province were developed autochthonously, having been deposited on older sialic basement (Thurston, 2002). Isotopic studies of the greenstones and plutons in the Abitibi Subprovince indicate predominantly juvenile magmatism (Carignan et al., 1993; Ludden and Hynes, 2000; Vervoort et al., 1994) with only minor (possible) contributions from older sialic crust (Davis et al., 2000). Together with the requirement for the presence of a sedimentary basin to the north of the Opatica domain, within which the rocks of the Opinaca-Nemiscau domain were deposited, the mainly juvenile nature of the magmatism in the Abitibi Subprovince points to a likely ocean basin setting for the Abitibi-Opatica terrane.

The presence of zircons of pre–2800 Ma age, documented in the tonalite orthogneiss of the northern Opatica domain (Davis et al., 1995) and in the volcanic rocks in the northern Abitibi Subprovince (Obatagamau Formation, in Quebec; Mortensen, 1993a), further suggests that the basin within which the Abitibi-Opatica terrane was built may have contained some rifted fragments of the older La Grande domain. In the Frotet-Evans greenstone belt (Fig. 2), Mg- and Fe-tholeiite volcanic rocks, which may be of an age similar to the oldest volcanic units in the northern Abitibi Subprovince (ca. 2755 Ma), were apparently deposited upon volcanic rocks as old as 2793 Ma that are reportedly of arc affinity (Boily and Dion, 2002) and that may also represent remnants of an older terrane to the north. The 2825 Ma Lac Rodayer pluton, within the northernmost Opatica domain, near the border with the Opinaca-Nemiscau Subprovince (Fig. 1), may also represent a remnant of the La Grande domain.

The southern boundary of the Abitibi-Opatica terrane, in Quebec, is a tectonic suture where the lower crust of an older terrane to the south wedged into the southern Abitibi crust. The wedge of older terrane is evident in the Lithoprobe deep seismic reflection profiles (Fig. 3) (Benn, 2006). The presence of the tectonically accreted older crust (which, to our knowledge, does not crop out) is also manifested in the isotopic compositions of late granitic intrusive rocks, in the Pontiac Subprovince (Carignan et al., 1993). The geochemical and isotopic compositions of metasedimentary rocks in the Pontiac Subprovince provide evidence for derivation of detritus from older, 3000 Ma to 3500 Ma crustal material (or recycled material), suggesting the Abitibi Subprovince was tectonically juxtaposed with an older crustal block (Feng et al., 1993). The plate collision along the southern margin of the Abitibi Subprovince involved the thrusting of the southern Abitibi Subprovince over the Pontiac sedimentary basin (Pontiac Subprovince) (Feng and Kerrich, 1991) and the resulting high-grade (kyanite facies) metamorphism in the metasedimentary rocks of the Pontiac Subprovince (Benn et al., 1994; Sawyer and Barnes, 1994). As yet, the southern margin of the Abitibi Subprovince has not been identified in Ontario, in the largely unmapped plutonic terrane to the south of the Kenogamissi complex.

The western and eastern boundaries of the Abitibi-Opatica terrane are both marked by structures of Proterozoic age, such that the original east-west dimensions of the terrane remain in question. If the Wawa granite-greenstone Subprovince, to the west of the Kapuskasing structure (Fig. 1), represented the western part of the Abitibi Subprovince (Polat and Kerrich, 2001; Wyman et al., 2002), it would indicate that the Abitibi-Opatica terrane was originally well over 1000 km in its east-west dimension.

## Magmatic Evolution

In the greenstones of the Abitibi Subprovince, the changes in the geochemical signatures of volcanic rocks, with time, point to important changes in magmatic sources and magmatic processes during the 60 m.y. magmatic history. The apparent petrogenetic dichotomy in the volcanic stratigraphy, where the products of plume-type and subduction-type magmatism are present in distinct parts of the stratigraphy and also are interstratified in other parts of the stratigraphy, has been recognized for some time (Wyman, 1999a, 1999b; Wyman and Kerrich, 2002; Sproule et al., 2002). The term "plume-arc interaction" has been employed when interpreting the geochemistry of the volcanic rocks. The synthesis provided above points to a specific succession of petrogenetic signatures in the volcanic rocks that is incorporated into the model, presented in the following section.

The new geochemical data for plutonic rocks, mainly from the Kenogamissi complex, allow us to identify the petrogenetic changes that were previously recognized in the volcanic rocks. In particular, it appears that the TTG suite must actually be subdivided into two "subseries" that most likely reflect different *P-T* conditions of melting (Moyen and Stevens, 2006; Moyen et al., 2007), probably corresponding to contrasting geodynamical sites. From 2744 Ma through ca. 2735 Ma, the tonalites were crystallized from TTG magmas that were most likely generated by the melting of the base of a thick mafic crust. That is consistent with the period of plume-related volcanism represented by the Pacaud Assemblage that lasted some 15 m.y. (Fig. 4) and that would have generated the thick oceanic plateau-type crust from which the tonalites were derived by partial melting. The later TTG magmatism, represented by the 2723 Ma Gogama orthogneiss, and 2713 Ma Regan tonalite were more likely produced by slab melting. The age of the Gogama TTG indicates that it can be correlated with the terminal stages of the first period of subduction-related volcanism, represented by the Deloro Assemblage. The age of the Regan TTG pluton shows that it was formed during the second period of subduction-related magmatism, corresponding to the eruption of the Kidd-Munro Assemblage (Fig. 4B).

In the Kenogamissi complex, the emplacement of the crustally derived granodioritic magmas began at ca. 2700 Ma, contemporaneously with the eruption of crustally derived lavas of the Blake River Assemblage.

## Geodynamic Origin and Tectonomagmatic Evolution

Any model for the tectonomagmatic evolution of the southeastern Superior Province must explain the presence of volcanic and plutonic rocks that were produced by subduction-related and plume-related magmatism. Previous models have proposed the sequential accretion of exotic terranes including oceanic plateaus, or hotspot-related rises, and magmatic arcs (Kimura et al., 1993; Desrochers et al., 1993; Wyman, 1999a). The presence of subduction and plume magmatic products is then explained as a tectonic collage. Common to those models is the requirement for tectonic sutures within the Abitibi greenstones. Another model suggests that one or more mantle plumes would have operated beneath an Abitibi magmatic arc to periodically supply komatiitic magmas to the autochthonous volcanic pile that also included the products of subduction magmatism (e.g., Sproule et al., 2002), relaxing the requirement for the presence of tectonic sutures between allochthonous terranes.

We propose an alternative model that explains the regional tectonomagmatic evolution, taking into account the following: (1) The volcanic stratigraphy in the Abitibi greenstones represents an autochthonous pile, with no tectonic sutures separating different parts of the stratigraphy (Ayer et al., 2002); (2) linear high-strain zones within the Abitibi-Opatica terrane do not represent tectonic sutures (Peschler et al., 2006); and thus (3) the Abitibi-Opatica terrane represents a single contiguous tectonic terrane (Benn, 2006); and (4) prior to tectonic accretion, deformation, and metamorphism, the Abitibi-Opatica terrane was bounded, to the north and south, by sedimentary basins (the Pontiac Subprovince and the Opinaca-Nemiscau domain). To account for those data and interpretations, we propose that the Abitibi-Opatica terrane represents a large, magmatically modified oceanic plateau.

Several authors have proposed that greenstone belts may represent the tectonically accreted remnants of oceanic plateaus (Stein and Goldstein, 1996; White et al., 1999; Kerr et al., 2000) or large igneous provinces formed on continental margins (Abbott, 1996). Generically, the model is interesting because it can explain the episodic growth of large regions of thick lithosphere that would be difficult to subduct and thus prone to tectonic accretion and preservation within Archean cratons. It is a particularly attractive model to explain the formation of thick crust during the superplume event that occurred in Late Archean time (Condie, 1998). In the specific case of the Abitibi-Opatica terrane, the oceanic plateau model explains the magmatic history recorded in the geochemistry of the volcanic and plutonic rocks and the lack of terrane sutures. Our model for the formation of the Abitibi-Opatica terrane as a magmatically modified oceanic plateau is similar, in some of its aspects, to the model proposed for modification of the southern Caribbean plateau, exposed on the island of Aruba, by subduction magmatism (White et al., 1999).

The geodynamic-tectonic model is explained with reference to Figure 9. In Figure 9A, we show the Abitibi-Opatica terrane as an oceanic plateau, ca. 2730 Ma, following the period of Pacaud-Obatagamau volcanism and the emplacement of early, ca. 2740 Ma tonalites. According to the geochemical signature of that first generation of TTGs, the magmas were most likely generated by melting of the base of the plateau crust, under granulite-grade conditions, very likely with heat input from the still-active plume from which the volcanic rocks were extracted. The crust is shown as 30 km in thickness, an approximation that is based on the need for a thick mafic crust and melting of its lower parts. Considering a mafic crust of $3 \times 10^4$ m thickness, formed by deposition of lavas and by basaltic underplating over a period of $25 \times 10^6$ yr, the plateau crust could have been built by the addition of 1 mm of volcanic stratigraphy, on average, per year.

Four lines of evidence suggest that the Abitibi-Opatica terrane was formed within a mature ocean basin, rather than within a plume-induced intracontinental rift basin. First, we point to the mainly juvenile isotopic signatures of rocks in the Abitibi-Opatica terrane, cited above. Second, there is no reported evidence from anywhere in the Abitibi Subprovince for extensional tectonic structures, such as faults or mafic dike complexes, that would be indicative of deposition during a prolonged period of rifting. The absence of sheeted dike complexes also indicates that the Abitibi-Opatica crust probably did not form at a divergent margin (Kerr et al., 2000). The pre-Abitibi (pre–2800 Ma) xenocrystic zircons, identified in plutonic and volcanic units, do indicate the presence of remnants of an older rifted margin within the ocean basin.

Third, evidence for subduction-related volcanism and plutonism, between 2730 Ma and 2702 Ma, suggests that a large ocean basin must have existed to the south of the Abitibi-Opatica terrane, prior to 2730 Ma. Assuming a constant convergence rate of ~10 cm/yr, 28 m.y. of subduction would have consumed 2800 km of oceanic crust. A convergence rate of even a few centimeters per year would have resulted in subduction of many hundreds of kilometers of oceanic crust. Fourth, a comparison of geochemical and isotopic compositions of metasedimentary rocks in the Abitibi and Pontiac Subprovinces indicates the presence of a tectonic suture, requiring the former presence of an ocean basin between the two (Feng et al., 1993).

The plateau crust in Figure 9A is bounded by thinner oceanic crust upon which detritus would have accumulated in the protobasins now preserved in the Pontiac Subprovince and the Opinaca-Nemiscau domain. We have no data that would allow us to constrain the dimensions of the Opinaca-Nemiscau basin; it is shown as a relatively small basin in Figure 9A for simplicity only.

Far-field stresses associated with plate reconfiguration could have initiated subduction in the thinner crust, along the southern margin of the thick Abitibi-Opatica plateau crust (Fig. 9A). We suggest that the 5 m.y. period of apparent volcanic quiescence in the Abitibi Subprovince (2735–2730 Ma, Fig. 4B) does not indicate the shutting down of the Abitibi-Opatica plume. That interpretation would require the presence of a second, younger plume, to produce the komatiitic rocks in the Kidd-Munro and Tisdale Assemblages. Instead, we suggest that the subduction of the relatively cold slab beneath the Abitibi-Opatica plateau may have been sufficient to curb plume volcanism for several million years, possibly deflecting the plume updip of the subducting slab (Fig. 9A).

Figure 9B shows the situation ca. 2720 Ma during renewed plume-type volcanism, corresponding to the Stoughton-Roquemaure Assemblage. We attribute the cessation of subduction volcanism and renewed plume-related volcanism to the creation of a slab window. The opening of the slab window could have resulted from the subduction of a divergent plate boundary (Fig. 9B, inset). An alternative interpretation would be that the subducting slab detached due to thermal weakening, arising from the presence of the underlying plume. In either case, the opening of the slab window would have allowed the renewed rise of the plume under the study area and the eruption of the komatiitic magmas whose geochemical signatures record the melting of highly refractory mantle, at relatively shallow depths (Sproule et al., 2002). The formation of the 2723 Ma Gogama TTG, with its slab-melting signature, might be explained by melting of the slab-window edges (Thorkelson and Breitsprecher, 2005) at this stage in the tectonic history.

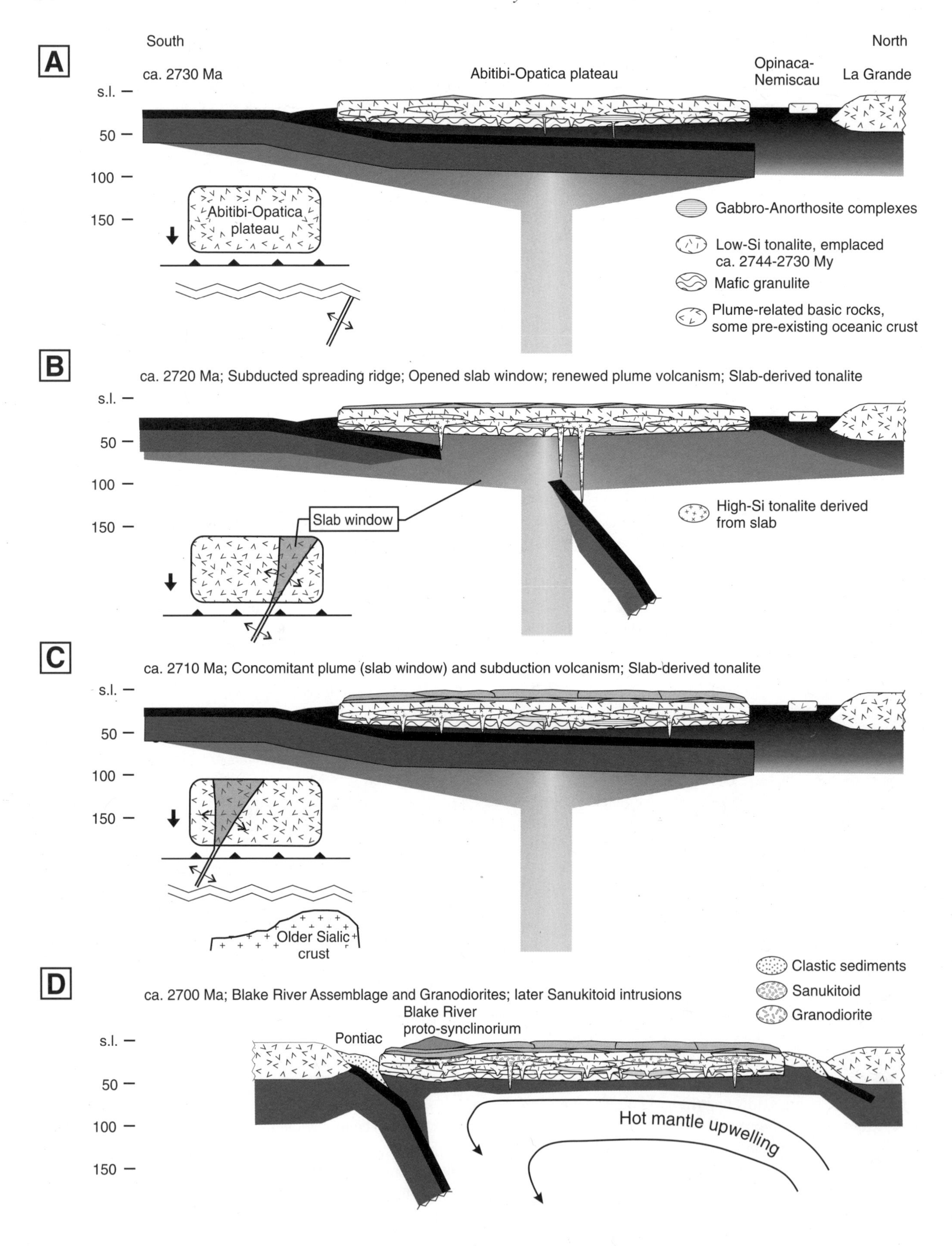
A
South
North
ca. 2730 Ma
Abitibi-Opatica plateau
Opinaca-
Nemiscau
La Grande
s.l.
50
100
150
Abitibi-Opatica
plateau
Gabbro-Anorthosite complexes
Low-Si tonalite, emplaced
ca. 2744-2730 My
Mafic granulite
Plume-related basic rocks,
some pre-existing oceanic crust
B
ca. 2720 Ma; Subducted spreading ridge; Opened slab window; renewed plume volcanism; Slab-derived tonalite
Slab window
High-Si tonalite derived
from slab
C
ca. 2710 Ma; Concomitant plume (slab window) and subduction volcanism; Slab-derived tonalite
Older Sialic
crust
D
ca. 2700 Ma; Blake River Assemblage and Granodiorites; later Sanukitoid intrusions
Clastic sediments
Sanukitoid
Granodiorite
Pontiac
Blake River
proto-synclinorium
Hot mantle upwelling

Figure 9. The proposed model for the precollisional, tectonomagmatic evolution of the Abitibi-Opatica terrane as a modified oceanic plateau. (A) Subduction has been initiated beneath the Abitibi-Opatica plateau, leading to the eruption of subduction-related lavas of the Deloro Assemblage that are deposited upon the plume-related volcanic rocks of the Pacaud Assemblage. The lower crust of the plateau underwent melting between 2744 Ma and 2730 Ma to produce the early tonalite suite and a granulitic lower crust. (B) Break-off of the subducting slab, possibly due to subduction of a spreading ridge (see inset), and renewed plume-related volcanism represented by the Stoughton-Roquemaure Assemblage. Generation and emplacement of high-Si tonalites that include a component of slab melt. (C) The slab window sweeps obliquely across the plateau, causing eruption of interstratified plume-related and subduction-related lavas of the Kidd-Munro Assemblage and the Tisdale Assemblage. Further generation and emplacement of high-Si tonalite plutons. (D) An older sialic block impinges on the southern margin of the Abitibi-Opatica plateau, causing the slab to sink and an upwelling of asthenospheric mantle. Crustal melting produces lavas of the Blake River Assemblage and granodiorite plutons. Sanukitoid intrusions (e.g., the Neville pluton in the Kenogamissi complex, Fig. 5) are derived by melting of the metasomatized mantle beneath the plateau in response to upwelling of hot asthenospheric mantle. Refer to Fig. 4 for ages of the volcanic assemblages and the plutonic suites.

The creation of a slab window can also explain the following period of interstratified plume-type and subduction-type volcanic units, recorded in the Kidd-Munro and Tisdale Assemblages (Fig. 4B). If a slab window formed beneath the Abitibi-Opatica plateau, allowing the renewed upwelling of the plume and eruption of komatiitic lavas at eruptive centers above the slab window, such low-viscosity lavas could flow laterally for tens of kilometers (Lesher et al., 1984) to be deposited upon recently erupted subduction-related lava flows, which may have been erupted from more numerous and widespread volcanoes. Figure 9C depicts an appropriate plate configuration ca. 2710 Ma. Further slab melting during this period would also have led to the generation of the 2713 Ma TTG of the Regan pluton. Slab melting was enhanced by the thermal interaction of the plume and the downgoing slab.

Figure 9D depicts the situation during the initial stages of the collisional event as an older sialic block (shown riding on the subducting plate, in Fig. 9C) entered the subduction zone. That older block is now buried beneath the overthrust Pontiac Subprovince (Fig. 3). The slowing of plate convergence associated with the onset of collision would cause the sinking and eventual detachment of the remnant of the subducting slab, the upwelling of hot mantle to replace it, and widespread crustal melting to produce components of the Blake River Assemblage and the 2700–2690 Ma granodiorites (Fig. 4).

We suggest that the influx of hot asthenosphere below the former mantle wedge (Fig. 9D) may also have resulted in the generation of the 2696–2680 Ma sanukitoid-type, hornblende-bearing diorites, monzodiorites, and granodiorites by melting of the previously metasomatized mantle wedge. It is worth noting that the sanukitoid-type plutons were emplaced ~10 m.y. earlier in the southern Opatica domain (Barlow pluton; see Table 1) than in the Kenogamissi complex (Neville pluton) in the southern Abitibi Subprovince (Fig. 4; Table 1). That would be consistent with progressive sinking of the slab during collision and the resulting influx of hot asthenosphere from north to south (Fig. 9D). That interpretation for the formation of the sanukitoid magmas in the Abitibi-Opatica terrane is similar to the model proposed for the production of high-Mg basalts and sanukitoids in the ca. 3000 Ma Mallina basin, Pilbara Craton (Smithies et al., 2004).

The proposed model provides a geodynamically feasible answer to an important question regarding the "plume-arc interaction" that is indicated by the interstratification of plume-type and subduction-type lavas. Specifically, it explains the occurrence together, in space and in time, of the cold lithospheric downwelling that is a subduction zone, with a hot asthenospheric upwelling that is a plume. The location of an oceanic plateau is dependent only upon the location of the plume from which the plateau lavas are generated. It is known that in modern Earth, a negative correlation exists between plumes (and hotspots) and convergent plate margins (Zhao, 2004), and there is no reason to think that a mantle plume would be more likely to form within a zone of cold downwelling in the Archean mantle than in the modern mantle. Once a plateau is formed, it may become unsubductable due to its thickness and strength (Abbott, 1996), and the thinner crust around the margins of the plateau should represent areas where stresses would be concentrated, and subduction initiated.

We suggest that far-field plate reconfigurations could provide the required stresses to begin subduction. According to this model, as long as the plume was still active below the plateau, plume-subduction interaction would be inevitable. Initially, the subduction of cold lithosphere into the plume mantle would tend to cool that region, possibly inhibiting plume magmatism. However, the subduction of an active plate margin, or the heating of the subducting slab by the plume, would make slab breakoff likely, leading to renewed plume magmatism. Subduction-related magmatism would therefore have been episodic, and the period of subduction magmatism that occurred between 2730 Ma and 2723 Ma (Deloro Assemblage, Fig. 4) would be explained by a failed subduction that was renewed later, ca. 2720 Ma, with the eruption of subduction-related components of the Kidd-Munro Assemblage.

Finally, with specific reference to the southeastern Superior Province, it is important to point out two limitations of the tectonic model that is shown in Figure 9. At present, we have little knowledge of the structural and metamorphic histories of the Opinaca-Nemiscau domain, which is an important element of the model. For example, the imbrication and partial subduction of the Opinaca-Nemiscau basin, in Figure 9D, remains to be tested. The interpretation that inherited zircons in rocks of the Opatica domain and Abitibi Subprovince are derived from the La Grande Subprovince crust also remains to be tested. However, further studies on those two questions are unlikely to invalidate our model of subduction below a magmatically active oceanic plateau, which may prove to be generally applicable to Archean terranes where evidence can be shown for mixed plume-related and subduction-related magmatic histories.

## CONCLUSIONS

This paper has summarized and synthesized the important data sets that must be considered in order to understand the geodynamic environment within which the Abitibi-Opatica terrane was formed, and its tectonomagmatic evolution, from initial plume magmatism, through modification of the plateau crust, plume-subduction interaction, and finally plate collision. We have also introduced a new set of geochemical data from several suites of granitic rocks that crop out within the Abitibi greenstones. The new data were shown to be consistent with the published interpretations of the geochemical data from the volcanic stratigraphy. Taken together, the abundant geological, geophysical, geochemical, and geochronological data for the Abitibi-Opatica terrane, and the less abundant data for surrounding regions, allowed us to propose a new plate tectonic model for the origin and evolution of the Abitibi-Opatica terrane.

The Abitibi-Opatica terrane is interpreted to represent a Late Archean oceanic plateau formed within a large ocean basin and subsequently modified by partial melting of its base and by subduction-related magmatism. The model explains the evolution in the petrogenetic signatures of the volcanic units and also of the plutonic units, with a long-lived plume magmatism that persisted during subduction magmatism. The model explains the plume-subduction interaction that is recorded in the volcanic and plutonic records by the formation of a slab window, due to subduction of a divergent plate margin or possibly due to thermally aided slab breakoff. The interaction of the plume with the subducting slab also resulted in the production of the TTG magmas with slab-melting geochemical signatures.

The geodynamic-tectonic model indicates that the pre- to syncollisional history of the Abitibi-Opatica terrane can be accounted for by an essentially actualistic plate tectonic model that calls for subduction initiation along the margin of a magmatically active oceanic plateau, and the compositional modification of the plateau due to subduction-related magmatism.

## ACKNOWLEDGMENTS

Fieldwork for this study was funded from a Natural Sciences and Engineering Research Council of Canada Discovery Grant awarded to K.B. J.F.M.'s postdoctoral fellowship at Stellenbosch University was funded by the South African National Research Foundation (NRF grant GUN 2053698) and by a scholarship from the Department of Geology, Stellenbosch University. The authors wish to thank the editors of this volume for the opportunity to publish the work. R. Kerrich and H. Smithies are thanked for their reviews, which greatly improved the quality of the paper.

## REFERENCES CITED

Abbott, D.H., 1996, Plumes and hotspots as sources of greenstone belts: Lithos, v. 37, p. 113–127, doi: 10.1016/0024-4937(95)00032-1.

Arth, J.G., and Hanson, G.N., 1975, Geochemistry and origin of the early Precambrian crust of northeastern Minnesota: Geochimica et Cosmochimica Acta, v. 39, p. 325–362, doi: 10.1016/0016-7037(75)90200-8.

Ayer, J., Amelin, Y., Corfu, F., Kamo, S., Ketchum, J., Kwok, K., and Trowell, N., 2002, Evolution of the southern Abitibi greenstone belt based on U-Pb geochronology: Autochthonous volcanic construction followed by plutonism, regional deformation and sedimentation: Precambrian Research, v. 115, p. 63–95, doi: 10.1016/S0301-9268(02)00006-2.

Ayer, J.A., Trowell, N.F., and Josey, S., 2004, Geological compilation of the Abitibi greenstone belt: Ontario Geological Survey, Miscellaneous Release, Data 143.

Barker, F., 1976, Trondhjemites: Definition, environment and hypothesis of origin, *in* Barker, F., ed., Trondhjemites, dacites and related rocks: Amsterdam, Elsevier, p. 1–12.

Beakhouse, G., and Davis, D.W., 2005, Evolution and tectonic significance of intermediate to felsic plutonism associated with the Hemlo greenstone belt, Superior Province, Canada: Precambrian Research, v. 137, p. 61–92, doi: 10.1016/j.precamres.2005.01.003.

Becker, J.K., and Benn, K., 2003, The Neoarchean Rice Lake batholith and its place in the tectonomagmatic evolution of the Swayze and Abitibi granite-greenstone belts, northeastern Ontario: Ontario Geological Survey Open File Report 6105, p. 42.

Bédard, J., 2006, A catalytic delamination-driven model for coupled genesis of Archaean crust and sub-continental lithospheric mantle: Geochimica et Cosmochimica Acta, v. 70, p. 1188–1214, doi: 10.1016/j.gca.2005.11.008.

Benn, K., 2004, Late Archean Kenogamissi complex, Abitibi Subprovince, Ontario: Doming, folding and deformation-assisted melt remobilization during syntectonic batholith emplacement: Transactions of the Royal Society of Edinburgh, Earth Sciences, v. 95, p. 297–307, doi: 10.1017/S0263593300001085.

Benn, K., 2006, Tectonic delamination of the lower crust during Late Archean collision of the Abitibi-Opatica and Pontiac terranes, Superior Province, Canada, *in* Benn, K., et al., eds., Archean geodynamics and environments: American Geophysical Union Geophysical Monograph 164, p. 267–282.

Benn, K., and Peschler, A.P., 2005, A detachment fold model for fault zones in the Late Archean Abitibi greenstone belt, Ontario: Tectonophysics, v. 400, p. 85–104, doi: 10.1016/j.tecto.2005.02.011.

Benn, K., Sawyer, E.W., and Bouchez, J.L., 1992, Orogen parallel and transverse shearing in the Opatica belt, Quebec: Implications for the structure of the Abitibi Subprovince: Canadian Journal of Earth Sciences, v. 29, p. 2429–2444.

Benn, K., Miles, W., Ghassemi, M.R., and Gillett, J., 1994, Crustal structure and kinematic framework of the northwestern Pontiac Subprovince, Quebec: An integrated structural and geophysical study: Canadian Journal of Earth Sciences, v. 31, p. 271–281.

Boily, M., and Dion, C., 2002, Geochemistry of boninite-type volcanic rocks in the Frotet-Evans greenstone belt, Opatica Subprovince, Quebec: Implications for the evolution of Archaean greenstone belts: Precambrian Research, v. 115, p. 349–371, doi: 10.1016/S0301-9268(02)00016-5.

Calvert, A.J., Sawyer, E.W., Davis, W.J., and Ludden, J.N., 1995, Archean subduction inferred from seismic images of a mantle suture in the Superior Province: Nature, v. 375, p. 670–674, doi: 10.1038/375670a0.

Card, K.D., and Poulsen, K.H., 1998, Geology and mineral deposits of the Superior Province of the Canadian Shield, *in* Lucas, S.B., and St.-Onge, M.R., eds., Geology of the Precambrian Superior and Grenville provinces and Precambrian fossils in North America: Geology of Canada, no. 7: Ottawa, Geological Survey of Canada, p. 13–204.

Carignan, J., Gariepy, C., Machado, N., and Rive, M., 1993, Pb isotopic geochemistry of granitoids and gneisses from the Late Archean Pontiac and Abitibi Subprovinces of Canada: Chemical Geology, v. 106, p. 299–316, doi: 10.1016/0009-2541(93)90033-F.

Champion, D.C., and Smithies, R.H., 1998, Archaean granites of the Yilgarn and Pilbara cratons, *in* The Bruce Chappell Symposium—Granites, island arcs, the mantle and ore deposits, Abstract Volume: Canberra, Australian Geological Survey Organization, p. 24–25.

Chown, E.H., Harrap, R., and Moukhsil, A., 2002, The role of granitic intrusions in the evolution of the Abitibi belt, Canada: Precambrian Research, v. 115, p. 291–310, doi: 10.1016/S0301-9268(02)00013-X.

Collins, W.J., Van Kranendonk, M.J., and Teyssier, C., 1998, Partial convective overturn of Archaean crust in the east Pilbara craton, Western Australia: Driving mechanisms and tectonic implications: Journal of Structural Geology, v. 20, p. 1405–1424.

Condie, K.C., 1981, Archean greenstone belts: Developments in Precambrian geology: Amsterdam, Elsevier, 434 p.

Condie, K.C., 1986, Origin and early growth rate of continents: Precambrian Research, v. 32, p. 261–278, doi: 10.1016/0301-9268(86)90032-X.

Condie, K.C., 1998, Episodic continental growth and supercontinents: A mantle avalanche connection?: Earth and Planetary Science Letters, v. 163, p. 97–108, doi: 10.1016/S0012-821X(98)00178-2.

Condie, K.C., and Benn, K., 2006, Archean geodynamics: Similar to or different from modern geodynamics?, *in* Benn, K., et al., eds., Archean geodynamics and environments: American Geophysical Union Geophysical Monograph 164, p. 47–59.

Corfu, F., and Noble, S.R., 1992, Genesis of the southern Abitibi greenstone belt, Superior Province, Canada: Evidence from zircon Hf isotope analyses using a single filament technique: Geochimica et Cosmochimica Acta, v. 56, p. 2081–2097, doi: 10.1016/0016-7037(92)90331-C.

Daigneault, R., St.-Julien, P., and Allard, G.O., 1990, Tectonic evolution of the northeast portion of the Archean Abitibi Greenstone Belt, Chibougamau area, Quebec: Canadian Journal of Earth Sciences, v. 27, p. 1714–1736.

Daigneault, R., Mueller, W.U., and Chown, E.H., 2002, Oblique Archean subduction: Accretion and exhumation of an oceanic arc during dextral transpression, Southern Volcanic Zone, Abitibi Subprovince, Canada: Precambrian Research, v. 115, p. 261–290, doi: 10.1016/S0301-9268(02)00012-8.

Daigneault, R., Mueller, W.U., and Chown, E.H., 2004, Abitibi greenstone belt plate tectonics: The diachronous history of arc development, accretion and collision, *in* Eriksson, P.G., et al., eds., Developments in Precambrian geology, volume 12: Amsterdam, Elsevier, p. 88–103.

Davis, D.W., 2002, U-Pb geochronology of Archean metasedimentary rocks in the Pontiac and Abitibi Subprovinces, Quebec, constraints on timing, provenance and regional tectonics: Precambrian Research, v. 115, p. 97–117, doi: 10.1016/S0301-9268(02)00007-4.

Davis, W.J., Machado, N., Gariépy, C., Sawyer, E.W., and Benn, K., 1995, U-Pb geochronology of the Opatica tonalite-gneiss belt and its relationship to the Abitibi greenstone belt, Superior Province, Quebec: Canadian Journal of Earth Sciences, v. 32, p. 113–127.

Davis, W.J., Lacroix, S., Gariepy, C., and Machado, N., 2000, Geochronology and radiogenic isotope geochemistry of plutonic rocks from the central Abitibi Subprovince: Significance to the internal subdivision and plutono-tectonic evolution of the Abitibi belt: Canadian Journal of Earth Sciences, v. 37, p. 117–133, doi: 10.1139/cjes-37-2-3-117.

Desrochers, J.-P., Hubert, C., Ludden, J.N., and Pilote, P., 1993, Accretion of Archean oceanic plateau fragments in the Abitibi greenstone belt, Canada: Geology, v. 21, p. 451–454, doi: 10.1130/0091-7613(1993)021<0451:AOAOPF>2.3.CO;2.

Dostal, J., and Mueller, W., 1996, An Archean oceanic felsic dyke swarm in a nascent arc: The Hunter Mine Group, Abitibi Greenstone Belt, Canada: Journal of Volcanology and Geothermal Research, v. 72, p. 37–57, doi: 10.1016/0377-0273(95)00082-8.

Dostal, J., and Mueller, W.U., 1997, Komatiite flooding of a rifted Archean rhyolitic arc complex: Geochemical signature and tectonic significance of the Stoughton-Roquemaure Group, Abitibi greenstone belt, Canada: Journal of Geology, v. 105, p. 545–563.

Eggins, S., 2003, Laser ablation ICP-MS analysis of geological materials prepared as lithium borate glasses: Geostandards Newsletter, v. 27, no. 2, p. 147–162, doi: 10.1111/j.1751-908X.2003.tb00642.x.

Feng, R., and Kerrich, R., 1991, Single zircon age constraints on the tectonic juxtaposition of the Archean Abitibi greenstone belt and Pontiac Subprovince, Quebec, Canada: Geochimica et Cosmochimica Acta, v. 55, p. 3437–3441, doi: 10.1016/0016-7037(91)90502-V.

Feng, R., and Kerrich, R., 1992a, Geochemical evolution of granitoids from the Archean Abitibi Southern Volcanic Zone and the Pontiac Subprovince, Superior Province, Canada: Implications for tectonic history and source regions: Chemical Geology, v. 98, p. 23–70, doi: 10.1016/0009-2541(92)90090-R.

Feng, R., and Kerrich, R., 1992b, Geodynamic evolution of the southern Abitibi and Pontiac terranes: Evidence from geochemistry of granitoid magma series (2700–2630 Ma): Canadian Journal of Earth Sciences, v. 29, p. 2266–2286.

Feng, R., Kerrich, R., and Maas, R., 1993, Geochemical, oxygen, and neodymium isotope compositions of metasediments from the Abitibi greenstone belt and Pontiac Subprovince, Canada: Evidence for ancient crust and Archean terrane juxtaposition: Geochimica et Cosmochimica Acta, v. 57, p. 641–658, doi: 10.1016/0016-7037(93)90375-7.

Foley, S.F., Tiepolo, M., and Vannucci, R., 2002, Growth of early continental crust controlled by melting of amphibolite in subduction zones: Nature, v. 417, p. 837–840, doi: 10.1038/nature00799.

Fourcade, S., Martin, H., and De Bremond D'Ars, J., 1992, Chemical exchange in migmatites during cooling: Lithos, v. 28, p. 43–53, doi: 10.1016/0024-4937(92)90022-Q.

Fowler, A.D., and Jensen, L.S., 1989, Quantitative trace-element modeling of the crystallization history of the Kinojevis and Blake River Groups, Abitibi Greenstone Belt, Ontario: Canadian Journal of Earth Sciences, v. 26, p. 1356–1367.

Frarey, M.J., and Krogh, T.E., 1986, U-Pb ages for late internal plutons of the Abitibi and eastern Wawa Subprovince, Ontario and Quebec: Geological Survey of Canada Current Research 86-1A, p. 43–48.

Gardien, V., Thompson, A.B., Grujic, D., and Ulmer, P., 1995, Experimental melting of biotite + plagioclase + quartz ± muscovite assemblages and implications for crustal melting: Journal of Geophysical Research, B, Solid Earth and Planets, v. 100, p. 15,581–15,591, doi: 10.1029/95JB00916.

Gariépy, C., Allègre, C.J., and Lajoie, J., 1984, U-Pb systematics in single zircons from the Pontiac metasediments, Abitibi greenstone belt: Canadian Journal of Earth Sciences, v. 21, p. 1296–1304.

Halls, H.C., and Zhang, B.X., 1998, Uplift structure of the southern Kapuskasing zone from 2.45 Ga dike swarm displacement: Geology, v. 26, p. 67–70, doi: 10.1130/0091-7613(1998)026<0067:USOTSK>2.3.CO;2.

Heather, K.B., and Shore, G.T., 1999, Geology of the Swayze greenstone belt, Ontario: Geological Survey of Canada Open File 3384.

Henry, P., Stevenson, R.K., and Gariépy, C., 1998, Late Archean mantle composition and crustal growth in the western Superior Province of Canada: Nd and Pb isotopic evidence from the Wawa, Quetico and Wabigoon Subprovinces: Geochimica et Cosmochimica Acta, v. 62, p. 143–157, doi: 10.1016/S0016-7037(97)00324-4.

Isnard, H., and Gariépy, C., 2004, Sm-Nd, Lu-Hf and Pb-Pb signatures of gneisses and granitoids from the La Grande belt: Extent of Late Archean crustal recycling in the northeastern Superior Province, Canada: Geochimica et Cosmochimica Acta, v. 68, p. 1099–1113, doi: 10.1016/j.gca.2003.08.004.

Jahn, B., Glikson, A.Y., Peucat, J.-J., and Hickman, A.H., 1981, REE geochemistry and isotopic data of Archaean silicic volcanics and granitoids from the Pilbara block, Western Australia: Implications for early crustal evolution: Geochimica et Cosmochimica Acta, v. 45, p. 1633–1652.

Kamber, B., Ewart, A., Collerson, K.D., Bruce, M.C., and McDonald, G.D., 2002, Fluid-mobile trace element constraints on the role of slab melting and implications for Archaean crustal growth models: Contributions to Mineralogy and Petrology, v. 144, p. 38–56.

Kay, R.W., and Kay, S.M., 1991, Creation and destruction of lower continental crust: Geologische Rundschau, v. 80, p. 259–278, doi: 10.1007/BF01829365.

Kerr, A.C., White, R.V., and Saunders, A.D., 2000, LIP reading: Recognizing oceanic plateaux in the geological record: Journal of Petrology, v. 41, p. 1041–1056, doi: 10.1093/petrology/41.7.1041.

Kerrich, R., and Feng, R., 1992, Archean geodynamics and the Abitibi-Pontiac collision: Implications for advection of fluids at transpressive collisional boundaries and the origin of giant quartz vein systems: Earth-Science Reviews, v. 32, p. 33–60, doi: 10.1016/0012-8252(92)90011-H.

Kerrich, R., and Polat, A., 2006, Archean greenstone-tonalite duality: Thermochemical mantle convection models or plate tectonics in the early Earth global dynamics?: Tectonophysics, v. 415, p. 141–165, doi: 10.1016/j.tecto.2005.12.004.

Kerrich, R., Wyman, D., Fan, J., and Bleeker, W., 1998, Boninite series: Low Ti-tholeiite associations from the 2.7 Ga Abitibi greenstone belt: Earth and Planetary Science Letters, v. 164, p. 303–316, doi: 10.1016/S0012-821X(98)00223-4.

Kerrich, R., Polat, A., Wyman, D., and Hollings, P., 1999, Trace element systematics of Mg-, to Fe-tholeiitic basalt suites of the Superior Province: Implications for Archean mantle reservoirs and greenstone belt genesis: Lithos, v. 46, p. 163–187, doi: 10.1016/S0024-4937(98)00059-0.

Kimura, G., Ludden, J.N., Desrochers, J.P., and Hori, R., 1993, A model of ocean-crust accretion for the Superior Province, Canada: Lithos, v. 30, p. 337–355, doi: 10.1016/0024-4937(93)90044-D.

King, E.M., Valley, J.W., Davis, D.W., and Edwards, G., 1998, Oxygen isotope ratios of Archean plutonic zircons from granite-greenstone belts of the Superior Province: Indicator of magmatic source: Precambrian Research, v. 92, p. 365–387, doi: 10.1016/S0301-9268(98)00082-5.

Kleinhanns, I.C., Kramers, J.D., and Kamber, B.S., 2003, Importance of water for Archaean granitoid petrology: A comparative study of TTG and potassic granitoids from Barberton Mountain Land, South Africa: Contributions to Mineralogy and Petrology, v. 145, p. 377–389, doi: 10.1007/s00410-003-0459-9.

Lafleche, M.R., Dupuy, C., and Bougault, H., 1992, Geochemistry and petrogenesis of Archean mafic volcanic rocks of the southern Abitibi Belt, Quebec: Precambrian Research, v. 57, p. 207–241, doi: 10.1016/0301-9268(92)90003-7.

Legault, M., Gauthier, M., Jebrak, M., Davis, D.W., and Baillargeon, F., 2002, Evolution of the subaqueous to near-emergent Joutel volcanic complex, Northern Volcanic Zone, Abitibi Subprovince, Quebec, Canada: Precambrian Research, v. 115, p. 187–221, doi: 10.1016/S0301-9268(02)00010-4.

Lesher, C.M., Arndt, N.T., and Groves, D.I., 1984, Genesis of komatiite-associated nickel sulphide deposits at Kambalda, Western Australia: A distal volcanic model, *in* Buchanan, D.L., and Jones, M.J., eds., Sulphide deposits in mafic and ultramafic rocks: London, Institution of Mining and Metallurgy, p. 70–80.

Ludden, J., and Hynes, A., 2000, The Lithoprobe Abitibi-Grenville transect: Two billion years of crust formation and recycling in the Precambrian Shield of Canada: Canadian Journal of Earth Sciences, v. 37, p. 459–476, doi: 10.1139/cjes-37-2-3-459.

Ludden, J., Hubert, C., and Gariépy, C., 1986, The tectonic evolution of the Abitibi greenstone belt of Canada: Geological Magazine, v. 123, p. 153–166.

Maaløe, S., 1982, Petrogenesis of Archaean tonalites: Geologische Rundschau, v. 71, p. 328–346, doi: 10.1007/BF01825045.

Martin, H., 1986, Effect of steeper Archean geothermal gradient on geochemistry of subduction-zone magmas: Geology, v. 14, p. 753–756, doi: 10.1130/0091-7613(1986)14<753:EOSAGG>2.0.CO;2.

Martin, H., 1994, The Archean grey gneisses and the genesis of the continental crust, *in* Condie, K.C., ed., Archean crustal evolution: Developments in Precambrian geology: Amsterdam, Elsevier, p. 205–259.

Martin, H., 1999, The adakitic magmas: Modern analogues of Archaean granitoids: Lithos, v. 46, p. 411–429, doi: 10.1016/S0024-4937(98)00076-0.

Martin, H., and Moyen, J.-F., 2002, Secular changes in TTG composition as markers of the progressive cooling of Earth: Geology, v. 30, p. 319–322, doi: 10.1130/0091-7613(2002)030<0319:SCITTG>2.0.CO;2.

Martin, H., Smithies, R.H., Rapp, R.P., Moyen, J.-F., and Champion, D.C., 2005, An overview of adakite, tonalite-trondhjemite-granodiorite (TTG) and sanukitoid: Relationships and some implications for crustal evolution: Lithos, v. 79, p. 1–24, doi: 10.1016/j.lithos.2004.04.048.

Moorbath, S., 1975, Evolution of Precambrian crust from strontium isotopic evidence: Nature, v. 254, p. 395–398, doi: 10.1038/254395a0.

Mortensen, J.K., 1987, U-Pb zircon ages for volcanic and plutonic rocks of the Noranda–Lake Abitibi area, Abitibi Subprovince, Quebec: Geological Survey of Canada Current Research 87-1A, p. 581–590.

Mortensen, J.K., 1993a, U-Pb geochronology of the eastern Abitibi Subprovince, part 1: Chibougamau-Matagami-Joutel region: Canadian Journal of Earth Sciences, v. 30, p. 11–28.

Mortensen, J.K., 1993b, U-Pb geochronology of the eastern Abitibi Subprovince, part 2: Noranda-Kirkland Lake area: Canadian Journal of Earth Sciences, v. 30, p. 29–41.

Mortensen, J.K., 1993c, U-Pb geochronology of the Lapparent massif, northeastern Abitibi Belt: Basement or synvolcanic pluton: Canadian Journal of Earth Sciences, v. 30, p. 42–47.

Mortensen, J.K., and Card, K.D., 1993, U-Pb age constraints for the magmatic and tectonic evolution of the Pontiac Subprovince, Quebec: Canadian Journal of Earth Sciences, v. 30, p. 1970–1980.

Moyen, J.-F., and Stevens, G., 2006, Experimental constraints on TTG petrogenesis: Implications for Archean geodynamics, *in* Benn, K., et al., eds., Archean geodynamics and environments: American Geophysical Union Geophysical Monograph 164, p. 149–175.

Moyen, J.-F., Martin, H., Jayananda, M., and Auvray, B., 2003, Late Archaean granites: A typology based on the Dharwar Craton (India): Precambrian Research, v. 127, p. 103–123, doi: 10.1016/S0301-9268(03)00183-9.

Moyen, J.-F., Stevens, G., Kisters, A.F.M., and Belcher, R.W., 2007, TTG plutons of the Barberton granitoid-greenstone terrain, South Africa, *in* Van Kranendonk, M.J., Smithies, R.H., and Bennet, V., eds., Earth's oldest rocks: Amsterdam, Elsevier, Developments in Precambrian Geology, v. 15, p. 607–668.

Mueller, W.U., Daigneault, R., Mortensen, J.K., and Chown, E.H., 1996, Archean terrane docking: Upper crust collision tectonics, Abitibi greenstone belt, Quebec, Canada: Tectonophysics, v. 265, p. 127–150, doi: 10.1016/S0040-1951(96)00149-7.

Oliver, H.S., Hughes, D.J., Hall, R.P., and Johns, G.W., 2000, Lithogeochemistry of the Shining Tree area, *in* Summary of field work and other activities: Ontario Geological Survey Open File Report 6032, p. 6.1–6.14.

Patiño-Douce, A.E., 2005, Vapor-absent melting of tonalite at 15–32 kbar: Journal of Petrology, v. 46, p. 275–290, doi: 10.1093/petrology/egh071.

Patiño-Douce, A.E., and Beard, J.S., 1995, Dehydration-melting of Bt gneiss and Qtz amphibolite from 3 to 15 kB: Journal of Petrology, v. 36, p. 707–738.

Percival, J.A., 1988, Deep geology out in the open: Nature, v. 335, p. 671, doi: 10.1038/335671a0.

Percival, J.A., and Krogh, T.E., 1983, U-Pb zircon geochronology of the Kapuskasing structural zone and vicinity in the Chapleau-Foleyet area, Ontario: Canadian Journal of Earth Sciences, v. 20, p. 830–843.

Percival, J.A., Stern, R.A., Skulski, T., Card, K.D., Mortensen, J.K., and Begin, N.J., 1994, Minto block, Superior Province: Missing link in deciphering assembly of the craton at 2.7 Ga: Geology, v. 22, p. 839–842, doi: 10.1130/0091-7613(1994)022<0839:MBSPML>2.3.CO;2.

Peschler, A.P., Benn, K., and Roest, W.R., 2004, Insights on Archean continental geodynamics from gravity modelling of granite-greenstone terranes: Journal of Geodynamics, v. 38, p. 185–207, doi: 10.1016/j.jog.2004.06.005.

Peschler, A.P., Benn, K., and Roest, W.R., 2006, Gold-bearing fault zones related to Late Archean orogenic folding of upper and middle crust in the Abitibi granite-greenstone belt, Ontario: Precambrian Research, v. 151, p. 143–159, doi: 10.1016/j.precamres.2006.08.013.

Polat, A., and Kerrich, R., 2001, Geodynamic processes, continental growth, and mantle evolution recorded in the Late Archean greenstone belts of the southern Superior Province, Canada: Precambrian Research, v. 112, p. 5–25, doi: 10.1016/S0301-9268(01)00168-1.

Polat, A., and Kerrich, R., 2006, Reading the geochemical fingerprints of Archean hot subduction volcanic rocks: Evidence for accretion and crustal recycling in a mobile tectonic regime, *in* Benn, K., et al., eds., Archean geodynamics and environments: American Geophysical Union Geophysical Monograph 164, p. 189–213.

Rapp, R.P., 2003, Experimental constraints on the origin of compositional variations in the adakite-TTG-sanukitoid-HMA family of granitoids, *in* EGS-AGU-EUG joint assembly, Nice, France, 6–11 April 2003, p. 8123.

Rapp, R.P., and Watson, E.B., 1995, Dehydration melting of metabasalt at 8–32 kbar: Implications for continental growth and crust-mantle recycling: Journal of Petrology, v. 36, p. 891–931.

Rapp, R.P., Watson, E.B., and Miller, C.F., 1991, Partial melting of amphibolite/eclogite and the origin of Archaean trondhjemites and tonalites: Precambrian Research, v. 51, p. 1–25, doi: 10.1016/0301-9268(91)90092-O.

Rapp, R.P., Shimizu, N., Norman, M.D., and Applegate, G.S., 1999, Reaction between slab-derived melts and peridotite in the mantle wedge: Experimental constraints at 3.8 GPa: Chemical Geology, v. 160, p. 335–356, doi: 10.1016/S0009-2541(99)00106-0.

Sawyer, E.W., 1998, Formation and evolution of granite magmas during crustal reworking: The significance of diatexites: Journal of Petrology, v. 39, p. 1147–1167, doi: 10.1093/petrology/39.6.1147.

Sawyer, E.W., and Barnes, S.J., 1994, Thrusting, magmatic intraplating, and metamorphic complex development in the Archean Belleterre-Angliers greenstone belt, Superior Province, Quebec, Canada: Precambrian Research, v. 68, p. 183–200, doi: 10.1016/0301-9268(94)90029-9.

Sawyer, E.W., and Benn, K., 1993, Structure of the high-grade Opatica Belt and adjacent low-grade Abitibi Subprovince, Canada: An Archean mountain front: Journal of Structural Geology, v. 15, p. 1443–1458, doi: 10.1016/0191-8141(93)90005-U.

Schmidt, M.W., 1993, Phase relations and composition in tonalite as a function of pressure: An experimental study at 650 °C: American Journal of Science, v. 293, p. 1011–1060.

Schmidt, M.W., and Thompson, A.B., 1996, Epidote in calc-alkaline magmas: An experimental study of stability, phase relationships and the role of epidote in magmatic evolution: American Mineralogist, v. 81, p. 462–474.

Shirey, S.B., and Hanson, G.N., 1984, Mantle-derived Archaean monzodiorites and trachyandesites: Nature, v. 310, p. 222–224, doi: 10.1038/310222a0.

Sigmarsson, O., Chmeleff, J., Morris, J.D., and Lopez-Escobar, L., 2002, Origin of $^{226}Ra$-$^{230}Th$ disequilibria in arc lavas from southern Chile and implications for magma transfer time: Earth and Planetary Science Letters, v. 196, p. 189–196, doi: 10.1016/S0012-821X(01)00611-2.

Smithies, R.H., and Champion, D.C., 2000, The Archaean high-Mg diorite suite: Links to tonalite-trondhjemite-granodiorite magmatism and implications for Early Archaean crustal growth: Journal of Petrology, v. 41, p. 1653–1671, doi: 10.1093/petrology/41.12.1653.

Smithies, R.H., Champion, D.C., and Sun, S.S., 2004, Evidence for early LREE-enriched mantle source regions: Diverse magmas from the c. 3.0 Ga Mallina Basin, Pilbara Craton, NW Australia: Journal of Petrology, v. 45, p. 1515–1537, doi: 10.1093/petrology/egh014.

Spear, F.S., 1993, Metamorphic phase equilibria and pressure-temperature-time paths: Mineralogical Society of America Monograph: Washington, D.C., Mineralogical Society of America, Monograph 1, 799 p.

Spiegelman, M., Kelemen, P., and Aharonov, E., 2001, Causes and consequences of flow organization during melt transport: The reaction infiltration instability in compactible media: Journal of Geophysical Research, v. 106, no. B2, p. 2061–2077, doi: 10.1029/2000JB900240.

Sproule, R.A., Lesher, C.M., Ayer, J.A., Thurston, P.C., and Herzberg, C.T., 2002, Spatial and temporal variations in the geochemistry of komatiitic basalts in the Abitibi greenstone belt: Precambrian Research, v. 115, p. 153–186, doi: 10.1016/S0301-9268(02)00009-8.

Stein, M., and Goldstein, S.L., 1996, From plume head to continental lithosphere in the Arabian-Nubian shield: Nature, v. 382, p. 773–778, doi: 10.1038/382773a0.

Stern, R.A., and Hanson, G.N., 1991, Archaean high-Mg granodiorites: A derivative of light rare earth enriched monzodiorite of mantle origin: Journal of Petrology, v. 32, p. 201–238.

Stern, R.A., Hanson, G.N., and Shirey, S.B., 1989, Petrogenesis of mantle-derived, LILE-enriched Archean monzodiorites and trachyandesites (sanukitoids) in southwestern Superior Province: Canadian Journal of Earth Sciences, v. 26, p. 1688–1712.

Stevenson, R.K., Henry, P., and Gariépy, C., 1999, Assimilation–fractional crystallization origin of Archean sanukitoid suites, western Superior Province, Canada: Precambrian Research, v. 96, p. 83–99, doi: 10.1016/S0301-9268(99)00009-1.

Sutcliffe, R., Smith, A., Doherty, W., and Barnett, R., 1990, Mantle derivation of Archaean amphibole-bearing granitoids and associated mafic rocks: Evidence from the southern Superior Province, Canada: Contributions to Mineralogy and Petrology, v. 105, p. 255–274, doi: 10.1007/BF00306538.

Thorkelson, D.J., and Breitsprecher, K., 2005, Partial melting of slab window margins: Genesis of adakitic and non-adakitic magmas: Lithos, v. 79, p. 25–41, doi: 10.1016/j.lithos.2004.04.049.

Thurston, P.C., 2002, Autochthonous development of Superior Province greenstone belts?: Precambrian Research, v. 115, p. 11–36, doi: 10.1016/S0301-9268(02)00004-9.

Van Kranendonk, M.J., Collins, W.J., Hickman, A.H., and Pawley, M.J., 2004, Critical tests of vertical vs. horizontal tectonic models for the Archaean East Pilbara granite-greenstone terrane, Pilbara Craton, Western Australia: Precambrian Research, v. 131, p. 173–211, doi: 10.1016/j.precamres.2003.12.015.

Vervoort, J.D., White, W.M., and Thorpe, R.I., 1994, Nd and Pb isotope ratios of the Abitibi Greenstone Belt: New evidence for very early differentiation of the Earth: Earth and Planetary Science Letters, v. 128, p. 215–229, doi: 10.1016/0012-821X(94)90146-5.

White, R.V., Tarney, J., Kerr, A.C., Saunders, A.D., Kempton, P.D., Pringle, M.S., and Klaver, G.T., 1999, Modification of an oceanic plateau, Aruba, Dutch Caribbean: Implications for the generation of continental crust: Lithos, v. 46, p. 43–68, doi: 10.1016/S0024-4937(98)00061-9.

Wyman, D.A., 1999a, A 2.7 Ga depleted tholeiite suite: Evidence of plume-arc interaction in the Abitibi Greenstone Belt, Canada: Precambrian Research, v. 97, p. 27–42, doi: 10.1016/S0301-9268(99)00018-2.

Wyman, D.A., 1999b, Lode gold deposits and Archean mantle plume–island arc interaction, Abitibi Subprovince, Canada: Journal of Geology, v. 107, p. 715–725, doi: 10.1086/314376.

Wyman, D.A., 2003, Upper mantle processes beneath the 2.7 Ga Abitibi belt, Canada: A trace element perspective: Precambrian Research, v. 127, p. 143–165, doi: 10.1016/S0301-9268(03)00185-2.

Wyman, D.A., and Kerrich, R., 2002, Formation of Archean continental lithospheric roots: The role of mantle plumes: Geology, v. 30, p. 543–546, doi: 10.1130/0091-7613(2002)030<0543:FOACLR>2.0.CO;2.

Wyman, D.A., Kerrich, R., and Polat, A., 2002, Assembly of Archean cratonic mantle lithosphere and crust: Plume-arc interaction in the Abitibi-Wawa subduction-accretion complex: Precambrian Research, v. 115, p. 37–62, doi: 10.1016/S0301-9268(02)00005-0.

Zegers, T.E., and Van Keken, P.E., 2001, Middle Archean continent formation by crustal delamination: Geology, v. 29, p. 1083–1086, doi: 10.1130/0091-7613(2001)029<1083:MACFBC>2.0.CO;2.

Zhao, D.P., 2004, Global tomographic images of mantle plumes and subducting slabs: Insight into deep Earth dynamics: Physics of the Earth and Planetary Interiors, v. 146, p. 3–34, doi: 10.1016/j.pepi.2003.07.032.

MANUSCRIPT ACCEPTED BY THE SOCIETY 14 AUGUST 2007

Printed in the USA

The Geological Society of America
Special Paper 440
2008

# When did plate tectonics begin? Evidence from the orogenic record

**Victoria Pease***
*Department of Geology & Geochemistry, Stockholm University, Stockholm, SE-106 91, Sweden*

**John Percival**
*Geological Survey of Canada, 601 Booth Street, Ottawa, Ontario K1A 0E8, Canada*

**Hugh Smithies**
*Geological Survey of Western Australia, 100 Plain Street, East Perth, Western Australia 6004, Australia*

**Gary Stevens**
*Department of Geology, Stellenbosch University, Stellenbosch, Western Cape 7602, South Africa*

**Martin Van Kranendonk**
*Geological Survey of Western Australia, 100 Plain Street, East Perth, Western Australia 6004, Australia*

## ABSTRACT

**Evidence of modern-style plate tectonics is preserved in the continental rock record as orogens and rifts; these orogens represent regions of mountain building resulting from compression between converging plates. Recognition of orogens in the ancient rock record can help identify when plate tectonics began on Earth. Evidence of Paleoproterozoic collisional orogeny is widely accepted. The development, however, of Archean collisional orogens is highly controversial, as is the operation of plate tectonics in general. We review the tectonic evolution of three well-studied Archean terranes—the Pilbara craton of Western Australia, the Barberton granite-greenstone terrane of South Africa, and the Superior Province of Canada—in terms of their geological development and evidence for Archean collisional and accretionary plate-tectonic processes in the context of secular evolution of the planet. The Pilbara craton preserves geological, geochemical, and geochronological evidence for continental rifting at 3.2 Ga, development of an oceanic-arc subduction complex at 3.12 Ga, and terrane accretion at 3.07 Ga. The Barberton granite-greenstone terrane of the Kaapvaal craton provides thermobarometric evidence for subduction-related high-pressure–low-temperature metamorphism juxtaposed against medium-pressure–high-temperature metamorphism associated with exhumation of high-grade rocks via orogenic collapse, which together are interpreted to represent a paired metamorphic belt. The Superior Province in the Canadian Shield records widespread accretionary and collisional assembly at ca. 2.7 Ga. This evidence argues for "modern-style" plate tectonics on Earth since at least 3.2 Ga.**

**Keywords:** Archean, accretion, orogeny, plate tectonics, metamorphism.

*vicky.pease@geo.su.se

Pease, V., Percival, J., Smithies, H., Stevens, G., and Van Kranendonk, M., 2008, When did plate tectonics begin? Evidence from the orogenic record, *in* Condie, K.C., and Pease, V., eds., When Did Plate Tectonics Begin on Planet Earth?: Geological Society of America Special Paper 440, p. 199–228, doi: 10.1130/2008.2440(10).

## INTRODUCTION

The evidence of plate-tectonic processes preserved in the continental rock record includes a wide range of geological features, such as accreted oceanic arcs, continental magmatic arcs, ophiolites, accretionary mélanges, turbiditic accretionary wedges, forearc and back-arc basin rocks, thrust fault–bounded panels of high-pressure metamorphic rocks, blueschists, large-scale nappes, igneous rock suites associated with extensional tectonics, and foredeep basins. Many of these features are found in areas of mountain building, or orogens, caused by subduction, terrane accretion, and the subsequent compression between converging tectonic plates. The three types of orogens recognized today are collisional, accretionary, and decretionary. Collisional orogens, such as the Himalayas, are formed by the collision between two continental plates (Dewey, 1969; Wilson, 1966) and are characterized by the addition of little or no juvenile material. Accretionary orogens, such as Japan, Indonesia, and Alaska, on the other hand, involve the consumption of oceanic crust and the addition of significant amounts of juvenile arc material (see review by Kusky and Bradley, 1999). Subduction-erosion in decretionary orogens, such as the Andes, results in the net loss of forearc material (e.g., von Huene and Scholl, 1991). While examples of accretionary and collisional orogens are preserved in the Precambrian record, examples of decretionary orogens have not been documented in the ancient record.

Regional studies of Phanerozoic orogenic belts have provided information indispensable to the development of our understanding of modern orogenic and tectonic processes (e.g., Miyashiro, 1961; Ernst, 1973, 1988; Smith, 1984; Chopin, 1987), yet relatively few studies have attempted to do the same for significantly older terranes (e.g., Williams and Currie, 1993; Collins and Van Kranendonk, 1999). Consequently, there is great potential for such studies to address the controversy regarding Paleo- to Mesoarchean geodynamic questions, particularly the operation of plate tectonics. Collisional orogens are recognized as far back as the Paleoproterozoic (e.g., Connelly and Mengel, 2000), whereas the earliest accretionary orogens date from the Mesoarchean (Smith et al., 1998; Van Kranendonk et al., 2002, 2007; Smithies et al., 2005b), or perhaps even the Paleoarchean (Nutman et al., 2002). Both types of orogens contain distinct terranes bounded by major faults (terrane boundaries). However, several studies have noted the lack of many of the most classical features of plate tectonics in Archean continental crust (cf. Hamilton, 1998, 2003; Stern, 2005; Brown, 2006). For example, characteristic rock associations such as blueshists or accretionary mélanges are not always present, although this is true even in some early Precambrian and Phanerozoic orogens (Şengör and Natal'in, 1996). Older orogens can also be highly reworked and fragmental, rendering evidence of their tectonic evolution even more cryptic. Consequently, in ancient orogens, it is the *convergence of a wide and independent range of evidence* (e.g., geological setting, geochemical characteristics of rock associations, structures, and metamorphism) that allows us to confidently infer a tectonic environment similar to that of modern orogens (e.g., Van Kranendonk, 2004a, and references therein).

We review the Pilbara craton of Western Australia, the Kaapvaal craton of South Africa, and the Superior Province in the Canadian Shield for evidence of plate tectonics in the Archean. Their development is considered in relation to the preserved orogenic record and secular changes in Earth evolution, such as differences in mantle temperatures, geochemical evolution of Earth's mantle and crust, and the advent of modern-style plate tectonics in Earth history.

## TECTONIC DEVELOPMENT OF THE 3.72–2.83 Ga PILBARA CRATON

The Pilbara craton, in the northwestern part of Western Australia (Fig. 1), is particularly important to Archean geoscientific studies because this well-exposed and relatively extensive (~122,000 km$^2$) region encompasses elements that collectively span virtually the entire Paleoarchean and Mesoarchean, from 3.72 Ga to 2.83 Ga, with relatively few and small temporal gaps. A detailed description of the extensive geological, geochronological, and geochemical data that have been collected from this craton is given in Van Kranendonk et al. (2002, 2004, 2007). The craton is divided into two contrasting domains, each of which preserves evidence for the operation of plate-tectonic processes in the Archean (Hickman, 2004; Van Kranendonk et al., 2007).

The largest of these domains is the 3.53–3.165 Ga East Pilbara terrane (Fig. 1), which has a structural style dominated by ovoid granite complexes mantled by synclinal greenstone belts (Hickman, 1983, 2004; Hickman and Van Kranendonk, 2004; Van Kranendonk et al., 2004), and represents the ancient nucleus of the craton. The second domain is the 3.27–3.11 Ga West Pilbara superterrane (Fig. 1). This is a collage of three lithostratigraphically and geochronologically distinct terranes—the Karratha, Regal, and Sholl terranes—that have a characteristic linear (belt-like) structural pattern, in which terranes are juxtaposed against one another across major structures, including the Regal thrust and the crustal-scale, predominantly strike-slip, Sholl shear zone.

The East Pilbara terrane and the West Pilbara superterrane are separated by dominantly coarse clastic sedimentary rocks of the 3.02–2.93 Ga De Grey Supergroup (Fig. 1), which were deposited in a series of late-tectonic basins. Syn- to late-tectonic intrusions occur across the craton but are concentrated close to the inferred suture between the East Pilbara terrane and the West Pilbara superterrane. The West Pilbara superterrane and the rocks of the De Grey Supergroup together form a belt that wraps around the northern and western edge of the East Pilbara terrane, and they, together with the late-tectonic intrusions, contain a wealth of evidence supporting the evolution of a modern-style convergent margin setting from ca. 3.13 to 2.91 Ga.

### The East Pilbara Terrane

The East Pilbara terrane is composed of the generally well-preserved Pilbara Supergroup and several granitic supersuites. The Pilbara Supergroup consists of four demonstrably autochthonous

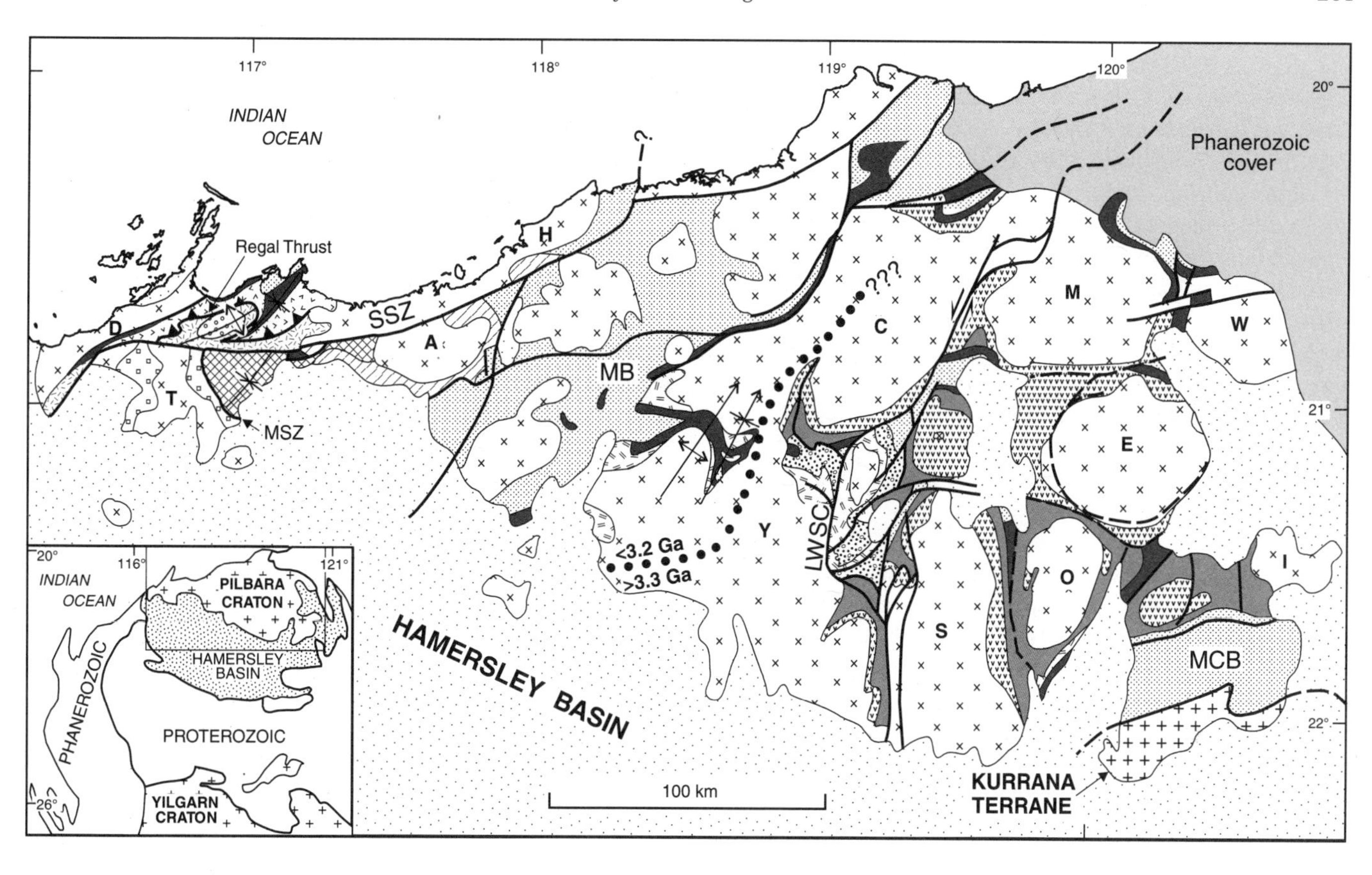


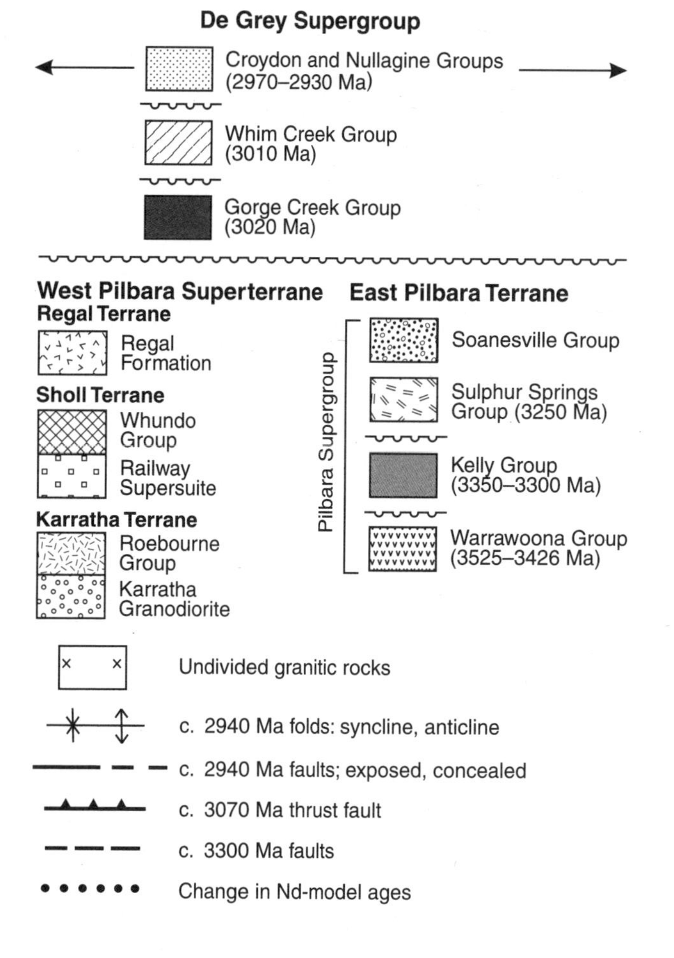


Figure 1. Simplified geological map of the Pilbara craton, showing Nd model age data plotted in 100 m.y. age groups (Van Kranendonk et al., 2007). A—Andover Granitic Complex; C—Carlindi Granitic Complex; D—Dampier Granitic Complex; E—Mount Edgar Granitic Complex; H—Harding Granitic Complex; LWSC—Lalla Rookh Western Shaw structural corridor; M—Muccan Granitic Complex; MB—Mallina Basin; MCB—Mosquito Creek Basin; MSZ—Maitland Shear Zone; O—Corunna Downs Granitic Complex; S—Shaw Granitic Complex; SSZ—Sholl shear zone; T—Cherrata Granitic Complex; Y—Yule Granitic Complex. Inset shows location of the craton in northwestern Australia; rectangle indicates location of the main figure.

and consistently upward-younging volcanic-sedimentary groups deposited from ca. 3.53 Ga to ca. 3.165 Ga (Van Kranendonk, 2000; Van Kranendonk et al., 2002, 2006, 2007). The lowermost, 3.53–3.42 Ga, Warrawoona Group contains up to 12 km of pillowed tholeiitic and komatiitic basalt, with subordinate komatiite. Locally voluminous felsic volcanic rocks cap each (ultra) mafic to felsic volcanic cycle formed during the almost 100 m.y. of continuous volcanism from 3.53 to 3.42 Ga (Hickman and Van Kranendonk, 2004). The unconformably overlying Kelly Group contains up to 8 km of tholeiitic basalt, basaltic komatiite, and komatiite, capped by rhyolite, which erupted between 3.35 and 3.315 Ga. This is in turn overlain by up to 4 km of komatiitic to rhyolitic volcanic rocks of the 3.27–3.235 Ga Sulfur Springs Group, and then by the dominantly sedimentary rocks of the Soanesville Group (ca. 3.235–3.165 Ga).

The maximum primary vertical thickness of the Pilbara Supergroup in any particular area is likely to have been less than 15 km as a result of accumulation in actively developing basins during periods of extension. Detailed geological mapping and extensive geochronology have revealed no stratigraphic repetitions in any of the greenstone belts. The discovery of abundant, old (i.e., older than 3.5 Ga) zircon xenocrysts in igneous rocks and detrital zircon grains in sedimentary rocks (e.g., Thorpe et al., 1992; Van Kranendonk, 2004b), as well as geochemical evidence for contamination of even the oldest rocks of the Pilbara Supergroup by felsic crust (Glikson and Hickman, 1981; Green et al., 2000) indicate that the supergroup was deposited on an older basement of at least partly sialic crust (e.g., Hickman, 1983, 1984; Van Kranendonk et al., 2002, 2007).

The voluminous basaltic rocks of the Pilbara Supergroup are derived from two sources, one that produced magmatic compositions very similar to modern plume-related magmas (e.g., Arndt et al., 2001; Van Kranendonk and Pirajno, 2004; Condie, 2005; Smithies et al., 2005a) and the other representing a lithospheric mantle source (Smithies et al., 2005a). The lithospheric mantle source appears to have become more depleted over time, whereas the primary composition of magmas from the plume-related source appears not to have changed significantly. The more primitive magmas erupted essentially free of significant input from felsic crust, although individual basalt units show a range in La/Sm and La/Yb ratios and in light rare earth element (LREE) concentrations that likely reflects at least some interaction with felsic crust (Smithies et al., 2007a, 2007b). La/Nb ratios in the more primitive magmas are low (0.97–1.19), and the rocks show no evidence of having sampled any source compositionally resembling a modern mantle wedge (Smithies et al., 2005a).

Over the same time period, geochemical data show that the composition of felsic volcanic units progressively evolved from tholeiitic to tonalite-trondhjemite-granodiorite (TTG)–like to intracrustal high-K melts (Smithies et al., 2007a, 2007b). These trends are also consistent with isotopic trends and age patterns from granites of the East Pilbara terrane (Champion and Smithies, 2001, 2007). Together, these data suggest that the felsic component of the terrane evolved mainly as a result of the progressive recycling of pre-existing crust, without any need for the significant addition of juvenile material through subduction processes.

### West Pilbara Superterrane

The oldest terrane of the West Pilbara superterrane is the Karratha terrane (Fig. 1), which consists of 3.27–3.25 Ga ultramafic to felsic volcanic and sedimentary rocks, with contemporaneous subvolcanic intrusions. These rocks are in thrust contact with 3.5-km-thick, mid-ocean-ridge basalt (MORB)–like basalts (Ohta et al., 1996) that comprise the Regal Formation of the Regal terrane (Hickman, 2004) (Fig. 1). Kiyokawa et al. (2002) suggested that the northwestern part of the Regal Formation represents a series of thrust-bound slices of basalts from different arc-related tectonic settings. However, major- and trace-element geochemical data show no significant compositional difference within the formation and suggest that it represents a single tholeiitic basalt package with MORB-like characteristics (Smithies et al., 2007a).

The Karratha and Regal terranes are separated from the Sholl terrane to the south by the kilometer-wide, long-lived Sholl shear zone (Smith et al., 1998) (Fig. 1). The Sholl terrane consists of the ca. 3.13–3.11 Ga Whundo Group and coeval intrusions.

The Whundo Group is a >10-km-thick succession of bimodal basaltic to felsic volcanic and volcaniclastic rocks (Fig. 2). It represents a lithostratigraphically and geochronologically exotic terrane: it is fault bounded and has no counterparts in any other part of the craton. Neither the base nor the top of the group is preserved; the former is intruded by granites, and the latter is unconformably overlain by younger rocks. Rocks of basaltic to andesitic composition dominate the Whundo Group but are compositionally distinct from the dominantly tholeiitic rocks of the Pilbara Supergroup. Moreover, the Whundo Group makes up a stratigraphically interlayered succession of a wide range of different magma types (Fig. 2), each of which has geochemical features closely reminiscent of modern subduction-related magmas (Smithies et al., 2005b). These include rocks with a strong boninite affinity near the base of the sequence that overlie LREE-enriched calc-alkaline andesites (CA1 and CA2 in Fig. 2), including lavas that preserve geochemical evidence for flux-melting (CA2 in Fig. 2), and an upper part that contains Nb-enriched basalts in close association with lavas of adakitic affinity. Smithies et al. (2005b) showed that the LREE enrichments in the boninites and calc-alkaline andesites were unlikely to be related to assimilation of any locally available "Pilbara crust" or to average Pilbara craton crust. In addition, the stratigraphic sequence contains no sedimentary material of continental provenance, and the volcanic rocks have Nd-isotopic compositions ($\varepsilon_{Nd}$ ~+2 to +3) indicative of juvenile crust (Sun and Hickman, 1998; Smithies et al., 2004a, 2005b).

### Postcollisional Sedimentary Basins: The De Grey Supergroup

Rocks of both the East Pilbara terrane and the West Pilbara superterrane are unconformably overlain by the 3.02–2.93 Ga De Grey Supergroup (Fig. 1), which is composed of predominantly coarse clastic rocks that consist of four groups deposited in separate basins across the Pilbara craton (Van Kranendonk et al., 2006, 2007). The lowermost succession is the ca. 3.02 Ga clastic sedimentary rocks, felsic tuffs and porphyries, and banded iron formation (Cleaverville Formation) of the Gorge Creek Group. The lateral continuity of this group across distinct lithotectonic terranes is the primary evidence indicating that accretion of the West Pilbara superterrane with the East Pilbara terrane occurred prior to this time, at ca. 3.07 Ga (see following discussion). The ca. 3.01–2.98 Ga Whim Creek Group is a complex association of coeval mafic to felsic volcanic, volcaniclastic, and intrusive rocks that unconformably overlies the Whundo and Gorge Creek Groups along the western margin of the Mallina Basin (MB in Fig. 1).

The Croydon Group forms the main fill of the Mallina Basin, which is interpreted to overlie the collisional suture between the East Pilbara terrane and the West Pilbara superterrane (Van Kranendonk et al., 2002). The main sedimentary package includes sandstone, siltstone, and conglomerate deposited on a series of submarine fans (Eriksson, 1982). These dominantly clastic rocks are overlain, with peperitic contacts, by locally spinifex-textured and variolitic basalts, which form the Louden and Mount Negri Volcanic Members of the Bookingarra Formation (Hickman, 1997; Pike and Cas, 2002; Van Kranendonk et al., 2006). Basalts very similar in composition to the Loudens and Negri Volcanic Members occur across the Mallina Basin interbedded with clastic sedimentary rocks, and also across the northern margin of the East Pilbara terrane (Salt Well Member) and on the northeastern margin of the East Pilbara terrane (Coonieena Basalt Member) (Fig. 3).

## Syn- to Late-Tectonic Intrusions

The ca. 2.955–2.92 Ga Sisters Supersuite is principally developed across the Mallina Basin and western part of the East Pilbara terrane, although small intrusions also occur within the West Pilbara superterrane. This supersuite is dominantly composed of leucocratic, K-rich monzogranites derived from remelting of older granitic components in the East Pilbara terrane and crust underlying the Mallina Basin (Champion and Smithies, 2001, 2007). Widespread monzogranite magmatism was earliest and most voluminous immediately adjacent to the Mallina Basin and becomes progressively younger to the east. However, other compositionally distinct suites of granitic rocks have also been recognized, the most important of which are, for the purposes of this discussion, the 2.955–2.945 Ga sanukitoids of the Indee Suite in the core of the Mallina Basin (Fig. 3; Smithies and Champion, 2000; Smithies et al., 2004b; Van Kranendonk et al., 2006) and the LREE-enriched gabbro rocks of the Langenbeck Suite in the Mallina Basin, which are the intrusive equivalents of basalts within the Croydon Group (Bookingarra Formation). Both of these are interpreted to contain geochemical evidence of derivation from a subduction-modified mantle source (Smithies et al., 2004b, 2007a).

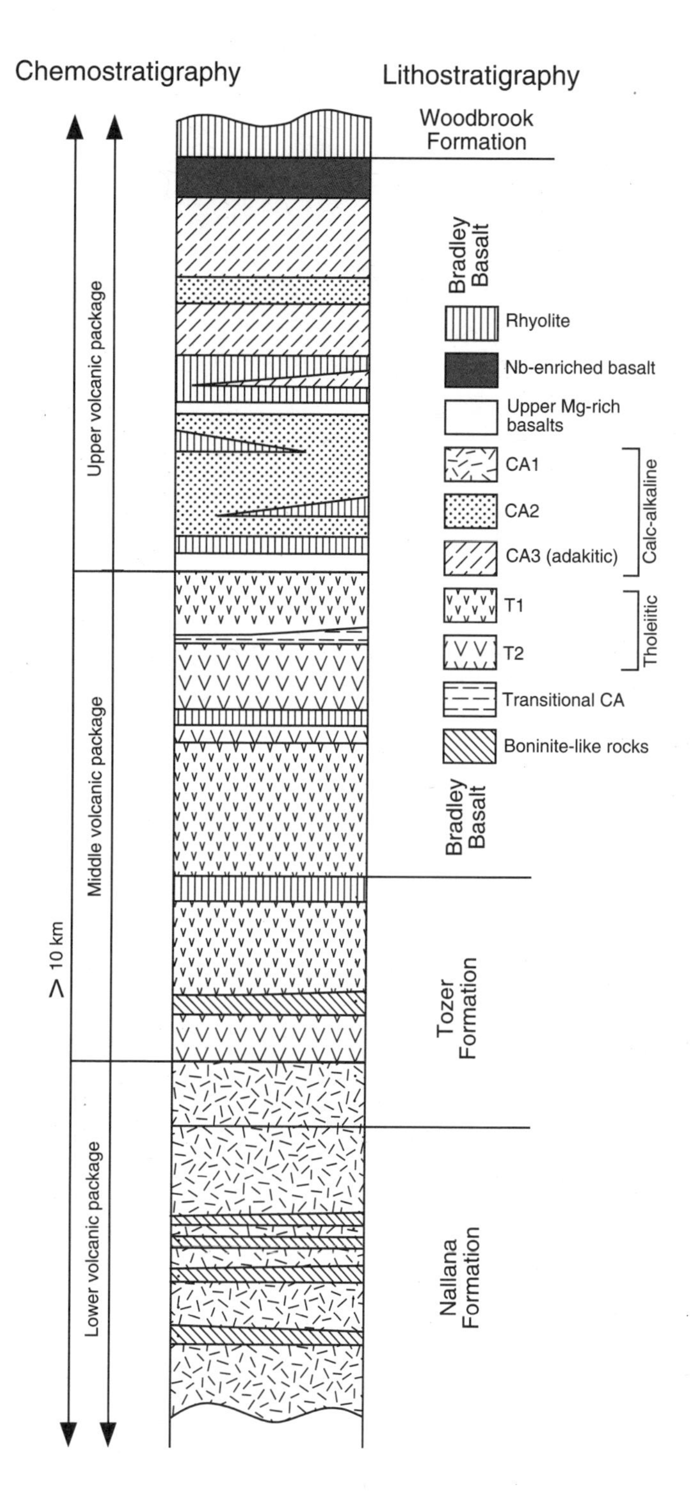


Figure 2. Lithogeochemical column of the ca. 3.12 Ga Whundo Group of the Sholl terrane (from Smithies et al., 2005a). Lithological subdivisions are on the right-hand side, and broad chemical subdivisions are on the left.

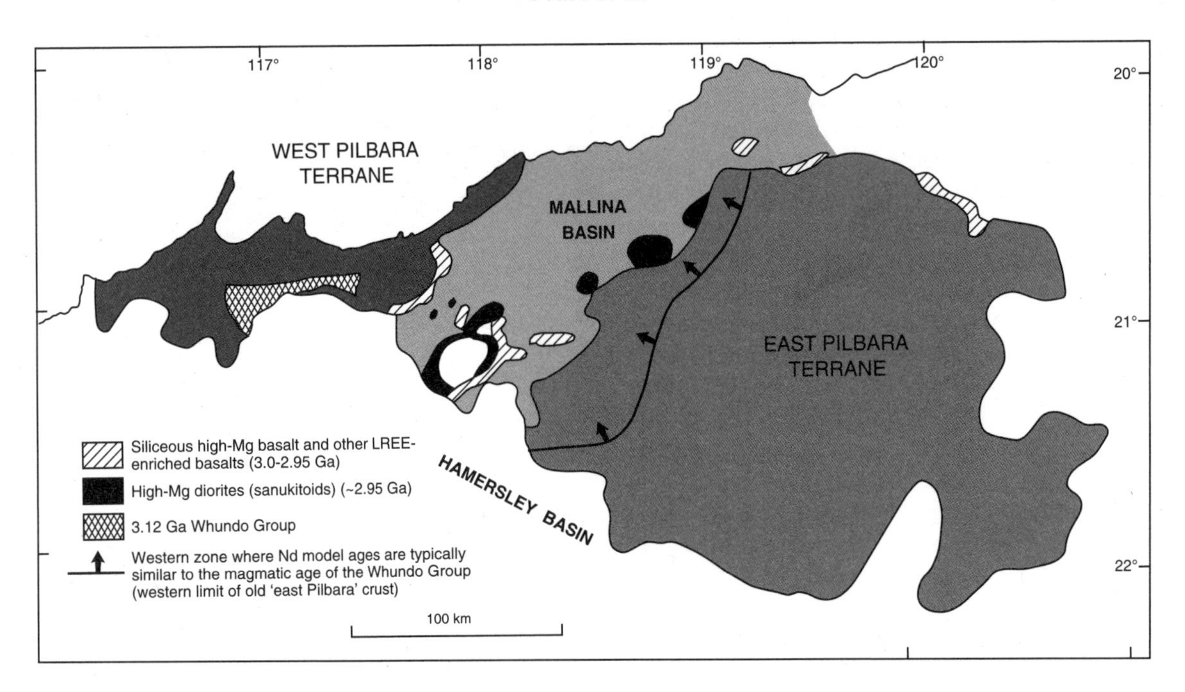


Figure 3. Map showing the distribution of various subduction-enriched magmatic units of the West Pilbara superterrane and the Mallina Basin, as well as the western limit of "old Pilbara crust" as defined by young (ca. 3.12 Ga or younger) Nd-depleted-mantle model ages (based on ~100 Nd isotopic analyses; Geological Survey of Western Australia and Geoscience Australia, unpublished data) (from Smithies et al., 2007a). LREE—light rare earth element.

## Tectonic Evolution of the Pilbara Craton

### *East Pilbara Terrane*

Several different models have been proposed for the tectonic evolution of the East Pilbara terrane. These include plate-tectonic–based models whereby formation of the Warrawoona Group is ascribed to a mid-ocean-ridge setting (Ueno et al., 2001; Kato and Nakamura, 2003) and felsic volcanic rocks of the group are interpreted to have formed in an arc or back-arc environment (Bickle et al., 1983, 1993; Barley et al., 1984, 1998). Other workers have suggested that the East Pilbara terrane was affected by periods of Alpine-style thrusting followed by episodes of extensional orogenic collapse and core complex formation at 3.47 Ga and 3.31 Ga (Bickle et al., 1980, 1985; Boulter et al., 1987; Zegers et al., 1996, 2001; van Haaften and White, 1998; Kloppenburg et al., 2001; Blewett, 2002) during Basin-and-Range–style extension (cf. Barley and Pickard, 1999). Nonuniformitarian models for East Pilbara tectonic evolution suggest that crust formation occurred as a result of plume magmatism (Van Kranendonk and Pirajno, 2004; Smithies et al., 2005a, 2007b; Van Kranendonk et al., 2007), and the characteristic dome-and-keel structural map pattern was a result of multiple episodes of partial convective overturn of the middle and upper crust at between 3.47 and 2.74 Ga (Hickman, 1975, 1984; Collins, 1989; Williams and Collins, 1990; Collins et al., 1998; Collins and Van Kranendonk, 1999; Van Kranendonk et al., 2001, 2004, 2007; Hickman and Van Kranendonk, 2004; Sandiford et al., 2004).

The new map, geochemical, and geochronological data outlined here (see Van Kranendonk et al., 2007, for greater detail) shed new light on the applicability of some or all of these models. Data that support a widespread, at least partly sialic, basement to the Pilbara Supergroup preclude formation of Warrawoona Group crust at a mid-ocean ridge. Models of ophiolite obduction (Ueno et al., 2001; Kato and Nakamura, 2003) conflict with geochemical evidence for crustal contamination of basalts, the right-way-up and upward-younging greenstone stratigraphy, the interbedding of basalts and felsic volcanic rocks, the geochemical data for progressive upward depletion of mantle-sourced magmas (Smithies et al., 2005a), and the low-pressure–high-temperature contact-metamorphic style, which are all uncharacteristic of accretionary orogens (Van Kranendonk et al., 2007).

A volcanic-arc or back-arc environment for the Warrawoona Group is incompatible with both the relatively homogeneous and tholeiitic geochemical characteristics of Warrawoona Group mafic volcanic rocks (Smithies et al., 2005a). Some felsic magmas of the lower Pilbara Supergroup share some geochemical features with Phanerozoic calc-alkaline rocks found in continental subduction settings, but they are typically more sodic (Smithies et al., 2007b). They are also stratigraphically bound, above and below, by thick sequences of basalt with low La/Nb and La/Sm ratios that show no evidence of having sampled any source compositionally resembling a modern mantle wedge. These felsic volcanic rocks can largely be explained as differentiated products of large tholeiitic magma chambers, with variable degree of mixing with, or assimilation of, TTG-like crust or magma (e.g., Smithies et al., 2007b).

There is evidence that many of the structures previously interpreted as the result of Alpine-style thrusting and extensional core complex formation at ca. 3.46 Ga (Bickle et al., 1980, 1985; Zegers et al., 2001) resulted from sinistral transpression at 2.94 Ga, *after* formation of the dome-and-keel pattern (Van Kranendonk et al., 2004). These authors also showed that the granitic domes have very little in common with core complexes and formed during radial extension of the cover off the rising domes (also see Hickman and Van Kranendonk, 2004). Periods of significant extension occurred during deposition of the Pilbara Supergroup, but these likely represent an integral component of the vertical tectonic process that dominated this terrane, rather than of processes similar to modern-style plate tectonics (Van Kranendonk et al., 2007). The stratigraphic and geochemical data from the Pilbara Supergroup show that it most likely formed in a hybrid setting between that of a typical modern-day oceanic plateau and modern large igneous provinces erupted onto stable continental crust, i.e., an oceanic-type plateau in terms of volume and setting, but erupted onto a basement of older continental crust, possibly with similarities to the modern Kerguelen Plateau (e.g., Operto and Charvis, 1995; Frey et al., 2000; Van Kranendonk and Pirajno, 2004).

***West Pilbara Superterrane***

In contrast with the East Pilbara terrane, the mainly younger tectonic units that comprise the West Pilbara superterrane and the basins that host the De Grey Supergroup preserve several strong lines of evidence that modern-style plate tectonic process were in operation, at least locally, by 3.12 Ga. The linear structural fabric of the superterrane, the presence of lithologically and geochronologically exotic belts juxtaposed across crustal-scale shear zones, and the presence of a broad belt of isotopically juvenile crust are consistent with some form of convergent margin setting and are difficult to account for within any other tectonic context (Smith et al., 1998; Smithies et al., 2004a, 2004b, 2005b; Van Kranendonk et al., 2007). Indeed, the operation of horizontal tectonics is now widely accepted for the West Pilbara superterrane, although the specifics of how and when this occurred are still debated (Ohta et al., 1996; Barley, 1997; Kiyokawa and Taira, 1998; Smith et al., 1998; Smithies and Champion, 2000; Kiyokawa et al., 2002; Van Kranendonk et al., 2002; Smith, 2003; Hickman, 2004; Smithies et al., 2004a, 2004b).

The Whundo Group, in particular, contains several geochemical features that, together with the isotopically juvenile composition of the rocks and the lack of any continental-derived sediment within the volcanic-sedimentary pile, point to an oceanic setting. The presence of boninites that have all of the compositional hallmarks of modern boninites, and the presence of calc-alkaline volcanics, Nb-enriched basalts, and adakites, strongly suggests an arc-like environment (Fig. 2; Smithies et al., 2005b). Indeed, enrichments in large ion lithophile elements (LILEs), high field strength elements (HFSEs), and LREE, and high ratios of LREE/HFSE and LILE/HFSE throughout the entire Whundo stratigraphy cannot be explained through assimilation of locally or regionally exposed, or average, Pilbara felsic crust (Smithies et al., 2005b). Unusual trends of increasing trace-element enrichment, i.e., higher La/Sm, with measures of *increasing* degrees of partial melting (e.g., increasing Cr, decreasing HFSE, HREE, etc.) in the calc-alkaline rocks are similar to trends seen in some arc and back-arc rocks (Taylor and Martinez, 2003), which can only be explained if the degree of melting is sympathetically tied to metasomatic enrichment, i.e., flux melting of a mantle source. Trends to higher Th/Ba stratigraphically upward are also consistent with an increasing slab-melt component. Additionally, the calc-alkaline basalts and andesites eventually give way to adakites and Nb-enriched basalts (Fig. 2), which *require* a major slab-melt component in their petrogenesis (Smithies et al., 2005b). Thus, the Whundo Group provides strong geochemical evidence for a subduction-enriched mantle source, strongly supporting independent geological relationships for modern-style subduction-accretion processes at 3.12 Ga (Smithies et al., 2005a, 2007b, 2007c). The Whundo Group likely represents Earth's oldest preserved arc, and this exotic terrane was accreted to the northwestern edge of the East Pilbara terrane at some stage between 3.12 and 2.95 Ga, probably at ca. 3.07 Ga (Van Kranendonk et al., 2007).

***De Grey Supergroup and Associated Intrusions***

Subduction modification of the mantle during deposition of the Whundo Group also had an influence on the composition of ca. 2.95 Ga Croydon Group basalts and granitic rocks of the Indee Suite (Sisters Supersuite), which together form a belt flanking the northern and western margin of the ancient cratonic nucleus represented by the East Pilbara terrane (Fig. 3). While there is no regional evidence for subduction contemporaneous with deposition of the ca. 3.02–2.93 Ga De Grey Supergroup and associated mafic magmas, or with emplacement of the ca. 2.95 Ga sanukitoids of the Indee Suite, the composition of these magmas points strongly to a highly enriched mantle source. Very high La/Yb and La/Sm ratios and strongly developed negative Nb anomalies suggest extensive contamination of the Croydon Group basalts (e.g., Arndt et al., 2001). However, the high La/Nb ratios and Nb concentrations in the basalt are inconsistent with this crustal component being simply average exposed Archean Pilbara crust, and the regional compositional homogeneity of these basalts cannot be explained through assimilation of crust as compositionally heterogeneous as the Pilbara craton (Smithies et al., 2004b). Likewise, while a mantle source component is required to explain the high Mg# and high Cr and Ni concentrations of the Indee Suite, even the most primitive diorites show extreme enrichment in Th and LREE and no apparent correlation to Mg#, $\varepsilon_{Nd}$, Cr, Ni, or $SiO_2$. A significantly LILE-enriched mantle source is the most reasonable explanation for the composition of these rocks (e.g., Shirey and Hanson, 1984; Evans and Hanson, 1997; Smithies and Champion, 2000).

The enriched component of the mantle source for the basalts of the Croydon Group and for the Indee Suite can be modeled as contrasting mixtures of the same two crustal components: "old-Pilbara crust" (>3.3 Ga, $\varepsilon_{Nd[2.95\ Ga]} < -2.3$, high La/Nb, [Sm, Zr]) and basaltic crust similar to the ca. 3.12 Ga Whundo Group ($\varepsilon_{Nd[2.95\ Ga]} > -0.4$, low La/Nb, (Sm, Zr]) (Smithies et al., 2004b). We suggest that prior to or during accretion of the Whundo Group to the East Pilbara craton, ca. 3.12 Ga Whundo mafic crust and homogeneous sediment derived from old (>3.3 Ga) Archean terranes were subducted to the southeast and that partial melts derived from the subducted sediments infiltrated the mantle wedge.

### A New Model for the Tectonic Evolution of the Pilbara Craton

Analysis of the available geological data for the Pilbara craton suggests a six-stage evolution for the Pilbara craton, from the early formation of ancient sialic crust, through plume events dominated by vertical tectonics, to horizontal tectonic events including rifting and collision (Fig. 4; Van Kranendonk et al., 2007). The 3.53–3.17 Ga East Pilbara terrane represents the ancient nucleus of the craton that formed through three distinct mantle plume events at 3.53–3.43 Ga, 3.35–3.29 Ga, and 3.27–3.24 Ga (stage 2 in Fig. 4). Each plume event resulted in the eruption of thick, dominantly basaltic volcanic successions on older crust to at least 3.72 Ga, and melting of crust to generate first TTG and then progressively more evolved granitic magmas. In each case, plume magmatism was accompanied by uplift and crustal extension. The combination of conductive heating from below, thermal blanketing from above, and internal heating of buried granitoids during these events led to episodes of partial convective overturn of upper and middle crust. These mantle melting events caused severe depletion of the subcontinental lithospheric mantle, making the East Pilbara terrane a stable, buoyant, unsubductable protocontinental nucleus by ca. 3.2 Ga. Extension accompanying the latest event led to rifting of the protocontinent margins between 3.2 and 3.17 Ga (stage 3 in Fig. 4).

After 3.2 Ga, horizontal tectonic forces dominated over vertical forces, as revealed by the geology of the three terranes (Karratha, Sholl, and Regal) of the West Pilbara superterrane. The ca. 3.12 Ga Whundo Group of the Sholl terrane is a fault-bounded, 10-km-thick volcanic succession that has geochemical characteristics of modern oceanic arcs (including boninites and evidence for flux melting), indicating steep Archean subduction (stage 4.1 in Fig. 4). At 3.07 Ga, the 3.12 Ga Sholl terrane, 3.27 Ga Karratha terrane, and ca. 3.2 Ga Regal terrane accreted together and onto the East Pilbara terrane during the Prinsep orogeny (stage 4.2 in Fig. 4). This was followed by development of an intracontinental sag basin, the De Grey Superbasin (stage 5 in Fig. 4), and widespread plutonism (2.99–2.93 Ga) as a result of orogenic relaxation and slab breakoff (late stage 5 in Fig. 4). Cratonwide compressional deformation at 2.95–2.93 Ga culminated with 2.91 Ga accretion between the 3.18 Ga Kurrana terrane and the East Pilbara terrane. This compression caused amplification of the dome-and-keel structure in the East Pilbara terrane. Final cratonization was affected by emplacement of 2.89–2.83 Ga post-tectonic granites.

## TECTONO-METAMORPHIC EVIDENCE FOR EARLY ARCHEAN (3.45–3.21 Ga) OROGENY IN THE BARBERTON GRANITE-GREENSTONE TERRANE

The tectonic evolution of the Barberton greenstone belt differs from that of the Pilbara craton in that the preserved history of the Barberton greenstone belt from 3570 to 3080 Ma has been interpreted within a lateral tectonic framework (e.g., de Wit et al., 1992; de Ronde and de Wit, 1994). The Barberton greenstone belt preserves a history that starts with the formation of oceanic crust in a seafloor spreading environment, evolves through two episodes of arc-related and trench-related processes, and culminates with a major episode of granite magmatic activity, marking craton stabilization toward the end of the second accretionary period. This history is reviewed briefly here and followed by a detailed examination of the second accretionary period for which the metamorphic evidence is best preserved. Recent investigations of this metamorphic event have provided constraints on the thermotectonic evolution of the Barberton granite-greenstone terrane during accretionary orogenesis at ca. 3230 Ma (e.g., Dziggel et al., 2002, 2006; Stevens et al., 2002; Kisters et al., 2003; Diener et al., 2005; Moyen et al., 2006).

Metamorphic constraints have contributed enormously to our understanding of the geodynamics of younger orogenies. However, in the Archean, perhaps because of a paucity of rock compositions useful for accurately recording pressure-temperature change, metamorphic studies have played relatively minor roles in advancing our understanding of tectonic evolution. Brown (2006) argued that paired metamorphic belts, i.e., two tectonically bounded metamorphic domains characterized by different metamorphic field gradients (typically a high-pressure, low-temperature block and a moderate-pressure, high-temperature block), constitute a defining characteristic for subduction-related orogeny, and that such features are absent from the Archean rock record. This may be so, but the question could be asked with some validity "have we looked properly?" in areas with potentially suitable exposure. The metamorphic studies from the Barberton granite-greenstone terrane summarized here provide evidence of allochthonous blocks recording different metamorphic field gradients, some low and others high. These are interpreted to reflect the prior existence of a paired metamorphic belt. As discussed later, the primary arguments for the creation of the composite Barberton granite-greenstone terrane by lateral tectonic processes rest on structural, geochronological, lithological, and geochemical evidence that allows components of the Barberton granite-greenstone terrane to be interpreted as arc terranes that were accreted at ca. 3230 Ma. The metamorphic signature of this event (reviewed later) documents the crustal response to accretionary orogenesis and demonstrates the following:

1. domains with metamorphic field gradients as low as 15 °C/km;
2. the existence of tectono-metamorphic domains characterized by different metamorphic field gradients, i.e., a high-pressure, low-temperature (15 °C/km) domain and

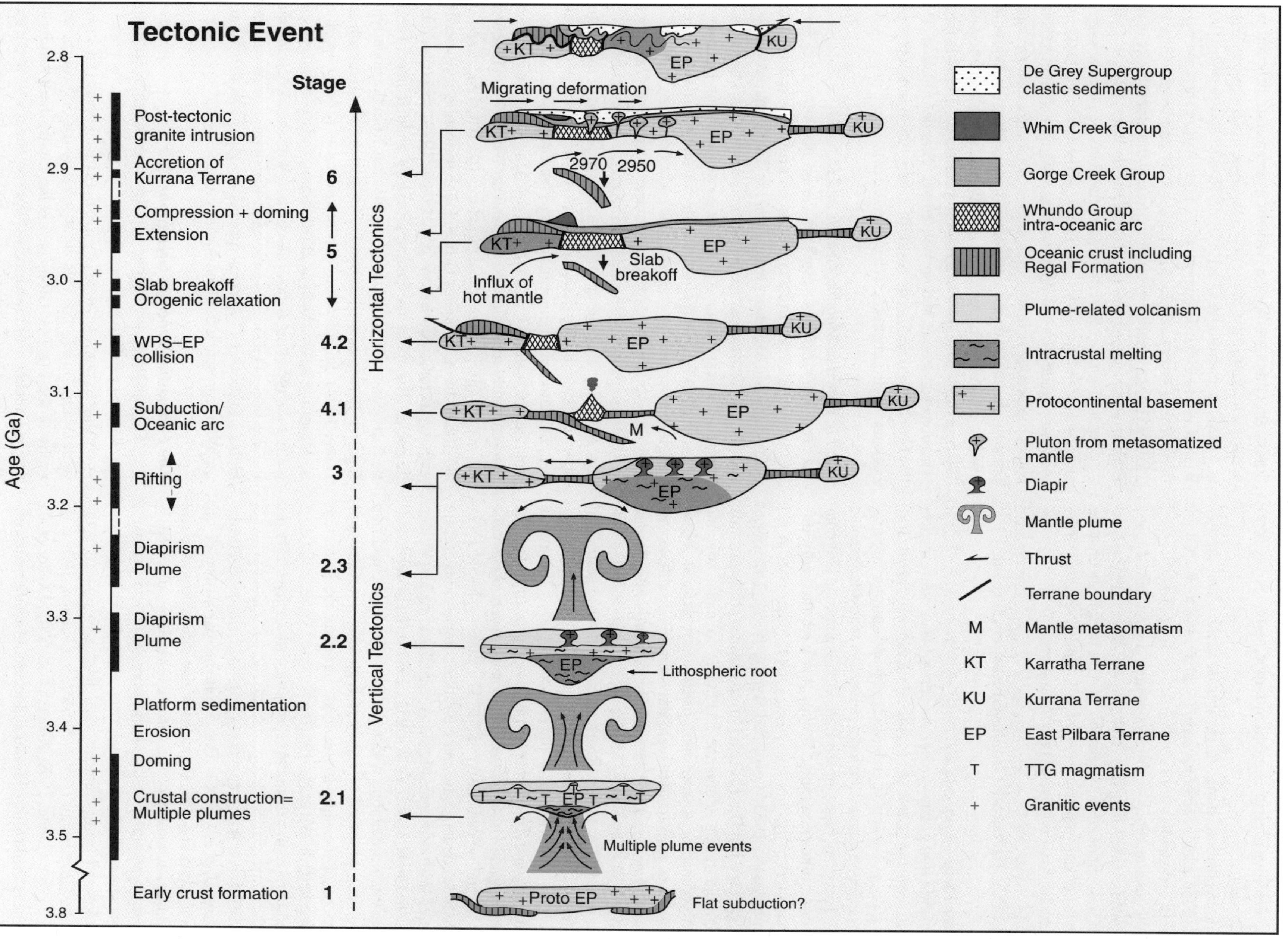


Figure 4. Diagrammatic sketch of events in the formation of the Pilbara craton (from Van Kranendonk et al., 2007). Change from vertical to horizontal tectonics occurred at ca. 3.2 Ga, coeval with inferred rifting of the margins of the East Pilbara terrane (stage 3). See text for details. WPS—West Pilbara superterrane.

a medium-pressure, medium-temperature (30 °C/km) domain;

3. the deposition, burial to midcrustal depths, and amphibolite-facies metamorphism (5 kbar and 650 °C) of metasedimentary rocks within ca. 15 m.y.; and
4. the exhumation of high-pressure, amphibolite-facies, lower-crustal domains (relative to lower-greenschist-facies supracrustal rocks of the Barberton greenstone belt) at rates on the order of 2–5 mm/yr resulting from extensional collapse of the D2 orogeny.

## Geology of the Barberton Granite-Greenstone Terrane

### *The Onverwacht Group and D1*

The Barberton granite-greenstone terrane is a composite terrane that forms the oldest nucleus to the Kaapvaal craton. It is composed of the Barberton greenstone belt and surrounding granitic rocks (Fig. 5), and it includes several tectono-stratigraphic domains. The stratigraphy of the Barberton greenstone belt consists of three lithostratigraphic groups that have markedly different characters (Viljoen and Viljoen, 1969a; Anhaeusser et al., 1981; Anhaeusser, 1983; Lowe and Byerly, 1999). The lowermost 3550–3250 Ma Onverwacht Group consists predominantly of amphibolite-facies submarine ultramafic to mafic volcanic rocks with komatiitic to komatiitic basaltic composition. These rocks have been interpreted to represent Archean ocean floor that experienced typical seafloor metamorphism shortly after formation between 3490 and 3450 Ma (de Ronde and de Wit, 1994; de Wit et al., 1987a). The mafic-ultramafic units are interbedded with felsic volcanic rocks, thin cherts, and BIF (banded iron formation). The stratigraphically lowest formations of the group, the Sandspruit and Theespruit Formations, differ from the rest of the Onverwacht Group in that they consist of a bimodal suite of volcanic and volcaniclastic rocks. The mafic volcanic rocks within these formations have more basaltic compositions and are interlayered with substantial felsic volcaniclastic rocks (Viljoen and Viljoen, 1969a, 1969b; de Wit et al., 1987b). These felsic volcaniclastic units are unique within the lower Onverwacht Group and have been proposed to represent high-level equivalents of TTG plutons of the same age as those intruding the Barberton greenstone belt (de Wit et al., 1987b).

In addition to the chemical sedimentary rocks that typify the rest of the Onverwacht Group, rare clastic sedimentary layers are also developed within the Theespruit and Sandspruit Formations (Dziggel et al., 2006). The tectonothermal history of the Theespruit and Sandspruit Formations also differs from the rest of the Onverwacht Group; the former have been subjected to high-pressure amphibolite-facies metamorphism and pervasive ductile shearing. This is in strong contrast to the low metamorphic grades and brittle deformational fabrics exhibited by the bulk of the Onverwacht Group (Viljoen and Viljoen, 1969a; de Wit et al., 1983; Cloete, 1999; Kisters et al., 2003; Diener et al., 2005). These marked lithological and tectono-metamorphic contrasts have led workers to conclude that the Theespruit Formation is allochthonous to the Barberton greenstone belt and was juxtaposed against the low-grade Onverwacht Group (de Wit et al., 1983; Armstrong et al., 1990; de Ronde and de Wit, 1994; Kisters et al., 2003; Diener et al., 2005). Within the lower portions of the Onverwacht Group, tectonically emplaced wedges of 3570–3511 Ma tonalitic gneisses represent the oldest rocks in the Barberton greenstone belt (de Ronde and de Wit, 1994).

Apart from the clearly faulted contact between the Theespruit and Sandspruit Formations and the rest of the Onverwacht Group, the lower-grade rocks of the Onverwacht Group contain tectonically emplaced allochthonous domains, some of which are interpreted to be the Archean equivalent of ophiolites (de Wit et al., 1987a), recumbent nappes, and downward-facing sequences (de Wit et al., 1987b) that formed prior to younger ca. 3450 Ma felsic intrusions (de Ronde and de Wit, 1994). This deformation (D1) has been interpreted as the result of compression in a forearc or back-arc environment (de Wit et al., 1987a).

### *The Fig Tree and Moodies Groups and D2*

The 3260–2250 Ma Fig Tree Group unconformably overlies the Onverwacht Group and is unconformably overlain by the 3225–3215 Ma Moodies Group. The Fig Tree Group consists of fluvial and shallow-marine sediments, as well as intermediate and felsic volcaniclastic rocks. The Moodies Group is made up of more coarse-grained clastic sediments and consists largely of conglomerates and quartzites. These rocks have the general character of flysch and molasse assemblages, respectively, and rocks of both groups were deposited during active deformation. In the case of the Fig Tree Group, a significant proportion of the stratigraphy consists of volcaniclastic rocks. In the central parts of the belt, D2 deformation imprints on Fig Tree Group rocks as tight to isoclinal folds and as extensive E-W–striking shear zones that are predominantly thrusts (de Ronde and de Wit, 1994). D2 refolds D1 nappes, thrusts, and other structures (de Wit, 1982; Lowe et al., 1985). In the Moodies Group, D2 deformation is associated with syntectonic deposition (e.g., Lowe, 1999) of coarse clastic rocks. The rocks are characterized by angular unconformities between units, abundant eroded detritus from the underlying Fig Tree Group, as well as basaltic lavas and tuffs (Heubeck and Lowe, 1994). Such features are interpreted to reflect deposition in deforming basins, such as allochthonous piggyback basins, as well as documenting a close association with an arc-terrane during the later stages of evolution (de Ronde and de Wit, 1994). Hubeck and Lowe (1994) interpreted the upper Moodies stratigraphy to represent a foreland-basin setting.

### *The Composite Barberton Greenstone Belt*

The Barberton greenstone belt was interpreted by de Ronde and de Wit (1994) to represent an accretionary belt composed of two major terranes separated by a major crustal break (the Inyoka fault system on Fig. 5). This is reflected by the differing stratigraphy of the Fig Tree Group across the Inyoka fault and by the varying thickness of the Moodies strata on either side of the fault (Heubeck and Lowe, 1994). Lowe (1994) developed this concept

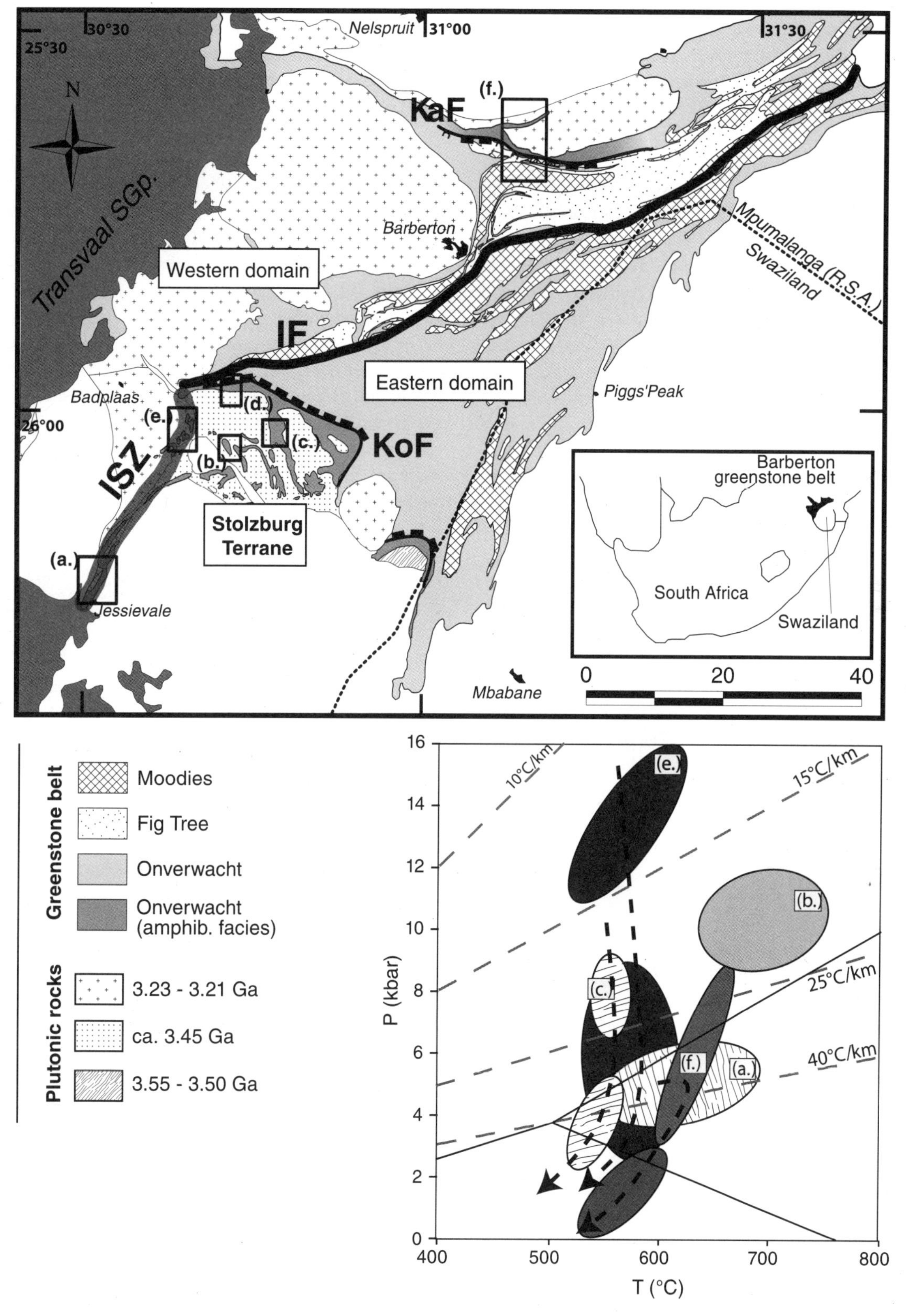


Figure 5. Geological map of the Barberton greenstone belt (after Anhaeusser et al., 1981). KaF—Kaap River fault; KoF—Komati fault; ISZ—Inyoni shear zone; IF—Inyoka-Saddleback fault. Boxes refer to study areas reviewed in the text. Western domain: box a—Schapenburg schist belt (Stevens et al., 2002); box f—Stentor Pluton (Otto et al., 2005; Dziggel et al., 2006). Eastern domain: box b—Central Stolzburg terrane (Dziggel et al., 2002); box c—Tjakastad schist belt (Diener et al., 2005); box d—Stolzburg schist belt (Kisters et al., 2003); box e—Inyoni shear zone (Dziggel et al., 2002; Moyen et al., 2006). The legend refers to shading on the map, not on the pressure-temperature (*P-T*) diagram, where shading denotes paired *P-T* ellipses from the same rock evolution path. The overall *P-T* trend is one of decompression.

further and demonstrated that the Barberton greenstone belt consists of four tectonically assembled, diachronous-stratigraphic blocks (Fig. 5). Thus, the Barberton granite-greenstone terrane records the formation of an Archean continental nucleus through several possible periods of orogeny. However, the merging of diachronous terranes along crustal-scale sutures (de Wit et al., 1987b, 1992; Armstrong et al., 1990; de Ronde and de Wit, 1994; Kamo and Davis, 1994; Lowe, 1994, 1999; de Ronde and Kamo, 2000; Kisters et al., 2003; Diener et al., 2005) appears to have been a function predominantly of D2 deformation. Thus, there is relatively strong evidence that the Barberton granite-greenstone terrane was shaped by lateral tectonic processes throughout its well-preserved geologic history. In this respect, although similar to the tectonic evolution proposed for the post–3.2 Ga Pilbara craton, this geodynamic style seems to have initiated earlier in the Barberton granite-greenstone terrane.

## Metamorphism in the Barberton Granite-Greenstone Terrane

From a metamorphic perspective the rocks of the Barberton granite-greenstone terrane consist of the following:

1. The generally low-grade (lower- to mid-greenschist facies) metamorphic supracrustal sequence of the Barberton greenstone belt;
2. an amphibolite-facies succession in the stratigraphically lowermost part of the Barberton greenstone belt (Theespruit Formation in the south), as well as metamorphosed ca. 3.50–3.45 Ga TTG plutons such as the Theespruit and Stolzburg plutons (Fig. 5) (e.g., Kisters et al., 2003);
3. syntectonic TTG plutons intruded during terrane amalgamation at ca. 3.21 Ga (e.g., Kaap Valley and Nelshoogte plutons, Fig. 5); and
4. younger, relatively potassic ca. 3.1 Ga granitoids (e.g., Mpuluzi batholith) emplaced syntectonically into active shear zones interpreted to have been associated with NW-SE–directed, subhorizontal shortening (Westraat et al., 2005; Belcher and Kisters, 2006).

The ca. 3.45 and 3.23 intrusive phases, in particular, have previously been regarded as the mechanism for amphibolite metamorphism (e.g., Anhaeusser, 1983; Kisters and Anhaeusser, 1995). Early studies saw amphibolite-facies metamorphism as a form of contact metamorphism driven by local magmatic heating, but during which deformation occurred as a result of diapiric intrusion of the granitoids. Dziggel et al. (2002) demonstrated that in greenstone remnants associated with the ca. 3.45 Ga Theespruit and Stolzburg plutons, peak amphibolite metamorphism occurred concurrently with D2 terrane amalgamation.

### *The Stolzburg Terrane*

High-grade crust to the south of the Barberton greenstone belt is known as the "Stolzburg terrane" (Fig. 5), and it is one of the best-studied high-grade regions in the Barberton granite-greenstone terrane (Kisters et al., 2003). The Stolzburg terrane corresponds to a portion of the "Songimvelo block" (Lowe, 1994); it is bounded by the Komati fault in the north and northeast, which defines a sharp metamorphic break between the amphibolite-facies Stolzburg terrane and the greenschist-facies domain represented by the main body of the Barberton greenstone belt (Kisters et al., 2003; Diener et al., 2005). To the west, the Stolzburg terrane is bounded by the Inyoni shear zone, which separates it from the 3.23–3.21 Ga Badplaas pluton. To the south, it is intruded by the 3.10 Ga Mpuluzi batholith and Boesmanskop syenite.

The Stolzburg terrane is composed of ca. 3.45 Ga TTG, including the Stolzburg and Theespruit plutons, other plutons, and areas of orthogneiss. The terrane contains greenstone material in the form of amphibolite-facies remnants along the pluton margins (Fig. 5), remnants of the Theespruit Formation along the southern margin of the Barberton greenstone belt, and remnants of the Sandspruit Formation within the granitoid terrane (Anhaeusser et al., 1981; Anahaeusser, 1983; Dziggel et al., 2002). The greenstone remnants consist of metamorphosed mafic and ultramafic metavolcanic rocks, with minor thin chert and BIF layers. This terrane also contains an up to 8-m-thick clastic sedimentary unit within which well-preserved primary sedimentary features such as trough cross-bedding occur. A minimum age of sedimentation is indicated by the age of crosscutting trondhjemite gneiss at 3431 Ma (Dziggel et al., 2002). The youngest detrital zircons contained within the sedimentary rocks are 3521 Ma, giving a maximum age of deposition and therefore constraining this unit to be part of the Onverwacht Group (Dziggel et al., 2002).

The metamorphic history of Onverwacht Group greenstone remnants exposed within the Stolzburg terrane (Fig. 5, box b) is recorded within two types of clastic metasediments, including a trough cross-bedded, meta-arkose and a planar-bedded, possibly more distal, metasedimentary unit of mafic geochemical affinity (Dziggel et al., 2002). The latter is characterized by the peak metamorphic mineral assemblage diopside + andesine + garnet + quartz. This assemblage, and garnet in particular, is extensively replaced by retrograde epidote. Peak metamorphic mineral assemblages of magnesio-hornblende + andesine + quartz, and quartz + ferrosilite + magnetite + grunerite have been recorded from adjacent amphibolites and interlayered BIF units, respectively. In these rocks, retrogression is marked by actinolitic rims around peak metamorphic magnesio-hornblende cores in the metamafic rocks, and by a second generation of grunerite that occurs as fibrous aggregates rimming orthopyroxene in the BIF. The peak metamorphic textures are typically post-tectonic and are texturally mature and well-equilibrated. Pressure-temperature (*P-T*) estimates, using a variety of barometers and geothermometers, yield pressures between 8 and 11 kbar and temperatures between 650 and 700 °C (Fig. 5). The relatively high pressures and low temperatures of peak metamorphism are interpreted to reflect a tectonic environment comparable to modern continent-continent collisional settings, and the Stolzburg terrane is interpreted to represent an exhumed mid- to lower-crustal fragment that formed "basement" to the Barberton greenstone belt at the time of the peak metamorphic event at ca. 3230 Ma (Dziggel et al., 2002).

### *High*-P–*Low*-T *Metamorphism*

The Inyoni shear zone is a complex structure extending in a southwesterly direction from the termination of the Stolzburg syncline into the granitoid-dominated terrane to the south (Fig. 5, box e). It forms the boundary between the Stolzburg terrane to the east and the Badplaas pluton to the west. The Inyoni shear zone contains a diverse assemblage of greenstone remnants (Dziggel et al., 2002; Moyen et al., 2006). These are mostly typical lower Onverwacht Group rocks, consisting of interlayered metamafic and meta-ultramafic units with occasional minor BIF horizons, but some clastic metasedimentary rocks also occur (Dziggel et al., 2002). The greenstone remnants are encapsulated within a matrix of TTG orthogneisses, components of which were intruded syntectonically during, or close to, peak metamorphism at 3229 Ma (Dziggel et al., 2006). Structures in the Inyoni shear zone are complex and the result of interference from (1) northeast-southwest shortening, which generated a predominantly vertical foliation with symmetrical folds and crenulation cleavage developed from meso- to macroscale, and (2) vertical extrusion of the Stolzburg terrane, resulting in the development of a synmelt vertical lineation and folds with vertical axes. Metamafic rocks contain a dominant foliation defined by hornblende. Metamorphic titanite, which formed in association with epidote through retrograde replacement of garnet and plagioclase, has an age of 3229 Ma (Dziggel et al., 2006).

Metasedimentary rocks within the Inyoni shear zone yield peak metamorphic *P-T* estimates between 8 and 11 kbar and 600–700 °C (Dziggel et al., 2002) (Fig. 5, box e). Metamafic samples preserve textural evidence of their prograde metamorphic evolution in the cores of rootless isoclinal folds, where growth-zoned garnets contain low-temperature mineral inclusion suites within their cores. Calculated *P-T* estimates for the prograde assemblage garnet, albitic plagioclase, clinopyroxene, and quartz in a number of sites range from 12 to 15 kbar and 600 to 650 °C (Moyen et al., 2006). Garnet in samples from higher-strain domains generally shows partial replacement by symplectitic coronas of epidote + Fe-tschermakite + quartz symplectite. Calculated *P-T* estimates from these assemblages produce retrograde conditions of 8–10 kbar at 580–650 °C (Moyen et al., 2006) (Fig. 5, box e). These decompression structures are in good agreement with peak metamorphic estimates from the nearby metavolcanic-sedimentary sequence. Locally, both the high- and low-strain domains record greenschist-facies chlorite + epidote + actinolite retrogression overprinting the amphibolite-facies assemblages.

### *Peak Metamorphism in Other Areas*

The Schapenburg schist belt (Fig. 5, box a) in the southern extremity of the terrane is one of several large (~3 × 12 km) greenstone remnants exposed in the granitoid-dominated terrane south of the Barberton greenstone belt. It is unique in having a well-developed metasedimetary sequence, in addition to the typical mafic-ultramafic volcanic rocks (Anhaeusser, 1983). Its metamorphic history has been studied by Stevens et al. (2002) and is summarized here. The metasedimentary sequence consists of two distinctly different units. In the southwestern portion of the belt, in low-strain domains, a meta-tuffaceous unit of granitoid composition contains minor agglomerate layers, as well as well-preserved cross-bedding and graded bedding. This unit underlies rhythmically banded metagraywackes consisting of alternating ~10-cm-thick metapelites and 1–2-cm-thick meta-psammites. On the basis of the graded bedding and trough cross-bedding preserved in the underlying meta-tuffaceous unit, the metasedimentary succession is shown to young to the east. In this direction, the succession is overlain by Onverwacht Group rocks. Detrital zircons within the metasediments have ages as young as 3240 Ma, and, consequently, they are correlated with lower-greenschist sediments of the Fig Tree Group exposed in the central portions of the Barberton greenstone belt, 60 km to the north. The Schapenburg metasediments are relatively $K_2O$-poor and preserve a peak metamorphic assemblage of garnet + cordierite + gedrite + biotite + quartz ± plagioclase. Post-tectonic peak metamorphic assemblages are texturally very well equilibrated, and the predominantly almandine garnets from all rock types show almost flat zonation patterns for Fe, Mg, Mn, and Ca. Consequently, there appears to be no preserved record of the prograde path. Peak metamorphic conditions are constrained to 640 ± 40 °C and 4.8 ± 1.0 kbar (Fig. 5, box a). The maximum age of metamorphism is defined by a syntectonic tonalite intrusion into the central portion of the schist belt at 3231 Ma. In combination with the age of the youngest detrital zircons in the metasediments, this age demonstrates that sedimentation, burial to midcrustal depths (~18 km), and equilibration under amphibolite-facies conditions were achieved in ~10 m.y., but no more than 20 m.y.

### *Exhumation of the High-Grade Domains*

The deformed and metamorphosed margins of the Stolzburg terrane in the north, where it abuts the lower-grade greenstone belt, have been studied in two separate areas. Detailed mapping of the contacts between the supracrustal and gneiss domains along the southern margin of the greenstone belt has identified an ~1-km-wide deformation zone (Fig. 5, box d) (Kisters et al., 2003). Amphibolite-facies rocks at, and below, the granite-greenstone contacts are characterized by rodded gneisses and strongly lineated amphibolite-facies mylonites. The prolate, coaxial fabrics are overprinted by greenschist-facies mylonites at higher structural levels, which cut progressively deeper into the underlying high-grade basement rocks. These mylonites developed during noncoaxial strain, and kinematic indicators consistently point to a top-to-the-NE sense of movement of the greenstone sequence with respect to the lower structural levels. This relationship between bulk coaxial NE-SW stretching of midcrustal basement rocks and noncoaxial, top-to-the-NE shearing along retrograde mylonites at upper-crustal levels is consistent with exhumation of deeper crustal levels (Kisters et al., 2003).

The dominant peak metamorphic assemblage preserved within amphibolite-facies domains throughout the study area is hornblende + plagioclase + sphene + quartz. Garnet-bearing assemblages are confined to specific narrow layers developed parallel to the compositional banding of the rocks ($S_0$). In all cases, retrogression is associated with the development of later shear fabrics ($S_1$ in retrograde mylonites), which postdate peak metamorphic porphyroblasts. These features suggest a primary bulk-compositional control (Fe/Fe + Mg ratios and the presence of carbonate) on the distribution of the garnet-bearing metamorphic assemblages and indicate that they are probably peak metamorphic grade equivalents of the more dominant amphibolites. Peak *P-T* conditions occurred at 5.5 ± 0.9 kbar and 491 ± 40 °C and 6.3 ± 1.5 kbar and 492 ± 40 °C (Kisters et al., 2003). Retrogression is marked by the development of actinolite + epidote + chlorite + quartz in the metamafic rocks and muscovite + chlorite + quartz in the metapelitic layer. These conditions are at lower grades than previously defined (Dziggel et al., 2002), but they are developed along a similarly low metamorphic field gradient.

In the Tjakastad schist belt (Fig. 5, box c), peak metamorphic and retrograde assemblages are syntectonic with fabrics developed during exhumation, and they illustrate the initiation of detachment formation at deep crustal levels and elevated temperatures (Diener et al., 2005). Peak metamorphic conditions occurred at 7.0 ± 1.2 kbar and 537 ± 45 °C, and 7.7 ± 0.9 kbar and 563 ± 14 °C (Fig. 5). The peak metamorphic assemblages are syntectonic, and peak metamorphic porphyroblasts (e.g., staurolite and plagioclase) are recrystallized and deformed within the exhumation fabric. Within rare low-strain domains in the garnet-bearing amphibolite, retrograde mineral assemblages can be seen to pseudomorph peak metamorphic garnet. In these sites, a new generation of garnet is developed within the assemblage garnet + chlorite + muscovite + plagioclase + quartz, resulting in *P-T* estimates of 3.8 ± 1.3 kbar and 543 ± 20 °C and indicating near isothermal decompression (Fig. 5). This is consistent with the presence of staurolite as part of the peak and retrograde assemblage, where the modeled staurolite stability field in relevant compositions is confined to a narrow temperature range between 580 and 650 °C over a pressure range between 10 and 3 kbar, and where the occurrence of sillimanite replaces kyanite within the staurolite-bearing rocks (Diener et al., 2005). Geochronological constraints, combined with the depths of burial, indicate that exhumation of the high-grade rocks occurred at 2–5 mm/yr. This is similar to the exhumation rates of crustal rocks in younger compressional orogenic environments. Coupled with low metamorphic field gradients of ~20 °C/km, it suggests that the Mesoarchean crust was cold and rigid enough to allow tectonic stacking, crustal overthickening, and an overall rheological response very similar to that displayed by modern doubly thickened continental crust (Diener et al., 2005).

Along the northern margin of the Barberton greenstone belt (Fig. 5, box f), the granitoid-greenstone contact is characterized by a shear zone that separates generally low-grade, greenschist-facies greenstone belt from midcrustal basement gneisses (Dziggel et al., 2006). The supracrustal rocks in the hanging wall are metamorphosed to upper-greenschist-facies conditions, whereas similar rocks and granitoid gneisses in the footwall are metamorphosed to amphibolite-facies grades. Peak *P-T* metamorphic conditions (Fig. 5) are calculated as 5 ± 1 kbar and 600–700 °C from amphibolite-facies assemblages and aluminous schists (Dziggel et al., 2006). This corresponds to an elevated metamorphic field gradient of ~30–40 °C/km. The peak metamorphic minerals in this area are also syntectonic; fabrics formed during exhumation of the high-grade rocks at ca. 3.23 Ga (Dziggel et al., 2006). Metamorphism replaced clinopyroxene and plagioclase in the metamafic rocks with coronitic epidote + quartz and actinolite + quartz symplectites, indicating retrograde *P-T* conditions of 1–3 kbar and 500–650 °C (Dziggel et al., 2006). This indicates that exhumation and decompression commenced under amphibolite-facies conditions, as supported by the synkinematic growth of peak metamorphic minerals during extensional shearing, followed by near-isobaric cooling to temperatures below 500 °C. The last stages of exhumation are characterized by solid-state doming of footwall gneisses and strain localization in contact-parallel greenschist-facies mylonites, which overprint the decompressed basement rocks. The southern margin of the Stentor domain is bounded by the Kaap River and Lily faults. These correspond to a major metamorphic break, from 6 to 8 kbar in the amphibolitic domain to nearly unmetamorphosed supracrustals in the Barberton greenstone belt immediately south of the faults (Dziggel et al. 2006). Schoene and Bowring (2007) demonstrated different cooling paths for granitoid rocks to the north and south of the Barberton greenstone belt. The southern basement underwent uplift in two stages, with final uplift coinciding with the 3.10 Ga event. In the north, uplift to high crustal levels occurred as part of the ca 3.23 Ga event. These findings are consistent with the differences observed in the *P-T* paths at 3.23 Ga for these two areas.

## The Late Thermotectonic Evolution of the Barberton Granite-Greenstone Terrane

With the rare exception of some ca 3.45 Ga metamorphic rocks developed within the Onverwacht Group within the upper-plate portion of the Barberton greenstone belt (e.g., Cloete, 1999), almost all the metamorphic rocks from the Barberton granite-greenstone terrane that have been studied in detail record the ca. 3.23 Ga metamorphic event. Based on this data, the following thermotectonic evolution is hypothesized (Fig. 6). At ca. 3.24 Ga, detritus from a volcanic arc was contributing to Fig Tree Group sediments, and TTG melts generated during the subduction process were filling magma chambers within the overriding plate. A pre-existing, old (≥3.45 Ga) and cold protocratonic domain, consisting of both older TTG plutons and a greenstone sequence, was converging with the subduction zone. The amphibolite-facies rocks discussed previously represent different domains within these two plates (e.g., C and E within metamafic rocks in the subducting plate and A within the Fig Tree Group in the overriding plate; Fig. 6A). At 3.23 Ga, the converging cratonic fragment accreted with the overriding plate (Fig. 6B). Relatively cold (~600 °C) granitic crust and associated remnants of greenstone material underwent high-pressure

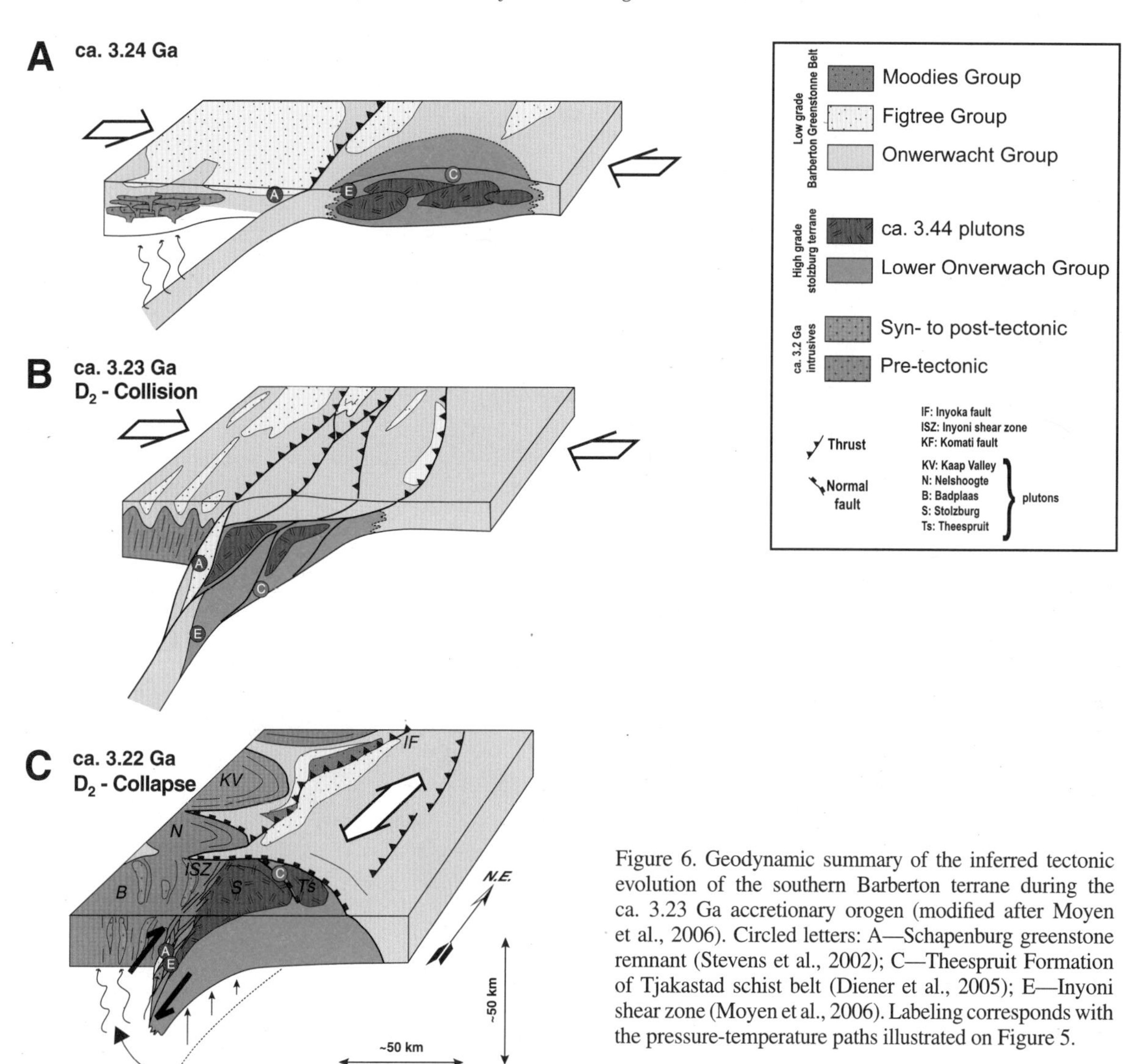


Figure 6. Geodynamic summary of the inferred tectonic evolution of the southern Barberton terrane during the ca. 3.23 Ga accretionary orogen (modified after Moyen et al., 2006). Circled letters: A—Schapenburg greenstone remnant (Stevens et al., 2002); C—Theespruit Formation of Tjakastad schist belt (Diener et al., 2005); E—Inyoni shear zone (Moyen et al., 2006). Labeling corresponds with the pressure-temperature paths illustrated on Figure 5.

amphibolite-facies metamorphism at ~40 km depth. Concurrent with this process, slivers of upper-plate rocks imbricated and entrained along the trace of the subduction zone underwent metamorphism along markedly lower-pressure metamorphic field gradients (e.g., ~650 °C at 18 km depth) than the rocks of the subducting slab. Postcollisional collapse in response to the resultant gravitational instability uplifted both the granitoid domains of the lower plate and the precollisional TTG plutons in the upper plate relative to the low-grade metamorphosed greenstone sequence (Fig. 6C). This created sharp metamorphic breaks along discrete structures bounding the low-grade sequence, which record as much as 18 km relative vertical displacement. The topographic manifestation of these processes produced intermontane basins into which Moodies Group deposition took place. Decompression melting of mafic rocks at the base of the crust produced further TTG magmatism.

## 2.7 Ga ACCRETIONARY AND COLLISIONAL ASSEMBLY OF THE SUPERIOR PROVINCE, CANADA

Views of the Superior Province have changed dramatically since Reymer and Schubert's (1984, 1986) estimate of a crustal growth rate of ~500 $km^3$ $km^{-1}$ $m.y.^{-1}$, 6–10 times the Mesozoic-Cenozoic average based on their assumption that ca. 2.7 Ga juvenile crust was characteristic of the whole province. Although a peak in production of juvenile crust at 2.7 Ga remains a robust observation on the global scale (Condie, 1998), during the past two decades recognition of numerous large fragments of ancient

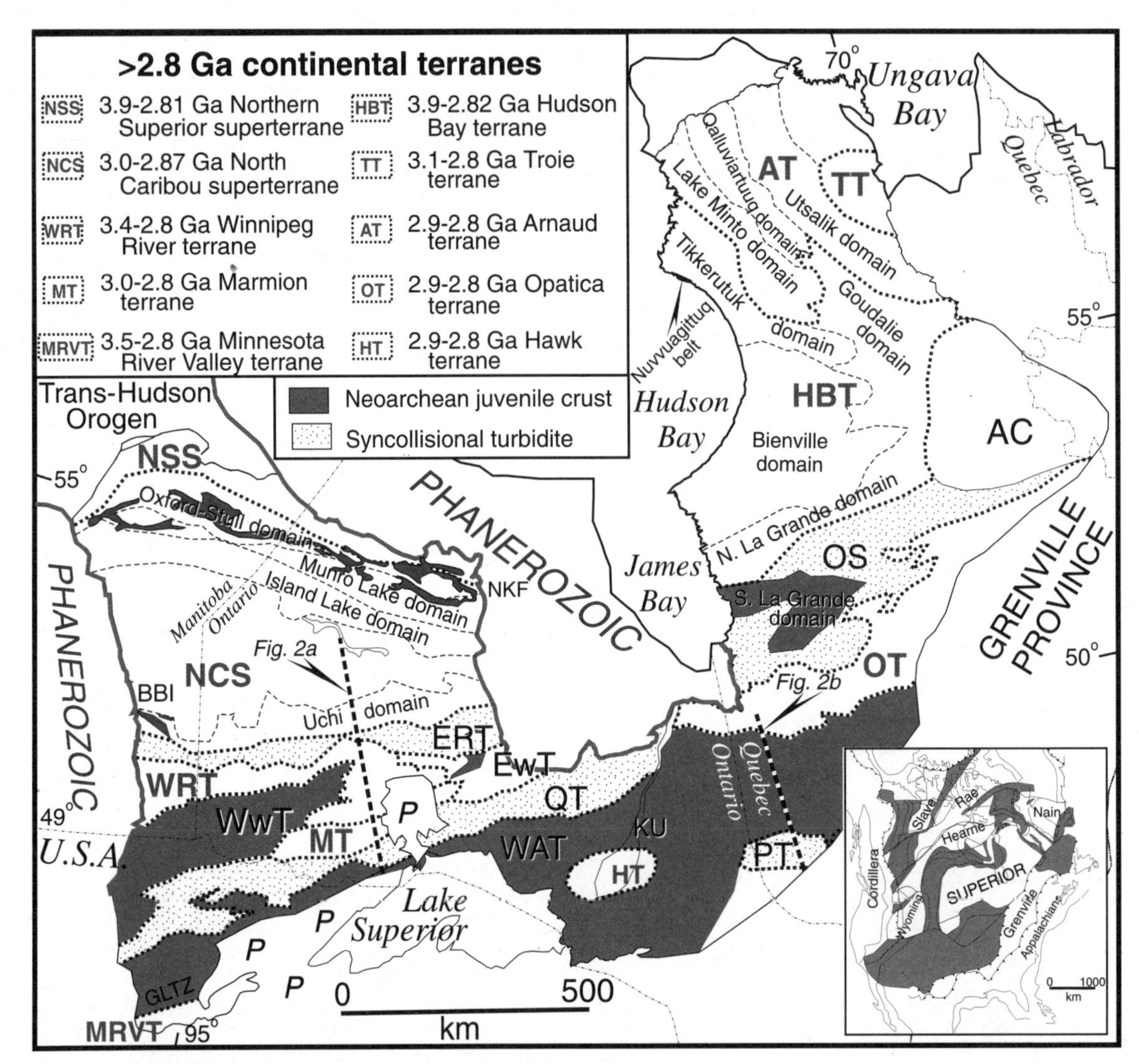


Figure 7. Tectonostratigraphic terrane map of the Superior Province showing distribution of Eo- to Mesoarchean continental fragments, Neoarchean juvenile crust, and synorogenic turbidite basins. Additional abbreviations not shown in legend: AC—Ashuanipi complex; BBI—Bidou–Black Island terrane; ERB—English River basin; EwT—eastern Wabigoon terrane; GLTZ—Great Lakes tectonic zone; KU—Kapuskasing uplift; NKF—North Kenyon fault; OB—Opinaca basin; P—flat-lying Proterozoic cover; PB—Pontiac basin; QB—Quetico basin; WAT—Wawa-Abitibi terrane; WwT—western Wabigoon terrane.

crust within the Superior Province has reduced the estimated volume of juvenile crust in the Superior Province, bringing growth rates into better agreement with Phanerozoic fluxes. We review the nature and style of interaction among Archean continental and oceanic fragments that led to the assembly of the Superior Province and that argue for establishment of modern-style plate tectonics on Earth by the end of the Archean.

Constituting the nucleus of the Canadian Shield (Fig. 7, inset), the Superior Province represents cool, refractory, 250-km-thick lithosphere (Van der Lee and Frederiksen, 2005). It has resisted major reworking by the ca. 1.8 Ga collision of the Trans-Hudson orogen, ca. 1.1 Ga impingement of the Grenville orogen and mid-continent rift system (Fig. 7), as well as emplacement of diabase dike swarms throughout the Proterozoic.

**Superior Province Eoarchean to Mesoarchean Terranes**

Sialic crustal fragments are the tectonic building blocks of the Superior Province (Percival et al., 2004a). Although generally poorly preserved due to Neoarchean plutonic reworking, distinct Nd model ages and tectonostratigraphic histories can be discerned for individual fragments (Fig. 8), which have consequently been interpreted as independent terranes (Percival et al., 2006a). Some terranes show evidence of assembly prior to their ca. 2.7 Ga incorporation into the Superior Province and hence are considered to be superterranes.

The poorly exposed Northern Superior superterrane (Skulski et al., 1999; Figs. 7 and 8) is recognized on the basis of isotopic evidence, including ca. 3.5 Ga orthogneiss (Böhm et

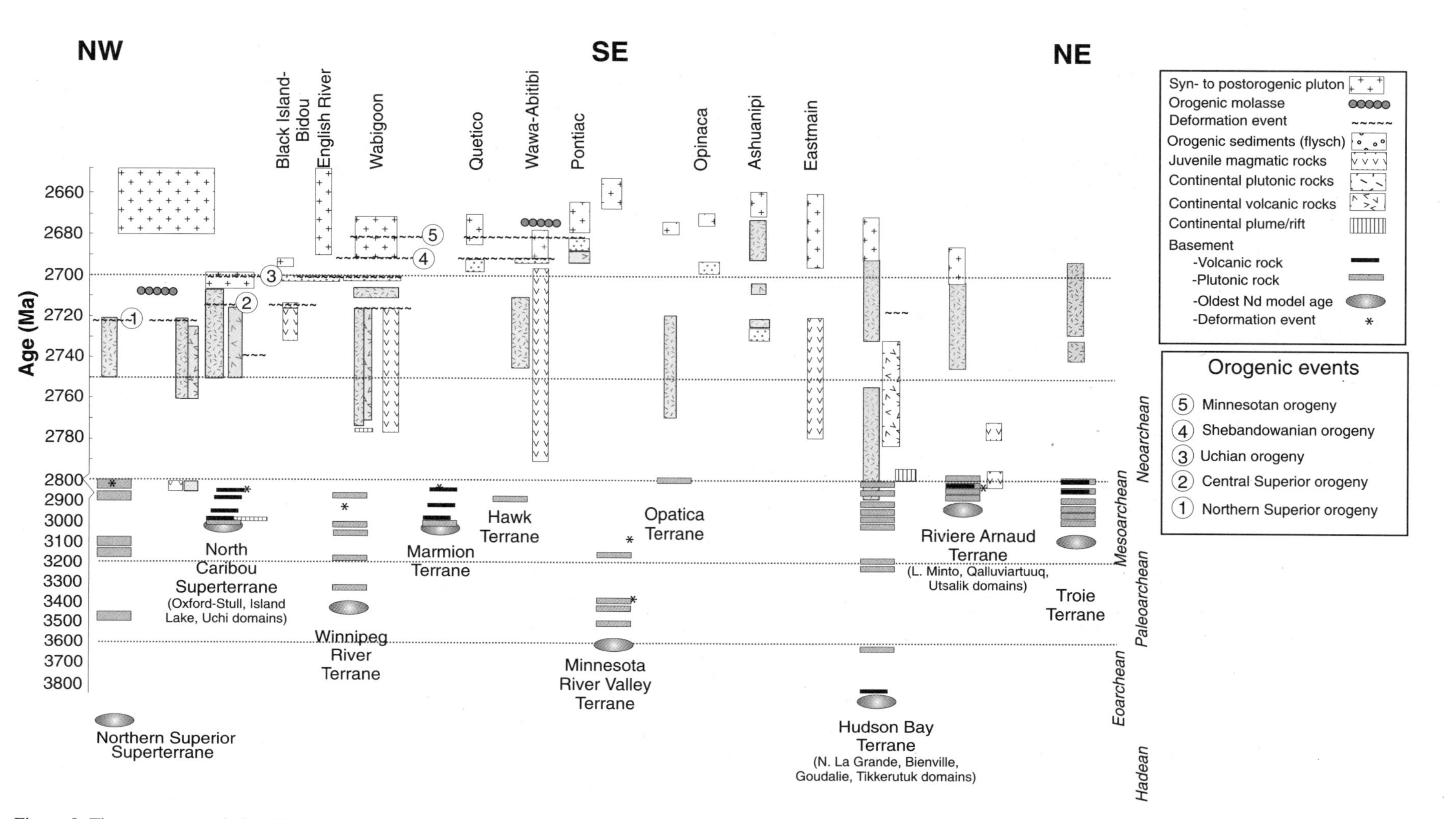


Figure 8. Time-space correlation diagram showing the age and internal relationships of continental and oceanic domains (modified after Percival et al., 2004a; Percival, 2007). Note the wide diversity in age and longevity of basement (pre–2.8 Ga) domains. The southward-younging progression of events between 2.72 and 2.68 Ga (orogenic events 1–5) is most evident in the northwestern Superior Province.

al., 2000), >3.5 Ga inherited zircon ages (Skulski et al., 2000), and 3.9 Ga detrital zircon ages (Böhm et al., 2003). These remnants were intruded by tonalite-trondhjemite-granodiorite (TTG) suite rocks at 3.2–3.1, 2.85–2.81, and 2.74–2.71 Ga, and were metamorphosed at 2.68 and 2.61 Ga (Skulski et al., 2000; Böhm et al., 2003).

To the east across Hudson and James Bays (Fig. 7), the Hudson Bay terrane (Leclair et al., 2006) may be correlative. It is characterized by ancient Nd model ages, >3.2 Ga zircon inheritance, and rare Meso- and Eoarchean rocks (Fig. 8). Narrow supracrustal belts (Percival et al., 1992, 1994; Skulski et al., 1996; Lin et al., 1996; Percival and Skulski, 2000; Leclair et al., 2006), including the ca. 3.8–3.6 Ga Nuvvuagittuq belt (Figs. 7 and 8; David et al., 2003; O'Neil et al., 2007) are separated by plutonic massifs, including widespread charnockitic intrusions (Stern et al., 1994; Skulski et al., 1996; Percival and Mortensen, 2002; Bédard, 2003; Stevenson et al., 2006; Boily et al., 2006). Several domains are recognized on the basis of age and lithology within the Hudson Bay terrane (Fig. 7): the northern La Grande (Roscoe and Donaldson, 1988; Goutier and Dion, 2004), Bienville (Skulski et al., 1998b; David, 2008; Isnard and Gariépy, 2004; Boily et al., 2006), Tikkerutuk (Percival et al., 2001; David, 2008), and Goudalie domains. The latter contains a Neoarchean suture that juxtaposes 2725 Ma continental arc-volcanic rocks against 2790 Ma juvenile oceanic-plateau rocks (Skulski et al., 1996; Lin et al., 1996).

The North Caribou superterrane (Figs. 7 and 8; Thurston et al., 1991) is the largest Mesoarchean domain of the Superior Province (Stott, 1997). It is characterized by widespread evidence of crust with ca. 3.0 Ga mantle extraction ages (Stevenson, 1995; Stevenson and Patchett, 1990; Corfu et al., 1998; Hollings et al., 1999; Henry et al., 2000). Petrogenetic studies suggest that crust at least 35 km thick was produced in a juvenile-arc setting (Whalen et al., 2003; Percival et al., 2006b). Subsequent rifting is recorded by ca. 2980 Ma quartzite–komatiite–iron formation rift sequences (Percival et al., 2006b; Sasseville et al., 2006), and an amalgamation event occurred prior to 2870 Ma (Stott et al., 1989; Thurston et al., 1991). Mesoarchean units have been variably reworked by Neoarchean magmatic and deformational events (Stone, 1998, 2005). The northern margin (Oxford-Stull domain, Figs. 7 and 8) represents a shortened continental margin collage (Syme et al., 1999; Corkery et al., 2000; Skulski et al., 2000; Stone et al., 2004). It is overlain by a 2722–2705 Ma successor arc of shoshonitic volcanic and associated sedimentary rocks (Corkery and Skulski, 1998; Corkery et al., 2000; Skulski et al., 2000; Lin et al., 2006), the latter of which record an influx of detrital zircon up to 3.65 Ga derived from the Northern Superior superterrane (Corkery et al., 1992, 2000; Lin et al., 2006). The southwestern margin is tectonically juxtaposed against juvenile Neoarchean rocks (Percival et al., 2006b). Elsewhere along this margin, a 300 m.y. record of tectonostratigraphic evolution includes early rifting (Tomlinson et al., 1998; Sanborn-Barrie et al., 2004), several episodes of continental-arc magmatism (Henry et al., 2000; Sanborn-Barrie et al., 2001, 2004), intra-arc rifting, and several phases of deformation and associated sedimentation (Dubé et al., 2004).

The Winnipeg River terrane (Figs. 7 and 8; Beakhouse, 1991) is composed of Neoarchean plutonic rocks with Paleo- to Mesoarchean (3.4–3.0 Ga) inheritance (Henry et al., 1998; Tomlinson et al., 2004; Whalen et al., 2002, 2004a), which distinguishes it from the Northern Superior and North Caribou superterranes to the north and the Marmion terrane (described later) to the south. Volcanic rocks (2930–2880 Ma; Sanborn-Barrie and Skulski, 1999) underlie "rift to drift" continental tholeiites, representing early rifting of the Winnipeg River basement (Skulski et al., 1998a; Sanborn-Barrie and Skulski, 2006). A complex Neoarchean structural-metamorphic history characterizes the Winnipeg River terrane, including 2717–2713 Ma folding, 2713–2707 Ma magmatism, horizontal extension, and post–2705 Ma upright folding (Brown, 2002; Percival et al., 2004b; Melnyk et al., 2006).

The Marmion terrane (Figs. 7 and 8) consists of 3010–2999 Ma juvenile tonalite basement (Davis and Jackson, 1988; Tomlinson et al., 2004), upon which several greenstone belts formed between 2990 and 2780 Ma (Hollings and Wyman, 1999; Stone et al., 2002; Tomlinson et al., 2003). The terrane may have been tectonically accreted to the Winnipeg River terrane by ca. 2.92 Ga (Tomlinson et al., 2004) or could have formed by magmatic addition of juvenile crust at the Winnipeg River margin.

Mesoarchean rocks within the predominantly juvenile, Neoarchean Wawa-Abitibi terrane include 2890–2880 Ma volcanic rocks of the Hawk assemblage (Figs. 7 and 8; Turek et al., 1992) and 2920 Ma tonalite (Moser, 1994). Inherited 2800–2900 Ma zircons in granitoid rocks to the east indicate a terrane at least 200 km long (J. Ketchum, 2005, personal commun.).

The poorly exposed Minnesota River Valley terrane (Figs. 7 and 8) consists predominantly of granitoid and gneissic rocks with zircon inheritance dating back to ca. 3.5 Ga and mantle extraction ages close to 3.6 Ga (Bickford et al., 2006, 2007). Its complex history includes Mesoarchean magmatic events at 3.48–3.50, 3.36–3.38, and 3.14 Ga, as well as metamorphism at 3.38 and 3.08 Ga, younger Neoarchean (2.62–2.60 Ga) magmatism and metamorphism (Bickford et al., 2006), and a low-grade overprint at 1.8 Ga (Goldich et al., 1984; Schmitz et al., 2006). The Great Lakes tectonic zone (Fig. 7), separating the Minnesota River Valley and Wawa-Abitibi terranes, dips northward based on isotopic inheritance in plutons cutting juvenile volcanic rocks to the north (Sims et al., 1997).

Bordering the Abitibi belt on the north, the Opatica terrane (Figs. 7 and 8) consists of 2.82 Ga tonalite, 2.77–2.70 tonalite-granodiorite, and 2.68 Ga granite and pegmatite (Benn et al., 1992; Sawyer and Benn, 1993; Davis et al., 1995; Benn and Moyen, this volume). Polyphase deformation results from west-directed thrusting overprinted by a fold-and-thrust belt (Sawyer and Benn, 1993).

The Arnaud terrane (Figs. 7 and 8; Leclair et al., 2006) consists of plutonic and sparse supracrustal rocks with mantle extraction ages in the 2.9–2.8 Ga range (Stern et al., 1994; Boily et al., 2006; Stevenson et al., 2006; David, 2008). It consists of several domains (Fig. 7) characterized by narrow supracrustal belts, paragneiss, and abundant 2.74–2.70 Ga hornblende and pyroxene-bearing granitic intrusions (Skulski et al., 1996; Percival et al., 2001; Parent et al., 2001).

The Troie terrane (Figs. 7 and 8; Leclair et al., 2006) is recognized on the basis of zircon inheritance and Nd model ages in excess of 3.0 Ga (Percival et al., 2001; Boily et al., 2006; Leclair et al., 2006; David, 2008). The Mesoarchean units resemble those of the northern La Grande domain, with which they have been linked on the basis of age, lithology, and isotopic characteristics (Leclair et al., 2006).

Ages of igneous and thermotectonic events provide templates that can be compared across terrane boundaries (Percival, 2007). Similarities between the Northern Superior and Hudson Bay terranes have been previously noted (e.g., Böhm et al., 2003), and their correlation is supported by detrital zircon ages and the proximity of the two terranes. Bickford et al. (2006, 2007) pointed out similarities between the Minnesota River Valley and Northern Superior terranes in terms of crystallization ages (cf. Böhm et al., 2003); however, the correlation is less compelling in the light of differences in crustal residence age and thermotectonic history (Fig. 8). Other continental fragments with common age features include the juvenile 3.0 Ga North Caribou and Marmion terranes. Their different rift histories appear to preclude origins from a common parent terrane (Tomlinson et al., 1999). Juvenile terranes with common 2.9–2.8 Ga histories include the Hawk, Opatica, and Arnaud terranes, allowing possible common origins. Matches for Superior Province microcontinental fragments may be found in other Archean cratons with "Superia" ancestry (Bleeker, 2003).

## Superior Province Neoarchean Juvenile (Oceanic) Terranes

The Oxford-Stull domain (Fig. 7) contains tectonically imbricated oceanic rocks with continent-margin rocks of North Caribou affinity, juxtaposed on D1 faults between 2820 and 2780 Ma (Corkery et al., 2000). The oceanic rocks are depleted tholeiitic pillow basalts that are consistent with an seafloor environment. They were accreted to the margin near the time of initiation of Neoarchean calc-alkaline magmatism (ca. 2780 Ma; Corkery et al., 2000).

Juvenile rocks of the Bidou and Black Island assemblages south of the North Caribou superterrane are interpreted to represent a paired arc–back-arc system deposited off the margin (Bailes and Percival, 2005; Percival et al., 2006b). They were juxtaposed with the North Caribou margin after 2707 Ma (Percival et al., 2006b).

The western Wabigoon terrane (Figs. 7 and 9A) is a classic granite-greenstone terrane consisting of sinuous belts of supracrustal rocks surrounding domiform granitoid batholiths (Blackburn et al., 1991). Volcanic rocks are predominantly juvenile, submarine tholeiitic basalts through calc-alkaline basalt-andesite-dacite sequences, erupted between 2775 and 2720 Ma. Continental basement has not been recognized, and the oldest volcanic sequence has oceanic-plateau affinity, possibly a product of plume-driven rifting. Many of the large batholith complexes are composed of juvenile tonalites emplaced synvolcanically during the peak of magmatic activity, between 2735 and 2730 Ma (Blackburn et al., 1991; Henry et al., 1998; Whalen et al., 2004b). The calc-alkaline volcanic rocks and coeval plutons are consistent with arc settings and have been attributed to increasing input of continental material through sediment subduction (Bernier et al., 1999; Whalen et al., 2004b). The western Wabigoon terrane likely evolved as a small ocean basin, possibly not far removed from a rifted margin, in a complex microplate environment (Skulski et al., 1998a; Percival et al., 2004b). Juvenile rocks of the eastern Wabigoon terrane have been interpreted as an oceanic back-arc assemblage (Stott et al., 2002).

The Wawa-Abitibi terrane (Figs. 7 and 9B) is the largest domain of juvenile rocks in the Superior Province, extending over an area of 200 × 1200 km. Restricted parts of the central terrane were built on ca. 2.9 Ga sialic crust of the Hawk terrane. Extensive zones of isotopically juvenile crust to the east and west are characterized by continuous stratigraphic sequences that extend over distances exceeding 100 km, suggesting autochthonous deposition rather than assembly through thrust stacking (Ayer et al., 2002). Seven submarine volcanic assemblages of calc-alkaline, tholeiitic, and komatiitic composition, ranging in age from 2770 to 2696 Ma, suggest a tectonic environment involving recurring interaction between plume and subduction-zone magmatism (Dostal and Mueller, 1997; Wyman, 1999; Ayer et al., 2002; Wyman et al., 2002; Sproule et al., 2002). Details of mantle processes surrounding this unusual style of magmatism are still under discussion. Unconformably overlying clastic sedimentary rocks of the 2687 Ma Porcupine Group (Ayer et al., 2004) are synorogenic with respect to D1 deformation, which may mark amalgamation of the Abitibi and Opatica terranes (Davis et al., 1994).

Greenstone belts of the Arnaud terrane include juvenile tholeiitic and calc-alkaline basalts of oceanic-plateau and arc affinity (Skulski et al., 1996). Zones of structural juxtaposition of oceanic and continental assemblages indicate tectonic assembly between 2718 and 2700 Ma (Skulski and Percival, 1996; Lin et al., 1996; Percival and Skulski, 2000). A wide belt of metasedimentary gneisses has been interpreted to represent a basin separating the Hudson Bay terrane, with its Eo- to Mesoarchean ancestry, from the adjacent <2.9 Ga Arnaud terrane (Leclair et al., 2006).

## Tectonic Assembly of the Superior Province

The ca. 2.72–2.71 Ga Northern Superior orogeny (Fig. 10) represents collision between the Northern Superior and North Caribou superterranes. Subduction polarity is inferred to have been southward, based on the presence of 2775–2733 Ma arc magmatic activity in the northern North Caribou superterrane, and south-over-north shear-zone movement (Corkery et al., 2000; Skulski et al., 2000; Lin et al., 2006; Parks et al., 2006). This geometry is mirrored by a steep slab of high resistivity in the mantle to 150 km depth (Craven et al., 2004; Percival et al.,

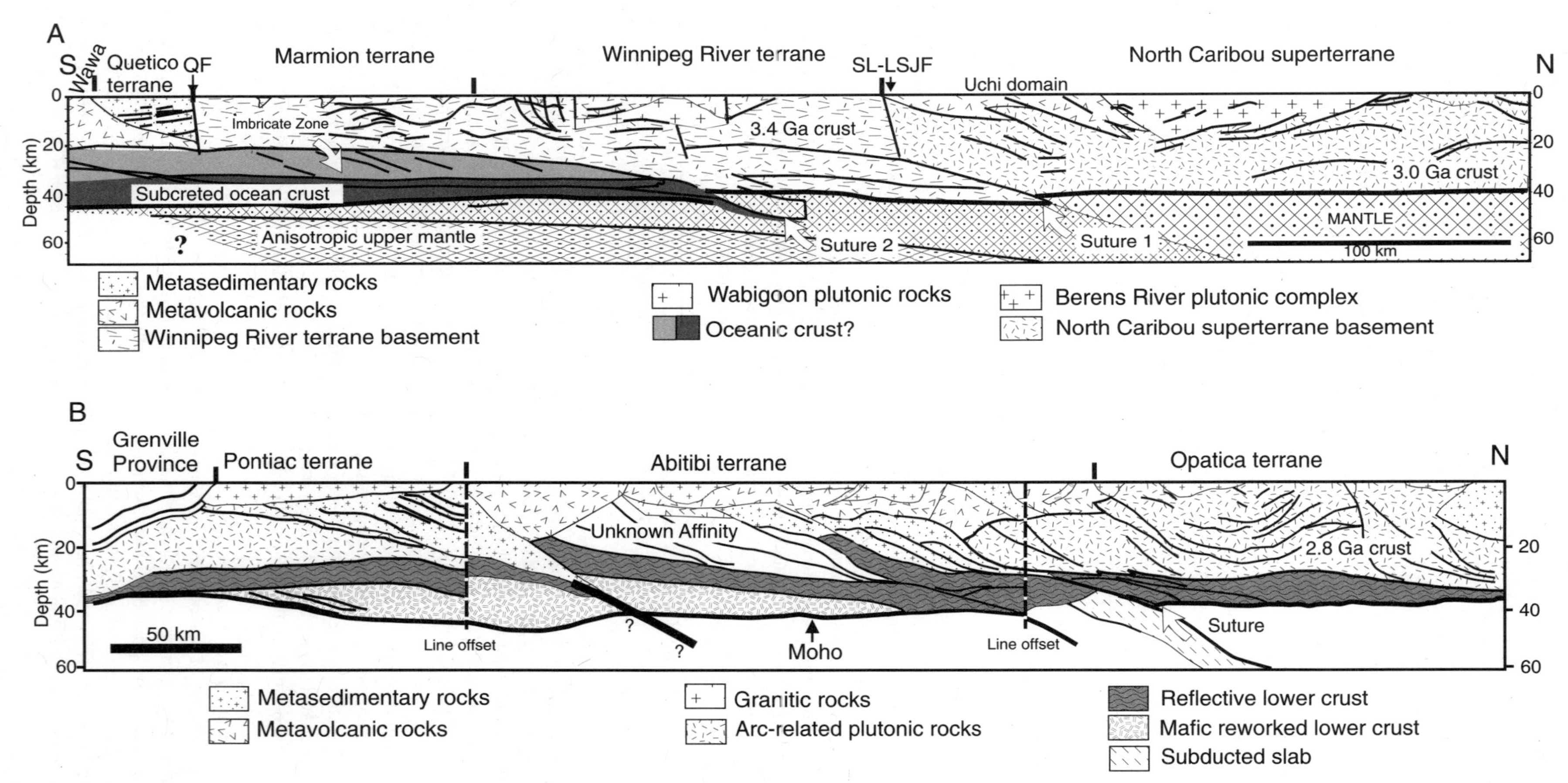


Figure 9. Regional cross sections of Superior Province based on Lithoprobe coincident seismic reflection-refraction profiles combined with surface geological, geochemical, and isotopic observations. (A) Western Superior line (after White et al., 2003; Musacchio et al., 2004). (B) Abitibi-Grenville line (after Calvert and Ludden, 1999). Abbreviations: QF—Quetico fault; SL-LSJF—Sydney Lake–Lake St. Joseph fault.

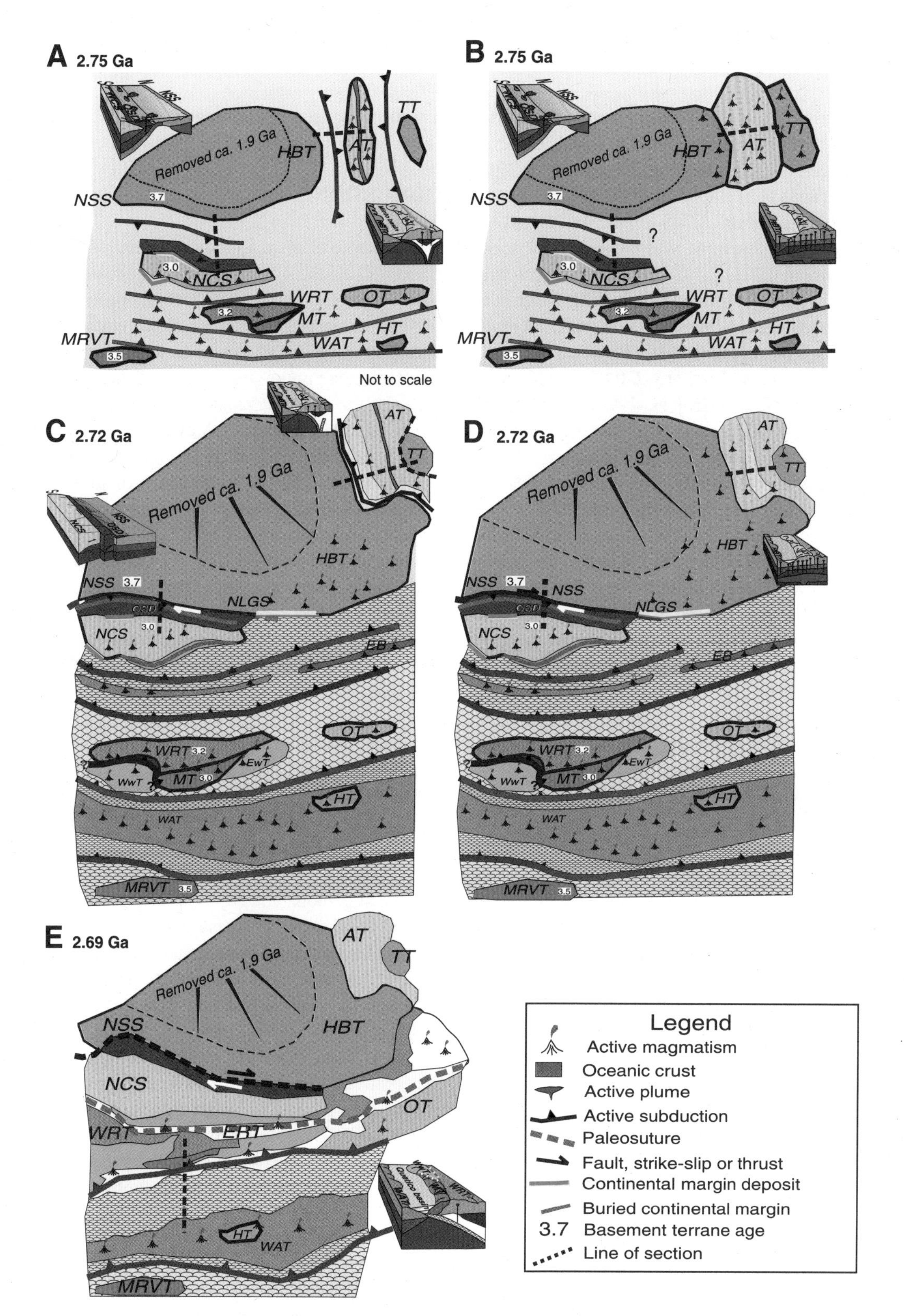


Figure 10. Schematic evolutionary model for growth of Superior Province. Abbreviations are as in Figure 6. (A) At 2.75 Ga, independent microcontinental fragments are separated by tracts of oceanic crust of unknown dimension. (B) Alternative interpretation in which the northeastern Superior Province is magmatically active as a result of megaplume impingement (see Boily et al., 2006). (C) By 2.72 Ga, the Northern Superior superterrane starts to collide with the northern margin of the North Caribou superterrane to initiate the composite Superior superterrane. The western Wabigoon terrane begins to impinge on the southwestern Winnipeg River terrane margin. NLGS—northern La Grande subprovince; EB—Eastmain belt. (D) Alternative interpretation in which the northeastern Superior Province continues to evolve within a megaplume environment. (E) Between 2.70 and 2.69 Ga, the Wawa-Abitibi terrane (WAT) docks with the composite Superior superterrane, accompanied by deposition of synorogenic Quetico flysch in the intervening trench, its burial, and metamorphism. Arc magmatism continues in the oceanic Wawa-Abitibi terrane, and postorogenic granitic magmatism is widespread across the composite Superior superterrane to the north.

2006a). Collision and uplift of the Northern Superior superterrane are recorded by the appearance of 3.5–3.6 Ga detrital zircon in <2711 Ma synorogenic sedimentary rocks deposited on North Caribou superterrane units (Corkery et al., 2000).

Widespread arc magmatism (2748–2708 Ma) across the southern North Caribou superterrane was the precursor to the Uchian orogeny, in which the ca. 3.4 Ga Winnipeg River terrane docked from the south (Fig. 10; Stott and Corfu, 1991; Corfu et al., 1995; Stott, 1997). The suture zone between the continental blocks is marked by the wide synorogenic English River basin (Figs. 7 and 10E), which was deposited after 2713–2704 Ma and overridden as the collision progressed (Corfu et al., 1995). Both the English River and Winnipeg River terranes were rapidly buried and heated by inferred southward overthrusting of the North Caribou superterrane, which is imaged by gently north-dipping reflectivity on seismic-reflection profiles (Fig. 9A; White et al., 2003).

Parts of the central Superior Province were assembled into a superterrane during the Central Superior orogeny, just prior to collision with the northern Superior collage (Figs. 10C and 10D). The Winnipeg River–Marmion terrane had rifted by ca. 2880 Ma, leading to formation of the 2775–2720 Ma western Wabigoon oceanic terrane. Both the amount of separation and nature of collision are debated. Blackburn et al. (1991) inferred an in situ rift, whereas Sanborn-Barrie and Skulski (1999, 2006) regarded the western Wabigoon as an oceanic terrane overridden by the Winnipeg River terrane at ca. 2700 Ma. Opposing subduction polarity and ca. 2710 Ma collision have been inferred by Davis and Smith (1991), Percival et al. (2004b), and Melnyk et al. (2006).

The Shebandowanian orogeny brought the oceanic Wawa-Abitibi terrane into juxtaposition with the amalgamated northern Superior collage at ca. 2.695 Ga (Fig. 10E; Corfu and Stott, 1986, 1998; Stott, 1997; Daigneault et al., 2006). Subduction polarity is inferred to have been to the north based on 2745–2700 Ma arc magmatism in the Wabigoon and Winnipeg River terranes to the north (Figs. 10C and 10D; Percival et al., 2006a). The suture zone is marked by the thick synorogenic turbiditic wedge of the Quetico basin. Seismic profiles across the boundary zone indicate gently northward-dipping reflectivity (Calvert and Ludden, 1999; White et al., 2003), but the Quetico wedge does not appear to project northward, consistent with a model of accretion rather than subduction of Quetico sedimentary material.

The ca. 2680 Ma Minnesotan orogeny is responsible for collision between the Minnesota River Valley terrane and the northern Superior collage. Northward subduction polarity is indicated by the geometry of seismic reflectors (Fig. 9A; White et al., 2003) and by the isotopic signature of old crust beneath the southern Wawa-Abitibi terrane (Sims et al., 1997). The Great Lakes tectonic zone (Fig. 7) is the probable surface expression of the suture (Sims and Day, 1993), and thick mafic lower crust to the north may be a subcreted slab of oceanic material (Musacchio et al., 2004; Percival et al., 2004b). In central Quebec, the <2685 Ma (Davis, 2002) Pontiac basin bounds the Wawa-Abitibi terrane to the south. There, the polarity of subduction also appears to be northward, based on north-dipping seismic reflectivity (Fig. 9B; Calvert et al., 1995) and the presence of peraluminous granite in the southern Abitibi (Feng and Kerrich, 1991, 1992; Chown et al., 2002).

## DISCUSSION

The role of plate tectonics on Archean Earth is highly controversial and the subject of ongoing debate. However, many elements of modern-style plate-tectonic processes are preserved in the ancient orogens of Earth's oldest craton and shield areas, as exemplified by our examples in the Pilbara craton of Australia, the Barberton granite-greenstone terrane of South Africa, and the Superior Province of Canada (Table 1). It is important to note that the evidence in support of Archean plate tectonics from each of these regions derives from independent and distinct observations. For the Pilbara craton, combined geological, geochemical, and geochronological evidence shows a marked contrast in tectonic styles between the older than 3.2 Ga East Pilbara terrane cratonic nucleus and the younger than 3.2 Ga West Pilbara superterrane. Prior to the accretion of these two disparate elements at 3.07 Ga, the former underwent continental rifting at 3.2 Ga, while the latter formed as a collage of rifted fragments of the East Pilbara terrane and a 3.12 Ga oceanic arc complex. Thermobarometric evidence from the Barberton granite-greenstone terrane, combined with a range of structural, geochemical, and geochronological data, indicates that between 3.4 and 3.2 Ga, high-pressure–low-temperature metamorphism associated with subduction was closely followed by exhumation of, and juxtaposition with, high-grade (medium-pressure–high-temperature metamorphism) rocks from depth via orogenic collapse. Evidence for the 2.7 Ga accretionary and collisional assembly of the Superior craton is also based on a range of geological, geophysical, geochemical, and geochronological data. Together, these data provide compelling evidence for the formation of continental crust through "modern-style" subduction-accretion processes (Table 1) since at least 3.2 Ga, although these processes may have operated differently on an older, relatively hotter, Earth.

It is clear from the evidence summarized here and from other studies that there are some important differences between Archean and "modern" plate tectonics (Hamilton, 1998, 2003; Stern, 2005; Brown, 2006, this volume). Many of these differences have long been known, such as the lack in Archean rocks of broken formations and large accretionary mélange units, blueschists, and classic ophiolite sections (Table 1; cf. Hamilton, 1998; McCall, 2003). These and other aspects of the differences between Archean and modern plate tectonics may be explained by a significantly greater heat flux associated with the thermal state of the planet in Archean times (cf. Franck, 1998; Brown, this volume; Hynes, this volume; van Hunen et al., this volume), which would have resulted in significantly weaker crust and lithosphere. This in turn would greatly affect how plates interacted and deformed along collision zones (e.g., Bailey, 1999; Cagnard et al., 2006). It also allows for a range of other crust-forming processes, i.e., stagnant lid convection (Korenaga, 2006), plume tectonics (Maruyama, 1994), and vertical tectonics (Collins et al., 1998; Bédard

TABLE 1. DISTINGUISHING FEATURES OF PLATE TECTONICS RELEVANT TO THE ARCHEAN

| Example | Evidence | Region |
|---|---|---|
| Continental drift, breakup, and/or seafloor spreading | Presence of discrete continental and juvenile oceanic domains | Superior, Barberton, Pilbara |
| | Seafloor magnetic anomalies | |
| | Dispersed faunal provinces across oceans | |
| | Split geological provinces across oceans | |
| | Rift-related dike swarms and associated continental flood basalt provinces along rift margins | Superior (locally) |
| | Rift valleys and rift-fill sequences, including thick clastic sedimentary successions and distinct volcanic lavas (typically alkaline) | |
| | Passive-margin mountain belts, caused by uplift over thinned continental crust underplated by inflowing mantle | |
| | Fit of continents; paleomagnetic data from (mostly) volcanic successions | |
| Subduction zones | Ocean bathymetry; fault-plane solutions and vector resolution | |
| Fault-bound exotic terranes in mountain belts, with different geological histories and provenances | Geology, geochronology, paleontology, paleomagnetic data | Superior, Barberton (locally), Pilbara |
| Transform faults | Offset of mid-oceanic-ridge segments in the opposite direction to the displacement vector across ocean-floor transform faults | |
| Curvilinear plate-margin processes and dimensions | Large scale (<1000 km), curvilinear orogenic belts with (mostly) univergent folds and thrust faults indicative of compression; gravity data of crustal thickening | Superior, Pilbara(?) |
| Lateral movements | Long strike-slip faults (e.g., Sleep, 1992) | Superior, Barberton |
| Accretionary mélange zones | Chaotic zones of tectonic mélange, with blocks of exotic rocks from many sources, including blueschists (e.g., Franciscan Complex, California) | Barberton |
| Fault-bound mantle slices at surface | Geology and mineral chemistry | |
| Hotspot chains | Age progression from site of current volcanism (e.g., Hawaii-Emperor seamount chain) | |
| Mantle enrichment by fluids or melts in suprasubduction-zone environments | Existence of isotopically juvenile, calc-alkaline basalt | Superior, Pilbara, |
| | High Mg#, Ni, Cr in slab-derived melts that have interacted with overlying mantle wedge (e.g., sanukitoids) | Superior |
| Stacking and crust-mantle imbrications on seismic data | Gently dipping crust and Moho offsets | Superior, Pilbara |
| Older over younger tectonic stacking | Fault-bound panels of older rock tectonically juxtaposed atop younger rocks (e.g., Moine thrust zone, Scotland; Franciscan Complex, California) | Superior (locally), Barberton (locally), Pilbara |
| Paired metamorphic belts (metamorphic duality) | Ultrahigh-temperature (arc and back-arc) versus eclogite high-pressure metamorphism (subduction zone) (e.g., Brown, 2006) | Barberton (locally) |

et al., 2003, Bédard, 2006), to have operated, operated more vigorously, or even to have been the dominant form of crust-mantle interaction. In fact, the difference in structural style preserved in the Pilbara craton, where pre–3.2 Ga crust is dominated by a relatively small-scale (tens to hundred kilometers) dome-and-keel geometry, and post–3.2 Ga terranes are dominated by large-scale (hundreds of kilometers) curvilinear belts and faults, suggests that Earth's Paleo- to Eoarchean tectonic regime may have been plume-driven, whereas modern-style steep subduction may have been initiated in the Mesoarchean (Van Kranendonk, 2004a, 2007; Smithies et al., 2005a; Van Kranendonk et al., 2007). The continental crust preserved in ancient cratons, the increasing preservation of continental crust through time (Hurley and Rand, 1969), and the appearance of cratons that have subduction imprints at, or younger than, 3.2 Ga suggest that Mesoarchean subduction operated in a similar fashion to present-day subduction and was capable of producing large cratons with thick mantle keels by the Neoarchean (see also discussion in Shirey et al., this volume). The widespread evidence of subduction-accretion in Neoarchean terranes and the large size of these cratons indicate that modern-style plate tectonics had evolved to become the dominant form of crust-mantle interaction at least by ca. 2.7 Ga.

## CONCLUSIONS

Evidence for modern-style plate tectonics is preserved in the ancient orogens of the Pilbara craton of Western Australia, the Barberton granite-greenstone terrane of the Kaapvaal craton in South Africa, and the Superior Province of the Canadian Shield.

The 3.53–2.83 Ga Pilbara craton documents the formation of ancient sialic crust. Subsequent mantle-plume events erupted thick volcanic successions and melted this crust to generate TTGs, followed by progressively more evolved granitic magmas. Plume magmatism was accompanied by uplift, crustal extension, and partial convective overturn of the upper and middle crust. Horizontal tectonic forces dominated over vertical forces after 3.2 Ga. Thick volcanic successions with the geochemical characteristics of modern oceanic arcs (including boninites and evidence for flux melting) indicate steep subduction in the Mesoarchean. Subduction was followed by terrane accretion at 3.07 Ga, the development of intracontinental basins, and widespread plutonism resulting from orogenic relaxation and slab breakoff.

The 3.45–3.21 Ga Barberton granite-greenstone terrane preserves tectono-metamorphic evidence for Paleoarchean subduction, orogeny, and extensional collapse. High-pressure–low-temperature and medium-pressure–high-temperature metamorphic terranes were juxtaposed at ca. 3230 Ma. A tectonic environment similar to modern ocean-continent collision, with subduction followed by later exhumation of high-grade terranes associated with orogenic collapse, suggests active plate tectonics. These terranes may be the Archean analogue of a paired metamorphic belt. Estimates of metamorphic field gradients (15–30 °C/km) and exhumation rates (2–5 mm/yr) agree with modern analogues.

The 2.7 Ga Superior Province records collision between, and accretion of, terranes of both continental and oceanic derivation. Continental-arc magmatism and accretion of oceanic crust document Neoarchean subduction and the operation of plate tectonics.

The evidence from the orogenic record indicates that plate tectonics have operated on Earth since at least 3.2 Ga, and other evidence (isotope geochemistry, ore geology, geodynamics, etc.) suggests that some form of plate tectonics may have operated as early as 3.6 Ga.

Meso- to Neoarchean plate tectonics operated in a similar way as modern plate tectonics, but with specific differences resulting from a hotter mantle.

## ACKNOWLEDGMENTS

Thanks are extended to the organizers of the stimulating Penrose Workshop that led to the production of this volume. We also thank R. Frost and an anonymous reviewer for constructive remarks, which improved the manuscript. Colleagues in the Geological Surveys of Canada, Manitoba, Ontario, and Quebec, as well as several universities, contributed to the synthesis of the Superior Province section through discussions and collaboration. This work was supported by the Swedish Research Council (Grant 629-2002-745) and is published with the permission of the Executive Director of the Geological Survey of Western Australia.

## REFERENCES CITED

Anhaeusser, C.R., 1983, The geology of the Schapenburg greenstone remnant and surrounding Archaean granitic terrane south of Badplaas, Eastern Transvaal, *in* Anhaeusser, C.R., ed., Contributions to the Geology of the Barberton Mountain Land, National Geodynamics Programme, Barberton Project: Geological Society of South Africa Special Publication 9, p. 31–44.

Anhaeusser, C.R., Robb, L.J., and Viljoen, M.J., 1981, Provisional Geological Map of the Barberton Greenstone Belt and Surrounding Granitic Terrane, Eastern Transvaal and Swaziland: Johannesburg, Geological Society of South Africa, scale 1:250,000.

Armstrong, R.A., Compston, W., de Wit, M.J., and Williams, I.S., 1990, The stratigraphy of the Barberton greenstone belt revisited: A single zircon ion microprobe study: Earth and Planetary Science Letters, v. 101, p. 90–106, doi: 10.1016/0012-821X(90)90127-J.

Arndt, N., Bruzak, G., and Reischmann, T., 2001, The oldest continental and oceanic plateaus: Geochemistry of basalts and komatiites of the Pilbara craton, Australia, *in* Ernst, R.E., and Buchan, K.L., eds., Mantle Plumes: Their Identification through Time: Geological Society of America Special Paper 352, p. 359–387.

Ayer, J., Amelin, Y., Corfu, F., Kamo, S., Ketchum, J., Kwok, K., and Trowell, N., 2002, Evolution of the southern Abitibi greenstone belt based on U-Pb geochronology: Autochthonous volcanic construction followed by plutonism, regional deformation and sedimentation: Precambrian Research, v. 115, p. 63–95, doi: 10.1016/S0301-9268(02)00006-2.

Ayer, J.A., Thurston, P.C., Dubé, B., Gibson, H.L., Hudak, G., Lafrance, B., Lesher, C.M., Piercey, S.J., Reed, L.E., and Thompson, P.H., 2004, Discover Abitibi greenstone architecture project: Overview of results and belt-scale implications, *in* Summary of Field Work and Other Activities 2004: Ontario Geological Survey Open-File Report 6145, p. 37–1 to 37–15.

Bailes, A.H., and Percival, J.A., 2005, Geology of the Black Island Area, Lake Winnipeg, Manitoba (Parts of NTS 62P1, 7 and 8): Manitoba Geological Survey Geoscientific Report GR2005–2, 33 p.

Bailey, R.C., 1999, Gravity-driven continental overflow and Archaean tectonics: Nature, v. 398, p. 413–415, doi: 10.1038/18866.

Barley, M.E., 1997, The Pilbara craton, *in* de Wit, M.J., and Ashwal, L., eds., Greenstone Belts: Oxford, UK, Oxford University Monographs on Geology and Geophysics 35, p. 657–664.

Barley, M.E., and Pickard, A.L., 1999, An extensive, crustally derived, 3325 to 3310 Ma silicic volcano-plutonic suite in the eastern Pilbara craton: Evidence from the Kelley belt, McPhee Dome, and Corunna Downs Batholith: Precambrian Research, v. 96, p. 41–62, doi: 10.1016/S0301-9268(99)00003-0.

Barley, M.E., Sylvester, G.C., and Groves, D.I., 1984, Archaean calc-alkaline volcanism in the Pilbara block, Western Australia: Precambrian Research, v. 24, p. 285–319, doi: 10.1016/0301-9268(84)90062-7.

Barley, M.E., Loader, S.E., and McNaughton, N.J., 1998, 3430 to 3417 Ma calc-alkaline volcanism in the McPhee Dome and Kelley belt, and growth of the eastern Pilbara craton: Precambrian Research, v. 88, p. 3–24, doi: 10.1016/S0301-9268(97)00061-2.

Beakhouse, G.P., 1991, Winnipeg River subprovince, *in* Thurston, P.C., Williams, H.R., Sutcliffe, R.H., and Stott, G.M., eds., Geology of Ontario: Ontario Geological Survey Special Volume 4, Part 1, p. 279–301.

Bédard, J.H., 2003, Evidence for regional-scale, pluton-driven, high-grade metamorphism in the Archaean Minto block, northern Superior Province, Canada: The Journal of Geology, v. 111, p. 183–205, doi: 10.1086/345842.

Bédard, J.H., 2006, A catalytic delamination-driven model for coupled genesis of Archaean crust and sub-continental lithospheric mantle: Geochimica et Cosmochimica Acta, v. 70, p. 1188–1214, doi: 10.1016/j.gca.2005.11.008.

Bédard, J.H., Brouillette, P., Madore, L., and Berclaz, A., 2003, Archean cratonization and deformation in the northern Superior Province, Canada: An evaluation of plate tectonic versus vertical tectonic models: Precambrian Research, v. 127, p. 61–87, doi: 10.1016/S0301-9268(03)00181-5.

Belcher, R.W., and Kisters, A.F.M., 2006, Syntectonic emplacement and deformation of the Heerenveen batholith: Conjectures on the structural setting of the 3.1 Ga granite magmatism in the Barberton granitoid-greenstone terrain, South Africa, *in* Reimold, W.U., and Gibson, R.L., Processes on the Early Earth: Geological Society of America Special Paper 405, p. 211–231.

Benn, K., and Moyen, J.-F., 2008, this volume, The late Archean Abitibi-Opatica terrane, Superior Province: A late Archean oceanic plateau modified by subduction and slab window magmatism, *in* Condie, K.C., and Pease, V., eds., When Did Plate Tectonics Begin on Planet Earth?: Geological Society of America Special Paper 440, doi: 10.1130/2008.2440(09).

Benn, K., Sawyer, E.W., and Bouchez, J.L., 1992, Orogen parallel and transverse shearing in the Opatica belt, Quebec: Implications for the structure of the Abitibi belt: Canadian Journal of Earth Sciences, v. 29, p. 2429–2444.

Bernier, F., Stevenson, R.K., Gariépy, C., and Franklin, J.M., 1999, Nd isotopic studies in the South Sturgeon Lake greenstone belt, north-west Ontario; A progress report, *in* Harrap, R.M., and Helmstaedt, H., eds., Western Superior Transect Sixth Annual Workshop: Lithoprobe Report 70, p. 117–121.

Bickford, M.E., Wooden, J.L., and Bauer, R.L., 2006, SHRIMP study of zircons from the Early Archean rocks in the Minnesota River Valley: Implications for the tectonic history of the Superior Province: Geological Society of America Bulletin, v. 118, p. 94–108, doi: 10.1130/B25741.1.

Bickford, M.E., Wooden, J.L., Bauer, R.L., and Schmitz, M.D., 2007, Paleoarchean gneisses in the Minnesota River Valley and northern Michigan, USA, *in* Van Kranendonk, M.J., Smithies, R.H., and Bennett, V., eds.,

Earth's Oldest Rocks: Amsterdam, Elsevier, Developments in Precambrian Geology, v. 15, p. 731–750.

Bickle, M.J., Bettenay, L.F., Boulter, C.A., Groves, D.I., and Morant, P., 1980, Horizontal tectonic intercalation of an Archaean gneiss belt and greenstones, Pilbara block, Western Australia: Geology, v. 8, p. 525–529, doi: 10.1130/0091-7613(1980)8<525:HTIOAA>2.0.CO;2.

Bickle, M.J., Bettenay, L.F., Barley, M.E., Chapman, H.J., Groves, D.I., Campbell, I.H., and de Laeter, J.R., 1983, A 3500 Ma plutonic and volcanic calc-alkaline province in the Archaean East Pilbara block: Contributions to Mineralogy and Petrology, v. 84, p. 25–35, doi: 10.1007/BF01132327.

Bickle, M.J., Morant, P., Bettenay, L.F., Boulter, C.A., Blake, T.S., and Groves, D.I., 1985, Archaean tectonics of the Shaw Batholith, Pilbara block, Western Australia: Structural and metamorphic tests of the batholith concept, *in* Ayers, L.D., Thurston, P.C., Card, K.D., and Weber, W., eds., Evolution of Archean Supracrustal Sequences: Geological Association of Canada Special Paper 28, p. 325–341.

Bickle, M.J., Bettenay, L.F., Chapman, H.J., Groves, D.I., McNaughton, N.J., Campbell, I.H., and de Laeter, J.R., 1993, Origin of the 3500–3300 Ma calc-alkaline rocks in the Pilbara Archaean: Isotopic and geochemical constraints from the Shaw Batholith: Precambrian Research, v. 60, p. 117–149, doi: 10.1016/0301-9268(93)90047-6.

Blackburn, C.E., John, G.W., Ayer, J., and Davis, D.W., 1991, Wabigoon subprovince, *in* Thurston, P.C., Williams, H.R., Sutcliffe, R.H., and Stott, G.M., eds., Geology of Ontario: Ontario Geological Survey Special Volume 4, Part 1, p. 303–381.

Bleeker, W., 2003, The late Archean record: A puzzle in ca. 35 pieces: Lithos, v. 71, p. 99–134, doi: 10.1016/j.lithos.2003.07.003.

Blewett, R., 2002, Archaean tectonic processes: A case for horizontal shortening in the North Pilbara granite-greenstone terrane, Western Australia: Precambrian Research, v. 113, p. 87–120, doi: 10.1016/S0301-9268(01)00204-2.

Böhm, C.O., Heaman, L.M., Creaser, R.A., and Corkery, M.T., 2000, Discovery of pre–3.5 Ga exotic crust at the northwestern Superior Province margin, Manitoba: Geology, v. 28, p. 75–78, doi: 10.1130/0091-7613(2000)28<75:DOPGEC>2.0.CO;2.

Böhm, C.O., Heaman, L.M., Stern, R.A., Corkery, M.T., and Creaser, R.A., 2003, Nature of Assean Lake ancient crust, Manitoba: A combined SHRIMP-ID-TIMS U-Pb geochronology and Sm-Nd isotope study: Precambrian Research, v. 126, p. 55–94, doi: 10.1016/S0301-9268(03)00127-X.

Boily, M., Leclair, A., Maurice, C., Berclaz, A., and David, J., 2006, Étude Géochimique et Isotopique du Nd des Assemblages Volcaniques et Plutoniques du Nord-Est de la Province du Supérieur (NEPS): Québec, Ministère des Ressources Naturelles, Report GM62031, 50 p.

Boulter, C.A., Bickle, M.J., Gibson, B., and Wright, R.K., 1987, Horizontal tectonics pre-dating upper Gorge Creek Group sedimentation, Pilbara block, Western Australia: Precambrian Research, v. 36, p. 241–258, doi: 10.1016/0301-9268(87)90023-4.

Brown, J.L., 2002, Neoarchean Evolution of the Western-Central Wabigoon Boundary Zone, Brightsand Forest Area, Ontario [M.Sc. thesis]: Ottawa, Ontario, University of Ottawa, 240 p.

Brown, M., 2006, Duality of thermal regimes is the distinctive characteristic of plate tectonics since the Neoarchean: Geology, v. 34, p. 961–964, doi: 10.1130/G22853A.1.

Brown, M., 2008, this volume, Characteristic thermal regimes of plate tectonics and their metamorphic imprint throughout Earth history: When did Earth first adopt a plate tectonics mode of behavior?, *in* Condie, K.C., and Pease, V., eds., When Did Plate Tectonics Begin on Planet Earth?: Geological Society of America Special Paper 440, doi: 10.1130/2008.2440(05).

Cagnard, F., Durrieu, N., Gapais, D., Brun, J.-P., and Ehlers, C., 2006, Crustal thickening and lateral flow during compression of hot lithospheres, with particular reference to Precambrian times: Terra Nova, v. 18, p. 72–78, doi: 10.1111/j.1365-3121.2005.00665.x.

Calvert, A., and Ludden, J.N., 1999, Archean continental assembly in the southeastern Superior Province of Canada: Tectonics, v. 18, p. 412–429, doi: 10.1029/1999TC900006.

Calvert, A.J., Sawyer, E.W., Davis, W.J., and Ludden, J.N., 1995, Archaean subduction inferred from seismic images of a mantle suture in the Superior Province: Nature, v. 375, p. 670–674, doi: 10.1038/375670a0.

Champion, D.C., and Smithies, R.H., 2001, Archean granites of the Yilgarn and Pilbara cratons, Western Australia, *in* Cassidy, K.F., Dunphy, J.M., and Van Kranendonk, M.J., eds., 4th International Archaean Symposium 2001, Extended Abstracts: Australian Geological Survey Organization (AGSO) Geoscience Australia Record 2001/37, p. 134–136.

Champion, D.C., and Smithies, R.H., 2007, Geochemistry of Paleoarchean granites of the East Pilbara terrane, Pilbara craton, Western Australia: Implications for early Archean crustal growth, *in* Van Kranendonk, M.J., Smithies, R.H., and Bennett, V., eds., Earth's Oldest Rocks: Amsterdam, Elsevier, Developments in Precambrian Geology, v. 15, p. 369–410.

Chopin, C., 1987, Very-high-pressure metamorphism in the Western Alps: Implications for subduction of continental crust: Philosophical Transactions of the Royal Society of London, v. 321, p. 183–197, doi: 10.1098/rsta.1987.0010.

Chown, E.H., Harrap, R., and Mouksil, A., 2002, The role of granitic intrusions in the evolution of the Abitibi belt, Canada: Precambrian Research, v. 115, p. 291–310, doi: 10.1016/S0301-9268(02)00013-X.

Cloete, M., 1999, Aspects of Volcanism and Metamorphism of the Onverwacht Group Lavas in the Southwestern Portion of the Barberton Greenstone Belt: Memoirs of the Geological Survey of South Africa 84, 232 p.

Collins, W.J., 1989, Polydiapirism of the Archaean Mt. Edgar batholith, Pilbara block, Western Australia: Precambrian Research, v. 43, p. 41–62, doi: 10.1016/0301-9268(89)90004-1.

Collins, W.J., and Van Kranendonk, M.J., 1999, The development of kyanite during partial convective overturn of Archaean granite-greenstone terranes: The Pilbara craton, Australia: Journal of Metamorphic Geology, v. 17, p. 145–156, doi: 10.1046/j.1525-1314.1999.00187.x.

Collins, W.J., Van Kranendonk, M.J., and Teyssier, C., 1998, Partial convective overturn of Archaean crust in the east Pilbara craton, Western Australia: Driving mechanisms and tectonic implications: Journal of Structural Geology, v. 20, p. 1405–1424, doi: 10.1016/S0191-8141(98)00073-X.

Condie, K.C., 1998, Episodic continental growth and supercontinents; A mantle avalanche connection?: Earth and Planetary Science Letters, v. 163, p. 97–108, doi: 10.1016/S0012-821X(98)00178-2.

Condie, K.C., 2005, High field strength element ratios in Archean basalts: A window to evolving sources of mantle plumes?: Lithos, v. 79, p. 491–504, doi: 10.1016/j.lithos.2004.09.014.

Connelly, J.N., and Mengel, F.C., 2000, Evolution of Archean components in the Paleoproterozoic Nagssugtoqidian orogen: West Greenland: Geological Society of America Bulletin, v. 112, p. 747–763.

Corfu, F., and Stott, G.M., 1986, U-Pb ages for late magmatism and regional deformation in the Shebandowan belt, Superior Province, Canada: Canadian Journal of Earth Sciences, v. 23, p. 1075–1082.

Corfu, F., and Stott, G.M., 1998, Shebandowan greenstone belt, western Superior Province; U-Pb ages, tectonic implications and correlations: Geological Society of America Bulletin, v. 110, p. 1467–1484, doi: 10.1130/0016-7606(1998)110<1467:SGBWSP>2.3.CO;2.

Corfu, F., Stott, G.M., and Breaks, F.W., 1995, U-Pb geochronology and evolution of the English River subprovince, an Archean low *P*–high *T* metasedimentary belt in the Superior Province: Tectonics, v. 14, p. 1220–1233, doi: 10.1029/95TC01452.

Corfu, F., Davis, D.W., Stone, D., and Moore, M., 1998, Chronostratigraphic constraints on the genesis of Archean greenstone belts, northwestern Superior Province, Ontario, Canada: Precambrian Research, v. 92, p. 277–295, doi: 10.1016/S0301-9268(98)00078-3.

Corkery, M.T., and Skulski, T., 1998, Geology of the Little Stull Lake Area (Part of NTS 53K/10 and /7): Report of Activities 1998: Manitoba Energy and Mines, Geological Services, p. 111–118.

Corkery, M.T., Davis, D.W., and Lenton, P.G., 1992, Geochronological constraints on the development of the Cross Lake greenstone belt: Canadian Journal of Earth Sciences, v. 29, p. 2171–2185.

Corkery, M.T., Cameron, H.D.M., Lin, S., Skulski, T., Whalen, J.B., and Stern, R.A., 2000, Geological Investigations in the Knee Lake Belt (Parts of NTS 53L): Report of Activities 2000: Manitoba Industry, Trade and Mines, Manitoba Geological Survey, p. 129–136.

Craven, J.A., Skulski, T., and White, D.W., 2004, Lateral and vertical growth of cratons: Seismic and magnetotelluric evidence from the western Superior transect: Lithoprobe Celebratory Conference: Vancouver, Lithoprobe Secretariat, University of British Columbia, Lithoprobe Report 86.

Daigneault, R., Mueller, W.U., and Chown, E.H., 2006, Abitibi greenstone belt plate tectonics: The diachronous history of arc development, accretion and collision, *in* Eriksson, P.G., Altermann, W., Nelson, D.R., Mueller, W.U., and Catuneanu, O., eds., The Precambrian Earth: Tempos and Events: Amsterdam, Elsevier, p. 88–103.

David, J., 2008, Compilation des Données Géochronologiques de la Partie Nord-Est de la Province du Supérieur: Québec, Ministère des Ressources Naturelles (in press).

David, J., Parent, M., Stevenson, R., Nadeau, P., and Godin, L., 2003, The Porpoise Cove supracrustal sequence, Inukjuak area: A unique example of Paleoarchean crust (ca. 3.8 Ga) in the Superior Province: Geological Association of Canada Program with Abstracts, v. 28 (online).

Davis, D.W., 2002, U-Pb geochronology of Archean metasedimentary rocks in the Pontiac and Abitibi subprovinces, Quebec; Constraints on timing, provenance and regional tectonics: Precambrian Research, v. 115, p. 97–117, doi: 10.1016/S0301-9268(02)00007-4.

Davis, D.W., and Jackson, M., 1988, Geochronology of the Lumby Lake greenstone belt: A 3 Ga complex within the Wabigoon subprovince, northwest Ontario: Geological Society of America Bulletin, v. 100, p. 818–824, doi: 10.1130/0016-7606(1988)100<0818:GOTLLG>2.3.CO;2.

Davis, D.W., and Smith, P.M., 1991, Archean gold mineralization in the Wabigoon subprovince, a product of crustal accretion: Evidence from U-Pb geochronology in the Lake of the Woods area, Superior Province, Canada: The Journal of Geology, v. 99, p. 337–353.

Davis, W.J., Gariepy, C., and Sawyer, E.W., 1994, Pre–2.8 Ga crust in the Opatica gneiss belt: A potential source of detrital zircons in the Abitibi and Pontiac subprovinces, Superior Province, Canada: Geology, v. 22, p. 1111–1114, doi: 10.1130/0091-7613(1994)022<1111:PGCITO>2.3.CO;2.

Davis, W.J., Machado, N., Gariepy, C., Sawyer, E.W., and Benn, K., 1995, U-Pb geochronology of the Opatica tonalite-gneiss belt and its relationship to the Abitibi greenstone belt, Superior Province, Quebec: Canadian Journal of Earth Sciences, v. 32, p. 113–127.

de Ronde, C.E.J., and de Wit, M.J., 1994, The tectonic history of the Barberton greenstone belt, South Africa: 490 million years of Archaean crustal evolution: Tectonics, v. 13, p. 983–1005, doi: 10.1029/94TC00353.

de Ronde, C.E.J., and Kamo, S.L., 2000, An Archaean arc–arc collisional event: A short-lived, (ca. 3 Myr) episode, Weltevreden area, Barberton greenstone belt, South Africa: Journal of African Earth Sciences, v. 30, p. 219–248, doi: 10.1016/S0899-5362(00)00017-8.

Dewey, J.F., 1969, Evolution of the Appalachian-Caledonian orogen: Nature, v. 222, p. 124–129, doi: 10.1038/222124a0.

de Wit, M.J., 1982, Gliding and overthrust nappe tectonics in the Barberton greenstone belt: Journal of Structural Geology, v. 4, p. 117–136, doi: 10.1016/0191-8141(82)90022-0.

de Wit, M.J., Fripp, R.E.P., and Stanistreet, I.G., 1983, Tectonic and stratigraphic implications of new field observations along the southern part of the Barberton greenstone belt, *in* Anhaeusser, C.R., ed., Contributions to the Geology of the Barberton Mountain Land, National Geodynamics Programme, Barberton Project: Geological Society of South Africa Special Publication 9, p. 21–29.

de Wit, M.J., Hart, R.A., and Hart, R.J., 1987a, The Jamestown ophiolite complex, Barberton Mountain Land: A section through 3.5 Ga oceanic crust: Journal of African Earth Sciences, v. 6, p. 681–730, doi: 10.1016/0899-5362(87)90007-8.

de Wit, M.J., Armstrong, R.A., Hart, R.J., and Wilson, A.H., 1987b, Felsic igneous rocks within the 3.3 to 3.5 Ga Barberton greenstone belt: High-level equivalents of the surrounding tonalite trondhjemite terrain, emplaced during thrusting: Tectonics, v. 6, p. 529–549.

de Wit, M.J., Roering, C., Hart, R.J., Armstrong, R.A., de Ronde, C.E.J., Green, R.W.E., Tredoux, M., Peberdy, E., and Hart, R.A., 1992, Formation of an Archaean continent: Nature, v. 357, p. 553–562, doi: 10.1038/357553a0.

Diener, J.F.A., Stevens, G., Kisters, A.F.M., and Poujol, M., 2005, Metamorphism and exhumation of the basal parts of the Barberton greenstone belt, South Africa: Constraining the rates of Mesoarchean tectonism: Precambrian Research, v. 143, p. 87–112.

Dostal, J., and Mueller, W., 1997, Komatiite flooding of a rifted Archean rhyolitic arc complex: Geochemical signature and tectonic significance of the Stoughton-Roquemaure Group, Abitibi greenstone belt, Canada: The Journal of Geology, v. 105, p. 545–563.

Dubé, B., Williamson, K., McNicoll, V., Malo, M., Skulski, T., Twomey, T., and Sanborn-Barrie, M., 2004, Timing of gold mineralization at Red Lake, northwestern Ontario, Canada: New constraints from U-Pb geochronology at the Goldcorp high-grade zone, Red Lake Mine and the Madsen Mine: Economic Geology and the Bulletin of the Society of Economic Geologists, v. 99, p. 1611–1641.

Dziggel, A., Stevens, G., Poujol, M., Anhaeusser, C.R., and Armstrong, R.A., 2002, Metamorphism of the granite-greenstone terrane south of the Barberton greenstone belt, South Africa: An insight into the tectono-thermal evolution of the "lower" portions of the Onverwacht Group: Precambrian Research, v. 114, p. 221–247, doi: 10.1016/S0301-9268(01)00225-X.

Dziggel, A., Knipfer, S., Kisters, A.F.M., and Meyer, F.M., 2006, *P-T* and structural evolution during exhumation of high-*T*, medium-*P* basement rocks in the Barberton Mountain Land, South Africa: Journal of Metamorphic Geology, v. 24, p. 535–551.

Eriksson, K.A., 1982, Geometry and internal characteristics of Archaean submarine channel deposits, Pilbara block, Western Australia: Journal of Sedimentary Petrology, v. 52, p. 383–393.

Ernst, W.G., 1973, Interpretative synthesis of metamorphism in the Alps: Geological Society of America Bulletin, v. 84, p. 2053–2078.

Ernst, W.G., 1988, Tectonic history of subduction zones inferred from retrograde blueschist *P-T* paths: Geology, v. 16, p. 1081–1084, doi: 10.1130/0091-7613(1988)016<1081:THOSZI>2.3.CO;2.

Evans, O.C., and Hanson, G.N., 1997, Late- to post-kinematic Archean granitoids of the S.W. Superior Province: Derivation through direct mantle melting, *in* de Wit, M.J., and Ashwal, L.D., eds., Greenstone Belts: Oxford, UK, Oxford University Press, p. 280–295.

Feng, R., and Kerrich, R., 1991, Single zircon age constraints on the tectonic juxtaposition of the Archean Abitibi greenstone belt and Pontiac subprovince, Quebec, Canada: Geochimica et Cosmochimica Acta, v. 55, p. 3437–3441, doi: 10.1016/0016-7037(91)90502-V.

Feng, R., and Kerrich, R., 1992, Geochemical evolution of granitoids from the Archean Abitibi southern volcanic zone and the Pontiac subprovince, Superior Province, Canada: Implications for tectonic history and source regions: Chemical Geology, v. 98, p. 23–70, doi: 10.1016/0009-2541(92)90090-R.

Franck, S., 1998, Evolution of the global mean heat flow over 4.6 Gyr: Tectonophysics, v. 291, p. 9–18, doi: 10.1016/S0040-1951(98)00027-4.

Frey, F.A., Coffin, M.F., Wallace, P.J., Weis, D., Zhao, X., Wise, S.W., Wähnert, V., Teagle, A.H., Saccocia, P.J., Reusch, D.N., Pringle, M.S., Nicolaysen, K.E., Neal, C.R., Müller, R.D., Moore, C.L., Mahoney, J.J., Keszthelyi, L., Inokuchi, H., Duncan, R.A., Delius, H., Damuth, J.E., Damasceno, D., Coxall, H.K., Borre, M.K., Boehm, F., Barling, J., Arndt, N.T., and Antretter, M., 2000, Origin and evolution of a submarine large igneous province: The Kerguelen Plateau and Broken Ridge, southern Indian Ocean: Earth and Planetary Science Letters, v. 176, p. 73–89, doi: 10.1016/S0012-821X(99)00315-5.

Glikson, A.Y., and Hickman, A.H., 1981, Geochemistry of Archaean Volcanic Successions, Eastern Pilbara Block, Western Australia: Australian Bureau of Mineral Resources, Geology and Geophysics Record 1981/36, 56 p.

Goldich, S.S., Wooden, J.L., Ankenbauer, G.A., Levy, T.M., and Suda, R.U., 1984, Origin of the Morton Gneiss, southwestern Minnesota: Part I. Lithology, *in* Morey, G.B., and Hanson, G.N., eds., Selected Studies of Archean Gneisses and Lower Proterozoic Rocks, Southern Canadian Shield: Geological Society of America Special Paper 182, p. 45–50.

Goutier, J., and Dion, C., 2004, Géologie et Minéralisation de la Sous-Province de La Grande, Baie-James. Québec Exploration 2004: Québec, Ministère des Ressources Naturelles, de la Faune et des Parcs, Report DV 2004–06, 21 p.

Green, M.G., Sylvester, P.J., and Buick, R., 2000, Growth and recycling of early Archaean continental crust: Geochemical evidence from the Coonterunah and Warrawoona Groups, Pilbara craton, Australia: Tectonophysics, v. 322, p. 69–88, doi: 10.1016/S0040-1951(00)00058-5.

Hamilton, W.B., 1998, Archaean magmatism and deformation were not the products of plate tectonics: Precambrian Research, v. 91, p. 143–179, doi: 10.1016/S0301-9268(98)00042-4.

Hamilton, W.B., 2003, An alternative Earth: GSA Today, v. 13, no. 11, p. 4–12, doi: 10.1130/1052-5173(2003)013<0004:AAE>2.0.CO;2.

Henry, P., Stevenson, R., and Gariépy, C., 1998, Late Archean mantle composition and crustal growth in the western Superior Province of Canada: Neodymium and lead isotopic evidence from the Wawa, Quetico, and Wabigoon subprovinces: Geochimica et Cosmochimica Acta, v. 62, p. 143–157, doi: 10.1016/S0016-7037(97)00324-4.

Henry, P., Stevenson, R., Larbi, Y., and Gariépy, C., 2000, Nd isotopic evidence for early to late Archean (3.4–2.7 Ga) crustal growth in the Western Superior Province (Ontario, Canada): Tectonophysics, v. 322, p. 135–151, doi: 10.1016/S0040-1951(00)00060-3.

Heubeck, C., and Lowe, D.R., 1994, Late syndepositional deformation and detachment tectonics in the Barberton greenstone belt, South Africa: Tectonics, v. 13, p. 1514–1536, doi: 10.1029/94TC01809.

Hickman, A.H., 1975, Precambrian structural geology of part of the Pilbara region: Western Australia Geological Survey Annual Report, v. 1974, p. 68–73.

Hickman, A.H., 1983, Geology of the Pilbara Block and Its Environs: Geological Survey of Western Australia Bulletin 127, 268 p.

Hickman, A.H., 1984, Archaean diapirism in the Pilbara block, Western Australia, *in* Kröner, A., and Greiling, R., eds., Precambrian Tectonics Illustrated: Stuttgart, E. Schweizerbarts'che Verlagsbuchhandlung, p. 113–127.

Hickman, A.H., 1997, A revision of the stratigraphy of Archaean greenstone successions in the Roebourne-Whundo area, west Pilbara: Western Australia Geological Survey Annual Review, v. 1996–97, p. 76–81.

Hickman, A.H., 2004, Two contrasting granite-greenstone terranes in the Pilbara craton, Australia: Evidence for vertical *and* horizontal tectonic regimes prior to 2900 Ma: Precambrian Research, v. 131, p. 153–172, doi: 10.1016/j.precamres.2003.12.009.

Hickman, A.H., and Van Kranendonk, M.J., 2004, Diapiric processes in the formation of Archaean continental crust, East Pilbara granite-greenstone terrane, Australia, *in* Eriksson, P.G., Altermann, W., Nelson, D.R., Mueller, W.U., and Catuneanu, O., eds., The Precambrian Earth: Tempos and Events: Amsterdam, Elsevier, p. 54–75.

Hollings, P., and Wyman, D.A., 1999, Trace element and Sm-Nd systematics of volcanic and intrusive rocks from the 3 Ga Lumby Lake greenstone belt, Superior Province; Evidence for Archean plume-arc interaction: Lithos, v. 46, p. 189–213, doi: 10.1016/S0024-4937(98)00062-0.

Hollings, P., Wyman, D.A., and Kerrich, R., 1999, Komatiite-basalt-rhyolite associations in northern Superior Province greenstone belts: Significance of plume-arc interaction in the generation of the protocontinental Superior Province: Lithos, v. 46, p. 137–161, doi: 10.1016/S0024-4937(98)00058-9.

Hurley, P.M., and Rand, J.R., 1969, Pre-drift continental nuclei: Science, v. 164, no. 3885, p. 1229–1242, doi: 10.1126/science.164.3885.1229.

Hynes, A., 2008, this volume, Some effects of a warmer Archean mantle on the geological record of plate tectonics, *in* Condie, K.C., and Pease, V., eds., When Did Plate Tectonics Begin on Planet Earth?: Geological Society of America Special Paper 440, doi: 10.1130/2008.2440(07).

Isnard, H., and Gariépy, C., 2004, Sm-Nd, Lu-Hf and Pb-Pb signatures of gneisses and granitoids from the La Grande belt; Extent of late Archean crustal recycling in the northeastern Superior Province, Canada: Geochimica et Cosmochimica Acta, v. 68, p. 1099–1113, doi: 10.1016/j.gca.2003.08.004.

Kamo, S.L., and Davis, D.W., 1994, Reassessment of Archaean crustal development in the Barberton Mountain Land, South Africa, based on U-Pb dating: Tectonics, v. 13, p. 167–192, doi: 10.1029/93TC02254.

Kato, Y., and Nakamura, K., 2003, Origin and global tectonic significance of early Archean cherts from the Marble Bar greenstone belt, Pilbara craton, Western Australia: Precambrian Research, v. 125, p. 191–243, doi: 10.1016/S0301-9268(03)00043-3.

Kisters, A.F.M., and Anhaeusser, C.R., 1995, Emplacement features of Archaean TTG plutons along the southern margin of the Barberton greenstone belt, South Africa: Precambrian Research, v. 75, p. 1–15, doi: 10.1016/0301-9268(95)00003-N.

Kisters, A.F.M., Stevens, G., Dziggel, A., and Armstrong, R.A., 2003, Extensional detachment faulting and core-complex formation in the southern Barberton granite-greenstone terrain, South Africa: Evidence for a 3.2 Ga orogenic collapse: Precambrian Research, v. 127, p. 355–378, doi: 10.1016/j.precamres.2003.08.002.

Kiyokawa, S., and Taira, A., 1998, The Cleaverville Group in the west Pilbara coastal granite-greenstone terrane of Western Australia: An example of a mid-Archaean immature oceanic island-arc succession: Precambrian Research, v. 88, p. 109–142, doi: 10.1016/S0301-9268(97)00066-1.

Kiyokawa, S., Taira, A., Byrne, T., Bowring, S., and Sano, Y., 2002, Structural evolution of the middle Archaean coastal Pilbara terrane, Western Australia: Tectonics, v. 21, p. 1044, doi: 10.1029/2001TC001296.

Kloppenburg, A., White, S.H., and Zegers, T.E., 2001, Structural evolution of the Warrawoona greenstone belt and adjoining granitoid complexes, Pilbara craton, Australia: Implications for Archaean tectonic processes: Precambrian Research, v. 112, p. 107–147, doi: 10.1016/S0301-9268(01)00172-3.

Korenaga, J., 2006, Archean geodynamics and the thermal evolution of Earth, *in* Benn, K., Mareschal, J.-C., and Condie, K. C., eds., Archean Geodynamics and Environments: Washington, D.C., American Geophysical Union, p. 7–32.

Kusky, T.M., and Bradley, D.C., 1999, Kinematics of melange fabrics: Examples and applications from the McHugh complex, Kenai Peninsula, Alaska: Journal of Structural Geology, v. 21, p. 1773–1796, doi: 10.1016/S0191-8141(99)00105-4.

Leclair, A., Labbé, J.-Y., Berclaz, A., David, J., Gosselin, C., Lacoste, P., Madore, L., Maurice, C., Roy, P., Sharma, K.N.M., and Simard, M., 2006, Government geoscience stimulates mineral exploration in the Superior Province, northern Quebec: Geoscience Canada, v. 22, p. 60–75.

Lin, S., Percival, J.A., and Skulski, T., 1996, Structural constraints on the tectonic evolution of a late Archean greenstone belt in the northeastern Superior Province, northern Quebec (Canada): Tectonophysics, v. 265, p. 151–167, doi: 10.1016/S0040-1951(96)00150-3.

Lin, S., Davis, D.W., Rotenberg, E., Corkery, M.T., and Bailes, A.H., 2006, Geological evolution of the northwestern Superior Province: Clues from geology, kinematics and geochronology in the Gods Lake Narrows area, Oxford Lake–Knee Lake–Gods Lake greenstone belt, Manitoba: Canadian Journal of Earth Sciences, v. 43, p. 749–765, doi: 10.1139/E06-068.

Lowe, D.R., 1994, Accretionary history of the Archaean Barberton greenstone belt (3.55–3.22 Ga), southern Africa: Geology, v. 22, p. 1099–1102, doi: 10.1130/0091-7613(1994)022<1099:AHOTAB>2.3.CO;2.

Lowe, D.R., 1999, Geological evolution of the Barberton greenstone belt and vicinity, *in* Lowe, D.R., and Byerly, G.R., eds., Geologic Evolution of the Barberton Greenstone Belt, South Africa: Geological Society of America Special Paper 329, p. 287–312.

Lowe, D.R., and Byerly, G.R., eds., 1999, Geologic Evolution of the Barberton Greenstone Belt, South Africa: Geological Society of America Special Paper 329, 319 p.

Lowe, D.R., Byerly, G.R., Ransom, B.L., and Nocita, B.R., 1985, Stratigraphic and sedimentological evidence bearing on structural repetition in early Archaean rocks of the Barberton greenstone belt, South Africa: Precambrian Research, v. 27, p. 165–186, doi: 10.1016/0301-9268(85)90011-7.

Maruyama, S., 1994, Plume tectonics: Geological Survey of Japan, v. 100, no. 1, p. 24–49.

McCall, J.G.H., 2003, A critique of the analogy between Archaean and Phanerozoic tectonics based on regional mapping of the Mesozoic-Cenozoic plate convergent zone in the Makran, Iran: Precambrian Research, v. 127, p. 5–18, doi: 10.1016/S0301-9268(03)00178-5.

Melnyk, M.J., Cruden, A.R., Davis, D.W., and Stern, R.A., 2006, U-Pb ages of magmatism constraining regional deformation in the Winnipeg River subprovince and Lake of the Woods greenstone belt: Evidence for Archean terrane accretion in the western Superior Province: Canadian Journal of Earth Sciences, v. 43, p. 967–993, doi: 10.1139/E06-035.

Miyashiro, A., 1961, Evolution of metamorphic belts: Journal of Petrology, v. 2, p. 277–311.

Moser, D., 1994, The geology and structure of the mid-crustal Wawa gneiss domain: A key to understanding tectonic variation with depth and time in the late Archean Abitibi-Wawa orogen: Canadian Journal of Earth Sciences, v. 31, p. 1064–1080.

Moyen, J.-F., Stevens, G., and Kisters, A.F.M., 2006, Record of mid-Archaean subduction from metamorphism in the Barberton terrain, South Africa: Nature, v. 442, p. 559–562, doi: 10.1038/nature04972.

Musacchio, G., White, D.J., Asudeh, I., and Thomson, C.J., 2004, Lithospheric structure and composition of the Archean western Superior Province from seismic refraction/wide-angle reflection and gravity modelling: Journal of Geophysical Research, v. 109, p. B03304, doi: 10.1029/2003JB002427.

Nutman, A.P., Friend, C.R.L., and Bennett, V.C., 2002, Evidence for 3650–3600 Ma assembly of the northern end of the Itsaq gneiss complex, Greenland: Implication for early Archaean tectonics: Tectonics, v. 21, 1005, doi: 10.1029/2000TC001203.

Ohta, H., Maruyama, S., Takahashi, E., Watanabe, Y., and Kato, Y., 1996, Field occurrence, geochemistry and petrogenesis of the Archaean mid-oceanic ridge basalts (A-MORBs) of the Cleaverville area, Pilbara craton, Western Australia: Lithos, v. 37, p. 199–221, doi: 10.1016/0024-4937(95)00037-2.

O'Neil, J., Maurice, C., Stevenson, R.K., Larocque, J., Cloquet, C., David, J., and Francis, D., 2007, The Geology of the 3.8 Ga Nuvvuagittuk (Porpoise Cove) greenstone belt, northeastern Superior Province, Canada, *in* Van Kranendonk, M.J., Smithies, R.H., and Bennett, V., eds., Earth's Oldest Rocks: Amsterdam, Elsevier, Developments in Precambrian Geology, v. 15, p. 219–250.

Operto, S., and Charvis, P., 1995, Kerguelen Plateau: A volcanic passive margin fragment?: Geology, v. 23, p. 137–140, doi: 10.1130/0091-7613(1995)023<0137:KPAVPM>2.3.CO;2.

Otto, A., Dziggel, A., Kisters, A.F.M., and Meyer, F.M., 2005, Polyphase gold mineralization in the structurally condensed granitoid-greenstone contact at the New Consort Gold Mine: Geological Society of America Abstracts with Programs, v. 37, no. 7, p. 96.

Parent, M., Leclair, A., David, J., and Sharma, K.N.M., 2001, Geology of the Lac Nedlouc area (NTS 34H et 24E): Québec, Ministère des Ressources Naturelles, Report RG 2000–09, 39 p.

Parks, J., Lin, S., Davis, D.W., and Corkery, M.T., 2006, Geochronological constraints on the history of the Island Lake greenstone belt and the relationship of terranes in the northwestern Superior Province: Canadian Journal of Earth Sciences, v. 43, p. 789–803, doi: 10.1139/E06-044.

Percival, J.A., 2007, Eo- to Mesoarchean terranes of the Superior Province and their tectonic context, *in* Van Kranendonk, M.J., Smithies, R.H., and Bennett, V., eds., Earth's Oldest Rocks: Amsterdam, Elsevier, Developments in Precambrian Geology, v. 15, p. 1065–1086.

Percival, J.A., and Mortensen, J.K., 2002, Water-deficient calc-alkaline plutonic rocks of northeastern Superior Province, Canada: Significance of charnockitic magmatism: Journal of Petrology, v. 43, p. 1617–1650, doi: 10.1093/petrology/43.9.1617.

Percival, J.A., and Skulski, T., 2000, Tectonothermal evolution of the northern Minto block, Superior Province, Québec, Canada: Canadian Mineralogist, v. 38, p. 345–378, doi: 10.2113/gscanmin.38.2.345.

Percival, J.A., Mortensen, J.K., Stern, R.A., Card, K.D., and Bégin, N.J., 1992, Giant granulite terranes of northeastern Superior Province: The Ashuanipi complex and Minto block: Canadian Journal of Earth Sciences, v. 29, p. 2287–2308.

Percival, J.A., Stern, R.A., Mortensen, J.K., Card, K.D., and Bégin, N.J., 1994, Minto block Superior Province: Missing link in deciphering tectonic assembly of the craton at 2.7 Ga: Geology, v. 22, p. 839–842, doi: 10.1130/0091-7613(1994)022<0839:MBSPML>2.3.CO;2.

Percival, J.A., Stern, R.A., and Skulski, T., 2001, Crustal growth through successive arc magmatism: Reconnaissance U-Pb SHRIMP data from the northeastern Superior Province, Canada: Precambrian Research, v. 109, p. 203–238, doi: 10.1016/S0301-9268(01)00148-6.

Percival, J.A., Bleeker, W., Cook, F.A., Rivers, T., Ross, G., and van Staal, C.R., 2004a, Panlithoprobe workshop IV: Intra-orogen correlations and comparative orogenic anatomy: Geoscience Canada, v. 31, p. 23–39.

Percival, J.A., McNicoll, V., Brown, J.L., and Whalen, J.B., 2004b, Convergent margin tectonics, central Wabigoon subprovince, Superior Province, Canada: Precambrian Research, v. 132, p. 213–244, doi: 10.1016/j.precamres.2003.12.016.

Percival, J.A., Sanborn-Barrie, M., Stott, G., Helmstaedt, H., Skulski, T., and White, D.J., 2006a, Tectonic evolution of the western Superior Province from NATMAP and Lithoprobe studies: Canadian Journal of Earth Sciences, v. 43, p. 1085–1117, doi: 10.1139/E06-062.

Percival, J.A., McNicoll, V., and Bailes, A.H., 2006b, Strike-slip juxtaposition of ca. 2.72 Ga juvenile arc and >2.98 Ga continent margin sequences and its implications for Archean terrane accretion, western Superior Province, Canada: Canadian Journal of Earth Sciences, v. 43, p. 895–927, doi: 10.1139/E06-039.

Pike, G., and Cas, R.A.F., 2002, Stratigraphic evolution of Archean volcanic rock–dominated rift basins from the Whim Creek belt, west Pilbara craton, Western Australia: International Association of Sedimentologists Special Publication 33, p. 213–234.

Reymer, A., and Schubert, G., 1984, Phanerozoic addition rates to the continental crust and crustal growth: Tectonics, v. 3, p. 63–77.

Reymer, A., and Schubert, G., 1986, Rapid growth of some major segments of continental crust: Geology, v. 14, p. 299–302, doi: 10.1130/0091-7613(1986)14<299:RGOSMS>2.0.CO;2.

Roscoe, S.M., and Donaldson, J.A., 1988, Uraniferous pyritic quartz pebble conglomerate and layered ultramafic intrusions in a sequence of quartzite, carbonate, iron formation and basalt of probable Archean age at Lac Sakami, Quebec: Geological Survey of Canada Paper 88-1C, p. 117–121.

Sanborn-Barrie, M., and Skulski, T., 1999, 2.7 Ga tectonic assembly of continental margin and oceanic terranes in the Savant Lake–Sturgeon Lake greenstone belt, Ontario: Geological Survey of Canada Current Research 1999-C, p. 209–220.

Sanborn-Barrie, M., and Skulski, T., 2006, Sedimentary and structural evidence for 2.7 Ga continental arc-oceanic arc collision in the Savant-Sturgeon greenstone belt, western Superior Province, Canada: Canadian Journal of Earth Sciences, v. 43, p. 995–1030, doi: 10.1139/E06-060.

Sanborn-Barrie, M., Skulski, T., and Parker, J.R., 2001, Three Hundred Million Years of Tectonic History Recorded by the Red Lake Greenstone Belt, Ontario: Geological Survey of Canada Current Research 2001–C19, 19 p.

Sanborn-Barrie, M., Rogers, N., Skulski, T., Parker, J.R., McNicoll, V., and Devaney, J., 2004, Geology and tectonostratigraphic assemblages, east Uchi, Red Lake and Birch-Uchi belts, Ontario: Geological Survey of Canada Open-File 4256 and Ontario Geological Survey Preliminary Map P-3460, scale 1:250,000.

Sandiford, M., Van Kranendonk, M.J., and Bodorkos, S., 2004, Conductive incubation and the origin of dome-and-keel structure in Archean granite-greenstone terrains: A model based on the eastern Pilbara craton, Western Australia: Tectonics, v. 23, p. TC1009, doi: 10.1029/2002TC001452.

Sasseville, C., Tomlinson, K.Y., Hynes, A., and McNicoll, V., 2006, Stratigraphy, structure, and geochronology of the 3.0–2.7 Ga Wallace Lake greenstone belt, western Superior Province, SE Manitoba, Canada: Canadian Journal of Earth Sciences, v. 43, p. 929–945, doi: 10.1139/E06-041.

Sawyer, E.W., and Benn, K., 1993, Structure of the high-grade Opatica belt and adjacent low-grade Abitibi subprovince, Canada; an Archean mountain front: Journal of Structural Geology, v. 15, p. 1443–1458, doi: 10.1016/0191-8141(93)90005-U.

Schmitz, M.D., Bowring, S.A., Southwick, D.L., Boerboom, T.J., and Wirth, K.R., 2006, High-precision U-Pb geochronology in the Minnesota River Valley subprovince and its bearing on Neoarchean to Paleoproterozoic evolution of the southern Superior Province: Geological Society of America Bulletin, v. 118, p. 82–93, doi: 10.1130/B25725.1.

Schoene, B., and Bowring, S.A., 2007, Determining accurate temperature-time paths from U-Pb thermochronology: An example from the Kaapvaal craton, southern Africa: Geochimica et Cosmochimica Acta, v. 71, p. 165–185, doi: 10.1016/j.gca.2006.08.029.

Şengör, A.C., and Natal'in, B.A., 1996, Turkic-type orogeny and its role in the making of continental crust: Annual Review of Earth and Planetary Sciences, v. 24, p. 263–337, doi: 10.1146/annurev.earth.24.1.263.

Shirey, S.B., and Hanson, G.N., 1984, Mantle-derived Archaean monzodiorites and trachyandesites: Nature, v. 310, p. 222–224, doi: 10.1038/310222a0.

Shirey, S., Kamber, B., Whitehouse, M., Mueller, P., and Basu, A., 2008, this volume, A review of the isotopic evidence for mantle and crustal processes in the Hadean and Archean: Implications for the onset plate tectonic subduction, *in* Condie, K.C., and Pease, V., eds., When Did Plate Tectonics Begin on Planet Earth?: Geological Society of America Special Paper 440, doi: 10.1130/2008.2440(01).

Sims, P.K., and Day, W.C., 1993, The Great Lakes tectonic zone—revisited: U.S. Geological Survey Bulletin 1904-S, p. 1–11.

Sims, P.K., Kotov, A.B., Neymark, L.A., and Peterman, Z.E., 1997, Nd isotopic evidence for middle and early Archean crust in the Wawa subprovince of Superior Province, Michigan, U.S.A.: Geological Association of Canada Abstract Volume, v. 23, p. A137.

Skulski, T., and Percival, J.A., 1996, Allochthonous 2.78 Ga oceanic plateau slivers in a 2.72 Ga continental arc sequence; Vizien greenstone belt, northeastern Superior Province, Canada: Lithos, v. 37, p. 163–179, doi: 10.1016/0024-4937(95)00035-6.

Skulski, T., Percival, J.A., and Stern, R.A., 1996, Archean crustal evolution in the central Minto block, northern Quebec, *in* Radiogenic Age and Isotopic Studies, Report 9: Geological Survey of Canada, Current Research 1995-F, p. 17–31.

Skulski, T., Sanborn-Barrie, M., and Stern, R.A., 1998a, Did the Sturgeon Lake belt form near a continental margin?: Vancouver, Lithoprobe Secretariat, University of British Columbia, Lithoprobe Report 65, p. 87–89.

Skulski, T., Stern, R.A., and Ciesielski, A., 1998b, Timing and sources of granitic magmatism, Bienville subprovince, northern Quebec [abstract]: Geological Association of Canada–Mineralogical Association of Canada Annual Meeting Abstract Volume, v. 23, p. A-175.

Skulski, T., Percival, J.A., Whalen, J.B., and Stern, R.A., 1999, Archean crustal evolution in the northern Superior Province: Tectonic and magmatic processes in crustal growth: A pan-Lithoprobe perspective: Vancouver, Lithoprobe Secretariat, University of British Columbia, Lithoprobe Report 75, p. 128–129.

Skulski, T., Corkery, M.T., Stone, D., Whalen, J.B., and Stern, R.A., 2000, Geological and geochronological investigations in the Stull Lake–Edmund Lake greenstone belt and granitoid rocks of the northwestern Superior Province: Report of Activities 2000: Manitoba Industry, Trade and Mines, Manitoba Geological Survey, p. 117–128.

Sleep, N., 1992, Archean plate tectonics: What can be learned from continental geology?: Canadian Journal of Earth Sciences, v. 29, no. 10, p. 2066–2071.

Smith, D., 1984, Coesite in clinopyroxene in the Caledonides and its implications for geodynamics: Nature, v. 310, no. 5979, p. 641–644, doi: 10.1038/310641a0.

Smith, J.B., 2003, The episodic development of intermediate to silicic volcano-plutonic suites in the Archaean West Pilbara, Australia: Chemical Geology, v. 194, p. 275–295, doi: 10.1016/S0009-2541(02)00243-7.

Smith, J.B., Barley, M.E., Groves, D.I., Krapez, B., McNaughton, N.J., Bickle, M.J., and Chapman, H.J., 1998, The Sholl shear zone, West Pilbara: Evidence for a domain boundary structure from integrated tectonic analyses, SHRIMP U-Pb dating and isotopic and geochemical data of granitoids: Precambrian Research, v. 88, p. 143–171, doi: 10.1016/S0301-9268(97)00067-3.

Smithies, R.H., and Champion, D.C., 2000, The Archaean high-Mg diorite suite: Links to tonalite-trondhjemite-granodiorite magmatism and implications for early Archaean crustal growth: Journal of Petrology, v. 41, p. 1653–1671, doi: 10.1093/petrology/41.12.1653.

Smithies, R.H., Champion, D.C., and Sun, S.-S., 2004a, The case for Archaean boninites: Contributions to Mineralogy and Petrology, v. 147, p. 705–721, doi: 10.1007/s00410-004-0579-x.

Smithies, R.H., Champion, D.C., and Sun, S.-S., 2004b, Evidence for early LREE-enriched mantle source regions: Diverse magmas from the ca. 3.0 Ga Mallina Basin, Pilbara craton, NW Australia: Journal of Petrology, v. 45, p. 1515–1537, doi: 10.1093/petrology/egh014.

Smithies, R.H., Van Kranendonk, M.J., and Champion, D.C., 2005a, It started with a plume—Early Archaean basaltic proto-continental crust: Earth and Planetary Science Letters, v. 238, p. 284–297, doi: 10.1016/j.epsl.2005.07.023.

Smithies, R.H., Champion, D.C., Van Kranendonk, M.J., Howard, H.M., and Hickman, A.H., 2005b, Modern-style subduction processes in the Mesoarchaean: Geochemical evidence from the 3.12 Ga Whundo intraoceanic arc: Earth and Planetary Science Letters, v. 231, p. 221–237, doi: 10.1016/j.epsl.2004.12.026.

Smithies, R.H., Champion, D.C., Van Kranendonk, M.J., and Hickman, A.H., 2007a, Geochemistry of volcanic rocks of the northern Pilbara craton: Western Australia Geological Survey, Report 104, 53 p.

Smithies, R.H., Champion, D.C., and Van Kranendonk, M.J., 2007b, The oldest well-preserved felsic volcanic rocks on Earth: Geochemical clues to the early evolution of the Pilbara Supergroup and implications for the growth of a Paleoarchean protocontinent, *in* Van Kranendonk, M.J., Smithies, R.H., and Bennett, V., eds., Earth's Oldest Rocks: Amsterdam, Elsevier, Developments in Precambrian Geology, v. 15, p. 339–367.

Smithies, R.H., Van Kranendonk, M.J., and Champion, D.C., 2007c, The Mesoarchaean emergence of modern style subduction, *in* Maruyama, S., and Santosh, M., eds., Island Arcs: Past and Present: Gondwana Research, v. 11, p. 50–68.

Sproule, R.A., Lesher, C.M., Ayer, J.A., Thurston, P.C., and Herzberg, C.T., 2002, Spatial and temporal variations in the geochemistry of komatiites and komatiitic basalts in the Abitibi greenstone belt: Precambrian Research, v. 115, p. 153–186, doi: 10.1016/S0301-9268(02)00009-8.

Stern, R.A., Percival, J.A., and Mortensen, J.K., 1994, Geochemical evolution of the Minto block: A 2.7 Ga continental magmatic arc built on the Superior proto-craton: Precambrian Research, v. 65, p. 115–153, doi: 10.1016/0301-9268(94)90102-3.

Stern, R.J., 2005, Evidence from ophiolites, blueschists, and ultrahigh-pressure metamorphic terranes that the modern episode of subduction tectonics began in Neoproterozoic time: Geology, v. 33, p. 557–560, doi: 10.1130/G21365.1.

Stevens, G., Droop, G.T.R., Armstrong, R.A., and Anhaeusser, C.R., 2002, Amphibolite-facies metamorphism in the Schapenburg schist belt: A record of the mid-crustal response to ~3.23 Ga terrane accretion in the Barberton greenstone belt: South African Journal of Geology, v. 105, p. 271–284, doi: 10.2113/1050271.

Stevenson, R.K., 1995, Crust and mantle evolution in the late Archean: Evidence from a Sm-Nd isotopic study of the North Spirit Lake greenstone belt, northwestern Ontario: Geological Society of America Bulletin, v. 107, p. 1458–1467, doi: 10.1130/0016-7606(1995)107<1458:CAMEIT>2.3.CO;2.

Stevenson, R.K., and Patchett, P.J., 1990, Implications for the evolution of continental crust from Hf isotope systematics of Archean detrital zircons: Geochimica et Cosmochimica Acta, v. 54, p. 1683–1697, doi: 10.1016/0016-7037(90)90400-F.

Stevenson, R.K., David, J., and Parent, M., 2006, Crustal evolution of the western Minto block, northern Superior Province, Canada: Precambrian Research, v. 145, p. 229–242, doi: 10.1016/j.precamres.2005.12.004.

Stone, D., 1998, Precambrian Geology of the Berens River Area, Northwest Ontario: Ontario Geological Survey, Open File Report 5963, 115 p.

Stone, D., 2005, Geology of Northern Superior Area, Ontario: Ontario Geological Survey Open-File Report 6140, 94 p.

Stone, D., Tomlinson, K.Y., Davis, D.W., Fralick, P., Hallé, J., Percival, J.A., and Pufahl, P., 2002, Geology and Tectonostratigraphic Assemblages, South-Central Wabigoon Subprovince: Geological Survey of Canada Open-File 4284 and Ontario Geological Survey Preliminary Map P-3448, scale 1:250,000.

Stone, D., Corkery, M.T., Hallé, J., Ketchum, J., Lange, M., Skulski, T., and Whalen, J., 2004, Geology and tectonostratigraphic assemblages, eastern Sachigo subprovince, Ontario and Manitoba: Manitoba Geological Survey Open-File OF2003-2 or Geological Survey of Canada Open-File 1582 and Ontario Geological Survey Preliminary Map P-3462, scale 1:250,000.

Stott, G.M., 1997, The Superior Province, Canada, *in* de Wit, M.J., and Ashwal, L.D., eds., Greenstone Belts: Oxford Monograph on Geology and Geophysics 35, p. 480–507.

Stott, G.M., and Corfu, F., 1991, Uchi subprovince, *in* Thurston, P.C., Williams, H.R., Sutcliffe, R.H., and Stott, G.M., eds., Geology of Ontario: Ontario Geological Survey Special Volume 4, Part 1, p. 145–238.

Stott, G.M., Corfu, F., Breaks, F.W., and Thurston, P.C., 1989, Multiple orogenesis in northwestern Superior Province: Geological Association of Canada Abstracts, v. 14, p. A56.

Stott, G.M., Davis, D.W., Parker, J.R., Straub, K.J., and Tomlinson, K.Y., 2002, Geology and tectonostratigraphic assemblages, eastern Wabigoon subprovince, Ontario: Geological Survey of Canada Open File 4285 and Ontario Geological Survey Map P-3449, scale 1:250,000.

Sun, S., and Hickman, A.H., 1998, New Nd-isotopic and geochemical data from the west Pilbara—Implications for Archaean crustal accretion and shear zone development: Australian Geological Survey Organisation, Research Newsletter, June 1998.

Syme, E.C., Corkery, M.T., Bailes, A.H., Lin, S., Skulski, T., and Stern, R.A., 1999, Towards a new tectonostratigraphy for the Knee Lake greenstone belt, Sachigo subprovince, Manitoba, *in* Harrap, R.M., and Helmstaedt, H., eds., Western Superior Transect Fifth Annual Workshop: Vancouver, Lithoprobe Secretariat, University of British Columbia, Lithoprobe Report 70, p. 124–131.

Taylor, B., and Martinez, F., 2003, Back-arc basin basalt systematics: Earth and Planetary Science Letters, v. 210, p. 481–497.

Thorpe, R.A., Hickman, A.H., Davis, D.W., Mortensen, J.K., and Trendall, A.F., 1992, U-Pb zircon geochronology of Archaean felsic units in the Marble Bar region, Pilbara craton, Western Australia: Precambrian Research, v. 56, p. 169–189, doi: 10.1016/0301-9268(92)90100-3.

Thurston, P.C., Osmani, I.A., and Stone, D., 1991, Northwestern Superior Province: Review and terrane analysis, *in* Thurston, P.C., Williams, H.R., Sutcliffe, R.H., and Stott, G.M., eds., Geology of Ontario: Ontario Geological Survey Special Volume 4, Part 1, p. 81–144.

Tomlinson, K.Y., Stevenson, R.K., Hughes, D.J., Hall, R.P., Thurston, P.C., and Henry, P., 1998, The Red Lake greenstone belt, Superior Province: Evidence of plume-related magmatism at 3 Ga and evidence of an older enriched source: Precambrian Research, v. 89, p. 59–76, doi: 10.1016/S0301-9268(97)00078-8.

Tomlinson, K.Y., Hughes, D.J., Thurston, P.C., and Hall, R.P., 1999, Plume magmatism and crustal growth at 2.9 to 3.0 Ga in the Steep Rock and Lumby Lake area, western Superior Province: Lithos, v. 46, p. 103–136, doi: 10.1016/S0024-4937(98)00057-7.

Tomlinson, K.Y., Davis, D.W., Stone, D., and Hart, T., 2003, New U-Pb and Nd isotopic evidence for crustal recycling and Archean terrane development in the south-central Wabigoon subprovince, Canada: Contributions to Mineralogy and Petrology, v. 144, p. 684–702.

Tomlinson, K.Y., Stott, G.M., Percival, J.A., and Stone, D., 2004, Basement terrane correlations and crustal recycling in the western Superior Province: Nd isotopic character of granitoid and felsic volcanic rocks in the Wabigoon subprovince, N. Ontario, Canada: Precambrian Research, v. 132, p. 245–274, doi: 10.1016/j.precamres.2003.12.017.

Turek, A., Sage, R.P., and Van Schmus, W.R., 1992, Advances in the U-Pb zircon geochronology of the Michipicoten greenstone belt, Superior Province, Ontario: Canadian Journal of Earth Sciences, v. 29, p. 1154–1165.

Ueno, Y., Maruyama, S., Isozaki, Y., and Yurimoto, Y., 2001, Early Archean (ca. 3.5 Ga) microfossils and $^{13}$C-depleted carbonaceous matter in the North Pole area, Western Australia: Field occurrence and geochemistry, *in* Nakashima, S., Maruyama, S., Brack, A., and Windley, B.F., eds., Geochemistry and the Origin of Life: Tokyo, Universal Academy Press, p. 203–236.

Van der Lee, S., and Frederiksen, A., 2005, Surface wave tomography applied to the North American upper mantle, *in* Levander, A., and Nolet, G., eds., Seismic Earth—Array Analysis of Broadband Seismograms: American Geophysical Union Geophysical Monograph 157, p. 67–80.

van Haaften, W.M., and White, S.H., 1998, Evidence for multiphase deformation in the Archean basal Warrawoona Group in the Marble Bar area, East Pilbara, Western Australia: Precambrian Research, v. 88, p. 53–66, doi: 10.1016/S0301-9268(97)00063-6.

van Hunen, J., Davies, G., Hynes, A., and van Keken, P., 2008, this volume, Tectonics of the early Earth: Some geodynamic considerations, *in* Condie, K.C., and Pease, V., eds., When Did Plate Tectonics Begin on Planet Earth?: Geological Society of America Special Paper 440, doi: 10.1130/2008.2440(08).

Van Kranendonk, M.J., 2000, Geology of the North Shaw 1:100,000 Sheet: Western Australia Geological Survey, Geological Series Explanatory Notes, scale 1:100,000, 86 p.

Van Kranendonk, M.J., 2004a, Archaean tectonics 2004: A review: Precambrian Research, v. 131, p. 143–151, doi: 10.1016/j.precamres.2003.12.008.

Van Kranendonk, M.J., 2004b, Geology of the Carlindie 1:100,000 Sheet: Western Australia Geological Survey, Geological Series Explanatory Notes, scale 1:100,000 45 p.

Van Kranendonk, M.J., and Pirajno, F., 2004, Geological setting and geochemistry of metabasalts and alteration zones associated with hydrothermal chert ± barite deposits in the ca. 3.45 Ga Warrawoona Group, Pilbara craton, Australia: Geochemistry, Exploration, Environment, Analysis, v. 4, p. 253–278.

Van Kranendonk, M.J., Hickman, A.H., and Collins, W.J., 2001, Comment on "Evidence for multiphase deformation in the Archaean basal Warrawoona Group in the Marble Bar area, East Pilbara, Western Australia": Precambrian Research, v. 105, p. 73–78, doi: 10.1016/S0301-9268(00)00094-2.

Van Kranendonk, M.J., Hickman, A.H., Smithies, R.H., Nelson, D.N., and Pike, G., 2002, Geology and tectonic evolution of the Archaean North Pilbara terrain, Pilbara craton, Western Australia: Economic Geology and the Bulletin of the Society of Economic Geologists, v. 97, p. 695–732.

Van Kranendonk, M.J., Collins, W.J., Hickman, A.H., and Pawley, M.J., 2004, Critical tests of vertical vs horizontal tectonic models for the Archaean East Pilbara granite-greenstone terrane, Pilbara craton, Western Australia: Precambrian Research, v. 131, p. 173–211, doi: 10.1016/j.precamres.2003.12.015.

Van Kranendonk, M.J., Hickman, A.H., Smithies, R.H., Williams, I.R., Bagas, L., and Farrell, T.R., 2006, Revised Lithostratigraphy of Archean Supracrustal and Intrusive Rocks in the Northern Pilbara Craton, Western Australia: Western Australia Geological Survey Record 2006/15, 57 p.

Van Kranendonk, M.J., Smithies, R.H., Hickman, A.H., and Champion, D.C., 2007, Secular tectonic evolution of Archean continental crust: Interplay between horizontal and vertical processes in the formation of the Pilbara craton, Australia: Terra Nova, v. 19, p. 1–38, doi: 10.1111/j.1365-3121.2006.00723.x.

Viljoen, M.J., and Viljoen, R.P., 1969a, An introduction to the geology of the Barberton granite-greenstone terrain: Geological Society of South Africa Special Publication 2, p. 9–28.

Viljoen, M.J., and Viljoen, R.P., 1969b, The geology and geochemistry of the lower ultramafic unit of the Onverwacht Group and a proposed new class of igneous rocks: Geological Society of South Africa Special Publication 2, p. 55–86.

von Huene, R., and Scholl, D.W., 1991, Observations at convergent margins concerning sediment subduction, subduction erosion, and the growth of continental crust: Reviews of Geophysics, v. 29, p. 279–316.

Westraat, J.D., Kisters, A.F.M., Poujol, M., and Stevens, G., 2005, Transcurrent shearing, granite sheeting and the incremental construction of the tabular 3.1 Ga Mpuluzi batholith, Barberton granite-greenstone terrain, South Africa: Journal of the Geological Society of London, v. 162, p. 373–388.

Whalen, J.B., Percival, J.A., McNicoll, V., and Longstaffe, F.J., 2002, A mainly crustal origin for tonalitic granitoid rocks, Superior Province, Canada: Implications for late Archean tectonomagmatic processes: Journal of Petrology, v. 43, p. 1551–1570, doi: 10.1093/petrology/43.8.1551.

Whalen, J.B., Percival, J.A., McNicoll, V., and Longstaffe, F.J., 2003, Intra-oceanic production of continental crust in a Th-depleted ca. 3.0 Ga arc complex, western Superior Province, Canada: Contributions to Mineralogy and Petrology, v. 146, p. 78–99, doi: 10.1007/s00410-003-0484-8.

Whalen, J.B., Percival, J.A., McNicoll, V., and Longstaffe, F.J., 2004a, Geochemical and isotopic (Nd-O) evidence bearing on the origin of late- to post-orogenic high-K granitoid rocks in the western Superior Province: Implications for late Archean tectonomagmatic processes: Precambrian Research, v. 132, p. 303–326, doi: 10.1016/j.precamres.2003.11.007.

Whalen, J.B., McNicoll, V., and Longstaffe, F.J., 2004b, Juvenile ca. 2.735–2.720 Ga high- and low-Al tonalitic plutons: Implications for TTG and VMS petrogenesis, western Superior Province, Canada: Precambrian Research, v. 132, p. 275–301, doi: 10.1016/j.precamres.2004.02.008.

White, D.J., Musacchio, G., Helmstaedt, H.H., Harrap, R.M., Thurston, P.C., van der Velden, A., and Hall, K., 2003, Images of a lower-crustal oceanic slab: Direct evidence for tectonic accretion in the Archean western Superior Province: Geology, v. 31, p. 997–1000, doi: 10.1130/G20014.1.

Williams, I.S., and Collins, W.J., 1990, Granite-greenstone terranes in the Pilbara block, Australia, as coeval volcano-plutonic complexes; evidence from U-Pb zircon dating of the Mount Edgar batholith: Earth and Planetary Science Letters, v. 97, p. 41–53, doi: 10.1016/0012-821X(90)90097-H.

Williams, P.R., and Currie, K.L., 1993, Character and regional significance of the sheared Archaean granite-greenstone contact near Leonora, Western Australia: Precambrian Research, v. 62, p. 343–365, doi: 10.1016/0301-9268(93)90029-2.

Wilson, J.T., 1966, Did the Atlantic close and then re-open?: Nature, v. 211, p. 676–681, doi: 10.1038/211676a0.

Wyman, D., 1999, A 2.7 Ga depleted tholeiite suite: Evidence for plume-arc interaction in the Abitibi greenstone belt, Canada: Precambrian Research, v. 97, p. 27–42, doi: 10.1016/S0301-9268(99)00018-2.

Wyman, D.A., Kerrich, R., and Polat, A., 2002, Assembly of Archean cratonic mantle lithosphere and crust: Plume-arc interaction in the Abitibi-Wawa subduction-accretion complex: Precambrian Research, v. 115, p. 37–62, doi: 10.1016/S0301-9268(02)00005-0.

Zegers, T.E., White, S.H., de Keijzer, M., and Dirks, P., 1996, Extensional structures during deposition of the 3460 Ma Warrawoona Group in the eastern Pilbara craton, Western Australia: Precambrian Research, v. 80, p. 89–105, doi: 10.1016/S0301-9268(96)00007-1.

Zegers, T.E., Nelson, D.R., Wijbrans, J.R., and White, S.H., 2001, SHRIMP U-Pb zircon dating of Archean core complex formation and pancratonic strike-slip deformation in the East Pilbara granite-greenstone terrain: Tectonics, v. 20, p. 883–908, doi: 10.1029/2000TC001210.

Manuscript Accepted by the Society 14 August 2007

Printed in the USA

The Geological Society of America
Special Paper 440
2008

# Seismic images of Paleoproterozoic microplate boundaries in the Fennoscandian Shield

**Annakaisa Korja***
**Pekka J. Heikkinen**
*Institute of Seismology, P.O. Box 68, FIN-00014 University of Helsinki, Finland*

## ABSTRACT

**Deep seismic-reflection data from the BABEL and FIRE profiles across the Fennoscandian Shield image Svecofennian crust that is made up of a collage of terranes. The data suggest sequential accretion of island arcs and microcontinents to the Karelian craton (1.9–1.8 Ga). These accretionary events may have caused temporary changes to arc geometries, accretionary episodes or collisional phases, and westward growth of the continent. The accreted terranes experienced gravitational collapse that stabilized the crust and exhumed the medium- to high-grade rocks to their present positions. The structures froze after isostatic balance was achieved and thus have been protected from later tectonic deformation. The accretionary growth period was terminated by continent-continent collision, after which it was possible for the Wilson cycle to operate at the margin of the newly formed continent.**

**As in modern accretionary systems, several tectonic environments are found. These include paleosubduction, obduction, continental transforms, collision of hot and cold terranes, and collapse of hot and cold collisions, which are all supportive of plate tectonics operating in the Paleoproterozoic. From the different collision zones, the following tectonic units can be recognized: hinterland-foreland fold-and-thrust belts, metamorphic cores, accreted arcs and basins, and foreland fold-and-thrust belts. The metamorphic cores are associated with granitoid complexes and/or core complexes. The plate-tectonic theory together with gravitational balancing is a viable model to explain the evolution of the Svecofennian orogen.**

**Keywords:** crustal structure, Paleoproterozoic, Fennoscandia, reflection seismics.

## INTRODUCTION

Plate-tectonic reconstructions (Rogers, 1996; Zhao et al., 2002; Stampfli and Borel, 2002) suggest that large-scale continent-continent collisions are episodic and related to supercontinent formation. The intervening periods are characterized by rifting, subduction, and accretionary orogenies. In the intervening periods, new crust is being formed, whereas in the supercontinent periods the crust is mostly reworked.

In modern plate-tectonic settings, new crust is formed at subduction zones at convergent margins. The convergent margin may be a long-lived one, like western North America, where subduction and accretion have continued for at least the last 650 m.y. (Coney et al., 1980; Scotese, 2001), or it may be more restricted in time and shut off by continent-continent collision, like the Alpine-Himalayan system (Stampfli and Borel, 2002). Small

*Annakaisa.Korja@helsinki.fi

Korja, A., and Heikkinen, P.J., 2008, Seismic images of Paleoproterozoic microplate boundaries in the Fennoscandian Shield, *in* Condie, K.C., and Pease, V., eds., When Did Plate Tectonics Begin on Planet Earth?: Geological Society of America Special Paper 440, p. 229–248, doi: 10.1130/2008.2440(11). For permission to copy, contact editing@geosociety.org. 

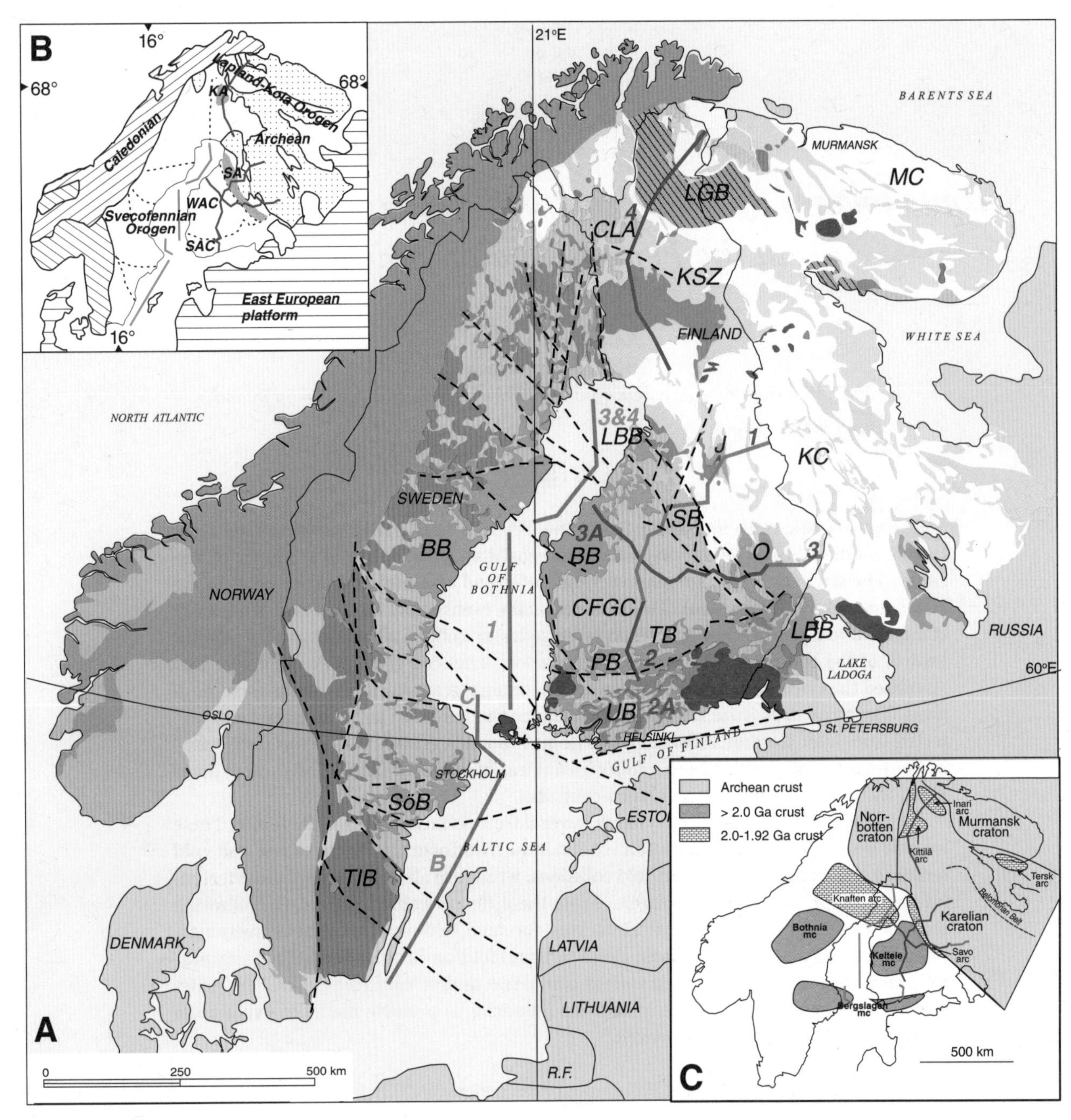
B
16°
68°
Caledonian
Lapland-Kola Orogen
KA
Archean
SA
WAC
Svecofennian Orogen
SAC
East European platform
A
21°E
BARENTS SEA
MURMANSK
MC
LGB
CLA
KSZ
FINLAND
WHITE SEA
NORTH ATLANTIC
3&4
LBB
J
KC
SWEDEN
SB
3A
BB
O
GULF OF BOTHNIA
NORWAY
CFGC
TB
LBB
RUSSIA
PB
LAKE LADOGA
60°E
OSLO
UB
2A
HELSINKI
St. PETERSBURG
GULF OF FINLAND
STOCKHOLM
SöB
BALTIC SEA
TIB
DENMARK
LATVIA
LITHUANIA
R.F.
0 250 500 km
C
Archean crust
> 2.0 Ga crust
2.0-1.92 Ga crust
Norrbotten craton
Inari arc
Murmansk craton
Kittilä arc
Tersk arc
Knaften arc
Karelian craton
Bothnia mc
Keitele mc
Savo arc
Bergslagen mc
500 km

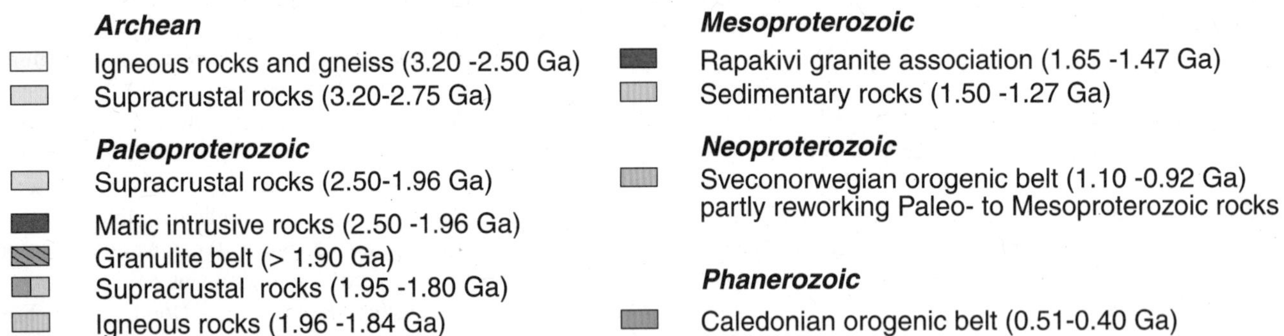
Archean
Igneous rocks and gneiss (3.20 -2.50 Ga)
Supracrustal rocks (3.20-2.75 Ga)
Paleoproterozoic
Supracrustal rocks (2.50-1.96 Ga)
Mafic intrusive rocks (2.50 -1.96 Ga)
Granulite belt (> 1.90 Ga)
Supracrustal rocks (1.95 -1.80 Ga)
Igneous rocks (1.96 -1.84 Ga)
Granite and migmatite (1.85 -1.75 Ga)
Igneous rocks (1.85 -1.66 Ga)
Mesoproterozoic
Rapakivi granite association (1.65 -1.47 Ga)
Sedimentary rocks (1.50 -1.27 Ga)
Neoproterozoic
Sveconorwegian orogenic belt (1.10 -0.92 Ga) partly reworking Paleo- to Mesoproterozoic rocks
Phanerozoic
Caledonian orogenic belt (0.51-0.40 Ga)
Alkaline intrusions
Phanerozoic sedimentary rocks

suspect terranes entering the system cause temporary changes in the plate movement geometries. The different tectonic histories of the accreting terranes and colliding continents result in a variety of tectono-magmatic events within one collisional system. For example, the Pyrenees display features of cold collision (Beaumont and Quinlan, 1994; Beaumont et al., 2000), the Tibetan Plateau is undergoing migmatite formation after hot orogeny (Vanderhaeghe and Teyssier, 2001a, 2001b; Jamieson et al., 2004), and the North American Cordilleras show signs of tectonic switching (Collins, 2002).

Hoffman (1988), Condie (2002), Cawood et al. (2006), and Artemieva (2006) all stressed the importance of the Paleoproterozoic as a time of rapid crustal growth and assembly of the Precambrian Shields—the cores of the modern continents. Plate-tectonic models involving small plates have been invoked to explain the lithological, metamorphic, and structural observations of the juvenile Paleoproterozoic orogens (Hietanen, 1975; Hoffman, 1988; Lahtinen et al., 2005; Korja et al., 2006b; Tyson et al., 2002; Santosh et al., 2006). Seismic images of the Paleoproterozoic orogens (BABEL Working Group, 1993; Hajnal et al., 1996; Cook et al., 1999; White et al., 2000) are similar to those found in modern orogens, indicating active plate-tectonic processes in the Paleoproterozoic. Minor differences in the geochemistry of the lithological associations and increased thickness of the crust and lithosphere have been attributed to higher heat flow in the Precambrian (Abbott and Mooney, 1995; Abbott et al., 2000). However, some doubts on the uniformitarian applicability of plate-tectonic theory for the Paleoproterozoic and Archean have been posed by Stern (2005), who pointed out that the Proterozoic differs from the Phanerozoic by the absence of eclogites and blueschists as well as the scarcity of ophiolites, and by Bédard (2006), who suggested that delamination and mantle turnovers characterized the early Precambrian.

The Svecofennian orogen in the Fennoscandian Shield is a typical example of a Paleoproterozoic juvenile domain. It is characterized by an abundance of arc-related granitoids and lesser amounts of volcanic rocks, low-pressure–high-temperature regional metamorphism (Korsman et al., 1999), thick crust, especially thick mafic lower crust, and deep electrical conductors marking terrane boundaries (Korja et al., 1993). In this paper, we will show seismic images of the Paleoproterozoic Svecofennian orogen, where juvenile crustal blocks have accreted to a continental margin. Like in modern accretionary systems, varying tectonic environments are found. These include subduction, obduction, transforms, collision of hot and cold terranes, and collapse of hot and cold collisions, all of which support the operation of plate tectonics in the Paleoproterozoic.

## GEOLOGICAL OUTLINE

The East European craton is a typical example of a stable Precambrian continental nucleus. It is composed of Fennoscandian, Ukrainian, and Volgo-Uralian parts that where joined together in Paleoproterozoic orogenies (Gorbatschev and Bogdanova, 1993; Lahtinen et al., 2005). The Fennoscandian Shield (Fig. 1A) is composed of Archean to Phanerozoic terranes. Its Precambrian nucleus was assembled from Archean and Paleoproterozoic terranes in the Svecofennian and Lapland-Kola orogenies, ca. 1.9–1.8 Ga (Gaál and Gorbatschev, 1987; Bogdanova et al., 2006; Daly et al., 2006).

The Svecofennian orogen is situated west of the Archean Karelian craton, and the Lapland-Kola orogen is situated between the Karelian and Murmansk cratons (Fig. 1B). The similarity of lithological associations, geochemical compositions, metamorphic histories, and geophysical structures to those of modern convergent-margin environments has inspired many plate-tectonic explanations for the formation of the Svecofennian and Lapland-Kola orogens (Hietanen, 1975; Gaál and Gorbatschev, 1987; Lahtinen, 1994; Korja, 1995; Gorbatschev and Bogdanova, 1993; Nironen, 1997; Daly et al., 2001, 2006). Recently, Lahtinen et al. (2005) and Korja et al. (2006b) explained the rapid crustal evolution and coexistence of multiple tectonic environments with a model where 2.0–1.9-b.y.-old arcs (Savo, Knaften, Kittilä), Paleoproterozoic (>2.0 Ga) microcontinents (Keitele, Bothnia, Bergslagen), and Archean continental fragments (Kola-Murmansk) arrived at the Karelian continental margins (Fig. 1C). The terrane accretion resulted in overlapping orogenic events and switching of subduction polarities between 1.92 and 1.88 Ga. The convergent margins changed into continent-continent collision zones upon the arrival of large continents (Sarmatia, Laurentia, Amazonia) between 1.87 and 1.79 Ga. Both collision phases were followed by collapse-stage reworking of the geometrical relationships. In these Paleoproterozoic collisional and extensional processes, the crust attained its characteristic large thickness (>56 km), thick high-velocity lower crust (*Vp* > 6.8 km/s at depths of 30–60 km), and thick high-velocity lithosphere (>200 km) (Korja et al., 1993; Sandoval et al., 2003; Peltonen and Brügman, 2006).

Figure 1. Geological setting of the Fennoscandian Shield. (A) FIRE (red and purple) and BABEL (blue) profiles on a simplified geological map of the Fennoscandian Shield after Koistinen et al. (2001). Major shear zones are marked with dashed lines. Abbreviations: BB—Bothnian belt; CLA—Central Lapland area; CFGC—Central Finland granitoid complex; J—Jormua ophiolite; KC—Karelian craton; KSZ—Kittilä shear zone; LBB—Ladoga–Bothnian Bay wrench fault; LGB—Lapland granulite belt; MC—Murmansk craton; O—Outokumpu ophiolite; PB—Pirkanmaa belt; SB—Savo belt; SöB—Sörmland belt; TB—Tampere belt; TIB—Transscandinavian Igneous Belt; U—Uusimaa belt. (B) Orogenic domains of the Fennoscandia and arc complexes within the Svecofennian orogen. Abbreviations: KA—Kittilä arc; SA—Savo arc; SAC—Southern Finland arc complex; WAC—Western Finland arc complex. (C) Crustal units older than 2.0 Ga within the Svecofennian and Lapland-Kola orogens (after Lahtinen et al., 2005); mc—microcontinent.

The central part of the Svecofennian orogen (Fig. 1B) consists of rocks formed within the Karelian craton, Savo arc, Western Finland arc complex, and Southern Finland arc complex (Korsman et al., 1997). The Lapland-Kola orogen consists of rocks formed within the Archean Karelian and Murmansk cratons and within juvenile Proterozoic terranes (Daly et al., 2001, 2006). Allochthonous units were thrust on the cratons, which underwent metamorphic overprinting and local shearing.

The Karelian craton is composed of Archean granitoid-gneiss complexes and greenstone belts (3.2–2.5 Ga), which are intruded by Paleoproterozoic layered mafic intrusions (2.5–2.1 Ga) and A-type granitoids (2.5–2.4 Ga), as well as numerous mafic dike swarms (2.4–1.97 Ga) (Sorjonen-Ward and Luukkonen, 2005). Autochthonous supracrustal rocks ranging from quartzites to pelites and mafic volcanic rocks were deposited on the Archean basement from 2.45 Ga onward. During the Svecofennian orogeny, allochthonous units were thrust from the west onto the Karelian craton, which underwent metamorphic overprinting, resetting of K-Ar isotopes, and localized shearing. The allochthonous units include meta-turbiditic units tectonically intercalated with 1.97–1.95-b.y.-old ophiolites (Jormua [J], Outokumpu [O]) and Archean slivers of various sizes (Sotkuma [S], Iisalmi [I]; Wegmann, 1928; Eskola, 1949; Koistinen, 1981; Kontinen, 1987; Peltonen et al., 1996). The reactivated areas were intruded by Paleoproterozoic granitoids (1.87–1.80 Ga) (Gaál and Gorbatschev, 1987; Korsman et al., 1997).

The SE-NW–trending Savo belt (SB, Fig. 1A) hosts rocks formed in the oceanic Savo arc and its vicinity (Lahtinen, 1994). It is composed of metavolcanic and meta-turbiditic rocks that are interlayered with gneissic tonalites (1.92 Ga). The heavily deformed volcano-sedimentary complex was intruded by calc-alkaline granodiorite, tonalite, and quartz-diorite plutons and overlain by volcanic rocks (1.89–1.88 Ga) during the peak of deformation and low pressure–high-temperature metamorphism (Korsman et al., 1999). The suture between the Savo belt and Karelian craton is overprinted by the major NW-SE–directed Ladoga–Bothnian Bay wrench fault (Korsman and Glebovitsky, 1999).

The Western Finland arc complex (WAC, Fig. 1B) is composed of supracrustal belts (Bothnian belt [BB], Tampere belt [TB], and Pirkanmaa belt [PB]) surrounding the Central Finland granitoid complex. The Bothnian belt (1.90–1.87 Ga) consists of meta-turbiditic sequences hosting mafic volcanic units as well as a large migmatite-granitoid complex (Korsman et al., 1997). The granitoid complex is composed of mainly postcollisional calc-alkaline granitoids (1.89–1.88 Ga) with minor amounts of mafic plutonic rocks and comagmatic metavolcanic sequences (Korsman et al., 1997; Nironen et al., 2000). The intrusion of the igneous rocks was simultaneous with the peak of the low pressure–high-temperature metamorphism ($T$ = 700–800 °C, $P$ = 400–500 MPa) and deformation (Korsman et al., 1999). Soon after or contemporaneously with the calc-alkaline granitoids, a series of alkaline, occasionally pyroxene-bearing granites with minor granodiorites and quartz monzonites intruded the Central Finland granitoid complex (between 1.88 Ga and 1.87 Ga). These, as well as a series of coeval gabbros and mafic dikes, intruded along NW-SE–trending shear zones in a transtensional environment (Elliott et al., 1998; Nironen et al., 2000).

The Tampere belt (1.90–1.89 Ga; TB, Fig. 1A) is made up of well-preserved metagraywacke and metavolcanic rocks formed in a mature arc, or close to a continental margin. The Pirkanmaa belt (PB, Fig. 1A) is made up of migmatitic graywackes, which are interpreted to be remnants of an accretionary prism (Kähkönen, 1987; Lahtinen, 1994; Korsman et al., 1999). Both southern supracrustal belts suffered from regional metamorphism and polyphase deformation. During the peak of the metamorphism, the belts were intruded by calc-alkaline and minor alkaline granodiorites and granites with ages of 1.89–1.87 Ga (Patchett and Kouvo, 1986; Nironen, 1989; Kilpeläinen, 1998).

The NE-SW–directed Southern Finland arc complex consists of volcanic belts (1.90–1.88 Ga), intervening metasedimentary units, as well as voluminous younger migmatites and granites (1.84–1.81 Ga). The volcanic belts are composed of volcanic rocks interlayered with pelitic to psammitic rocks and occasional carbonate and quartzite layers. The supracrustal rocks were intruded by calc-alkaline granitoids during the first deformation and metamorphic peak at 1.88 Ga (Mouri et al., 1999). The supracrustal units were juxtaposed before the second deformation cycle and associated low pressure–high-temperature regional metamorphism and granite and migmatite formation between 1.85 and 1.82 Ga (Ehlers et al., 1993; Korsman et al., 1999; Kurhila et al., 2005).

## SEISMIC DATA

The seismic data used in this paper (Figs. 2–11) are from the deep-reflection lines BABEL (BABEL Working Group, 1993; Korja and Heikkinen, 2005) and FIRE (Kukkonen et al., 2006), which transect the Paleoproterozoic Svecofennian and Lapland-Kola orogens (Fig. 1). The BABEL profiles B, C, 1, and 3–4 span 1200 km of marine seismic profiles in the Baltic Sea and the Gulf of Bothnia. FIRE profiles 1, 2, 3, and 4 span 2100 km of land seismic profiles in areas with only minor Quaternary sedimentary cover. BABEL B, C, 1, and FIRE 2 cover the southern part of the Svecofennian orogen; BABEL 1, 3–4, and FIRE 1 and 3 cover the Western Finland arc complex. BABEL 3–4 and FIRE 1 and 3 image the suture zone between the Savo arc and Karelian craton. FIRE 4 studies the northern part of the Svecofennian orogen and the Lapland-Kola orogen.

Most of the seismic sections shown here are migrated NMO (normal move out) stacks. We also used migrated DMO (dip move out) stacks in near-surface sections on FIRE profiles. The sections imaging the whole crust (Figs. 2, 3, 4, 6, 9, 10, and 11) are displayed as automatic line drawings and/or as instantaneous amplitudes. The instantaneous-amplitude sections are averaged both horizontally and vertically, and they are plotted as gray-scale intensities (for technical details, see Korja and Heikkinen, 2005; Kukkonen et al., 2006). The coherency-filtered DMO sections imaging only the upper crust are displayed either in variable intensity (Fig. 8) or variable area (Fig. 9) mode.

In seismic-reflection sections, gently dipping surfaces are imaged better than structures with steep dips. Because of this, it is possible that the tops and crests of synforms and antiforms are imaged as piece-wise continuous subhorizontal reflectors, leading to biased interpretation (Ji and Long, 2006). These problems can be partly avoided by displaying the data as gray-shaded instantaneous-amplitude sections, which enhance differences in reflectivity between adjacent blocks and help to define block boundaries. In the upper crust, the DMO sections can be used to image the steeply dipping structures.

In the geological interpretation, it is assumed that reflections arise mainly from lithological contacts with large velocity and density contrasts. In a highly metamorphosed and deformed area such as the Svecofennian orogen, lithological contacts can be either primary contacts like dikes or igneous batholiths intruding deformed supracrustal rocks or they may represent a tectonic contact formed during folding and faulting. Weak reflectivity generally indicates a rock mass with little or no density and velocity contrasts, e.g., monotonous intrusions and older crustal pieces in which the internal structure has been homogenized in the scale of reflectivity prior to deformation.

Tectonic block boundaries are either thrust faults, normal faults, or strike slip-faults that separate blocks with different reflection properties. On seismic sections, compression produces concave (thrusts) reflections and extension produces convex (listric) reflections in general (Meissner, 1996). Detachment zones and décollements are continuous subhorizontal reflections toward which other reflections truncate. The detachment surfaces likely form at the rheological boundaries where the lithologies change from dominantly wet-quartz– to dry-quartz–dominated rheology and from dry-quartz– to plagioclase-dominated rheology (Kusznir and Park, 1987; Meissner, 1996). Décollements are mechanically weak layers along which thrust faulting takes place.

Subvertical strike-slip zones are observed indirectly because they are associated with transparent zones, decreased reflectivity, and displacement of continuous reflections (Harding, 1985). The decreased reflectivity is probably related to sets of closely spaced fracture zones, which destroy the continuity of reflecting boundaries. A well-known example is the San Andreas Fault, from which reflections have disappeared (Zhu, 2000).

## SEISMIC IMAGES OF PLATE-TECTONIC ENVIRONMENTS

The Svecofennian orogen evolved in a convergent-margin setting, beginning with subduction, accretion of microcontinents and arc terranes, and finally terminating in a continent-continent collision (Lahtinen et al., 2005; Korja et al., 2006b). The crustal formation finalized in gravitational collapse, during which the crustal blocks moved to their present positions. In the following, we will analyze the seismic-reflection images and try to separate the traces left by the different tectonic processes of convergent margins (subduction, transforms, arc accretion, continent-continent collision, and gravitational collapse).

### Subduction

Seismic profile BABEL 3–4 (Figs. 1 and 2) images a frozen paleosubduction zone, associated accretionary prism (S), and an older crustal indentor (I) under which crust is subducting (Fig. 2; BABEL Working Group, 1990, 1993; Korja and Heikkinen, 2005). The lower crust of the subducting plate (P) has seismic P-wave velocities larger than 6.8 km/s, suggesting mafic lithologies (Gohl and Pedersen, 1995). This, together with the high reflectivity, has been interpreted to indicate stacking of thin oceanic crust (Gohl and Pedersen, 1995). Beaumont and Quinlan (1994) interpreted the structure to image pro- and retrowedge deformation of a collision zone and concluded that subduction had come to a terminal phase and that collision had already begun. The reflection image is similar to that found underneath New Zealand and the Pyrenees (Beaumont et al., 2000), where collision is taking place.

The survival of dipping reflections in the lower crust implies that the crustal material has behaved rigidly (Meissner, 1996). Recently, Maggi et al. (2000) and Jackson et al. (2004) suggested that if the lower crust is in granulite-facies conditions, then it is rigid and not plastic, as previously believed. Based on the geophysical properties of the crust, Henkel et al. (1991) and Korja (1995) suggested that the high-velocity lower crust of the Svecofennian orogen is composed of metastable mafic granulites (Ellis and Maboko, 1992; Austrheim, 1990; Bjørnerud and Austrheim, 2004). Recent analysis of lower-crustal xenoliths from kimberlite pipes (Hölttä et al., 2000) in the eastern part of the Svecofennian orogen has provided further support for this interpretation (Kuusisto et al., 2006).

Because mafic granulites are lighter than the underlying mantle, they are not easily subducted, and because they are heavier than average continental crust (Christensen and Mooney, 1995), they are not easily obducted either. This might be the reason why subduction geometry could survive in a nascent collision zone. We reinterpret the BABEL 3–4 image as a failed subduction of mafic metastable granulitic lower crust of the Bothnia microcontinent beneath the Keitele microcontinent.

### Transform

In map view (Fig. 1), the Ladoga–Bothnian Bay wrench fault is a series of anastomosing shear zones subparallel to the Savo belt and the western edge of the exposed Karelian craton. The crust is split into small blocks that display complex lithological relationships. High-grade metamorphic blocks with pyroxene-bearing postkinematic granitoids, schollen migmatitic rocks intruded by nickel-sulfide–bearing gabbros, and metavolcanic rocks are exposed within the Ladoga–Bothnian Bay wrench fault zone (Korsman and Glebovitsky, 1999).

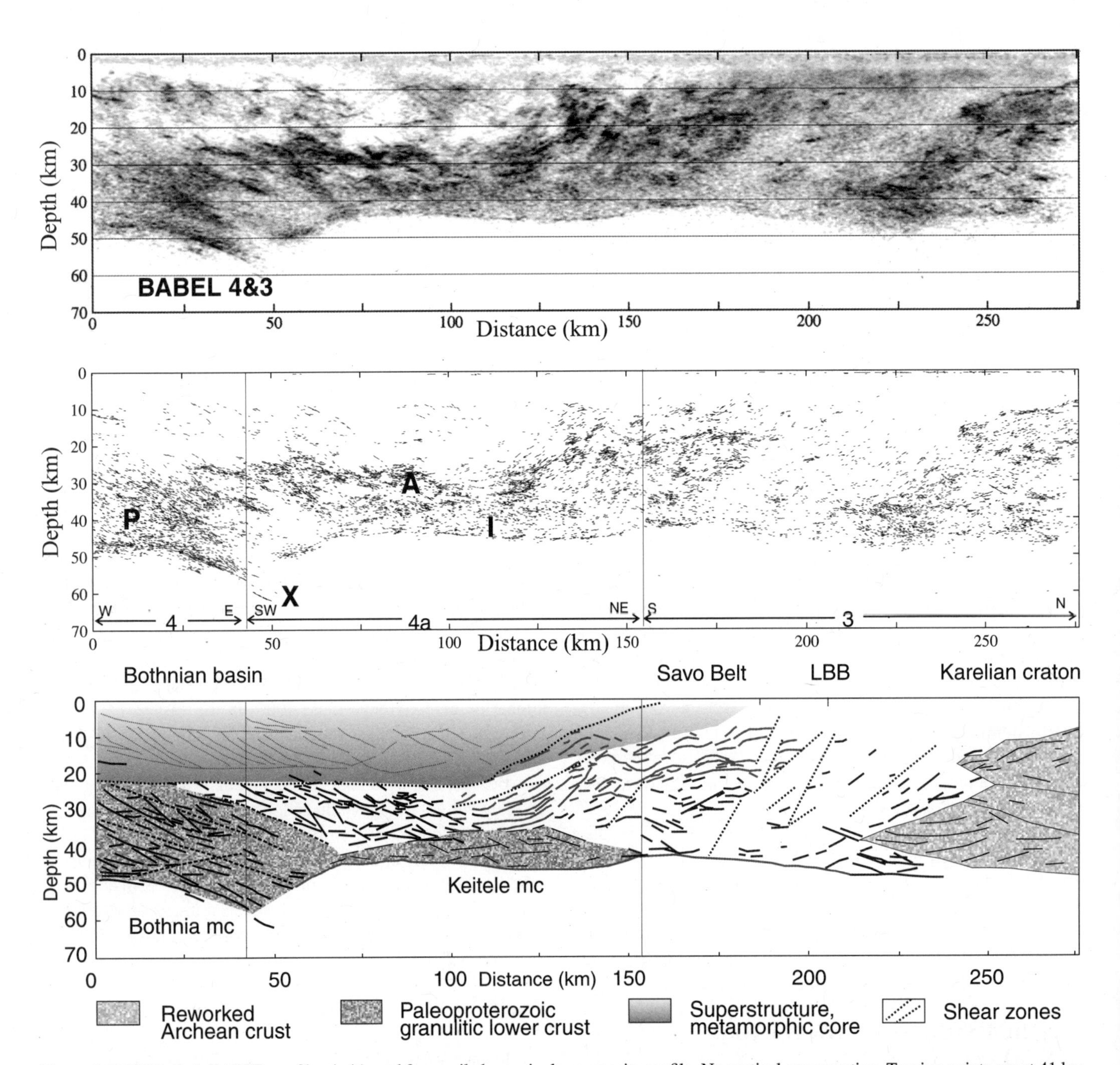


Figure 2. BABEL 3–4, BABEL profiles 4, 4A, and 3 compiled as a single composite profile. No vertical exaggeration. Turning points are at 41 km and 156 km. Upper panel: An averaged instantaneous-amplitude section after Korja and Heikkinen (2005). Middle panel: An automatic line drawing after Korja and Heikkinen (2005). Lower panel: A geological interpretation modified from Korja and Heikkinen (2005). Abbreviations: A—accretion prism; I—older indentor; P—subducting plate; X—paleosubduction zone; LBB—Ladoga–Bothnian Bay wrench fault; mc—microcontinent.

The Ladoga–Bothnian Bay wrench fault is crossed by three reflection lines: BABEL 3–4 and FIRE 1 and 3 (Fig. 1). In the seismic sections (Figs. 2, 3, and 4), it is displayed as a series of subvertical transparent zones penetrating the entire crust. Between the major faults, the reflection amplitudes decrease (Figs. 3–5). BABEL profiles 3–4 (Fig. 2) show that the fault zone has developed on the continental side of the subduction zone (S). On FIRE 3a (Fig. 3), two of the well-studied late-stage granites (Suvasvesi and Kermajärvi; CMPs 3800–2800) displaying dextral motion (Halden, 1982) are observed as poorly reflective upper-crustal bodies bounded by steep transparent zones. Together, these observations point to transfer motion and oblique convergence.

In a detailed upper crustal section along FIRE 1 (Fig. 5), a synformal structure (CMP 11400–10250) is disrupted by steep reflections toward which the eastward-dipping reflections are bending. At the surface, the subvertical disruptive structures correlate with the shear zones transecting the area. The reflection section is interpreted to image upward movement of the well-reflective supracrustal units along the steep structures (Korja et al., 2006a).

All together, the features suggest a transform boundary zone where granites intrude in transtensional basins, and high-grade blocks move upward in transpressive zones as suggested by Schulmann et al. (2003) and Thompson et al. (2001). Both the seismic image and the map view of the Ladoga–Bothnian Bay wrench fault zone show that it is similar to the San Andreas transform fault (Langenheim et al., 2005), where crustal wedging is taking place currently along strike-slip duplexes (Woodcock and Fischer, 1986).

Recently, Pattison et al. (2006) identified another large-scale, strike-slip fault, the Kittilä shear zone (KSZ) (Fig. 1), on FIRE 4 seismic sections (Figs. 6 and 12). Gaál et al. (1989) had previously suggested an upper-crustal fault zone in the area. The reflection patterns suggest a positive flower structure (Harding, 1985) forming in a transpressive environment. Because the shear zone separates two crustal blocks with different reflection properties, it indicates large displacements along the fault in a major suture zone (Pattison et al., 2006).

## Obduction

At the western margin of the Karelian craton, a detailed upper-crustal seismic section along FIRE 3 (Fig. 7) images fine-scale layering, folding, and thrusting, as well as faulting. The thrusts observed in the allochthonous formations at the surface (Wegmann, 1928; Koistinen, 1981, Park et al., 1984) can be correlated with the upper-crustal low-angle reflections on FIRE 3 (Figs. 3 and 7), and more steeply dipping tectonic contacts correlate with basin margins (Sorjonen-Ward, 2006). The reflection structures imply thin-skin thrusting and complicated basement cover relationships reminiscent of the Appalachians (Cook et al., 1979). We interpret the subhorizontal surface separating the highly reflective layered units from the more monotonous units to be a décollement surface separating a foreland fold-and-thrust belt from the relatively undeformed basement.

On the whole, the upper-crustal structure, including parautochthonous sedimentary basins (Koli, Höytiäinen; Ward, 1987; Sorjonen-Ward, 2006), allochthonous units including ophiolites (Outokumpu association; Koistinen, 1981), and a complex basement cover interface (Park and Bowes, 1983) is typical of foreland fold-and-thrust belts (Coward, 1994), where basement, autochthonous, and allochthonous units are duplexed by thrusting. The mantled gneiss domes (e.g., Sotkuma; Eskola, 1949) appear to be allochthonous basement thrust wedges, as previously suggested by Park et al. (1984) and Kohonen (1995). Because ophiolites are usually obducted onto the arriving lower plate, the oceanic part of the Karelian plate was once moving toward the west and subducting.

Similar, although more complicated, structural relationships are found on FIRE 1 (Figs. 1, 5, and 8), where Archean crustal blocks are bounded by Paleoproterozoic schist belt remnants. On a more detailed, upper-crustal section along FIRE 1 (Fig. 8), the allochthonous Archean blocks are imaged to have been thrust on allochthonous supracrustal formations composed of thin-skin thrust wedges. The subhorizontal, thin-skinned part of the thrust belt (CMPs 5250–6250) is thin or has been completely exhumed.

We interpret the upper-crustal sections along FIRE 1 and FIRE 3 as displaying complementary images of the foreland fold-and-thrust belt. FIRE 1 displays a deeper erosion level with thick-skin stacking at surface a higher metamorphic grade and large amounts of migmatitic granites. FIRE 3 on the other hand displays a shallower erosion level with thin-skin thrusting, lower metamorphic grade, and minor amounts of partial melting.

## Collisional Plate Boundaries

Continental collision zones are characterized by bivergent thrust and stack structures. The bivergent structures are found in schist belts, which represent volcano-sedimentary units squeezed between accreting crustal blocks during collision. In Fennoscandia, the downward continuation of the schist belts has been traced with seismic and geoelectrical methods. These units are highly reflective (P in Figs. 2, 3, and 11), and they have low P-wave velocity values and *Vp*/*Vs* ratios (Hyvönen et al., 2006). Some of the formations have high conductivity values (Korja et al., 2002), indicating graphite- and sulfide-bearing supracrustal sequences typically deposited in shallow, closed basins. On FIRE profiles, bivergent zones are found beneath the Savo belt (FIRE 3 CMP 2000–8000; Figs. 3 and 9), Pirkanmaa belt (FIRE 2 CMP 3000–7000; Fig. 11), Central Lapland area (FIRE 4 CMP 9000–11000; Fig. 6), and Lapland granulite belt (FIRE 4; 325–525 km in distance; Fig. 12); on BABEL lines, bivergent zones are found beneath Södermanland (SöB; BABEL B; Fig. 12; 50–250 km in distance) and beneath Bothnian Bay (BABEL 3–4; Figs. 2 and 12).

Profile FIRE 3 (Fig. 3) displays westward thickening of the crust by thick-skin stacking and formation of anticlinal ramps (A) in the middle crust. Within the Savo belt, the middle crust displays a series of ramp anticlines that gradually change to thrust wedges, grow steeper, and become subvertical close to the suture

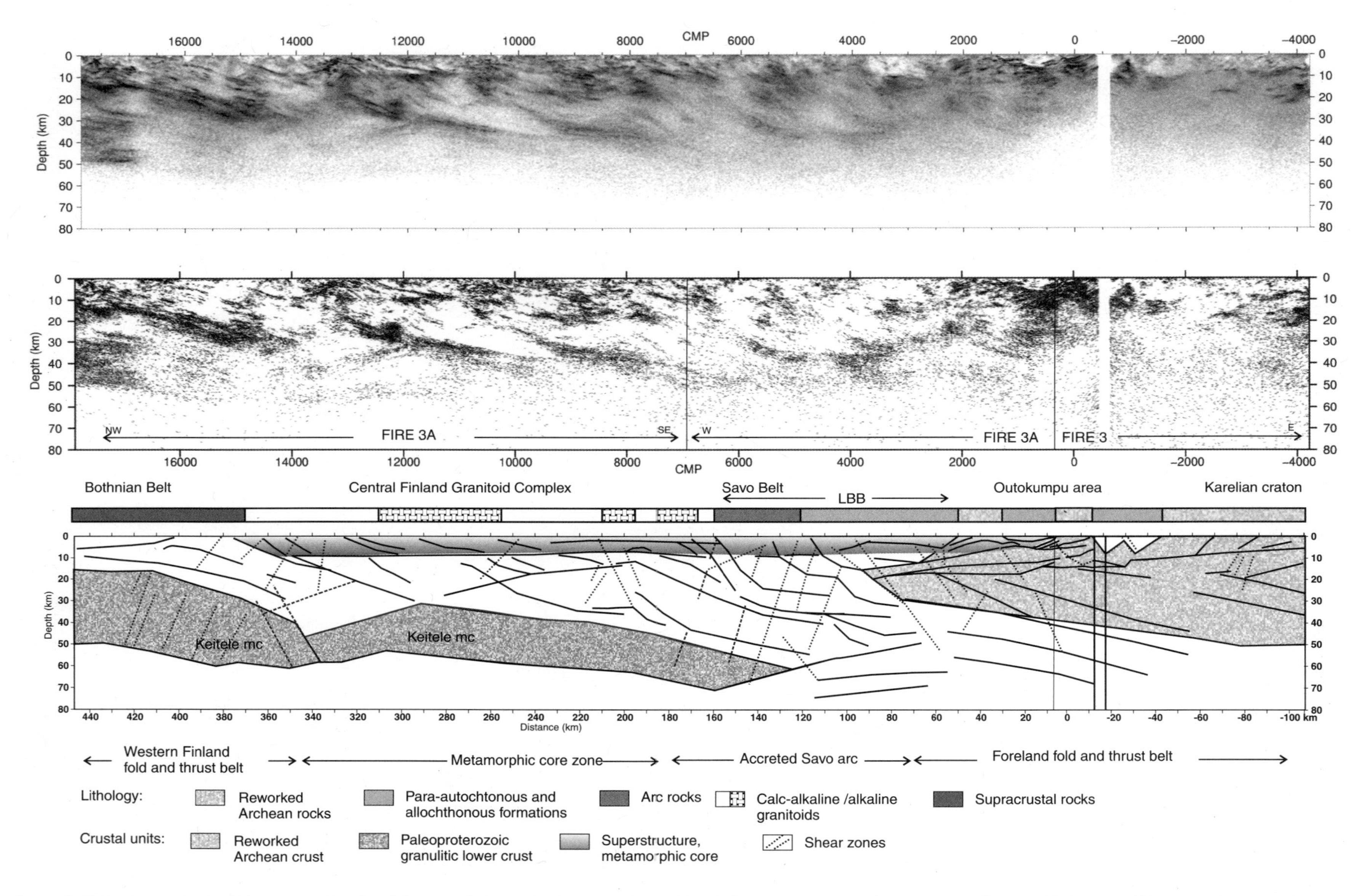


Figure 3. FIRE deep seismic-reflection profiles 3 and 3a compiled as a single composite profile FIRE 3–3a. Distance is in common midpoints (CMPs) and kilometers. No vertical exaggeration. Upper panel: An averaged instantaneous-amplitude section after Kukkonen et al. (2006). Middle panel: An automatic line drawing. Major turning points are marked with a vertical line. Lower panel: Lithology at surface after Korsman et al. (1997) and a geological interpretation. Cross point of line FIRE 1 is marked with a vertical line. LBB—Ladoga–Bothnian Bay wrench fault.

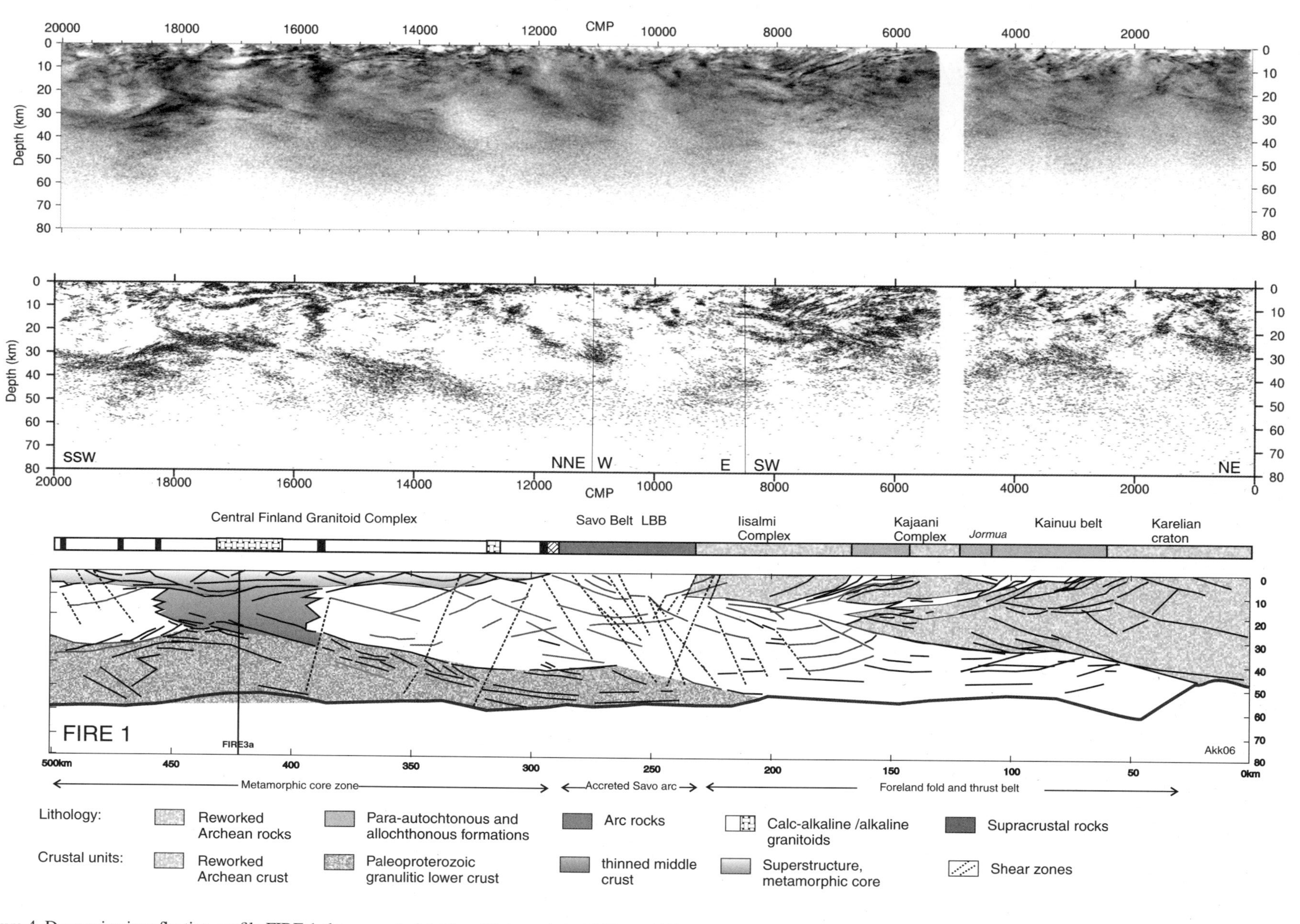


Figure 4. Deep seismic-reflection profile FIRE 1 shown as straight line. Distance is in CMPs and kilometers. No vertical exaggeration. Upper panel: An averaged instantaneous-amplitude section after Kukkonen et al. (2006). Middle panel: An automatic line drawing. Major turning points are marked with vertical lines. Lower panel: Lithology at surface after Korsman et al. (1997) and Korja et al. (2006a) and a geological interpretation. Cross point of line FIRE 3a–3 is marked with a vertical line. LBB—Ladoga–Bothnian Bay wrench fault.

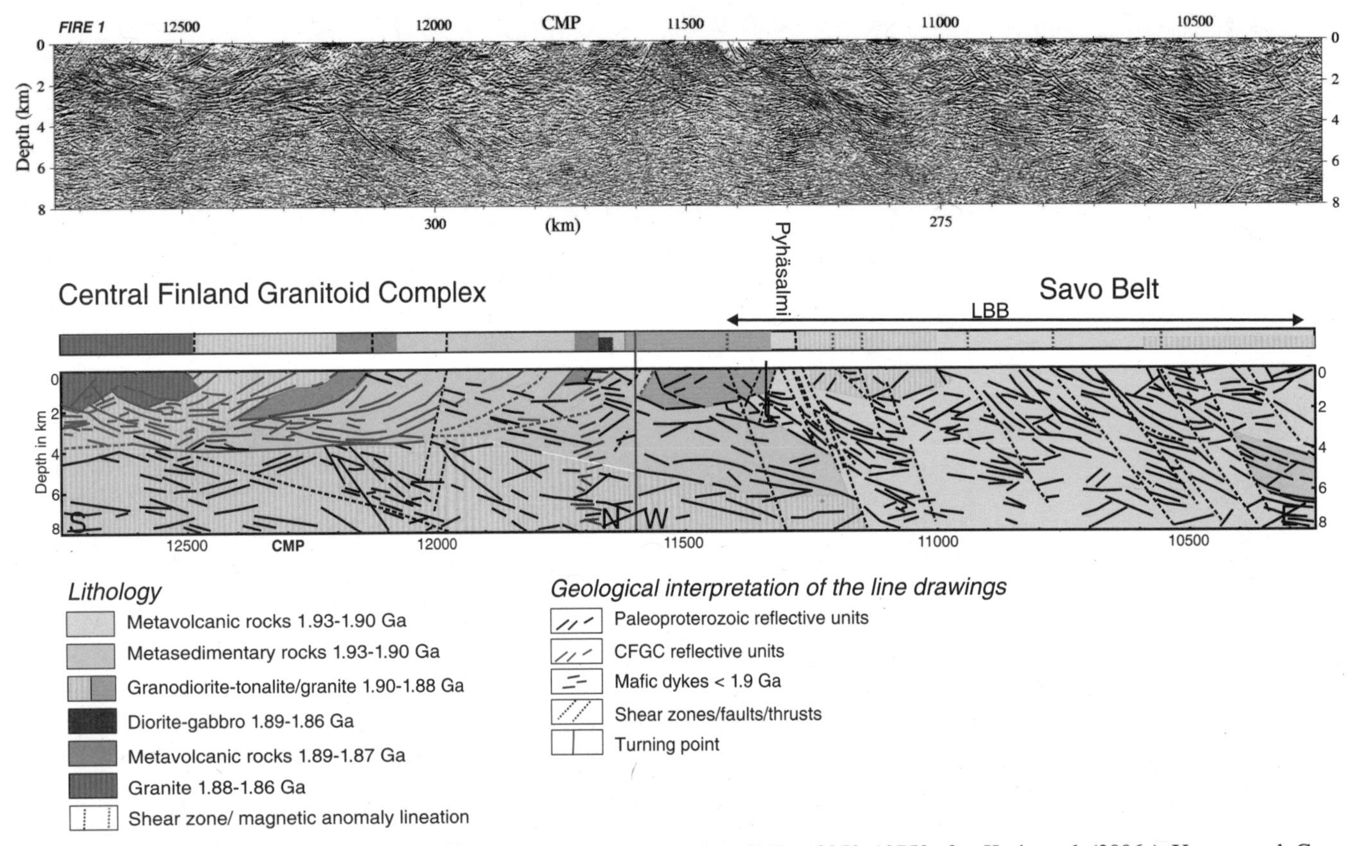


Figure 5. A geological interpretation of the uppermost 8 km of FIRE 1 between CMPs 10250–12750 after Korja et al. (2006a). Upper panel: Gray-scale variable-intensity DMO (dip move out) section. Lower panel: Lithology at the surface and a geological interpretation. Major turning point is marked with a red line on the lower panel. Note the turning of the reflections toward subvertical transparent zones disrupting the seismic pattern between CMPs. LBB—Ladoga–Bothnian Bay wrench fault.

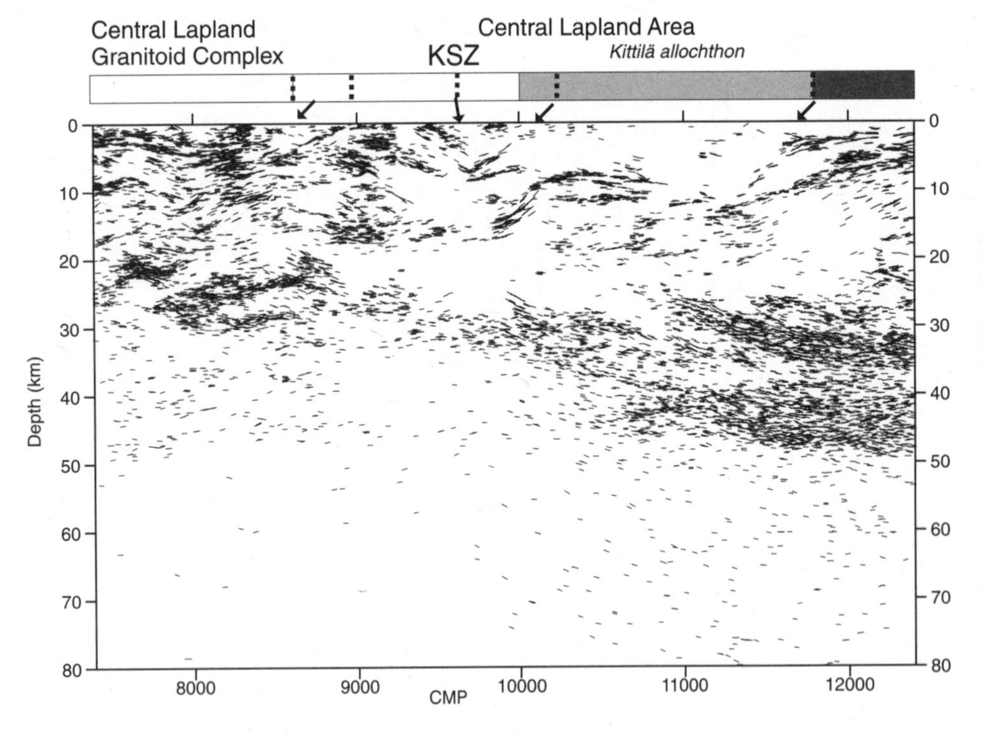


Figure 6. An automatic line drawing along FIRE 4 between CMPs 7500 and 12500. 1000 CMPs correspond to 25 km in distance. Lithological units and shear zones at surface are after Pattison et al. (2006). Crustal-scale reflectivity patterns change over the Kittilä shear zone (KSZ), which is a subvertical transparent zone dipping to the apparent north.

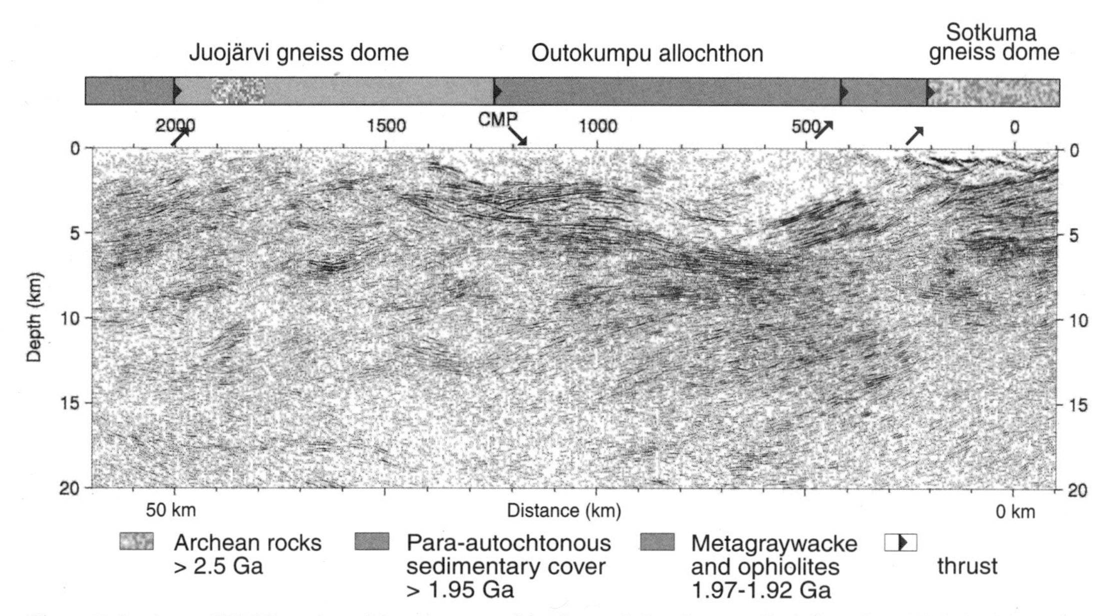


Figure 7. A migrated DMO section of the uppermost 8 km beneath Outokumpu allochthon along FIRE 3a-3 between CMPs -10 and 2150. Lithology at the surface is after Korsman et al. (1997). Note the thin-skin thrusting and gentle folding of the allochthonous units. 1000 CMPs correspond to 25 km in distance.

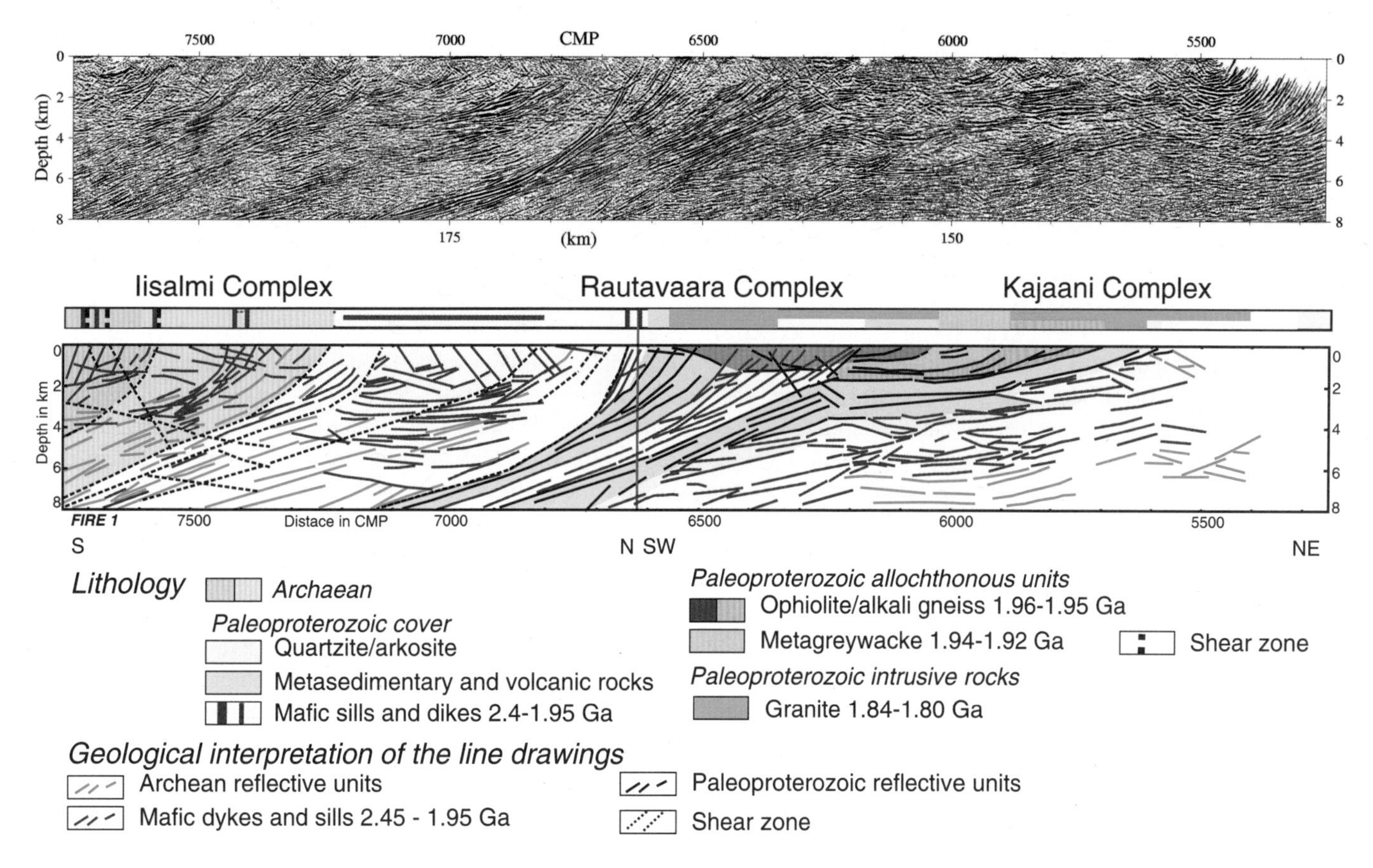


Figure 8. A geological interpretation of the uppermost 8 km of FIRE 1 between CMPs 5250–7750 after Korja et al. (2006a). Upper panel: Gray-scale variable-intensity DMO (dip move out) section. Lower panel: Lithology at the surface and a line drawing. A major turning point is marked with a red line on the lower panel. Note the thin-skin thrusting of the Paleoproterozoic supracrustal units.

(Fig. 9). The seismic image displays the sequential shortening of the middle- and upper-crustal flakes of the Savo arc. Some of the shortening is taken up by subvertical shear zones of the Ladoga–Bothnian Bay wrench fault, described previously. The lowermost crust, however, is poorly reflective and displays subhorizontal reflectivity. This is interpreted to indicate that the lower crust of the arc was detached and that it behaved differently during accretion. This may indicate that during collision, the lower crust of the Savo arc continued to subduct, whereas upper and middle crust was added to the western continent or continental arc (Keitele). This is analogous to the Izu-Bonin-Japan arc collision, where the upper and middle crust of the Izu-Bonin has accreted to the Japan arc and the lower crust continues to subduct (Sato et al., 2005; Takahashi et al., 2007).

On FIRE 1, the lower crust beneath the Karelian craton stands out as a distinct unit because it shows a reflective Moho with a 15 km step and few moderately westward-dipping reflections in its upper parts (Figs. 4 and 10). On FIRE 3 (Fig. 3), the step in Moho is less well imaged. The step is directed north-south, and it can be followed by refraction profiles and on the Moho map for a few hundred kilometers (Korja et al., 1993). Xenoliths coming from the lower crust indicate that the crustal material is Paleoproterozoic mafic granulite (Hölttä et al., 2000). We interpret the material to be trapped oceanic crust that is rigid enough to preserve structures other than subhorizontal ones (Meissner, 1996) and the upper boundary of which is characterized by normal faults. This suggests that during the final collision, the Archean crust, along with its allochthonous cover, overrode oceanic crust attached either to the Savo arc or Karelian craton.

FIRE 2-A (Figs. 1 and 11) images the collision structures produced by accretion in the southern part at 1.89 Ga (Lahtinen et al., 2005). The upper crust displays bivergent structures below the Pirkanmaa migmatite belt and Tampere volcanic belt (Nironen et al., 2006), whereas moderately north-dipping convex structures and occasional ramp anticlines (A) are found in the middle crust. The Pirkanmaa migmatite belt has previously been interpreted as an accretionary prism (Kilpeläinen, 1998; Lahtinen, 1994; Kähkönen, 2005). Recently, Nironen et al. (2006) interpreted it to be a crustal scale pop-up structure. The complex has been squeezed between poorly reflective blocks, here interpreted as the Keitele and Bergslagen microcontinents (Fig. 1C). The Tampere belt, which has been thrust northward on the southern margin of the Keitele microcontinent, may be part of a foreland fold-and-thrust belt. In the lower crust, a poorly reflective body from which reflections spin off is found. The body has high velocities and high *Vp*/*Vs* ratios (Hyvönen et al., 2006), and it is associated with a Bouguer anomaly high (Kozlovskaya et al., 2004). It is interpreted as another piece of trapped oceanic crust.

## Collapse Structures

During collapse of the overthickened crust, an extensional environment develops on top of the collisional structure in the thickened crust and displaces reflections formed during collision. Since the structures are three-dimensional, they are imaged differently from different directions. Along the southern part of FIRE 1 (CMP 12000–20000; Fig. 4) and northern part of FIRE 2 (CMP 0–3000; Fig. 11), the crust is layered, and there are two major layer boundaries across which seismic-reflection properties change. Above the upper subhorizontal highly reflective boundary (8–12 km in depth), the reflections mimic graben and horst structures. The apparent subhorizontal reflections are in reality perpendicular cross sections of listric reflections flattening out on detachment surface between 8 and 12 km in depth as seen on the perpendicular line FIRE 3 (Fig. 3). On FIRE 1, the middle crust has more patchy reflectivity that is disrupted by reflective zigzag patterns and transparent subvertical zones. It shows considerable symmetrical thinning in the southern part of the profile (CMP 14000–19000). The lower boundary of the middle crust is quite reflective as bright reflections flatten out on its undulating surface. The lower crust has patchy reflectivity, and the Moho boundary is gradational and observed primarily as a decrease in reflectivity. Both the upper to middle crust boundary and the Moho boundary are upwarping (convex).

Symmetrical thinning of the middle crust, upwarping of the lower crust, and subvertical crustal faults dipping symmetrically toward the center give the crust a structure reminiscent of McKenzie-type pure shear rifting. We interpret the layered crustal structure to have formed in a gravitational collapse. The collapse phase is best developed in the Central Finland granitoid complex, beneath which the upper-crustal detachment zone is formed, middle crust is thinned, and the lower crust is upwarping. At the surface, the area is characterized by calc-alkaline magmatism (1.89–1.88 Ga) followed by bimodal magmatism (1.88–1.87 Ga) with volcanic rocks, alkaline granitoids, and gabbroic components (Nironen et al., 2000). Kinematic analysis suggests a transtensional environment (Nironen, 2003).

The middle and lower crust in southernmost Finland is weakly reflective and displays only minor reflective events, mimicking the large-scale horst and graben structure on FIRE 2A (Fig. 11; CMP1900–5000). It is interpreted to image an older nucleus (Bergslagen microcontinent), where contractional structures have been inverted to become an extensional core complex structure like in the Cordilleras (Wernicke, 1985). The exposed core is characterized by a granulite complex, migmatites, granites, and layered mafic intrusions (Nironen et al., 2006). It is likely that gravitational collapse also took place here and produced the extensional structures and exposed the core complex structure. In the southernmost part of the profile, the collapse structure is overlain by an allochthonous sheet of migmatitic rocks belonging to a younger orogeny that took place in the south (see BABEL B; Fig. 12).

## DISCUSSION

Juvenile continental crust is formed at island arcs. In terms of seismic-reflectivity characteristics, arcs can be described as poorly reflective crustal blocks that are capped and surrounded

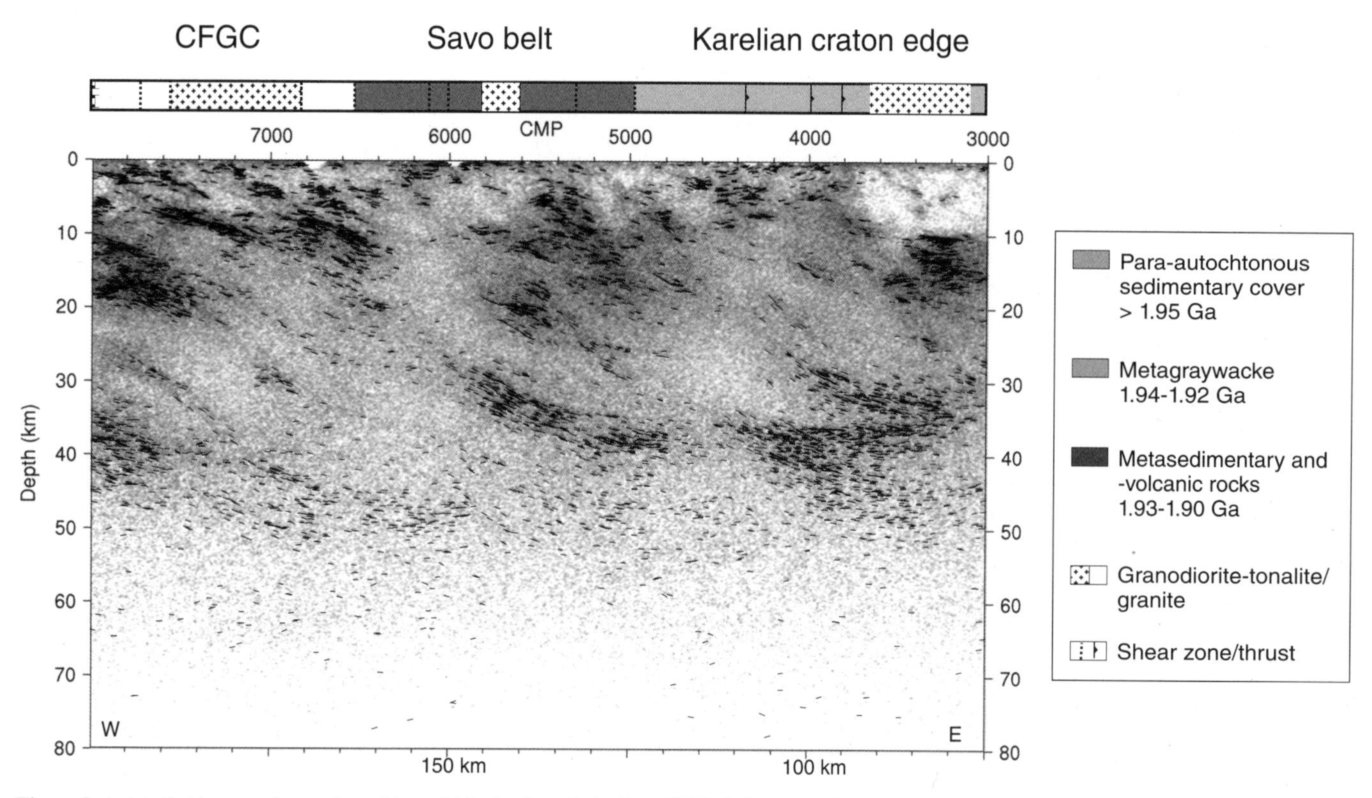


Figure 9. A detailed image of crustal stacking within the Savo belt along FIRE 3a between CMPs 3000 and 8000 on an averaged instantaneous-amplitude section combined with automatic line drawing. Surface lithology is after Korsman et al. (1997). CFGC—Central Finland granitoid complex.

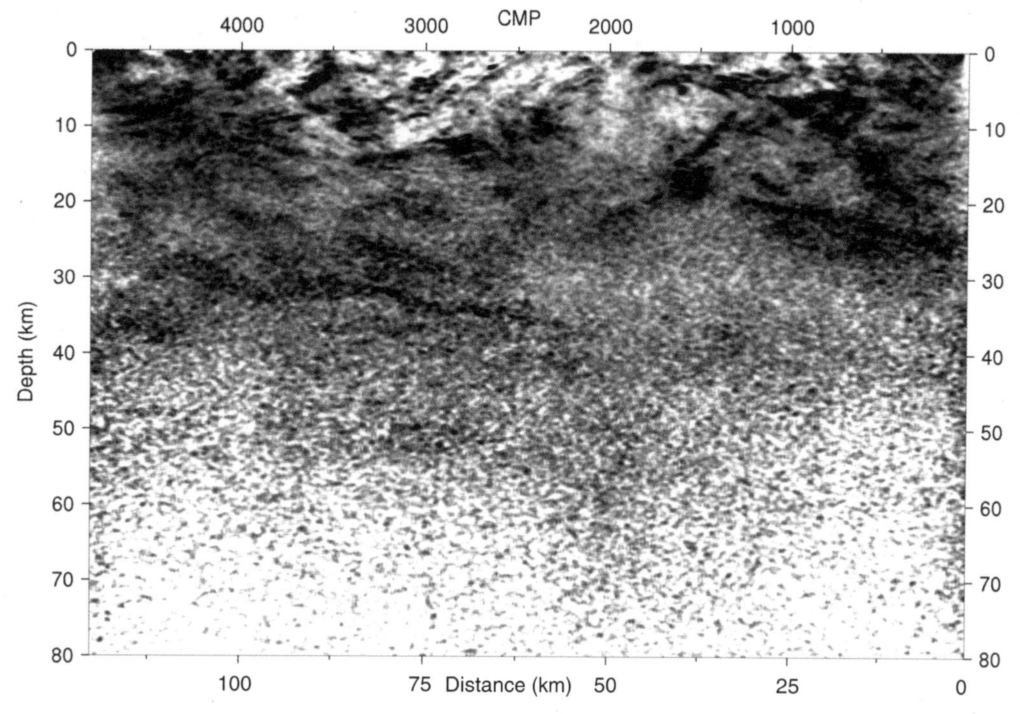


Figure 10. An averaged instantaneous-amplitude section of the lower crust beneath the western margin of the Karelian craton along FIRE 1 between CMPs 0 and 4700. Note the step in the Moho boundary and thick-skin stacking of the lower-crustal bocks. 1000 CMPs correspond to 25 km in distance.

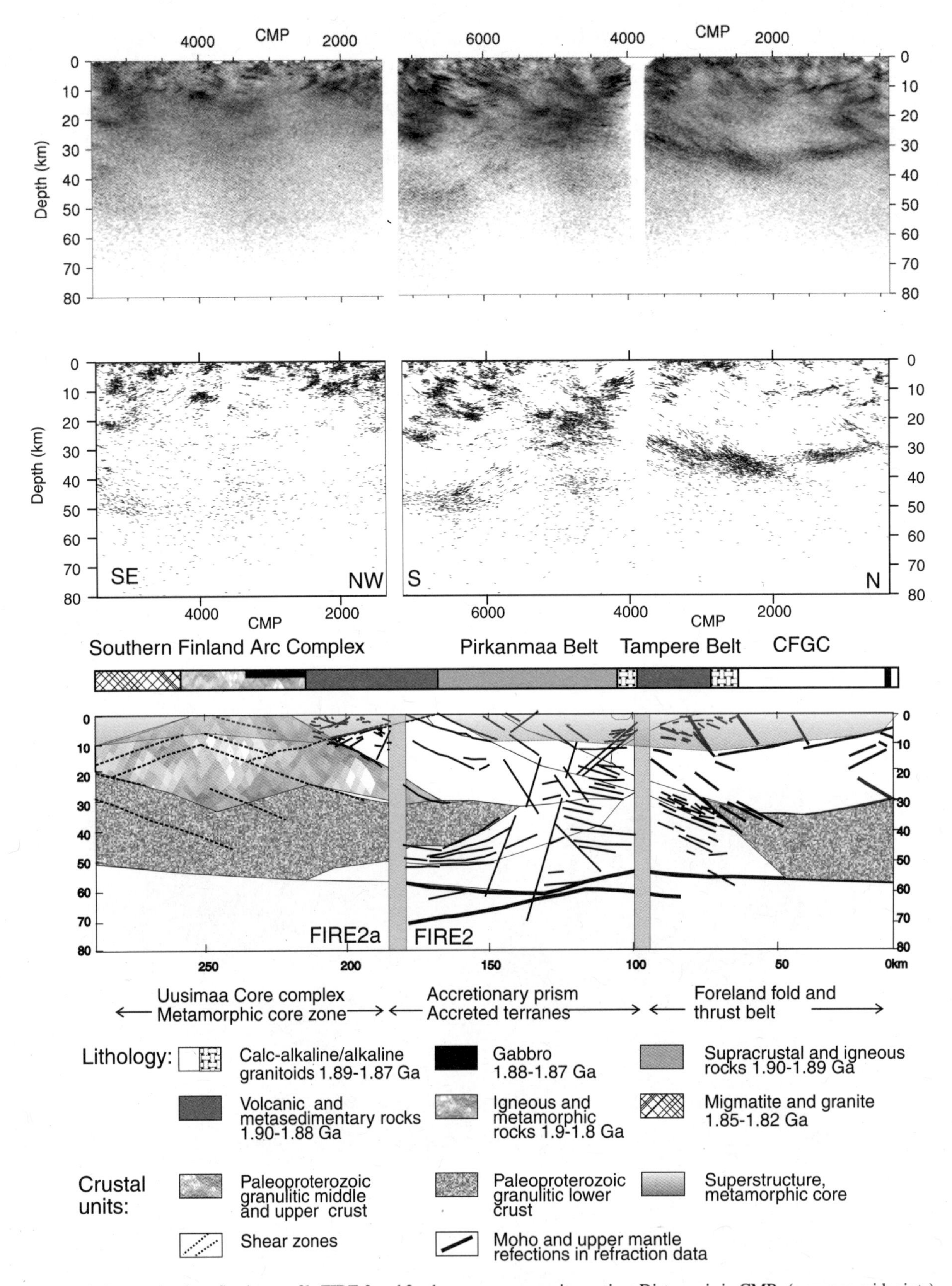


Figure 11. Deep seismic-reflection profile FIRE 2 and 2a shown as a composite section. Distance is in CMPs (common midpoints) and kilometers. No vertical exaggeration. Upper panel: An averaged instantaneous-amplitude section after Kukkonen et al. (2006). Middle panel: An automatic line drawing. Major turning points are marked with vertical lines. Lower panel: Lithology at surface after Korsman et al. (1997) and Nironen et al. (2006) and a geological interpretation. CFGC—Central Finland granitoid complex.

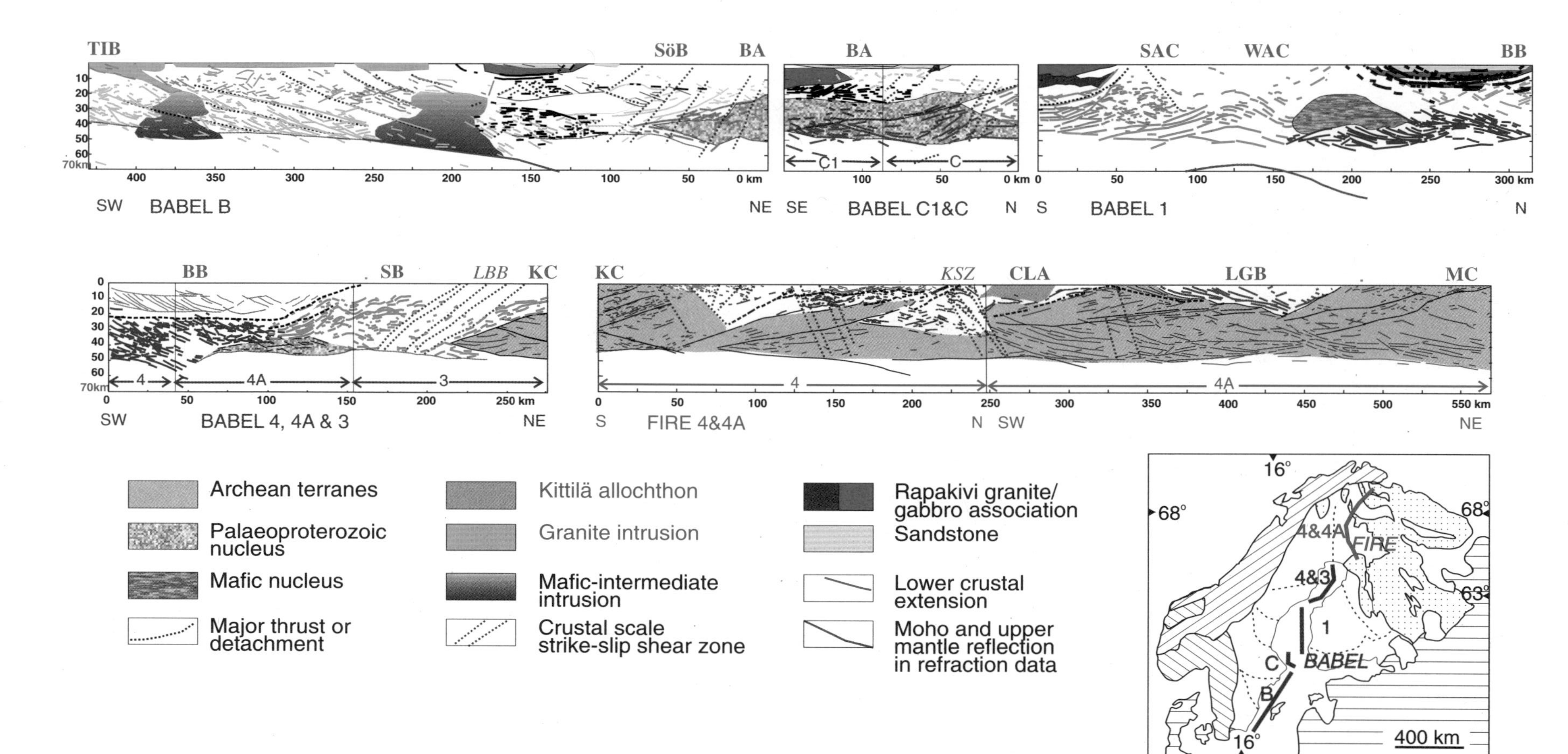


Figure 12. A geological interpretation of the BABEL lines B, C, 1, 3–4, and FIRE 4 compiled from Korja and Heikkinen (2005) and Pattison et al. (2006). The line drawing has no vertical exaggeration. Geologic units in the map are as in Figure 1. Colored lines denote reflections arising from units with different reflection properties; mafic sills are in black. Mantle reflectors (MR) are from Abramovitz et al. (1997) and Heikkinen and Luosto (2000). TIB—Transscandinavian igneous belt; BA—Bergslagen area.

by more reflective, layered supracrustal units (Suyehiro et al., 1996; Shillington et al., 2004; Sato et al., 2005). The crustal thickness of arcs varies between 15 km and 30 km, and their structure can be described simply using a three-layer model with upper-granitic, middle intermediate, and mafic lower crust (Mooney et al., 1998). This can also be described as rheological layering of wet quartz–, dry quartz–, and plagioclase-dominated lithologies in a geothermal environment (Kusznir and Park, 1987). When such rheologically heterogeneous crust is accreted, the stronger middle and upper crustal layers detach from the dense and rheologically weak lower crust. The more competent upper and middle crust is obducted as "flakes" and the lower crust undergoes continental subduction (e.g., Taiwan orogeny; Wu et al., 1997; Sato et al., 2005). Similar flakes are seen in FIRE 1 and 3 beneath the Savo belt. This suggests to us that reflective units described from meta-arcs (Avalonia, Hottah; van der Velden and Cook, 1999) might be associated with upper-crustal flakes composed of volcanic units interlayered with sedimentary units and layer-parallel granitoids.

When more evolved continental crust with a granulitic lower crust enters a subduction zone, subduction is no longer possible. Consequently, the system freezes, and another subduction zone develops on the far side of the colliding plate, as shown in BABEL line 3–4 (Figs. 3 and 11). In the Fennoscandian example seen on the compilation profile, a new subduction system has developed on the southern margin of the accreting system in front of a mafic indentor (Korja et al., 2001). Similar jumps of subduction zones have been described at modern accretionary margins (Stampfli and Borel, 2002).

Both the BABEL and FIRE data sets support the model that the Svecofennian orogen is composed of accreted terranes and intervening basins as earlier suggested by Korja et al. (1993). The seismic data also corroborate the idea of sequential arrival of terranes to the Karelian craton margin and westward growth of the continent between 1.9 and 1.8 Ga. The accretion likely caused temporary changes in the arc geometries. In the north, the continent grew by accretion of several smaller Archean crustal fragments, possibly pieces of stretched continental margin (Gaál and Gorbatschev, 1987; Lahtinen et al., 2005), to the Karelian margin (Pattison et al., 2006).

In the last Svecofennian orogenic episode, the Nordic orogen at 1800 Ma (Lahtinen et al., 2005; Korja et al., 2006b), the system changed from an accretionary margin to a continent-continent collision zone. This margin has since experienced several Wilson cycles. The Nordic orogen was first split to form the Grenvillian Sea before the Grenvillian orogen, then Tethys Sea before the Caledonian orogen, and the latest reactivation involved the opening of the Atlantic Ocean. We suggest that the present-day type of plate tectonics, with large cratonic continental plates, started to operate after the Paleoproterozoic accretionary stage, which may have ended after formation of the Columbia supercontinent (Rogers and Santosh, 2002; Santosh et al., 2006). Attempts to form supercontinents had already occurred in the Archean (Bleeker, 2003), but they were mostly broken into smaller pieces that drifted apart and formed the cores of Precambrian Shields that stabilized in the Paleoproterozoic. Microplate tectonics or accretionary tectonics were thus complemented by large plate tectonics from the Mesoproterozoic onward. In the younger geological record, microplates are found rimming the larger plates in Phanerozoic orogens and in active accretionary margins like the Indonesian archipelago northwest of the Australian continental margin.

The growth of the Fennoscandian continental plate took place by the accretion of juvenile Paleoproterozoic arcs and microplates. The rapid change from juvenile arc crust to stable continental crust seems to be linked to thick mafic granulitic lower crust underlying the area (Korja, 1995; Korsman et al., 1999; Kuusisto et al., 2006). The granulites may have formed either as lower parts of island arcs, like the granulites recently found in the Kohistan arc (Garrido et al., 2006), or they might have formed in the hot Svecofennian orogeny characterized by low pressure–high-temperature metamorphism (Korsman et al., 1999; Vanderhaege and Teyssier, 2001a, 2001b; Jamieson et al., 2004). Because mafic granulitic lower crust is less dense than the mantle, it cannot subduct. If it is pushed to greater depths by overthrusting of overriding crustal slices, it will rise isostatically. Once isostasy is retained, the uplift ceases, strain hardening occurs, and the crustal structures can be preserved (Kusznir and Park, 1987).

A Tibetan plateau–type upper-crustal extensional area, the Central Finland granitoid complex, developed on top of a collisional structure as a result of stacking of hot juvenile crustal pieces that began to melt within the thickened crust. At depths and temperatures where granulite facies are formed, anatexis begins (20 km, 750 °C) and migmatites are formed. A major change in rheological properties takes place at the migmatite front, where lithologies change from wet to dry or from amphibole- and mica-bearing to pyroxene-bearing lithologies (Vanderhaege and Teyssier, 2001b; Jamieson et al., 2004). Partial melting weakens the zone along which upper-crustal detachment surfaces form (Beaumont et al., 2004). Below the detachment, the crust produces granitoid suites from different melt portions at different crustal levels. These melts are observed as the postcollisional calc-alkaline granitoids intruding the Central Finland granitoid complex and Ladoga–Bothnian Bay wrench fault area. The temperature increase is, however, buffered by energy-consuming melting reactions, and, thus, the temperature may not increase beyond 900 °C in the underlying crustal layers, which attain granulite-facies mineralogy (Vielzeuf et al., 1990). The formation of the migmatite front further emphasizes the layered nature of the crust (Culshaw et al., 2006).

The upper-crustal detachment zone is the migmatite front below which anatexis takes place and synorogenic magmatism is initiated by partial melting of the thickened arc crust. In Central Finland, postkinematic granitoids formed in response to the lower-crustal uplift, associated decompressional melting, magmatic underplating, and partial melting of the lower and middle crust. The granitoids were emplaced above the migmatite front, and granulite-facies rocks are found below the front. In an orogenic context, the Central Finland granitoid complex is the metamorphic core of the orogen.

In the Kola-Lapland orogeny, the Archean Murmansk and composite Karelian cratons were welded together (Gaál and Gorbatschev, 1987; Lahtinen et al., 2005). New crust was formed only in intervening basins like the Lapland granulite belt, which is now exposed as a medium- to high-pressure block between the Murmansk and Karelian cratons (Tuisku and Huhma, 2006). The seismic image suggests a symmetrical head-on collision of the two continents. The Lapland granulite belt forms a symmetrical, large-scale flower structure or a pop-up structure (Pattison et al., 2006). Although the structure looks symmetrical, the lithology and metamorphism show asymmetry and increasing amounts of deformation, shearing, and higher pressures toward the south. The structure could have formed in a two-sided cold collision, where a high degree of deformation and metamorphism should be observed on the retrowedge side of the orogen (Beaumont and Quinlan, 1994). According to Kusznir and Park (1987), the development of subhorizontal reflectivity and detachment surfaces in the lower crust suggests cold collision. Pattison et al. (2006) associated the bright subhorizontal lower-crustal reflections with the extensional collapse of the Lapland-Kola orogen. High-pressure granulite complexes of similar age are found elsewhere in the world, and they have been linked to the formation of the supercontinent Columbia (Santosh et al., 2006).

## CONCLUSIONS

The reflection structures and lithological assemblages observed in the Fennoscandian Shield are similar to modern ones. There are remnants of ophiolites recording paleo-oceanic crust, transform faults recording lateral motion, traces of subduction zones, and modern-style bivergent collision zones.

The reflectivity patterns suggest that the Svecofennian crust was compiled from small terranes with differentiated crust that accreted to the continental margin. It is suggested that the upper parts of frozen subduction zones might be preserved elsewhere where granulitic lower crust fails to subduct.

During thermal maturation and gravitational stabilization, the hot arc crust underwent extensive partial melting and gravitational collapse, whereas the colder plate margins were better preserved in the suture zones. The stabilization processes led to highly variable three-dimensional structures, where the thickest parts are found underneath the suture zones and slightly thinner areas occur in the stacked continental-arc pieces.

We suggest that modern stable crust is potentially formed and preserved in similar environments where small plates with well-developed crustal structure and lithosphere are accreted to continental margins. Strain hardening in an extensional setting finalizes the process and stabilizes the crust.

The plate-tectonic theory, together with gravitational-balancing processes (gravitational collapse), is a viable model to explain the geological and geophysical structures observed in the Paleoproterozoic of Fennoscandia.

## ACKNOWLEDGMENTS

We are grateful for lively and fruitful discussions with the FIRE Working Group, especially to Raimo Lahtinen, Mikko Nironen, and Nicole Pattison. We also thank Kati Karkkulainen for her help in drafting the figures. The thorough reviews by Arie van der Velden and Walter Mooney greatly improved the manuscript.

## REFERENCES CITED

Abbott, D.H., and Mooney, W.D., 1995, The structural and geochemical evolution of the continental crust: Support for the oceanic plateau model of continental growth: Reviews of Geophysics, v. 99, supplement, p. 231–242.

Abbott, D.H., Herzberg, C., Mooney, W., Sparks, D., and Zhang, Y.S., 2000, Quantifying Precambrian crustal extraction: The root is the answer: Tectonophysics, v. 322, p. 163–190, doi: 10.1016/S0040-1951(00)00062-7.

Abramovitz, T., Berthelsen, A., and Thybo, H., 1997, Proterozoic sutures and terranes in the southeastern Baltic Shield interpreted from BABEL seismic data: Tectonophysics, v. 270, p. 259–277, doi: 10.1016/S0040-1951(96)00213-2.

Artemieva, I.M., 2006, Global 1×1 thermal model TC1 for the continental lithosphere: Implications for lithosphere secular evolution: Tectonophysics, v. 416, p. 245–277, doi: 10.1016/j.tecto.2005.11.022.

Austrheim, H., 1990, The granulite-eclogite facies transition: A comparison of experimental work and a natural occurrence in the Bergen Arcs, western Norway: Lithos, v. 25, p. 163–169, doi: 10.1016/0024-4937(90)90012-P.

BABEL Working Group, 1990, Evidence for early Proterozoic plate tectonics from seismic reflection profiles in the Baltic Shield: Nature, v. 348, p. 34–38, doi: 10.1038/348034a0.

BABEL Working Group, 1993, Integrated seismic studies of the Baltic Shield using data in the Gulf of Bothnia region: Geophysical Journal International, v. 112, p. 305–324, doi: 10.1111/j.1365-246X.1993.tb01172.x.

Beaumont, C., and Quinlan, G., 1994, A geodynamic framework for interpreting crustal-scale seismic-reflectivity patterns in compressional orogens: Geophysical Journal International, v. 116, p. 754–783, doi: 10.1111/j.1365-246X.1994.tb03295.x.

Beaumont, C., Munoz, J.A., Hamilton, J., and Fullsack, P., 2000, Factors controlling the Alpine evolution of the central Pyrenees inferred from a comparison of observations and geodynamical models: Journal of Geophysical Research–Solid Earth, v. 105, p. 8121–8145.

Beaumont, C., Jamieson, R.A., Nguyen, M.H., and Medvedev, S., 2004, Crustal channel flows: 1. Numerical models with applications to the tectonics of the Himalayan-Tibetan orogen: Journal of Geophysical Research–Solid Earth, v. 109, doi: 10.1029/2003JB002809.

Bédard, J.H., 2006, A catalytic delamination-driven model for coupled genesis of Archean crust and sub-continental lithospheric mantle: Geochimica et Cosmochimica Acta, v. 70, p. 1188–1214, doi: 10.1016/j.gca.2005.11.008.

Bjørnerud, M.G., and Austrheim, H., 2004, Inhibited eclogite formation: The key to the rapid growth of strong and buoyant Archean continental crust: Geology, v. 32, p. 765–768, doi: 10.1130/G20590.1.

Bleeker, W., 2003, The late Archean record: A puzzle in ca. 35 pieces: Lithos, v. 71, p. 99–134, doi: 10.1016/j.lithos.2003.07.003.

Bogdanova, S.V., Gorbatschev, R., Grad, M., Janik, T., Guterch, A., Kozlovskaya, E., Motuza, G., Skridlaite, G., Starostenko, V., Taran, L., and EUROBRIDGE and POLONAISE Working Groups, 2006, EUROBRIDGE: New insight into the geodynamic evolution of the East European craton, *in* Gee, D.G., and Stephenson, R.A., eds., European Lithosphere Dynamics: Geological Society of London Memoir 32, p. 599–626.

Cawood, P.A., Kröner, A., and Pisarev, S., 2006, Precambrian plate tectonics: Criteria and evidence: GSA Today, v. 16, no. 7, p. 4–11, doi: 10.1130/GSAT01607.1.

Christensen, N.I., and Mooney, W.D., 1995, Seismic velocity structure and composition of the continental crust: A global view: Journal of Geophysical Research, v. 100, p. B9761–B9766, doi: 10.1029/95JB00259.

Collins, W.J., 2002, Hot orogens, tectonic switching, and creation of continental crust: Geology, v. 30, p. 535–538, doi: 10.1130/0091-7613(2002)030<0535:HOTSAC>2.0.CO;2.

Condie, K.C., 2002, Continental growth during 1.9 Ga superplume event: Journal of Geodynamics, v. 34, p. 249–264, doi: 10.1016/S0264-3707(02)00023-6.

Coney, P.J., Jones, D.L., and Monger, J.W.H., 1980, Cordilleran suspect terranes: Nature, v. 288, p. 329–333.

Cook, F.A., Albaugh, D.S., Brown, L.D., Kaufman, S., Oliver, J.E., and Hatcher, R.D., 1979, Thin-skinned tectonics in the crystalline southern Appalachians; COCORP seismic-reflection profiling of the Blue Ridge and Piedmont: Geology, v. 7, p. 563–567, doi: 10.1130/0091-7613(1979)7<563:TTITCS>2.0.CO;2.

Cook, F.A., van der Velden, A.J., Hall, K.W., and Roberts, B.J., 1999, Frozen subduction in Canada's Northwest Territories: Lithoprobe deep lithospheric reflection profiling of the western Canadian Shield: Tectonics, v. 18, p. 1–24, doi: 10.1029/1998TC900016.

Coward, M., 1994, Continental collision, *in* Hancock, P.L., ed., Continental Deformation: Oxford, Pergamon Press, p. 264–288.

Culshaw, N.G., Beaumont, C., and Jamieson, R.A., 2006, The orogenic superstructure-infrastructure concept: Revisited, quantified, and revived: Geology, v. 34, p. 733–736, doi: 10.1130/G22793.1.

Daly, J.S., Balagansky, V.V., Timmerman, M.J., Whitehouse, M.J., de Jong, K., Guise, P., Bogdanova, S., Gorbatschev, R., and Bridgwater, D., 2001, Ion microprobe U-Pb zircon geochronology and isotopic evidence for a trans-crustal suture in the Lapland-Kola orogen, northern Fennoscandian Shield: Precambrian Research, v. 105, p. 289–314, doi: 10.1016/S0301-9268(00)00116-9.

Daly, J.S., Balagansky, V.V., Timmerman, M.J., and Whitehouse, M.J., 2006, The Lapland-Kola orogen: Palaeoproterozoic collision and accretion of the northern Fennoscandian lithosphere, *in* Gee, D.G., and Stephenson, R.A., eds., European Lithosphere Dynamics: Geological Society of London Memoir 32, p. 561–578.

Ehlers, C., Lindroos, A., and Selonen, O., 1993, The late Svecofennian granite-migmatite zone of southern Finland—A belt of transpressive deformation and granite emplacement: Precambrian Research, v. 64, p. 295–309, doi: 10.1016/0301-9268(93)90083-E.

Elliott, B.A., Rämö, O.T., and Nironen, M., 1998, Mineral chemistry constraints on the evolution of the 1.88–1.87 Ga post-kinematic granite plutons in the Central Finland granitoid complex: Lithos, v. 45, p. 109–129, doi: 10.1016/S0024-4937(98)00028-0.

Ellis, D.J., and Maboko, M.A.H., 1992, Precambrian tectonics and the physicochemical evolution of the continental crust. I. The gabbro-eclogite transition revisited: Precambrian Research, v. 55, p. 491–506, doi: 10.1016/0301-9268(92)90041-L.

Eskola, P., 1949, The problem of mantled gneiss domes: Quarterly Journal of the Geological Society of London, v. 104, p. 461–476.

Gaál, G., and Gorbatschev, R., 1987, An outline of the Precambrian evolution of the Baltic Shield: Precambrian Research, v. 35, p. 15–52, doi: 10.1016/0301-9268(87)90044-1.

Gaál, G., Berthelsen, A., Gorbatschev, R., Kesola, R., Lehtonen, M.I., Marker, M., and Raase, P., 1989, Structure and composition of the Precambrian crust along the POLAR profile in the northern Baltic Shield: Tectonophysics, v. 162, p. 1–25, doi: 10.1016/0040-1951(89)90354-5.

Garrido, C.J., Bodinier, J.-L., Burg, J.-P., Zeilinger, G., Hussain, S., Dawood, H., Nawaz Chaudhry, M., and Gervilla, F., 2006, Petrogenesis of mafic garnet granulite in the lower crust of Kohistan paleoarc complex (Northern Pakistan): Implications for infracrustal differentiation of island arcs and generation of continental crust: Journal of Petrology, v. 47, p. 1873–1914, doi: 10.1093/petrology/egl030.

Gohl, K., and Pedersen, L.B., 1995, Collisional tectonics of the Baltic Shield in the northern Gulf of Bothnia from seismic data of the BABEL project: Geophysical Journal International, v. 120, p. 209–226, doi: 10.1111/j.1365-246X.1995.tb05921.x.

Gorbatschev, R., and Bogdanova, S., 1993, Frontiers in the Baltic Shield: Precambrian Research, v. 64, p. 3–21, doi: 10.1016/0301-9268(93)90066-B.

Hajnal, Z., Lucas, S., White, D., Lewry, J., Bezdan, S., and Stauffer, M., R. and Thomas M., D., 1996, Seismic reflection images of high-angle faults and linked detachments in the Trans-Hudson orogen: Tectonics, v. 15, p. 427–439.

Halden, N.M., 1982, Structural, metamorphic and igneous history of the migmatites in the deep levels of wrench fault regime, Savonranta, eastern Finland: Transactions of the Royal Society of Edinburgh, Earth Sciences, v. 73, p. 17–30.

Harding, T.P., 1985, Seismic characteristics and identification of negative flower structures, positive flower structures and positive structural inversions: American Association of Petroleum Geologists Bulletin, v. 69, p. 582–600.

Heikkinen, P., and Luosto, U., 2000, Review of some features of the seismic velocity models in Finland, *in* Pesonen, L.J., Korja, A., and Hjelt, S.-E., eds., Lithosphere 2000: A Symposium on the Structure, Composition And Evolution of the Lithosphere in Finland (4–5 October 2000): Espoo, Otaniemi, Institute of Seismology, University of Helsinki, Report S-41, p. 35–41.

Henkel, H., Lee, M.K., Lund, C.-E., and Rasmussen, T.M., 1991, An integrated geophysical interpretation of the 2000 km FENNOLORA section of the Baltic Shield, *in* Freeman, R., Giese P., and Mueller, St., eds., The European Geotraverse: Integrative Studies, Results from the Fifth Earth Science Study Centre, Rauischholzhausen Germany, 26.3.–7.4., 1990: Strasbourg, European Science Foundation, p. 1–48.

Hietanen, A., 1975, Generation of potassium-poor magmas in the northern Sierra Nevada and the Svecofennian in Finland: Journal of Research of the U.S. Geological Survey, v. 3, p. 631–645.

Hoffman, P.F., 1988, United plates of America, the birth of a craton: Early Proterozoic assembly and growth of Laurentia: Annual Review of Earth and Planetary Sciences, v. 16, p. 543–603, doi: 10.1146/annurev.ea.16.050188.002551.

Hölttä, P., Huhma, H., Mänttäri, I., Peltonen, P., and Juhanoja, J., 2000, Petrology and geochemistry of mafic granulite xenoliths from the Lahtojoki kimberlite pipe, eastern Finland: Lithos, v. 51, p. 109–133, doi: 10.1016/S0024-4937(99)00077-8.

Hyvönen, T., Tiira, T., Korja, A., Heikkinen, P., Rautioaho, E., and the SVEKALAPKO Seismic Tomography Working Group, 2006, A tomographic crustal velocity model of the central Fennoscandian Shield: Geophysical Journal International, v. 168, p. 1210–1226.

Jackson, J.A., Austrheim, H., McKenzie, D., and Priestley, K., 2004, Metastability, mechanical strength, and the support of mountain belts: Geology, v. 32, p. 625–628, doi: 10.1130/G20397.1.

Jamieson, R.A., Beaumont, C., Medvedev, S., and Nguyen, M.H., 2004, Crustal channel flows: 2. Numerical models with implications for metamorphism in the Himalayan-Tibetan orogen: Journal of Geophysical Research–Solid Earth, v. 109, doi: 10.1029/2003JB002811.

Ji, S., and Long, Ch., 2006, Seismic reflection response of folded structures and implications for the interpretations of deep seismic reflection profiles: Journal of Structural Geology, v. 28, p. 1380–1387, doi: 10.1016/j.jsg.2006.05.003.

Kähkönen, Y., 1987, Geochemistry and tectonomagmatic affinities of the metavolcanic rocks of the early Proterozoic Tampere schist belt, southern Finland: Precambrian Research, v. 35, p. 295–311, doi: 10.1016/0301-9268(87)90060-X.

Kähkönen, Y., 2005, Svecofennian supracrustal rocks, *in* Lehtinen, M., Nurmi, P.A., and Rämö, O.T., eds., Precambrian Geology of Finland: Key to the Evolution of the Fennoscandian Shield: Amsterdam, Elsevier, Developments in Precambrian Geology, v. 14, p. 343–405.

Kilpeläinen, T., 1998, Evolution and 3D Modelling of Structural and Metamorphic Patterns of the Palaeoproterozoic Crust in the Tampere-Vammala Area, Southern Finland: Geological Survey of Finland Bulletin 397, 124 p. + 2 app.

Kohonen, J., 1995, From Continental Rifting to Collisional Shortening—Paleoproterozoic Kaleva Metasediments of the Höytiäinen Area in North Karelia, Finland: Geological Survey of Finland Bulletin 380, 79 p.

Koistinen, T.J., 1981, Structural evolution of an early Proterozoic strata-bound Cu-Co-Zn deposit, Outokumpu, Finland: Transactions of the Royal Society of Edinburgh, Earth Sciences, v. 72, p. 115–158.

Koistinen, T., Stephens, M.B., Bogatchev, V., Nordgulen, Ø., Wennerström, M., and Korhonen, J., 2001, Geological map of the Fennoscandian Shield: Espoo, Geological Survey of Finland; Trondheim, Geological Survey of Norway; Uppsala, Geological Survey of Sweden; and Moscow, Ministry of Natural Resources of Russia, scale 1:2,000,000.

Kontinen, A., 1987, An early Proterozoic ophiolite—The Jormua mafic-ultramafic complex, northeastern Finland: Precambrian Research, v. 35, p. 313–341, doi: 10.1016/0301-9268(87)90061-1.

Korja, A., 1995, Structure of the Svecofennian Crust—Growth and Destruction of the Svecofennian Orogen [Ph.D. thesis]: Helsinki, Institute of Seismology, University of Helsinki, 36 p.

Korja, A., and Heikkinen, P., 2005, The accretionary Svecofennian orogen—Insight from the BABEL profiles: Precambrian Research, v. 136, p. 241–268, doi: 10.1016/j.precamres.2004.10.007.

Korja, A., Korja, T., Luosto, U., and Heikkinen, P., 1993, Seismic and geoelectric evidence for collisional and extensional events in the Fennoscandian Shield—Implications for Precambrian crustal evolution: Tectonophysics, v. 219, p. 129–152, doi: 10.1016/0040-1951(93)90292-R.

Korja, A., Heikkinen, P., and Aaro, S., 2001, Crustal structure of the northern Baltic Sea paleorift: Tectonophysics, v. 331, p. 341–358, doi: 10.1016/S0040-1951(00)00290-0.

Korja, A., Lahtinen, R., Heikkinen, P., Kukkonen, I.T., and FIRE Working Group, 2006a, A geological interpretation of the upper crust along FIRE 1, *in* Kukkonen, I.T., and Lahtinen, R., eds., Finnish Reflection Experiment FIRE 2001–2005: Geological Survey of Finland Special Paper 43, p. 45–76.

Korja, A., Lahtinen, R., and Nironen, M., 2006b, The Svecofennian orogen: A collage of microcontinents and island arcs, *in* Gee, D., and Stephenson, R., eds., European Lithosphere Dynamics: Geological Society of London Memoir 32, p. 561–578.

Korja, T., Engels, M., Zhamaletdinov, A.A., Kovtun, A.A., Palshin, N.A. Smirnov, M.Yu., Tokarev, D.A., Asming, V.E., Vanyan, L.L., Vardaniants, I.L., and the BEAR Working Group, 2002, Crustal conductivity in Fennoscandia—A compilation of a database on crustal conductance in the Fennoscandian Shield: Earth, Planets, Space, v. 54, p. 535–558.

Korsman, K., and Glebovitsky, V., eds., 1999, Raahe-Ladoga zone structure-lithology, metamorphism and metallogeny: A Finnish-Russian cooperation project 1996–1999. Map 2: Metamorphism of the Raahe-Ladoga Zone: Espoo, Geological Survey of Finland, scale 1:1,000,000.

Korsman, K., Koistinen, T., Kohonen, J., Wennerström, M., Ekdahl, E., Honkamo, M., Idman, H., and Pekkala, Y., eds., 1997, Suomen kallioperäkartta-Berggrundskarta över Finland (Bedrock map of Finland): Espoo, Finland, Geological Survey of Finland, scale 1:1,000,000.

Korsman, K., Korja, T., Pajunen, M., Virransalo, P., and GGT/SVEKA Working Group, 1999, The GGT/SVEKA transect: Structure and evolution of the continental crust in the Paleoproterozoic Svecofennian orogen in Finland: International Geological Review, v. 41, p. 287–333.

Kozlovskaya, E., Elo, S., Hjelt, S.E., Yliniemi, J., Pirttijärvi, M., and the SVEKALAPKO Seismic Tomography Working Group, 2004, 3-D density model of the crust of southern and central Finland obtained from joint interpretation of the SVEKALAPKO crustal P wave velocity models and gravity data: Geophysical Journal International, v. 158, p. 827–848.

Kukkonen, I.T., Heikkinen, P., Ekdahl, E., Hjelt, S.-E., Yliniemi, J., Jalkanen, E., and FIRE Working Group, 2006, Acquisition and geophysical characteristics of reflection seismic data on FIRE transects, Fennoscandian Shield, *in* Kukkonen, I.T., and Lahtinen, R., eds., Finnish Reflection Experiment FIRE 2001–2005: Geological Survey of Finland Special Paper 43, p. 13–43.

Kurhila, M., Vaasjoki, M., Mänttäri, I., Rämö, T., and Nironen, M., 2005, U-Pb ages and Nd isotope characteristics of the lateorogenic, migmatizing microcline granites in southwestern Finland: Bulletin of the Geological Society of Finland, v. 77, p. 105–128.

Kusznir, N.J., and Park, R.G., 1987, The extensional strength of the continental lithosphere: Its dependence on geothermal gradient, and crustal composition and thickness, *in* Coward, M.P., Dewey, J.F., and Hancock, P.L., eds., Continental Extension Tectonics: Geological Society of London Special Publication 28, p. 35–52.

Kuusisto, M., Kukkonen, I.T., Heikkinen, P., and Pesonen, L.J., 2006, Lithological interpretation of crustal composition in the Fennoscandian Shield with seismic velocity data: Tectonophysics, v. 420, p. 283–299, doi: 10.1016/j.tecto.2006.01.014.

Lahtinen, R., 1994, Crustal Evolution of the Svecofennian and Karelian Domains during 2.1–1.79 Ga, with Special Emphasis on the Geochemistry and Origin of 1.93–1.91 Ga Gneisses, Tonalites and Associated Supracrustal Rocks in the Rautalampi Area, Central Finland: Geological Survey of Finland Bulletin 378, 128 p.

Lahtinen, R., Korja, A., and Nironen, M., 2005, Palaeoproterozoic tectonic evolution, *in* Lehtinen, M., Nurmi, P., and Rämö, T., eds., The Precambrian Geology of Finland—Key to the Evolution of the Fennoscandian Shield: Amsterdam, Elsevier, p. 418–532.

Langenheim, V.E., Jachens, R.C.C., Matti, J.C., Hauksson, E., Morton, D.M., and Christensen, A., 2005, Geophysical evidence for wedging in the San Gorgonio Pass structural knot, southern San Andreas fault zone, southern California: Geological Society of America Bulletin, v. 117, p. 1554–1572, doi: 10.1130/B25760.1.

Maggi, A., Jackson, J.A., McKenzie, D., and Priestly, K., 2000, Earthquake focal depths, effective thickness, and the strength of the continental lithosphere: Geology, v. 28, p. 495–498, doi: 10.1130/0091-7613(2000)28<495:EFDEET>2.0.CO;2.

Meissner, R., 1996, Faults, and folds, fact and fiction: Tectonophysics, v. 264, p. 279–293, doi: 10.1016/S0040-1951(96)00132-1.

Mooney, W.D., Laske, G., and Masters, G.T., 1998, CRUST 5.1; a global crustal model at 5 degrees × 5 degrees: Journal of Geophysical Research, v. 103, p. 727–747, doi: 10.1029/97JB02122.

Mouri, H., Korsman, K., Huhma, H., 1999, Tectono-metamorphic evolution and timing of the melting processes in the Svecofennian Tonalite-Trondhjemite Migmatite Belt: An example from Luopioinen, Tampere area, southern Finland, *in* Kähkönen, Y., and Lindqvist, K., eds., Studies related to the Global Geoscience Transects/SVEKA Project in Finland: Bulletin of the Geological Society of Finland, v. 71, no. 1, p. 31–56.

Nironen, M., 1989, Emplacement and Structural Setting of Granitoids in the Early Proterozoic Tampere and Savo Schist Belts, Finland—Implications for Contrasting Crustal Evolution: Geological Survey of Finland Bulletin 346, 83 p.

Nironen, M., 1997, The Svecofennian orogen: A tectonic model: Precambrian Research, v. 86, p. 21–44, doi: 10.1016/S0301-9268(97)00039-9.

Nironen, M., 2003, Keski-Suomen Granitoidikompleksi: Karttaselitys (Summary: Central Finland Granitoid Complex—Explanation to a Map): Geological Survey of Finland Report 157, 45 p. + 1 app. map.

Nironen, M., Elliott, B.A., and Rämö, O.T., 2000, 1.88–1.87 Ga post-kinematic intrusions of the Central Finland granitoid complex: A shift from C-type to A-type magmatism during lithospheric convergence: Lithos, v. 53, no. 1, p. 37–58, doi: 10.1016/S0024-4937(00)00007-4.

Nironen, M., Korja, A., Heikkinen, P., and the FIRE Working Group, 2006, A geological interpretation of the upper crust along FIRE 2 and FIRE 2A, *in* Kukkonen, I.T., and Lahtinen, R., eds., Finnish Reflection Experiment FIRE 2001–2005: Geological Survey of Finland Special Paper 43, p. 77–103.

Park, A.F., and Bowes, D.R., 1983, Basement-cover relationships during polyphase deformation in the Svecokarelides of the Kaavi district, eastern Finland: Transactions of the Royal Society of Edinburgh, Earth Sciences, v. 74, p. 95–118.

Park, A.F., Bowes, D.R., Halden, N.M., and Koistinen, T.J., 1984, Tectonic evolution at an early Proterozoic continental margin: The Svecokarelides of eastern Finland: Journal of Geodynamics, v. 1, p. 359–386, doi: 10.1016/0264-3707(84)90016-4.

Patchett, P.J., and Kouvo, O., 1986, Origin of continental crust of 1.9–1.7 Ga age: Nd isotopes and U-Pb ages in the Svecokarelian terrain of South Finland: Contributions to Mineralogy and Petrology, v. 92, p. 1–12, doi: 10.1007/BF00373959.

Pattison, N.L., Korja, A., Lahtinen, R., Ojala, V.J., and the FIRE Working Group, 2006, FIRE seismic reflection profiles 4, 4A and 4B: Insights into the crustal structure of northern Finland from Ranua to Näätämö, *in* Kukkonen, I.T., and Lahtinen, R., eds., Finnish Reflection Experiment FIRE 2001–2005: Geological Survey of Finland Special Paper 43, p. 161–222.

Peltonen, P., and Brügmann, G., 2006, Origin of layered continental mantle (Karelian craton, Finland): Geochemical and Re-Os isotope constraints: Lithos, v. 89, p. 405–423, doi: 10.1016/j.lithos.2005.12.013.

Peltonen, P., Kontinen, A., and Huhma, H., 1996, Petrology and geochemistry of metabasalts from the 1.95 Ga Jormua ophiolite, northeastern Finland: Journal of Petrology, v. 37, p. 1359–1383, doi: 10.1093/petrology/37.6.1359.

Rogers, J.J.W., 1996, A history of continents in the past three billion years: The Journal of Geology, v. 104, p. 91–107.

Rogers, J.J.W., and Santosh, M., 2002, Configuration of Columbia, a Mesoproterozoic supercontinent: Gondwana Research, v. 5, p. 5–22, doi: 10.1016/S1342-937X(05)70883-2.

Sandoval, S., Kissling, E., and Ansorge, J., 2003, High-resolution body wave tomography beneath the SVEKALAPKO array: I. A priori three-dimensional crustal model and associated traveltime effects on teleseismic wave fronts: Geophysical Journal International, v. 153, p. 75–87, doi: 10.1046/j.1365-246X.2003.01888.x.

Santosh, M., Sajeev, K., and Li, J.H., 2006, Extreme crustal metamorphism during Columbia supercontinent assembly: Evidence from North China craton: Gondwana Research, v. 10, p. 256–266, doi: 10.1016/j.gr.2006.06.005.

Sato, H., Hirata, N., Koketsu, K., Okaya, D., Abe, S., Kobayashi, R., Matsubara, M., Iwasaki, T., Ito, T., Ikawa, T., Kawanaka, T., Kasahara, K., and Harder, H., 2005, Earthquake source fault beneath Tokyo: Science, v. 309, p. 462–464, doi: 10.1126/science.1110489.

Schulmann, K., Thompson, A.B., Lexa, O., and Jezek, J., 2003, Strain distribution and fabric development modelled in active and ancient transpressive zones: Journal of Geophysical Research–Solid Earth, v. 108, p. 2023, doi: 10.1029/2001/JB000632.

Scotese, C.R., 2001, Atlas of Earth History: Volume 1. Paleogeography: Arlington, Texas, PALEOMAP Project, 52 p.

Shillington, D.J., Van Avendonk, H., Holbrook, W.S., Kelemen, P., and Hornbach, M., 2004, Composition and structure of the central Aleutian island arc from arc-parallel wide-angle seismic data: Geochemistry, Geophysics, Geosystems, v. 5, p. Q10006, doi: 10.1029/2004GC000715.

Sorjonen-Ward, P., 2006, Geological and structural framework and preliminary interpretation of the FIRE 3 and FIRE 3A reflection seismic profiles, central Finland, *in* Kukkonen, I.T., and Lahtinen, R., eds., Finnish Reflection Experiment FIRE 2001–2005: Geological Survey of Finland Special Paper 43, p. 105–160.

Sorjonen-Ward, P., and Luukkonen, E., 2005, Archean rocks, *in* Lehtinen, M., Nurmi, P., and Rämö, T., eds., The Precambrian Geology of Finland—Key to the Evolution of the Fennoscandian Shield: Amsterdam, Elsevier, p. 19–99.

Stampfli, G.M., and Borel, G.D., 2002, A plate tectonic model for the Paleozoic and Mesozoic constrained by dynamic plate boundaries and restored synthetic oceanic isochrones: Earth and Planetary Science Letters, v. 196, p. 17–33, doi: 10.1016/S0012-821X(01)00588-X.

Stern, R.J., 2005, Evidence from ophiolites, blueschists, and ultrahigh-pressure metamorphic terranes that the modern episode of subduction tectonics began in Neoproterozoic time: Geology, v. 33, p. 557–560, doi: 10.1130/G21365.1.

Suyehiro, K., Takahashi, N., Ariie, Y., Yokoi, Y., Hino, R., Shinohara, M., Kanazawa, T., Hirata, N., Tokuyama, H., and Taira, A., 1996, Continental crust, crustal underplating, and low-Q upper mantle beneath an oceanic island arc: Science, v. 272, p. 390–392, doi: 10.1126/science.272.5260.390.

Takahashi, N., Kodaira, S., Klemperer, S.L., Tatsumi, Y., Kaneda, Y., and Suyehiro, K., 2007, Crustal structure and evolution of the Mariana intra-oceanic island arc: Geology, v. 35, p. 203–206, doi: 10.1130/G23212A.1.

Thompson, A.B., Schulmann, K., Jezek, J., and Tolar, V., 2001, Thermally softened continental extensional zones (arcs and rifts) as precursors to thickened orogenic belts: Tectonophysics, v. 332, p. 115–141, doi: 10.1016/S0040-1951(00)00252-3.

Tuisku, P., and Huhma, H., 2006, Evolution of migmatitic granulite complexes: Implications from Lapland granulite belt, Part II: Isotopic dating: Bulletin of the Geological Society of Finland, v. 78, p. 143–175.

Tyson, A.R., Morozova, E.A., Karlstrom, K.E., Chamberlain, K.R., Smithson, S.B., Dueker, K.G., and Foster, C.T., 2002, Proterozoic Farwell Mountain–Lester Mountain suture zone, northern Colorado: Subduction flip and progressive assembly of arcs: Geology, v. 30, p. 943–946, doi: 10.1130/0091-7613(2002)030<0943:PFMLMS>2.0.CO;2.

Vanderhaeghe, O., and Teyssier, C., 2001a, Crustal-scale rheological transitions during late-orogenic collapse: Tectonophysics, v. 335, p. 211–228, doi: 10.1016/S0040-1951(01)00053-1.

Vanderhaeghe, O., and Teyssier, C., 2001b, Partial melting and flow of orogens: Tectonophysics, v. 342, p. 451–472, doi: 10.1016/S0040-1951(01)00175-5.

van der Velden, A.J., and Cook, F.A., 1999, Proterozoic and Cenozoic subduction complexes: A comparison of geometric features: Tectonics, v. 18, p. 575–581, doi: 10.1029/1999TC900011.

Vielzeuf, D., Clemens, J.D., Pin, C., and Moinet, E., 1990, Granites, granulites and crustal differentiation, *in* Vielzeuf, D., and Vidal, Ph., eds., Granulites and Crustal Evolution: Dordrecht, Netherlands, Kluwer Academic Publishers, p. 59–85.

Ward, P., 1987, Early Proterozoic deposition and deformation at the Karelian craton margin in southeastern Finland: Precambrian Research, v. 35, p. 71–93, doi: 10.1016/0301-9268(87)90046-5.

Wegmann, C.E., 1928, Ueber die Tektonik der jüngeren Faltung in Ostfinnland: Fennia, v. 50, p. 1–22.

Wernicke, B., 1985, Theory of large-scale, uniform-sense normal simple shear of the continental lithosphere: Canadian Journal of Earth Sciences, v. 22, p. 108–125.

White, D.J., Zwanzig, H.V., and Hajnal, Z., 2000, Crustal suture preserved in the Paleoproterozoic Trans-Hudson orogen, Canada: Geology, v. 28, p. 527–530, doi: 10.1130/0091-7613(2000)28<527:CSPITP>2.0.CO;2.

Woodcock, N.H., and Fischer, M., 1986, Strike-slip duplexes: Journal of Structural Geology, v. 8, p. 725–735, doi: 10.1016/0191-8141(86)90021-0.

Wu, F.T., Rau, R.J., and Salzberg, D., 1997, Taiwan orogeny: Thin-skinned or lithospheric collision?: Tectonophysics, v. 274, p. 191–220, doi: 10.1016/S0040-1951(96)00304-6.

Zhao, G.C., Cawood, P.A., Wilde, S.A., and Sun, M., 2002, Review of global 2.1–1.8 Ga orogens: Implications for a pre-Rodinia supercontinent: Earth-Science Reviews, v. 59, p. 125–162, doi: 10.1016/S0012-8252(02)00073-9.

Zhu, L., 2000, Crustal structure across the San Andreas fault, southern California, from teleseismic converted waves: Earth and Planetary Science Letters, v. 179, p. 183–190, doi: 10.1016/S0012-821X(00)00101-1.

Manuscript Accepted by the Society 14 August 2007

Printed in the USA

The Geological Society of America
Special Paper 440
2008

# *Plate tectonics on early Earth? Weighing the paleomagnetic evidence*

**David A.D. Evans***
*Department of Geology & Geophysics, Yale University, New Haven, Connecticut 06520-8109, USA*

**Sergei A. Pisarevsky***
*School of GeoSciences, University of Edinburgh, Edinburgh EH9 3JW, Scotland, UK*

## ABSTRACT

**Paleomagnetism is the only quantitative method available to test for lateral motions by tectonic plates across the surface of ancient Earth. Here, we present several analyses of such motions using strict quality criteria from the global paleomagnetic database of pre–800 Ma rocks. Extensive surface motion of cratons can be documented confidently to older than ca. 2775 Ma, but considering only the most reliable Archean data, we cannot discern differential motion from true polar wander (which can also generate surface motions relative to the geomagnetic reference frame). In order to find evidence for differential motions between pairs of Precambrian cratons, we compared distances between paleomagnetic poles through precisely isochronous intervals for pairs of cratons. The existing database yields several such comparisons with ages ranging from ca. 1110 to ca. 2775 Ma. Only one pair of these ages, 1110–1880 Ma, brackets significantly different apparent polar wander path lengths between the same two cratons and thus demonstrates differential surface motions. If slightly less reliable paleomagnetic results are considered, however, the number of comparisons increases dramatically, and an example is illustrated for which a single additional pole could constrain differential cratonic motion into the earliest Paleoproterozoic and late Neoarchean (in the interval 2445–2680 Ma). In a separate analysis based in part upon moderately reliable paleomagnetic poles, if a specific reconstruction is chosen for Laurentia and Baltica between ca. 1265 and 1750 Ma, then those cratons' rotated apparent polar wander paths show convergence and divergence patterns that accord with regional tectonics and appear to be remarkably similar to predictions from a plate-tectonic conceptual model. Carefully targeted and executed future paleomagnetic studies of the increasingly well-dated Precambrian rock record can imminently extend these tests to ca. 2700 Ma, and with substantially more effort, to perhaps as old as ca. 3500 Ma.**

**Keywords:** plate tectonics, paleomagnetism, Precambrian, Archean, Proterozoic.

*Evans, corresponding author: dai.evans@yale.edu; Pisarevsky: sergei.pisarevsky@ed.ac.uk.

Evans, D.A.D., and Pisarevsky, S.A., 2008, Plate tectonics on early Earth? Weighing the paleomagnetic evidence, *in* Condie, K.C., and Pease, V., eds., When Did Plate Tectonics Begin on Planet Earth?: Geological Society of America Special Paper 440, p. 249–263, doi: 10.1130/2008.2440(12). For permission to copy, contact editing@geosociety.org. 

## PRINCIPLES AND PRACTICALITIES

Most definitions of plate tectonics include as a major component, if not a complete basis, the lateral motion by large blocks of rigid lithosphere across the surface of the planet. Whereas many of the papers in this volume focus on indirect petrological, geochemical, or isotopic evidence for (or against) early plate tectonics on Earth, this contribution considers the most direct form of evidence for lateral surface motions, consulting the global paleomagnetic record from Archean and Proterozoic cratons. Within a single craton, shifts in paleomagnetic directions from rocks of different ages demonstrate rotations or latitudinal changes of that block relative to the geomagnetic reference frame. When coeval paleomagnetic records are compared between two or more cratons, differential block motion may be distinguished. If paleomagnetic directions vary across a large craton in a manner that is consistent with some assumed geomagnetic field geometry (e.g., a geocentric axial dipole, "GAD"), then internal rigidity of that craton may be inferred. In theory, then, the paleomagnetic method can be used to test the most fundamental kinematic elements of plate tectonics in early Earth history.

In practice, several limitations of the geological record prevent application of paleomagnetic techniques to solve these problems. First, in order to infer geographic movement from observed temporal variations in paleomagnetic direction, one must rely on a specified model for the geomagnetic reference frame. For over half a century, the GAD model has served as the starting point for all paleomagnetic investigations of phenomena occurring at time scales longer than that of typical geomagnetic secular variation (~$10^3$–$10^4$ yr; Merrill et al., 1996). The GAD model has proved to be consistent, or very nearly so, for time-averaged paleomagnetic directions from rocks formed during the last five million years, of both normal and reverse polarity, within typical errors of ~5° for individual or mean paleomagnetic results (McElhinny et al., 1996; Tauxe, 2005, and references therein).

Determinations of any subsidiary departure from a GAD field in ancient times are limited by the spatial distribution of paleomagnetic data in the context of plate reconstruction models. Pre-Jurassic data cannot be transferred across plate boundaries, for the sake of comparisons and checks on internal consistency, due to the lack of intact oceanic lithosphere from those ages; instead, potential departures from a GAD model must be investigated using single large plates (i.e., supercontinents), presumed episodes of true polar wander (TPW), or latitude-based statistical tests. Inasmuch as supercontinent reconstructions and hypothesized TPW events are based heavily, if not entirely, on paleomagnetic data, the danger of circularity in these tests of field geometry is apparent.

Global statistical tests of pre-Mesozoic nondipole components are limited to zonal field harmonics, and they are further restricted, due to symmetry considerations in the reversing geodynamo, to the geocentric axial octupole component (g3) and its related family of higher harmonics. The magnitude of the octupole component has been proposed to vary between negligible levels and as much as ~20% relative to the dipole during the past 300 m.y. (Torsvik and Van der Voo, 2002). Following the method developed by M.E. Evans (1976) and several subsequent studies, Kent and Smethurst (1998) found a database-wide bias toward shallow inclinations of the entire pre-Mesozoic paleomagnetic record. An obvious candidate for systematic error could be the shallowing of inclinations within sedimentary rocks due to post-depositional compaction, but the bias was also observed within an igneous-only subset of results. The method assumes random sampling of latitudes by the continents as they migrate across the globe, and some recent attention has been paid to this requirement (Meert et al., 2003; M.E. Evans, 2005). As more high-quality data accumulate, there is certainly opportunity to refine this analysis in the future. In the meantime, D. Evans (2006) showed the consistency of paleomagnetic latitudes determined for large evaporite basins from modern times through most of the last two billion years, permitting a GAD field geometry over that interval. Ancient geomagnetic field strength is not well agreed upon, but variations throughout the entire geological record almost always fall within the range of 10%–150% of the present value (Biggin and Thomas, 2003; Macouin et al., 2003; Valet et al., 2005; Tarduno et al., 2006).

In summary, there is diverse evidence from the paleomagnetic record that a modern-like GAD geomagnetic field, whether generally weaker or comparable in strength to that of today, existed throughout the Phanerozoic and Proterozoic Eons, to first approximation. This uniformitarian geodynamo therefore provides a stable frame of reference by which we may measure and compare cratonic motions to assess the likelihood of plate tectonics on early Earth. However, there are more difficult obstacles presented by the geological record, especially for the Archean.

The stability of paleomagnetic remanence in ancient rocks is governed primarily by original ferromagnetic mineralogy and grain size (Butler and Banerjee, 1975) and by subsequent metamorphic conditions. The Curie temperatures of the two most common primary ferromagnetic minerals in crustal rocks, magnetite and hematite, are 580 °C and 675 °C, respectively, roughly within the amphibolite metamorphic facies. However, practical experience from the last half-century of paleomagnetic work has demonstrated that even lower-middle greenschist metamorphism is likely to erase most or all of the earlier magnetic remanence history. This is largely due to the integrated time-temperature history of prolonged tectonothermal activity, which is able to cause superparamagnetic relaxation (resetting) of ferromagnetic mineral grains well below their Curie temperatures (Pullaiah et al., 1975). In a global perspective, subgreenschist stratified Archean rocks are rare. (Unlayered intrusive rocks pose difficulties for paleomagnetic study due to the possibility of unquantifiable tilting during exhumation.) Notable exceptions are the low-grade cover successions of the Kaapvaal and Pilbara cratons, which will be described later, and isolated prehnite-pumpellyite domains within Neoarchean greenstone belts. For a successful paleomagnetic test of relative motions between two Archean crustal blocks, one must not only find the rare example of a subgreenschist stratified unit on the first block, but also the equally rare example of the same age on the second block. Such possibilities become vanishingly small within the preserved geological record on Earth older than ca. 3 Ga. Indeed, as will be shown later, even the relatively rich Proterozoic paleomagnetic record yields few valid tests of relative motions using this technique.

The predictive power of plate kinematics is founded on the internal rigidity of large regions of lithosphere; kinematic data obtained from one side of a plate can then be extrapolated to the remainder of its area. Large sedimentary basins and extensive dike swarms with internally coherent geographical outcrop patterns can demonstrate lithospheric rigidity of cratons. Such features are common in the Proterozoic rock record but dwindle abruptly further back into the Archean record, especially older than 2.7 Ga (e.g., Eriksson and Fedo, 1994; Ernst and Buchan, 2001). Other papers in this volume address whether this change represents secular evolution of tectonic style or merely a degree of preservation of the oldest crust. For our purposes, tectonic coherence of adjacent crustal blocks must be treated as "suspect" in the manner of terranes in accretionary orogens, and the lateral extents of cratons subjected to our paleomagnetic tests steadily decrease with age. For example, rocks of the Abitibi greenstone belt, Canada, can be considered *a priori* as rigidly connected to all of Laurussia during 0.2–0.4 Ga, or to the successively accreting portions of Laurentia at 0.4–1.8 Ga, but only as far as the Superior craton at 1.8–2.0 Ga, only the eastern Superior craton at ca. 2.0–2.65 Ga, and only the Abitibi subprovince itself for the pre–2.65 Ga history (see Hoffman, 1996; Pesonen et al., 2003; Halls and Davis, 2004; Buchan et al., 2007a; Percival, 2007). These areal constraints further limit our ability to apply quantitative paleomagnetic tests to the most ancient geological record.

## METHODS

The present analysis of differential motions between Precambrian cratons employs three independent techniques, all of which require pairs of precisely coeval paleomagnetic poles from the same two cratonic blocks, as well as the assumptions of a pure geocentric-axial-dipole magnetic field and a constant planetary radius. The first method relies on an intact amalgamation of cratonic blocks that has persisted to the present time. As long as those blocks have traveled together rigidly, their paleomagnetic apparent polar wander (APW) paths should align perfectly without applying any relative rotations. With increasing age, a complete paleomagnetic data set would document the timing of amalgamation of the individual tectonic elements by way of discordant paleomagnetic poles from ages prior to that assembly. Timing of assembly would be bracketed between the oldest concordant poles and youngest discordant poles when viewed in present coordinates. In the conceptual diagram of Figure 1, the current amalgamation of cratons A + B persists back to the age of their suturing at time t3. This same age should mark the initial convergence of APW paths from each of the blocks A and B in their present positions. A modification of this technique holds true if the now-separated cratons and their APW paths can be restored confidently into a prior period of

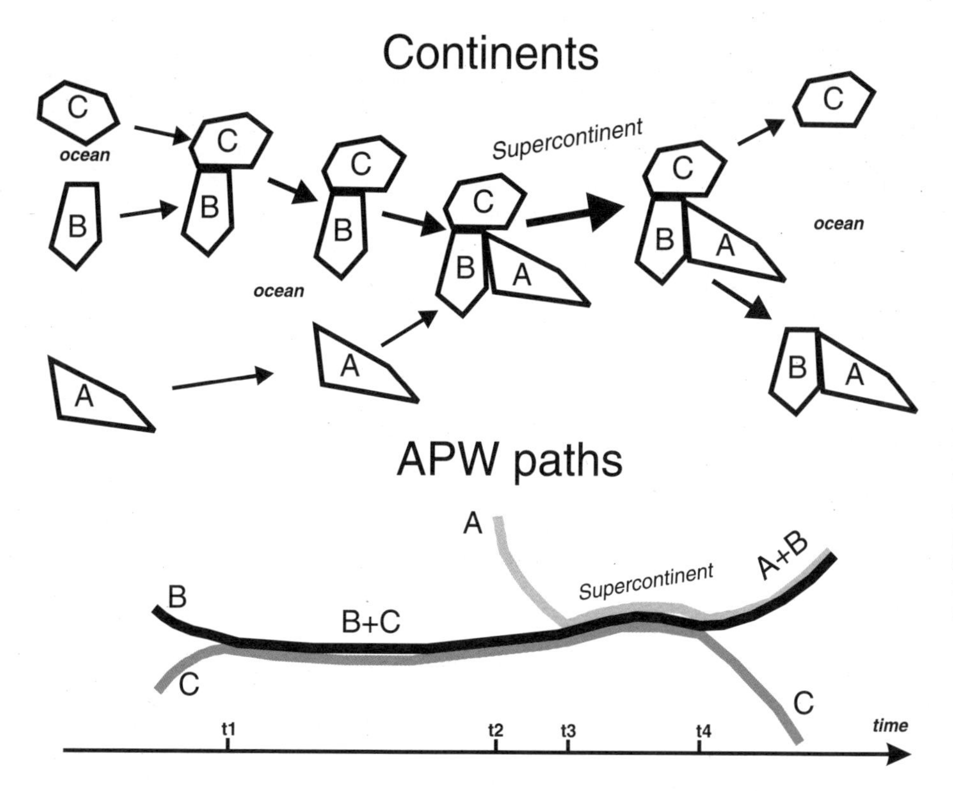


Figure 1. Schematic cartoon of the paleomagnetic record of supercontinental amalgamation and dispersal, involving cratons A, B, and C. Time increases from left to right, with collisions at ages t1 and t3, rifting at age t4, and an arbitrary reference age for discussion in the text at t2. The degree of relative separation is indicated vertically, and this simplified illustration does not discriminate between translational versus local-rotational components of relative motion between cratons. In this idealized example, each apparent polar wander (APW) path is considered to consist of a dense data set of reliable paleomagnetic poles. This figure was inspired by McGlynn and Irving (1975).

assembly such as the Gondwanaland or Pangea configurations of major continental blocks in the early Mesozoic, which have been precisely reconstructed by seafloor-spreading histories (e.g., Besse and Courtillot, 2002). In this case, the APW path of craton C can be rotated to A + B via the same total-reconstruction parameters that reunified the supercontinent, and the age of craton C's initial suturing (t1) is similarly indicated by its converging APW path at that time (Fig. 1).

The remaining technique can be used on cratonic elements for which a specific ancient tectonic association has not yet been established. By this method, great-circle distances between paleomagnetic poles of differing ages are calculated for a given craton, and these are compared with distances between poles from the isochronous interval on another craton. If the angular distances are significantly distinct between blocks, over precisely the same interval of time, then the blocks must have been in relative motion at some time during that interval. Using the schematic diagrams of Figure 1, suppose that cratons A and C, now on separate plates, are constrained by paleomagnetic data at the precise ages of t2 and t4. The difference in total APW path length between those cratons (that of A being significantly longer) proves that they could not have traveled together as a rigid block for that entire interval of time (t2 to t4). In general, this method cannot distinguish between relative translations and vertical-axis rotations; it simply determines whether we can reject the null hypothesis that the two blocks were part of the same kinematic entity. If for those same cratons, paleomagnetic data are available only at times t3 and t4, such a null hypothesis cannot be rejected; however, existence of the supercontinent is not required by these data alone because independently moving plates can generate APW paths of equal length. This test is quantified by the precisions of ages and paleomagnetic pole locations.

To begin our analyses, we considered all ~1500 entries for rocks older than 800 Ma, as listed in the global paleomagnetic database (version 4.6; Pisarevsky, 2005, and available at http://dragon.ngu.no; supplemented by ~20 subsequently published results). We chose this minimum age limit because of near-universal agreement among geoscientists that the hallmarks of modern plate tectonics can be seen at numerous places in the geological record back to at least ca. 800 Ma (Stern, 2005), and also because of the practicality in treating a relatively small proportion of the nearly 10,000 published paleomagnetic results in the global database. Because the implications of our tests are so far-reaching, i.e., a positive result *demands* differential motion between cratons, we chose to limit our discussion to only the data that (we suspect) nearly all paleomagnetists would agree are reliable, providing a definitive basis for our conclusions. To achieve this goal objectively, we queried the global paleomagnetic database with strict quality filters.

Our selection of paleomagnetic poles required satisfying *all* of the following criteria. First, we required precise age of the rock within limits of ±15 m.y. For relative plate translation rates of ~5 cm/yr (typical for continents since breakup of Pangea), these maximal error bounds of 30 m.y. of motion will cause 1500 km of separation, or ~15° of arc at Earth's surface. Typical angular uncertainties on the most reliable paleomagnetic poles are ~5°–10°, and linear addition of two of these uncertainties results in an equivalent amount of error as provided by the age limits. Ages of the most precisely dated paleomagnetic poles have precisions less than a few m.y., afforded by recent advances in U-Pb geochronology. We excluded ages from K-Ar, Rb-Sr, and Pb-Pb on sedimentary rocks, which, in some instances, may have tight precision but dubious accuracy. Second, we required reasonably precise statistics on the mean direction, with radius of 95% confidence cone ($A_{95}$) less than 20°. Although this is moderately lax, our analysis considers the angular uncertainties quantitatively, and we merely want to distinguish results that likely represent a single directional mode rather than mixed or highly scattered results. Third, we selected only data treated by modern demagnetization techniques including least-squares principal component analysis or, at the very least, vector subtraction. Results from stable-endpoint analysis or "blanket" cleaning were included only when the rocks have been subjected to at least a pilot restudy using more advanced methods. Fourth, at least one positive field test must have been performed to confirm the antiquity of paleomagnetic remanence. Not all field stability tests guarantee primary remanences, and we have included some results in Table 1 that conceivably could be (ancient) remagnetizations. Nevertheless, two-thirds of the selected results in Table 1 have passed tests that imply a primary origin of remanence (baked-contact test, syndepositional fold test, or intraformational conglomerate test), and *all* of the final subset of key poles in Table 2 are among these. Fifth, the rocks must be firmly attached to a craton, with little possibility for local vertical-axis rotation (which could greatly affect pole distances). Also in this general category, plutonic rocks must be assessable for structural tilt. This requirement accepts parallel dike swarms and layered plutonic complexes, but it excludes most granites, gneisses, and deformed mafic lenses within orogenic belts. The discerning reader will have identified elements similar to Van der Voo's (1990) criteria 1–5 in the preceding discussion. We did not require dual polarity of remanence (his criterion 6) or dissimilarity to younger portions of the apparent polar wander path (his criterion 7). Our selection criteria are also similar in spirit to the "key pole" approach of Buchan et al. (2000, 2001).

We selected about fifty results that passed all five criteria, representing only ~3% of the published poles from the pre–800 Ma time interval (Table 1). Our conclusions are therefore on the conservative side for interpretations of differential surface kinematics from paleomagnetic data. It should be noted that despite this low rate for acceptance of results among the entire paleomagnetic database, more than half of the entries on our high-quality list were generated within the past ten years; thus, the rate of high-quality paleomagnetic pole generation has been increasing, in part due to greater recognition among paleomagnetists of the need for adequate statistics and field stability tests, and in part due to the greater precision of isotopic age determinations (principally U-Pb) from the Precambrian geologic record. Also, on an optimistic note, almost 400 of the results that we evaluated were

TABLE 1. PRECISELY DATED PRE–800 Ma PALEOMAGNETIC POLES FROM STRATIFIED ROCKS OR UNDEFORMED DIKE SWARMS IN STABLE CRATONIC BLOCKS

| Rock unit (craton) | Age (Ma) | Test* | Pole (°N,°E) | $A_{95}$° | Pole or age reference |
|---|---|---|---|---|---|
| Kandyk suite (Siberia) | 1005 ± 4 | f | –03, 177 | 4 | Pavlov et al. (2002) |
| Bangemall sills (Australia) | 1070 ± 6 | C | 34, 095 | 8 | Wingate et al. (2002) |
| Lake Shore traps (Laurentia) | 1087 ± 2 | f, C | 22, 181 | 7 | Diehl and Haig (1994) |
| Portage Lake volc. (Laurentia) | ca. 1095 | f, G | 27, 178 | 5 | Hnat et al. (2006) |
| **Logan sills mean (Laurentia)** | **1109 +4/–2** | **C** | **49, 220** | **4** | **Buchan et al. (2000)** |
| **Umkondo mean (Kalahari)** | **ca. 1110** | **C** | **64, 039** | **4** | **Gose et al. (2006)** |
| Abitibi dikes (Laurentia) | 1141 ± 2 | C | 43, 209 | 14 | Ernst and Buchan (1993) |
| Sudbury dikes (Laurentia) | 1235 +7/–3 | C | –03, 192 | 3 | Palmer et al. (1977); Stupavsky and Symons (1982) |
| **Post-Jotnian intr. (Baltica)** | **ca. 1265** | **C** | **04, 158** | **4** | **Buchan et al. (2000)** |
| **Mackenzie mean (Laurentia)** | **1267 ± 2** | **C** | **04, 190** | **5** | **Buchan and Halls (1990)** |
| McNamara Fm (Laurentia) | 1401 ± 6 | f | –14, 208 | 7 | Elston et al. (2002) |
| Purcell lava (Laurentia) | ca. 1440 | f | –24, 216 | 5 | Elston et al. (2002) |
| Snowslip Fm (Laurentia) | 1450 ± 14 | f | –25, 210 | 4 | Elston et al. (2002) |
| St. Francois Mtns (Laurentia) | ca. 1476 | g, c, F | –13, 219 | 6 | Meert and Stuckey (2002) |
| Western Channel (Laurentia) | ca. 1590 | C | 09, 245 | 7 | Irving et al. (1972, 2004); Hamilton and Buchan (2007) |
| Emmerugga Dol. (N. Australia) | ca. 1645 | f | –79, 203 | 6 | Idnurm et al. (1995) |
| Tatoola Sandstone (N. Australia) | 1650 ± 3 | f | –61, 187 | 6 | Idnurm et al. (1995) |
| West Branch volc. (N. Australia) | 1709 ± 3 | G | –16, 201 | 11 | Idnurm (2000) |
| Peters Creek volc. (N. Australia) | ca. 1725 | g | –26, 221 | 5 | Idnurm (2000) |
| Cleaver dikes (Laurentia) | 1740 +5/–4 | c, C | 19, 277 | 6 | Irving et al. (2004) |
| Taihang dikes (N. China) | 1769 ± 3 | f, C | 36, 247 | 3 | Halls et al. (2000) |
| **Post-Waterberg (Kalahari)** | **ca. 1875** | **C** | **09, 015** | **17** | **Hanson et al. (2004); de Kock (2007)†** |
| **Molson dikes B (Superior)** | **ca. 1880** | **C** | **27, 219** | **4** | **Halls and Heaman (2000)** |
| Minto dikes (E. Superior) | 1998 ± 2 | C | 38, 174 | 10 | Buchan et al. (1998) |
| Bushveld mean (Kaapvaal) | ca. 2050 | f | 12, 027 | 4 | Evans et al. (2002) |
| **Waterberg sequence I (Kaapvaal)** | **2054 ± 4** | **c, F, G** | **37, 051** | **11** | **de Kock et al. (2006)** |
| **Kuetsyarvi lavas (Karelia)** | **2058 ± 6** | **G** | **23, 298** | **7** | **Torsvik and Meert (1995); Melezhik et al. (2007)** |
| Cauchon dikes (W. Superior) | 2091 ± 2 | C | 53, 180 | 9 | Halls and Heaman (2000) |
| Marathon R pol. (W. Superior) | ca. 2105 | C | 54, 180 | 7 | Buchan et al. (1996); Hamilton et al. (2002); Halls et al. (2005) |
| Marathon N pol. (W. Superior) | 2126 ± 1 | C | 45, 199 | 7 | Buchan et al. (1996); Halls et al. (2005) |
| Biscotasing dikes (E. Superior) | 2167 ± 2 | C | 28, 223 | 11 | Buchan et al. (1993) |
| **Nipissing sills N1 (E. Superior)** | **2217 ± 4** | **C** | **–17, 272** | **10** | **Buchan et al. (2000)** |
| **Ongeluk lava (Kaapvaal)** | **2222 ± 13** | **G** | **–01, 101** | **5** | **Evans et al. (1997)** |
| Dharwar dikes (Dharwar) | 2366 ± 1 | C | 16, 057 | 6 | Halls et al. (2007) |
| Widgiemooltha (Yilgarn) | ca. 2415 | c | 08, 337 | 8 | Evans (1968); Smirnov and Evans (2006) |
| **Karelian D comp (Karelia)** | **ca. 2445** | **C** | **–12, 244** | **15** | **Mertanen et al. (2006)** |
| **Matachewan N pol. (E. Superior)** | **2446 ± 3** | **C** | **–50, 244** | **5** | **Halls and Davis (2004)** |
| Matachewan R pol. (W. Superior) | 2459 ± 5 | C | –51, 257 | 9 | Halls and Davis (2004); Halls et al. (2005) |
| Great Dike mean (Zimbabwe) | 2575 ± 1 | c | 21, 058 | 6 | Jones et al. (1976); Wilson et al. (1987); Mushayandebvu et al. (1994); Oberthür et al. (2002) |
| **Nyanzian lava (Tanzania)** | **ca. 2680** | **g, f** | **–14, 330** | **6** | **Meert et al. (1994)** |
| **Otto Stock (E. Superior, Abitibi)** | **2680 ± 1** | **c** | **–69, 047** | **5** | **Pullaiah and Irving (1975); Buchan et al. (1990)** |
| Stillwater complex (Wyoming) | 2705 ± 4 | c | –67, 292 | 17 | Xu et al. (1997) |
| **Fortescue Package 1 (Pilbara)** | **2772 ± 2** | **f, g** | **–41, 160** | **4** | **Schmidt and Embleton (1985); Strik et al. (2003)** |
| **Derdepoort lavas (Kaapvaal)** | **ca. 2782** | **G** | **–40, 005** | **18** | **Wingate (1998)** |
| Duffer M component (Pilbara) | 3467 ± 5 | f | 44, 086 | 7 | McElhinny and Senanayake (1980); Van Kranendonk et al. (2002) |

*Field stability test abbreviations: f—fold test, F—folding penecontemporaneous with rock formation, c—inverse contact test, C—baked-contact test, g—conglomerate test, G—intraformational conglomerate test. Note: capitalized symbols indicate primary magnetization, whereas lowercase symbols indicate merely ancient remanence relative to the geological feature of the test.

†de Kock (2007) has shown that the ca. 1875 Ma post-Waterberg sills are distinctly older than the Sibasa lavas, so only the mean from the sills, rather than the combined pole (Hanson et al., 2004), is listed here.

TABLE 2. ANGULAR DISTANCES OF PRECISELY COEVAL PAIRS OF PALEOMAGNETIC POLES BETWEEN TWO CRATONS (PRE–800 Ma)

| Ages (Ma) | Craton 1 | Pole dist. (°) | Craton 2 | Pole dist. (°) |
|---|---|---|---|---|
| Key poles only | | | | |
| **1110–1880** | **Superior** | **22 ± 8** | **Kalahari** | **58 ± 21** |
| 1110–2220 | Superior | 81 ± 14 | Kaapvaal | 79 ± 9 |
| 1265–2445 | Superior | 71 ± 10 | Karelia | 87 ± 19 |
| 1880–2220 | Superior | 68 ± 14 | Kaapvaal | 86 ± 22 |
| Adding a single non-key pole (2680 ± 3 Ma Varpaisjärvi basement, Karelia; –64°N, 133°E, $A_{95}$ = 8°; Neuvonen et al., 1981) | | | | |
| 1265–2680 | Superior | 70 ± 10 | Karelia | 71 ± 12 |
| **2445–2680** | **Superior** | **60 ± 10** | **Karelia** | **88 ± 23** |
| Adding another non-key pole (2686 ± 28 Ma Allanridge lavas, Kaapvaal; –68°N, 356°E, $A_{95}$ = 6°; de Kock, 2007) | | | | |
| **1110–2680** | **Superior** | **20 ± 9** | **Kaapvaal** | **45 ± 10** |
| **1880–2680** | **Superior** | **42 ± 9** | **Kaapvaal** | **78 ± 23** |
| 2220–2680 | Superior | 88 ± 15 | Kaapvaal | 86 ± 11 |

*Note*: Boldface type indicates significantly distinguishable values between the two cratons, requiring their differential motion at some time within the given age interval (according to the inherent assumptions of our methods elaborated in the text).

lacking in only one of the first four reliability criteria, including ~40 in which the remanences have been demonstrated to be primary but merely lack adequate statistics or modern demagnetization techniques. Therefore, even if the global paleomagnetic community were merely to revisit those same rocks to bolster the existing published results, we would anticipate a near-doubling of highly reliable data that could be achieved within a few years.

Among the presently available subset listed in Table 1, we next identify groups of poles from different cratons that are coeval within ten million years of rock age (and presumably primary magnetic remanence, as indicated by the field stability tests). Wherever two cratons share coeval poles from more than one age, those poles are listed in Table 2. This is the subset of results that may be considered by the tests for relative motion as described previously.

Paleomagnetic data from Archean rocks are particularly relevant to a discussion on the origins of plate tectonics on Earth, and we note that only a few Archean results (seven, to be precise) pass the stringent reliability criteria we have set forth. Some of the unreliable or unusable results deserve mention here. A substantial portion of the Archean data are determined from unstratified plutonic rocks (some recent examples include the Kaap Valley pluton, Layer et al., 1996; Mbabane pluton, Layer et al., 1989; and Pikwitonei granulite, Zhai et al., 1994) and so are not considered in this analysis. Although figuring prominently in discussions on possible connections between the Kaapvaal and Pilbara cratons (Wingate, 1998; Zegers et al., 1998), poles from the Millindinna and Usushwana complexes (Schmidt and Embleton, 1985; Layer et al., 1988) are excluded because of poor age constraints and lack of field stability tests on the ages of remanence. The Superior craton has yielded two groups of poles with ages near 2710 Ma, with the most reliable examples from each being the Ghost Range complex (Geissman et al., 1982) and the Red Lake granites (Costanzo-Alvarez and Dunlop, 1993; several distinct plutons showed the same remanence direction, thus decreasing the likelihood of regional tilt biases). The two groups of poles are widely separated, however, and each contains members from various Superior craton subprovinces. No preference is made in this analysis regarding these two groups, among which no single paleomagnetic pole satisfies all five reliability criteria. Reliable data are available from the Tanzanian craton at ca. 2680 Ma (Meert et al., 1994), precisely the same age as results from Superior (Table 1), but, unfortunately, no other age comparisons on reliable paleomagnetic poles can be made between those two cratons. Finally, despite a series of well-dated and moderately to highly reliable poles from the Pilbara craton spanning most of geon 27 (Strik et al., 2003), only a single result of comparable age and reliability is to be found anywhere else in the world, from the Kaapvaal craton (Wingate, 1998), which lacks the additional highly reliable poles needed to test for Archean differential motion of those two cratons. Although two of the eight precise age pairs (highlighted in boldface on Table 1) are Archean, those two most ancient pairs are solitary with regard to the two cratons they represent, and they will be unusable for these tests until they are joined by complementary data.

The distribution of our accepted data is skewed toward a few well-studied cratons: Superior (merged with other cratons in Laurentia at ca. 1750 Ma), Karelia (the core of Fennoscandia and merged with the other components of Baltica at ca. 1800–1700 Ma), and Kaapvaal (merged with Zimbabwe to form Kalahari at least by ca. 2000 Ma, if not earlier). Our conclusions, therefore, might not be globally representative. However, the substantial changes in paleolatitude by these few blocks, including some cases of demonstrably differential motion (see following), indicate that the motions we describe are representative of broad regions across the globe.

## RESULTS

Perhaps surprisingly, from the entire database of more than 1500 entries, we obtained only ten paleomagnetic poles from five distinct ages that may be used in our analysis (Table 2). It is important to remember that these ten data points are not the only "key" poles from the pre–800 Ma paleomagnetic database, which number approximately fifty by our evaluation. The present data set is limited primarily by the requirement of two pairs of precisely coeval poles from the same two cratons. Ironically, there are several instances of a well-defined APW path segment for one craton that disappears at the same age that another craton's APW path comes into focus. For example, the Superior/Laurentian craton APW path is reasonably well-constrained between 2200 and 1750 Ma, but is unknown for immediately younger segments, when for instance the North Australian cratonic APW path becomes well-defined between 1725 and 1640 Ma. As the North Australian path dwindles in reliability with younger ages, the Laurentian path regains precision at 1590 Ma and younger intervals. Just a few well-targeted studies could fill these gaps.

From the five precise pairings of "key" poles listed in Table 1, four age comparisons emerge, all involving the Superior (or Laurentian) craton: Kalahari at 1110–1880 Ma, Kaapvaal at 1110–2220 Ma and 1880–2220 Ma, and Karelia at 1267–2445 Ma (Table 2). These poles and cratons are plotted in present coordinates (Fig. 2) for the sake of clarity, because subsequent diagrams show them rotated into other reference frames. The four key pole pairs are illustrated in Figure 3, where all elements have been reconstructed to the paleogeographic grid at the younger of the two ages. The only demonstrably differential motion among these four examples involves Superior and Kalahari at 1110–1880 Ma (Fig. 3A), as seen by the lack of overlap between the ca. 1880 Ma poles when rotated into the common 1110 Ma paleogeographic reference frame. In this instance, the cratonic reconstruction shown in Figure 3A is tectonically meaningless, but the cratons are illustrated nonetheless to assist understanding of the concepts. This analysis does not specify when during the 1110–1880 Ma interval the differential motion occurred, so the conservative interpretation would constrain the antiquity of differential cratonic motion, by this method, merely to somewhat older than 1110 Ma.

In the remaining three tests, the lengths of APW are indistinguishable. At face value, this would appear to permit the hypothesis that the two cratons traveled together throughout each entire interval of time, whether as part of a supercontinent, or experiencing true polar wander, or both. However, such a model is untenable for Superior-Kaapvaal between 1110 and 2220 Ma because it has already been demonstrated (Fig. 3A) that those cratons were in relative motion between 1110 and 1880 Ma. The common APW path length in Figure 3B, therefore, is fortuitous. Superior and Kaapvaal could have been members of a supercontinent between 1880 and 2220 Ma (Fig. 3C), but if so, then about four similarly sized cratons will need to be found to fill the ~60° gap between them. Superior and Karelia reconstruct directly adjacent to each other for the interval 1265–2445 Ma (Fig. 3D), but this direct juxtaposition over a billion years is not permitted by the known assembly of eastern Laurentia, occupying the same space as Karelia, during that interval.

Throughout the 1110–1880 Ma age interval, in which differential cratonic motion has now been established, most of the world's cratons can be considered as internally rigid blocks, as evidenced by large, geometrically coherent dike swarms and epicratonic or platformal sedimentary basins (e.g., Ernst and Buchan, 2001; Zhao et al., 2004). By demonstrating differential lateral motion between internally rigid lithospheric blocks older than 1110 Ma, we can quantitatively refute the hypothesis of Stern (2005) that plate tectonics began at ca. 800 Ma. The present comparisons cannot refute Stern's (this volume) modified hypothesis of stagnant-lid tectonics between 1800 and 800 Ma, because it is possible that all of the demonstrated differential motion (shown here to be within 1110–1880 Ma) could have occurred entirely during the 1800–2000 Ma "proto–plate tectonic" regime proposed in Stern's revised model.

To demonstrate the rapid expansion of this kind of analysis allowed by even a modestly growing paleomagnetic database, we will consider two "non-key" poles that, if considered to be reliable or verified as such in the near future, could push back the global record of relative cratonic motion to the interval 2450–2680 Ma, and prove additional between-craton motions in Proterozoic time. The first result is that from the 2680 ± 3 Ma Varpaisjarvi basement complex in east-central Finland; a positive baked-contact test with younger than ca. 2450 Ma intrusions demonstrates a possibly primary remanence (Neuvonen et al., 1981), but the lack of paleohorizontal control requires omission of this result due to our strict quality criteria. However, if we were to include this single result, then comparison with the venerable yet high-quality and precisely coeval Otto Stock lamprophyre dikes pole from 2680 ± 1 Ma in Superior (Pullaiah and Irving, 1975) allows two additional estimates of APW path length: from 1265 Ma and from 2445 Ma (Table 2; Fig. 4). The first test is negative; great-circle distances between 1265 Ma and 2680 Ma poles are nearly identical between the two cratons (Fig. 4A). The second test, between 2445 Ma and 2680 Ma, demonstrates significantly different APW path lengths, even with the poles stretched to their uncertainty limits (Fig. 4B). This result would appear to indicate that Superior and Karelia could not have lain on the same plate through the Neoarchean to earliest Paleoproterozoic interval.

The second example is from the Allanridge lavas on the Kaapvaal craton in South Africa. These rocks stratigraphically separate the Ventersdorp Group, with ages as young as ca. 2710 Ma (Armstrong et al., 1991), from the overlying Transvaal Supergroup, in which protobasins have ages as old as ca. 2665 Ma (Barton et al., 1995). The lavas have been studied paleomagnetically by Strik et al. (2007) and de Kock (2007), the latter of whom documented a positive intraformational conglomerate test and computed a new, combined mean paleomagnetic pole. The only criterion lacking in this combined result is a precise age determination, but the best estimate of ca. 2686 Ma is

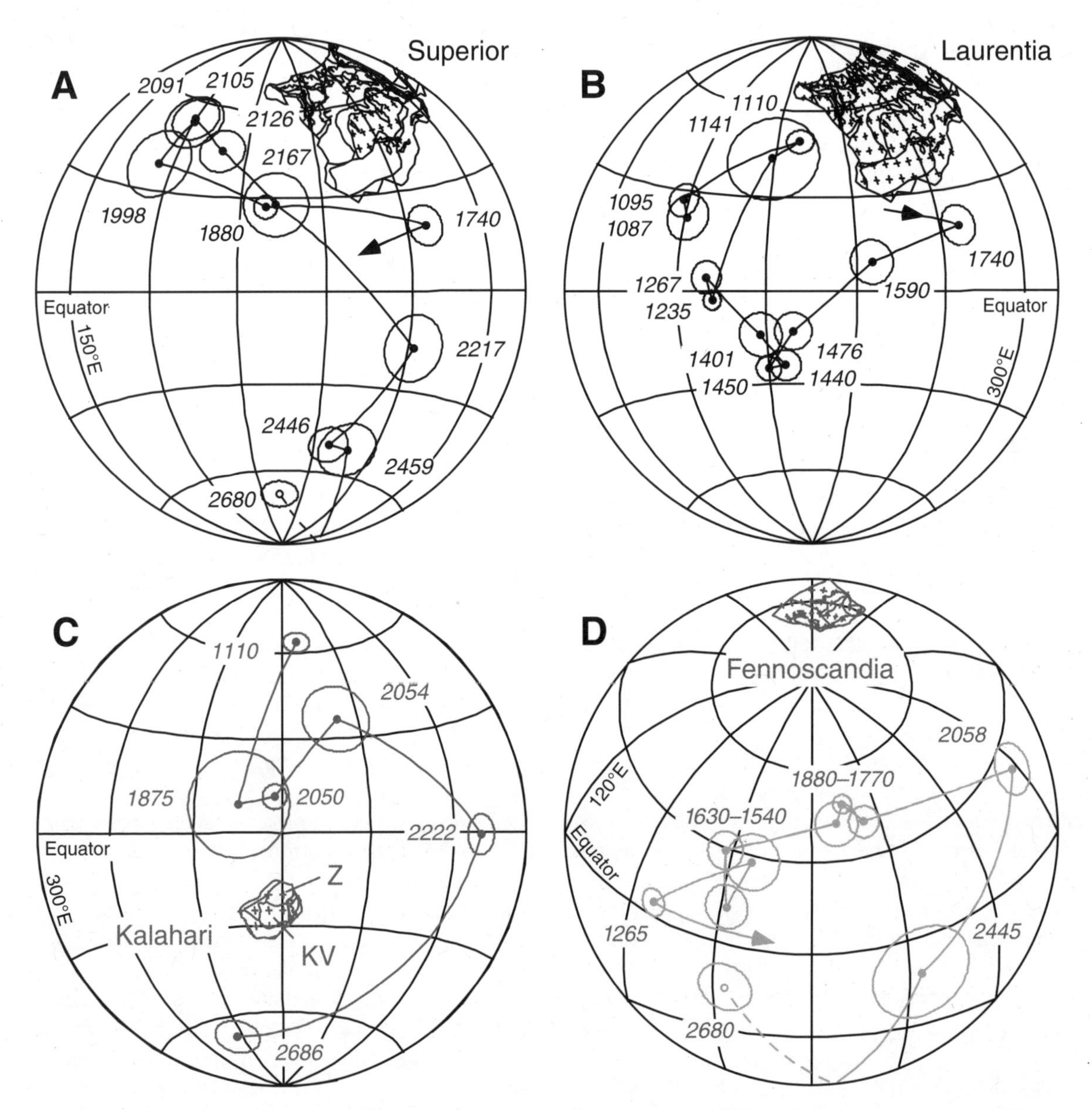


Figure 2. Paleomagnetic poles amenable to precise apparent polar wander (APW) length tests between pre–800 Ma cratons, in each craton's present coordinate reference frame. Greater complexities in the APW paths are evident from a broader consideration of non-key poles, but these are omitted here for clarity. As in subsequent global equal-area projection figures, black color represents Laurentia or the Superior craton, whereas gray color represents other cratons; all ages are in Ma. (A) Superior APW path. (B) Laurentia APW path, continuing in age progression from the Superior path. (C) Kaapvaal–Kalahari APW path (KV—Kaapvaal craton area; Z—Zimbabwe craton area). (D) Karelia–Fennoscandia APW path. All poles are listed in Tables 1 and 2, with the exception of the 1880–1540 Ma Fennoscandian poles (non-key; listed in Buchan et al., 2000; or Pesonen et al., 2003), which are included for comparison with Figure 5.

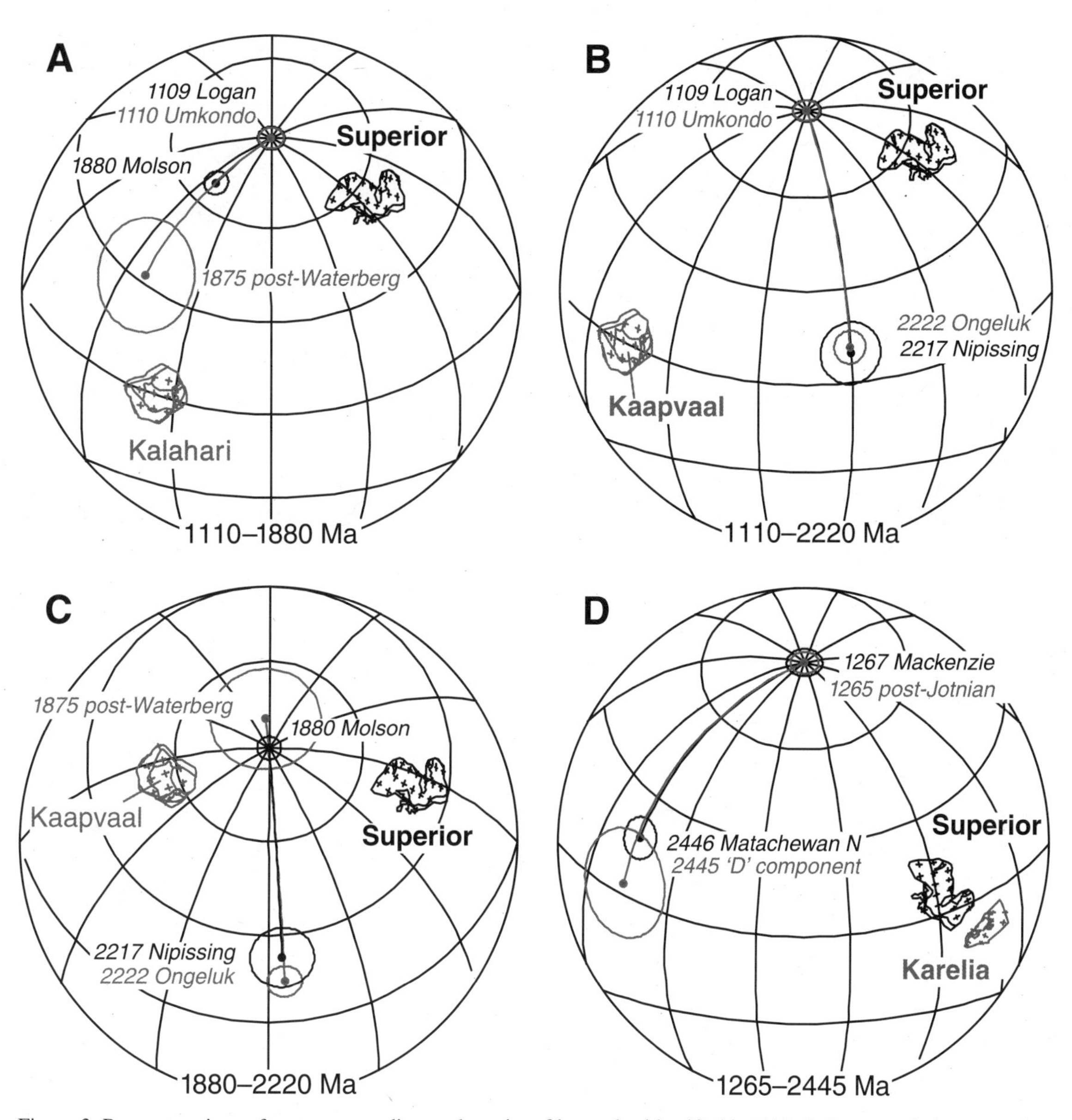


Figure 3. Reconstructions of cratons according to the pairs of key poles identified in Table 2. Due to polarity uncertainties, these reconstructions are nonunique and represent only half of the viable solutions for each pole pair. Coordinate system is the reconstructed paleolatitude grid of the younger pole in each pair. (A) Superior-Kalahari, 1110–1880 Ma; (B) Superior-Kaapvaal, 1110–2220 Ma; (C) Superior-Kaapvaal, 1880–2220 Ma; (D) Superior-Karelia, 1265–2445 Ma. Note that only the first panel (A) illustrates demonstrably independent motion between those two cratons (indicated by the different apparent polar wander [APW] lengths).

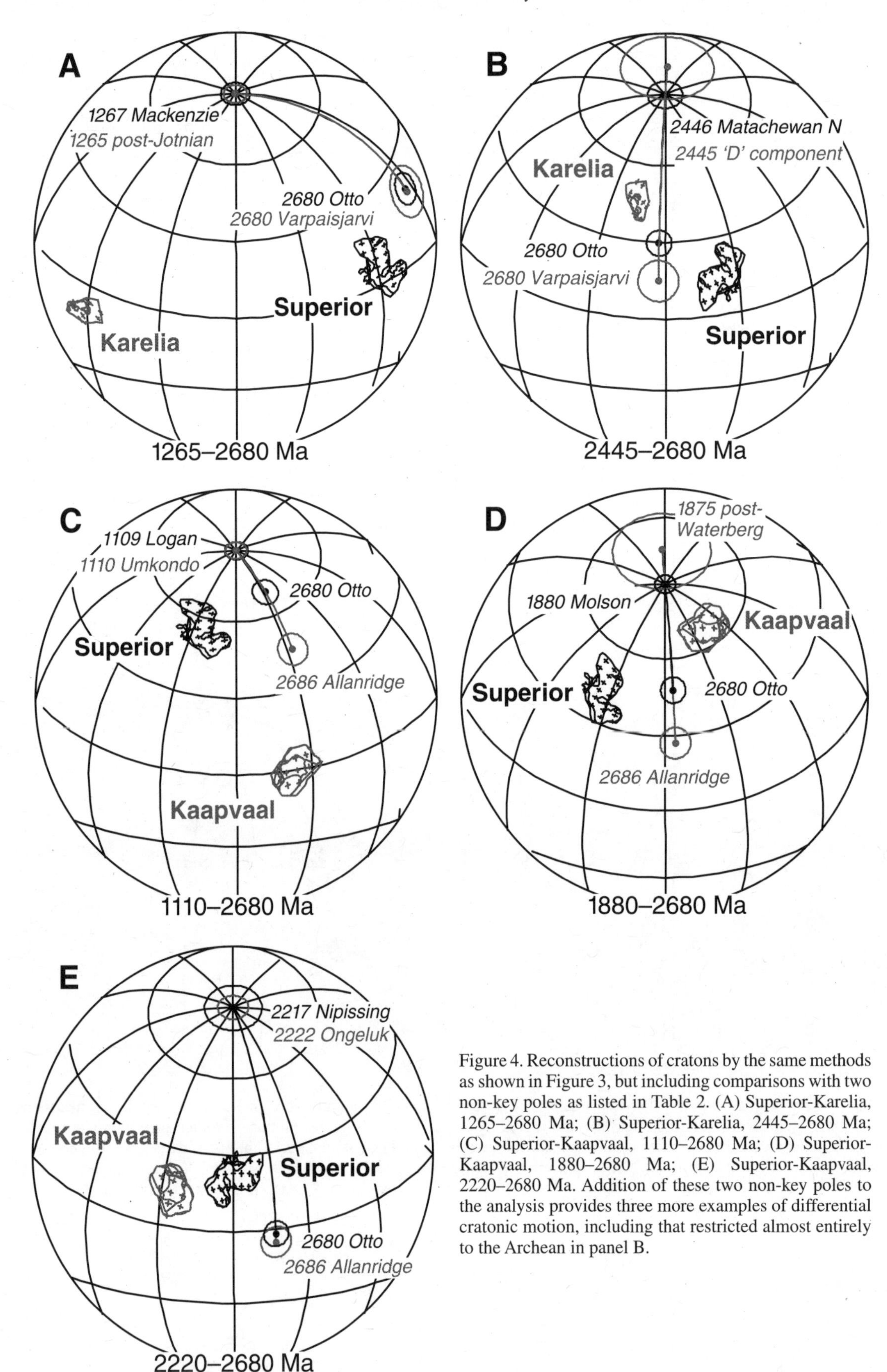


Figure 4. Reconstructions of cratons by the same methods as shown in Figure 3, but including comparisons with two non-key poles as listed in Table 2. (A) Superior-Karelia, 1265–2680 Ma; (B) Superior-Karelia, 2445–2680 Ma; (C) Superior-Kaapvaal, 1110–2680 Ma; (D) Superior-Kaapvaal, 1880–2680 Ma; (E) Superior-Kaapvaal, 2220–2680 Ma. Addition of these two non-key poles to the analysis provides three more examples of differential cratonic motion, including that restricted almost entirely to the Archean in panel B.

similar to that of the Otto Stock as described already. Because of the three key pole comparisons between Superior/Laurentia and Kaapvaal/Kalahari already noted, addition of this single fourth pole from Africa, in comparison with the Otto Stock result from Superior, provides three new APW length tests. Shown in Figure 4 (panels C, D, and E), the younger two tests are positive, requiring differential motion between the cratons, whereas the oldest test is negative. In that last example (Fig. 4E), an intriguing juxtaposition is presented that could indicate a supercraton connection between the two blocks (sensu Bleeker, 2003), as first suggested by Evans and Powell (2000). Although the test is negative in terms of identifying differential cratonic motion and early plate tectonics on Earth, it could instead lead the way toward insights into Archean-Paleoproterozoic paleogeography.

Recognizing that these two examples are not yet of high reliability, we discuss them merely to demonstrate how tractable the APW length-comparison method can be: a few well-chosen subject rocks for paleomagnetic study in the near future may provide convincing evidence for Archean differential motions between cratons, or alternatively, produce potential supercraton juxtapositions of similar antiquity.

Our analysis using APW lengths has been highly conservative, in that we have selected only the most reliable paleomagnetic data. There are many published results that we suspect indicate primary and accurate magnetizations but that cannot be included in this study due to the strict quality criteria outlined already. If we were to include the entire database of moderately reliable results, we would certainly be reporting many additional examples of between-craton motions. Cawood et al. (2006) considered some of these and concluded that differential cratonic motion was demonstrable as old as 2070 and 2680 Ma. Those constituent data, however, suffer from uncertainties of paleohorizontal (e.g., the Kaapvaal ca. 2680 Ma pole from Mbabane pluton) or not precisely equal ages (e.g., ca. 2060–2050 Ma slowly cooling Bushveld complex versus the ca. 2075 Ma Fort Frances and slightly older dikes).

The second kind of paleomagnetic test we employed to determine possible differential cratonic motions utilizes known or hypothesized paleogeographic reconstructions and the logic developed in the discussion of Figure 1. If two cratons can be demonstrated to be joined in a supercontinent for a certain interval of time, then discordant older paleomagnetic poles in that reference frame will indicate differential cratonic motions in assemblage of the supercontinent. This type of test has been attempted in previous studies, although not using the precisely age-comparable and reliable subset of data selected herein. One example is the conclusion by Meert et al. (1995), based on discrepant ca. 800–600 Ma poles and well supported by tectonic syntheses (e.g., Collins and Pisarevsky, 2005), that Gondwanaland did not assemble until after that time. Another is the less confident hypothesis of Elming et al. (2001) that Baltica was not yet assembled at 1750 Ma, based on discrepant results from Sarmatia versus Fennoscandia. The tectonic assembly of Laurentia, better studied than any other Paleoproterozoic craton (Hoffman, 1988, and hundreds of subsequent publications), is a prime target for paleomagnetic analysis, and indeed, alternating fixist versus mobilist interpretations have been made through the past few decades (e.g., Christie et al., 1975; Cavanaugh and Nairn, 1980; Dunsmore and Symons, 1990; Irving et al., 2004; Symons and Harris, 2005). Recent work in western Canada, as yet published only in abstract (Buchan et al., 2007b; Evans and Raub, 2007) promises to clarify this debate with several new, key paleomagnetic poles. Initial results from those studies suggest that differential motion of Laurentian cratons can be documented between ca. 1880 and 1750 Ma. A final example, subtle and elegant, concerns a minor (10°–20°) vertical-axis rotation between the western and eastern halves of the Superior craton, as documented by dike trends and paleomagnetic data older than ca. 2000 Ma (Halls and Davis, 2004; Buchan et al., 2007a). The Kapuskasing zone serves as the locus of this deformation, which is constrained in age to ca. 1900–2000 Ma (Halls and Davis, 2004; Buchan et al., 2007a). Differential motion between (mostly) rigid blocks with narrow intervening boundary zones of deformation could well be considered as *bona fide* plate tectonics, but the continuation of Archean greenstone belts across the Kapuskasing zone implies that the total strain was minor (West and Ernst, 1991). The Superior example thus demonstrates a rheological form of plate-like tectonics, but not on a global scale.

An additional constraint may be provided by a particularly favorable cratonic reconstruction between Laurentia and Baltica for ca. 1265–1750 Ma. This reconstruction, named NENA by Gower et al. (1990), has been confirmed paleomagnetically by directional data from the Mackenzie and post-Jotnian mafic large igneous provinces (Buchan et al., 2000; Pesonen et al., 2003), as well as orientations of maximum magnetic susceptibility axes in rocks of those suites (Elming and Mattsson, 2001). Paleomagnetic results of lesser reliability, from rocks as old as ca. 1830 Ma, also support this connection (Buchan et al., 2000). The 1880 Ma poles from Superior and Fennoscandia, however, are discordant if rotated into the NENA reconstruction (Fig. 5). If NENA is valid to ages as old as ca. 1830, as supported by both paleomagnetic and tectonostratigraphic data (Gower et al., 1990; Buchan et al., 2000), then these data indicate relative motion between Superior and Karelia to achieve the assembly between 1880 and 1830 Ma. Unfortunately, the 1830 Ma poles are not reliable enough to support this conclusion definitively, but future work on rocks of this age has the potential to strengthen the argument considerably.

The demise of NENA is indicated by diverging APW paths of Laurentia and Baltica for ages younger than 1265 Ma. There is no precise estimate of the separation age; the oldest discrepant poles are ca. 1050 Ma, and those are not of "key" status (Fig. 5; poles are discussed in Buchan et al., 2000). Nonetheless, the overall APW pattern is intriguingly similar to that predicted by a plate-tectonic model of supercontinental assembly and dispersal, as illustrated in Figure 1. Concordance of the paleomagnetic reconstruction with a tectonically based juxtaposition (Gower et al., 1990) is the most compelling attribute of the aggregate

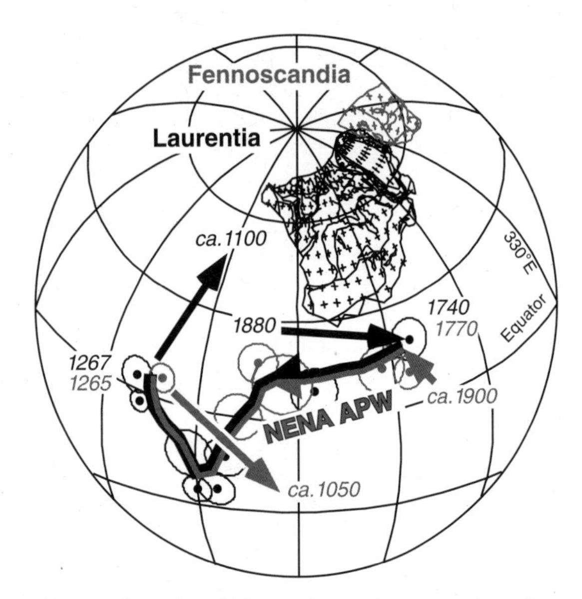


Figure 5. 1880–1100 Ma key paleomagnetic poles from Laurentia/Superior (black) in its present coordinate reference frame, compared with mostly non-key Baltica/Fennoscandia poles (gray) rotated to a similar configuration as "NENA" by Gower et al. (1990). Rotation parameters: (47.5°, 001.5°, +49.0°). References are in Table 1, or in Buchan et al. (2000), or Pesonen et al. (2003). Apparent polar wander (APW) paths converge between 1880 and ca. 1750 Ma, during the time of NENA assembly, and they diverge after 1265 Ma, coincident with NENA breakup.

data set. According to this evidence, modern-style plate tectonics appears to have been in operation throughout the most of the Proterozoic, negating the Mesoproterozoic stagnant-lid model of Stern (this volume).

## CONCLUSIONS AND FUTURE PROSPECTS

Limitations of the Archean geological record, particularly the dearth of subgreenschist stratified volcano-sedimentary successions older than 3000 Ma—as well as our currently poor knowledge of the Archean geomagnetic field—substantially hinder paleomagnetic methods to test whether plate tectonics existed at those times. The available highest-quality paleomagnetic data provide only a single example of differential cratonic motions within the Paleo- and Mesoproterozoic Eras, when at least some portions of Earth's surface were rigid blocks moving laterally relative to each other. Lithospheric rigidity and lateral motion are two important aspects of a tectonic regime that is "plate-like," and, therefore, the most robust paleomagnetic record argues for at least one instance of plate tectonics prior to 1110 Ma (Table 2).

We noted herein that if certain "non-key" poles were used in the analysis, then the differential motions could be better constrained, by various paleomagnetic methods, to as old as the earliest Proterozoic Eon. In the longer term, the present analysis might be able to extend Earth's earliest record of greenschist- or lower-grade stratified rocks, primarily known from the Barberton area of Kaapvaal in southern Africa, and the Pilbara region of Western Australia. These two terranes have been hypothesized to constitute our planet's oldest "supercontinent" (Zegers et al., 1998; extending the work of Cheney, 1996), but paleomagnetic studies of stratified rocks from the 3500–2800 Ma interval are few in number (e.g., McElhinny and Senanayake, 1980; Yoshihara and Hamano, 2004), and only one result satisfied our quality criteria. Even that lone result, the magnetite-borne remanence from the Duffer Formation in the Pilbara craton (McElhinny and Senanayake, 1980), is only constrained to be prefolding in age, and the folding could have occurred more than 100 m.y. following deposition (Van Kranendonk et al., 2002). Establishing the 700-m.y.-long APW paths for the Kaapvaal and Pilbara cratons will be challenging at the least, and might even be impossible given the moderate metamorphic grade and difficulties in testing whether Earth's primitive geodynamo can serve as a stable kinematic reference frame. Nevertheless, these two successions hold the greatest promise for paleomagnetism to address whether plate tectonics existed deep within Archean time.

We showed here that incorporation of just two merely moderately reliable sets of paleomagnetic data allowed us to identify differential motion of cratons as old as 2445–2680 Ma. If there is interest within the paleomagnetic community, those and similarly ancient data can be "upgraded" to the highest level of quality within a matter of years. A more extensive research program on the 2800–3500 Ma volcano-sedimentary successions of Kaapvaal and Pilbara holds the best promise for detecting any relative motions between cratons in the older part of the Archean Eon.

## ACKNOWLEDGMENTS

The authors gratefully acknowledge discussions with Wouter Bleeker, Ken Buchan, Michiel de Kock, Henry Halls, Larry Heaman, Lauri Pesonen, and Theresa Raub. The manuscript benefited from insightful anonymous reviews and the suggestions of editor Vicky Pease. Evans was supported by National Science Foundation grant EAR-0310922 and a fellowship from the David & Lucile Packard Foundation. This paper contributes to UNESCO-IGCP (United Nations Educational, Scientific and Cultural Organisation–International Geoscience Programme) Project 509, "Palaeoproterozoic Supercontinents and Global Evolution."

## REFERENCES CITED

Armstrong, R.A., Compston, W., Retief, E.A., Williams, I.S., and Welke, H.J., 1991, Zircon ion microprobe studies bearing on the age and evolution of the Witwatersrand triad: Precambrian Research, v. 53, p. 243–266, doi: 10.1016/0301-9268(91)90074-K.

Barton, J.M., Blignaut, E., Salnikova, E.B., and Kotov, A.B., 1995, The stratigraphical position of the Buffelsfontein Group based on field relationships and chemical and geochronological data: South African Journal of Geology, v. 98, p. 386–392.

Besse, J., and Courtillot, V., 2002, Apparent and true polar wander and the geometry of the geomagnetic field over the last 200 Myr: Journal of Geophysical Research, v. 107, no. B11, doi: 10.1029/2000JB000050.

Biggin, A.J., and Thomas, D.N., 2003, Analysis of long-term variations in the geomagnetic poloidal field intensity and evaluation of their relationship with global geodynamics: Geophysical Journal International, v. 152, p. 392–415, doi: 10.1046/j.1365-246X.2003.01849.x.

Bleeker, W., 2003, The late Archean record: A puzzle in ca. 35 pieces: Lithos, v. 71, p. 99–134, doi: 10.1016/j.lithos.2003.07.003.

Buchan, K.L., and Halls, H.C., 1990, Paleomagnetism of Proterozoic mafic dyke swarms of the Canadian Shield, *in* Parker, A.J., Rickwood, P.C., and Tucker, D.H., eds., Mafic Dykes and Emplacement Mechanisms: Rotterdam, Balkema, p. 209–230.

Buchan, K.L., Neilson, D.J., and Hale, C.J., 1990, Relative age of Otto Stock and Matchewan dykes from paleomagnetism and implications for the Precambrian polar wander path: Canadian Journal of Earth Sciences, v. 27, p. 915–922.

Buchan, K.L., Mortensen, J.K., and Card, K.D., 1993, Northeast-trending Early Proterozoic dykes of southern Superior Province: Multiple episodes of emplacement recognized from integrated paleomagnetism and U-Pb geochronology: Canadian Journal of Earth Sciences, v. 30, p. 1286–1296.

Buchan, K.L., Halls, H.C., and Mortensen, J.K., 1996, Paleomagnetism, U-Pb geochronology, and geochemistry of Marathon dykes, Superior Province, and comparison with the Fort Francis swarm: Canadian Journal of Earth Sciences, v. 33, p. 1583–1595.

Buchan, K.L., Mortensen, J.K., Card, K.D., and Percival, J.A., 1998, Paleomagnetism and U-Pb geochronology of diabase dyke swarms of Minto block, Superior Province, Canada: Canadian Journal of Earth Sciences, v. 35, p. 1054–1069, doi: 10.1139/cjes-35-9-1054.

Buchan, K.L., Mertanen, S., Park, R.G., Pesonen, L.J., Elming, S.-Å., Abrahamsen, N., and Bylund, G., 2000, Comparing the drift of Laurentia and Baltica in the Proterozoic: The importance of key palaeomagnetic poles: Tectonophysics, v. 319, p. 167–198, doi: 10.1016/S0040-1951(00)00032-9.

Buchan, K.L., Ernst, R.E., Hamilton, M.A., Mertanen, S., Pesonen, L.J., and Elming, S.-A., 2001, Rodinia: The evidence from integrated palaeomagnetism and U-Pb geochronology: Precambrian Research, v. 110, p. 9–32, doi: 10.1016/S0301-9268(01)00178-4.

Buchan, K.L., Goutier, J., Hamilton, M.A., Ernst, R.E., and Matthews, W.A., 2007a, Paleomagnetism, U-Pb geochronology, and geochemistry of Lac Esprit and other dyke swarms, James Bay area, Quebec, and implications for Paleoproterozoic deformation of the Superior Province: Canadian Journal of Earth Sciences, v. 44, p. 643–664, doi: 10.1139/E06-124.

Buchan, K.L., van Breemen, O., and LeCheminant, A.N., 2007b, Towards a Paleoproterozoic apparent polar wander path (APWP) for the Slave Province: Paleomagnetism of precisely-dated mafic dyke swarms: Geological Association of Canada Abstracts, v. 32, p. 11.

Butler, R.F., and Banerjee, S.K., 1975, Theoretical single-domain grain size range in magnetite and titanomagnetite: Journal of Geophysical Research, v. 80, p. 4049–4058.

Cavanaugh, M.D., and Nairn, A.E.M., 1980, The role of the geologic province in Precambrian paleomagnetism: Earth-Science Reviews, v. 16, p. 257–276, doi: 10.1016/0012-8252(80)90046-X.

Cawood, P.A., Kröner, A., and Pisarevsky, S., 2006, Precambrian plate tectonics: Criteria and evidence: GSA Today, v. 16, no. 7, p. 4–11, doi: 10.1130/GSAT01607.1.

Cheney, E.S., 1996, Sequence stratigraphy and plate tectonic significance of the Transvaal succession of southern Africa and its equivalent in Western Australia: Precambrian Research, v. 79, p. 3–24, doi: 10.1016/0301-9268(95)00085-2.

Christie, K.W., Davidson, A., and Fahrig, W.F., 1975, The paleomagnetism of Kaminak dikes—No evidence of significant Hudsonian plate motion: Canadian Journal of Earth Sciences, v. 12, p. 2048–2064.

Collins, A.S., and Pisarevsky, S.A., 2005, Amalgamating eastern Gondwana: The evolution of the circum-Indian orogens: Earth-Science Reviews, v. 71, p. 229–270, doi: 10.1016/j.earscirev.2005.02.004.

Costanzo-Alvarez, V., and Dunlop, D.J., 1993, Paleomagnetism of the Red Lake greenstone belt, northwestern Ontario: Possible evidence for the timing of gold mineralization: Earth and Planetary Science Letters, v. 119, p. 599–615, doi: 10.1016/0012-821X(93)90065-H.

de Kock, M.O., 2007, Paleomagnetism of Selected Neoarchean-Paleoproterozoic Cover Sequences on the Kaapvaal Craton [Ph.D. thesis]: Johannesburg, University of Johannesburg, 275 p.

de Kock, M.O., Evans, D.A.D., Dorland, H.C., Beukes, N.J., and Gutzmer, J., 2006, Paleomagnetism of the lower two unconformity bounded sequences of the Waterberg Group, South Africa: Towards a better-defined apparent polar wander path for the Paleoproterozoic Kaapvaal craton: South African Journal of Geology, v. 109, p. 157–182, doi: 10.2113/gssajg.109.1-2.157.

Diehl, J.F., and Haig, T.D., 1994, A paleomagnetic study of the lava flows within the Copper Harbor Conglomerate, Michigan: New results and implications: Canadian Journal of Earth Sciences, v. 31, p. 369–380.

Dunsmore, D.J., and Symons, D.T.A., 1990, Paleomagnetism of the Lynn Lake gabbros in the Trans-Hudson orogen and closure of the Superior and Slave cratons, *in* Lewry, J.F., and Stauffer, M.R., eds., The Early Proterozoic Trans-Hudson Orogen of North America: Geological Association of Canada Special Paper 37, p. 215–228.

Elming, S.-A., and Mattsson, H., 2001, Post-Jotnian basic intrusions in the Fennoscandian Shield, and the break up of Baltica from Laurentia: A palaeomagnetic and AMS study: Precambrian Research, v. 108, p. 215–236, doi: 10.1016/S0301-9268(01)00131-0.

Elming, S.-A., Mikhailova, N.P., and Kravchenko, S., 2001, Palaeomagnetism of Proterozoic rocks from the Ukrainian Shield: New tectonic reconstructions of the Ukrainian and Fennoscandian shields: Tectonophysics, v. 339, p. 19–38, doi: 10.1016/S0040-1951(01)00032-4.

Elston, D.P., Enkin, R.J., Baker, J., and Kisilevsky, D.K., 2002, Tightening the belt: Paleomagnetic-stratigraphic constraints on deposition, correlation, and deformation of the Middle Proterozoic (ca. 1.4 Ga) Belt-Purcell Supergroup, United States and Canada: Geological Society of America Bulletin, v. 114, p. 619–638, doi: 10.1130/0016-7606(2002)114<0619:TTBPSC>2.0.CO;2.

Eriksson, K.A., and Fedo, C.M., 1994, Archean synrift and stable-shelf sedimentary successions, *in* Condie, K.C., ed., Developments in Precambrian Geology: Volume 11. Archean Crustal Evolution: Amsterdam, Elsevier, p. 171–204.

Ernst, R.E., and Buchan, K.L., 1993, Paleomagnetism of the Abitibi dyke swarm, southern Superior Province, and implications for the Logan Loop: Canadian Journal of Earth Sciences, v. 30, p. 1886–1897.

Ernst, R.E., and Buchan, K.L., 2001, Large mafic magmatic events through time and links to mantle plume heads, *in* Ernst, R.E., and Buchan, K.L., eds., Mantle Plumes: Their Identification through Time: Geological Society of America Special Paper 352, p. 483–575.

Evans, D.A.D., 2006, Proterozoic low orbital obliquity and axial-dipolar geomagnetic field from evaporite palaeolatitudes: Nature, v. 444, p. 51–55, doi: 10.1038/nature05203.

Evans, D.A.D., and Powell, C.McA., 2000, Toward a quantitative reconstruction of Kenorland, the Archean-Paleoproterozoic supercontinent: Geological Society of Australia Abstracts, v. 59, p. 144.

Evans, D.A.D., and Raub, T.M.D., 2007, Growth of Laurentia from a paleomagnetic perspective: Critical review and update: Geological Association of Canada Abstracts, v. 32, p. 26.

Evans, D.A.D., Beukes, N.J., and Kirschvink, J.L., 1997, Low-latitude glaciation in the Palaeoproterozoic era: Nature, v. 386, p. 262–266, doi: 10.1038/386262a0.

Evans, D.A.D., Beukes, N.J., and Kirschvink, J.L., 2002, Paleomagnetism of a lateritic paleo-weathering horizon and overlying Paleoproterozoic redbeds from South Africa: Implications for the Kaapvaal apparent polar wander path and a confirmation of atmospheric oxygen enrichment: Journal of Geophysical Research, v. 107, no. B12, doi: 10.1029/2001JB000432.

Evans, M.E., 1968, Magnetization of dikes: A study of the paleomagnetism of the Widgiemooltha dike suite, Western Australia: Journal of Geophysical Research, v. 73, p. 3261–3270.

Evans, M.E., 1976, Test of the dipolar nature of the geomagnetic field throughout Phanerozoic time: Nature, v. 262, p. 676–677, doi: 10.1038/262676a0.

Evans, M.E., 2005, Testing the geomagnetic dipole hypothesis: Palaeolatitudes sampled by large continents: Geophysical Journal International, v. 161, p. 266–267, doi: 10.1111/j.1365-246X.2005.02628.x.

Geissman, J.W., Strangway, D.W., Tasillo-Hirt, A.M., and Jensen, L.J., 1982, Paleomagnetism and structural history of the Ghost Range intrusive complex, central Abitibi belt, Ontario: Further evidence for the late Archean geomagnetic field of North America: Canadian Journal of Earth Sciences, v. 19, p. 2085–2099.

Gose, W.A., Hanson, R.E., Dalziel, I.W.D., Pancake, J.A., and Seidel, E.K., 2006, Paleomagnetism of the 1.1 Ga Umkondo large igneous province in southern Africa: Journal of Geophysical Research, v. 111, doi: 10.1029/2005JB003897.

Gower, C.F., Ryan, A.B., and Rivers, T., 1990, Mid-Proterozoic Laurentia-Baltica: An overview of its geological evolution and a summary of the contributions made by this volume, *in* Gower, C.F., Rivers, T., and Ryan,

A.B., eds., Mid-Proterozoic Laurentia-Baltica: Geological Association of Canada Special Paper 38, p. 1–20.

Halls, H.C., and Davis, D.W., 2004, Paleomagnetism and U-Pb geochronology of the 2.17 Ga Biscotasing dyke swarm, Ontario, Canada: Evidence for vertical-axis crustal rotation across the Kapuskasing zone: Canadian Journal of Earth Sciences, v. 41, p. 255–269, doi: 10.1139/e03-093.

Halls, H.C., and Heaman, L.M., 2000, The paleomagnetic significance of new U-Pb age data from the Molson dyke swarm, Cauchon Lake area, Manitoba: Canadian Journal of Earth Sciences, v. 37, p. 957–966, doi: 10.1139/cjes-37-6-957.

Halls, H.C., Li, J., Davis, D., Hou, G., Zhang, B., and Qian, X., 2000, A precisely dated Proterozoic palaeomagnetic pole for the North China craton, and its relevance to palaeocontinental reconstruction: Geophysical Journal International, v. 143, p. 185–203, doi: 10.1046/j.1365-246x.2000.00231.x.

Halls, H.C., Stott, G.M., and Davis, D.W., 2005, Paleomagnetism, Geochronology and Geochemistry of Several Proterozoic Mafic Dike Swarms in Northwestern Ontario: Ontario Geological Survey Open-File Report 6171, 59 p.

Halls, H.C., Kumar, A., Srinivasan, R., and Hamilton, M.A., 2007, Paleomagnetism and U-Pb geochronology of easterly trending dykes in the Dharwar craton, India: Feldspar clouding, radiating dyke swarms and the position of India at 2.37 Ga: Precambrian Research, v. 155, p. 47–68, doi: 10.1016/j.precamres.2007.01.007.

Hamilton, M.A., and Buchan, K.L., 2007, U-Pb geochronology of the Western Channel diabase, Wopmay orogen: Implications for the APWP for Laurentia in the earliest Mesoproterozoic: Geological Association of Canada Abstracts, v. 32, p. 35–36.

Hamilton, M.A., Davis, D.W., Buchan, K.L., and Halls, H.C., 2002, Precise U-Pb dating of reversely magnetized Marathon diabase dykes and implications for emplacement of giant dyke swarms along the southern margin of the Superior Province, Ontario: Geological Survey of Canada Current Research, v. 2002–F6, 10 p.

Hanson, R.E., Gose, W.A., Crowley, J.L., Ramezani, J., Bowring, S.A., Bullen, D.S., Hall, R.P., Pancake, J.A., and Mukwakwami, J., 2004, Paleoproterozoic intraplate magmatism and basin development on the Kaapvaal craton: Age, paleomagnetism and geochemistry of ~1.93 to ~1.87 Ga post-Waterberg dolerites: South African Journal of Geology, v. 107, p. 233–254, doi: 10.2113/107.1-2.233.

Hnat, J.S., van der Pluijm, B.A., and Van der Voo, R., 2006, Primary curvature in the Mid-Continent Rift: Paleomagnetism of the Portage Lake volcanics (northern Michigan, USA): Tectonophysics, v. 425, p. 71–82, doi: 10.1016/j.tecto.2006.07.006.

Hoffman, P.F., 1988, United plates of America, the birth of a craton: Early Proterozoic assembly and growth of Laurentia: Annual Review of Earth and Planetary Sciences, v. 16, p. 543–603, doi: 10.1146/annurev.ea.16.050188.002551.

Hoffman, P.F., 1996, Tectonic genealogy of North America, *in* van der Pluijm, B.A., and Marshak, S., eds., Earth Structure: An Introduction to Structural Geology and Tectonics: New York, McGraw-Hill, p. 459–464.

Idnurm, M., 2000, Towards a high resolution Late Palaeoproterozoic–earliest Mesoproterozoic apparent polar wander path for northern Australia: Australian Journal of Earth Sciences, v. 47, p. 405–429, doi: 10.1046/j.1440-0952.2000.00788.x.

Idnurm, M., Giddings, J.W., and Plumb, K.A., 1995, Apparent polar wander and reversal stratigraphy of the Palaeo-Mesoproterozoic southeastern McArthur Basin, Australia: Precambrian Research, v. 72, p. 1–41, doi: 10.1016/0301-9268(94)00051-R.

Irving, E., Donaldson, J.A., and Park, J.K., 1972, Paleomagnetism of the Western Channel diabase and associated rocks, Northwest Territories: Canadian Journal of Earth Sciences, v. 9, p. 960–971.

Irving, E., Baker, J., Hamilton, M., and Wynne, P.J., 2004, Early Proterozoic geomagnetic field in western Laurentia: Implications for paleolatitudes, local rotations and stratigraphy: Precambrian Research, v. 129, p. 251–270, doi: 10.1016/j.precamres.2003.10.002.

Jones, D.L., Robertson, I.D.M., and McFadden, P.L., 1976, A palaeomagnetic study of Precambrian dyke swarms associated with the Great Dyke of Rhodesia: Transactions of the Geological Society of South Africa, v. 78, p. 57–65.

Kent, D.V., and Smethurst, M.A., 1998, Shallow bias of paleomagnetic inclinations in the Paleozoic and Precambrian: Earth and Planetary Science Letters, v. 160, p. 391–402, doi: 10.1016/S0012-821X(98)00099-5.

Layer, P.W., Kröner, A., McWilliams, M., and Burghele, A., 1988, Paleomagnetism and age of the Archean Usushwana complex, southern Africa: Journal of Geophysical Research, v. 93, p. 449–457.

Layer, P.W., Kröner, A., McWilliams, M., and York, D., 1989, Elements of the Archean thermal history and apparent polar wander of the eastern Kaapvaal craton, Swaziland, from single grain dating and paleomagnetism: Earth and Planetary Science Letters, v. 93, p. 23–34, doi: 10.1016/0012-821X(89)90181-7.

Layer, P.W., Kröner, A., and McWilliams, M., 1996, An Archean geomagnetic reversal in the Kaap Valley pluton, South Africa: Science, v. 273, p. 943–946, doi: 10.1126/science.273.5277.943.

Macouin, M., Valet, J.P., Besse, J., Buchan, K., Ernst, R., LeGoff, M., and Scharer, U., 2003, Low paleointensities recorded in 1 to 2.4 Ga Proterozoic dykes, Superior Province, Canada: Earth and Planetary Science Letters, v. 213, p. 79–95, doi: 10.1016/S0012-821X(03)00243-7.

McElhinny, M.W., and Senanayake, W.E., 1980, Paleomagnetic evidence for the existence of the geomagnetic field 3.5 Ga ago: Journal of Geophysical Research, v. 85, p. 3523–3528.

McElhinny, M.W., McFadden, P.L., and Merrill, R.T., 1996, The time-averaged paleomagnetic field 0–5 Ma: Journal of Geophysical Research, v. 101, p. 25,007–25,027, doi: 10.1029/96JB01911.

McGlynn, J.C., and Irving, E., 1975, Paleomagnetism of early Aphebian diabase dykes from the Slave structural province, Canada: Tectonophysics, v. 26, p. 23–38, doi: 10.1016/0040-1951(75)90111-0.

Meert, J.G., and Stuckey, W., 2002, Revisiting the paleomagnetism of the 1.476 Ga St. Francois Mountains igneous province, Missouri: Tectonics, v. 21, no. 2, doi: 10.1029/2000TC001265.

Meert, J.G., Van der Voo, R., and Patel, J., 1994, Paleomagnetism of the late Archean Nyanzian system, western Kenya: Precambrian Research, v. 69, p. 113–131, doi: 10.1016/0301-9268(94)90082-5.

Meert, J.G., Van der Voo, R., and Ayub, S., 1995, Paleomagnetic investigation of the Neoproterozoic Gagwe lavas and Mbozi complex, Tanzania, and the assembly of Gondwana: Precambrian Research, v. 74, p. 225–244, doi: 10.1016/0301-9268(95)00012-T.

Meert, J.G., Tamrat, E., and Spearman, J., 2003, Non-dipole fields and inclination bias: Insights from a random walk analysis: Earth and Planetary Science Letters, v. 214, p. 395–408, doi: 10.1016/S0012-821X(03)00417-5.

Melezhik, V.A., Huhma, H., Condon, D.J., Fallick, A.E., and Whitehouse, M.J., 2007, Temporal constraints on the Paleoproterozoic Lomagundi-Jatuli carbon isotopic event: Geology, v. 35, p. 655–658, doi: 10.1130/G23764A.1.

Merrill, R.T., McElhinny, M.W., and McFadden, P.L., 1996, The Magnetic Field of the Earth: Paleomagnetism, the Core, and the Deep Mantle: New York, Academic Press, 531 p.

Mertanen, S., Vuollo, J.I., Huhma, H., Arestova, N.A., and Kovalenko, A., 2006, Early Paleoproterozoic-Archean dykes and gneisses in Russian Karelia of the Fennoscandian Shield—New paleomagnetic, isotope age and geochemical investigations: Precambrian Research, v. 144, p. 239–260, doi: 10.1016/j.precamres.2005.11.005.

Mushayandebvu, M.F., Jones, D.L., and Briden, J.C., 1994, A palaeomagnetic study of the Umvimeela dyke, Zimbabwe: Evidence for a Mesoproterozoic overprint: Precambrian Research, v. 69, p. 269–280, doi: 10.1016/0301-9268(94)90091-4.

Neuvonen, K.J., Korsman, K., Kouvo, O., and Paavola, J., 1981, Paleomagnetism and age relations of the rocks in the main sulphide ore belt in central Finland: Bulletin of the Geological Society of Finland, v. 53, p. 109–133.

Oberthür, T., Davis, D.W., Blenkinsop, T.G., and Höhndorf, A., 2002, Precise U-Pb mineral ages, Rb-Sr and Sm-Nd systematics for the Great Dyke, Zimbabwe—Constraints on late Archean events in the Zimbabwe craton and Limpopo belt: Precambrian Research, v. 113, p. 293–305, doi: 10.1016/S0301-9268(01)00215-7.

Palmer, H.C., Merz, B.A., and Hayatsu, A., 1977, The Sudbury dikes of the Grenville Front region: Paleomagnetism, petrochemistry, and K-Ar age studies: Canadian Journal of Earth Sciences, v. 14, p. 1867–1887.

Pavlov, V.E., Gallet, Y., Petrov, P.Yu., and Zhuravlev, D., 2002, The Ui Group and late Riphean sills in the Uchur-Maya area: Isotope and paleomagnetic data and the problem of the Rodinia supercontinent: Geotectonics, v. 36, p. 278–292.

Percival, J.A., 2007, Geology and metallogeny of the Superior Province, Canada, *in* Goodfellow, W.D., ed., Mineral Resources of Canada: A Synthesis of Major Deposit Types, District Metallogeny, the Evolution of Geological Provinces, and Exploration Methods: Geological Association of Canada, Mineral Deposits Division, Special Publication No. 5, p. 903–928.

Pesonen, L.J., Elming, S.-Å., Mertanen, S., Pisarevsky, S., D'Agrella-Filho, M.S., Meert, J.G., Schmidt, P.W., Abrahamsen, N., and Bylund, G., 2003,

Palaeomagnetic configuration of continents during the Proterozoic: Tectonophysics, v. 375, p. 289–324, doi: 10.1016/S0040-1951(03)00343-3.
Pisarevsky, S.A., 2005, New edition of the global paleomagnetic database: Eos (Transactions, American Geophysical Union), v. 86, no. 17, p. 170, doi: 10.1029/2005EO170004.
Pullaiah, G., and Irving, E., 1975, Paleomagnetism of the contact aureole and late dikes of the Otto Stock, Ontario, and its application to Early Proterozoic apparent polar wandering: Canadian Journal of Earth Sciences, v. 12, p. 1609–1618.
Pullaiah, G., Irving, E., Buchan, K.L., and Dunlop, D.J., 1975, Magnetization changes caused by burial and uplift: Earth and Planetary Science Letters, v. 28, p. 133–143, doi: 10.1016/0012-821X(75)90221-6.
Schmidt, P.W., and Embleton, B.J.J., 1985, Prefolding and overprint magnetic signatures in Precambrian (~2.9–2.7 Ga) igneous rocks from the Pilbara craton and Hamersley basin, NW Australia: Journal of Geophysical Research, v. 90, p. 2967–2984.
Smirnov, A.V., and Evans, D.A.D., 2006, Paleomagnetism of the ~2.4 Ga Widgiemooltha dike swarm (Western Australia): Preliminary results: Eos (Transactions, American Geophysical Union), v. 87, no. 36, abstract no. GP23A-01.
Stern, R.J., 2005, Evidence from ophiolites, blueschists, and ultrahigh-pressure metamorphic terranes that the modern episode of subduction tectonics began in Neoproterozoic time: Geology, v. 33, p. 557–560, doi: 10.1130/G21365.1.
Stern, R.J., 2008, this volume, Modern-style plate tectonics began in Neoproterozoic time: An alternate interpretation of Earth dynamics and history, *in* Condie, K.C., and Pease, V., eds., When Did Plate Tectonics Begin on Planet Earth?: Geological Society of America Special Paper 440, doi: 10.1130/2008.2440(13).
Strik, G., Blake, T.S., Zegers, T.E., White, S.H., and Langereis, C.G., 2003, Palaeomagnetism of flood basalts in the Pilbara craton, Western Australia: Late Archaean continental drift and the oldest known reversal of the geomagnetic field: Journal of Geophysical Research, v. 108, doi: 10.1029/2003JB002475.
Strik, G., de Wit, M.J., and Langereis, C.G., 2007, Palaeomagnetism of the Neoarchaean Pongola and Ventersdorp Supergroups and an appraisal of the 3.0–1.9 Ga apparent polar wander path of the Kaapvaal craton, southern Africa: Precambrian Research, v. 153, p. 96–115, doi: 10.1016/j.precamres.2006.11.006.
Stupavsky, M., and Symons, D.T.A., 1982, Extent of Grenvillian remanence components in rocks of the Southern Province: Canadian Journal of Earth Sciences, v. 19, p. 698–708.
Symons, D.T.A., and Harris, M.J., 2005, Accretion history of the Trans-Hudson orogen in Manitoba and Saskatchewan from paleomagnetism: Canadian Journal of Earth Sciences, v. 42, p. 723–740, doi: 10.1139/e04-090.
Tarduno, J.A., Cottrell, R.D., and Smirnov, A.V., 2006, The paleomagnetism of single silicate crystals: Recording geomagnetic field strength during mixed polarity intervals, superchrons, and inner core growth: Reviews of Geophysics, v. 44, doi: 10.1029/2005RG000189.
Tauxe, L., 2005, Inclination flattening and the geocentric axial dipole hypothesis: Earth and Planetary Science Letters, v. 233, p. 247–261, doi: 10.1016/j.epsl.2005.01.027.
Torsvik, T.H., and Meert, J.G., 1995, Early Proterozoic palaeomagnetic data from the Pechenga zone (north-west Russia) and their bearing on Early Proterozoic palaeogeography: Geophysical Journal International, v. 122, p. 520–536, doi: 10.1111/j.1365-246X.1995.tb07011.x.
Torsvik, T.H., and Van der Voo, R., 2002, Refining Gondwana and Pangea palaeogeography: Estimates of Phanerozoic non-dipole (octupole) fields: Geophysical Journal International, v. 151, p. 771–794, doi: 10.1046/j.1365-246X.2002.01799.x.
Valet, J.-P., Meynadier, L., and Guyodo, Y., 2005, Geomagnetic dipole strength and reversal rate over the past two million years: Nature, v. 435, p. 802–805, doi: 10.1038/nature03674.
Van der Voo, R., 1990, The reliability of paleomagnetic data: Tectonophysics, v. 184, p. 1–9, doi: 10.1016/0040-1951(90)90116-P.
Van Kranendonk, M.J., Hickman, A.H., Smithies, R.H., and Nelson, D.R., 2002, Geology and tectonic evolution of the Archean North Pilbara terrain, Pilbara craton, Western Australia: Economic Geology and the Bulletin of the Society of Economic Geologists, v. 97, p. 695–732.
West, G.F., and Ernst, R.E., 1991, Evidence from aeromagnetics on the configuration of Matachewan dykes and the tectonic evolution of the Kapuskasing structural zone, Ontario, Canada: Canadian Journal of Earth Sciences, v. 28, p. 1797–1811.
Wilson, J.F., Jones, D.L., and Kramers, J.D., 1987, Mafic dyke swarms in Zimbabwe, *in* Halls, H.C., and Fahrig, W.F., eds., Mafic Dyke Swarms: Geological Association of Canada Special Paper 34, p. 433–444.
Wingate, M.T.D., 1998, A palaeomagnetic test of the Kaapvaal-Pilbara (Vaalbara) connection at 2.78 Ga: South African Journal of Geology, v. 101, p. 257–274.
Wingate, M.T.D., Pisarevsky, S.A., and Evans, D.A.D., 2002, Rodinia connections between Australia and Laurentia: No SWEAT, no AUSWUS?: Terra Nova, v. 14, p. 121–129, doi: 10.1046/j.1365-3121.2002.00401.x.
Xu, W., Geissman, J.W., Van der Voo, R., and Peacor, D.R., 1997, Electron microscopy of iron oxides and implications for the origin of magnetizations and rock magnetic properties of Banded Series rocks of the Stillwater complex, Montana: Journal of Geophysical Research, v. 102, no. B6, p. 12,139–12,157, doi: 10.1029/97JB00303.
Yoshihara, A., and Hamano, Y., 2004, Paleomagnetic constraints on the Archean geomagnetic field intensity obtained from komatiites of the Barberton and Belingwe greenstone belts, South Africa and Zimbabwe: Precambrian Research, v. 131, p. 111–142, doi: 10.1016/j.precamres.2004.01.003.
Zegers, T.E., de Wit, M.J., Dann, J., and White, S.H., 1998, Vaalbara, Earth's oldest supercontinent: A combined structural, geochronologic, and paleomagnetic test: Terra Nova, v. 10, p. 250–259, doi: 10.1046/j.1365-3121.1998.00199.x.
Zhai, Y., Halls, H.C., and Bates, M.P., 1994, Multiple episodes of dike emplacement along the western margin of the Superior Province, Manitoba: Journal of Geophysical Research, v. 99, p. 21,717–21,732, doi: 10.1029/94JB01851.
Zhao, G., Sun, M., Wilde, S.A., and Li, S., 2004, A Paleo-Mesoproterozoic supercontinent: Assembly, growth and breakup: Earth-Science Reviews, v. 67, p. 91–123, doi: 10.1016/j.earscirev.2004.02.003.

Manuscript Accepted by the Society 14 August 2007

Printed in the USA

The Geological Society of America
Special Paper 440
2008

# *Modern-style plate tectonics began in Neoproterozoic time: An alternative interpretation of Earth's tectonic history*

**Robert J. Stern***
*Geosciences Department, University of Texas at Dallas, Richardson, Texas 75080-3021, USA*

## ABSTRACT

**Modern-style plate tectonics are mostly driven by the excess density of oceanic lithosphere sinking deeply in subduction zones and can be sustained as long as melt is produced at mid-ocean ridges. Among the silicate planets, the mechanism of plate tectonics is unique to Earth, indicating that special circumstances are required. Given that the potential temperature of Earth's mantle has decreased by several hundred degrees Celsius since Archean time, the density of oceanic lithosphere must have systematically increased, which has profound implications for the viability of plate tectonics through time. Two things must be done to advance our understanding of Earth's tectonic history: (1) uncritical uniformitarianism should be avoided; and (2) the geologic record must be thoughtfully and objectively interrogated. Theoretical considerations should motivate the exploration, but geologic evidence will provide the answers. The debate needs to address the criteria for identifying tectonic style in ancient rocks, whether this evidence is likely to be preserved, and what the record indicates. The most important criteria are the temporal distribution of ophiolites, blueschists, ultrahigh-pressure terranes, eclogites, paired metamorphic belts, passive margins, subduction-related batholiths, arc igneous rocks, isotopic evidence of recycling, and paleomagnetic constraints. This list of criteria should evolve; objective redefinitions and reviews of, especially, the eclogite paired metamorphic belt and subduction-related batholith records are needed. Also, the likely effects of major tectonic changes on other Earth systems should be considered, such as true polar wander, climate change, and biosphere changes. The modern episode of plate tectonics began in Neoproterozoic time, <1.0 Ga ago, with earlier alternating episodes of proto–plate tectonics (1.8–2.0 and 2.5–2.7 Ga); unstable stagnant-lid tectonics dominated the rest of Proterozoic time and an unknown part of Archean time.**

**Keywords:** plate tectonics, Precambrian, tectonics.

## INTRODUCTION

The controversy about when and how modern-style plate tectonics on Earth began presents many opportunities to better understand our planet. The question invites the broadest possible inquiry into Earth's tectonic history, opening a richly interdisciplinary opportunity to discover how it and other silicate planets have evolved. We must engage this controversy because we cannot understand the operation of the Earth system until we understand when modern-style subduction/plate tectonics began and what Earth's tectonic style was before this.

The geoscientific subdiscipline of tectonics is concerned with deformation of and heat loss through the outer conductive shell—the lithosphere—of Earth. Other silicate planets have

*rjstern@utdallas.edu.

Stern, R.J., 2008, Modern-style plate tectonics began in Neoproterozoic time: An alternative interpretation of Earth's tectonic history, *in* Condie, K.C., and Pease, V., eds., When Did Plate Tectonics Begin on Planet Earth?: Geological Society of America Special Paper 440, p. 265–280, doi: 10.1130/2008.2440(13). For permission to copy, contact editing@geosociety.org. 

distinctive tectonic styles as well, if not plate tectonics. Tectonic style reflects how the planet loses its internal heat, and plate tectonics represent only one of at least three tectonothermal regimes known for silicate planets (Sleep, 2000). Indeed, plate tectonics are an unusual way for a silicate planet to cool. This encourages us to explore the possibility that Earth's tectonic style has changed as it cooled over the past 4.6 b.y. For this exploration to be successful, a philosophical "map" or strategy is needed. This should address all of the salient considerations, including: (1) a definition of modern-style plate tectonics; (2) an examination of how silicate planets lose heat and deform today; (3) a discussion of the physical requirements for plate tectonics; (4) an outline of the geologic criteria for recognizing the past operation of plate tectonics; and (5) an outline of Earth's tectonic history based on the previous considerations. These considerations should evolve as the exploration proceeds.

In the following sections, the geodynamic consensus that subduction of negatively buoyant lithosphere powers modern-style plate tectonics is leveraged to explore when in the history of cooling Earth this is likely to have started. First, the tectonic styles of the other silicate planets are used to note that plate tectonics are unusual from a planetary perspective, suggesting that modern-style plate tectonics may also be unusual in Earth history. Hutton's admonition "The present is the key to the past" may often be true, but it cannot be said that Earth's present tectonic style is the key to understanding the tectonic style of other planets. For this reason, the Huttonian approach will not help us understand earliest Earth history (Rollinson, 2007), and we cannot know how far back in time this approach might be useful. Instead, a non-uniformitarian perspective (discussed further in Criteria section) is needed for effective exploration of Earth's tectonic evolution. Next, consideration is given to when in our planet's history the lithosphere would have been sufficiently dense to power plate motions. Finally, a modified list of empirical geologic criteria for recognizing the products of modern-style plate tectonics (Stern, 2007) is presented and discussed. This leads to the conclusion that modern-style plate tectonics began relatively late in Earth history, sometime in the Neoproterozoic era (1000–542 Ma). This conclusion is surely controversial, but there is likely to be unanimous agreement that a better understanding of Earth's tectonic history is key for understanding the total Earth system.

## WHAT IS MODERN-STYLE PLATE TECTONICS?

We must define "modern-style plate tectonics" for this exploration to be productive. A useful definition may be obtained by considering what drives the plates today—mostly deep subduction—and how this is likely to have evolved over Earth history.

Most geoscientists have been immersed in the basics of plate tectonics throughout their careers: Earth's lithosphere is a mosaic of several plates, each of which moves relative to the others in ways that can be described with a rotation pole and angular velocity (Cox and Hart, 1986). Plate boundaries can be convergent (subduction), divergent (spreading), or transform (strike-slip). The plate-driving force is noticeably absent from this classic description; although allusions in classrooms and introductory texts are invariably made to mantle convection, these are generally not useful because there are several types of convection (Anderson, 2001). It is now clear that for modern-style plate tectonics to occur, the plates drive themselves as they organize mantle convection (Davies and Richards, 1992; Lithgow-Bertelloni and Richards, 1998). Plate motions result mostly from the negative buoyancy of oceanic lithosphere, which slides down and away from mid-ocean ridges (ridge push) toward the trench and sinks in subduction zones (Forsyth and Uyeda, 1975). The thermal boundary layer, which is the useful definition of lithosphere (Anderson, 1995), is buoyant when it is created. Zero-age oceanic lithosphere consists only of basaltic crust, which is much less dense than asthenosphere. As lithosphere ages and moves away from its creation at a mid-ocean ridge, it cools and thickens, first by hydrothermal circulation near the ridge and then by thermal conduction of the underlying mantle away from the ridge. The mantle part of lithosphere is colder and therefore denser than asthenosphere; consequently, the lithosphere inexorably increases in density as it thickens. This results in a buoyancy crossover, when oceanic lithosphere becomes denser than the asthenosphere beneath it. For today's Earth, this "buoyancy crossover time" takes 20–40 m.y. (Hynes, 2005; Oxburgh and Parmentier, 1977; van Hunen et al., this volume), but in a hotter early Earth, it probably took much longer, as discussed in the Physical Requirements section.

Ridge push provides a small part of the driving force (Conrad and Lithgow-Bertelloni, 2002; Schellart, 2004), and gravitational descent of lithosphere in subduction zones provides most of the rest; because of this, modern plate tectonics may more aptly be called "subduction tectonics" (Stern, 2005). The sinking of slabs deep into the mantle is especially important for powering plate motions. Mantle tomographic images imply that some subducted slabs can be followed to 1100–1300 km depth, perhaps down 1700 km, but it is not clear yet whether such slabs sink all the way to the core-mantle boundary (Fukao et al., 2001; van der Hilst et al., 1997). Subducting lithosphere controls plate motions in two ways, by (1) pulling on the plate on the surface (slab pull) and (2) causing a regional mantle downwelling (slab suction). Observed motions of the plates are best predicted if slab pull and slab suction forces each account for about half of the total driving force (Conrad and Lithgow-Bertelloni, 2002, 2004). A small amount of plate-driving force is applied by basal drag on continental lithosphere (Eaton and Fredericksen, 2007; Forsyth and Uyeda, 1975). This can retard as well as propel continents but is not significant for powering the fast-moving oceanic plates.

These considerations offer perspectives about how to effectively distinguish between modern-style plate tectonics and earlier tectonic modes that may be similar in some respects. For example, Ernst (2007) subdivided Earth's plate tectonic history into four stages. In the first three stages, plate motions were controlled by asthenospheric convection dragging buoyant lithosphere along and down (Fig. 1, B1). This may be analogous to the "delamination and shallow subduction" mode of Foley et al. (2003, p. 249)

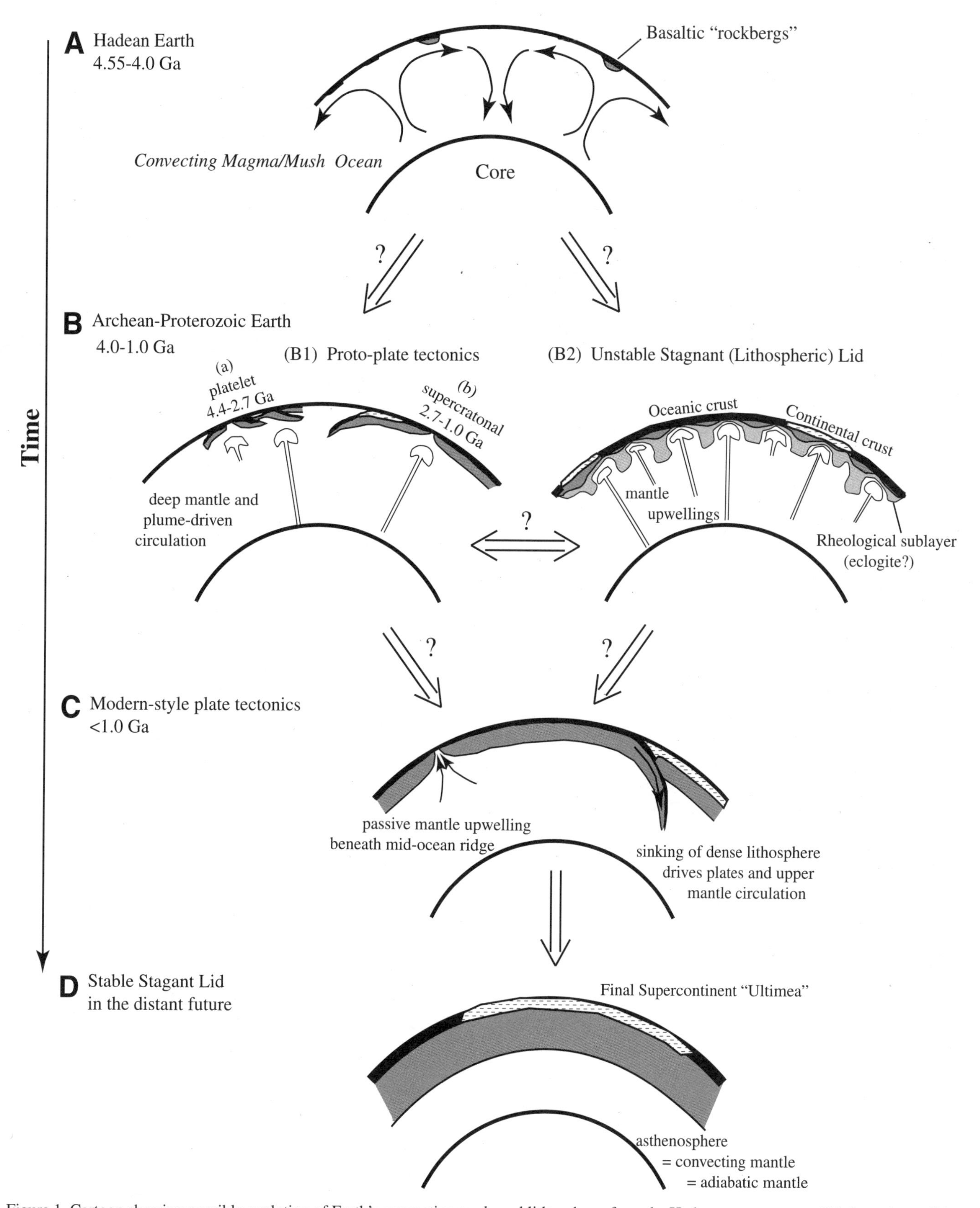


Figure 1. Cartoon showing possible evolution of Earth's convective mode and lithosphere, from the Hadean magma ocean (A) through possible pre–plate tectonic and unstable stagnant-lid regimes (B) to modern-style plate tectonics (C) to future stable stagnant lid (D). This figure was modified after Ernst (2007) and Solomatov and Moresi (1996). Tectonic scenarios A, B1, B2, C, and D are labeled on Figures 2 and 3.

or "Proterozoic plate tectonics regime" of Brown (2007, p. 194). This possible tectonic mode is called "proto–plate tectonics" here. Only for the youngest stage does Ernst (2007) infer that subducting lithosphere caused plate motions. This mechanism drives the plates today, as discussed later. Because the driving forces were fundamentally different from those of today, Ernst's (2007) first three stages are not modern-style plate tectonics, only the last stage. With the problem thus defined, we can usefully consider how these forces might have changed as Earth cooled.

## COMPARATIVE TECTONICS OF THE SILICATE PLANETS AND EARTH'S MOON

Modern-style plate tectonics represent only one of at least three and perhaps four tectonic modes. Of the five largest silicate bodies in the solar system (Mercury, Venus, Earth, Moon, and Mars), only Earth has plate tectonics (Sleep, 2000; Stevenson, 2003). Because larger bodies retain heat longer than small ones, it may not be surprising that the largest silicate planet—Earth—has a different tectonic style than its planetary brethren. Water is another important consideration. The fact that most active plate boundaries on Earth are under water may be an important reason that we have plate tectonics and other planets do not. Water weakens rocks and lowers the strength of the lithosphere, viscosity, and melting point of the mantle (Mei and Kohlstedt, 2000; Regenauer-Lieb et al., 2001). The following discussion is about how the unusual cooling style of the largest silicate planet might have been established. These modes of heat loss are also summarized in Figures 1 and 2.

All the silicate planets had infancies characterized by partially to largely molten "magma oceans" (Abe, 1997, p. 27) due to the tremendous early heat sources: planetary accretion, core formation, abundance of radioactive U, Th, and K, presence of now-extinct radionuclides, continued bolide impacts, and perhaps the Sun's T-tauri phase (Anderson, 1989; Rollinson, 2007). Radiation from a molten surface is a very effective way for a planet to lose heat, but once this solidifies and a thermal boundary layer forms and thickens, the planet cools more slowly. The mode of heat loss and thus tectonic mode must change on a cooling planet (Sleep, 2000). The "stagnant-lid" tectonothermal regime, whereby a single lithospheric plate encompasses the entire planetary surface and any mantle convection occurs beneath this lid (Solomatov and Moresi, 1996), is the dominant heat loss mode for the large silicate bodies, and all but Earth are presently in this mode. There are many variants of stagnant-lid tectonothermal regimes, from very thick, cold, stable (Luna-style) lids to hot, unstable (Venus-style) lids. Stable stagnant lids are characterized by weakly convecting mantle beneath them, whereas unstable stagnant lids can lie above vigorously convecting mantle. A range of deformational and magmatic activity is correspondingly expected across this spectrum.

We have much to learn about stagnant-lid planetary tectonothermal regimes, and, unfortunately, we must infer from dry, hot

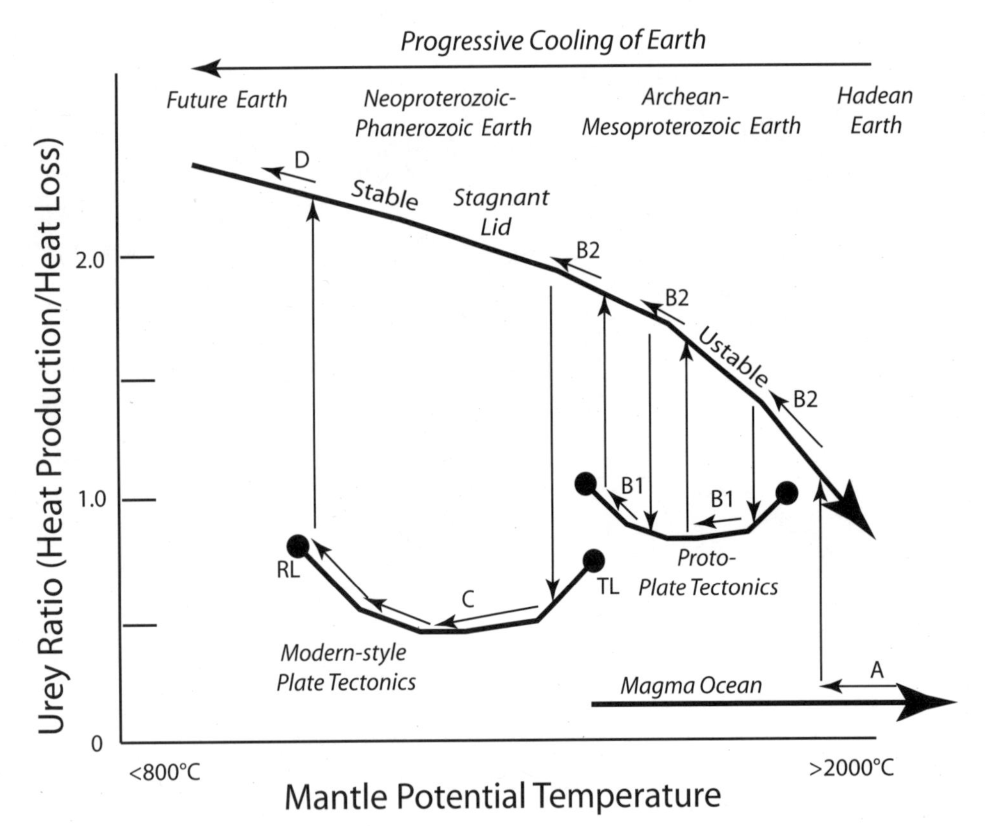


Figure 2. Diagram showing evolution of cooling silicate bodies through four modes of planetary heat loss—magma ocean, proto–plate tectonics, modern-style plate tectonics, and stagnant lid—as a function of Urey ratio (heat production/heat loss) and $T_p$ (mantle potential temperature) (modified from Sleep, 2000). Stagnant lid encompasses a wide range of magmatic and tectonic styles for planets with only one lithospheric plate, from that of essentially "dead" bodies (Mercury, Earth's moon) to magmatically and tectonically active planets such as Venus and probably early Earth; this is simplified as a stable (cold) and unstable (hot) stagnant lid. Arrowed line segments illustrate the possible tectonic evolution of a cooling Earth as schematized in Figure 1. RL—ridge lock, TL—trench lock. Line segments labeled A, B1, B2, C, and D follow tectonic scenarios in Figure 1.

Venus how wet, hot Earth might have behaved. We especially do not understand the unstable stagnant-lid mode, where eclogitic lithosphere delamination can recycle crust deep into the interior. For thin unstable lithosphere, conductive cooling is increasingly supplemented by convective heat loss due to periodic advective breakthroughs such as hotspot or heat pipe volcanism. Warmer planets like Venus have very unstable stagnant lids. The Martian stagnant lid is moderately unstable and is disrupted by the large Tharsis volcanoes and the Valles Marineris rift. Venusian resurfacings (Strom et al., 1994) may mark times of runaway delamination, when mantle melting was stimulated by foundered portions of the lower lithosphere, perhaps similar in some respects to what Bédard (2006) inferred for Archean Earth. Similar lithospheric failure on Mars probably caused that planet's hemispheric discontinuity, where a densely cratered, ancient lithosphere is preserved in the south and a much younger lithosphere is preserved in the north (Nimmo, 2005). Ultimately, it is a planet's fate to cool by conduction alone; when this happens, the planet becomes tectonically and magmatically inactive, either dead or hibernating. This is the present situation of Mercury and Earth's moon, both of which lose heat only by conduction and have strong stagnant lids that are very cold, thick, and stable.

Given that stagnant-lid tectonics is the dominant mode of silicate bodies today, it seems likely that Earth also experienced one or more stagnant-lid episodes during its 4.5 b.y. history. Stagnant-lid tectonics on the hot early Earth was surely very unstable, accompanied by vigorous mantle convection and abundant tectonic and magmatic activity, as explored by Hamilton (2007).

The principal tectonic alternatives inferred from other silicate planets are compared to the evolutionary tectonic history model of Ernst (2007) in Figure 1. Figure 2 is modified after Sleep (2000) to show some of the modes of planetary heat loss discussed previously as a function of Urey ratio (relative importance of heat production to heat loss) and mantle potential temperature ($T_p$, a measure of mantle temperature adiabatically decompressed to pressures corresponding to the planetary surface). Earth's Urey ratio has been estimated to range from 0.16 (Korenaga, 2006) to 0.65–0.85 (Schubert et al., 1980); most geodynamicists accept a value of ~0.4 (Butler and Peltier, 2002). The three main tectonothermal modes are associated with interior processes that are independent of tectonic style, including intraplate volcanism, large igneous provinces, and lithospheric delamination (Anderson, 2005). These curves amplify the conclusion that modern-style plate tectonics can only occur when appropriate mantle thermal conditions exist (van Hunen et al., this volume).

Plate tectonics are a very effective mode of planetary heat loss, as reflected by estimates of Earth's Urey ratio. This is because spreading at mid-ocean ridges efficiently delivers interior heat to the surface, where it is dissipated by seafloor hydrothermal activity, at the same time that subduction mixes cold surface matter with Earth's hot interior. It is such an effective mode of planetary heat loss that conditions favoring plate tectonics may only exist for relatively brief intervals in planetary evolution. In the future, Earth's plate tectonic regime will shut down by ridge lock, when the mantle becomes too cool to melt by adiabatic decompression, and the mid-ocean ridges will no longer be weak enough to act as plate boundaries (Sleep, 2000). Sometime in the past, plate tectonics could not occur because of trench lock. Trench lock results when decompression of Earth's hotter mantle generates oceanic crust that is too thick and buoyant to subduct. It is not known how long conditions favoring plate tectonics might have existed on Earth, but these considerations deem it very unlikely that these should persist for the life of the planet. These considerations also suggest that conditions favorable for plate tectonics will occur relatively late in planetary evolution.

Planetary heat-flow considerations favor a recent start for modern-style plate tectonics. A Urey ratio ~0.3 extrapolated back in time yields a "thermal catastrophe" ca. 1.0–1.5 Ga (Korenaga, 2006). The term "thermal catastrophe" is a misnomer, because nothing of the sort happened; instead, it reflects the fact that for much of its history, Earth lost heat slowly, especially during stagnant-lid episodes. The "thermal catastrophe" actually marks a dramatic increase in heat loss due to the relatively recent onset of the modern plate-tectonic regime.

## PHYSICAL REQUIREMENTS FOR INITIATION OF MODERN-STYLE PLATE TECTONICS

Given that the gravitational instability of old oceanic lithosphere mostly powers modern plate motions, an understanding of when modern-style plate tectonics began requires an understanding of how lithospheric density has changed as Earth cooled over the past 4.5 b.y. Modern-style plate tectonics could not have started until large parts of the oceanic lithosphere became gravitationally unstable. As noted earlier, the density of oceanic lithosphere today is controlled by the thickness of its mantle portion, but oceanic crustal thickness is also important. Oceanic crustal thickness is controlled by $T_p$, which determines the amount of mantle melting for a given amount of upwelling (McKenzie and Bickle, 1988). The mantle potential temperature for Archean Earth was perhaps 300–500 °C higher than today (Nisbet et al., 1993). Decompression melting of the hotter Archean mantle must have generated thicker oceanic crust (Davies, 1992) and thus more buoyant lithosphere. However, others have argued that the upper mantle of a hotter early Earth would have been more depleted than the present upper mantle and thus at higher temperature could have generated oceanic crustal thickness similar to that of today (Davies, 2006).

Another important question concerns the conditions under which modern-style plate tectonics can be sustained, because subduction quickly removes negatively buoyant lithosphere. This requires tracking the buoyancy crossover time through Earth history. Conductive cooling and thickening are functions of age; consequently, oceanic lithosphere should ultimately become gravitationally unstable relative to asthenosphere, although for a hotter early Earth, this probably required many tens or even hundreds of millions of years compared to 20–40 m.y. today. If so, negatively buoyant lithosphere would have been removed by subduction much more rapidly than it could form, and subduction would only have been possible intermittently, if at all. Today, the mean age of oceanic lithosphere is ~100 m.y. old (Parsons, 1982), much greater than the buoyancy crossover time, so that there is an inexhaustible supply of negatively buoyant lithosphere, and subduction can be sustained indefinitely. This is an important reason why modern-style plate tectonics can continue indefinitely, at least until ridge lock occurs sometime in the future.

The foregoing considerations alone cannot tell when modern-style plate tectonics began. To start subduction, the lithosphere must rupture and subside so that asthenosphere can flow over it, allowing the lithosphere to sink (Stern, 2004). The great strength of the lithosphere, which increases with age, is the fundamental problem facing all models of subduction initiation (Gurnis et al., 2004). For this reason, many researchers conclude that subduction zones form where the lithosphere is already ruptured, such as at major transform faults or fracture zones (Stern, 2004; Toth and Gurnis, 1998). The transition from stagnant-lid to plate tectonics also requires lithospheric rupture to start the subduction process, but formation of major transform faults requires plate tectonics. Perhaps a meteorite impact or major episode of lithospheric delamination caused the initial rupture that ultimately evolved into the first subduction zone?

Van Hunen et al. (this volume) also stress that for modern-style subduction to occur, the lithosphere must be strong enough to resist internal deformation while remaining weak enough to bend into the subduction zone, and that plate edges must be weak enough to decouple from neighboring plates. The strength of the asthenosphere is also an important control; the higher $T_p$ of the Archean would have greatly decreased asthenospheric viscous resistance, allowing plates to sink more easily back into the mantle. However, without a way to create negatively buoyant lithosphere faster than it can be subducted, continuous subduction and thus modern-style plate tectonics are impossible.

In the past, other sources of lithospheric density could have helped drive plates, such as shallower eclogitization of subducted oceanic crust (van Hunen and van den Berg, 2008). Eclogite is much denser than cold oceanic lithosphere (3.45 g/cm$^3$ vs. 3.3 g/cm$^3$; Anderson, 2005), and if the much thicker oceanic crust that probably was generated in hotter early Earth was converted to eclogite, this might have provided the negative density to power a different style of plate tectonics. Conversion to eclogite occurs at depths of ~40–50 km today, so development of eclogite-driven self-sustaining subduction would require a plate to somehow be forced down to such a depth.

## CRITERIA FOR RECOGNIZING THE START OF PLATE TECTONICS IN THE GEOLOGIC RECORD

The theoretical considerations outlined already motivate critical examination of the geologic record for information about when plate tectonics began and what happened before. The strictly uniformitarianist view—that because plate tectonics operates today, it has existed continuously and without modification since Earth accreted—is absurd and should be rejected. This leaves the question of "When did plate tectonics begin?" unanswered. The theoretical considerations outlined here suggest that it was later in Earth history rather than earlier, but only the geologic record has the answer. To get this information from the geologic record, we must first identify the most reliable fossil indicators of past plate-tectonic activity, especially those that are likely to be preserved; these indicators are also considered by other contributions in this volume. We should also entertain the possibility that plate tectonics have operated episodically on this planet. Above all, we should keep our minds open and favor approaches based on multiple working hypotheses (Chamberlain, 1897) rather than the straight-jacket of substantive uniformitarianism (Gould, 1965). Gould distinguished substantive uniformitarianism (i.e., uniform geologic processes through time) from methodological uniformitarianism (i.e., constancy of natural laws), and noted that the latter was obvious and the former ridiculous. This perspective is essential for a successful exploration of the initiation of plate tectonics.

We should take special care to avoid circular reasoning and not confuse observations with interpretations (e.g., the observation is pillowed basalt with Nb depletion, the interpretation is formation in a back-arc basin). The mere existence of continental crust, igneous and metamorphic rocks, and deformation of a certain age does not necessarily require plate tectonics to form them. We know that Venus and Mars do not have plate tectonics, yet they enjoy igneous activity and deformation, as must have pre–plate tectonic Earth. We should also keep in mind that the transition to modern-style plate tectonics on Earth may have been protracted, with multiple episodes of proto–plate tectonics or plate-like tectonics driven by different forces interspersed with episodes of stagnant-lid behavior.

There are many criteria for modern-style plate tectonics, but the clearest of these preserve evidence of protracted, deep subduction. Stern (2007) identified criteria for recognizing the operation of plate tectonics in the geologic record. This includes, in no particular order: ophiolites, blueschist, and ultrahigh-pressure metamorphic belts, eclogites, passive margins, transform faults, paleomagnetic evidence, igneous rocks with subduction-related geochemistry, and isotopic evidence for recycling. This list is modified here to broaden the eclogite discussion and to include paired metamorphic belts and subduction-related batholiths as important criteria. While transform faults are essential to plate tectonics, long strike-slip faults can occur in other tectonic modes and on other planets; for example, strike-slip faulting has been inferred for Venus from radar imagery (Koenig and Aydin, 1998; Tuckwell and Ghail, 2003). For this reason, long strike-slip faults are not listed here as an important criterion.

Geologic indicators may be incompletely preserved but should be preserved somewhere on the planet if they were produced. Today, plate tectonics are associated with 55,000 km of convergent and collisional margins, and a similar global scale should have characterized past episodes. If modern-style plate tectonics operated in the past, then abundant evidence should be preserved, even if only a small fraction of these ancient margins is

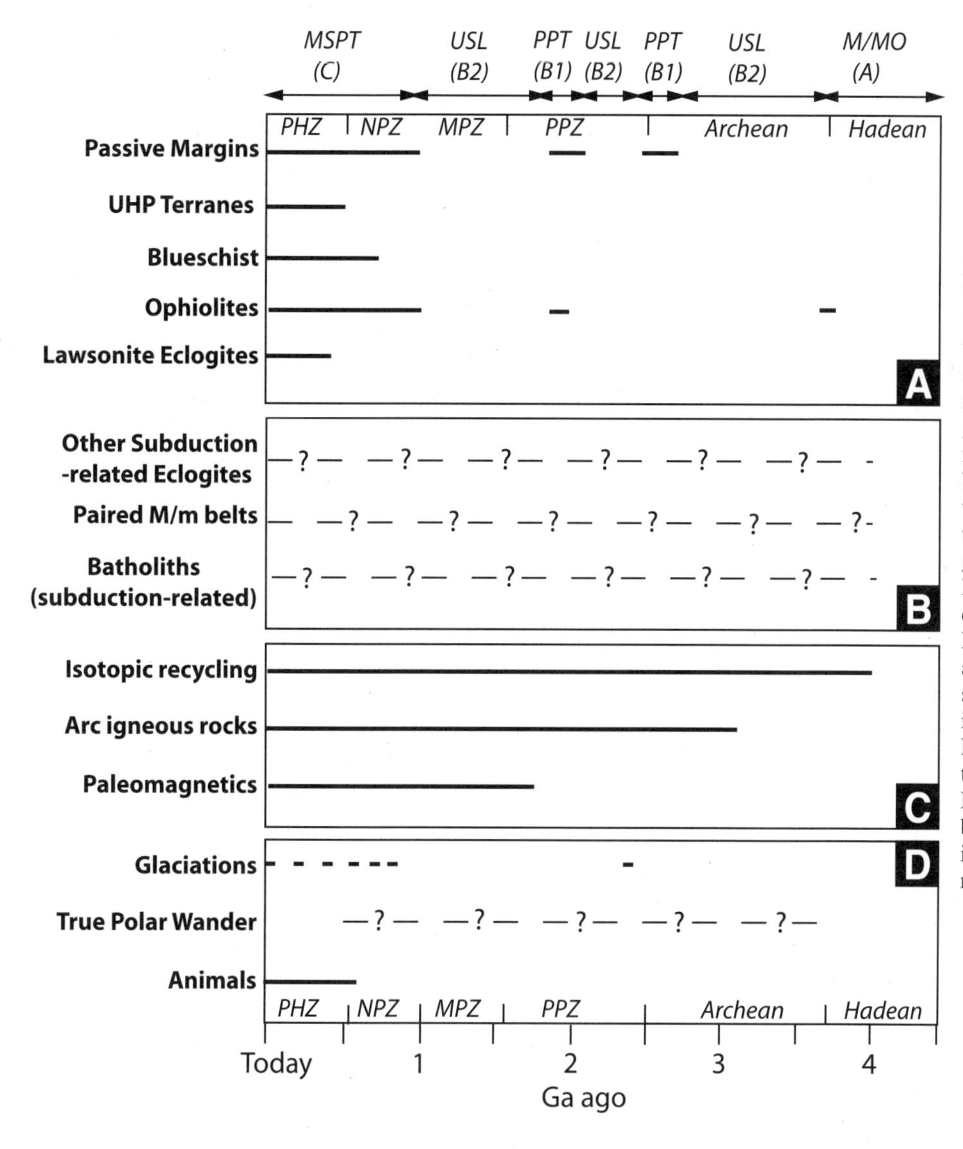


Figure 3. Criteria for the operation of plate tectonics back through Earth history. (A) "Geologic" evidence (passive margins, ultrahigh-pressure [UHP] terranes, blueschists, ophiolites, and lawsonite eclogites). (B) Three additional lines of potential geologic evidence (other eclogites, paired metamorphic belts, and subduction-related batholiths), which are not yet well enough understood to be temporally resolved on this figure. (C) Temporal distribution of geochemical, isotopic, and paleomagnetic lines of evidence. (D) Possible effects of changes in terrestrial tectonics on other Earth systems. Double-arrowed segments at top labeled A, B1, B2, and C show possible timing of tectonic scenarios presented in Figures 1 and 2. PPZ—Paleoproterozoic, MPZ—Mesoproterozoic, NPZ—Neoproterozoic, PHZ—Phanerozoic; MSPT—Modern-style plate tectonics; USL—unstable stagnant lid; PPT—Proto–plate tectonics; M/MO—mush/magma ocean; M/m—metamorphism.

preserved. The distribution of these lines of evidence through time is summarized in Figure 3, using four panels: Figure 3A shows "geologic" evidence (passive margins, ultrahigh-pressure terranes, blueschists, ophiolites, and lawsonite eclogites); Figure 3B presents three additional lines of potential geologic evidence (other subduction-related eclogites, paired metamorphic belts, and subduction-related batholiths), which are not yet well enough understood to be temporally resolved on this figure; Figure 3C shows the temporal distribution of geochemical, isotopic, and paleomagnetic lines of evidence; and Figure 3D shows when tectonic transitions might have triggered secondary effects on other Earth systems. There are different ways that these lines of geologic evidence can be considered, not only when the assemblage or property first appeared but also when it first became common. It is perhaps not surprising that the various lines of evidence suggest different times for the start of plate tectonics. Others present a different list of criteria and reach different conclusions (Condie and Kröner, this volume). Discussion about the best criteria should be encouraged at this stage of the exploration.

## Ophiolites

Ophiolites are fragments of oceanic lithosphere tectonically emplaced on continental crust, and they are reliable indicators of plate-tectonic activity. Ophiolites manifest two modes of lithospheric motion expected from plate tectonics: seafloor spreading to form the crustal section and plate convergence to emplace this on the continent. The location(s) of ophiolite formation is controversial; fortunately, this is not pertinent here because all tectonic settings for ophiolites require subduction for formation and emplacement (Stern, 2004). Complete ophiolites, with pelagic sediments, pillowed basalts, sheeted dikes, gabbros, and tectonized ultramafic residues (harzburgite and lherzolite), are rare. The search for this evidence of past plate tectonics should take fragmentary ophiolites into account, but essential components should be present, if disrupted. Sheeted dikes are especially scarce in otherwise complete ophiolites. Pillowed tholeiitic basalts with concave-downward rare earth element (REE) patterns and associated gabbros and harzburgitic ultramafics containing Cr-rich spinels in an allochthonous thrust sheet are minimum requirements for inferring that a disrupted mafic-ultramafic suite represents an ophiolite. Using these criteria, the 1.73 Ga Payson "ophiolite" (Dann, 1997) is not an ophiolite because it lacks tectonized peridotite and is demonstrably autochthonous, with screens of older continental crust between dike swarms. Ophiolites may be readily removed by erosion because the best-preserved examples are emplaced as the uppermost unit of a nappe stack, and these are likely to be eroded first. On the other hand, ophiolites are very thick and, if deformed, are likely to be at least partially preserved in orogenic belts.

Ophiolites of Neoproterozoic and younger age are common (Dilek, 2003; Stern, 2005), but pre-Neoproterozoic ophiolites are rare and often controversial. There are good ophiolites at ca. 2.0 Ga (Purtuniq, Canada, Scott et al., 1992; Jormua, Finland, Peltonen et al., 1996). Paleoproterozoic ophiolites show interesting differences from Neoproterozoic and younger ophiolites, which generally have "suprasubduction-zone" (SSZ) affinities. Jormua metabasalts are enriched, have no suprasubduction-zone characteristics, and are interpreted to have formed in a small Red Sea–like oceanic rift. The 2.5 Ga Dongwanzi ophiolite of NE China was thought to be the oldest ophiolite (Kusky et al., 2001), but this claim has since been challenged (Zhai et al., 2002; Zhao et al., 2007). The oldest of the purported ancient ophiolites may be the 3.8 Ga Isua example (Furnes et al., 2007). It is not surprising that early perimobile Earth witnessed seafloor spreading.

Condie and Kröner (this volume) concluded that the general absence of ophiolites from the pre-Neoproterozoic record is because the middle- to lower-crustal sections and mantle lithosphere were generally lost during collisional tectonics, so that only the upper, pillow-lava–rich volcanic crust was emplaced, and the rest of the oceanic crust was disposed of by delamination or other means. Wakabayashi and Dilek (2003, p. 435) noted that "ophiolite obduction mechanisms must be spatially associated with subduction"; it should be added that obduction requires the ophiolite-to-be to lie on the overriding plate, usually the forearc. Stern (2004) argued that most Phanerozoic ophiolites were formed in forearcs during subduction initiation and were emplaced when buoyant crust was subducted, the subduction zone failed, and the buoyant crust isostatically rebounded. With modern, subduction-related ophiolite emplacement mechanisms, it is very difficult to see how the lower part of an ophiolite, thick or thin, could be so effectively removed.

## Blueschist

Blueschist is a metamorphosed mafic rock containing abundant sodic amphibole that forms as a result of high-pressure (*P*), low-temperature (*T*) metamorphism (Maruyama et al., 1996). Blueschists are synonymous with "B-type" ultrahigh-pressure terranes (Maruyama et al., 1996) and are characteristic of Pacific-type orogenic belts (Ernst, 2003). The perception that blueschist only forms in subduction zones is based on their association with ancient subduction mélanges and is confirmed by studies of active subduction zones (Abers et al., 2006; Maekawa et al., 1995; Zhang et al., 2004). The best-studied blueschists, those of the Franciscan and Sanbagawa terranes, appear to have been subducted 15–70 km deep before returning to the surface (Ernst, 2003).

It has been known for a half-century that blueschists are not found in very ancient rocks (Roever, 1964), most recently confirmed by Brown (2008). It is now widely acknowledged that the oldest blueschists date from Neoproterozoic time, ca. 800–700 Ma (Maruyama et al., 1996); pre-Neoproterozoic blueschists are unknown. This does not seem to reflect poor preservation of pre-Neoproterozoic blueschists. Because blueschists are preserved in fossil subduction zones which reach deep into the Earth, they are not likely to be removed by erosion. The preservation of ca. 3.2 Ga high-*P*, low-*T* metamorphic rocks in South Africa (Moyen et al., 2006) further indicates that blueschists would be preserved somewhere if they had been produced and exhumed in pre-Neoproterozoic time. This assessment is confirmed in Brown's (2006) review of how terrestrial metamorphism has changed with time: "...there is a clear distinction between the Archean, where the data are minimal, and the Proterozoic and Phanerozoic, where the data are abundant, and between the pre-Neoproterozoic, where [blueschist] metamorphism does not occur, and Neoproterozoic and younger belts, where [blueschist] metamorphism is common. These observations are inconsistent with a progressively degraded record with increasing age" (Brown 2006, p. 963).

Many geoscientists think that the absence of pre-Neoproterozoic blueschist indicates that Earth was hotter so that subduction did not generate blueschist. Early Earth surely had significantly higher mantle potential temperature but this may not have affected the thermal structure of subduction zones and, thus, whether or not blueschist was produced. The thermal structure of subduction zones mostly reflects the age of the subducted lithosphere and the convergence rate (Peacock, 2002), not mantle potential temperature. In fact, the thicker oceanic crust expected for early Earth might be expected to result in a rate of production of blueschist greater than that of the recent past if Archean subduction zones existed. This controversy might be resolved with robust geodynamic models for subduction zones in a hotter Earth.

## Eclogites

Eclogites are metabasalts and metagabbros, typically recording pressures >1.2 GPa (45 km depth) and temperatures >600–650 °C. They are generally coarse-grained but can be fine-grained if deformed. Eclogite is often thought to reflect plate-tectonic processes—especially subduction—but there are many ways to generate garnet-clinopyroxene rocks other than in subduction zones. Coleman et al. (1965) considered the diversity of eclogites at the dawn of plate-tectonic theory and recommended "elimination of the single eclogite metamorphic facies concept"

(p. 483). They divided eclogites into three broad groups based on the proportion of pyrope (MgAl garnet) in garnet: Eclogite xenolith garnets in kimberlites, basalts, or ultramafic rocks (>55% pyrope); in eclogite lenses within migmatite gneissic terrains (30%–55% pyrope); and in lenses of eclogite within alpine-type metamorphic rocks (<30% pyrope). Jadeite (NaAl clinopyroxene) content in pyroxene progressively increases in the first two groups, whereas diopside decreases. The difference in Ca-Mg partitioning between coexisting garnet-pyroxene in eclogites of the same bulk composition indicates a broad range of pressure-temperature conditions during crystallization.

Certainly some eclogites are diagnostic of subduction, especially Coleman's group C (associated with glaucophane schists) and the "low-*T*" eclogites of Carswell (1990), but others probably do not require subduction to form. Some eclogites may form during subduction initiation, such as Franciscan eclogites (Anczkiewicz et al., 2004); other eclogites (medium-temperature eclogite–high-pressure granulite [EHP-GM] suite of Brown, 2006) may be related to continental collision. Some eclogites are probably lower-crustal delaminates, as confirmed by integrated petrologic-geophysical studies in the United States, Europe, and Asia (Ruppert et al., 1998; Tilmann et al., 2003). Ducea (2002) demonstrated that the Sierra Nevada Batholith was affected by foundering of 35–50 km of lower-crustal pyroxenite residues. Instability of lower-crustal eclogite may be responsible for the near constant 45 km depth to the Moho. In proof of this shallow recycling, the reprocessed remnants of Gondwana lithosphere, lost during the terminal collision that formed the East African orogen and other Neoproterozoic orogens, can be recognized in the isotopic composition of Indian Ocean basalts (Janney et al., 2005); other enriched oceanic basalts, especially in the Atlantic, may have a similar source (Kamenetsky et al., 2001). These considerations suggest that, whereas eclogite-driven delamination today occurs contemporaneous with plate tectonics, this style of surficial recycling may have been especially important during unstable stagnant-lid episodes (Fig. 1, B2).

As noted previously, eclogites are so much denser than other crustal or upper-mantle rocks that they are sure to provide important gravitational forces for downwelling, whether as part of a subducted plate, delaminated crustal root, or otherwise. These eclogites are likely to be preserved in the mantle and returned to the surface from time to time as well as melt. We need clear, diagnostic criteria to distinguish subduction-related eclogites from those that form in other tectono-metamorphic environments, similar to what has been discussed by Jacob (2004). Lawsonite-bearing eclogites are a useful example; they clearly manifest metamorphism in a subduction zone and are only found in Phanerozoic terranes (Tsujimori et al., 2006). Further efforts to distinguish subduction-related versus other kinds of eclogites need to be articulated, perhaps combining the pioneering approach of Coleman et al. (1965) with the detailed mineral-chemical approach of Tsujimori et al. (2006) and geochemical-isotopic approach of Jacob (2004).

## Ultrahigh-Pressure Terranes

Ultrahigh-pressure (UHP) metamorphic terranes are important indicators of ancient subduction zones. These form when continental crust is subducted to depths >100 km and returns to the surface. Metamorphic assemblages in ultrahigh-pressure terranes include coesite and/or diamond and indicate peak metamorphic conditions of ~700–900 °C and 3–4 GPa or more (Liou et al., 2004). In contrast to the long-appreciated and well-understood blueschist record, ultrahigh-pressure terranes have been recognized for only the past 15 yr or so, and new localities will surely be found in the future. The oldest known ultrahigh-pressure locality is in Mali, where coesite-bearing gneiss was metamorphosed at ca. 620 Ma (Jahn et al., 2001). The oldest diamond-bearing ultrahigh-pressure terrane is found in Kazakhstan, where diamond- and coesite-bearing paragneiss of the Kokchetav massif was subducted >120 km deep at ca. 530 Ma (Maruyama and Liou, 1998). The first evidence for deep subduction of continental crust is thus found in late Neoproterozoic and early Cambrian rocks; more evidence is likely to be found because the geological community's ability to recognize ultrahigh-pressure terranes is improving but still limited.

## Paired Metamorphic Belts

Paired metamorphic belts are diagnostic of modern-style plate tectonics; they always form on the overriding plate, with the high-*P*, low-*T* (blueschist-bearing) belt lying next to the trench and the low-*P*, high-*T* (greenschist) belt near the arc, which itself is often preserved as a batholith. The paired metamorphic belt concept evolved early in the development of plate-tectonic theory (Miyashiro, 1972). This is the characteristic thermal structure of a subduction zone (Kelemen et al., 2003; van Keken et al., 2002), so paired metamorphic belts are simple, clear manifestations of a mature subduction zone, with very low heat flow near the trench and high heat flow near the magmatic axis. Paired metamorphic belts with a mélange zone, forearc basin, and magmatic arc like Mesozoic California or late Paleozoic–Mesozoic Japan (Banno and Nakajima, 1992) are compelling evidence of a mature and long-lived subduction zone. These belts are tens to hundreds of kilometers wide and thus are robust features that can be expected to partially persist through episodes of orogeny and erosion.

Brown (2006) argued that the definitive hallmark of plate tectonics is a duality of thermal environments (requiring convergent plate margins), and that the imprint of this hallmark is found in the rock record as the contemporary occurrence of spatially distinct contrasting types of metamorphism. This is not the same as paired metamorphic belt, because any silicate planet convection will develop dual pressure-temperature metamorphic regimes. A duality of thermal environments indicates convection, but this does not require modern-style plate tectonics.

Paired metamorphic belt models need to be revised because only ~30% of convergent plate margins are accretionary; a much greater proportion of convergent plate margins is erosional (Clift and Vannucchi, 2004; von Huene and Scholl, 1993). Accretionary convergent margins are likely to preserve the high-*P*, high-*T* belt as a broad zone of mélange that is subducted to depth and returned to the surface (Cloos and Shreve, 1988a, 1988b). Erosional convergent margins are likely to preserve a much narrower belt of high-*P*, high-*T* rocks, but in this case, the proximity of ophiolite or forearc basement is clear.

### Batholiths

Evidence of a magmatic "island" arc strongly suggests modern-style plate tectonics because it demonstrates that a subduction zone existed at a depth of 100–250 km. Both intra-oceanic and Andean-type convergent margins should develop a felsic plutonic middle crust (Stern, 2002). Volcanic rocks are readily removed by erosion, but underlying plutonic rocks are much more difficult to erase (Hamilton and Myers, 1967), and in Precambrian rocks, we may need to rely on the plutonic record of arc history. Batholiths are dense "forests" of plutons and are dominated by felsic plutonic lithologies that often include tonalite-trondhjemite-granodiorite (TTG) suites common in the Archean. Field studies and U-Pb zircon dating reveal remarkable complexity in the evolution of batholiths and plutons (Glazner et al., 2004).

Subduction-related batholiths can also help to resolve subduction polarity (through relations with paired metamorphic belts and using the K-h relationship; Dickinson, 1975). The sequence from the trench to the back-arc should be high-*P*, low-*T* metamorphic belt, low-*P*, high-*T* metamorphic belt, batholith. Arcs and batholiths are huge features, hemispheric in scale. Subduction-related batholiths, with clear across-arc variations in magma chemical and isotopic compositions, are strong evidence for modern-style plate tectonics, especially where associated with a paired metamorphic belt. The arc and back-arc region is invariably characterized by high heat flow (Hyndman et al., 2005), although it may be a zone of rifting, thrusting, or be strain-neutral. One problem is that not all batholiths require subduction to form; all that is needed is for the crust to heat to the melting point and/or to extensively fractionate mafic melts. The point that batholiths unrelated to plate tectonics can and have been generated was stressed by Hamilton (1998).

Characteristics of subduction-related batholiths include: (1) an aspect ratio ≈ 10–100 (hundreds to thousands of kilometers long, tens to hundreds of kilometers wide); (2) alignment parallel to and ~200 km inboard from pertinent trench/suture; (3) life span of tens to a few hundred millions of years; and (4) across-batholith compositional variations, including higher K and other large ion lithophile elements (LILEs) with greater distance from trench/suture (K-h relationship). Such batholiths are commonly preserved around the Pacific Rim today

### Chemical Composition of Igneous Rocks

The chemical composition of igneous rocks can help to reveal tectonic setting; this line of evidence of course complements inference from batholiths. In particular, compositions of ancient igneous rocks like those produced in Cenozoic island arcs are strong evidence for subduction and thus plate tectonics. Basalts from modern intra-oceanic arc systems show distinctive enrichments or "spikes" in fluid-mobile elements (e.g., K, Sr, Pb) relative to neighboring elements. These spikes are associated with strong relative depletions in incompatible high field strength elements (HFSE), especially Nb and Ta (Stern, 2002). These sorts of studies benefit greatly from the increasing availability of high-quality trace-element analyses produced from proliferating inductively coupled plasma–mass spectrometry (ICP-MS) analytical facilities. These characteristics can be inferred from normalized plots of trace elements ("spider diagrams") and are also the basis of a wide range of trace-element "discriminant diagrams," which have been widely used to infer the tectonic setting of ancient igneous rocks (Pearce and Cann, 1973; Shervais, 1982). Interpretations of ancient basalts using this approach are useful for inferring the existence of ancient magmatic arcs that formed by subduction, and there are many papers that do this (Cawood et al. [2006] offer several examples). Kerrich and Polat (2006) inferred Archean examples of convergent margin associations in Archean time, focusing on boninites, Mg-andesites, and adakites.

As is the case for other lines of inference regarding the operation of plate tectonics, care should be taken to ensure that important chemical characteristics are not the result of alteration or crustal contamination, which can result in trace-element patterns that mimic those of arc igneous rocks. The first concern in particular requires petrographic studies before samples are selected for chemical analysis. The fact that spikes of fluid-mobile elements seen on multi-element diagrams for ancient rocks may reflect greenschist-facies alteration can be partly overcome by emphasizing immobile incompatible trace elements, especially HFSE cations (e.g., Th, Nb, La). This approach applied to ancient rocks is particularly useful for primitive basalts because these are likely to be less contaminated by preexisting continental crust, and any contamination that does occur is likely to be revealed by isotopic and geochronologic studies. However, even this approach must recognize that many igneous rocks unrelated to subduction show the Nb-Ta depletions typical of subduction zones, such as Phanerozoic continental flood basalts (http://www.geokem.com/flood-basalts-2.html). Foley (this volume) also cautioned against overinterpreting trace-element spikes seen in geochemical data for Archean igneous rocks, particularly because the causes of the distinctive "subduction signature" in modern arc igneous rocks are not well understood.

Studies of granitic rocks have shown some success (Pearce et al., 1984), but the fact that felsic igneous rocks are often crustal melts means that their trace-element compositions can be more reflective of crustal source than tectonic process. It should be noted that the composition of the continental crust as a whole shows fractionation between fluid-mobile versus high field strength element cations that is consistent with formation in a subduction-related environment (Rudnick and Gao, 2003).

These fractionations strongly suggest control due to subduction, but the other tectonic modes discussed here should also be capable of recycling water into the upper mantle to generate these signatures.

**Passive-Margin Record**

The passive-margin record is another powerful constraint. Passive continental margins such as those around the Gulf of Mexico and the Atlantic Ocean form when continents rift apart and produce an intervening, long-lived ocean basin. Passive margins form as an integral part of plate tectonics on a planet—like Earth—with buoyant, subaerially exposed continents and submerged oceans. Passive margins take tens to hundreds of millions of years to develop fully, allowing hundreds of meters of thermal subsidence that creates even greater accommodation space for sediments. The proximity of passive margins to continental sediment sources such as rivers and glaciers ensures that they will contain enormous thicknesses of sediments. These sedimentary sources are built on very different crusts: continental crust is on the landward side, oceanic crust is on the seaward side, and transitional crust is beneath. Because passive margins are associated with thick continental crust, they are difficult to subduct entirely and, although deformed, should be preserved in ancient suture zones. Examination of the geologic record indicates that the oldest passive margins date to ca. 2.7 Ga, but they become common at ca. 2.0 Ga and again ca. 0.9 Ga (Bradley, 2007, personal commun.). Hynes (this volume) noted, however, that the hotter Archean mantle would have resulted in less total accommodation space for passive-margin sediments than is observed for modern passive margins, perhaps by a factor of two or more, and this may explain the scarcity of passive-margin sequences in the Archean record. Clearly, the passive-margin record is important for understanding Earth's tectonic history.

**Paleomagnetic Measurements**

Paleomagnetic measurements help to constrain the initiation of significant differential motion between crustal blocks (proxies for independent plates). Cawood et al. (2006) concluded that the paleomagnetic database demonstrates differential movements of continents in pre-Neoproterozoic times. They noted that comparisons of coeval paleopoles yield ambiguous longitudes and polarities but nevertheless indicate that Australia and Baltica drifted independently between 1770 and 1500 Ma, requiring independent plate motions and thus plate tectonics (Pesonen et al., 2003). Other examples of differential movement of continental blocks in pre-Neoproterozoic times are given in Pesonen et al. (2003), Pisarevsky and McElhinny (2003), and Powell et al. (2001). The latest critical assessment of the paleomagnetic record for evidence of relative motion found strong evidence for differential cratonic motions in Paleo- to Mesoproterozoic time but not earlier (Evans and Pisarevsky, this volume). This may reflect the fact that uncertainties resulting from the overprinting of primary magnetization by diagenesis or metamorphism increase with age, while opportunities to compare suitable units on different blocks decrease with age. Additional serious complications arise for interpretation of paleomagnetic data if Earth's magnetic field has not always been a dipole approximately collinear with the geographic poles. There is also a substantial body of evidence that the Precambrian witnessed episodes of true polar wander (Goldreich and Toomre, 1969; Maloof et al., 2006). Nevertheless, these issues should be resolvable with a carefully planned effort to identify suitable suites to paleomagnetically sample on different cratons. Paleomagnetic results are important for answering the question of "When did plate tectonics begin?," and I enthusiastically endorse the call for a coordinated, international paleomagnetic effort to better understand when various cratons moved relative to each other and when they did not (Evans and Pisarevsky, this volume).

**Isotopic Compositions**

Isotopic compositions are uniquely powerful for inferring times of crust-mantle differentiation and for identifying deeply recycled surface materials. There is abundant evidence for early recycling of crust and sediments into the mantle from as far back as the early Archean (Shirey et al., this volume). Nd isotopic compositions are consistent with the formation of an early depleted reservoir in the mantle, indicating very early differentiation of Earth's crust and mantle (Boyet and Carlson, 2005). Isotopic studies of inclusions in diamonds from eclogites derived from the mantle beneath Archean cratons give isotopic compositions of O, Sr, and Pb that indicate seafloor alteration and recycling into the mantle (MacGregor and Manton, 1986; Schulze et al., 2003), and many of these eclogites themselves show isotopic evidence of low-$T$ seawater alteration (Jacob, 2004; MacGregor and Manton, 1986). Mass-independent $^{33}S$ isotope anomalies in diamonds are also strong evidence that these materials were originally at the surface early in Earth history and were recycled into the mantle (Farquhar et al., 2002). The isotopic record clearly indicates that the mantle differentiated very early and that crust and sediments were also recycled very early. Shirey et al. (this volume) interpreted the isotopic record to indicate subduction and plate tectonics, but surface materials can descend into the mantle as a result of proto–plate tectonics, delamination associated with an unstable stagnant lid, or even without any tectonics at all, as a result of being deeply injected by meteorite impacts. Objective assessment of the isotopic data concludes strong evidence for early recycling but cannot distinguish between subduction and other convective modes.

**Effects on Other Earth Systems**

The exploration of the initiation of modern-style plate tectonics has important implications for the functioning of other Earth systems, an approach stressed by Rollinson (2007). We should consider these, as well as the direct evidence articulated already. Three possible secondary effects are shown in Figure 3D, including glaciations, true polar wander, and the rise of animals. This is not to imply that all such changes require major tectonic changes. For example, the Paleoproterozoic glaciation may have had a biologic cause in the proliferation of photosynthetic cyanobacteria and drawdown of $CO_2$ (Kopp et al., 2005), but major tectonic changes may have led to significant changes in these systems, as suggested by Fairchild and Kennedy (2007, p. 914): "The most plausible root causes for long-term change in the surface Earth System are deep earth processes and biological innovation, for example, via their effect on master reactions I and II. [Note: (I) $CO_2 + H_2O = CH_2O$ (organic C) + $O_2$ ; (II) Silicates + $CO_2$ = secondary silicates + aqueous silica + cations + $HCO_3$.] However, although reorganizations of mantle circulation might have resulted in stepwise reductions in outgassing and/or changes in tectonic style, as proposed for example at 2.5–2.0 Ga and around 1 Ga, the evidence is disputed, and so links are unclear."

The links between tectonic changes and explosive volcanism (leading to atmospheric cooling) and volcanic $CO_2$ emissions (leading to atmospheric warming) are straightforward. Several episodes of Neoproterozoic glaciation, articulated as "Snowball Earth" (Hoffman et al., 2002), may have had a tectonic trigger. Stern et al. (2008) related Neoproterozoic cooling leading to glaciation, at least partly, to a significant increase in explosive volcanism, using total Neoproterozoic crustal growth as a proxy for explosive volcanism. The start of modern-style plate tectonics may also have affected Earth's moment of inertia and rotational behavior (perhaps triggering true polar wander; Goldreich and Toomre, 1969; Maloof et al., 2006). Other Earth systems may also have been affected by the beginning of modern-style plate tectonics, including the biosphere. There is a strong temporal similarity between the beginning of modern-style plate tectonics and the development of animals. Links between Neoproterozoic tectonics and biosphere are less obvious than between tectonics and climate. Valentine and Moores (1970) suggested that continental breakup stimulated evolution because this increased both the isolation of species and the area of shallow continental shelves; this tectonic control would have been important for biosphere development after Rodinia began to break up in early Neoproterozoic (Cryogenian) time (Bogdanova et al., 2008). Changing patterns of tectonics and magmatism may have also impacted life by increasing the supply of nutrients (P, Fe, etc.) to the ocean as a result of increased orogeny, erosion, and run-off and by increased explosive volcanism (Anbar and Knoll, 2002).

## A PLATE TECTONIC AND PRE–PLATE TECTONIC HISTORY OF EARTH

These geologic criteria for modern-style plate tectonics are only a beginning. The list of criteria should be evaluated and modified accordingly. Similarly, the time line for each criterion should be updated frequently and individually. In any case, presentation of the criteria and time lines in this fashion is an excellent way to move the discussion forward. Each of us should weigh and interpret the evidence individually and from the perspective of our own expertise. Given the criteria and time lines shown in Figure 3, how are these best interpreted? Clearly, the isotope record shows that there has been material moving from Earth's surface and mixing into the mantle; this permits but does not require modern-style plate tectonics because this could also be accomplished by unstable stagnant-lid and proto–plate tectonic modes. The evidence for arc-like geochemical signatures also dates from early in Earth history, strongly suggesting the presence of subduction zones and thus the operation of plate tectonics, but this could also be accomplished by other tectonic modes. Paleomagnetic data indicate differential plate motions at least as far back as Paleoproterozoic time, and the ophiolite record starts about this time. A very important perspective is revealed by metamorphic rocks, especially blueschists, ultrahigh-pressure terranes, and lawsonite eclogites, because these demonstrate deep subduction and the operation of modern-style plate tectonics as opposed to proto–plate tectonics or unstable stagnant-lid tectonics. There is no record of deep subduction prior to Neoproterozoic time preserved in metamorphic rocks, although Pease et al. (this volume) reached a very different conclusion.

Just as Earth history can be divided into Precambrian and Phanerozoic time, so perhaps can its tectonic history be divided into the past 1 b.y. of modern-style plate tectonics, and a preceding 3.5 b.y. of pre–modern-style plate tectonics (Figs. 1, 2, and 3). This first phase was itself complex, involving a Hadean magma/mush ocean becoming an Archean to Mesoproterozoic mode of alternating unstable stagnant-lid and proto–plate tectonics. Something akin to plate tectonics began early in order to form the Isua ophiolite, but there is no record that this style of magmato-tectonics continued. Modern-style plate tectonics began in the last 25% of Earth history. Earth's early tectonic style may have been similar in some ways to plate tectonics—especially in being able to deeply recycle surface materials to depth in the mantle and to generate arc-like magmas, but it was not likely driven by lithospheric negative buoyancy or associated with deep subduction. The presence of ophiolites and paleomagnetic evidence for independent plate motions indicates that a short but intense episode of proto–plate tectonics occurred ca. 1.8–2.0 Ga during Paleoproterozoic time—perhaps as a result of collapse of ancient oceanic basins that had finally become negatively buoyant with respect to underlying asthenosphere, but, again, there is no evidence for deep subduction. The Paleoproterozoic episode might not have had deep subduction and was short-lived because of a long buoyancy crossover time, whereby subduction quickly exhausted the supply of negatively buoyant oceanic lithosphere. After the Paleoproterozoic, the proto–plate tectonics pulse ended, and several hundreds of millions of years elapsed before modern-style plate tectonics began in Neoproterozoic time. Establishment of self-sustaining subduction and plate tectonics late in Earth history reflects the fact that Earth had cooled sufficiently to allow a relatively short buoyancy crossover time, so that oceanic lithosphere became negatively buoyant after a few tens of millions of years. Clearly, the evidence for deep subduction and the first great ophiolite graveyards date from Neoproterozoic time.

## CONCLUDING REMARKS

We cannot pretend to understand the operation of the Earth system until we understand when modern-style subduction and plate tectonics began and what came before this. These grand questions offer richly interdisciplinary avenues of solid Earth research, ones that can strengthen ties between many strands of earth science research and at the same time yield new insights into our planet's unique tectonics. Making progress will require critical reexamination of a wide range of fundamental issues. We should reject the straitjacket of substantive uniformitarianism and embrace the more flexible approach of multiple working hypotheses. The exploration requires that we more fully engage our imaginations and consult the full range of geoscientific and related disciplines. We should also examine other silicate planets for insights as we consider more fully the range of tectonic styles that Earth might have experienced as it cooled and differentiated.

Ultimately, we must recognize the magmatic and tectonic fingerprints of a pre–plate tectonic Earth in the geologic record. As discussed herein, it is easy to imagine pre–plate tectonic Earth but much more difficult to convincingly demonstrate it. This requires identification of a geologic feature that could not have formed by modern-style plate-tectonic processes. Where should we start to look for such rocks? There is increasing evidence that the latest Paleoproterozoic and Mesoproterozoic might be a good time period for some of this effort to be concentrated. These rocks are abundantly preserved, and problems with strict interpretation of uniformitarian plate-tectonic models have been noted (Bickford and Hill, 2007; Hoffman, 1989).

A final comment: the exploration of the initiation of plate tectonics must be broadly interdisciplinary if it is to be successful. The overall effort, focused on critical examination of the Precambrian geologic record and development of refined geodynamic models, promises to lead to an improved understanding of how the Earth system developed through time.

## ACKNOWLEDGMENTS

My research into modern and ancient crust formation has been funded by the U.S. National Science Foundation. Much of my understanding resulted from discussions with colleagues at Stanford (especially G. Ernst, S. Klemperer, J.-G. Liou, D. Scholl, N. Sleep, and T. Tsujimori) while I was Blaustein Fellow during fall 2005 and with colleagues at Caltech (especially D. Anderson, P. Asimow, M. Gurnis, D. Stevenson, and R. Workman) while I was a Tectonics Observatory Fellow in spring 2006. Insights provided by participants of the Landers 2006 Penrose conference are also greatly appreciated, as are the thoughts of C. Barnes, M. Brown, M. Ducea, D.A.D. Evans, A. Glazner, W. Hamilton, J.J.W. Rogers, and S. Shirey. I also greatly appreciate the constructive reviews of Jan Kramers, A. Kröner, Eldridge Moores, Victoria Pease, and Hugh Rollinson. I especially thank my co-convenors K. Condie and A. Kröner for helping to organize this fascinating Penrose conference. This is University of Texas at Dallas Geosciences contribution 1120.

## REFERENCES CITED

Abe, Y., 1997, Thermal and chemical evolution of the terrestrial magma ocean: Physics of the Earth and Planetary Interiors, v. 100, p. 27–39, doi: 10.1016/S0031-9201(96)03229-3.

Abers, G.A., van Keken, P.E., Kneller, E.A., Ferris, A., and Stachnik, J.C., 2006, The thermal structure of subduction zones constrained by seismic imaging: Implications for slab dehydration and wedge flow: Earth and Planetary Science Letters, v. 241, p. 387–397, doi: 10.1016/j.epsl.2005.11.055.

Anbar, A.A., and Knoll, A.H., 2002, Proterozoic ocean chemistry and evolution: A bioinorganic bridge?: Science, v. 297, p. 1137–1142, doi: 10.1126/science.1069651.

Anczkiewicz, R., Platt, J.P., Thirlwall, M.F., and Wakabayashi, J., 2004, Franciscan subduction off to a slow start: Evidence from high-precision Lu-Hf garnet ages on high grade-blocks: Earth and Planetary Science Letters, v. 225, p. 147–161, doi: 10.1016/j.epsl.2004.06.003.

Anderson, D.L., 1989, Theory of the Earth: Oxford, Blackwell, 382 p.

Anderson, D.L., 1995, Lithosphere, asthenosphere, and perisphere: Reviews of Geophysics, v. 33, p. 125–149, doi: 10.1029/94RG02785.

Anderson, D.L., 2001, Top-down tectonics?: Science, v. 293, p. 2016–2018, doi: 10.1126/science.1065448.

Anderson, D.L., 2005, Large igneous provinces, delamination, and fertile mantle: Elements, v. 1, p. 271–275.

Banno, S., and Nakajima, T., 1992, Metamorphic belts of Japanese Islands: Annual Review of Earth and Planetary Sciences, v. 20, p. 159–179, doi: 10.1146/annurev.ea.20.050192.001111.

Bédard, J.H., 2006, A catalytic delamination-driven model for coupled genesis of Archean crust and sub-continental lithospheric mantle: Geochimica et Cosmochimica Acta, v. 70, p. 1188–1214, doi: 10.1016/j.gca.2005.11.008.

Bickford, M.E., and Hill, B.M., 2007, Does the arc accretion model adequately explain the Paleoproterozoic evolution of southern Laurentia? An expanded interpretation: Geology, v. 35, p. 167–170, doi: 10.1130/G23174A.1.

Bogdanova, S., Li, Z., Moores, E.M., and Pisarevsky, S., 2008, Testing the Rodinia hypothesis: Records in its building blocks: Precambrian Research, v. 160, p. 1–4.

Boyet, M., and Carlson, R.W., 2005, $^{142}$Nd evidence for early (>4.53 Ga) global differentiation of the silicate Earth: Science, v. 309, p. 576–581, doi: 10.1126/science.1113634.

Brown, M., 2006, A duality of thermal regimes is the hallmark of plate tectonics since the Neoarchean: Geology, v. 34, p. 961–964, doi: 10.1130/G22853A.1.

Brown, M., 2007, Metamorphic conditions in orogenic belts: A record of secular change: International Geology Review, v. 49, p. 193–234.

Butler, S.L., and Peltier, W.R., 2002, Thermal evolution of Earth: Models with time-dependent layering of mantle convection which satisfy the Urey ratio constraint: Journal of Geophysical Research, v. 107, no. B6, doi: 10.1029/2000JB000018.

Carswell, D.A., 1990, Eclogites and the eclogite facies; definitions and classification, *in* Carswell, D.A., ed., Eclogite Facies Rocks: Glasgow, Blackie, p. 1–13.

Cawood, P.A., Kröner, A., and Pisarevsky, S., 2006, Precambrian plate tectonics: Criteria and evidence: GSA Today, v. 16, no. 7, p. 4–11, doi: 10.1130/GSAT01607.1.

Chamberlain, T.C., 1897, The method of multiple working hypotheses: The Journal of Geology, v. 6, no. 5, p. 837–848.

Clift, P., and Vannucchi, P., 2004, Controls on tectonic accretion versus erosion in subduction zones: Implications for the origins and recycling of the continental crust: Reviews of Geophysics, v. 42, doi: 10.1029/2003RG000127.

Cloos, M., and Shreve, R.L., 1988a, Subduction-channel model of prism accretion, mélange formation, sediment subduction, and subduction erosion at convergent plate margins: 1. Background and description: Pure and Applied Geophysics, v. 128, p. 455–500, doi: 10.1007/BF00874548.

Cloos, M., and Shreve, R.L., 1988b, Subduction-channel model of prism accretion, mélange formation, sediment subduction, and subduction erosion at convergent plate margins: 2. Implications and discussion: Pure and Applied Geophysics, v. 128, p. 501–545, doi: 10.1007/BF00874549.

Coleman, R.G., Lee, D.E., Beatty, L.B., and Brannock, W.W., 1965, Eclogites and eclogites: Their differences and similarities: Geological Society of America Bulletin, v. 76, p. 483–508, doi: 10.1130/0016-7606(1965)76[483:EAETDA]2.0.CO;2.

Condie, K.C., and Kröner, A., 2008, this volume, When did plate tectonics begin? Evidence from the geologic record, *in* Condie, K.C., and Pease, V., eds., When Did Plate Tectonics Begin on Planet Earth?: Geological Society of America Special Paper 440, doi: 10.1130/2008.2440(14).

Conrad, C.P., and Lithgow-Bertelloni, C., 2002, How mantle slabs drive plate tectonics: Science, v. 298, no. 5591, p. 207–209, doi: 10.1126/science.1074161.

Conrad, C.P., and Lithgow-Bertelloni, C., 2004, The temporal evolution of plate driving forces: Importance of "slab suction" versus "slab pull" during the Cenozoic: Journal of Geophysical Research, v. 109, no. B10407, doi: 10.1029/2004JB0022991.

Cox, A., and Hart, R.B., 1986, Plate Tectonics: How it works: Oxford, Blackwell, 416 p.

Dann, J.C., 1997, Pseudostratigraphy and origin of the early Proterozoic Payson ophiolite, central Arizona: Geological Society of America Bulletin, v. 109, no. 3, p. 347–365, doi: 10.1130/0016-7606(1997)109<0347:PAOOTE>2.3.CO;2.

Davies, G.F., 1992, On the emergence of plate tectonics: Geology, v. 20, p. 963–966, doi: 10.1130/0091-7613(1992)020<0963:OTEOPT>2.3.CO;2.

Davies, G.F., 2006, Gravitational depletion of the early Earth's upper mantle and the viability of plate tectonics: Earth and Planetary Science Letters, v. 243, p. 376–382, doi: 10.1016/j.epsl.2006.01.053.

Davies, G.F., and Richards, M.A., 1992, Mantle convection: The Journal of Geology, v. 100, p. 151–206.

de Roever, W.P., d., 1964, On the cause of the preferential distribution of certain metamorphic minerals in orogenic belts of different age: Geologische Rundschau, v. 54, p. 933–941.

Dickinson, W.R., 1975, Potash-Depth (K-h) Relations in Continental Margins and Intra-Oceanic Magmatic Arcs: Geology, v. 3, p. 53–56.

Dilek, Y., 2003, Ophiolite concept and its evolution, *in* Dilek, Y., and Newcomb, S., eds., Ophiolite Concept and the Evolution of Geological Thought: Geological Society of America Special Paper 373, p. 1–16.

Ducea, M., 2002, Constraints on the bulk composition and root foundering rates of continental arcs: A California arc perspective: Journal of Geophysical Research, v. 107, no. B11, p. 2304, doi: 10.1029/2001JB000643.

Eaton, D.W., and Fredericksen, A., 2007, Seismic evidence for convection-driven motion of the North American plate: Nature, v. 446, p. 428–431, doi: 10.1038/nature05675.

Ernst, W.G., 2003, High-pressure and ultrahigh-pressure metamorphic belts—Subduction, recrystallization, exhumation, and significance for ophiolite studies, *in* Dilek, Y., and Newcombe, S., eds., Ophiolite Concept and Evolution of Geological Thought: Geological Society of America Special Paper 373, p. 365–384.

Ernst, W.G., 2007, Speculations on evolution of the terrestrial lithosphere-asthenosphere system—Plumes and plates: Gondwana Research, v. 11, p. 38–49, doi: 10.1016/j.gr.2006.02.007.

Evans, D.A.D., and Pisarevsky, S.A., 2008, this volume, Plate tectonics on the early Earth?—Weighing the paleomagnetic evidence, *in* Condie, K.C., and Pease, V., eds., When Did Plate Tectonics Begin on Planet Earth?: Geological Society of America Special Paper 440, doi: 10.1130/2008.2440(12).

Fairchild, I.J., and Kennedy, M.J., 2007, Neoproterozoic glaciation in the Earth System: J. Geol. Society London, v. 164, p. 895–921.

Farquhar, J., Wing, B.A., McKeegan, K.D., Harris, J.W., Cartigny, P., and Thiemens, M.H., 2002, Mass-independent sulfur of inclusions in diamond and sulfur recycling on the early Earth: Science, v. 298, p. 2369–2372, doi: 10.1126/science.1078617.

Foley, S., 2008, this volume, A trace element perspective on Archean crust formation and on the presence or absence of Archean subduction, *in* Condie, K.C., and Pease, V., eds., When Did Plate Tectonics Begin on Planet Earth?: Geological Society of America Special Paper 440, doi: 10.1130/2008.2440(02).

Foley, S., Buhre, S., and Jacob, D.E., 2003, Evolution of the Archean crust by delamination and shallow subduction: Nature, v. 421, p. 249–252.

Forsyth, D., and Uyeda, S., 1975, On the relative importance of the driving forces of plate motions: Geophysical Journal International, v. 43, p. 163–200.

Fukao, Y., Widiyantoro, S., and Obayashi, M., 2001, Stagnant slabs in the upper and lower mantle transition region: Reviews of Geophysics, v. 39, no. 3, p. 291–323, doi: 10.1029/1999RG000068.

Furnes, H., de Wit, M., Staudigel, H., Rosing, M., and Muehlenbachs, K., 2007, A vestige of Earth's oldest ophiolite: Science, v. 315, p. 1704–1707.

Glazner, A.F., Bartley, J.M., Coleman, D.S., Gray, W., and Taylor, R.Z., 2004, Are plutons assembled over millions of years by amalgamation from small magma chambers?: GSA Today, v. 14, no. 4, p. 4–11, doi: 10.1130/1052-5173(2004)014<0004:APAOMO>2.0.CO;2.

Goldreich, P., and Toomre, A., 1969, Some remarks on polar wandering: Journal of Geophysical Research, v. 74, no. 10, p. 2555–2567.

Gould, S.J., 1965, Is uniformitarianism necessary?: American Journal of Science, v. 263, p. 223–228.

Gurnis, M., Hall, C., and Lavier, L., 2004, Evolving force balance during incipient subduction: Geochemistry, Geophysics, Geosystems, v. 5, p. Q07001, doi: 10.1029/2003GC000681.

Hamilton, W.B., 1998, Archean tectonics and magmatism were not products of plate tectonics: Precambrian Research, v. 91, p. 143–179, doi: 10.1016/S0301-9268(98)00042-4.

Hamilton, W.B., 2007, Earth's first two billion years—The era of internally mobile crust, *in* Hatcher, R.D., Jr., Carlson, M.P., McBride, J.H., and Martínez, J.R., eds., 4-D Framework of Continental Crust: Geological Society of America Memoir 200, p. 233–296.

Hamilton, W.B., and Myers, W.B., 1967, The Nature of Batholiths: U.S. Geological Survey Report 554–C, p. C1–C30.

Hoffman, P.F., 1989, Speculations on Laurentia's first gigayear (2.0 to 1.0 Ga): Geology, v. 17, p. 135–138, doi: 10.1130/0091-7613(1989)017<0135:SOLSFG>2.3.CO;2.

Hoffman, P.F., Kaufman, A.J., Halverson, G.P., and Schrag, D.P., 2002, The snowball Earth hypothesis: Testing the limits of global change: Terra Nova, v. 14, p. 129–155, doi: 10.1046/j.1365-3121.2002.00408.x.

Hyndman, R.D., Currie, C.A., and Mazzotti, S.P., 2005, Subduction zone backarcs, mobile belts, orogenic heat: GSA Today, v. 15, no. 2, p. 4–10.

Hynes, A., 2005, Buoyancy of the oceanic lithosphere and subduction initiation: International Geology Review, v. 47, p. 938–951.

Hynes, A., 2008, this volume, Effects of a warmer mantle on the characteristics of Archean passive margins, *in* Condie, K.C., and Pease, V., eds., When Did Plate Tectonics Begin on Planet Earth?: Geological Society of America Special Paper 440, doi: 10.1130/2008.2440(07).

Jacob, D.E., 2004, Nature and origin of eclogite xenoliths from kimberlites: Lithos, v. 77, p. 295–316, doi: 10.1016/j.lithos.2004.03.038.

Jahn, B.-M., Caby, R., and Monie, P., 2001, The oldest UHP eclogites of the world: Age of UHP metamorphism, nature of protoliths and tectonic implications: Chemical Geology, v. 178, p. 143–158, doi: 10.1016/S0009-2541(01)00264-9.

Janney, P.E., LeRoex, A.P., and Carlson, R.W., 2005, Hafnium isotope and trace element constraints on the nature of mantle heterogeneity beneath the central Southwest Indian Ridge (13°E to 47°E): Journal of Petrology, v. 46, no. 12, p. 2427–2464, doi: 10.1093/petrology/egi060.

Kamenetsky, V.S., Maas, R., Sushchevskaya, N.M., Norman, M.D., Cartwright, I., and Peyve, A.A., 2001, Remnants of Gondwanan continental lithosphere in oceanic upper mantle: Evidence from the South Indian Ridge: Geology, v. 29, p. 243–246, doi: 10.1130/0091-7613(2001)029<0243:ROGCLI>2.0.CO;2.

Kelemen, P.B., Rilling, J.L., Parmentier, E.M., Mehl, L., and Hacker, B.R., 2003, Thermal structure due to solid-state flow in the mantle wedge beneath arcs, *in* Eiler, J., and Hirschmann, M., eds., Inside the Subduction Factory: Washington, D.C., American Geophysical Union, p. 293–311.

Kerrich, R., and Polat, A., 2006, Archean greenstone-tonalite duality: Thermochemical mantle convection models or plate tectonics in the early Earth global dynamics?: Tectonophysics, v. 415, p. 141–164, doi: 10.1016/j.tecto.2005.12.004.

Koenig, E., and Aydin, A., 1998, Evidence for large-scale strike-slip faulting on Venus: Geology, v. 26, no. 6, p. 551–554, doi: 10.1130/0091-7613(1998)026<0551:EFLSSS>2.3.CO;2.

Kopp, R.E., Kirschvink, J.L., Hilburn, I.A., and Nash, C.Z., 2005, The Paleoproterozoic snowball Earth: A climate disaster triggered by the evolution of oxygenic photosynthesis: Proceedings of the National Academy of Sciences of the United States of America, v. 102, no. 32, p. 11,131–11,136, doi: 10.1073/pnas.0504878102.

Korenaga, J., 2006, Archean geodynamics and the thermal evolution of Earth, *in* Benn, K., Mareschal, J.-C., and Condie, K., eds., Archean Geodynamic Processes: American Geophysical Union Geophysical Monograph 164, p. 7–32.

Kusky, T.M., Li, J.H., and Tucker, R.T., 2001, The Dongwanzi ophiolite: Complete Archean ophiolite with extensive sheeted dike complex, North China craton: Science, v. 292, p. 1142–1145, doi: 10.1126/science.1059426.
Liou, J.G., Tsujimori, T., Zhang, R.Y., Katayama, I., and Maruyama, S., 2004, Global UHP metamorphism and continental subduction/collision: The Himalayan model: International Geology Review, v. 46, p. 1–27.
Lithgow-Bertelloni, C., and Richards, M.A., 1998, The dynamics of Cenozoic and Mesozoic plate motions: Reviews of Geophysics, v. 36, p. 27–78, doi: 10.1029/97RG02282.
MacGregor, I.D., and Manton, W.I., 1986, Roberts Victor eclogites: Ancient oceanic crust: Journal of Geophysical Research, v. B91, p. 14,063–14,079.
Maekawa, H., Fryer, P., and Ozaki, A., 1995, Incipient blueschist-facies metamorphism in the active subduction zone beneath the Mariana forearc, *in* Taylor, B., and Natland, J., eds., Active Margins and Marginal Basins of the Western Pacific: Washington, D.C., American Geophysical Union, p. 281–289.
Maloof, A.C., Halverson, G.P., Kirschvink, J.L., Shrag, D.P., Weiss, B.P., and Hoffman, P.F., 2006, Combined paleomagnetic, isotopic, and stratigraphic evidence for true polar wander from the Neoproterozoic Akademikerbreen Group, Svalbard, Norway: Geological Society of America Bulletin, v. 118, p. 1099–1124, doi: 10.1130/B25892.1.
Maruyama, S., and Liou, J.G., 1998, Ultrahigh-pressure metamorphism and its significance on the Proterozoic-Phanerozoic boundary: The Island Arc, v. 7, p. 6–35, doi: 10.1046/j.1440-1738.1998.00181.x.
Maruyama, S., Liou, J.G., and Terabayashi, M., 1996, Blueschists and eclogites of the world and their exhumation: International Geology Review, v. 38, p. 490–596.
McKenzie, D., and Bickle, M.J., 1988, The volume and composition of melt generated by extension of the lithosphere: Journal of Petrology, v. 29, p. 625–679.
Mei, S., and Kohlstedt, D.L., 2000, Influence of water on plastic deformation of olivine aggregates: 1. Diffusion creep regime: Journal of Geophysical Research, v. 105, no. B9, p. 21,457–21,470, doi: 10.1029/2000JB900179.
Miyashiro, A., 1972, Metamorphism and related magmatism in plate tectonics: American Journal of Science, v. 272, p. 629–656.
Moyen, J.-F., Stevens, G., and Kisters, A., 2006, Record of mid-Archean subduction from metamorphism in the Barberton terrain, South Africa: Nature, v. 442, p. 559–562, doi: 10.1038/nature04972.
Nimmo, F., 2005, Tectonic consequences of Martian dichotomy modification by lower-crustal flow and erosion: Geology, v. 33, no. 7, p. 533–536, doi: 10.1130/G21342.1.
Nisbet, E.G., Cheadle, M.J., Arndt, N.T., and Bickle, M.J., 1993, Constraining the potential temperature of the Archean mantle: A review of the evidence from komatiites: Lithos, v. 30, p. 291–307, doi: 10.1016/0024-4937(93)90042-B.
Oxburgh, E.R., and Parmentier, E.M., 1977, Compositional and density stratification in oceanic lithosphere—Causes and consequences: Journal of the Geological Society of London, v. 133, p. 343–355.
Parsons, B., 1982, Causes and consequences of the relation between area and age of the sea floor: Journal of Geophysical Research, v. 87, p. 289–302.
Peacock, S.M., 2002, Thermal structure and metamorphic evolution of subducting slabs, *in* Eiler, J., and Hirschman, M., eds., Subduction Factory: Washington, D.C., American Geophysical Union, p. 7–22.
Pearce, J.A., and Cann, J.R., 1973, Tectonic setting of basic volcanic rocks determined using trace element analyses: Earth and Planetary Science Letters, v. 19, p. 290–300, doi: 10.1016/0012-821X(73)90129-5.
Pearce, J.A., Harris, N.B.W., and Tindle, A.G., 1984, Trace element discrimination diagrams for the tectonic interpretation of granitic rocks: Journal of Petrology, v. 25, p. 956–983.
Pease, V., Percival, J., Stevens, G., and Van Kranendonk, M., 2008, this volume, When did plate tectonics begin? Evidence from the orogenic record, *in* Condie, K.C., and Pease, V., eds., When Did Plate Tectonics Begin on Planet Earth?: Geological Society of America Special Paper 440, doi: 10.1130/2008.2440(10).
Peltonen, P., Kontinen, A., and Huhma, H., 1996, Petrology and Geochemistry of Metabasalts from the 1.95 Ga Jormua Ophiolite, Northeastern Finland: Journal of Petrology, v. 37, no. 6, p. 1359–1383.
Pesonen, L.J., Elming, S.-A., Mertanen, S., Pisarevsky, S.A., D'Agrella-Filho, M.S., Meert, J., Schmidt, P.W., Abrahmsen, N., and Bylund, G., 2003, Palaeomagnetic configuration of continents during the Proterozoic: Tectonophysics, v. 375, p. 289–324, doi: 10.1016/S0040-1951(03)00343-3.
Pisarevsky, S., and McElhinny, M.W., 2003, Global Paleomagnetic Visual Database: EOS, Transactions of the American Geophysical Union, v. 84, no. 20, p. 192.
Powell, C.M., Jones, D.L., Pisarevsky, S.A., and Wingate, M.T.D., 2001, Paleomagnetic constraints on the position of the Kalahari craton in Rodinia: Precambrian Research, v. 110, p. 33–46.
Regenauer-Lieb, K., Yuen, D.A., and Branlund, J., 2001, The initiation of subduction: Criticality by addition of water?: Science, v. 294, no. 5542, p. 578–580, doi: 10.1126/science.1063891.
Rollinson, H., 2007, Early Earth Systems: A Geochemical Approach: Malden, Massachusetts, Blackwell, 285 p.
Rudnick, R.L., and Gao, S., 2003, 3.01 Composition of the continental crust, *in* Rudnick, R.L., ed., The Crust, Volume 3 of Treatise on Geochemistry (Holland, H.D., and Turekian, K.K., eds.): Amsterdam, Elsevier, p. 1–64, doi:10.1016/B0-08-043751-6/03016-4.
Ruppert, S., Fliedner, M., and Zandt, G., 1998, Thin crust and active upper mantle beneath the southern Sierra Nevada in the western United States: Tectonophysics, v. 286, p. 237–252, doi: 10.1016/S0040-1951(97)00268-0.
Schellart, W.P., 2004, Quantifying the net slab pull force as a driving mechanism for plate tectonics: Geophysical Research Letters, v. 31, no. L07611, doi: 10.1029/2004GL019528.
Schubert, G.D., Stevenson, D., and Cassen, P., 1980, Whole planet cooling and the radiogenic heat source contents of the Earth and Moon: Journal of Geophysical Research, v. 85, p. 2531–2538.
Schulze, D.E., Harte, B., Valley, J.W., Brenan, J.M., and Channer, D.M.R., 2003, Extreme crustal oxygen isotope signatures preserved in coesite in diamond: Nature, v. 423, p. 68–70, doi: 10.1038/nature01615.
Scott, D.J., Helmstaedt, H., and Bickle, M.J., 1992, Purtuniq Ophiolite, Cape Smith Belt, northern Quebec, Canada: A reconstructed section of early Proterozoic oceanic crust: Geology, v. 20, p. 173–176,
Shervais, J.W., 1982, *T-V* plots and the petrogenesis of modern and ophiolitic lavas: Earth and Planetary Science Letters, v. 59, p. 101–118, doi: 10.1016/0012-821X(82)90120-0.
Shirey, S.B., Kamber, B.S., Whitehouse, M.J., Mueller, P.A., and Basu, A.R., 2008, this volume, A review of the isotopic evidence for mantle and crustal processes in the Hadean and Archean: Implications for the onset of plate tectonic subduction, *in* Condie, K.C., and Pease, V., eds., When Did Plate Tectonics Begin on Planet Earth?: Geological Society of America Special Paper 440, doi: 10.1130/2008.2440(01).
Sleep, N.L., 2000, Evolution of the mode of convection within terrestrial planets: Journal of Geophysical Research, v. 105, no. E7, p. 17,563–17,578, doi: 10.1029/2000JE001240.
Solomatov, V.S., and Moresi, L.-N., 1996, Stagnant lid convection on Venus: Journal of Geophysical Research, v. 101, no. E2, p. 4737–4754, doi: 10.1029/95JE03361.
Stern, R.J., 2002, Subduction zones: Reviews of Geophysics, v. 40, p. 1012, doi: 10.1029/2001RG000108.
Stern, R.J., 2004, Subduction initiation: Spontaneous and induced: Earth and Planetary Science Letters, v. 226, p. 275–292, doi: 10.1016/S0012–821X(04)00498–4.
Stern, R.J., 2005, Evidence from ophiolites, blueschists, and ultrahigh-pressure metamorphic terranes that the modern episode of subduction tectonics began in Neoproterozoic time: Geology, v. 33, p. 557–560, doi: 10.1130/G21365.1.
Stern, R.J., 2007, When did plate tectonics begin on Earth? Theoretical and empirical constraints: Chinese Bulletin of Science, v. 52, no. 5, p. 578–591.
Stern, R.J., Avigad, D., Miller, N., and Beyth, M., 2008, From volcanic winter to snowball Earth: An alternative explanation for Neoproterozoic biosphere stress, *in* Dilek, Y., and Muehlenbachs, H.F.K., eds., Links Among Geological Processes, Microbial Activities, and Evolution of Life: Heidelberg, Springer (in press)
Stevenson, D.J., 2003, Styles of mantle convection and their influence on planetary evolution: Comptes Rendus Geoscience, v. 335, p. 99–111, doi: 10.1016/S1631-0713(03)00009-9.
Strom, R.G., Schaber, G.G., and Dawson, D.D., 1994, The global resurfacing of Venus: Journal of Geophysical Research, v. 99, p. 10,899–10,926, doi: 10.1029/94JE00388.
Tilmann, F., Ni, J., and Team, I.I.S., 2003, Seismic imaging of the downwelling Indian lithosphere beneath central Tibet: Science, v. 300, p. 1424–1427, doi: 10.1126/science.1082777.
Toth, J., and Gurnis, M., 1998, Dynamics of subduction initiation at preexisting fault zones: Journal of Geophysical Research, v. 103, no. B8, p. 18,053–18,067, doi: 10.1029/98JB01076.
Tsujimori, T., Sisson, V.B., Liou, J.G., Harlow, G.E., and Sorenson, S.S., 2006, Very-low-temperature record of the subduction process: A review of

worldwide lawsonite eclogites: Lithos, v. 92, p. 609–624, doi:10.1016/j.lithos.2006.03.054.

Tuckwell, G.W., and Ghail, R.C., 2003, A 400-km-scale strike-slip zone near the boundary of Thetis Regio, Venus: Earth and Planetary Science Letters, v. 211, p. 45–55, doi: 10.1016/S0012-821X(03)00128-6.

Valentine, J.W., and Moores, E.M., 1970, Plate-tectonic regulation of faunal diversity and sea level: A model: Nature, v. 228, p. 657–659, doi: 10.1038/228657a0.

van der Hilst, R.D., Widyantoro, S., and Engdahl, E.R., 1997, Evidence for deep mantle circulation from global tomography: Nature, v. 386, p. 578–584, doi: 10.1038/386578a0.

van Hunen, J., and van den Berg, A. P., 2008, Plate tectonics on the early Earth: Limitations imposed by strength and buoyancy of subducted lithosphere: Lithos (in press).

van Hunen, J., van Keken, P., Hynes, A., and Davies, G.F., 2008, this volume, Tectonics of the early Earth: Some geodynamic considerations, *in* Condie, K.C., and Pease, V., eds., When Did Plate Tectonics Begin on Planet Earth?: Geological Society of America Special Paper 440, doi: 10.1130/2008.2440(08).

van Keken, P.E., Kiefer, B., and Peacock, S.M., 2002, High-resolution models of subduction zones: Implications for mineral dehydration reactions and the transport of water into the deep mantle: Geochemistry, Geophysics, Geosystems, v. 3, no. 10, p. 1056, doi: 10.1029/2001GC000256.

von Huene, R., and Scholl, D.W., 1993, The return of sialic material to the mantle indicated by terrigenous material subducted at convergent margins: Tectonophysics, v. 219, p. 163–175, doi: 10.1016/0040-1951(93)90294-T.

Wakabayashi, J., and Dilek, Y., 2003, What constitutes 'emplacement' of an ophiolite? Mechanisms and relationship to subduction initiation and formation of metamorphic soles, *in* Dilek, Y., and Robinson, P.T., eds., Ophiolites in Earth History: Geological Society of London Special Publication 218, p. 427–447.

Zhai, M., Zhao, G., Zhang, Q., Kusky, T.M., and Li, J.-H., 2002, Is the Dongwanzi complex an Archean ophiolite?: Science, V. 295, p. 923.

Zhang, H., Thurber, C.H., Shelly, D., Ide, S., Beroza, G.C., and Hasegawa, A., 2004, High-resolution subducting-slab structure beneath northern Honshu, revealed by double-difference tomography: Geology, v. 32, no. 4, p. 361–364, doi: 10.1130/G20261.2.

Zhao, G., Wilde, S.A., Li, S., Sun, M., Grant, M.L., and Li, X., 2007, U-Pb Zircon age constraints on the Dongwanzi ultramafic-mafic body, North China, confirm it is not an Archean ophiolite: Earth and Planetary Science Letters, v. 255, p. 85–93.

Manuscript Accepted by the Society 14 August 2007

Printed in the USA

The Geological Society of America
Special Paper 440
2008

# When did plate tectonics begin? Evidence from the geologic record

**Kent C. Condie**
*Department of Earth and Environmental Science, New Mexico Institute of Mining & Technology, Socorro, New Mexico 87801, USA*

**Alfred Kröner**
*Institut für Geowissenschaften, Universität Mainz, 55099 Mainz, Germany*

## ABSTRACT

**Modern-style plate tectonics can be tracked into the geologic past with petrotectonic assemblages and other plate-tectonic indicators. These indicators suggest that modern plate tectonics were operational, at least in some places on the planet, by 3.0 Ga, or even earlier, and that they became widespread by 2.7 Ga. The scarcity of complete ophiolites before 1 Ga may be explained by thicker oceanic crust and preservation of only the upper, basaltic unit. The apparent absence of blueschists and ultrahigh-pressure metamorphic rocks before ca. 1 Ga may reflect steeper subduction geotherms and slower rates of uplift at convergent margins. It is unlikely that plate tectonics began on Earth as a single global "event" at a distinct time, but rather it is probable that it began locally and progressively became more widespread from the early to the late Archean.**

**Keywords:** plate tectonics, Archean tectonics, plate tectonic assemblages, mantle recycling.

## INTRODUCTION

Geodynamic modeling and isotopic data from mantle xenoliths, kimberlitic diamonds, and inherited or detrital zircons clearly indicate that materials that formed near or close to Earth's surface were recycled into the deep mantle prior to 2.5 Ga (Kamber et al., 2003; Boyet and Carlson, 2005; van Thienen et al., 2004; Shirey et al., 2004, this volume; van Hunen et al., this volume). Although plate tectonics are one way to accomplish such recycling, it may not be the only way. If new lithosphere is formed on a planetary surface, it requires "returns" where older lithosphere sinks into the mantle. For instance, one model proposed for recycling of lithosphere on Venus involves delamination of the deep roots of thickened lithosphere and settling into the mantle (Turcotte, 1995). Some recent models for early Earth call upon a similar return mechanism during the Archean. Both Zegers and van Keken (2001) and Bédard (2006) have proposed that thick Archean oceanic plateaus delaminated near their base where mafic rocks inverted to dense eclogite, which sunk into the mantle, carrying part of the lithosphere with it. This model does not require modern subduction zones. Other probable contributors to recycling of crust and upper mantle into the deep mantle during the Hadean are large impacts on the surface of Earth (Grieve et al., 2006).

If we want to identify the onset of "modern-style" plate tectonics, we need to track evidences for this style of plate tectonics in the geologic record. Modern-style plate tectonics must have spreading centers to create new oceanic lithosphere, depositional systems generating "ocean-plate stratigraphy" (e.g., Matsuda and Isozaki, 1991), and subduction zones where this lithosphere is returned to the mantle. Transform and transcurrent faults must also exist. To track modern-style plate tectonics (hereafter referred to simply as plate tectonics), we must determine the most reliable indicators of plate tectonics to track this process into the geologic past. We must also recognize that Earth's crustal record

Condie, K.C., and Kröner, A., 2008, When did plate tectonics begin? Evidence from the geologic record, *in* Condie, K.C., and Pease, V., eds., When Did Plate Tectonics Begin on Planet Earth?: Geological Society of America Special Paper 440, p. 281–294, doi: 10.1130/2008.2440(14). For permission to copy, contact editing@geosociety.org. 

is constantly being obliterated by surface erosion, sediment subduction, and subduction-zone erosion (Clift and Vannucchi, 2004), as well as being overprinted by deformation and metamorphism in the deep crust. Although the preserved geologic record is incomplete, it seems likely that evidence for the operation of plate tectonics is preserved. Because young continental crust is produced by subduction-zone processes, we should be able to track subduction into the past using older continental crust. However, we need to use caution, since in the earliest part of Earth history, continental crust may have been produced by other mechanisms, provided these mechanisms account for depletion in heavy rare earth elements (REEs) and in Nb and Ta (Bédard, 2006; Grieve et al., 2006). Given the uncertainties about preservation, any estimate as to when plate tectonics began must be taken as a minimum age.

It may not be easy to recognize the igneous and tectonic fingerprints of a pre–plate tectonic phase in Earth history in the geologic record. In the following discussion, we summarize lines of evidence for plate tectonics that are preserved in the geologic record. These are divided into two groups: petrotectonic assemblages and other indicators. Although few, if any, single rock types are diagnostic of plate tectonics, various rock assemblages together with other geologic features are characteristic of plate tectonics. Petrotectonic assemblages, as first proposed by Dickinson (1970), include rock "packages" that originate in specific tectonic settings and are characteristic of those settings. Even though some petrotectonic assemblages form on continents and others form in ocean basins, they all have one thing in common: they form in response to plate tectonics. Of course, one always has to allow for the remote possibility that some of these assemblages may have formed in early Archean tectonic settings that are no longer in existence. There are different ways in which various lines of evidence can be considered (Cawood et al., 2006), not only when an assemblage or a property of plate tectonics first appeared, but also when it first became common. First appearances are minimum ages only since earlier examples may not have been preserved or discovered yet. These lines of evidence are listed in Tables 1 and 2, and their significance is discussed in the following sections. It is important to point out at the onset that "single lines of evidence" (including single petrotectonic assemblages) may not be definitive of modern-style plate tectonics, but it is the *convergence* of evidence at any period of time that is most useful in tracking plate tectonics into the past.

TABLE 1. PETROTECTONIC ASSEMBLAGES CHARACTERISTIC OF PLATE TECTONICS

| Assemblage | Widespread distribution (Ga) | First appearance (Ga) |
|---|---|---|
| Ophiolites | ≤1.0 | 3.8? |
| Arc–back-arc | 2.7 | 3.1 |
| Accretionary prisms & OPS | ≤1.0 | 2.7 (3.8?) |
| Forearc basins | ≤2.0 | 2.7 (3.25?) |
| Blueschists & UHP rocks | ≤0.1 | 0.85 (1.0?) |
| Passive margins | ≤2.0 | 2.7 (2.9?) |
| Continental rift | ≤2.0 | 3.0 |
| Metallic mineral deposits | ≤2.7 | 3.5–3.4 |

*Note:* OPS—ocean plate stratigraphy; UHP—ultrahigh pressure.

TABLE 2. OTHER INDICATORS OF PLATE TECTONICS

| Indicator | Widespread distribution (Ga) | First appearance (Ga) |
|---|---|---|
| UHP metamorphism | ≤0.1 | 0.6 |
| Paired metamorphic belts | ≤2.7 | 3.3 |
| Transcurrent faults & sutures | ≤2.7 | 3.6 (?) |
| Collisional orogens | ≤2.0 | 2.2 |
| Accretionary orogens | ≤2.7 | 3.8–3.7 (?) |
| Paleomagnetism | ≤2.7 | ≥3.2 (?) |
| Geochemistry | ≤2.7 | 3.1 |
| Isotopes | ≤3.0 | ≥4.0 |
| Continents | ≤2.7 | ≥3.0 (?) |

*Note:* UHP—ultrahigh pressure.

## TRACKING PETROTECTONIC ASSEMBLAGES

### Ophiolites

Ophiolites are a reliable indicator of plate tectonics. They manifest two modes of lithospheric motion expected from plate tectonics: seafloor spreading to form oceanic crust and plate convergence to emplace the ophiolite. The exact tectonic setting of ophiolites is not so important because all tectonic settings for ophiolites require subduction for formation and emplacement (Stern, 2004, 2005). Complete ophiolites, with pelagic sediments, pillowed basalts, sheeted dikes, gabbros, and tectonized ultramafic rocks (dunite, harzburgite, and lherzolite) are rare in crust of any age. The evidence of plate tectonics should include fragmentary and disrupted ophiolites, but some of the ophiolite components should be preserved in single orogenic belts if they were obducted onto continental crust or incorporated into accretionary prisms along active plate margins. Pillowed tholeiitic basalts with mid-ocean-ridge basalt (MORB) or marginal-basin chemical characteristics, layered gabbros, sheared harzburgites containing Cr-rich spinels, and dunites with podiform chromites are some requirements for the identification of disrupted ophiolites.

The oldest well-preserved complete or almost complete ophiolites are the Jormua ophiolite in Finland (1.95 Ga; Peltonen et al., 1996), the Purtuniq ophiolite in the northern part of the Trans-Hudson orogen in Canada (ca. 2.0 Ga; Scott et al., 1992), and perhaps the Payson ophiolite in Arizona, USA (1.73 Ga; Dann, 1997). Although the Dongwanzi greenstones in northern China have been interpreted as an Archean ophiolite (2.5 Ga; Kusky et al., 2001), this interpretation is no longer viable (Zhai et al., 2002; Zhao et al., 2007). However, Furnes et al. (2007) have reported sheeted dikes and associated pillow basalts in the 3.8 Ga Isua greenstone belt in SW Greenland, although not all investigators that have studied the Isua succession agree with this interpretation. Ophiolites do not become widespread in the geologic record until ca. 1.0 Ga (Stern, 2005, this volume).

### Arc and Back-Arc Assemblages

Among the most distinctive of the plate-tectonic petrotectonic assemblages are arc and back-arc rock assemblages, both of which provide evidence of subduction zones. The most definitive and widespread assemblages are the tholeiite-andesite-dacite-rhyolite and tonalite-trondhjemite-granodiorite (TTG) suites, which characterize arcs. The relative proportion of rock types varies: oceanic arcs contain mostly tholeiites, basaltic andesites, and high-Mg andesites, whereas continental-margin arcs contain greater proportions of intermediate and felsic end members (Condie, 1994). At deeper erosion levels, plutonic equivalents of these rocks dominate. Other, less widespread and less frequent arc-related volcanics include adakites, boninites, shoshonites, sanukitoids and high-Mg andesites. In addition, back-arc basins contain large amounts of volcaniclastic and epiclastic sediment as well as hydrothermal vent deposits. How far back in time can these assemblages be tracked?

Archean greenstones can roughly be divided into two major types: those that consist chiefly of pillow basalt and komatiite, the "mafic plain type," and those that also contain calc-alkaline volcanics and related sediments, the "arc type" (Thurston and Chivers, 1990; Condie, 1994; Condie and Benn, 2006). Arc-type greenstones are widespread in late Archean cratons such as the Superior, Slave, Yilgarn, and Eastern and Southern Africa cratons (Condie and Harrison, 1976; Davis and Condie, 1977; Hallberg et al., 1976; Giles and Hallberg, 1982; Cousens, 2000; Polat and Kerrich, 2001; Percival et al., 2004; Kerrich and Polat, 2006). They appear to be the most widespread type of greenstone in the late Archean of Zimbabwe and Kenya (Condie and Harrison, 1976), where andesite is an important component. Although composed dominantly of pillow basalts, arc-type greenstones include up to 50% calc-alkaline volcanics, dominantly andesites and dacites (Boily and Dion, 2002). In addition, these greenstones contain appreciable amounts of graywacke and various volcaniclastic rocks, suggestive of deposition in an arc system. The Raquette Lake Formation in the Archean Slave Province is a superb example of a volcanic-sedimentary succession that appears to represent a fragment of a continental-margin arc system (Mueller and Corcoran, 2001). Detailed mapping and sedimentologic studies show facies of high-energy clastic sedimentation contemporary with explosive volcanism, much like that found along modern continental-margin arcs like Japan and the Andes. In recent years, the occurrences of boninite, shoshonite, and high-Mg andesite have been documented in several Archean greenstones, some as old as 3.4 Ga (Parman et al., 2001; Polat and Kerrich, 2001; Boily and Dion, 2002; Smithies et al., 2004).

Most Archean terranes are bimodal in composition, as pointed out long ago by Barker and Arth (1976). Some recent investigators have cited this bimodality as another argument against plate tectonics in the Archean (Hamilton, 1998). However, bimodal magmatism is widespread in the Phanerozoic record, both in continental rift systems and in many island arcs (Yoder, 1973; Feiss, 1982; Nakajima and Arima, 1998; Smith et al., 2003; Leat et al., 2003). Thus, bimodal magmatism is clearly not an Archean phenomenon, and it occurs in at least two major plate-tectonic settings.

The striking similarity of lithologic associations in arc-type Archean greenstones, including boninite, shoshonite, adakite, and other minor volcanic rock types, argues strongly for a subduction-related origin for these greenstones. Although arc-type greenstones were widespread by 2.7 Ga, the oldest well-documented example is the Whundo Group in the Pilbara craton in Western Australia, which dates back to 3.12 Ga (Smithies et al., 2005b, 2007; Pease et al., this volume).

### Accretionary Prisms and Ocean Plate Stratigraphy

Mélange complexes formed in accretionary prisms are the result of tectonic transfer of oceanic crust and pelagic sediments to an arc or active continental margin, and even after high-grade metamorphism, many such deposits can still be recognized. The apparent absence of accretionary prisms in the Archean rock record has been cited as evidence against the existence of Archean subduction (Hamilton, 1998). However, as pointed out by Stern (this volume), most Archean arc systems were probably intra-oceanic, where subduction erosion dominated, and thus, robust accretionary prisms are not expected to have formed. There are a few mélange-like terranes that have been identified in the Archean and interpreted as accretionary prisms (Komiya et al., 1999; Kitajima et al., 2001; Shervais, 2006), but not all investigators agree with this interpretation (Fedo et al., 2001; Rollinson, 2002).

A tectonic mélange resembling a classic Mesozoic-style accretionary mélange has been described from the Schreiber greenstone belt (2.75–2.70 Ga) in the Superior Province of Canada (Polat and Kerrich, 1999). Like Phanerozoic counterparts, this Archean mélange is characterized by varying degrees of fragmentation and mixing, ranging from intact sedimentary layers and volcanic flows to intensely sheared, transposed, and mixed tectonic mélange. It differs from Phanerozoic examples in the composition of the blocks that are included within the mélange, and also in metamorphic facies. The total package of rocks in the Schreiber greenstone, however, is comparable to those found in young arc successions that evolved through compressional to transpressional deformation. In addition, there are several late Archean greenstones that have been described that contain large blocks ranging from dismembered to intact, which are similar to blocks in young accretionary prisms (Thurston, 1994; Mueller et al., 1996; Stott, 1997; de Wit and Ashwal, 1997; de Wit, 1998; Kusky, 1998; Shervais, 2006). Although controversial, the Isua supracrustal sequence in SW Greenland, with an age of ca. 3.8 Ga, may be the oldest known example of an accretionary prism (Komiya et al., 1999; Shervais, 2006).

The identification of ocean-plate stratigraphy is also a key issue in identifying accretionary complexes since ocean-plate stratigraphy retains a travel log of the subducted oceanic plate from its birth at an oceanic ridge to its demise in a subduction zone (Cawood, 1982; Matsuda and Isozaki, 1991). This was demonstrated in the Mesozoic-Cenozoic accretion complexes of Japan, and details on how ocean-plate stratigraphy can be recognized in accretionary prisms were provided by Isozaki (1997).

There is controversy whether ocean-plate stratigraphy and associated structural features suggesting incorporation in an accretionary prism have been recognized in early Archean greenstone belts. Komiya et al. (1999) interpreted the Isua belt as the oldest known example of ocean-plate stratigraphy and accretion tectonics, whereas others contest this interpretation and argue that all original stratigraphic relationships have been destroyed by strong ductile deformation (Myers, 2004). Alteration in the Ivisaartoq greenstone belt in West Greenland has been interpreted as seafloor hydrothermal alteration associated with suprasubduction oceanic crust (Polat et al., 2007, 2008). Kitajima et al. (2001), Komiya et al. (2002a), and Kato and Nakamura (2003) interpreted part of the ca. 3.4 Ga Pilbara "stratigraphy" as representing ocean-plate stratigraphy and claim to have recognized structural features resembling those in modern accretionary prisms. This view is opposed by Smithies et al. (2005a) and Van Kranendonk et al. (2002, 2007), who maintain that the Eastern Pilbara terrane is a largely uninterrupted volcanic-sedimentary sequence. Further investigations are needed to resolve this issue.

## Foreland Basins

There are two types of foreland basins: retroarc foreland basins that form behind continental-arc systems, and peripheral foreland basins associated with collisional orogens (Condie, 2005b). Both are filled largely with detritus derived from advancing fold-and-thrust belts. The key tectonic feature of foreland basins is the syntectonic character of the sediments. Often, the proximal basin margin becomes involved with the advancing thrusts. Clastic sediments shed from the thrusts are uplifted and redeposited in the evolving foreland basin. Coarse-grained immature sediments, such as conglomerates and arkosic sandstones, characterize proximal facies, and fine-grained clastic sediments, often with marine carbonates, characterize more distal facies.

Both retroarc and peripheral foreland basin successions are widespread in the Paleoproterozoic. Classic examples are described from the Trans-Hudson and Wopmay orogens in Canada dating to 1.9–1.85 Ga (Bowring and Grotzinger, 1992; Ansdell et al., 1995). The Capricorn orogen in Western Australia also contains both retroarc and peripheral foreland basins, the latter of which are associated with the final collision of the Pilbara and Yilgarn cratons (Betts et al., 2002). The Padbury Group was deposited in a peripheral foreland basin along the southern margin of the Capricorn orogen ca. 1.96 Ga (Pirajno and Occhipinti, 2000). Several major volcanic-sedimentary basins in NE Australia appear to represent foreland basin deposits associated with major tectonic events related to a convergent plate boundary in what is now central Australia ca. 1.8–1.7 Ga (Giles et al., 2002).

In the late Archean, numerous examples of retroarc foreland basins have been described in various Archean cratons (Eriksson et al., 1994). Classic examples occur in the Superior craton in SE Canada, especially those associated with the Abitibi greenstone belt at 2.7 Ga (Mueller and Donaldson, 1992; Camire et al., 1993; Mueller et al., 1994). The oldest reported example of a possible foreland basin (retroarc type) is in the southern facies of the Fig Tree Group in the Barberton greenstone in South Africa at ca. 3.25 Ga (Lowe and Nocita, 1999).

Hence, it appears that foreland basins were clearly operational by the mid-Archean and widespread in the Paleoproterozoic.

## Blueschists

Blueschists are metamorphosed mafic rocks or metasediments containing sodic amphibole, which is stable under high-pressure (*P*) and low-temperature (*T*) conditions (Maruyama et al., 1996; Ernst, 2003). The conclusion that blueschists form only in subduction zones is based on their association with ancient subduction mélanges and is confirmed by studies of active subduction zones (Abers et al., 2006; Maekawa et al., 1995; Zhang et al., 2004). Blueschists of the Franciscan and Sanbagawa terranes appear to have been subducted to 15–70 km depths before returning to the surface (Ernst, 2003).

It is now accepted that the oldest known blueschists are Neoproterozoic in age (ca. 850–700 Ma; Maruyama ct al., 1996). These occur in West Africa, India, southern China (Shu et al., 1994), northern Mongolia (Sklyarov, 1990), and Anglesey, England (Kawai et al., 2006). Older examples are unknown. This apparent absence may not be entirely a preservation problem, since blueschists are preserved in fragments of tectonically uplifted subduction zones. However, hotter upper mantle during the Archean may have impeded exhumation of blueschists. The recent discovery of ca. 3.2 Ga, high-*P*, low-*T* metamorphic rocks in the southern Barberton greenstone terrane, South Africa (Moyen et al., 2006) suggests that Archean blueschists may be preserved somewhere in the Archean crust.

## Passive Continental Margins

Passive continental margins such as those around the Atlantic Ocean form when continents rift apart and an ocean basin forms outboard of the continents. Passive margins take tens to hundreds of millions of years to form, allowing hundreds of meters of thermal subsidence, which creates accommodation space for sediment accumulation. As pointed out by Hynes (this volume), most passive margins may develop by stretching continental crust with cool mantle roots. The proximity of many passive margins to continental river and glacier mouths ensures that they will contain an enormous thickness of sediments. Passive-margin sedimentary sequences are built on crusts of very different types: continental crust on the landward side, oceanic crust on the seaward side, and transitional crust beneath. Because passive-margin deposits are associated with thick continental crust, they are difficult to subduct and are expected to be partly preserved in ancient suture zones.

Examination of the geologic record indicates that passive-margin deposits are first preserved in the late Archean but do not become widespread until the Paleoproterozoic, beginning ca, 2.5 Ga. The oldest well-documented passive-margin deposits are in the Pilbara craton at 2685–2645 Ma and in the Kaapvaal craton at 2640–2470 Ma (Bradley, 2005; Schröder et al., 2006). A possible passive-margin sequence has been described in the SW Superior craton at Steeprock Lake at ca. 2900 Ma (Wilks and Nisbet, 1988). Also, some metapelite-marble-quartzite associations in Archean gneiss terranes are candidates for passive-margin sequences.

### Continental Rifts

The onset of continental fragmentation is recorded by rifting of continental crust, which is shown, for example, by the active rift systems in the Red Sea, Gulf of Aden, and East Africa today. Hence, continental rifts are part of the supercontinent cycle and thus part of plate tectonics. Continental rifts have distinct petrotectonic assemblages related to passive or active rifts (Condie, 2005b). Passive rifts are characterized by large volumes of immature clastic sediments such as arkosic sandstones and conglomerates, whereas active rifts have associated tholeiitic and bimodal magmatism. In tracking continental rifts into the geologic past, caution must be used because rift assemblages associated with flood basalts may be difficult to distinguish from non–"flood basalt"–type rift assemblages. Flood basalt rifts do not necessarily require plate tectonics if they are formed by mantle plumes (Condie, 2005b).

Although continental rift assemblages have been widely preserved since the Paleoproterozoic (Blake and Stewart, 1992; Betts et al., 1998; Jackson et al., 2000; Page et al., 2000), the earliest occurrence of rift assemblages is in the mid- to late Archean (Ayres and Thurston, 1985; Blake and Barley, 1992; Percival et al., 1994; Bleeker, 2002). The oldest continental rift assemblages are in the Kaapvaal craton in southern Africa. One of the first recognized continental rift deposits is the Nsuze Group in South Africa, which was deposited on older continental crust ca. 3 Ga (Burke et al., 1985; Weilers, 1990; de Villarroel and Lowe, 1993). The Dominion Group deposited in the proto-Witwatersrand basin at about the same time (3.1 Ga) is also probably a continental rift assemblage associated with a large igneous province event (Jackson, 1991; Schmitz et al., 2004; McCarthy and Rubridge, 2005).

### Mineral Deposits

Mineral deposits have heterogeneous distributions, and each major deposit type shows distinctive temporal patterns and relationships to specific tectonic settings (Barley et al., 1998; Groves et al., 2005; Kerrich et al., 2005). These reflect a complex interplay between formational and preservational forces that largely reflects changes in tectonic processes and environmental conditions in an evolving Earth. Mineral deposits most closely tied to subduction zones include orogenic gold, Cu-porphyry and epithermal deposits, and volcanic-hosted massive sulfide (VHMS) deposits (Groves et al., 2005; Barley et al., 1998).

The earliest orogenic gold deposits were formed in the Pilbara and Kaapvaal cratons between ca. 3.4 and 3.1 Ga (Zegers et al., 2002). The ages of later deposits define two major Precambrian peaks at 2.75–2.55 Ga and 2.1–1.75 Ga, a marked scarcity of deposits between 1.7 Ga and 600 Ma, and a more cyclic distribution from ca. 600 Ma to 50 Ma (Goldfarb et al., 2001). Lode gold deposits in the Superior Province and Kaapvaal cratons provide evidence in support of recycling of fluids through modern-style subduction zones by 3.1 Ga (Wyman et al., this volume).

Porphyry Cu–Au and epithermal-type Au-Ag deposits have a strong tectonic control in convergent-margin settings (Sillitoe, 1976); they form at high crustal levels in arc and back-arc environments with high to extreme uplift rates. The oldest examples of porphyry-type mineral deposits at ca. 3.3 Ga occur in greenstones in Western Australia (Barley, 1982).

VHMS systems form at ocean ridges (e.g., East Pacific Rise) and in back-arc basins (e.g., the Lau Basin). They are accreted onto convergent margins as oceanic terranes, or they may form directly in back-arc basins (Solomon and Quesada, 2003). There are numerous examples of VHMS deposits in late Archean greenstones (Cawood et al., 2006). Major peaks in VHMS deposits at 2.7 Ga and 1.9 Ga also correspond to peaks in both orogenic gold deposits and crustal growth (Condie, 2000). The oldest VHMS deposits are found in the eastern Pilbara and Barberton terranes at 3.50–3.25 Ga and correspond broadly with the oldest global orogenic gold events (Zegers et al., 2002).

## OTHER PLATE-TECTONIC INDICATORS

### Ultrahigh-Pressure Metamorphism

Ultrahigh-pressure (UHP) metamorphic terrains are another important indicator of subduction. UHP rocks form when continental crust is subducted to depths >100 km and then returns to the surface. Metamorphic assemblages in UHP terrains include coesite and/or diamond and indicate peak metamorphic conditions of ~700–900 °C and 3–4 GPa or more (Liou et al., 2004). In contrast to blueschists, the significance of which has long been appreciated, UHP terrains have been recognized for only the past 15 yr or so. The oldest reliably dated UHP locality is in Mali, where coesite-bearing gneiss was metamorphosed at ca. 620 Ma (Jahn et al., 2001). The oldest diamond-bearing UHP terrain is in Kazakhstan, where diamond- and coesite-bearing paragneiss of the Kokchetav massif was subducted >120 km deep at ca. 530 Ma (Maruyama and Liou, 1998). The first recognized evidence for deep subduction of continental crust is thus found in the late Neoproterozoic and early Cambrian.

Eclogites are often thought to reflect plate-tectonic processes, but there are several ways to generate high-pressure garnet-clinopyroxene metamorphic rocks. Lawsonite-bearing eclogites clearly manifest metamorphism in a subduction zone, but these are only found in Phanerozoic terranes (Tsujimori et al., 2006). Medium-*T*, high-*P* eclogites and granulites are now known from the late Archean, but these represent significantly higher geothermal gradients than are found in modern subduction zones (Brown, 2006a, 2006b, this volume). Brown (2006a, 2006b) concluded that low-*T*, high-*P* metamorphic rocks characteristic of modern subduction zones are not common before ca. 1 Ga. Eclogites at ca. 2 Ga are well documented from southern Tanzania and appear to be derived from MORB-type oceanic crust (Möller et al., 1995). Re/Os ages from inclusions in diamonds contained in MORB-like eclogite xenoliths yield much older ages suggestive of recycling of oceanic crust into mantle by at least 3 Ga (Richardson et al., 2001).

### Paired Metamorphic Belts

Modern plate tectonics are characterized by a duality of thermal regimes; this is characterized by two principal types of regional-scale metamorphic belts herein referred to as paired metamorphic belts (Miyashiro, 1961; Brown, 2006a, 2008). This duality is the hallmark of plate tectonics. The imprint of plate tectonics in the geologic record is the contemporaneous occurrence of these two contrasting types of metamorphic belts, reflecting the duality of thermal regimes. The first type, characterized by low temperature gradients, results in low-*T*, high-*P* metamorphism, corresponding to subduction zones. The second is characterized by high thermal gradients and gives rise to high-*T*, low-*P* metamorphism, corresponding to the arc and back-arc region.

Prior to ca. 3 Ga, greenstone belts record *P-T* conditions characteristic of low- to moderate-pressure and moderate- to high-temperature metamorphism (Brown, 2006a, 2006b). Examples of this *P-T* regime are the 3.8 Ga metamorphism in the Isua supracrustal belt of SW Greenland (Appel et al., 2001; Komiya et al., 2002b) and the 3.2 Ga metamorphism recorded in the Barberton greenstone in South Africa (Moyen et al., 2006). This duality of low-*T*, high-*P* and high-*T*, low-*P* metamorphic belts is common in late Archean and younger greenstones, and the oldest occurrence may be at ca. 3.3 Ga (Brown, 2006b).

### Large Transcurrent Faults and Suture Zones

Many large transcurrent faults are plate-bounding strike-slip faults like the San Andreas fault in North America. The identification of ancient large transcurrent faults clearly indicates rigid plates and weak plate boundaries and, hence, the probable operation of plate tectonics (Sleep, 2005). Well-documented major late Archean strike-slip faults occur in the Yilgarn and Pilbara cratons of Australia and in the Superior craton in Canada (Krapez and Barley, 1987; Van Kranendonk and Collins, 1998; Van Kranendonk et al., 2002; Percival et al., 2006). Sleep (2005) argued that the Inyoka fault in the Barberton greenstone belt of South Africa is the oldest major transcurrent fault at 3.2 Ga, but others interpret the fault as a thrust (Moyen et al., 2006).

Until we have a more systematic treatment of strike-slip fault kinematics through time, identification of ancient transcurrent faults must be done with caution. Not all strike-slip faults are transcurrent faults, and relatively short strike-slip faults with small offsets form in a wide range of tectonic settings, which may or may not reflect plate-tectonic activity. For example, non–plate-tectonic strike-slip faulting has been inferred for Venus from radar imagery (Koenig and Aydin, 1998; Tuckwell and Ghail, 2003). This faulting may reflect differential stresses due to downwelling mantle, which is coupled to the lithosphere to produce horizontal compressive stresses and horizontal shearing of Venus' stagnant lid.

Suture zones are sites of intense tectonism where plates or terranes collided. Although often broad and well defined in orogenic belts and associated with tectonic mélanges, these zones become narrow in the lower crust, where they are mostly represented by intense ductile deformation and mylonite belts. The true nature of many ductile shear zones is difficult to establish, and they have therefore been interpreted as cryptic sutures. However, many are considered as terrane boundaries, particularly when the terranes they separate have contrasting ages and tectono-metamorphic evolution (e.g., Nutman et al., 2002). Such terrane boundaries have been identified in many Paleoproterozoic domains but also in Archean cratons such as the Kaapvaal of southern Africa (de Wit et al., 1992; see recent review by Anhaeusser, 2006) and the Superior of Canada (Desrochers and Hubert, 1996). The oldest suture defining a terrane boundary is probably in the 3.6 Ga Itsaq Gneiss Complex of West Greenland (Nutman et al., 2002).

### Orogens

Two kinds of orogens result from the operation of plate tectonics (Dewey, 1977). Collisional orogens form during collision of two or more plates. If the angle of collision is significant, the crust is greatly thickened, and thrusting, metamorphism, and partial melting may rework the colliding plate margins. A relatively minor amount of juvenile crust is produced or tectonically "captured" during collisional orogeny (Windley, 1992; Pease et al., this volume). In contrast, accretionary orogens involve accretion and suturing of juvenile terranes and small crustal blocks (ophiolites, island arcs, oceanic plateaus, microcontinents, etc.) and the formation of mixed sedimentary-magmatic accretionary prisms. Accretionary orogens contain variable amounts of older crust, mostly interpreted as exotic continental slivers.

Although collisional orogens are not well documented before ca. 2 Ga, accretionary orogens may have been in existence since the early Archean (Condie, 2007). The oldest well-documented collisional orogen is the Capricorn orogen in Western Australia. This orogen began to develop at ca. 2.2 Ga as the Yilgarn and Pilbara cratons collided (Betts et al., 2002; Cawood and Tyler, 2004). Major phases of deformation and plutonism are recorded at 2.2, 2.0–1.96, and 1.83–1.78 Ga. The major collision between the Zimbabwe and Kaapvaal cratons also occurred at ca. 2 Ga, forming the Limpopo orogen (Kamber et al., 1995; Holzer et al., 1998; Kröner et al., 1999). Probably the most widely studied and best-understood Proterozoic collisional belt is the Trans-Hudson orogen in Canada, which formed as the Superior and Hearne cratons collided ca. 1.9 Ga (Bickford et al., 1990; Lucas et al., 1994).

Accretionary orogens have been recognized in the Paleoproterozoic and late Archean. Most, if not all, Archean granite-greenstone terrains appear to represent fragments of accretionary orogens (Condie, 1994, 2005b). Well-documented examples are found in the Superior, Zimbabwe, Slave, Yilgarn, and Kaapvaal cratons (Bowring and Grotzinger, 1992; Percival et al., 1994; Jelsma et al., 1996; Bleeker, 2002; Schmitz et al., 2004). Although not agreed on by all investigators, the oldest accretionary orogens may be represented by some of the oldest preserved rocks. For instance, the Itsaq Gneiss Complex in SW Greenland may be an example of an accretionary orogen that formed between 3.86 and 3.7 Ga (Nutman et al., 2002; Friend and Nutman, 2005). Reflection seismology studies in Canada and the Baltic Shield have also reported deep sub-Moho reflectors with shallow dips that may represent fossil subduction zones associated with Archean accretionary orogens (Percival et al., 2004). In addition, terranes in the Archean Yilgarn craton in Western Australia exhibit distinct seismic structures, which appear to have been acquired before craton assembly in the late Archean (Reading et al., 2007).

## Paleomagnetism

Paleomagnetism can potentially constrain the timing of significant differential motion between lithospheric blocks in the geologic past. However, uncertainty increases with time, and reconstructions of Precambrian continental motions are often uncertain, ambiguous, or both (Scotese, 2004; Pisarevsky, 2005). These uncertainties result from the possibility that Earth's magnetic field may not always have been a dipole approximating the geographic poles, the probable existence of true polar wander, and the overprinting of primary magnetization by remagnetization associated with deformation and metamorphism.

Cawood et al. (2006) concluded that the existing paleomagnetic database demonstrates differential movements between continents in pre-Neoproterozoic time. As an example, the positions of Australia and Baltica are paleomagnetically well defined at both 1770 and 1500 Ma. Although comparison of coeval paleopoles yields ambiguous longitudes and polarities, the paleomagnetic results nevertheless indicate that Australia and Baltica drifted independently between 1770 and 1500 Ma, requiring independent plate motions and thus, plate tectonics (Pesonen et al., 2003). Paleomagnetic data from the Archean of the Kaapvaal craton indicate that there has been significant polar wander during the late Archean (Layer et al., 1989; Kröner and Layer, 1992). Results suggest that between 3450 and 2700 Ma, the Kaapvaal craton moved from polar to equatorial latitudes at least twice. An apparent polar wander path from the Pilbara craton implies a latitudinal drift of ~21°, with an estimated drift rate of 12–112 cm/yr, which is large compared with current plate velocities, suggesting that continents might have moved faster during the Archean than in the Phanerozoic (Suganuma et al., 2006). Other examples of differential movement of continental blocks prior to 1 Ga are given in Pesonen et al. (2003), Pisarevsky and McElhinny (2003), and Powell et al. (2001). Evans and Pisarevsky (this volume), however, make a strong case that the most reliable pre–1880 Ma paleomagnetic results do not allow a distinction between differential plate motion and true polar wander.

Although paleomagnetic studies are clearly an important method that can be used to identify ancient plate motions, only well-dated and reliable pole positions are useful. At this point in time, the most important conclusion from paleomagnetic results is that continental plates were probably moving about on Earth's surface by the late Archean (and probably earlier), and hence plate tectonics in some form must have been operative by this time.

## Geochemical Constraints on Tectonic Setting

The temporal record of igneous rocks with arc-like chemical compositions can be used to infer when arcs, subduction zones, and, hence, plate tectonics appeared in the geologic record (Condie, 1994). Basalts from modern intra-oceanic arc systems show distinctive enrichments or "spikes" in fluid-mobile elements (e.g., K, Sr, Pb) relative to neighboring elements. These spikes are associated with strong relative depletions in incompatible high field strength element (HFSE) cations, especially Nb and Ta (Stern, 2002). Arc-like affinities can be inferred from normalized plots of trace elements ("spider diagrams") and are also the basis of a wide range of trace-element "discriminant diagrams," which have been used to infer the tectonic setting of ancient igneous rocks (Pearce and Cann, 1973; Shervais, 1982; Condie, 1994, 2005a). As is the case for other lines of inference regarding the operation of plate tectonics, care should be taken to ensure that such characteristics are not the result of crustal contamination or inheritance from the subcontinental lithosphere, as is seen for many flood basalts and continental-margin arc basalts. This approach applied to ancient rocks is particularly useful for oceanic basalts because these are likely to be less contaminated by preexisting continental crust, and any contamination that does occur is likely to be revealed by isotopic and geochronologic studies. The most convincing demonstration that a given suite of ancient igneous rocks formed in association with a subduction zone is given by plots of Th/Yb versus Nb/Yb (or Ta/Yb) or Nb/Th versus Zr/Nb (Condie, 1994; Pearce and Peate, 1995; Condie, 2005a). These diagrams are especially useful because the element ratios are not significantly affected by alteration and because it is possible to identify crustal contamination trajectories.

Basalt geochemistry has been widely used to track arc-related volcanics into the Archean, and there are numerous examples of late Archean greenstones with arc affinities (Kerrich et al., 1998; Puchtel et al., 1999; Percival et al., 2004; Polat et al., 2005; Sandeman et al., 2006; Kerrich and Polat, 2006). Arc-type greenstones are the most important greenstones in the late Archean terranes of Zimbabwe and Kenya (Condie and Harrison, 1976). In recent years, the occurrences of boninite, high-Mg andesite, adakite, and Nb-enriched andesites, all with arc geochemistry, have been documented in several late Archean greenstones (Boily and Dion, 2002; Polat and Kerrich, 2001; Kerrich and Polat, 2006). The oldest well-documented example of basalts with arc geochemistry, with an age of 3.12 Ga, is in the Whundo greenstone in the Pilbara craton of Western Australia (Smithies et al., 2005a). Boninites have been identified from the 3.8 Ga Isua greenstone in Southwest Greenland (Polat et al., 2008), although Smithies et al. (2004) questioned whether these rocks are true boninites.

### Continental Crust

The existence of continents is also evidence for past subduction. Arc-type geochemical signatures in granitoids can be traced into the early Archean (Kamber et al., 2002; Moyen and Stevens, 2005; Foley, this volume). Subduction is necessary to produce the distinct geochemical signatures of continental crust (i.e., enrichment in large ion lithophile elements [LILE] and relative depletion in Nb and Ta), and hence it must have been present in some form by 3.5 Ga. By 3 Ga, significant amounts of continental crust had formed in the Pilbara, Kaapvaal, and Superior cratons (Pease et al., this volume). The bits and pieces of early continental crust older than 3 Ga may not necessarily require modern-style subduction, but they do require some mechanism whereby Nb and Ta were sequestered in the source, whereas LILEs were concentrated in TTG magmas. It is also necessary for pre–3 Ga continental crust to have been recycled into the mantle to explain some of the geochemical and isotopic anomalies in the mantle today (Kleinhanns et al., 2003; Kamber et al., 2003; Shirey et al., this volume). Lastly, continents contain orogens that formed by collision and/or accretion and are often composed of terranes (see Pease et al., this volume). All these require lateral transport of the lithosphere, and this suggests that modern-style plate tectonics were in existence by at least 3 Ga.

### Isotopic Evidence for Mantle Recycling

Isotopic compositions of modern and ancient igneous rocks are powerful indices for inferring times of crust-mantle differentiation and for identifying surficial materials that have been mixed back into the deep mantle (Macdougall and Haggerty, 1999; Richardson et al., 2001; Eisele et al., 2002; Frei et al., 2004). There is abundant evidence for early recycling of crust and sediments into the mantle from as far back as the early Archean, including the Pb-Pb array for MORB and some oceanic-island basalts (Hart and Gaetani, 2006). Among these evidences are the following, which are extensively discussed by Shirey et al. (this volume). Isotopic ratios of $^{142}Nd/^{144}Nd$ in terrestrial rocks are consistent with the formation of an early depleted reservoir in the mantle, indicating very early differentiation of Earth's crust and probable recycling into the mantle (Boyet and Carlson, 2005; Caro et al., 2006). Isotopic compositions (O, Sr, Pb, Os) of inclusions in diamonds from eclogites derived from the mantle beneath Archean cratons have characteristics that reflect seafloor alteration and recycling of oceanic crust into the mantle (MacGregor and Manton, 1986; Richardson et al., 2001; Schulze et al., 2003). Ages of these inclusions are chiefly between 3.8 and 3.0 Ga, although some range to over 4.0 Ga. Negative Eu anomalies in majoritic garnet inclusions in diamond also require a crustal origin for the protolith of these inclusions (Tappert et al., 2005) and thus recycling into the mantle. The carbon isotopic composition of the host diamonds indicates a source from organic matter probably recycled into the mantle in subducting slabs. Mass-independent $^{33}S$ isotope anomalies in diamonds are also strong evidence that these materials were originally at the surface early in Earth history and were recycled into the mantle (Farquhar et al., 2002). Also, the Hadean Hf isotope record provided by detrital zircons from sediments shows that felsic components in the crust were produced by at least 4.3 Ga (Watson and Harrison, 2005; Shirey et al., this volume).

Thus, the isotopic record in diamond inclusions, the $^{142}Nd/^{144}Nd$ ration in terrestrial rocks, and the Hf isotopic composition of detrital zircons all clearly suggest that the mantle differentiated very early and that crust and sediments were also recycled into the mantle very early. Although such recycling is consistent with the early onset of subduction on Earth, other mantle recycling processes may explain the isotopic data equally well during the early Archean and the Hadean (Tolstikhin and Hofmann, 2005; Shirey et al., this volume).

## WHEN DID MODERN-STYLE PLATE TECTONICS BEGIN?

Most plate-tectonic indicators given in Tables 1 and 2 suggest that modern plate tectonics were operational, at least in some places on the planet, around 3 Ga and that they became widespread by 2.7 Ga. However, we are faced with three indicators (ophiolites, UHP metamorphism, and blueschists) that suggest a much later starting date around 1.0 Ga (Stern, 2005). Does this mean that the other petrotectonic assemblages and plate-tectonic indicators are invalid or does it mean that the three anomalous indicators have other explanations?

There are several problems with the recognition and preservation of ophiolite successions in the Archean. If Archean oceanic crust was significantly thicker than modern oceanic crust, as seems likely (Sleep and Windley, 1982; Foley et al., 2003), the deeper layers of oceanic crust that contain layered gabbros and ultramafic cumulate rocks may not have been accreted to the continents in subduction zones, but, instead, they may have been recycled into the mantle (Condie and Benn, 2006). Obduction of thick oceanic crust could have resulted in delamination of the middle- to lower-crustal sections and mantle lithosphere during collisional tectonics and the emplacement and preservation of only the upper, pillow-lava rich volcanic crust ("oceanic flake tectonics" as first described by Hoffman and Ranalli [1988]). One of the great challenges of the future is to recognize remnants of Archean oceanic crust, when and if it is preserved in greenstones.

Perhaps many of the mafic plain-type greenstones are remnants of oceanic crust rather than of oceanic plateaus (Condie, 1994; Tomlinson and Condie, 2001).

Stern (2005) concluded that the apparent absence of blueschists and ultrahigh-pressure (UHP) metamorphic rocks before ca. 1 Ga is important evidence for when the modern episode of subduction began. Others think that the absence of blueschists and UHP rocks from the Archean record results from the fact that the Archean Earth was hotter, and that subduction geotherms did not pass into the blueschist stability field (Martin, 1994; Pollack, 1997; Peacock, 2003; Ernst, 2003). However, it is not yet certain whether a higher mantle potential temperature results in a corresponding rise in the temperature of subduction zones. The thermal structure of subduction zones appears to reflect chiefly the age of the subducted lithosphere and the convergence rate (Peacock, 2003). Although earlier paleomagnetic data have suggested that minimum plate velocities in the Archean were similar to average plate velocities today (Layer et al., 1989), recent paleomagnetic data indicate that Archean plate motions were significantly faster than at present (Strik et al., 2003). Also the average age of subducting lithosphere in the Archean may have been younger than at present (Davies, 1992). The most convincing evidence of steeper subduction geotherms in the Archean, however, comes from thermobarometric studies of Archean metamorphic rocks (Komiya et al., 2002b; Moyen et al., 2006). Collectively, these lines of evidence lead us to the simplest and most straightforward explanation for the apparent absence of pre–1 Ga blueschists and UHP rocks in the geologic record: most Archean subduction zones were too hot. In addition to steeper Archean subduction geotherms, the rate of uplift of UHP rocks may have been so slow in the Archean that the UHP mineral assemblages recrystallized and can no longer be recognized.

It is very unlikely that plate tectonics began on Earth as a single global "event" at a distinct time in the early Archean. Rather, it is probable that a few mantle "returns" became cool enough that localized steep subduction became possible. The Whundo greenstones in the Western Pilbara may contain one such example around 3.1 Ga (Smithies et al., 2005a, 2007). As subduction zones continued to cool, more and more slabs were able to descend at steep angles, until by the late Archean, "steep-mode" subduction similar to modern subduction was widespread on the planet.

An intriguing question is whether a proposed global mantle melting event at 2.7 Ga (Condie, 1998, 2001, 2005b) had any effect on the evolving subduction systems. Could this event have provided a "push" that spread modern-style plate tectonics over the entire globe, thus accounting for the sudden widespread occurrence of plate-tectonic indicators at 2.7 Ga? If so, what is this telling us about the nature of the 2.7 Ga event? O'Neill et al. (2007) recently suggested that subduction may have been episodic during the Precambrian, thus explaining peaks in crustal production. In their model, higher mantle temperatures result in lower lithospheric stresses, causing rapid pulses of subduction interspersed with periods of relative quiescence. Stern (this volume) also suggested that episodic plate tectonics may have operated in the Precambrian. Clearly, there is much left to understand about the cooling history of planet Earth and why it is so different from Venus and Mars.

## ACKNOWLEDGMENTS

The manuscript was substantially improved by comments from Brian Windley, Peter Cawood, Hugh Smithies, Jan Kramers, Hugh Rollinson, and Eldridge Moores. This is Mainz Excellence-Cluster Geocycles contribution 23.

## REFERENCES CITED

Abers, G.A., van Keken, P.E., Keller, E.A., Ferris, A., and Stachnik, J.C., 2006, The thermal structure of subduction zones constrained by seismic imaging: Implications for slab dehydration and wedge flow: Earth and Planetary Science Letters, v. 241, p. 387–397, doi: 10.1016/j.epsl.2005.11.055.

Anhaeusser, C.R., 2006, A reevaluation of Archean intracratonic terrane boundaries on the Kaapvaal craton, South Africa: Collisional suture zones?, *in* Reimold, W.U., and Gibson, R.L., eds., Processes on the Early Earth: Geological Society of America Special Paper 405, p. 193–210.

Ansdell, K.M., Lucas, S.B., Connors, K., and Stern, R.A., 1995, Kisseynew metasedimentary gneiss belt, Trans-Hudson orogen: Back-arc origin and collisional inversion: Geology, v. 23, p. 1039–1043, doi: 10.1130/0091-7613(1995)023<1039:KMGBTH>2.3.CO;2.

Appel, P., Rollinson, H.R., and Touret, J., 2001, 3.8 Ga hydrothermal fluids from fluid inclusions in a volcanic breccia from the Isua greenstone belt, West Greenland: Precambrian Research, v. 112, p. 27–49, doi: 10.1016/S0301-9268(01)00169-3.

Ayres, L.D., and Thurston, P.C., 1985, Archean supracrustal sequences in the Canadian Shield: An overview: Geological Association of Canada Special Paper 28, p. 343–379.

Barker, F., and Arth, J.G., 1976, Generation of trondhjemitic-tonalitic liquids and Archean bimodal trondhjemite-basalt suites: Geology, v. 4, p. 596–600, doi: 10.1130/0091-7613(1976)4<596:GOTLAA>2.0.CO;2.

Barley, M.E., 1982, Porphyry style mineralization associated with early Archean calc-alkaline igneous activity, Eastern Pilbara, Western Australia: Economic Geology and the Bulletin of the Society of Economic Geologists, v. 77, p. 1230–1236.

Barley, M.E., Krapez, B., Groves, D.I., and Kerrich, R., 1998, The late Archean bonanza: Metallogenic and environmental consequences of the interaction between mantle plumes, lithospheric tectonics and global cyclicity: Precambrian Research, v. 91, p. 65–90, doi: 10.1016/S0301-9268(98)00039-4.

Bédard, J.H., 2006, A catalytic delamination-driven model for coupled genesis of Archean crust and sub-continental lithospheric mantle: Geochimica et Cosmochimica Acta, v. 70, p. 1188–1214, doi: 10.1016/j.gca.2005.11.008.

Betts, P.G., Lister, G.S., and O'Dea, M.G., 1998, Asymmetric extension of the middle Proterozoic lithosphere, Mount Isa terrane, Queensland, Australia: Tectonophysics, v. 296, p. 293–316, doi: 10.1016/S0040-1951(98)00144-9.

Betts, P.G., Giles, D., Lister, G.S., and Frick, L.R., 2002, Evolution of the Australian lithosphere: Australian Journal of Earth Sciences, v. 49, p. 661–695, doi: 10.1046/j.1440-0952.2002.00948.x.

Bickford, M., Collerson, K., Lewry, J., Van Schmus, W., and Chiarenzelli, J., 1990, Proterozoic collisional tectonism in the Trans-Hudson orogen, Saskatchewan: Geology, v. 18, p. 14–18, doi: 10.1130/0091-7613(1990)018<0014:PCTITT>2.3.CO;2.

Blake, D.H., and Barley, M.E., 1992, Tectonic evolution of the late Archean to early Proterozoic Mount Bruce Megasequence set, Western Australia: Tectonics, v. 11, p. 1415–1425.

Blake, D.H., and Stewart, A.J., 1992, Detailed Studies of the Mount Isa Inlier: Australian Bureau of Mineral Resources Bulletin 243, 135 p.

Bleeker, W., 2002, Archean tectonics: A review, with illustrations from the Slave craton: Geological Society of London Special Publication 199, p. 151–181.

Boily, M., and Dion, C., 2002, Geochemistry of boninite-type volcanic rocks in the Frotet-Evens greenstone belt, Opatica subprovince, Quebec: Implications for the evolution of Archean greenstone belts: Precambrian Research, v. 115, p. 349–371, doi: 10.1016/S0301-9268(02)00016-5.

Bowring, S.A., and Grotzinger, J.P., 1992, Implications of new chronostratigraphy for tectonic evolution of Wopmay orogen, NW Canadian Shield: American Journal of Science, v. 292, p. 1–20.

Boyet, M., and Carlson, R.W., 2005, $^{142}$Nd evidence for early (>4.53 Ga) global differentiation of the silicate Earth: Science, v. 309, p. 576–581, doi: 10.1126/science.1113634.

Bradley, D., 2005, Life spans of passive margins prior to arc collision, late Archean to present: Geological Society of America Abstracts with Programs, v. 37, no. 7, p. 493.

Brown, M., 2006a, A duality of thermal regimes is the hallmark of plate tectonics since the Neoarchean: Geology, v. 34, p. 961–964, doi: 10.1130/G22853A.1.

Brown, M., 2006b, Metamorphic conditions in orogenic belts: A record of secular change: International Geology Review, v. 49, no. 3, p. 193–234.

Brown, M., 2008, this volume, Characteristic thermal regimes of plate tectonics and their metamorphic imprint through Earth history, *in* Condie, K.C., and Pease, V., eds., When Did Plate Tectonics Begin on Planet Earth?: Geological Society of America Special Paper 440, doi: 10.1130/2008.2440(05).

Burke, K., Kidd, W.S.F., and Kusky, T.M., 1985, The Pongola structure of southeastern Africa: The world's oldest preserved rift?: Journal of Geodynamics, v. 2, p. 35–49, doi: 10.1016/0264-3707(85)90031-6.

Camire, G.E., Lafleche, M.R., and Ludden, J.N., 1993, Archean metasedimentary rocks from the NW Pontiac subprovince of the Canadian Shield: Chemical characterization, weathering and modelling of the source areas: Precambrian Research, v. 62, p. 285–305, doi: 10.1016/0301-9268(93)90026-X.

Caro, G., Bourdon, B., Birck, J.L., and Moorbath, S., 2006, High-precision $^{142}$Nd/$^{144}$Nd measurements in terrestrial rocks: Constraints on the early differentiation of the Earth's mantle: Geochimica et Cosmochimica Acta, v. 70, p. 164–191, doi: 10.1016/j.gca.2005.08.015.

Cawood, P.A., 1982, Structural relations in the subduction complex of the Paleozoic New England fold belt, eastern Australia: The Journal of Geology, v. 90, p. 381–392.

Cawood, P.A., and Tyler, I.M., 2004, Assembling and reactivating the Proterozoic Capricorn orogen: Lithotectonic elements, orogenies, and significance: Precambrian Research, v. 128, p. 201–218, doi: 10.1016/j.precamres.2003.09.001.

Cawood, P.A., Kröner, A., and Pisarevsky, S., 2006, Precambrian plate tectonics: Criteria and evidence: GSA Today, v. 16, no.7, p. 4–11, doi: 10.1130/GSAT01607.1.

Clift, P., and Vannucchi, P., 2004, Controls on tectonic accretion versus erosion in subduction zones: Implications for the origins and recycling of the continental crust: Reviews of Geophysics, v. 42, doi: 10.1029/2003RG000127.

Condie, K.C., 1994, Greenstones through time, *in* Condie, K.C., ed., Archean Crustal Evolution: Amsterdam, Elsevier, p. 85–120.

Condie, K.C., 1998, Episodic continental growth and supercontinents: A mantle avalanche connection?: Earth and Planetary Science Letters, v. 163, p. 97–108, doi: 10.1016/S0012-821X(98)00178-2.

Condie, K.C., 2000, Episodic continental growth models: Afterthoughts and extensions: Tectonophysics, v. 322, p. 153–162, doi: 10.1016/S0040-1951(00)00061-5.

Condie, K.C., 2001, Mantle Plumes and Their Record in Earth History: Cambridge, UK, Cambridge University Press, 305 p.

Condie, K.C., 2005a, High field strength element ratios in Archean basalts: A window to evolving sources of mantle plumes?: Lithos, v. 79, p. 491–504, doi: 10.1016/j.lithos.2004.09.014.

Condie, K.C., 2005b, Earth as an Evolving Planetary System: Amsterdam, Elsevier Academic Press, 447 p.

Condie, K.C., 2007, Accretionary orogens in space and time, *in* Hatcher, R.D., Jr., Carlson, M.P., McBride, J.H., and Martínez Catalán, J.R., 4-D Framework of Continental Crust: Geological Society of America Memoir 200, p. 145–158.

Condie, K.C., and Benn, K., 2006, Archean geodynamics: Similar to or different from modern geodynamics?: American Geophysical Union Monograph 164, p. 47–59.

Condie, K.C., and Harrison, N.M., 1976, Geochemistry of the Archean Bulawayan Group, Midlands greenstone belt, Rhodesia: Precambrian Research, v. 3, p. 253–271, doi: 10.1016/0301-9268(76)90012-7.

Cousens, B.L., 2000, Geochemistry of the Archean Kam Group, Yellowknife greenstone belt, Slave Province, Canada: The Journal of Geology, v. 108, p. 181–197, doi: 10.1086/314397.

Dann, J.C., 1997, Pseudostratigraphy and origin of the early Proterozoic Payson ophiolite, central Arizona: Geological Society of America Bulletin, v. 109, p. 347–365, doi: 10.1130/0016-7606(1997)109<0347:PAOOTE>2.3.CO;2.

Davies, G.F., 1992, On the emergence of plate tectonics: Geology, v. 20, p. 963–966, doi: 10.1130/0091-7613(1992)020<0963:OTEOPT>2.3.CO;2.

Davis, P.A., Jr., and Condie, K.C., 1977, Trace element model studies of Nyanzian greenstone belts, western Kenya: Geochimica et Cosmochimica Acta, v. 41, p. 271–277, doi: 10.1016/0016-7037(77)90235-6.

Desrochers, J.P., and Hubert, C., 1996, Structural evolution and early accretion of the Archean Malarctic composite block, southern Abitibi greenstone belt, Quebec, Canada: Canadian Journal of Earth Sciences, v. 33, p. 156–1569.

de Villarroel, G.H., and Lowe, D.R., 1993, Stratigraphy, petrography and provenance of Archean sedimentary rocks of the Nsuze Group, Pongola Supergroup, in the Wit M'folozi Inlier, South Africa: American Association of Petroleum Geologists Bulletin, v. 77, p. 320.

Dewey, J.F., 1977, Suture zone complexities: A review: Tectonophysics, v. 40, p. 53–67, doi: 10.1016/0040-1951(77)90029-4.

de Wit, M.J., 1998, On Archean granites, greenstones, cratons and tectonics: Does the evidence demand a verdict?: Precambrian Research, v. 91, p. 181–226, doi: 10.1016/S0301-9268(98)00043-6.

de Wit, M.J., and Ashwal, L.D., 1997, Greenstone Belts: Oxford, Oxford University Press, 809 p.

de Wit, M.J., Roering, C., Hart, R.J., Armstrong, R.A., de Ronde, C.E.J., Green, R.W.E., Tredoux, M., Peperdy, E., and Hart, R.A., 1992, Formation of an Archaean continent: Nature, v. 357, p. 553–562, doi: 10.1038/357553a0.

Dickinson, W.R., 1970, Relations of andesites, granites, and derivative sandstones to arc-trench tectonics: Reviews of Geophysics, v. 8, p. 813–860.

Eisele, J., Sharma, M., Galer, S.J.G., Blichert-Toft, J., Devey, C.W., and Hofmann, A.W., 2002, The role of sediment recycling in EM-1 inferred from Os, Pb, Hf, Nd, Sr isotope and trace element systematics of the Pitcairn hotspot: Earth and Planetary Science Letters, v. 196, p. 197–212, doi: 10.1016/S0012-821X(01)00601-X.

Eriksson, P.G., Krapez, B., and Fralick, P.W., 1994, Sedimentology of Archean greenstone belts: Signatures of tectonic evolution: Earth-Science Reviews, v. 37, p. 1–88, doi: 10.1016/0012-8252(94)90025-6.

Ernst, W.G., 2003, High-pressure and ultrahigh-pressure metamorphic belts—Subduction, recrystallization, exhumation, and significance for ophiolite studies, *in* Dilek, Y., and Newcombe, S., eds., Ophiolite Concept and Evolution of Geological Thought: Geological Society of America Special Paper 373, p. 365–384.

Evans, D.A.D., and Pisarevsky, S.A., 2008 , this volume, Plate tectonics on the early Earth? Weighing the paleomagnetic evidence, *in* Condie, K.C., and Pease, V., eds., When Did Plate Tectonics Begin on Planet Earth?: Geological Society of America Special Paper 440, doi: 10.1130/2008.2440(12).

Farquhar, J., Wing, B.A., McKeegan, K.D., Harris, J.W., Cartigny, P., and Thiemens, M.H., 2002, Mass-independent sulfur of inclusions in diamond and sulfur recycling on the early Earth: Science, v. 298, p. 2369–2372, doi: 10.1126/science.1078617.

Fedo, C.M., Myers, J.S., and Appel, P.W.U., 2001, Depositional setting and paleogeographic implications of earth's oldest supracrustal rocks, the >3.7 Ga Isua Greenstone belt, West Greenland: Sedimentary Geology, v. 141–142, p. 61–77.

Feiss, P.G., 1982, Geochemistry and tectonic setting of the volcanics of the Carolina slate belt: Economic Geology and the Bulletin of the Society of Economic Geologists, v. 77, p. 273–293.

Foley, S.F., 2008 , this volume, Trace element constraints on Archean magmatic and metamorphic processes, *in* Condie, K.C., and Pease, V., eds., When Did Plate Tectonics Begin on Planet Earth?: Geological Society of America Special Paper 440, doi: 10.1130/2008.2440(02).

Foley, S.F., Buhre, S., and Jacob, D.E., 2003, Evolution of the Archean crust by shallow subduction and recycling: Nature, v. 421, p. 249–252.

Frei, R., Polat, A., and Meibom, A., 2004, The Hadean upper mantle conundrum: Evidence for source depletion and enrichment from Sm-Nd, Re-Os and Pb isotopic compositions in 3.71 Ga boninite-like metabasalts from the Isua supracrustal belt, Greenland: Geochimica et Cosmochimica Acta, v. 68, p. 1645–1660, doi: 10.1016/j.gca.2003.10.009.

Friend, C.R.L., and Nutman, A.P., 2005, Complex 3670–3500 Ma orogenic episodes superimposed on juvenile crust accreted between 3850 and 3690 Ma, Itsaq Gneiss Complex, SW Greenland: The Journal of Geology, v. 113, p. 375–397, doi: 10.1086/430239.

Furnes, H., de Wit, M., Staudigel, H., Rosing, M., and Muehlenbachs, K., 2007, A vestige of Earth's oldest ophiolite: Science, v. 315, p. 1704–1707, doi: 10.1126/science.1139170.

Giles, C.W., and Hallberg, J.A., 1982, The genesis of the Archean Welcome Well volcanic complex, Western Australia: Contributions to Mineralogy and Petrology, v. 80, p. 307–318, doi: 10.1007/BF00378003.

Giles, D., Betts, P., and Lister, G., 2002, Far-field continental backarc setting for the 1.80–1.67 Ga basins of NE Australia: Geology, v. 30, p. 823–826, doi: 10.1130/0091-7613(2002)030<0823:FFCBSF>2.0.CO;2.

Goldfarb, R.J., Groves, D.I., and Gardoll, S., 2001, Orogenic gold and geologic time: A global synthesis: Ore Geology Reviews, v. 18, p. 1–75, doi: 10.1016/S0169-1368(01)00016-6.

Grieve, R.A.F., Cintala, M.J., and Therriault, A.M., 2006, Large-scale impacts and the evolution of the Earth's crust: The early years, *in* Reimold, W.U., and Gibson, R.L., eds., Processes on Early Earth: Geological Society of America Special Paper 405, doi: 10.1130/2006.2405(02).

Groves, D.I., Vielreicher, R.M., Goldfarb, R.J., and Condie, K.C., 2005, Controls on the heterogeneous distribution of mineral deposits through time: Geological Society of London Special Publication 248, p. 71–101.

Hallberg, J.A., Carter, D.N., and West, K.N., 1976, Archean volcanism and sedimentation near Meekatharra, Western Australia: Precambrian Research, v. 3, p. 577–595, doi: 10.1016/0301-9268(76)90020-6.

Hamilton, W.B., 1998, Archean magmatism and deformation were not products of plate tectonics: Precambrian Research, v. 91, p. 143–179, doi: 10.1016/S0301-9268(98)00042-4.

Hart, S.R., and Gaetani, G.A., 2006, Mantle paradoxes, the sulfide solution: Contributions to Mineralogy and Petrology, v. 152, p. 295–308, doi: 10.1007/s00410-006-0108-1.

Hoffman, P.F., and Ranalli, G., 1988, Archean oceanic flake tectonics: Geophysical Research Letters, v. 15, p. 1077–1080.

Holzer, L., Frei, R., Barton, J. M., and Kramers, J.D., 1998, Unraveling the record of successive high grade events in the central zone of the Limpopo belt using single phase dating of metamorphic mineral: Precambrian Research, v. 87, p. 87–115.

Hynes, A., 2008 , this volume, Effects of a warmer mantle on the characteristics of Archean passive margins, *in* Condie, K.C., and Pease, V., eds., When Did Plate Tectonics Begin on Planet Earth?: Geological Society of America Special Paper 440, doi: 10.1130/2008.2440(07).

Isozaki, Y., 1997, Jurassic accretion tectonics of Japan: Island Arc, v. 6, no. 1, p. 25–51.

Jackson, M.C., 1991, The Dominion Group: A Review of the Late Archean Volcano-Sedimentary Sequence and Implications for the Tectonic Setting of the Witwatersrand Supergroup, South Africa: Johannesburg, South Africa, Economic Research Unit, University of Witwatersrand, Information Circular 240, 40 p.

Jackson, M.J., Scott, D.L., and Rawlings, D.J., 2000, Stratigraphic framework for the Leichhardt and Calvert Superbasins: Review and correlations of the pre–1700 Ma successions between Mt. Isa and McArthur River: Australian Journal of Earth Sciences, v. 47, p. 381–403, doi: 10.1046/j.1440-0952.2000.00789.x.

Jahn, B.-M., Caby, R., and Monie, P., 2001, The oldest UHP eclogites of the world: Age of UHP metamorphism, nature of protoliths and tectonic implications: Chemical Geology, v. 178, p. 143–158, doi: 10.1016/S0009-2541(01)00264-9.

Jelsma, H.A., Vinyu, M.L., Valbrach, P.J., Davies, G.R., Wijbrans, J.R., and Verdurmen, E.A.T., 1996, Constraints on Archean crustal evolution of the Zimbabwe craton: A U-Pb zircon, Sm-Nd and Pb-Pb whole-rock study: Contributions to Mineralogy and Petrology, v. 124, p. 55–70, doi: 10.1007/s004100050173.

Kamber, B.S., Kramers, J.D., Napier, R., Cliff, R.A., and Rollinson, H.R., 1995, The Triangle shear zone, Zimbabwe, revisited: New data document an important event at 2.0 Ga in the Limpopo belt: Precambrian Research, v. 70, p. 191–213, doi: 10.1016/0301-9268(94)00039-T.

Kamber, B.S., Ewart, A., Collerson, K.D., Bruce, M.C., and McDonald, G.D., 2002, Fluid-mobile trace element constraints on the role of slab melting and implications for Archean crust growth models: Contributions to Mineralogy and Petrology, v. 144, p. 38–56.

Kamber, B.S., Collerson, K.D., Moorbath, S., and Whitehouse, M.J., 2003, Inheritance of early Archean Pb-isotope variability from long-lived Hadean protocrust: Contributions to Mineralogy and Petrology, v. 145, p. 25–46.

Kato, Y., and Nakamura, K., 2003, Origin and global tectonic significance of early Archean cherts from the Marble Bar greenstone belt, Pilbara craton, Western Australia: Precambrian Research, v. 125, p. 191–243, doi: 10.1016/S0301-9268(03)00043-3.

Kawai, T., Windley, B.F., Terabayashi, M., Yamamoto, H., Maruyama, S., and Isozaki, Y., 2006, Mineral isograds and metamorphic zones of the Anglesey blueschist belt, UK: Implications for the metamorphic development of a Neoproterozoic subduction-accretion complex: Journal of Metamorphic Geology, v. 24, p. 591–602.

Kerrich, R., and Polat, A., 2006, Archean greenstone-tonalite duality: Thermochemical mantle convection models or plate tectonics in the early Earth global dynamics?: Tectonophysics, v. 415, p. 141–164, doi: 10.1016/j.tecto.2005.12.004.

Kerrich, R., Wyman, D., Fan, J., and Bleeker, W., 1998, Boninite series: Low Ti-tholeiite associations from the 2.7 Ga Abitibi greenstone belt: Earth and Planetary Science Letters, v. 164, p. 303–316, doi: 10.1016/S0012-821X(98)00223-4.

Kerrich, R., Goldfarb, R., and Richards, J.P., 2005, Metallogenic provinces in an evolving geodynamic framework: Economic Geology and the Bulletin of the Society of Economic Geologists, v. 100, p. 1097–1136.

Kitajima, K., Maruyama, S., Utsunomiya, S., and Liou, J.G., 2001, Seafloor hydrothermal alteration at an Archaean mid-ocean ridge: Journal of Metamorphic Geology, v. 19, p. 583–599, doi: 10.1046/j.0263-4929.2001.00330.x.

Kleinhanns, I.C., Kramers, J.D., and Kamber, B.S., 2003, Importance of water for Archean granitoid petrology: A comparative study of TTG and potassic granitoids from Barberton Mountain Land, South Africa: Contributions to Mineralogy and Petrology, v. 145, p. 377–389, doi: 10.1007/s00410-003-0459-9.

Koenig, E., and Aydin, A., 1998, Evidence for large-scale strike-slip faulting on Venus: Geology, v. 26, p. 551–554, doi: 10.1130/0091-7613(1998)026<0551:EFLSSS>2.3.CO;2.

Komiya, T., Maruyama, S., Nohda, S., Masuda, M., Hayashi, H., and Okamoto, S., 1999, Plate tectonics at 3.8–3.7 Ga: Field evidence from the Isua accretionary complex, southern West Greenland: The Journal of Geology, v. 107, p. 515–554, doi: 10.1086/314371.

Komiya, T., Maruyama, S., Hirata, T., and Yurimoto, H., 2002a, Petrology and geochemistry of MORB and OIB in the mid-Archean North Pole region, Pilbara craton, Western Australia: Implications for the composition and temperature of the upper mantle at 3.5 Ga: International Geology Review, v. 44, p. 988–1016.

Komiya, T., Hayahi, M., Maruyama, S., and Yurimoto, H., 2002b, Intermediate-P–type Archean metamorphism of the Isua supracrustal belt: Implications for secular change of geothermal gradients at subduction zones and for Archean plate tectonics: American Journal of Science, v. 302, p. 806–826, doi: 10.2475/ajs.302.9.806.

Krapez, B., and Barley, M.E., 1987, Archaean strike-slip faulting and related ensialic basins: Evidence from the Pilbara block, Australia: Geological Magazine, v. 124, p. 555–567.

Kröner, A., and Layer, P.W., 1992, Crust formation and plate motion in the early Archean: Science, v. 256, p. 1405–1411, doi: 10.1126/science.256.5062.1405.

Kröner, A., Jaeckel, P., Brandl, G., Nemchin, A.A., and Pidgeon, R.T., 1999, Single zircon ages for granitoid gneisses in the Central zone of the Limpopo belt, southern Africa, and geodynamic significance: Precambrian Research, v. 93, p. 299–337, doi: 10.1016/S0301-9268(98)00102-8.

Kusky, T.M., 1998, Tectonic setting and terrane accretion of the Archean Zimbabwe craton: Geology, v. 26, p. 163–166, doi: 10.1130/0091-7613(1998)026<0163:TSATAO>2.3.CO;2.

Kusky, T.M., Li, J.H., and Tucker, R.T., 2001, The Dongwanzi ophiolite: Complete Archean ophiolite with extensive sheeted dike complex, North China craton: Science, v. 292, p. 1142–1145, doi: 10.1126/science.1059426.

Layer, P.W., Kröner, A., McWilliams, M., and York, D., 1989, Elements of the Archean thermal history and apparent polar wander of the eastern Kaapvaal craton, Swaziland, from single grain dating and paleomagnetism: Earth and Planetary Science Letters, v. 93, p. 23–34, doi: 10.1016/0012-821X(89)90181-7.

Leat, P.T., Smellie, J.L., Millar, I.L., and Larter, R.D., 2003, Magmatism in the South Sandwich arc: Geological Society of London Special Publication 219, p. 285–313.

Liou, J.G., Tsujimori, T., Zhang, R.Y., Katayama, I., and Maruyama, S., 2004, Global UHP metamorphism and continental subduction/collision: The Himalayan model: International Geology Review, v. 46, p. 1–27.

Lowe, D.R., and Nocita, B.W., 1999, Foreland basin sedimentation in the Mapepe Formation, southern-facies Fig Tree Group, *in* Lowe, D.R., and Byerly, G.R., eds., Geologic Evolution of the Barberton Greenstone Belt, South Africa: Geological Society of America Special Paper 329, p. 233–258.

Lucas, S., White, D., Hajnal, Z., Lewry, J., Green, A., Clowes, R., Zwanzig, H., Ashton, K., Schledewitz, D., Stauffer, M., Norman, A., Williams, P.F., and Spence, G., 1994, Three-dimensional collisional structure of the Trans-Hudson orogen, Canada: Tectonophysics, v. 232, p. 161–178, doi: 10.1016/0040-1951(94)90082-5.

Macdougall, J.D., and Haggerty, S.E., 1999, Ultradeep xenoliths from African kimberlites: Sr and Nd isotopic compositions suggest complex history: Earth and Planetary Science Letters, v. 170, p. 73–82, doi: 10.1016/S0012-821X(99)00091-6.

MacGregor, I.D., and Manton, W.I., 1986, Roberts Victor eclogites: Ancient oceanic crust: Journal of Geophysical Research, v. B91, p. 14,063–14,079.

Maekawa, H., Fryer, P., and Ozaki, A., 1995, Incipient blueschist-facies metamorphism in the active subduction zone beneath the Mariana forearc, *in* Taylor, B., and Natland, J., eds., Active Margins and Marginal Basins of the Western Pacific: American Geophysical Union Monograph 88, p. 281–289.

Martin, H., 1994, The Archean grey gneisses and the genesis of the continental crust, *in* Condie, K.C., ed., Archean Crustal Evolution: Amsterdam, Elsevier, p. 205–259.

Maruyama, S., and Liou, J.G., 1998, Ultrahigh-pressure metamorphism and its significance on the Proterozoic-Phanerozoic boundary: The Island Arc, v. 7, p. 6–35, doi: 10.1046/j.1440-1738.1998.00181.x.

Maruyama, S., Liou, J.G., and Terabayashi, M., 1996, Blueschists and eclogites of the world and their exhumation: International Geology Review, v. 38, p. 485–594.

Matsuda, T., and Isozaki, Y., 1991, Well-documented travel history of Mesozoic pelagic chert in Japan: From remote ocean to subduction zone: Tectonics, v. 10, p. 475–499.

McCarthy, T., and Rubridge, B., eds., 2005, The Story of Earth and Life—A Southern African Perspective on a 4.6-Billion-Year Journey: Cape Town, South Africa, Struik Publishers, 333 p.

Miyashiro, A., 1961, Evolution of metamorphic belts: Journal of Petrology, v. 2, no. 3, p. 277–311.

Möller, A., Appel, P., Mezger, K., and Schenk, V., 1995, Evidence for the 2 Ga subduction zone: Eclogites in the Usagaran belt of Tanzania: Geology, v. 23, p. 1067–1070, doi: 10.1130/0091-7613(1995)023<1067:EFAGSZ>2.3.CO;2.

Moyen, J.-F., and Stevens, G., 2005, Experimental constraints on TTG petrogenesis: Implications for Archean geodynamics: American Geophysical Union Monograph 164, p. 47–59.

Moyen, J.-F., Stevens, G., and Kisters, A., 2006, Record of mid-Archean subduction from metamorphism in the Barberton terrain, South Africa: Nature, v. 442, p. 559–562, doi: 10.1038/nature04972.

Mueller, W.U., and Corcoran, P.L., 2001, Volcano-sedimentary processes operating on a marginal continental arc: The Archean Raquette Lake Formation, Slave Province, Canada: Sedimentary Geology, v. 141–142, p. 169–204, doi: 10.1016/S0037-0738(01)00074-4.

Mueller, W., and Donaldson, J.A., 1992, Development of sedimentary basins in the Archean Abitibi belt, Canada: An overview: Canadian Journal of Earth Sciences, v. 29, p. 2249–2265.

Mueller, W., Donaldson, J.A., and Doucet, P., 1994, Volcanic and tectono-plutonic influences on sedimentation in the Archean Kirkland basin, Abitibi greenstone belt, Canada: Precambrian Research, v. 68, p. 201–230, doi: 10.1016/0301-9268(94)90030-2.

Mueller, W.U., Daigneault, R., Mortensen, J.K., and Chown, E.H., 1996, Archean terrane docking: Upper crust collision tectonics, Abitibi greenstone belt, Quebec, Canada: Tectonophysics, v. 265, p. 127–150, doi: 10.1016/S0040-1951(96)00149-7.

Myers, J.S., 2004, Isua enigmas: Illusive tectonic, sedimentary, volcanic and organic features of the >3.7 Ga Isua greenstone belt, SW Greenland, *in* Eriksson, P.G., Altermann, W., Nelson, D.R., Mueller, W.U., and Catuneanu, O., eds., The Precambrian Earth: Tempos and Events: Amsterdam, The Netherlands, Elsevier Publishers, p. 66–73.

Nakajima, K., and Arima, M., 1998, Melting experiments on hydrous low-K tholeiite: Implications for the genesis of tonalitic crust in the Izu-Bonin-Mariana arc: The Island Arc, v. 7, p. 359–373.

Nutman, A.P., Friend, C.R., and Bennett, V.C., 2002, Evidence for 3650–3600 Ma assembly of the northern end of the Itsaq Gneiss Complex, Greenland: implication for early Archean tectonics: Tectonics, v. 21, doi: 10.1029/2000TC001203.

O'Neill, C., Lenardic, A., Moresi, L., Torsvik, T.H., and Lee, C.T.A., 2007, Episodic Precambrian subduction: EOS Transactions, American Geophysical Union v. 88, no. 23, Joint Assembly Supplement, Abstract U44A-04.

Page, R.W., Jackson, M.J., and Krassay, A.A., 2000, Constraining sequence stratigraphy in north Australian basins: SHRIMP U-Pb zircon geochronology between Mt. Isa and McArthur River: Australian Journal of Earth Sciences, v. 47, p. 431–459, doi: 10.1046/j.1440-0952.2000.00797.x.

Parman, S.W., Grove, T.L., and Dann, J.C., 2001, The production of Barberton komatiites in an Archean subduction zone: Geophysical Research Letters, v. 28, p. 2513–2516, doi: 10.1029/2000GL012713.

Peacock, S.M., 2003, Thermal structure and metamorphic evolution of subducting slabs, *in* Eiler, J., and Hirschman, M., eds., Subduction Factory: American Geophysical Union Monograph 138, p. 7–22.

Pearce, J.A., and Cann, J.R., 1973, Tectonic setting of basic volcanic rocks determined using trace element analyses: Earth and Planetary Science Letters, v. 19, p. 290–300, doi: 10.1016/0012-821X(73)90129-5.

Pearce, J.A., and Peate, D.W., 1995, Tectonic implications of the composition of volcanic arc magmas: Annual Review of Earth and Planetary Sciences, v. 23, p. 251–285, doi: 10.1146/annurev.ea.23.050195.001343.

Pease, V., Percival, J., Smithies, H., Stevens, G., and van Kranendonk, M., 2008, this volume, When did plate tectonics begin? Evidence from the orogenic record, *in* Condie, K.C., and Pease, V., eds., When Did Plate Tectonics Begin on Planet Earth?: Geological Society of America Special Paper 440, doi: 10.1130/2008.2440(10).

Peltonen, P., Kontinen, A., and Huhma, H., 1996, Petrology and geochemistry of metabasalts from the 1.95 Ga Jormua ophiolite, northeastern Finland: Journal of Petrology, v. 37, p. 1359–1383, doi: 10.1093/petrology/37.6.1359.

Percival, J.A., Stern, R.A., Skulski, T., Card, K.D., Mortensen, J.K., and Begin, N.J., 1994, Minto block, Superior Province: Missing link in deciphering assembly of the craton at 2.7 Ga: Geology, v. 22, p. 839–842, doi: 10.1130/0091-7613(1994)022<0839:MBSPML>2.3.CO;2.

Percival, J.A., Bleeker, W., Cook, F.A., Rivers, T., Ross, G., and van Staal, C., 2004, Pan-Lithoprobe intra-orogen correlations and comparative orogenic anatomy: Geoscience Canada, v. 31, p. 23–39.

Percival, J.A., McNicoll, V., and Bailes, A.H., 2006, Strike-slip juxtaposition of ca. 2.72 Ga juvenile arc and >2.98 Ga continent margin sequences and its implications for Archean terrane accretion, western Superior Province, Canada: Canadian Journal of Earth Sciences, v. 43, p. 895–927, doi: 10.1139/E06-039.

Pesonen, L.J., Elming, S.-A., Mertanen, S., Pisarevsky, S.A., D'Agrella-Filho, M.S., Meert, J., Schmidt, P.W., Abrahmsen, N., and Bylund, G., 2003, Palaeomagnetic configuration of continents during the Proterozoic: Tectonophysics, v. 375, p. 289–324, doi: 10.1016/S0040-1951(03)00343-3.

Pirajno, F., and Occhipinti, S.A., 2000, Three Paleoproterozoic basins—Yerrida, Bryah and Padbury—Capricorn orogen, Western Australia: Australian Journal of Earth Sciences, v. 47, p. 675–688, doi: 10.1046/j.1440-0952.2000.00800.x.

Pisarevsky, S.A., 2005, New edition of the global paleomagnetic database: Eos (Transactions, American Geophysical Union), v. 86, p. 170, doi: 10.1029/2005EO170004.

Pisarevsky, S.A., and McElhinny, M.W., 2003, Global paleomagnetic database developed into its visual form: Eos (Transactions, American Geophysical Union), v. 84, p. 192, doi: 10.1029/2003EO200007.

Polat, A., and Kerrich, R., 1999, Formation of an Archean tectonic mélange in the Schreiber-Hemlo greenstone belt, Superior Province, Canada: Implications for Archean subduction-accretion process: Tectonics, v. 18, p. 733–755, doi: 10.1029/1999TC900032.

Polat, A., and Kerrich, R., 2001, Magnesian andesites, Nb-enriched basalt-andesites, and adakites from late Archean 2.7 Ga Wawa greenstone belts, Superior Province, Canada: Implications for late Archean subduction zone petrogenetic processes: Contributions to Mineralogy and Petrology, v. 141, p. 36–52.

Polat, A., Kusky, T., Li, J., Fryer, B., Kerrich, R., and Patrick, K., 2005, Geochemistry of Neoarchean volcanic and ophiolitic rocks in the Wutaishan greenstone belt, Central orogenic belt, North China craton: Implications for geodynamic setting and continental growth: Geological Society of America Bulletin, v. 117, p. 1387–1399, doi: 10.1130/B25724.1.

Polat, A., Appel, P., Frei, W.U.R., Pan, Y., Dilek, Y., Ordóñez-Calderón, J.C., Fryer, B., Hollis, J.A., and Raith, J.G., 2007, Field and geochemical characteristics of the Mesoarchean (3075 Ma) Ivisaartoq greenstone belt, southern West Greenland: Evidence for seafloor hydrothermal alteration in supra-subduction oceanic crust: Gondwana Research, v. 11, p. 69–91.

Pollack, H.N., 1997, Thermal characteristics of the Archean, *in* de Wit, M., and Ashwal, L.D., eds., Greenstone Belts: Oxford, Oxford University Press, p. 223–233.

Powell, C.M., Jones, D.L., Pisarevsky, S.A., and Wingate, M.T.D., 2001, Paleomagnetic constraints on the position of the Kalahari craton in Rodinia: Precambrian Research, v. 110, p. 33–46, doi: 10.1016/S0301-9268(01)00179-6.

Puchtel, I.S., Hofmann, A.W., Amelin, Y.V., Garbe-Schonberg, C.D., Samsonov, A.V., and Shchipansky, A.A., 1999, Combined mantle plume–island arc model for the formation of the 2.9 Ga Sumozero-Kenozero greenstone belt, SE Baltic Shield: Isotope and trace element constraints: Geochimica et Cosmochimica Acta, v. 63, p. 3579–3595, doi: 10.1016/S0016-7037(99)00111-8.

Reading, A.M., Kennett, B.L.N., and Goleby, B., 2007, New constraints on the seismic structure of West Australia: Evidence from terrane stabilization prior to the assembly of an ancient continent?: Geology, v. 35, p. 379–382, doi: 10.1130/G23341A.1.

Richardson, S.H., Shirey, S.B., Harris, J.W., and Carlson, R.W., 2001, Archean subduction recorded by Re-Os isotopes in eclogitic sulfide inclusions in Kimberley diamonds: Earth and Planetary Science Letters, v. 191, p. 257–266, doi: 10.1016/S0012-821X(01)00419-8.

Rollinson, H.R., 2002, The metamorphic history of the Isua greenstone belt, West Greenland, *in* Fowler, C.M.R., Ebinger, C.J., and Hawkesworth, C.J., eds., The Early Earth: Physical, Chemical and Biological Development: Geological Society of London Special Publication 199, p. 329–350.

Sandeman, H.A., Hanmer, S., Tella, S., Armitage, A.A., Davis, W.J., and Ryan, J.J., 2006, Petrogenesis of Neoarchean volcanic rocks of the MacQuoid supracrustal belt: A back-arc setting for the NW Hearne subdomain, western Churchill Province, Canada: Precambrian Research, v. 144, p. 140–165, doi: 10.1016/j.precamres.2005.11.001.

Schmitz, M.D., Bowring, S.A., de Wit, M.J., and Gartz, V., 2004, Subduction and terrane collision stabilize the western Kaapvaal craton tectosphere 2.9 Ga: Earth and Planetary Science Letters, v. 222, p. 363–376, doi: 10.1016/j.epsl.2004.03.036.

Schröder, S., Lacassie, J., and Beukes, N., 2006, Stratigraphic and geochemical framework of the Agouron drill cores, Transvaal Supergroup (Neoarchean-Paleoproterozoic, South Africa): South African Journal of Geology, v. 109, p. 23–54, doi: 10.2113/gssajg.109.1-2.23.

Schulze, D.E., Harte, B., Valley, J.W., Brenan, J.M., and Channer, D.M.R., 2003, Extreme crustal oxygen isotope signatures preserved in coesite in diamond: Nature, v. 423, p. 68–70, doi: 10.1038/nature01615.

Scotese, C.R., 2004, A continental drift flipbook: The Journal of Geology, v. 112, p. 729–741, doi: 10.1086/424867.

Scott, D.J., Helmstaedt, H., and Bickle, M.J., 1992, Purtuniq ophiolite, Cape Smith belt, northern Quebec, Canada: A reconstructed section of early Proterozoic oceanic crust: Geology, v. 20, p. 173–176, doi: 10.1130/0091-7613(1992)020<0173:POCSBN>2.3.CO;2.

Shervais, J.W., 1982, *T-V* plots and the petrogenesis of modern and ophiolitic lavas: Earth and Planetary Science Letters, v. 59, p. 101–118, doi: 10.1016/0012-821X(82)90120-0.

Shervais, J.W., 2006, The significance of subduction-related accretionary complexes in early Earth processes, *in* Reimold, W.U., and Gibson, R.L., eds., Processes on Early Earth: Geological Society of America Special Paper 405, p. 173–192.

Shirey, S.B., Richardson, S.H., and Harris, J.W., 2004, Integrated models of diamond formation and craton evolution: Lithos, v. 77, p. 923–944, doi: 10.1016/j.lithos.2004.04.018.

Shirey, S.B., Kamber, B.S., Whitehouse, M.J., Mueller, P.A., and Basu, A.R., 2008, this volume, A review of the geochemical evidence for mantle and crustal processes in the Hadean and Archean: Implications for the onset of plate tectonic subduction, *in* Condie, K.C., and Pease, V., eds., When Did Plate Tectonics Begin on Planet Earth?: Geological Society of America Special Paper 440, doi: 10.1130/2008.2440(01).

Shu, L.S., Zhou, G.Q., Shi, Y.S., and Yin, J., 1994, Study of the high pressure metamorphic blueschist and its late Proterozoic age in the Eastern Jiangnan belt: Chinese Science Bulletin, v. 39, p. 1200–1204.

Sillitoe, R.H., 1976, Andean mineralization: A model for the metallogeny of convergent plate margins: Geological Association of Canada Special Paper 14, p. 59–100.

Sklyarov, E.V., 1990, Ophiolites and blueschists of southeast Sayan, *in* First International Symposium on Geodynamic Evolution and Main Sutures of Central Asia: Ulan-Ude, Russia, IGCP, Project 283 Guidebook for excursion, 55 p.

Sleep, N.H., 2005, Evolution of the continental lithosphere: Annual Review of Earth and Planetary Sciences, v. 33, p. 369–393, doi: 10.1146/annurev.earth.33.092203.122643.

Sleep, N.H., and Windley, B.F., 1982, Archean plate tectonics: Constraints and inferences: The Journal of Geology, v. 90, p. 363–379.

Smith, I.E.M., Worthington, T.J., Stewart, R.B., Price, R.C., and Gamble, J.A., 2003, Felsic volcanism in the Kermadec arc, SW Pacific: Crustal recycling in an oceanic setting: Geological Society of London Special Publication 219, p. 99–118, doi: 10.1144/GSL.SP.2003.219.01.05.

Smithies, R.H., Champion, D.C., and Sun, S.-S., 2004, The case for Archean boninites: Contributions to Mineralogy and Petrology, v. 147, p. 705–721, doi: 10.1007/s00410-004-0579-x.

Smithies, R.H., Champion, D.C., Van Kranendonk, M.J., Howard, H.M., and Hickman, A.H., 2005a, Modern-style subduction processes in the Mesoarchean: Geochemical evidence from the 3.12 Ga Whundo intra-oceanic arc: Earth and Planetary Science Letters, v. 231, p. 221–237, doi: 10.1016/j.epsl.2004.12.026.

Smithies, R.H., Van Kranendonk, M.J., and Champion, D.C., 2005b, It started with a plume—Early Archaean basaltic proto-continental crust: Earth and Planetary Science Letters, v. 238, p. 284–297, doi: 10.1016/j.epsl.2005.07.023.

Smithies, R.H., Van Kranendonk, M.J., and Champion, D.C., 2007, The Mesoarchean emergence of modern-style subduction: Gondwana Research, v. 11, p. 50–68, doi: 10.1016/j.gr.2006.02.001.

Solomon, M., and Quesada, C., 2003, Zn-Pb-Cu massive sulfide deposits: Brine-pool types occur in collisional orogens, black smoker types occur in backarc and/or arc basins: Geology, v. 31, p. 1029–1032, doi: 10.1130/G19904.1.

Stern, R.A., Syme, E.C., Bailes, A.H., and Lucas, S.B., 1995, Paleoproterozoic arc volcanism in the Flin Flon belt, Trans-Hudson orogen, Canada: Contributions to Mineralogy and Petrology, v. 119, p. 117–141.

Stern, R.J., 2002, Subduction zones: Reviews of Geophysics, v. 40, 1012, doi: 10.1029/2001RG000108.

Stern, R.J., 2004, Subduction initiation: Spontaneous and induced: Earth and Planetary Science Letters, v. 226, p. 275–292, doi: 10.1016/S0012-821X(04)00498-4.

Stern, R.J., 2005, Evidence from ophiolites, blueschists, and ultrahigh-pressure metamorphic terranes that the modern episode of subduction tectonics began in Neoproterozoic time: Geology, v. 33, p. 557–560, doi: 10.1130/G21365.1.

Stern, R.J., 2008, this volume, Modern-style plate tectonics began in Neoproterozoic time: An alternative interpretation of Earth's tectonic history, *in* Condie, K.C., and Pease, V., eds., When Did Plate Tectonics Begin on Planet Earth?: Geological Society of America Special Paper 440, doi: 10.1130/2008.2440(13).

Stott, G.M., 1997, The Superior Province, Canada, *in* de Wit M.J., and Ashwal, L.D., eds., Greenstone Belts: Oxford, Clarendon Press, p. 480–507.

Strik, G., Blake, T.S., Zergers, T.E., White, S.H., and Langereis, C.G., 2003, Paleomagnetism of flood basalts in the Pilbara craton, Western Australia: Late Archean continental drift and the oldest known reversal of the geomagnetic field: Journal of Geophysical Research, v. 108, doi: 10.1029/2003JB002475.

Suganuma, Y., Hamano, Y., Niitsuma, S., Hoashi, M., Hisamitsu, T., Niitsuma, N., Kodama, K., and Nedachic, M., 2006, Paleomagnetism of the Marble Bar Chert Member, Western Australia: Implications for apparent polar wander path for Pilbara craton during Archean time: Earth and Planetary Science Letters, v. 252, p. 360–371, doi: 10.1016/j.epsl.2006.10.003.

Tappert, R., Stachel, T., Harris, J.W., Muehlenbachs, K., Ludwig, T., and Brey, G.P., 2005, Subducting oceanic crust: The source of deep diamonds: Geology, v. 33, p. 565–568, doi: 10.1130/G21637.1.

Thurston, P.C., 1994, Archean volcanic patterns, *in* Condie, K.C., ed., Archean Crustal Evolution: Amsterdam, Elsevier, p. 45–84.

Thurston, P.C., and Chivers, K.M., 1990, Secular variation in greenstone sequence development emphasizing Superior Province, Canada: Precambrian Research, v. 46, p. 21–58, doi: 10.1016/0301-9268(90)90065-X.

Tolstikhin, I.N., and Hofmann, A.W., 2005, Early crust on top of the Earth's core: Physics of the Earth and Planetary Interiors, v. 148, p. 109–130, doi: 10.1016/j.pepi.2004.05.011.

Tomlinson, K.Y., and Condie, K.C., 2001, Archean mantle plumes: Evidence from greenstone belt geochemistry, *in* Ernst, R.E., and Buchan, K.L., eds., Mantle Plumes: Their Identification through Time: Geological Society of America Memoir 352, p. 341–357.

Tsujimori, T., Sisson, V.B., Liou, J.G., Harlow, G.E., and Sorenson, S.S., 2006, Very-low-temperature record of the subduction process: A review of worldwide lawsonite eclogites: Lithos, v. 92, no. 3-4, p. 809–624, doi:10.1016/j.lithos.2006.03.054.

Tuckwell, G.W., and Ghail, R.C., 2003, A 400-km-scale strike-slip zone near the boundary of Thetis Regio, Venus: Earth and Planetary Science Letters, v. 211, p. 45–55, doi: 10.1016/S0012-821X(03)00128-6.

Turcotte, D.L., 1995, How does Venus lose heat?: Journal of Geophysical Research, v. 100, p. 16,931–16,940, doi: 10.1029/95JE01621.

van Hunen, J., Davies, G., Hynes, A., and van Keken, P.E., 2008, this volume, Tectonics of the early Earth: Some geodynamic considerations, *in* Condie, K.C., and Pease, V., eds., When Did Plate Tectonics Begin on Planet Earth?: Geological Society of America Special Paper 440, doi: 10.1130/2008.2440(08).

Van Kranendonk, M.J., and Collins, W.J., 1998, Timing and tectonic significance of late Archaean, sinistral strike-slip deformation in the Central Pilbara structural corridor, Pilbara craton, Western Australia: Precambrian Research, v. 88, p. 207–232, doi: 10.1016/S0301-9268(97)00069-7.

Van Kranendonk, M.J., Hickman, A.H., Smithies, R.H., Nelson, D.N., and Pike, G., 2002, Geology and tectonic evolution of the Archaean North Pilbara terrain, Pilbara craton, Western Australia: Economic Geology and the Bulletin of the Society of Economic Geologists, v. 97, p. 695–732.

Van Kranendonk, M.J., Smithies, R.H., Hickman, A.H., and Champion, D.C., 2007, Secular tectonic evolution of Archean continental crust: Interplay between horizontal and vertical processes in the formation of the Pilbara craton, Australia: Terra Nova, v. 19, p. 1–38, doi: 10.1111/j.1365-3121.2006.00723.x.

van Thienen, P., van den Berg, A.P., and Vlaar, N.J., 2004, Production and recycling of oceanic crust in the early Earth: Tectonophysics, v. 386, p. 41–65, doi: 10.1016/j.tecto.2004.04.027.

Watson, E.B., and Harrison, T.M., 2005, Zircon thermometer reveals minimum melting conditions on earliest Earth: Science, v. 308, p. 841–844, doi: 10.1126/science.1110873.

Weilers, B.F., 1990, A Review of the Pongola Supergroup and Its Setting on the Kaapvaal Craton: Johannesburg, South Africa, Economic Research Unit, University of Witwatersrand, Information Circular 228, 69 p.

Wilks, M.E., and Nisbet, E.G., 1988, Stratigraphy of the Steep Rock Group, NW Ontario: A major Archean unconformity and Archean stromatolites: Canadian Journal of Earth Sciences, v. 25, p. 370–391.

Windley, B.F., 1992, Proterozoic collision and accretionary orogens, *in* Condie, K.C., ed., Proterozoic Crustal Evolution: Amsterdam, Elsevier, p. 419–446.

Yoder, H.S., Jr., 1973, Contemporaneous basaltic and rhyolitic magmas: The American Mineralogist, v. 58, p. 153–171.

Zegers, T.E., and van Keken, P.E., 2001, Middle Archean continent formation by crustal delamination: Geology, v. 29, p. 1083–1086, doi: 10.1130/0091-7613(2001)029<1083:MACFBC>2.0.CO;2.

Zegers, T.E., Barley, M.E., Groves, D.I., McNaughton, N.J., and White, W.J., 2002, Oldest gold: Deformation and hydrothermal alteration in the early Archean shear-zone hosted Bamboo Creek deposit, Pilbara, Western Australia: Economic Geology and the Bulletin of the Society of Economic Geologists, v. 97, p. 757–776.

Zhai, M., Zhao, G., Zhang, Q., Kusky, T.M., and Li, J.-H., 2002, Is the Dongwanzi complex an Archean ophiolite?: Science, v. 295, p. 923, doi: 10.1126/science.295.5557.923a.

Zhang, H., Thurber, C.H., Shelly, D., Ide, S., Beroza, G.C., and Hasegawa, A., 2004, High-resolution subducting-slab structure beneath northern Honshu, revealed by double-difference tomography: Geology, v. 32, p. 361–364, doi: 10.1130/G20261.2.

Zhao, G., Wilde, S.A., Li, S., Sun, M., Grant, M.L., and Li, X., 2007, U-Pb zircon age constraints on the Dongwanzi ultramafic-mafic body, North China, confirm it is not an Archean ophiolite: Earth and Planetary Science Letters, v. 255, p. 85–93, doi: 10.1016/j.epsl.2006.12.007.

Manuscript Accepted by the Society 14 August 2007

Printed in the USA